GASEOUS DIELECTRICS VII

GASEOUS DIELECTRICS VII

Edited by

Loucas G. Christophorou

and

David R. James

Oak Ridge National Laboratory
Oak Ridge, Tennessee

SPRINGER SCIENCE+BUSINESS MEDIA, LLC

Library of Congress Cataloging-in-Publication Data

On file

Proceedings of the Seventh International Symposium on Gaseous Dielectrics, held April 24–28, 1994, in Knoxville, Tennessee

ISBN 978-1-4899-1297-8 ISBN 978-1-4899-1295-4 (eBook)
DOI 10.1007/978-1-4899-1295-4

Originally published by Plenum Press, New York in 1994
Softcover reprint of the hardcover 1st edition 1994

PREFACE

The Seventh International Symposium on Gaseous Dielectrics was held in Knoxville, Tennessee, U.S.A., on April 24-28, 1994. The symposium continued the interdisciplinary character and comprehensive approach of the preceding six symposia.

Gaseous Dielectrics VII is a detailed record of the symposium proceedings. It covers recent advances and developments in a wide range of basic, applied and industrial areas of gaseous dielectrics. It is hoped that *Gaseous Dielectrics VII* will aid future research and development in, and encourage wider industrial use of, gaseous dielectrics.

The Organizing Committee of the Seventh International Symposium on Gaseous Dielectrics consisted of G. Addis (U.S.A.), L. G. Christophorou (U.S.A.), F. Y. Chu (Canada), A. H. Cookson (U.S.A.), O. Farish (U.K.), I. Gallimberti (Italy), A. Garscadden (U.S.A.), D. R. James (U. S. A.), E. Marode (France), T. Nitta (Japan), W. Pfeiffer (Germany), Y. Qiu (China), I. Sauers (U.S.A.), R. J. Van Brunt (U.S.A.), and W. Zaengl (Switzerland). The local arrangements committee consisted of members of the Health Sciences Research Division and personnel of the Conference Office of the Oak Ridge National Laboratory, and staff of the University of Tennessee (UTK). The contributions of each member of these committees, the work of the Session Chairmen, the interest of the participants, and the advice of innumerable colleagues are gratefully acknowledged. I am especially indebted to Dr. Isidor Sauers, Dr. David R. James, Mrs. Joan E. Carrington, Ms. Jo Ann Cripps, and Mrs. Doris Crowell for their assistance during the symposium and for their help with the manuscripts.

The symposium was hosted by the Oak Ridge National Laboratory and the University of Tennessee and was sponsored by the U.S. Department of Energy, the UTK/ORNL Science Alliance, the Aero Propulsion and Power Laboratories of the Wright Research and Development Center and the Tennessee Valley Authority; it was organized in cooperation with the Institute of Electrical and Electronics Engineers, Inc., the Power Engineering Society, and the Dielectrics and Electrical Insulation Society. The continued support of the Oak Ridge National Laboratory and the University of Tennessee and the financial assistance of the sponsors are acknowledged with gratitude.

L. G. Christophorou, Symposium Chairman

Oak Ridge, Tennessee
July, 1994

CONTENTS

SECTION 1: BASIC PHYSICS OF GASEOUS DIELECTRICS

SECTION 4: PARTIAL DISCHARGES/DIAGNOSTICS

SECTION 5: BREAKDOWN CHARACTERISTICS UNDER SHORT TIME FRONTS AND FAST TRANSIENTS

SECTION 6: SPECIAL APPLICATIONS OF GASEOUS DIELECTRICS

SECTION 7: PULSED POWER SWITCHING/GAS RECOVERY

SECTION 8: GAS DECOMPOSITION/PLASMA CHEMISTRY

SECTION 9: DC GIS/CHARGING OF SOLID INSULATORS

SECTION 10: DEVELOPMENTS IN GAS-INSULATED EQUIPMENT

SECTION 11: RELIABILITY TESTING OF GIS

SECTION 12: CORONA/SURFACE DISCHARGES

SECTION 1: BASIC PHYSICS OF GASEOUS DIELECTRICS

ANION FORMATION IN MOLECULES AND CLUSTERS: ELEMENTARY PROCESSES INDUCED BY SLOW ELECTRONS

Eugen Illenberger

Institut für Physikalische und Theoretische Chemie
Freie Universität Berlin
Takustrasse 3
D-14195 Berlin, Germany

INTRODUCTION

We will discuss here some basic aspects in the interaction of low energy electrons (0-10 eV) with molecules and weakly bound van der Waals clusters with particular emphasis to processes resulting in negative ions. We shall illustrate in a case study that (i) the primary step of electron localization und (ii) the evolution of a temporary negative ion differs substantially on going from isolated molecules under single collision conditions ($\leq 10^{-5}$ mbar) to higher pressures (beyond single collision conditions , $\geq 10^{-4}$ mbar) and finally to the corresponding molecular clusters.

Consider a neutral molecule M (isolated or in close proximity to other molecules) interacting with free electrons. At certain specific energies resonant electron capture can directly form a "temporary negative ion" (TNI):[1-3]

$$e^- + M \quad \rightarrow \quad M^{-(*)} \qquad (1)$$

which can principally react via the following scheme

$$M^{-(*)} \quad \xrightarrow{\tau_a} \quad M^{(*)} + e^- \qquad (2)$$

$$M^{-(*)} \quad \xrightarrow{\tau_d} \quad R + X^- \qquad (3)$$

$$M^{-(*)} \quad \xrightarrow{\tau_s} \quad M^- + \text{Energy} \qquad (4)$$

Process (2) is autodetachment recovering the neutral molecule (eventually in an excited state), (3) is dissociative electron attachment leading to stable fragments, and (4) is stabilization into the thermodynamically stable parent anion. This last process can only occur for molecules possessing a *positive electron affinity*.

Gaseous Dielectrics VII, Edited by L.G. Christophorou and D.R. James, Plenum Press, New York, 1994

It is evident that the *formation* and *decomposition* of a TNI strongly depends on whether the overall process occurs on a single and isolated molecule ("unimolecularly") or within molecules at higher pressure or some stage of aggregation.

In the gas phase under single collision conditions the only stabilization mechanism (4) is radiative cooling which is generally much slower than the competing processes (2) and (3), so that in a mass spectrometric experiment one usually detects fragment ions. Decomposition into stable fragments (3) occurs on a time scale between $\tau_d \approx 10^{-14}$ s (direct electronic dissociation along repulsive energy surfaces) to $\tau_d \geq 10^{-12}$ s (electronic and vibrational predissociation).

On proceeding to van der Waals clusters,[4] the situation may change with respect to the *formation* of a TNI, its *energy* and its *decomposition*. Generally, the *energy* of the TNI in the cluster is shifted to lower energies, due to the polarization interaction of the negative charge with the surrounding molecules. Some of the various aspects concerning negative ion formation in clusters are illustrated in Fig. 1.

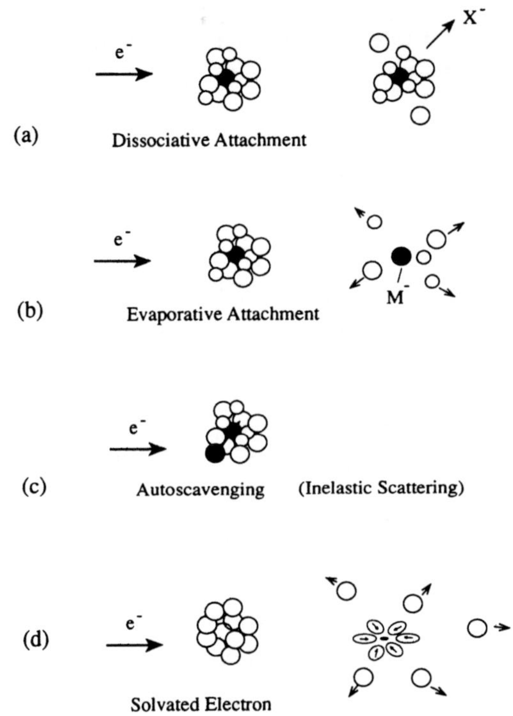

Figure 1. Schematic representation of some of the various possibilities of electron localization and evolution of a molecular anion in clusters.

In addition to dissociative attachment (Fig. 1a) stable anions M^- can be formed by collisional stabilization, thereby evaporating the target cluster ("evaporative electron attachment", Fig 1b) or the formation of solvated ions of the form $M_n \cdot X^-$ in the course of scattering processes within the ionized target cluster. Futhermore, electron localization can proceed in two steps in that the incoming electron is first *inelastically scattered* by one molecule and the slowed down electron *captured* by another molecule within one cluster (self or auto-scavenging, Fig. 1c). These inelastic scattering processes can considerably contribute to anion formation in clusters.[4] Finally electron localization and stabilization may be provided by the collective response of dipolar molecules creating a solvated electron, Fig. 1d.[5] In that case the excess electron is not bound to an individual molecule.

EXPERIMENTAL CONSIDERATIONS

Negative ion formation from molecules and clusters can be studied with a crossed beam **arrangement** schematically shown in Fig. 2. A beam of nearly monoenergetic electrons is crossed with a molecular beam and the resulting ions are extracted from the interaction volume and analyzed with a mass spectrometer.[4] The formation of a particular negative ion is then recorded as a function of the electron energy. Time-of-flight (TOF) measurements of the products allow information on the energy distribution of the particular reaction studied. The molecular beam is either a supersonic beam containing a distribution of clusters generated by adiabatic expansion or simply an effusive beam consisting of isolated molecules at a certain pressure in the interaction volume.

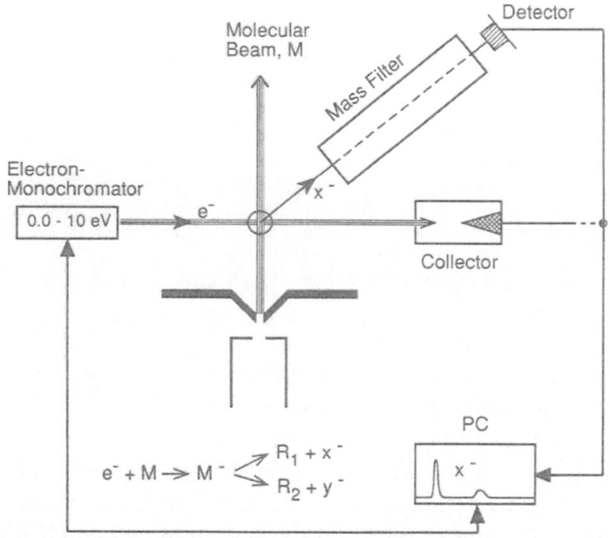

Figure 2 Schematic representation of the experimental arrangement for the study of electron attachment to molecules and clusters.

CASE STUDY: CF_3I

Figure 3 shows the energy profiles of fragment ions observed from **gas phase** CF_3I under **single collision conditions**. Two prominent resonances due to dissociative electron attachment are observed, one near 0 eV and another around 3.8 eV[6]. The spectra in Fig. 3 were recorded at different target gas pressures for the different ions to ensure *single collision conditions*.

The formation of I^- and CF_3^- exhibit a particular behavior in that they are formed via the release of remarkable amounts of excess energy as evident from the TOF measurements (not shown here).

The resonance at 0 eV can be described as the ground state of CF_3I^- formed by the accommodation of the extra electron into the first normally unoccupied MO with a strongly localized σ^*(C-I) character directly decomposing into I^- + $CF_3^\cdot$. The electronic state at 3.8 eV is most likely a *core excited* resonance: the incoming electron concomitantly excites one

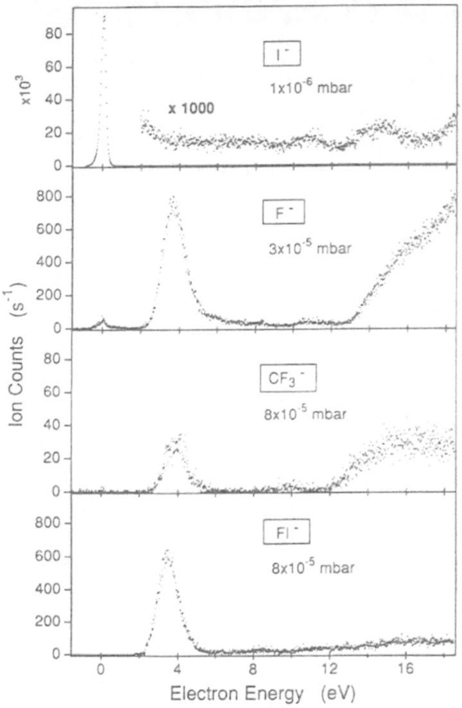

Figure 3. Negative ion formation from CF_3I under single collision conditions. The different products are recorded under different target gas pressure to ensure single collision conditions. From ref. 6.

of the molecule's electrons and we have two electrons in normally unfilled MOs. This state decomposes into CF_3^- + I with high kinetic energy.

The observation that no particular I^- contribution appears near 3.8 eV can be rationalized by the fact that the resonance involves promotion of a 5p electron located principally on the iodine atom to the localized σ*(C-I) MO. This leaves a hole or a positive charge on I in the initially formed molecular anion thus making the decomposition into I^- unlikely.

The situation is summarized in the potential energy diagram (Fig. 4). The zero eV resonance is attributed to a one electron process. Although the associated potential energy surface has a minimum CF_3I^- is not formed under single collision conditions.

In electron attachment to ***clusters*** the following new features appear:

1. Formation of ionic products of the form M_n^- ($n \geq 1$) and $M_n{\cdot}I^-$ ($n \geq 1$) (M = CF_3I) with M_n^- considerably less abundant than $M_n{\cdot}I^-$.
2. The appearance of distinct new features in the energy profiles of I^-, $M_n{\cdot}I^-$ and CF_3^-.

Figure (5a) shows that the molecular anion M^- appears near zero eV. It is generated via formation of ground state CF_3I^- within the target cluster and subsequent collisional stabilization thereby evaporating the target cluster ("evaporative electron attachment") according to

$$e^- (\approx 0\ eV) + (CF_3I)_m \rightarrow CF_3I^- \cdot (CF_3I)_{m-1} \rightarrow CF_3I^- + (m-1)\ CF_3I$$

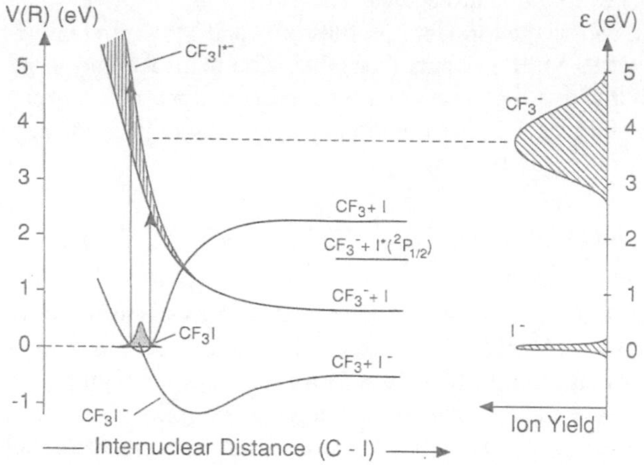

Figure 4. Schematic potential energy curves for the electronic states relevant in dissociative electron attachment to CF3I. From ref. 6.

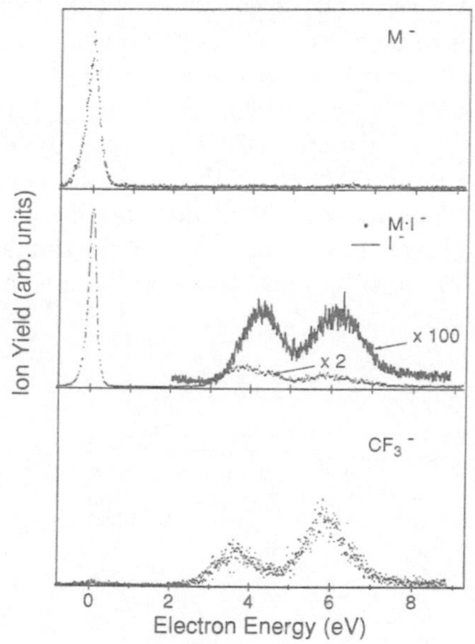

Figure 5. Ion yield curve of some products in electron attachment to CF3I clusters . From ref. 6

Of course, the stabilization energy may not in every case be sufficient to produce the maximum of (m-1) evaporated molecules.

Figure (5b) shows the ion yield profiles of I^- and $M{\cdot}I^-$ (the features remain the same in the homologous series $M_n{\cdot}I^-$). Apart from their dominant formation near zero eV, distinct new features with maxima near 4 eV and 6 eV become apparent. At zero eV all product ions must have the ground state TNI as precursor which dissociates in the target cluster. In the case of $M{\cdot}I^-$ the reaction can be expressed as

$$e^- (\approx 0\ eV) + (CF_3I)_m \rightarrow CF_3I^- \cdot (CF_3I)_{m-1} \rightarrow I^- \cdot CF_3I + CF_3(CF_3I)_{m-2}$$

In Fig. (5b) the intensity of I^- and $M{\cdot}I^-$ at zero eV is normalized to the same height. Above 2 eV the scale for I^- and $M{\cdot}I^-$ is then multiplied by 100 and 2, respectively.

For the products I^- and $M_n{\cdot}I^-$ ($n \geq 1$) we interpret the features at 4 eV and 6 eV as a result of *inelastic electron scattering* in the target cluster prior to electron localization: In the first step the electron is inelastically scattered from one molecule and the slowed down electron is captured by a second molecule *within the same cluster*. Such a mechanism was first proposed by Klots and Compton[7] in CH_3I clusters and was called *self scavenging* to emphasize the involvement of a single van der Waals cluster in contrast to the usual scavenging mechanism which is observed at higher pressures between two uncorrelated molecules (see below).

The cluster results presented above can be compared with anion formation using effusive gas inlet but under higher pressure, i. e, ***beyond single collision conditions.***

We have recorded ions at target gas pressures up to $8x10^{-4}$ mbar. Under these conditions, there is no indication of M^- and $M_n{\cdot}I^-$ formation and the ion yield curves of F^-, CF_3^- and FI^- remain virtually unaffected. However, for the I^- signal we observe the evolution of new features above 3 eV. Figure 6 compares the energy profile of I^- arising from the cluster beam at 1 bar stagnation pressure to that from the effusive beam at $4x10^{-4}$ mbar on an arbitrary scale for the intensities. Above 3 eV the effusive signal now increases with the *square* of the pressure as expected for a *secondary* reaction. The zero eV peak is not presented in Fig. 5, since the large amount of I^- ions causes saturation of the detector and the counting electronics. The "effusive" I^- signal above 3 eV arises from inelastic (either direct or

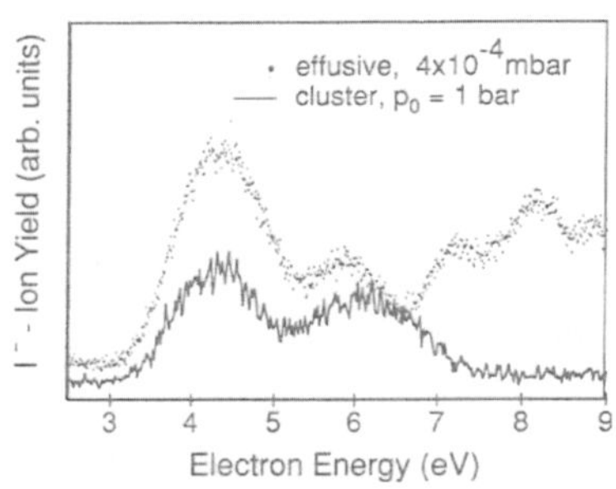

Figure 6. Comparison of the scavenger features on the channel I^- in clusters and isolated molecules beyond single collision conditions. From ref. 6.

resonant) scattering of the primary electron and capture of the slow electron by another molecule. In contrast to the process in clusters, this occurs between two *uncorrelated* monomers.

To establish this interpretation, we have added of a few % SF_6 to the CF_3I gas and recorded SF_6^-. This results in an SF_6^- energy profile which completely coincides with the "effusive" spectrum of I^- in Fig. 6.

The effusive high pressure spectrum of I^- thus represents the threshold excitation spectrum (TEES) of CF_3I. The first two peaks of the TEES resemble the features of I^- and $M{\cdot}I^-$ formation from clusters with the 6 eV feature blue shifted in clusters. Interestingly, the TEES spectrum exhibits distinct further structures above 7 eV while this is not seen for I^- and $M_n{\cdot}I^-$ formation from clusters.

Molecules of the type CH_3X show many further optical transitions above 6 eV which are usually assigned to members of Rydberg series[8] and which contribute to the TEES spectrum. In clusters these higher excited states do obviously *not* result in the formation of stable negative ions. A similar observation was made by Klots and Compton in CH_3I clusters[7] and by our laboratory in the case of O_2 clusters.[4] In both cases only the *lower* part of the TEES spectrum was imaged in anion formation from clusters. These higher excited states are either not existent in clusters or, more likely, the delocalization of the excitation destroys the ability of the rest of the cluster to capture an electron.

Acknowledgments

This work has been supported by the Deutsche Forschungsgemeinschaft and Fonds der Chemischen Industrie.

References

1. L.G. Christophorou, "Atomic and Molecular Radiation Physics", Wiley Interscience, London (1971).

2. E. Illenberger, J. Momigny, "Gaseous Molecular Ions. An Introduction to Elementary Processes Induced by Ionization", Steinkopff Verlag, Weinheim; Springer Verlag, New York (1992).

3. T. Oster, A. Kühn and E. Illenberger, Gas Phase Negative Ion Chemistry, *Int. J. Mass Spectrom. Ion Proc. 89:1 (1989).*

4. E. Illenberger, Electron Attachment Reactions in Molecular Clusters, *Chem. Rev. 92:1589 (1993).*

5. H. Haberland, Solvated Electron Clusters, in "The Chemical Physics of Atomic and Molecular Clusters", G. Scoles (Ed.), North Holland, Amsterdam (1990).

6. T. Oster, O. Ingolfsson, M. Meinke, T. Jaffke and E. Illenberger, Anion Formation from Gaseous and Condensed CF_3I on Low Energy Electron Impact, *J. Chem. Phys. 99: 5141 (1993).*

7. C.E. Klots and R.N. Compton, Self-Scavenging of Electrons in van der Waals Molecules of Methyl Iodide, *Chem. Phys. Lett. 73:589 (1980).*

8. M.B. Robin, "Higher Excited States of Polyatomic Moleculcs", Vol I, Academic, New York (1974).

DISCUSSION

L. G. CHRISTOPHOROU: In a gas-insulated apparatus slow electrons can be attached to molecules in the gas phase itself and also to molecules adsorbed on surfaces. How do the two processes differ?

E. ILLENBERGER: This is an extremely important point. Electron attachment to condensed or adsorbed molecules is characterized by a shift of the resonance to lower energies - typically 1-2 eV with respect to the gas phase - due to polarization of the surrounding medium. This can lead to a dramatic increase of the attachment cross section.

M. GOLDMAN: In general we could think that a surface acts as a third body. Are there many cases where dissociative attachment can take place?

E. ILLENBERGER: Yes. Although a surface represents an effective sink for the energy deposited onto an adsorbed molecule, dissociative electron attachment reactions on adsorbed molecules are well established in recent electron scattering experiments from surfaces.

R. J. VAN BRUNT: Please comment about the structures observed in the electron attachment cross sections for C_{60} and C_{70}. What is the source (resonance type) of these structures, and why is more structure observed for C_{60} than for C_{70}?

E. ILLENBERGER: The structures in the C_{60}^- attachment spectrum above 1 eV correlate with the C_{60} electron energy loss spectrum. Consequently, we interpret these structures as core-excited resonances, while the peak near zero eV is due to C_{60}^- in its electronic ground state. The less structured C_{70}^- attachment spectrum is most likely due to stronger electron-phonon coupling.

A. GARSCADDEN: Your autodetachment lifetimes of C_{60}^- are very long compared to the transition state lifetimes - by 6 or 7 orders of magnitude. Does the energy dispersal model (statistical) adequately explain this effect when SF_6^- and C_{60}^- autodetachment lifetimes are similar but the number of modes is different?

E. ILLENBERGER: The long lifetimes are related to the large number of internal degrees of freedom (for example, vibrational modes). So C_{60}^- is expected to have a much longer lifetime than SF_6^-; the lifetime of C_{60}^- is $\sim$ 1 s while that of SF_6^- is only a few μs. Given this, the statistical model can describe the autodetachment lifetimes satisfactorily.

R. N. COMPTON: Is the position of the $^2\Pi_g$ in the $(O_2)_nN_2$ cluster at the same position as that in the free N_2^- resonance? One would expect the N_2^- resonance in the cluster to be shifted downward in energy by approximately the electron affinity of O_2.

E. ILLENBERGER: The energy of the N_2^- resonance is slightly shifted to lower energy.

R. N. COMPTON: Do you see the vibrational structure in the $N_2^-(O_2)_n$ resonance corresponding to the free N_2^- resonance?

E. ILLENBERGER: The question is whether vibrational structure of N_2^- can be observed--regardless of the experimental energy resolution--on the product O_2^- or O_4^-. While the attachment profile of N_2 clearly shows vibrational structure, the subsequent formation of O_2^- or O_4^- is effected by (1) the energy distribution of the autodetached electrons, (2) the attachment profile of O_2 within the particular cluster and (3) the stabilization process leading to O_2^- or O_4^- by evaporation of the rest of the aggregate.

E. MARODE: What is the lifetime of the molecular nitrogen negative ion?

E. ILLENBERGER: The lifetime of the $N_2^-(^2\Pi_g)$ is on the order of a few vibrational periods, i.e., in the 10^{-14}-10^{-13}s range.

E. MARODE: The angular distribution of the outgoing electron after elastic collisions is well known and corresponds to an equidistributed angular distribution. Could the inelastic collision forming this temporary negative molecular ion, which actually ends like an elastic collision, change this final distribution? This question is actually relevant to Monte-Carlo simulation of gas discharges, since in such simulations this distribution is taken fully into account.

E. ILLENBERGER: The angular distribution of the resonantly scattered electrons is anisotropic, reflecting the π^* molecular orbital.

R. J. VAN BRUNT: Could you comment on the pressure dependence (expected and observed) for $(O_2)_2^-$ formation in N_2/O_2 mixtures?

E. ILLENBERGER: The usual pressure dependence to identify secondary and primary processes cannot be applied in the case of a supersonic beam expansion. By introducing N_2 as background gas, however, it is established that the reported processes occur within one single N_2/O_2 cluster.

R. NAGPAL: In the experiments with N_2/O_2 clusters, how do you ensure that the 2.0-2.5 eV N_2^- peak is from electron interactions with N_2 in the cluster and not with free N_2 in the gas phase? What is the ratio of the electron collision frequency with N_2 in the cluster to that with free N_2?

E. ILLENBERGER: This problem was carefully examined by adding N_2 as background gas into the supersonic beam apparatus. These experiments showed that under the present experimental conditions we are too far away (i.e., by orders of magnitude) to pick up slow electrons from isolated N_2.

EFFECT OF TEMPERATURE ON THE ELECTRON ATTACHMENT AND DETACHMENT PROPERTIES OF c-C_4F_6

P. G. Datskos[#], L. G. Christophorou[#], and J. G. Carter

Atomic, Molecular, and High Voltage Physics Group
Health Sciences Research Division
Oak Ridge National Laboratory
Oak Ridge, Tennessee 37831-6122, U.S.A.

ABSTRACT

The temperature dependence of the low-energy electron attachment and autodetachment processes for c-C_4F_6 in a N_2 buffer gas has been studied in the temperature, T, range of 300 to 600 K and the mean electron energy, $<\varepsilon>$, range from 0.19 to 1.0 eV. The low-energy electron attachment rate constant for c-C_4F_6 shows only a slight dependence on gas temperature. In contrast, the autodetachment frequency increases by more than four orders of magnitude when T is increased from 300 to 600 K. This increase in autodetachment is due to the increase in the internal energy content of the c-$C_4F_6^-$ anion with increasing T. The autodetachment process under consideration is a heat-activated process and has an activation energy E^* of 0.24 eV. The significance of these results to gaseous dielectrics is indicated.

INTRODUCTION

The electron attachment and detachment properties of dielectric gas molecules play an important role in determining their dielectric properties[1-3]. For example, at room temperature the large electron attachment cross sections[4] for hexafluorobutene (c-C_4F_6), hexafluorobenzene (C_6F_6) and sulfur hexafluoride (SF_6) which at low electron energies are due to both parent and fragment anions are primarily responsible for their high dielectric strengths[2,3]. As high electron attachment leads to high dielectric strength, the presence of

---

[#]Also at The Department of Physics and Astronomy, The University of Tennessee, Knoxville, Tennessee 37996-1200.

Gaseous Dielectrics VII, Edited by L.G. Christophorou
and D.R. James, Plenum Press, New York, 1994

electron detachment can lead to the production of avalanche initiating electrons[5-7] and high probability of detachment can result in a reduction of the dielectric strength of a gaseous medium (e. g., see Ref. 8 and later in this paper).

Changes in the internal energy of a molecule can drastically affect its electron attaching and detaching properties. Such changes in the internal energy content of a molecule can be effected by gas heating or by laser irradiation. Modifying the ability of an electronegative gas to capture and to release electrons results in alteration of its dielectric properties. For example, the decline in the limiting electric field strength, $(E/N)_{lim}$, with increasing T reported earlier[8] for dielectric gases which at low electron energies form parent anions can be attributed to increased autodetachment, while the increase in $(E/N)_{lim}$ for gases which capture electrons only dissociatively, can be attributed to the increase in the negative ion formation due to enhancement of dissociative electron attachment with increasing T[8].

In this paper we report on the effect of temperature, T, on the electron attachment rate constant, $k_a(<\varepsilon>,T)$, and the electron detachment frequency, $t_d^{-1}(T)$ for c-C_4F_6 and c-$C_4F_6^-$, respectively, in a buffer gas of N_2 for electron energies ≤1 eV. In this electron energy range c-C_4F_6 captures electrons nondissociatively, very efficiently[9], forming long-lived ($\tau_a > 6\mu s$[10]) parent negative ions. At energies above 2 eV fragment anions (F^-, $C_3F_3^-$, and $C_4F_5^-$) are produced with much lower cross sections[10].

The electron attachment and detachment processes envisioned are

$$e + c\text{-}C_4F_6 \xrightarrow{k_c} c\text{-}C_4F_6^{-*} + (N_2) \xrightarrow{p} c\text{-}C_4F_6^- \tag{1}$$

$$c\text{-}C_4F_6^- \xrightarrow{t_d^{-1}} c\text{-}C_4F_6 + e \tag{2}$$

namely, electrons are captured by c-C_4F_6 molecules with a rate constant $k_c(T)$ forming transient c-$C_4F_6^{-*}$ anions which then are stabilized via collisions with, primarily, the buffer gas (N_2) molecules with a probability of stabilization per colission *p* yielding stable c-$C_4F_6^-$ anions. The captured electrons can then thermally autodetach from the stable anion [reaction (2)] with a detachment frequency t_d^{-1}.

EXPERIMENTAL TECHNIQUE

In Fig. 1 is shown a schematic of the experimental arrangement employed. The cell consisted of a six-way stainless-steel cube with one sapphire window and two electrical feedthroughs. The two parallel stainless steel electrodes were held at a distance of ~0.42 cm. The electron swarm was produced by a fast N_2 laser pulse (FWHM ~6×10^{-10} s) which strikes the cathode through a 0.1 cm-diameter hole in the anode electrode.

The cell was filled with small quantities of electron attaching gas c-C_4F_6 in a buffer gas of N_2. The N_2 gas number density, N_T, was varied from 1.61×10^{19} to 9.66×10^{19} molecules cm^{-3} and the attaching gas number density N_a was varied from 1.6×10^{13} to 64.4×10^{13} molecules cm^{-3}. Under the influence of an applied uniform electric field the electron swarm drifts towards the anode . The electron energy distribution $f(\varepsilon,E/N,T)$ is characteristic of N_2 since $N_a \ll N_T$. As the electrons drift a fraction of them is captured by the c-C_4F_6 molecules via reaction (1). Heat-activated detachment [reaction (2)] gives rise to delayed electrons which can be in the drift gap long after the initial (prompt) electrons have reached the anode. The induced electron current monitored as a function

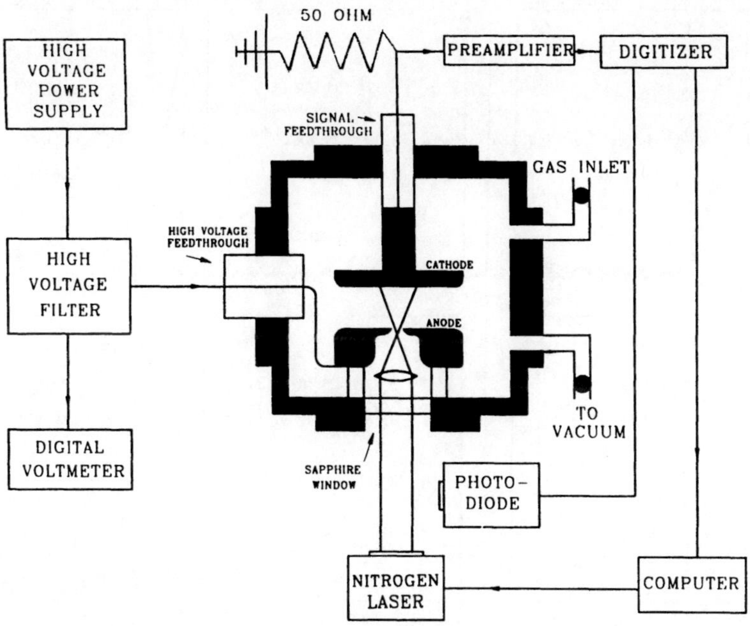

Figure 1. Schematic diagram of the experimental apparatus used in the present studies.

of time can be described by[11,12]

$$i_e(t) = \frac{e w_e}{d} \int_{w_i t}^{\min[w_e t,\, d]} \rho_e(x,t)\, dx \tag{3}$$

where $\rho_e(x,t)$ is the electron number density; w_e and w_i are the electron and ion drift velocities ($w_e \gg w_i$) and d is the drift distance. When the initial electrons are produced by a short duration laser pulse, *i.e.*, for initial conditions $\rho_e(x,0)=n_0\delta(x)$ the electron number density is given by[11-15]

$$\rho_e(x,t) = \frac{n_0}{w_e - w_i} \exp\left[-\frac{1}{t_a}\left(\frac{x-w_i t}{w_e - w_i}\right) - \frac{1}{t_d}\left(\frac{w_e t - x}{w_e - w_i}\right)\right] \times \left[\delta\left(\frac{w_e t - x}{w_e - w_i}\right) + \sqrt{\frac{1}{t_a t_d}\frac{x - w_i t}{w_e t - x}} \times I_1\left(\frac{2}{(w_e - w_i)}\sqrt{\frac{1}{t_a t_d}(w_e t - x)(x - w_i t)}\right)\right] \tag{4}$$

where t_a^{-1} and t_d^{-1} are, respectively, the electron attachment and detachment frequencies, n_0 is the initial number of electrons in the swarm (i.e. the number of electrons at $x=0$, $t=0$), and I_1 is the first order modified Bessel function.

The change in the electron current due to the removal (electron attachment) and gain (autodetachment) of electrons as T is increased and thus as the internal energy content of the neutral molecules and anions is increased can be viewed in the typical experimental waveforms shown in Fig. 2. The evolution of the autodetachment process and its effect on the electron current as T is increased is clearly seen for this gas in Fig. 2.

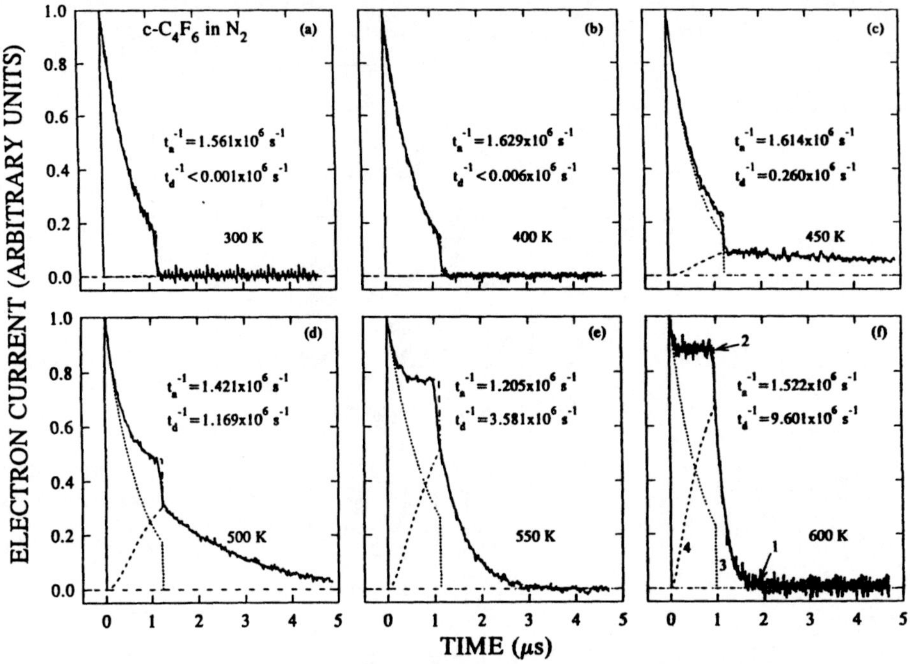

Figure 2. Electron current waveforms for c-C_4F_6 in N_2 at T=300, 400, 450, 500, 550, and 600 K. All waveforms are for E/N=1.24×10^{-17} V cm^2, $N_T = 6.44 \times 10^{19}$ molecules cm^{-3}, $N_a = 6.44 \times 10^{13}$ molecules cm^{-3}. The solid curves (–; curve 1 in Fig. 2f) are the experimentally measured total electron currents as a function of time; the dash-dot (-·-·-) curves (curve 2; Fig. 2f) are the calculated electron current waveforms for the t_a^{-1} and t_d^{-1} values obtained from the fitting of Eqn (3) (see the text and the values of t_a^{-1} and t_d^{-1} given in the figure) to curves 1; the dotted (···) curves (curve 3; Fig. 2f) represent the contribution to the electron current of the initial (prompt) electron swarm when only electron attachment occurs, and the broken (---) curves (curve 4; Fig. 2f) represent the contribution to the electron current from the autodetached electrons.

RESULTS AND DISCUSSION

Electron Attachment Rate Constant $k_a(<\varepsilon>,T)$

From waveforms such as those in Fig. 2 we determined the $t_a^{-1}(<\varepsilon>,T)$ and $t_d^{-1}(<\varepsilon>,T)$ simultaneously via a nonlinear least squares fit procedure to Eqn (3) for a number of mean electron energies and gas temperatures. The rate constant for electron attachment was then determined as $k_a = t_a^{-1}/N_a$ where N_a is the attaching gas number density. The $k_a(<\varepsilon>,T)$ for c-C_4F_6 was found to be independent of N_a at all T studied. A small increase in $k_a(<\varepsilon>,T)$ with increasing N_T was observed indicating a more efficient collisional stabilization of the transient c-$C_4F_6^{-*}$ anion at higher N_T. In order to account for this increase we determined the values of $\lim_{N_T \to \infty} [k_a(<\varepsilon>,T)]$ by plotting the measured $1/k_a$ at each N_T as a function of $1/N_T$ and extrapolating $1/N_T$ to 0, at each T. These values of $k_a(<\varepsilon>,T)$ are plotted in Fig. 3 as a function of $<\varepsilon>$ for a number of temperatures. It can be seen that the effect of T on $k_a(<\varepsilon>,T)$ is small.

Autodetachment Frequency $t_d^{-1}(<\varepsilon>,T)$

The gas temperature has a profound effect on reaction (2) for c-C_4F_6. This is clearly

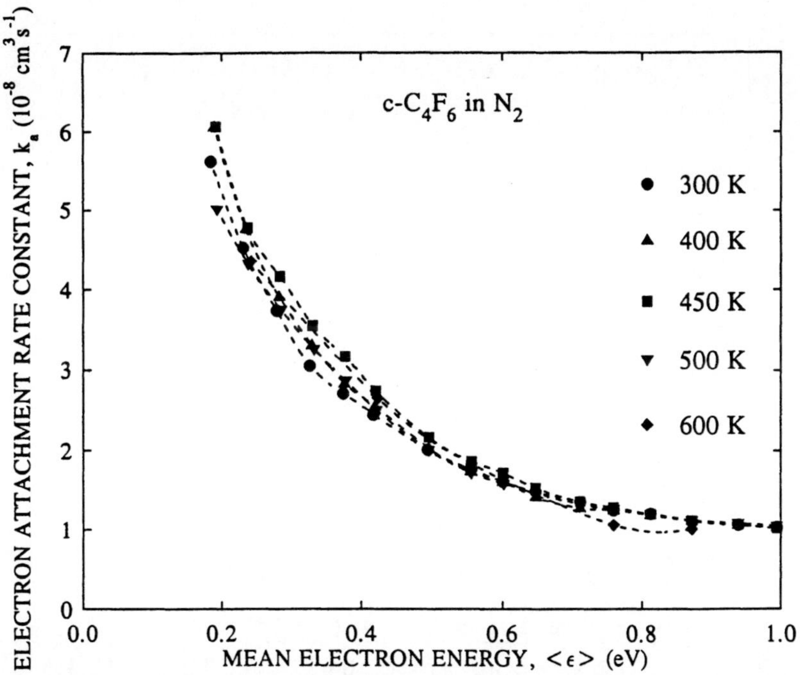

Figure 3. k_a vs. $<\varepsilon>$ for c-C_4F_6 at the indicated temperatures.

shown in Fig. 2 where the measured electron current waveforms change significantly with gas temperature. The t_d^{-1} we determined for c-$C_4F_6^-$ are plotted in Fig. 4 as a function of $<\varepsilon>$ for T= 450, 500, 550, and 600 K. Below 450 K the electron detachment is very small and only an upper limit can be placed on t_d^{-1}; t_d^{-1} is $< 10^3$ s^{-1} at T=300 K and $<$ 6×10^3 s^{-1} at T=400 K. For the N_2 gas number densities used, the c-$C_4F_6^{-*}$ anions undergo a large number of collisions within the lifetime of the "isolated" c-$C_4F_6^{-*}$ anions[10,12]. We can then assume that the preponderance of the c-$C_4F_6^-$ ions are in their lowest vibrational/rotational energy state allowed at each T when the electron is thermally detached from them. At each T employed we estimated the vibrational energy, $<\varepsilon>_{int}(T)$ of the c-C_4F_6 molecules in excess of the zero-point energy using the vibrational frequencies of Ref. 16. We then assumed that the total internal energy content for the c-$C_4F_6^-$ anions is the same as for the neutral c-C_4F_6 molecule and plotted (Fig. 5) t_d^{-1} as a function of $<\varepsilon>_{int}$. It can be seen that $t_d^{-1}(<\varepsilon>_{int})$ increases by more than three orders of magnitude as $<\varepsilon>_{int}$ changes from ~0.16 to ~0.56 eV; $t_d^{-1} < 10^3$ s^{-1} at $<\varepsilon>$ ~0.16 eV (see Fig. 2).

Dependence of the Anion Autodetachment on the Molecular Electron Affinity

If we assume that all the c-$C_4F_6^-$ ions are in their lowest vibrational/rotational state allowed at each T and that when they autodetach they reach the same energy state for the neutral molecule, then t_d^{-1} may be related to T by

$$t_d^{-1} = t_{d0}^{-1} e^{-E^*/kT} \tag{5}$$

where E^* is the activation energy of process (2) and t_{d0}^{-1} is the autodetachment frequency when T→∞. When our measured autodetachment frequencies are plotted as $\ln(t_d^{-1})$ vs. 1/T a straight line is obtained (Fig. 6) from which we estimated E^* to be 0.237 eV. A similar

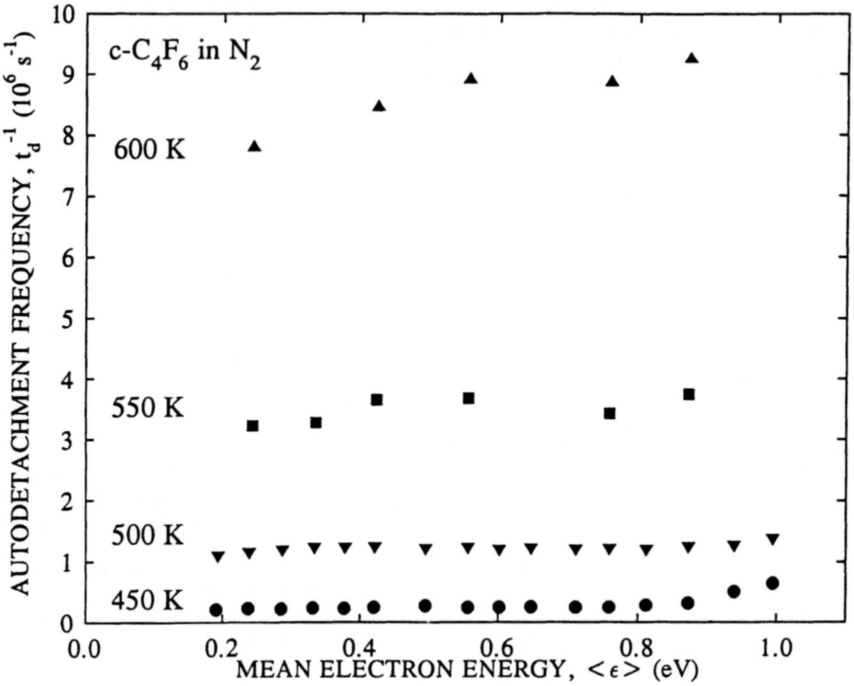

Figure 4. $t_d^{-1}(<\varepsilon>,T)$ vs. $<\varepsilon>$ for c-$C_4F_6^-$ at the indicated temperatures.

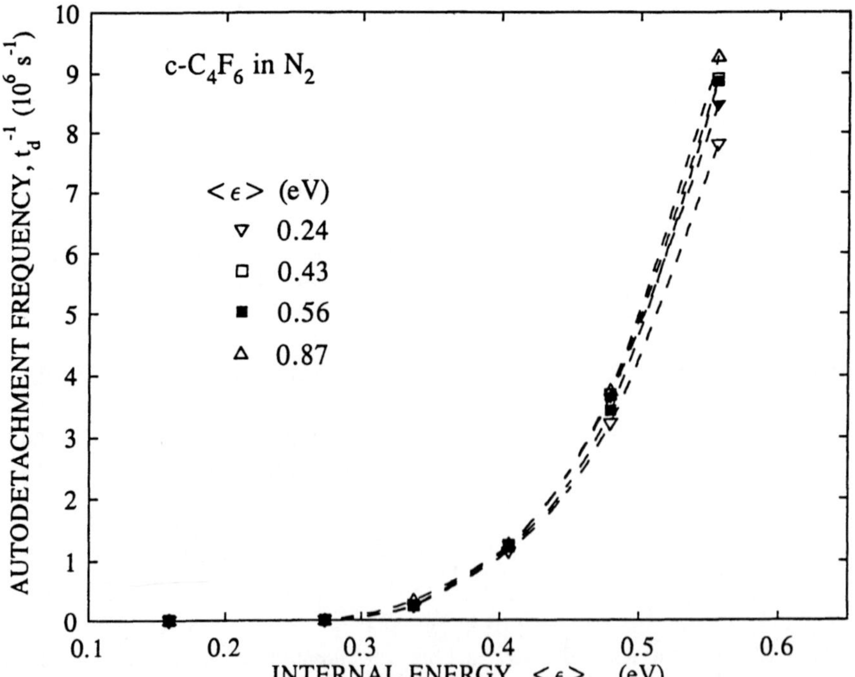

Figure 5. t_d^{-1} vs. $<\varepsilon>_{int}$ for c-$C_4F_6^-$ for four values of $<\varepsilon>$.

study[11] for $C_6F_6^-$ gave a value for E^* equal to 0.477 eV. If we take E^* to be an estimate of the electron affinity (EA) of the molecule, then $E^*(C_6F_6^-)$ compares well with the reported[17] $EA(C_6F_6)$ of 0.52 eV. There are no reported values for the EA(c-C_4F_6) but we can take E^*(c-$C_4F_6^-$)=0.24 eV to be a lower estimate. The large electron detachment can then be understood on the basis of the small values of EA for these molecules since the smaller the E^* the larger the exponential term in Eqn (5). Consistent with this, for SF_6 for which the EA is 1.05 eV[18] we observed[13] no electron detachment from SF_6^- below 600 K.

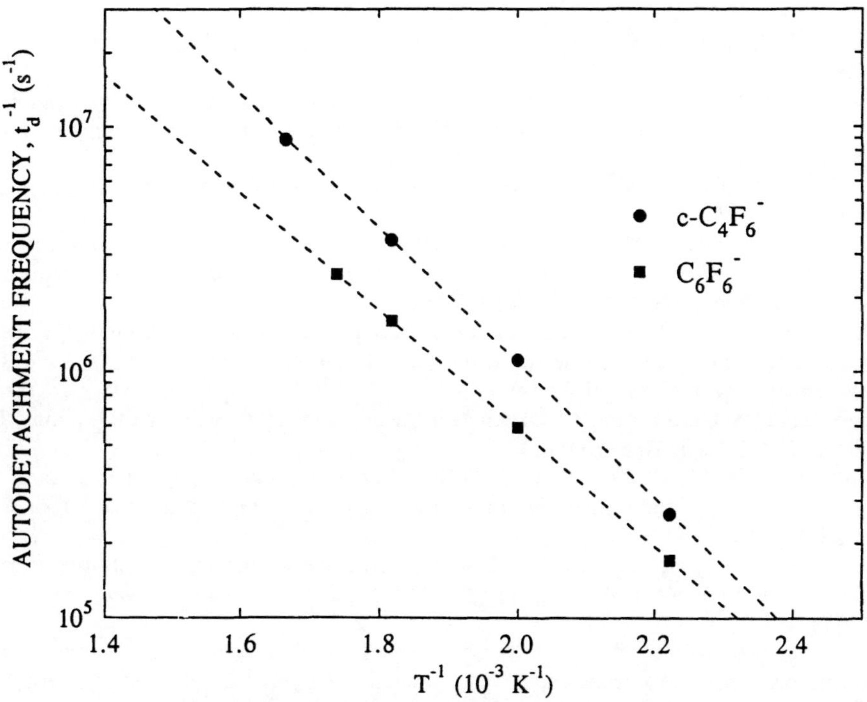

Figure 6. t_d^{-1} vs. 1/T for c-$C_4F_6^-$ and $C_6F_6^-$.

Dependence of the Dielectric Strength on Electron Attachment and Detachment

Earlier studies[8] have shown that the $(E/N)_{lim}$ for electronegative gases which at low electron energies form parent negative ions decreases with increasing gas temperature and for molecules that capture electrons only dissociatively the $(E/N)_{lim}$ increases with increasing T. For example, the $(E/N)_{lim}$ for c-C_4F_8 declined by as much as 20% as T was raised from 297 to 573 K and the $(E/N)_{lim}$ for $CClF_3$ increased by ~ 3% in the T-range 294 to 573 K[8]. A similar but larger increase in $(E/N)_{lim}$ (~14%) was recently observed[12] for SF_6 which at low electron energies forms predominantly parent negative ions and at higher energies only fragment negative ions. The decrease in $(E/N)_{lim}$ with increasing T is related to increases in the heat-activated electron detachment which is likely for dielectric gases which capture electrons forming parent anions while the increase in the dielectric strength with increasing gas temperature follows the T-dependence of the dissociative electron attachment.

ACKNOWLEDGEMENT

This work was sponsored by the Wright Laboratory, U.S. Department of the Air Force, under contract No. AF 33615-92-C-2221 with the University of Tennessee and by the Office of Health and Environmental Research, U.S. Department of Energy under contract No. DE-AC05-84OR21400 with Martin Marietta Energy Systems, Inc.

REFERENCES

1. J. M. Meek and J. D. Craggs (eds.), "*Electrical Breakdown in Gases*", Wiley, New York (1978).
2. L. G. Christophorou and S. R. Hunter, From basic reseacrh to application, in: "*Electron-Molecule Interactions and Their Applications*", L. G. Christophorou (ed.), Academic Press, Orlando, Vol 2, Chap. 5 (1984).
3. L. G. Christophorou and L. A. Pinnaduwage, Basic physics of gaseous dielectrics, *IEEE Trans. Electr. Insul.* 25:55 (1990).
4. L. G. Christophorou, D. L. McCorkle, and A. A. Christodoulides, Electron attachment processes, in: "*Electron-Molecule Interactions and Their Applications*", L. G. Christophorou (ed.), Academic Press, Orlando, Vol 1, Chap. 6, (1984).
5. Pinnekamp and N. Wiegart, Breakdown characteristics of inhomogeneous gaps in high pressure SF_6, in: "*Gaseous Dielectrics IV*", L. G. Christophorou and M. O. Pace (Eds.) Pergamon, New York, pp 91 (1984).
6. N. Wiegart, A model for the production of initial electrons by detachment of SF_6^- ions, *IEEE Trans. Electr. Insul.* 20:587 (1985).
7. J. K. Olthoff and R. J. Van Brunt, Collisional detachment and decomposition rates of SF_6^-, SF_5^- and F^- in SF_6: Implications for ion transport and electrical discharges, *J. Chem. Phys.* 91:2261 (1989).
8. L. G. Christophorou, R. A. Mathis, S. R. Hunter, and J. G. Carter, Effect of temperature on the uniform field breakdown strength of electronegative gases, *J. Appl. Phys.* 63:52 (1988).
9. A. A. Christodoulides, L. G. Christophorou, R. Y. Pai, and C. M. Tung, Electron attachment to perfluorocarbon compounds. I. c-C_4F_6, 2-C_4F_6, 1,2-C_4F_6, C_4F_8 and 2-C_4F_8, *J. Chem. Phys.* 70:1156 (1979).
10. I. Sauers, L. G. Christophorou, and J. G. Carter, Electron attachment to perfluorocarbon compounds. III. Fragmentation of aliphatic perfluorocarbons of interest to gaseous dielectrics, *J. Chem. Phys.* 71:3016 (1979).
11. P. G. Datskos, L. G. Christophorou, and J. G. Carter, Temperature-enhanced electron detachment from $C_6F_6^-$, *J. Chem. Phys.* 98:7875 (1993).
12. P. G. Datskos, L. G. Christophorou, and J. G. Carter, Temperature dependence of electron attachment and detachment in SF_6 and c-C_4F_6, *J. Chem. Phys.* 99:8607 (1993).
13. C. Wen and J. M. Wetzer, Electron avalanches influenced by detachment and conversion processes, *IEEE Trans. Electr. Insul.* 23:999 (1988).
14. C. Wen and J. M. Wetzer, Time-resolved avalanche waveforms in octafluorocyclobutane, *IEEE Trans. Electr. Insul.* 24:143 (1989).
15. T. H. Teich, Detachment of electrons from negative ions in electrical discharges, in: *Gaseous Dielectrics VI*, L. G. Christophorou and I. Sauers (Eds.), Plenum Press, New York p. 215 (1991).
16. J. R. Nielsen, M. Z. El-Sabban, and M. Alpert, Vibrational spectra of cyclobutforene (hexafluorobutene), *J. Chem. Phys.* 23:324 (1955).
17. S. Chowdhury, E. P. Grimsrud, T. Heinis, and P. Kebarle, Electron affinities of perfluorobenzene and perfuorophenyl compounds, *J. Am. Chem. Soc.* 108:3630 (1986).
18. E. P. Grimsrud, S. Chowdhury, and P. Kebarle, Electron affinity of SF_6 and perfluoromethylcyclohexane. The unusual kinetics of electron transfer reactions $A^- + B = A + B^-$ where $A = SF_6$ or perfluorinated cyclo-alkanes, *J. Chem. Phys.* 83:1059 (1985).

DISCUSSION

R. NAGPAL: How stable is c-C_4F_6 towards the increase in temperature from 300 to 600 K?

P. G. DATSKOS: It seems to be very stable. We checked this by measuring the electron attachment rate constant as a function of time and found no time dependence. Furthermore, the electron current waveforms recorded ~ 2 hours after the introduction of the mixture in the hot chamber were identical to those recorded right after the mixture was introduced.

E. ILLENBERGER: Is the activation energy for delayed autodetachment identical to the electron affinity of the respective molecule?

P. G. DATSKOS: We do not think it is identical to the electron affinity, because the exponential dependence $(\sim e^{-E^*/kT})$ of t^{-1}_d is based on the assumption that when the electron is detached, it reaches the same energy state. However, the activation energy should be at least a lower limit to the electron affinity.

E. KUFFEL: The measurements you described were conducted with N_2 as the buffer gas with a small admixture of attaching gas. What would the attachment be in practical mixtures containing approximately equal proportions of buffer and attaching gases?

P. G. DATSKOS: The electron attachment cross section (or rate constant) for c-C_4F_6 and SF_6 is independent of the attaching gas pressure. However, for a given E/N, the mean electron energy will depend on the relative concentration of attaching gas and buffer gas. For the same mean electron energy, the rate constant will be the same and independent of the attaching gas/buffer gas concentration.

ATTACHMENT OF LOW ENERGY ELECTRONS TO "HOT" SF_6 MOLECULES

P. G. Datskos[#], L. G. Christophorou[#], and J. G. Carter

Atomic, Molecular, and High Voltage Physics Group
Health Sciences Research Division
Oak Ridge National Laboratory
Oak Ridge, Tennessee 37831-6122, U.S.A.

ABSTRACT

The low energy electron attachment rate constant, k_a, for SF_6 has been measured in dilute mixtures of SF_6 with N_2 buffer gas in the temperature, T, range of 300 to 550 K and over a range of E/N values corresponding to mean electron energies, $<\varepsilon>$, ranging from 0.19 to 1.0 eV. As the gas temperature increases, the magnitude and energy dependence of $k_a(<\varepsilon>,T)$ remain unaffected. In the same T-range the collisionally stabilized SF_6^- is very stable with respect to autodetachment; its lifetime is > 1 ms. The limiting electric field strength $(E/N)_{lim}$ for SF_6 was found to increase by ~14% as T is raised from 300 to 600 K. The significance of these findings in gaseous dielectrics is indicated.

INTRODUCTION

Electron attachment to sulfur hexafluoride (SF_6) has drawn large interest over the past several years. The large electron attachment cross section[1-5] for SF_6 due to the formation of parent and fragment negative ions is primarily responsible for its high dielectric strength (compared to air) and for its use in several technologies. For example, SF_6 has been extensively used as a high dielectric strength medium[1], as a fluorine donor in excimer lasers[6] and in semiconductor technologies in plasma etching of GaAs layers[7] and silicon wafers[8].

Electron beam studies[1,4,5,9] have shown that at room temperature SF_6 attaches low energy electrons very efficiently forming SF_6^{-*} parent anions and also fragment anions such

---

[#]Also at The Department of Physics and Astronomy, The University of Tennessee, Knoxville, Tennessee 37996-1200.

Gaseous Dielectrics VII, Edited by L.G. Christophorou
and D.R. James, Plenum Press, New York, 1994

as F^-, F_2^-, SF_2^-, SF_3^-, SF_4^-, and SF_5^-. The "isolated" SF_6^{-*} parent anion is long-lived with an autodetachment lifetime $\tau_a > 1$ μs[10]. For gas temperatures above ambient electron beam studies[4] have shown that the cross section for forming the fragment anion SF_5^-, resulting from SF_6, increases by a few orders of magnitude when T is increased from 300 to 880 K. Another electron beam study[2] has shown that the energy-integrated total electron attachment cross section remains unaffected by increases in T from 300 to 1200 K.

Electron swarm studies[5] have shown that at room temperature the thermal electron attachment rate constant is very large $[(k_a)_{th} \sim 2.3 \times 10^{-7} cm^3\ s^{-1}]$ and that $k_a(<\varepsilon>,T)$ decreases monotonically with increasing mean electron energy. For T higher than 300 K other swarm studies[3,11] have shown that $(k_a)_{th}$ is virtually independent of T.

In this paper we report on the effect of temperature on the electron attachment rate constant, $k_a(<\varepsilon>,T)$, for SF_6 in a buffer gas of N_2 for mean electron energies below 1 eV. Changes in the gas temperature result in changes in the internal energy of a molecule which in turn can affect its dielectric properties. The limiting electric field strength $(E/N)_{lim}$ for SF_6 was measured as a function of T and found to increase with increasing T.

EXPERIMENTAL TECHNIQUE

The experimental technique we employed has been described in detail elsewhere[12, 13]. Briefly, photoelectrons generated by a short laser pulse (FWHM ~ 6×10^{-10} s) at the cathode of a parallel-plate electrode arrangement are depleted by attachment to SF_6 molecules mixed with N_2 buffer gas in the interelectrode region as they drift to the anode under an external applied electric field. The SF_6 gas number density, N_a, was varied from 1.93×10^{13} to 16.1×10^{13} molecules cm^{-3} and the N_2 gas number density, N_T, from 1.61×10^{19} to 9.66×10^{19} molecules cm^{-3}.

We recorded the induced electron current due to the motion of the electrons in the gap as a function of time. The depletion of electrons due to electron attachment and the gain due to electron detachment from the collisionally stabilized SF_6^- ions influences the magnitude and time dependence of the electron current and an analysis[12,13] of such electron current wave forms yields simultaneously the electron attachment $t_a^{-1}(<\varepsilon>,T)$ and electron detachment $t_d^{-1}(<\varepsilon>,T)$ frequencies. The evolution of the electron current as T is increased from 300 to 550 K can be viewed from the typical wave forms in Fig. 1.

RESULTS AND DISCUSSION

Electron Attachment Rate Constant $k_a(<\varepsilon>,T)$ for SF_6

From wave forms such as those in Fig. 1 we determined the $t_a^{-1}(<\varepsilon>,T)$ and $t_d^{-1}(<\varepsilon>,T)$ simultaneously via a nonlinear least squares fit procedure[12,13] for a number of E/N values (corresponding to $<\varepsilon>$ in the range 0.19 to 1.0 eV) and T (300-600 K). The rate constant for electron attachment to SF_6 was then determined as $k_a(<\varepsilon>,T)=t_a^{-1}(<\varepsilon>,T)/N_a$. The $k_a(<\varepsilon>,T)$ was found to be independent of N_a and N_T at all T studied. At the N_T employed the time between collisions of SF_6^{-*} and N_2 is less than 10^{-10} s (Ref 13). Our measured values of $k_a(<\varepsilon>,T)$ are plotted in Fig. 2 as a function of $<\varepsilon>$ for all the temperatures studied. It can be seen that the effect of T on $k_a(<\varepsilon>,T)$ is negligible.

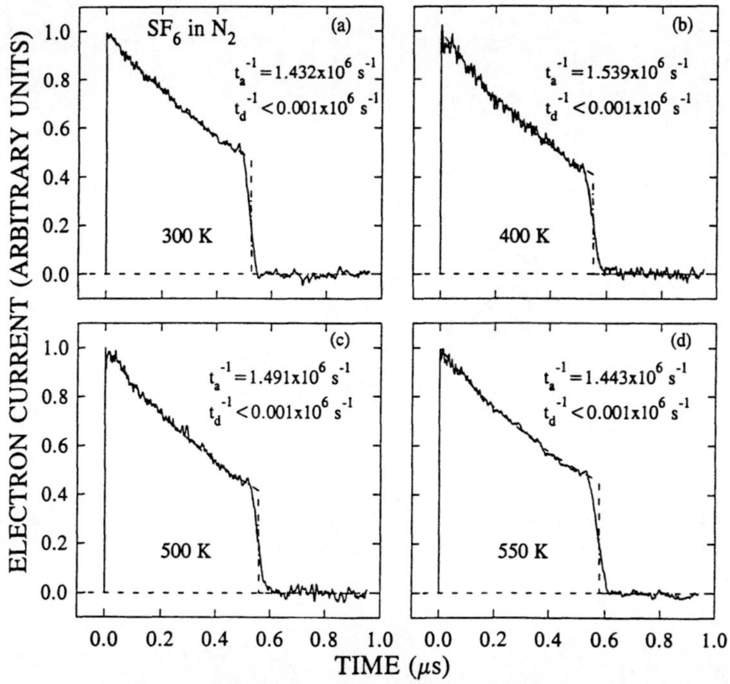

Figure 1. Electron current wave forms for SF_6 in N_2 at T=300, 400, 500, and 550 K. All wave forms are for $E/N=3.10x10^{-17}$ V cm^2, $N_T=6.44x10^{19}$ molecules cm^{-3}, and $N_a=9.66x10^{13}$ molecules cm^{-3}. The solid curves are the experimentally measured total electron current wave forms; the dash-dot (-·-·-) curves are the calculated electron current wave forms using the $t_a^{-1}(<\varepsilon>,T)$ and $t_d^{-1}(<\varepsilon>,T)$ values obtained from the analysis (see Refs 12 and 13 for an extensive discussion) and shown in the figure. Note that due to the negligible electron detachment from SF_6^- only an upper limit on $t_d^{-1}(T)$ was obtained.

Autodetachment Frequency $t_d^{-1}(<\varepsilon>,T)$ for SF_6^-

The stabilized SF_6^- ion was found to be very stable with respect to autdodetachment for T ≤ 550 K. This is clearly demonstrated by the wave forms in Fig. 1. The electron current is due to only the prompt electrons and there is no evidence for any contribution from thermally autodetached (from SF_6^-) electrons. In Fig. 3 are shown the electron current wave forms at T=550 K for concentrations of SF_6 ranging from $3.22x10^{13}$ to $16.10x10^{13}$ molecules cm^{-3}. Even for the largest N_a we employed, there is still no noticeable contribution to the total electron current from any delayed electrons. These measurements put an upper limit to $t_d^{-1}(T)$ for SF_6^-; $t_d^{-1}(T)$ is less than 10^3 s^{-1} at T=550 K. This corresponds to a lifetime t_d of SF_6^- of more than 10^{-3} s. This is the lifetime for SF_6^- to autodetach thermally from its lowest allowable, at each T, energy state (which should not be confused with the lifetime of the "isolated" SF_6^{-*} anion measured in the electron beam studies).

The negligible electron detachment from SF_6^- can be understood on the basis of its large electron affinity (EA). While reported EA values for this molecule vary from ~0.5 to ~1.5 eV (Ref 14) the currently accepted value of the EA(SF_6) is 1.05 eV (Ref 15).

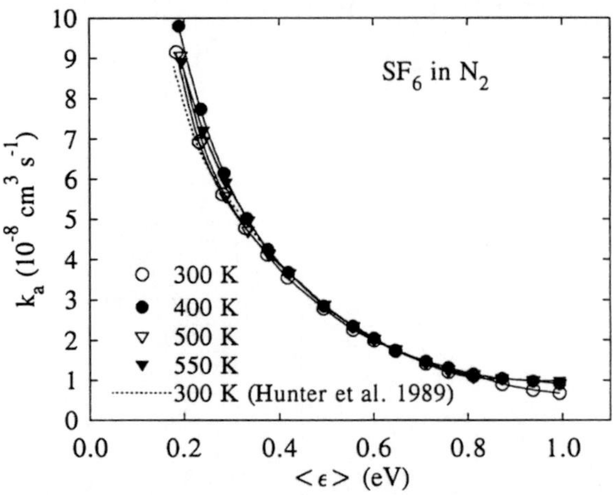

Figure 2. $k_a(<\varepsilon>,T)$ vs. $<\varepsilon>$ for SF_6 at the indicated temperatures. The broken curve is the total electron attachment rate constant reported in Ref 5.

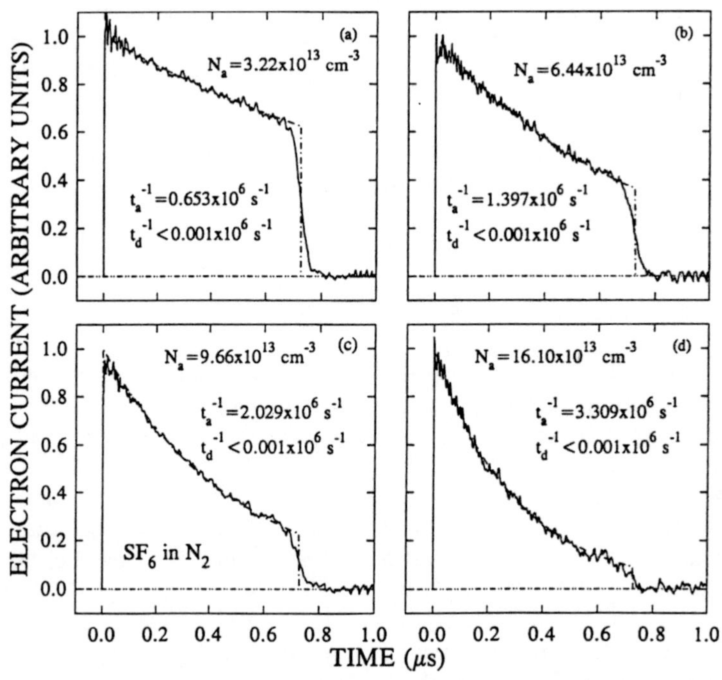

Figure 3. Electron current wave forms for SF_6 in N_2 for SF_6 concentrations $N_a=3.22\times10^{13}$, 6.44×10^{13}, 9.66×10^{13}, and 16.1×10^{13} molecules cm^{-3}. The wave forms are for $T=550$ K, $E/N=2.17\times10^{-17}$ V cm^2, and $N_T=6.44\times10^{19}$ molecules cm^{-3}. In each case the solid curve (—) is the experimentally measured total electron current wave form; the dash-dot (-·-·-) curve is the calculated electron current wave form for the $t_a^{-1}(<\varepsilon>,T)$ and $t_d^{-1}(<\varepsilon>,T)$ values obtained from the analysis[12,13] and listed in the figure (see the text).

Limiting electric field strength $(E/N)_{lim}$ of SF_6 as a function of T

From our experimental current wave forms as a function of E/N we were able to determine the "limiting breakdown field strength", $(E/N)_{lim}$, for pure SF_6 as a function of gas temperature. We define the $(E/N)_{lim}$ as the maximum E/N value -- at each temperature -- for which the number of electrons reaching the anode electrode is the same as the number of photoelectrons initially ejected from the cathode into the drift region. That is, the net gain (loss) of electronic charge during the drift time is zero and thus electrical breakdown is not possible. For the case of an electronegative gas for which no electron detachment processes are present this represents the situation for which the electron attachment frequency, $t_a^{-1}[(E/N)_{lim}]$, equals the electron ionization frequency, $t_i^{-1}[(E/N)_{lim}]$.

Earlier studies[16] have shown that the $(E/N)_{lim}$ for electronegative gases which at low electron energies form predominantly parent negative ions decreases with increasing gas temperature and for molecules that capture electrons predominantly dissociatively the $(E/N)_{lim}$ increases with increasing T. The decrease in $(E/N)_{lim}$ with increasing T is related to increases in the heat-activated electron detachment which is likely for dielectric gas molecules which capture electrons forming parent anions; the increase in the dielectric strength with increasing gas temperature follows the T-dependence of the dissociative electron attachment.

We determined the $(E/N)_{lim}$ for pure SF_6 for which we observed no electron detachment under our experimental conditions (see Figs 1 and 2) from the measured total current wave forms such as those in Fig. 4. In Fig. 4 the initial electron current is designated by I_1 and the total current at the end of the electron drift time is designated by I_2; I_2 is made up of two components, one due to ions (I_3) and another due to electrons (I_2-I_3). From a determination of I_1 and (I_2-I_3) we calculated the quantity $R=[(I_2-I_3)-I_1]/I_1$ and plotted it as a function of E/N. We took the E/N value for which $R=0$ to determine the value of $(E/N)_{lim}$. In Fig. 5 we plot R vs E/N for two temperatures (300 and 550 K). The lines through the experimental points are linear least squares fits to these points and the intersection of the fitted line with the E/N-axis is the $(E/N)_{lim}$ value for that temperature.

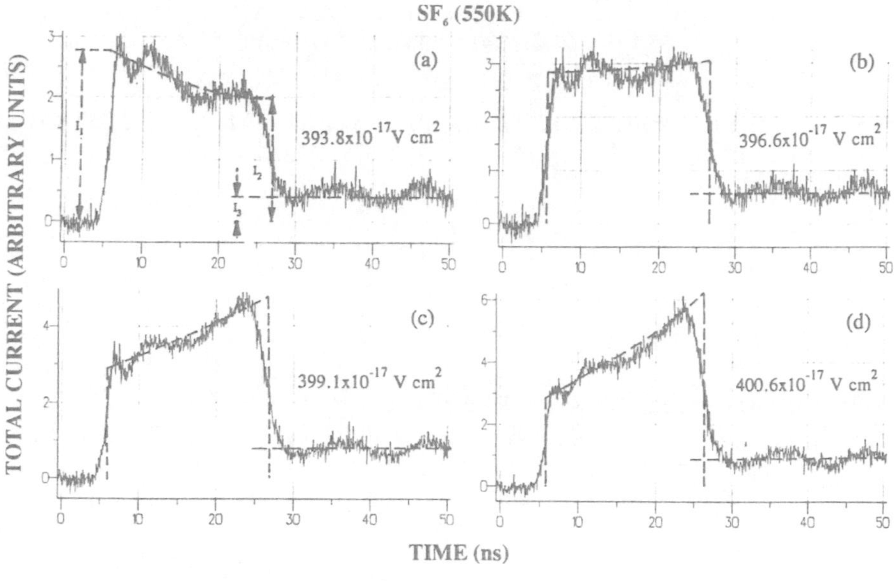

Figure 4. Total current wave forms for SF_6 for four values of E/N around the $(E/N)_{lim}$ at fixed T (550 K) (see the text).

The $(E/N)_{lim}$ values we determined in this manner for SF_6 at 300, 400, 500, 550, and 600 K are plotted in Fig. 6. The room temperature value of $(E/N)_{lim}$ determined in this study agrees very well with that reported earlier[17]. From Fig. 6 it can be seen that the $(E/N)_{lim}$ for SF_6 increases monotonically with increasing T. This we attribute to the increase in the dissociative electron attachment to SF_6 with increasing T (which is the dominant electron attachment process at high E/N). The effect of the presence of any electron detachment processes (autodetachment or collisional detachment) would tend to decrease the value of $(E/N)_{lim}$ for a fixed T since there would be more electrons ejected back into the medium released from the negative ions. In this investigation for all the temperatures studied and for E/N values up to $(E/N)_{lim}$ there was no evidence of electron detachment occurring in SF_6.

It is also interesting to note that for temperatures higher than 500 K we observed loss of SF_6 molecules from the chamber. This was probably due to some reaction[18] of SF_6 with the hot copper gaskets we used in the cell forming CuF_2. This problem was eliminated when we used gold plated copper gaskets to seal the cell.

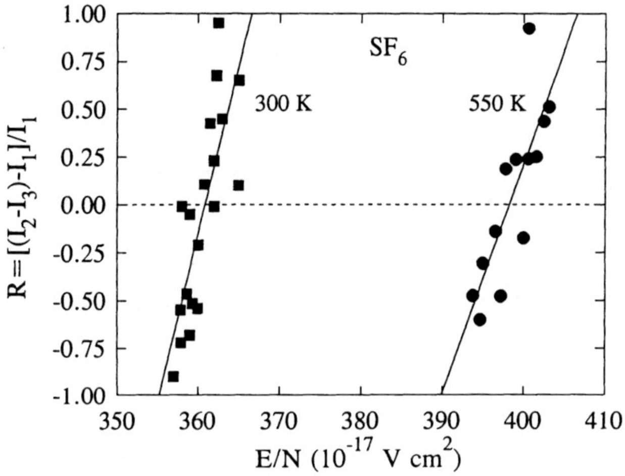

Figure 5. The ratio $R=[(I_2-I_3)-I_1]/I_1$ for SF_6 as a function of E/N for 300 and 550 K (see the text).

ACKNOWLEDGEMENT

This work was sponsored by the Wright Laboratory, U.S. Department of the Air Force, under contract No. AF 33615-92-C-2221 with the University of Tennessee and by the Office of Health and Environmental Research, U.S. Department of Energy under contract No. DE-AC05-84OR21400 with Martin Marietta Energy Systems, Inc.

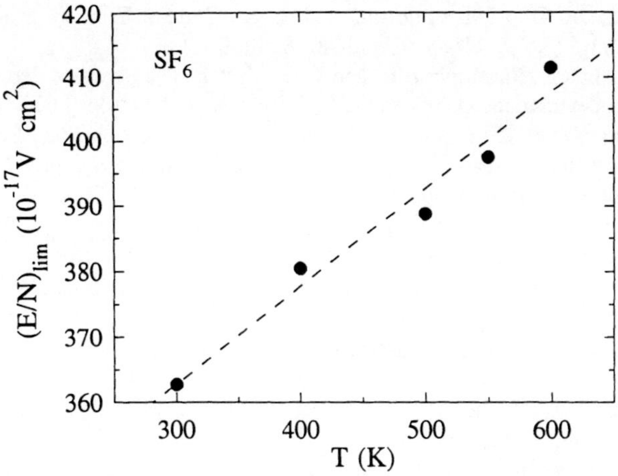

Figure 6. $(E/N)_{lim}$ for SF_6 as a function of T. The dashed line through the experimental points is a linear least squares fit to the data.

REFERENCES

1. L. G. Christophorou and S. R. Hunter, From basic research to application, in: "*Electron-Molecule Interactions and Their Applications*", L. G. Christophorou (ed.), Academic Press, Orlando, Vol 2, Chap. 5 (1984).
2. D. Spence and G. J. Schulz, Temperature dependence of electron attachment at low energies for polyatomic molecules, *J. Chem. Phys.*, 58:1800 (1973).
3. Z. Lj Petrović, R. W. Crompton, Thermal-electron attachment to SF_6 at room temperature and 500 K, *J. Phys. B: At. Mol. Phys.*, 17:2777 (1985).
4. C. L. Chen and P. J. Chantry, Photon-enhanced dissociative electron attachment in SF_6 and its isotopic selectivity, *J. Chem. Phys.*, 71:3897 (1979).
5. S. R. Hunter, J. G. Carter, and J. G. Christophorou, Low energy electron attachment to SF_6 in N_2, Ar, and Xe buffer gases, *J. Chem. Phys.*, 90:4879 (1989).
6. F. Kannari, H. Inagaki, and M. Obara, High efficiency multikilojoule HF chemical lasers using an electron beam initiated low-pressure mixture of H_2/F_2,NF_3 or H_2/F_2,SF_6, *Appl. Phys. Lett.*, 48:266 (1986).
7. S. Salimian, C. B. Cooper, III, R. Norton, and J. Bacon, Reactive ion etch process with highly controllable GaAs-to-AlGaAs selectivity using SF_6 and $SiCl_4$, *Appl. Phys. Lett.*, 51:1083 (1987).
8. W. W. Brandt and T. Honda, Mass spectrometric transient study of dc plasma etching of Si in SF_6/O_2 mixtures, *J. Appl. Phys.*, 60:1595 (1986).
9. M. Fenzlaff, R. Gerhard, and E. Illenberger, Associative and dissociative electron attachment to SF_6 and SF_5Cl, *J. Chem. Phys.*, 71:3016 (1979).
10. L. G. Christophorou, D. L. McCorkle, and A. A. Christodoulides, Electron attachment processes, in: "*Electron-Molecule Interactions and Their Applications*", L. G. Christophorou (ed.), Academic Press, Orlando, Vol 1, Chap. 6, (1984).
11. F. C. Fehsenfeld, Electron attachment to SF_6, *J. Chem. Phys.*, 53:2000 (1970).

12. P. G. Datskos, L. G. Christophorou, and J. G. Carter, Temperature-enhanced electron detachment from $C_6F_6^-$, *J. Chem. Phys.* 98:7875 (1993).
13. P. G. Datskos, L. G. Christophorou, and J. G. Carter, Temperature dependence of electron attachment and detachment in SF_6 and c-C_4F_6, *J. Chem. Phys.* 99:8607 (1993).
14. A. A. Christodoulides, D. L. McCorkle, and L. G. Christophorou, Electron affinities of atoms, molecules and radicals, in: "*Electron-Molecule Interactions and Their Applications*", L. G. Christophorou (ed.), Academic Press, Orlando, Vol 2, Chap. 6, (1984).
15. E. P. Grimsrud, S. Chowdhury, and P. Kebarle, Electron affinity of SF_6 and perfluoromethylcyclohexane. The unusual kinetics of electron transfer reactions $A^- + B = A + B^-$ where $A = SF_6$ or perfluorinated cyclo-alkanes, *J. Chem. Phys.* 83:1059 (1985).
16. L. G. Christophorou, R. A. Mathis, S. R. Hunter, and J. G. Carter, Effect of temperature on the uniform field breakdown strength of electronegative gases, *J. Appl. Phys.* 63:52 (1988).
17. L. G. Christophorou and L. A. Pinnaduwage, Basic physics of gaseous dielectrics, *IEEE Trans. Electr. Insul.* 25, 55 (1990).
18. See discussion at the end of: L. G. Christophorou, R. A. Mathis, S. R. Hunter, and J. G. Carter, Effect of temperature on the uniform field breakdown strength of electronegative gases, in: "*Gaseous Dielectrics V*", L. G. Christophorou and D. W. Bouldin (Eds), Pergamon Press, New York, pp 88 (1988).

MEASUREMENTS OF THE APPARENT IONIZATION COEFFICIENT FOR SYNTHETIC MIXTURES OF DECOMPOSITION GASES FROM POLYMERIC INSULATING MATERIALS

Daniel L. Schweickart and Alan Garscadden

Wright Laboratory
Aero Propulsion and Power Directorate
Wright-Patterson Air Force Base, OH 45433

INTRODUCTION

Realistic issues must be considered when solid electrical insulation is used in a low pressure environment. An important one from a reliability viewpoint is the physical basis for the electrical breakdown mechanisms in the gaseous medium, which inevitably accompanies the solid insulation. Whether the volume of the breakdown is macroscopic or microscopic, the Townsend mechanism can be dominant if the geometry and surface conditions are appropriate. In other cases, surface quenching or enhancements can modify the ionization growth.

The gases investigated are related to a range of polymers typically used in a low pressure (or aerospace) environment. They were not limited to any specific insulating material. The selection process was accomplished by surveying the gaseous decomposition products reported to evolve from a number of polymers in published pyrolysis studies. Relevant gas mixtures were then synthesized from pure molecular gas species and characterized in a series of experiments.

To further the objective of understanding the electrical failure modes of gases which evolve from insulation at low pressures, a methodology was established to enable the measurement of the apparent ionization coefficients. An original version of a pulsed-Townsend drift tube was designed, constructed and instrumented to this end.[1] The transport and growth parameters were measured in a broad range of synthesized binary mixtures chosen from the three molecular gases CO_2, CO and N_2O, as well as in the pure molecular species. In general, these measurements were made within the E/n range from 200 to 500 townsends, i.e. above the critical E/n value.

MEASUREMENT TECHNIQUE

This investigation relies on measurement of the spatial development of charge growth rather than curve fitting to temporal waveforms. In this technique a charge sensitive preamp integrates the total charge growth as a photo-initiated electron swarm traverses the space from the cathode to the anode. The anode, at a fixed distance from the cathode, acts as a charge collector for the external circuit. Charge growth depends upon ionizing collisions and attachment between electrons and molecules as the swarm crosses the gap.

This pulsed, charge growth technique incorporating total charge measurement works best when the apparent ionization coefficient λ is large and positive, i.e. $\lambda = (\alpha - \eta) > 0$. This condition is consistent with the focus of this investigation, which is primarily the range

Gaseous Dielectrics VII, Edited by L.G. Christophorou and D.R. James, Plenum Press, New York, 1994

of E/n for incipient breakdown in species with minimal attachment. (While the molecular gases used in this investigation are not considered highly attaching, they have been shown to exhibit some electronegative behavior: particularly CO_2 [2,3], and N_2O [4,5].) The measurement is based on the logarithmic relationship given by

$$\lambda = \frac{1}{g} \ln\left(\frac{Q}{Q_0}\right) \tag{1}$$

where Q is the total charge collected per photoemission event at a given E/n value, Q_0 is the total initial charge and g is the gap spacing. It should be noted that Q is measured only over E/n ranges for which secondary electron emission from the anode surface is negligible. (This is accomplished in part by a special meshed-anode design that helps to inhibit impact stimulated secondary electron emission by "trapping" any such secondary electrons in a zero field region.[6])

APPARENT IONIZATION COEFFICIENT DATA FOR PURE GASES

Before comparing the various mixture ratios of the three binary mixtures, it is useful to compare the measured ionization coefficient values for the pure gases (CO_2, CO, N_2O) to empirical values published in the literature. All measurements were made at a pressure of 1 torr with fill gases of at least 99.99% specified purity. At this pressure, the apparent

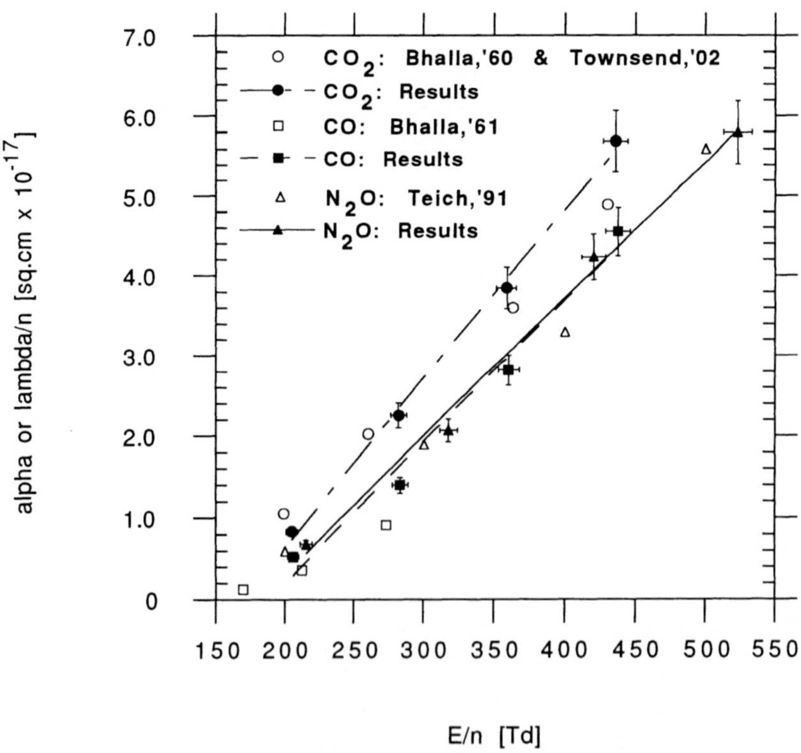

Fig. 1. Ionization Coefficients for CO_2, CO and N_2O with Random Errors

ionization coefficient data was obtained from about 200 to 500 townsends for the selected gases and mixtures. Over this limited range of E/n for these gases, the ionization coefficients are essentially linear with E/n.

The results shown in Fig. 1 are a comparison of published, empirical data for CO_2, CO and N_2O to the values obtained in these measurements. The vertical and horizontal error bars shown are the calculated random errors for the dependent and independent parameters, respectively, of our data. (The worst case [Δx,Δy] errors of [2.0%, 6.7%] are approximately the same for the binary mixtures discussed later.) The vertical axis of Fig. 1 has a dual label to recognize the fact that λ reduces to α, when attachment is lacking. As was discussed by Dutton [7], attachment under low energy swarm conditions does occur in CO_2 and N_2O. Hence, the referenced experiments for these two gases measured λ/n.

The published data shown for CO is of α/n, as it does not show significant attachment below 300 townsends. Christophorou and co-workers [8] indicate that the electron energy threshold is about 9.62 eV for the primary dissociative attachment process in CO, resulting in the negative ion O^- and C. The highest E/n value for which the measurements were made is about 440 townsends. A slight extrapolation (within the same order of magnitude) of the data in Dutton [9] shows the characteristic energy at 440 townsends to be about 6 eV. Hence, the measurements made in CO above 300 townsends are unlikely to be affected by attachment processes, especially since they were made at only 1 torr. Additionally, detachment of the O^- electron in pure CO is very efficient (forming CO_2 and an electron)[10]. Thus, the data shown is taken as α/n.

The published data shown in Fig. 1 agree reasonably well with the measured values for all three gases. The measured values for CO_2 seem to plot a line with a slightly higher slope than the combined values of Bhalla and Townsend [11]. The measured N_2O values have almost the same slope as the Teich data [12]. Since these values are for λ/n, some variation can be expected as attachment effects are pressure dependent. A meaningful comparison of the α/n measurements for CO to the Bhalla data [11], taken at low (< 300 Td) E/n values, is difficult. Other α/n values for CO, above 300 townsends, were not found in the literature.

BREAKDOWN CRITERIA IN GAS MIXTURES

Previous work by Wieland [13], Chantry and Wooton [14], and Van Brunt [15] have shown the usefulness (within limits) of one technique for predicting the uniform field breakdown limiting value $(E/n)_{lim}$ for gas mixtures with at least one strongly electronegative component. It uses partial pressure weighted averaging of the density-normalized ionization coefficient α/n and attachment coefficient η/n, for each component gas, to determine the functions $\alpha/n(E/n)_{mix}$ and $\eta/n(E/n)_{mix}$ for the mixture. The value of $(E/n)_{lim,mix}$ is then determined by the condition when $\alpha/n(E/n)_{mix} = \eta/n(E/n)_{mix}$. Since none of the constituent gases (CO_2, CO, N_2O) for the binary mixtures studied in this work are strongly electronegative, determination of a breakdown limit for the mixture by this criterion is immaterial, specifically in the context of the Townsend discharge.

The Townsend criterion for self sustaining breakdown, in a medium with minimal attachment ($\alpha >> \eta$), can be expressed as $\gamma \exp(\lambda d) > 1$, where the typical α has been replaced by λ. Since the secondary emission coefficient γ values are largely functions of geometry, materials and electrode surface conditions, λ is the only parameter directly dependent upon the gaseous medium properties. Hence, it is of interest to determine the validity of linear scaling to the λ/n measurements. The focus is to determine if partial pressure weighted averaging of λ/n holds for the binary gas mixtures and E/n range studied in this work. If not, are synergistic effects on the function $\lambda/n(p_{xx},E/n)$ apparent? In this context the term "negative" synergism for λ/n will apply when the λ/n values fall below the predicted, linearly-scaled values, since this occurrence lessens the likelihood of a self sustaining discharge. "Positive" λ/n synergism will be for points above the line.

APPARENT IONIZATION COEFFICIENT DATA FOR BINARY MIXTURES

The binary mixtures under consideration are CO_2/CO, CO_2/N_2O and CO/N_2O. In order to characterize λ/n according to mixture composition, it was sufficient to parameterize it with respect to "p_{xx}" (the percentage of buffer gas XX in the mixture), while holding E/n at a constant value. To accomplish this, a linear fit to the experimental data was used. The parameterizations lead to a "family of curves" for various E/n values. (This form can be useful since in applications, operational ranges of E/n will often be predetermined by external design constraints.)

CO_2/CO Mixtures

Fig. 2 shows the λ/n data for five gas mixtures plotted as a function of the parameter p_{CO2} for the selected E/n values. A linear, least-squares fitting routine was used to generate

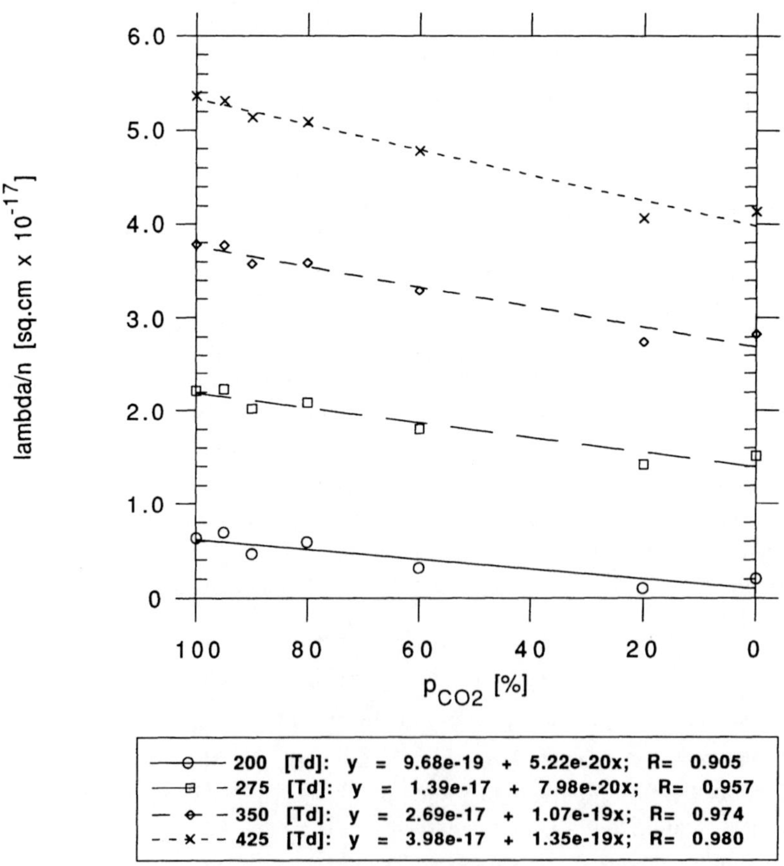

Fig. 2. Apparent Ionization Coefficient Data vs. p_{CO2} in CO_2/CO Mixtures

the "curves" for each of the selected E/n values. The linear, least-squares method used yields a linear correlation coefficient designated as R, where a perfect fit is indicated by R = 1. The fit equation for each line is shown, where the x-parameter represents p_{CO2} and the y-parameter represents λ/n. The value of the linear correlation coefficient R is also shown for each fit equation. The linear plots for all four selected E/n values are within 10% of a perfect fit. The potential for negative synergism appears at the point where $p_{CO2} = 20$ for all four E/n values. However, considering the random errors previously shown and the lack of any other p_{CO2} data points between 0 and 60, there is insufficient information to draw such a conclusion. Hence, partial pressure weighted averaging of λ/n seems to be reasonable for CO_2/CO mixtures.

CO_2/N_2O Mixtures

Fig. 3 shows the λ/n data for five gas mixtures plotted as a function of the parameter p_{CO2} for the selected E/n values. With the exception of the 200 townsend plot, the lines generated from the linear, least-squares fitting routine are within 6% of a perfect fit for all four selected E/n values. There appears to be a slight negative synergism for the λ/n values from $p_{CO2} = 80$ to 20. Nevertheless, the magnitude of calculated error for the data, coupled

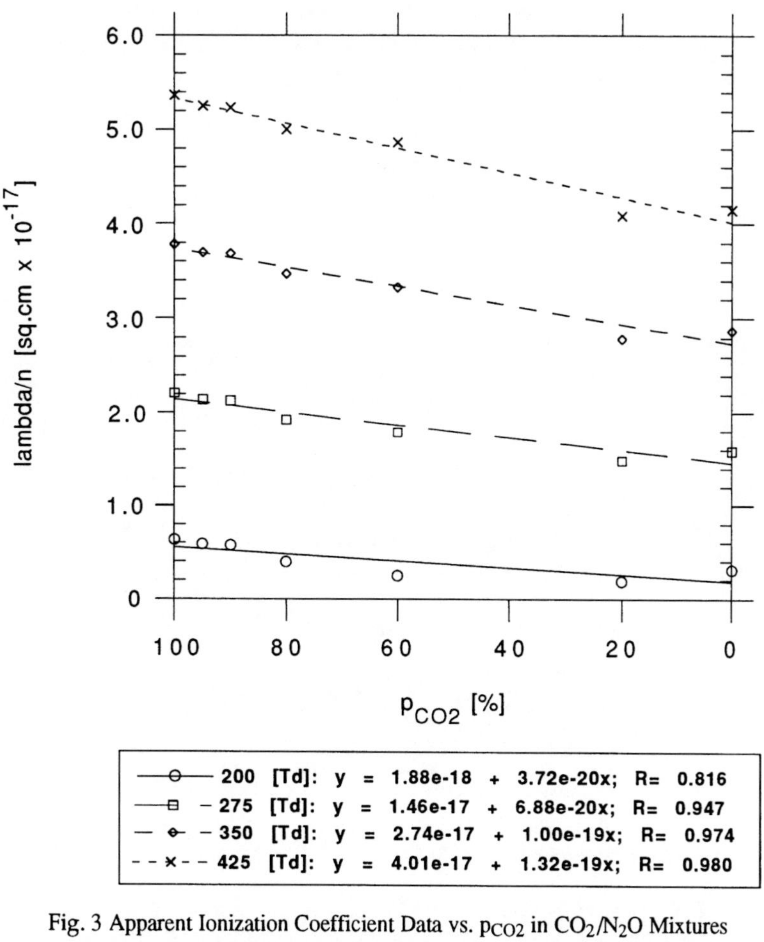

Fig. 3 Apparent Ionization Coefficient Data vs. p_{CO2} in CO_2/N_2O Mixtures

with the factor of only two difference between the λ/n values for the pure gases at low E/n, makes this deviation from linear behavior somewhat insignificant. At the least, the partial pressure weighted averaging appears to hold for CO_2/N_2O mixtures from 275 to 425 townsends.

CO/N_2O Mixtures

Fig. 4 shows the λ/n data for five mixtures plotted as a function of the parameter p_{CO} for the selected E/n values. For reference, a thin straight line between the two p_{CO} limits is included to represent the "ideal" partial pressure weighted values. None of the four lines generated from the linear, least-squares fitting routine are considered to be a good fit to the data, i.e. R = 0.145 to 0.400. While the negative synergistic effect is not a large one, it is evidenced in the CO/N_2O mixtures from 275 to 425 townsends. The effect is best exhibited by mixtures with p_{CO} values between 90 and 60, which show extrema at about 80. The maximum deviation from the pure gas λ/n values is about 10% for the 425 townsend data. The synergism appears to be real since this deviation is greater than the worst case, calculated random error of 6.9%. To a lesser extent, the same trend can be seen in the 275 and 300

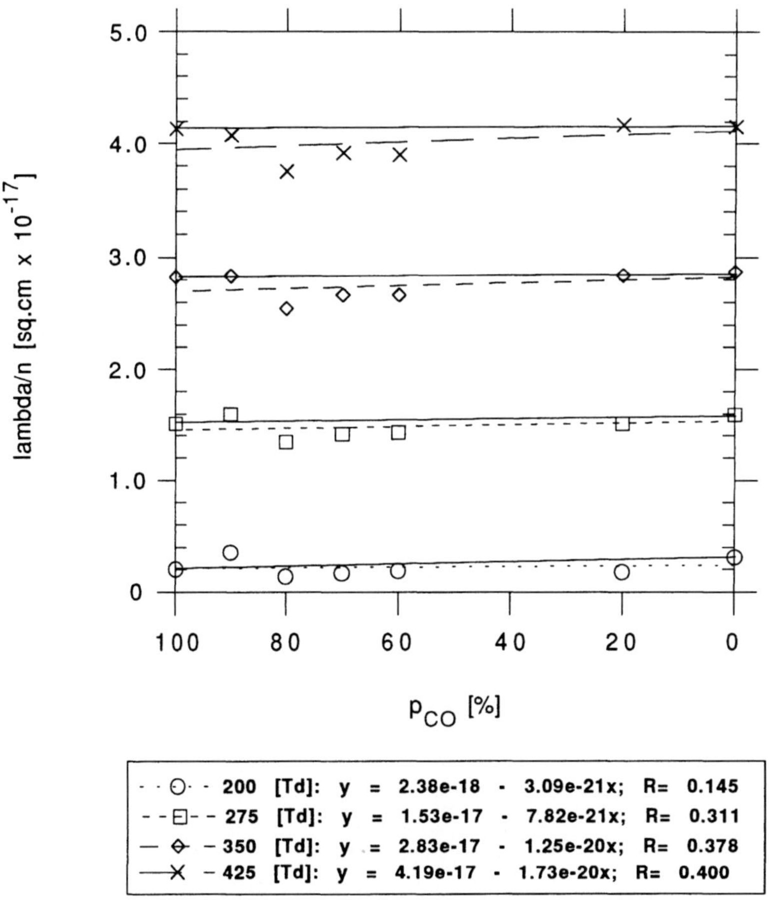

Fig. 4. Apparent Ionization Coefficient Data vs. p_{CO} in CO/N_2O Mixtures

townsend data. Theoretical calculations of the apparent ionization coefficient for CO/N_2O mixtures were recently conducted by R. Nagpal [16]. The preliminary results show qualitative agreement with the negative synergistic effect at higher E/n values. (The modeling technique utilizes a numerical solution to the Boltzmann equation for charge transport and scales the electron energy distribution functions of the component gases according to the mixture percentages.) However, under the specified conditions, partial pressure weighting can still give a rough prediction of apparent ionization coefficient values for CO/N_2O mixtures.

CONCLUSIONS

This investigation shows that partial pressure weighted averaging provides a good "rule of thumb" in determining λ/n values as functions of the partial pressure percentages (p_{xx}) for discrete E/n values, for the molecular gas mixtures studied. This can be useful information for selection of an economic insulating mixture of these gases. It is speculated that this "rule of thumb" might also be useful as a "first cut" for other molecular gas mixtures, provided they possess a similar non-electronegative structure. Consequently, real time partial pressure measurements may provide a useful diagnostic to estimate the overall integrity of an insulation system in a low pressure environment of both reactive and non-reactive gases.

REFERENCES

1. D. L. Schweickart, "The Measurement and Characterization of Charge Transport and Growth in Decomposition Gases from Electrical Insulation Used in Low Pressure Environments", Dissertation, University of Texas at Arlington (1993).
2. J. Dutton, "A Survey of Electron Swarm Data," Reprint No. 69 from *Journal of Physical and Chemical Reference Data* 4: 577 (1975).
3. H. Raether, "Electron Avalanches and Breakdown in Gases", Butterworth, Washington (1964).
4. G. R. G. Raju and R. Hackam, "Breakdown Field Strength of SF6, N2O, SF6+N2, SF6+N2O," *J. Appl. Phys.* 52: 3912 (1981).
5. T. M. Teich, "Electron Drift and Attachment in Nitrous Oxide," in "Proc. Twentieth International Conference on Phenomena in Ionized Gases", Pisa, Italy [Il Cioco] (1991).
6. D. L. Schweickart, "The Measurement and Characterization of Charge Transport and Growth in Decomposition Gases from Electrical Insulation Used in Low Pressure Environments", Dissertation, University of Texas at Arlington (1993).
7. J. Dutton, "A Survey of Electron Swarm Data," Reprint No. 69 from *Journal of Physical and Chemical Reference Data* 4: 577 (1975).
8. L. G. Christophorou, D. L. McCorkle, and A. A. Christodoulides, "Electron Attachment Processes," in "Electron-Molecule Interactions and Their Applications", Vol. 1, ed. L. G. Christophorou, Academic Press, Orlando (1984).
9. J. Dutton, "A Survey of Electron Swarm Data," Reprint No. 69 from *Journal of Physical and Chemical Reference Data* 4: 577 (1975).
10. J. L. Mauer and G. J. Schulz, "Associative Detachment of O^- with CO, H_2 and O_2," *Phys. Rev. A* 7: 593 (1973).
11. J. Dutton, "A Survey of Electron Swarm Data," Reprint No. 69 from *Journal of Physical and Chemical Reference Data* 4: 577 (1975).
12. T. M. Teich, "Electron Drift and Attachment in Nitrous Oxide," in "Proc. Twentieth International Conference on Phenomena in Ionized Gases", Pisa, Italy [Il Cioco] (1991).
13. A. Wieland, "Gasdurchschlagmechanismen in Elektronegativen Gasen (SF6) und in Gasgemischen," *ETZ-A* 94: 370 (1973).
14. P. J. Chantry and R. E. Wooton, "A Critique of Methods for Calculating the Dielectric Strength of Gas Mixtures," *J. Appl. Phys.* 52: 2731 (1981).
15. R. J. Van Brunt, "Common Parameterizations of Electron Transport, Collision Cross Section, and Dielectric Strength Data for Binary Gas Mixtures," *J. Appl. Phys.* 61: 1773 (1987).
16. R. Nagpal, Private Communication, Wright Laboratory, October 1992.

DISCUSSION

R. J. VAN BRUNT: Please comment on the calculations that have been performed by Nagpal. Are the electron energy distributions extracted from these calculations consistent with the requirements of the Wieland approximation?

D. L. SCHWEICKART: The calculations done by Nagpal employ a time-dependent solution of the Boltzmann transport equation using comprehensive electron collision cross section sets of Ar, CO, and N_2O. The calculations were first benchmarked to the electron transport data of the pure gases and were then extended to the mixtures for the studied experimental conditions. It should be noted that the electron energy distributions are different for CO and N_2O over the E/N range of interest. Also, neither gas is strongly electronegative. Hence, the Wieland approximation is not strictly applicable to the electron transport data in these calculations. Nevertheless, linear partial-pressure-weighted averaging of the apparent ionization coefficient in CO/N_2O mixtures is applicable for "rule-of-thumb" approximations. In an engineering applications content, the 10% maximum deviation from linearity should be covered by breakdown safety margins.

M. GOLDMAN: Can you assume that at the low pressures where you work there are no chemical reactions between the primary species created by the decomposition of the different polymers?

D. L. SCHWEICKART: It depends on the environmental conditions of the application. If the volume is initially evacuated, the rise in number density due to decomposition by-products will be small. Hence, subsequent by-product reactions will be minimal. But if the volume has not been "conditioned", residual species such as O_2, H_2, N_2, and H_2O may be present at higher concentrations, providing the opportunity for reactions which can result in secondary by-products.

M. GOLDMAN: If you find CO or CO_2, why do you not always find H_2O associated with these products?

D. L. SCHWEICKART: In the construction of most long polymer molecular chains some bond energies are lower than others. Low ones typically break at lower temperatures. The pyrolysis temperatures of interest are typically $<400°C$, as the insulation materials will tend to show detrimental mechanical effects if operated above this temperature. Hence, even if a material contains both H and O atoms, it may not evolve H_2O at lower pyrolysis (or operating) temperatures. Alternatively, the inherent composition of the polymer may preclude the production of H_2O (it may carbonize first). With polycarbonate, for instance, H_2O is not a major decomposition by-product when the material is pyrolyzed in an oxygen-free environment [A. Davis and J. H. Golden, "The Application of Gas Chromatography to the Characterization and Thermal Decomposition of a Polycarbonate," *J. Gas Chromatogr.* Vol. 5, No. 2, pp. 81-85, 1975].

L. G. CHRISTOPHOROU: What is the basis of the assumption that "all gases have identical electron energy distribution functions"? Under what conditions is this the case?

D. L. SCHWEICKART: The assumption was a conditional one, required for validity of the Wieland approximation for gas mixtures with at least one strongly electronegative component. It obviously does not apply to pure gases or to many gas mixtures with components whose cross sections and ionization thresholds significantly differ. However, if the electron energy distributions are close, over a limited E/N region of interest, the Wieland approximation is reasonably accurate and useful for engineering calculations.

A NEW COLLISION CROSS SECTION SET FOR SILANE

Rajesh Nagpal and Alan Garscadden

Plasma Research Section, Wright Laboratory
Wright-Patterson Air Force Base, Ohio 45433

INTRODUCTION

Plasma chemistry of monosilane (SiH_4) has provided rich sources of technological applications through the synthesis of amorphous silicon thin films. In order to identify the important precursors and understand the mechanisms/morphology of film growth, low energy electron collisions with SiH_4 have received considerable attention in the literature.[1] Besides being technologically important, fundamentally, SiH_4 is also of theoretical interest (as analogous to CH_4) as a prototype polyatomic molecule for electron scattering calculations.[2]

To date, almost all the information about the low energy ($\leq$ 10 eV) vibrational excitation and momentum transfer cross sections in SiH_4 has been been unfolded from the swarm data in pure SiH_4 and SiH_4 -rare gas mixtures.[3] However, there is a considerable scatter in the reported cross sections and the existing data base is far from *converged*. The reason for these discrepancies is twofold:

(a) Due to the large magnitude of the inelastic vibrational excitation cross sections as compared to those of momentum transfer (σ_m), the conventional two-term approximation for solving the Boltzmann transport equation (BTE) in the cross section unfolding procedure is grossly invalid. Therefore either a multi-term or a Monte-Carlo solution of electron kinetics is required to deduce a consistent set of SiH_4 cross sections to enable comparison with experimental beam or theoretical data.

(b) There has been insufficient parametrization in the swarm analyses. This problem was relieved to some extent by the work of Kurachi and Nakamura,[4,5] where the swarm analyses of SiH_4 -Ar,Kr mixtures, and of pure SiH_4 enabled them to significantly improve the pre-existing SiH_4 cross section data base. However, consistency in the cross sections of the above work was limited by the use of two-term approximation in the analysis. Further, these cross sections also fail more critical tests against the higher order electron transport coefficients in multi-term calculations.[6]

The present study reports a cross section set of SiH_4 which is consistent with all the swarm data to date. The inconsistency in the cross sections due to (a) and (b) above has been removed by the use of a Monte-Carlo solution of electron kinetics in the cross section unfolding procedure, and an additional analysis of SiH_4 -He mixtures.[7] We show that the non-Ramsauer buffer gas, Helium is a valuable tool for achieving consistency in the cross sections derived from swarm data.

Gaseous Dielectrics VII, Edited by L.G. Christophorou
and D.R. James, Plenum Press, New York, 1994

ANALYSIS

The approach adopted in the present work is a comparison between two-term and Monte-Carlo calculated electron transport coefficients versus experimental data in pure SiH_4 and SiH_4 -rare gas mixtures. The two-term solution of the BTE is well documented in the literature.[8] Briefly, the electron velocity distribution $f(v)$ is expanded in spherical harmonics of the angle with respect to the electric field vector.

$$f(\vec{v}) = \sum_{n=0} f_n(v) P_n(cos\theta) \tag{1}$$

where v is the electron velocity and P_n are the zeroth order Legendre polynomials of degree n. In the two-term approximation, it is assumed that the electron collision frequency for momentum transfer is much greater than the inelastic collision frequency and hence it is adequate to truncate the expansion to the the first two terms.

$$f(\vec{v}) = f_0(v) + \frac{\vec{v}}{v}\vec{f_1}(v) \tag{2}$$

where f_0 and f_1 are the isotropic and anisotropic components of the electron distribution respectively. Whereas in pure rare gases the above approximation is usually valid, in molecular gases or in molecular gas-rare gas mixtures, the original assumptions can be violated. The two-term derived cross sections[5] of SiH_4 (henceforth referred to as TTCS) show that σ_m has a Ramsauer minimum and the vibrational cross sections near their thresholds are larger than σ_m. We find that f_1 is actually larger than f_0 for the electron energies near the inelastic thresholds leading to a complete breakdown of the two-term approximation. The above effect also occurs in SiH_4 -Ramsauer rare gas mixtures (Ar,Kr). Therefore for an accurate description of the electron transport and unfolding of the cross sections, we have used a Monte Carlo method[9] for these and similar mixtures.

The starting inputs to the calculations are the TTCS. Comparison of discrepancies between two-term and Monte-Carlo calculations versus large relative anisotropy (f_1/f_0) in the electron energy distribution leads to a mapping of the electron energy range where there is inconsistency in the cross sections. The cross sections were then modified in the mapped energy range by trial and error until the Monte-Carlo calculated electron transport coefficients were in satisfactory agreement with experimental data. Further details of our analysis have been presented in a recent publication.[10]

RESULTS AND DISCUSSION

Vibrational Excitation and Momentum Transfer Cross Sections

Figure 1 shows the relative anisotropy in the energy distribution for pure SiH_4 . It is clear that the two-term approximation has broken down completely. The sudden onset of the vibrational inelastic processes shows up as spikes at their respective thresholds (0.113 and 0.271 eV), and the relative anisotropy is the highest for the near threshold electron energies. Similar results are also obtained for higher concentrations of SiH_4 in SiH_4 -Ar and SiH_4 -Kr mixtures. Therefore, the electron transport coefficients in pure SiH_4 and the above gas mixtures, calculated with the TTCS, using a Monte-Carlo or a multi-term calculation will not give good agreement with experimental data. This is indeed the case as illustrated in Figure 2 where the results using TTCS for electron drift velocities (w_d) and characteristic energies (ϵ_k) in pure SiH_4 are presented. While there is, as required, a consistency for the two-term calculations for w_d data, there is a rather poor agreement with the experimental[11] ϵ_k values (maximum error $\approx$ 30 %). More importantly, there is a significant discrepancy ($\approx$ 10 %) of the two-term calculated w_d data with the Monte-Carlo calculations at the E/N values where the relative anisotropy is the highest. The magnitudes of the errors obtained above are well above the most conservative estimates in swarm experiments. Therefore, the most significant changes were made to σ_m for SiH_4 in the energy range 0-1 eV and to the first vibrational excitation cross section (σ_{24}) near its threshold to achieve satisfactory agreement ($\leq 4\%$) with the experimental data using Monte-Carlo calculations in the fitting procedure. Figure 3 shows the calculated w_d and ϵ_k in pure SiH_4 versus experimental data[11,12] using the modified cross sections. The improved fits, particularly to the ϵ_k data, as compared to figure 2 illustrate consistency in the data base. The new cross sections are also compatible with the experimental swarm data of SiH_4 -Ar,Kr mixtures.[5]

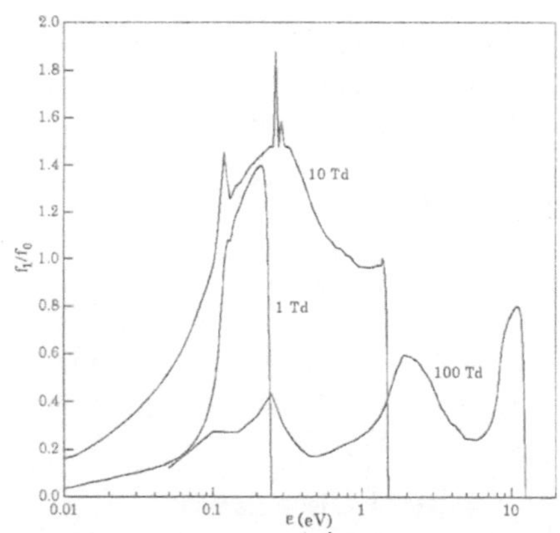

Figure 1 Relative anisotropy in pure SiH_4

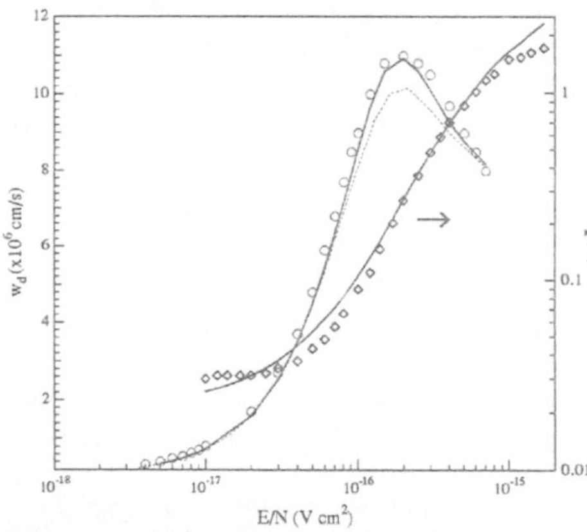

Figure 2 Electron drift velocities (o) and characteristic energies (◇) in pure SiH_4 . Points represent experimental data. Full lines and dotted lines represent two-term and Monte-Carlo calculations respectively using two-term derived cross sections.[5]

In SiH_4 -He mixtures, the comparisons are quite poor with the TTCS. Since Helium does not have a Ramsauer minimum in σ_m, the two-term calculation is valid in SiH_4 -He even at relatively high concentrations of SiH_4 in the gas mixture (3.5 %). Therefore as compared to more complicated multi-term or Monte-Carlo analyses, the relatively simpler two-term analysis can be used to unfold accurately the SiH_4 cross sections from SiH_4 -He mixtures. Non-uniqueness of the TTCS is shown clearly in Figure 4 where the calculated w_d differ from the experimental data[7] by up to 15 %.

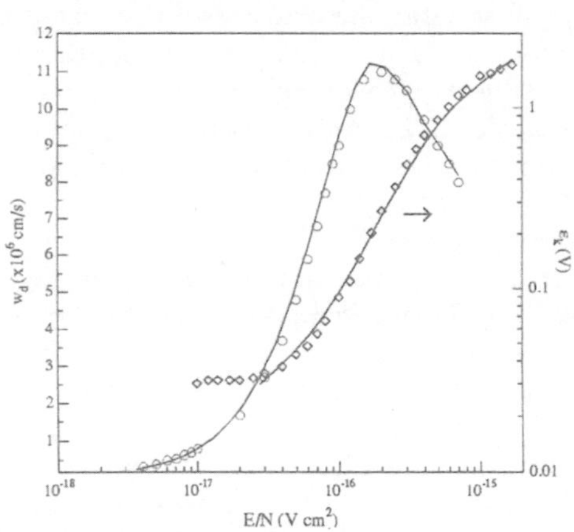

Figure 3 Monte-Carlo calculated (lines) electron drift velocities (o) and characteristics energies (◇) in pure SiH_4 versus experimental data (points) with the new derived cross sections

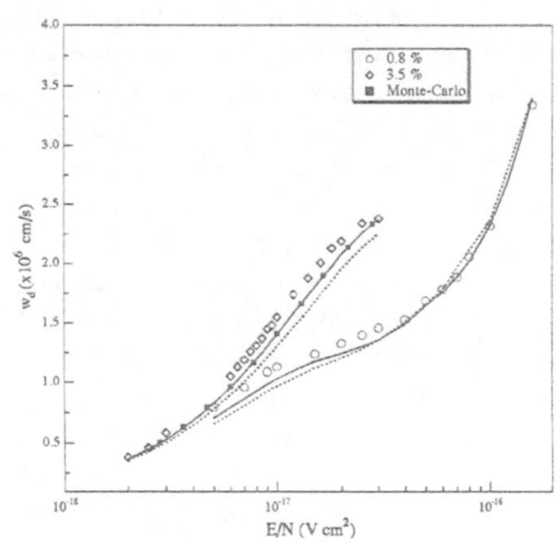

Figure 4 Electron drift velocities in SiH_4 -He mixtures. o,◇ represent experimental data. Full and dotted lines represent two-term results using the new and two-term derived[5] cross sections respectively.

The above discrepancies can be interpreted as following. There is an imperfection in the two-term cross section unfolding procedure due to the breakdown of the approximation in pure SiH_4 and SiH_4 -Ar,Kr mixtures. Therefore as long as the cross sections with the above imperfection are used for the mixtures from which they were derived, there would always appear consistent. However, removal of the imperfection in the two-term unfolding procedure by the use of SiH_4 -He mixtures brings out the non-uniqueness of the TTCS. The use of SiH_4 -He mixtures is therefore the critical test of consistency in the cross sections. Otherwise, the data base can only be considered as a limited fit to the experimental data from which it was derived. Figure 4 also shows the improved fit to SiH_4 -He mixtures with the new derived cross sections. An almost perfect agreement between the two term amd Monte-Carlo calculated w_d data for 3.5 % SiH_4 -He mixtures (full squares in Figure 4) demonstrates consistency in the derived cross sections from the Helium buffer gas swarm data.

Attachment, Dissociation and Ionization Cross Sections

For completeness, the cross section data base derived in the previous section needs to be extended towards higher electron energies to include dissociation and ionization. This is very important because in real deposition reactors, it is the silane radical/ion chemistry that determines the morphology and the growth rate of plasma deposited films.[13] The principal electron impact channels of radical/ion production in SiH_4 plasmas are:

$$e + SiH_4 \rightarrow e + SiH_{4-n} + nH \text{ or } \frac{n}{2}H_2 \qquad \textbf{Neutral Dissociation}$$

$$e + SiH_4 \rightarrow e + SiH_{4-n}^{-} + nH \qquad \textbf{Dissociative Attachment}$$

$$e + SiH_4 \rightarrow e + SiH_{4-n}^{+} + nH + 2e \qquad \textbf{Dissociative Ionization}$$

As compared to low energy cross sections of momentum transfer and vibrational excitation processes, there exist quite a few experimental measurements for dissociation,[14] attachment[15-17] and ionization[18,19] cross sections in SiH_4 to assist the swarm analyses.[20] Further, since the calculations are now towards high E/N ($\geq$ 100 Td, 1 Td = 1×10^{-17} Vcm^2), the two-term calculations can be used to unfold accurately the cross sections. This is because towards high E/N, the vibrational excitation processes (which caused the breakdown of the approximation) are no longer the dominant channels of inelastic losses in the calculations.

While the measured attachment cross sections in SiH_4 are quite small ($\leq 2 \times 10^{-18}$ cm^2), electron transport calculations show that for E/N $\leq$ 150 Td, the electron attachment rates (k_a) are larger than the ionization rates (k_i). This is due to the presence of relatively large dissociation cross sections (threshold 8.4 eV) between the attachment cross sections ($\geq$ 7 eV) and the ionization threshold (11.6 eV). The large dissociation inelastic losses then cause a rapid decrease of the electron energy distribution near the ionization threshold resulting in lower ionization rates as compared to those of attachment.[21] Following the above argument, the attachment, dissociation and ionization processes in SiH_4 are closely inter-related and hence the swarm analysis in conjunction with the experimental data can be used to accurately derive their respective cross sections.

Figure 5 presents the calculated ionization coefficients (α/N = k_i/w_d) in SiH_4 versus experimental data. For these calculations, cross sections for the total ionization processes were taken from Chatham et al,[18] those for total attachment were taken from Srivastava et al[17] and those for total dissociation were taken from Perrin et al.[14] The dissociative ionization cross sections were subtracted from those of total dissociation. The above cross sections are the same as in TTCS except for those of the attachment process. However, the above change does not affect the overall results since the total attachment cross sections of Srivastava et al[17] are only slightly lower in magnitude than the attachment cross sections for only SiH_3^- in the data of Ebinghaus et al[15] which was used in the TTCS. Due to a very large scatter in the attachment coefficient (η/N) measurements,[20] we have not attempted to adjust the attachment cross sections to enforce a fit to the η/N data. The inter-relationship between the attachment, dissociation and ionization processes in SiH_4, and consistency in their respective cross sections are confirmed in figure 5 where there is a good agreement between the calculated and experimental effective ionization coefficients ($\bar{\alpha} = \frac{\alpha-\eta}{p}$). More importantly, the calculations predict correctly the limiting value of E/N at which $\bar{\alpha} = 0$. The calculated limiting E/N=180 Td corresponds closely to the experimentally determined[20] value of E/p=65 $Vcm^{-1}Torr^{-1}$ (183.7 Td for p=1 torr at 0° C).

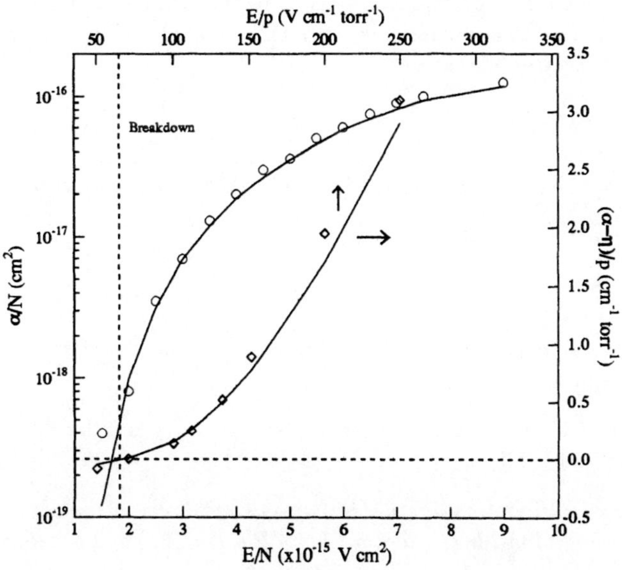

Figure:5 Ionization (○) and effective ionization (◇) coefficients in SiH_4 versus E/N and E/p respectively. Points represent experimental data.[20] Lines represent two-term calculations using the cross sections of Figure 6.

Finally, figure 6 presents the new derived cross sections of SiH_4 . The legend subscripts *T, m, 24, 13, a, d, i* correspond to the total (elastic+inelastic), momentum transfer, first vibrational, second vibrational, attachment, dissociation and ionization cross sections respectively.

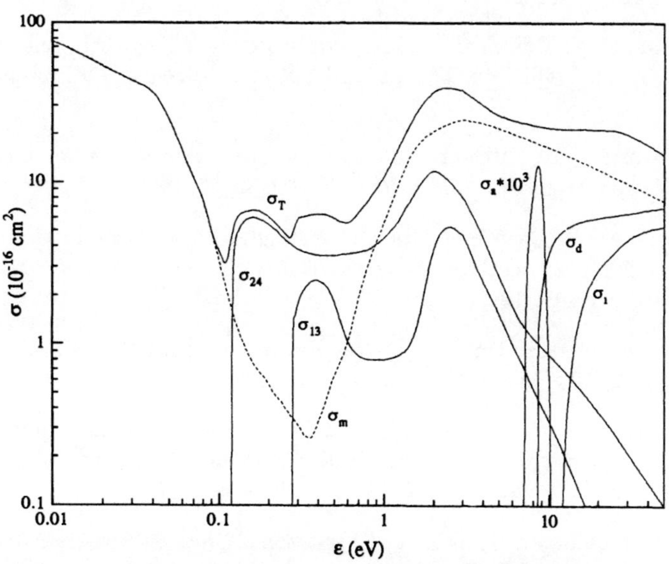

Figure:6 Electron collision cross sections for SiH_4

CONCLUSIONS

A recommended cross section set for monosilane over the energy range 0.01 to 50 eV is presented. The inconsistency in the previously derived cross sections by the two-term method is removed in the present work by the use of a Monte-Carlo calculation and an additional swarm analysis of SiH_4 -He mixtures. We have shown that the non-Ramsauer buffer gas Helium is a valuable tool for checking consistency in the swarm derived cross sections. The presented cross section set is also consistent with the attachment, dissociation and ionization data towards higher electron energies.

ACKNOWLEDGEMENT

The work of R. Nagpal has been supported by the Wright Laboratory through the University of Southwestern Louisiana.

REFERENCES

[1] A. Garscadden, G. L. Duke, and W. F. Bailey. Electron Kinetics of Silane Discharges. *Appl. Phys. Lett.*, 43:1012–14, 1983.

[2] C. Winstead and V. McKoy. Low-energy Electron Scattering by Silane (SiH_4). *Phys. Rev. A*, 42:5357–61, 1990.

[3] W. L. Morgan. A Critical Evaluation of Low-energy Electron Impact Cross Sections for Plasma Processing Modelling. *Plasma Chem. Plasma Proc.*, 12:477–93, 1992.

[4] M. Kurachi and Y. Nakamura. Electron Swarm Parameters in $SiH_4 - Ar$ Mixtures. *J. Phys. D: Appl. Phys.*, 21:602–06, 1988.

[5] M. Kurachi and Y. Nakamura. Electron Swarm Parameters in SiH_4-Rare Gas Mixtures and Collision Cross Sections for Monosilane Molecules. *IEEE Trans. Plasma Sc.*, 19:262–69, 1991.

[6] J. Wang, P. Ségur, and M. C. Bordage. Swarm Determination of Low Energy Electron-Molecule Cross Sections in Silane. In *International Conference on Phenomena in Ionized Gases*, pages 446–47, 1991.

[7] M. L. Andrews, K. A. Kirkendall, and A. Garscadden. Electron Drift Velocities in Silane-Helium and Silane-Argon Mixtures. In *36th Annual Gaseous Electronics Conference*, 1983.

[8] A. V. Phelps and L. C. Pitchford. Anisotropic Scattering of Electrons by N_2 and its Effect on Electron Transport. *Phys. Rev. A*, 31:2932–49, 1985.

[9] D. O. Ramos. A Monte Carlo Simulation of Drifting Electrons in a Gas Mixture Subjected to a Uniform DC Electric Field Using the Null Collision Technique. Master's thesis, Wright State University, 1990.

[10] R. Nagpal and A. Garscadden. Low Energy Electron Collision Cross Sections for Silane. *J. Appl. Phys.*, 75:703–09, 1994.

[11] P. G. Millican and I. C. Walker. Electron Swarm Characteristic Energies (D_r/μ) in Methane, Perdeuteromethane, Silane, Perdeuterosilane, Phosphine and Hydrogen Sulphide at low E/N. *J. Phys. D: Appl. Phys.*, 20:193–96, 1987.

[12] W. J. Pollock. Momentum Transfer and Vibrational Cross Sections in Non-Polar Gases. *Trans. Farad, Soc.*, 64:2919–26, 1968.

[13] T. Shirafuji, Y. Hayashi, and K. Tachibana. In-Situ Ellipsometric Monitoring of Growth of Amorphous and Polycrystalline Silicon Thin Films Deposited by RF Plasma CVD. In *XI International Symposium of Plasma Chemistry, Loughborough UK*, pages 817–22, 1993.

[14] J. Perrin, J. P. M. Schmitt, G. DE Rosny, B. Drevillon, J. Huc, and A. Lloret. Dissociation Cross Sections of Silane and Disilane by Electron Impact. *Chem. Phys.*, 73:383–94, 1982.

[15] H. Ebinghaus, K. Kraus, W. Muller-Duysing, and H. Neuert. Negative Ionen Durch Elektronenresonanzeinfang in PH_3, AsH_3, and SiH_4. *Z. Naturforsch. A*, 19A:732–36, 1964. in German.

[16] P. Haaland. Dissociative Attachment in Silane. *J. Chem. Phys.*, 93:4066–72, 1990.

[17] S. K. Srivastava, E. Krishnakumar, and A. C. de A. e Souza. Cross Sections for the Formation of Negative Ions by Electron Impact on Silane. *Int. J. Mass Spec. Ion Proc.*, 107:83–89, 1991.

[18] H. Chatham, D. Hils, R. Robertson, and A. Gallagher. Total and Partial Electron Collisional Ionization Cross Sections for CH_4, C_2H_6, SiH_4, and Si_2H_6. *J. Chem. Phys.*, 81:1770–77, 1984.

[19] P. Haaland. Dissociative Ionization of Silane. *Chem. Phys. Lett.*, 170:146–52, 1990.

[20] M. Shimozuma and H. Tagashira. Measurement of the Ionization and Attachment Coefficients in Monosilane and Disilane. *J. Phys. D: Appl. Phys.*, 19:L179–82, 1986.

[21] A. Garscadden. Effects due to Negative Ions and Particles in Plasmas. In M. Capitelli and J. N. Bardsley, editors, *Non-Equilibrium Processes in Partially Ionized Gases*, pages 541–50. Plenum Press, New York, 1990.

IONIZATION PROCESSES IN FULLERENES (C_{60}/C_{70})

J. Huang, D. Ding, C. E. Klots, and R. N. Compton

Oak Ridge National Laboratory
P.O.Box 2008
Oak Ridge, TN 37831-6125

ABSTRACT

The formation of positive and negative ions in the all-carbon-cage fullerene molecules C_{60} and C_{70} has been studied using both electron beams and laser beams. Structures in the low energy electron attachment cross section for C_{60} occur from the near zero energy threshold all the way up to ~10eV. This behavior can be ascribed to angular momentum $[\ell(\ell+1)/r^2]$ barrier shape resonances (the electronic ground state of C_{60}^- corresponds to ℓ=5). The long lifetimes for C_{60}^- with large internal energy (~17eV) are attributed to equipartition of energy among the large number of available vibrational modes. Delayed photoionization of C_{60} by visible pulsed lasers and cw-Ar ion lasers is ascribed to "thermionic emission" from highly vibrationally excited C_{60} molecules. Some possible implications of the unusual ionic properties of C_{60}/C_{70} to the dielectric strength of liquids and gases will be discussed.

INTRODUCTION

The all carbon cage fullerene molecules have received enormous theoretical and experimental attention since their production in macroscopic quantities[1]. Although, the C_{60} fullerene molecule was theoretically predicted in the late 1950's, it was only after Smalley and co-workers emphasized its existence[2] that the field of fullerenes exploded. The electronic and ionic properties of C_{60} have been of much recent interest. Further interest has been highlighted by the possible applications of fullerenes to areas such as high-T_c superconductivity, hydrogen storage, and drugs.

Gaseous Dielectrics VII, Edited by L.G. Christophorou
and D.R. James, Plenum Press, New York, 1994

In the first part of this paper, we describe the cw-Ar ion laser ionization of C_{60} ; in the second part, we discuss the electron attachment properties of C_{60} and suggest possible applications of fullerenes in the dielectric strength of materials.

C_{60} has previously been shown to attach electrons in the range from 0 to 15eV[3,4]. A later electron swarm experiment suggested that an activation energy[5] was required for free electron attachment to C_{60}. The authors of the beam experiment suggested that the resonances at low energy were a result of nuclear and electron-excited Feshbach resonances. In this study, we have performed high resolution electron beam experiments which show a threshold (~0.15eV) for the electron attachment to C_{60} and further peaks at 1, 5.5, and 8eV which we attribute to shape resonances resulting from angular momentum barriers in the e-C_{60} interaction.

EXPERIMENTAL

The laser ionization apparatus consists of a cw-Ar ion laser and a time-of-flight mass spectrometer (TOFMS) as shown in Fig. 1. The Ar ion laser has nine wavelengths and can be operated under all- or single-line mode. The maximum photon flux density can be 10^3 or 10^5 W/cm^2 for all-line output by using a 35 or 12.5cm focal length lens, respectively. The photons interact with C_{60} either in a cross beam or colinear beam setup.

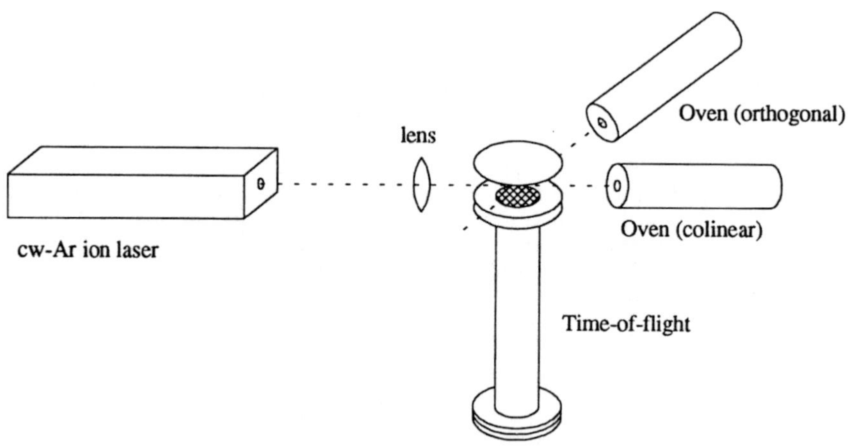

Fig. 1. The geometrical setup of the cw laser ionization

In the electron attachment experiment, an electron gun with an electron energy monochromator is used in place of the cw laser(Fig. 2). The electron beam was pulsed into the TOF interaction region with 5μs width for each pulse. After the electron bunch passes the C_{60} beam, a 10μs draw-out voltage(-40V) pulse to the TOF repeller plate pushes the negative ions into the TOF drift tube. The typical flight time for C_{60}^- is ~35μs. The electron energy resolution is measured by the zero energy attachment resonance of SF_6.

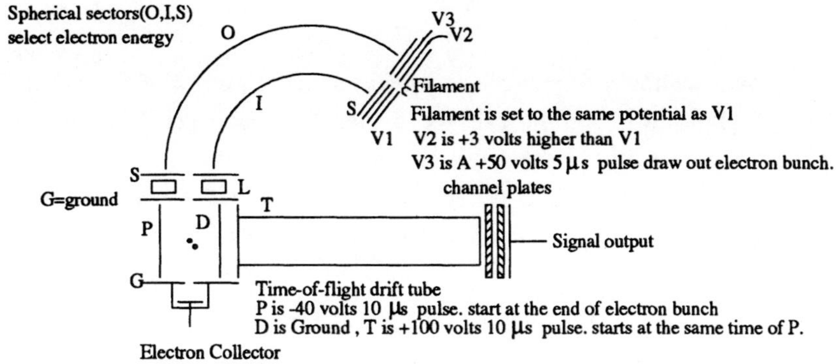

Fig. 2. Electron attachment apparatus

RESULTS AND DISCUSSION

CW laser ionization of C_{60}

The C_{60}^+ signal intensity shows a laser power dependence of order 4.2±0.5. A typical power dependence (ion counts as a function of laser flux in a double logarithmic plot) is shown in Fig. 3 for C_{60}^+ ions with an orthogonal beam setup and an all-line output laser. In the colinear configuration it was possible to obtain data using a single 488.0 nm or 514.5 nm line (having an even lower laser power density of $1\sim4\times10^3$ W/cm^2). The observed slopes of the power dependence for the C_{60}^+ and C_{70}^+ peaks are approximately the same at the higher temperature (T~500 °C), with a colinear beam arrangement, but saturation of the ion counts is observed for C_{70} in the region of higher power density. Roughly the same slopes of the power dependencies for both laser lines are obtained.

The photoexcitation behavior of fullerenes described here is reminiscent of the IR laser multiphoton dissociation previously observed for polyatomic molecules (*e.g.*, SF_6).[6] Multiphoton dissociation through the "quasicontinuum" has been exhaustively studied for many years. In the case of the fullerenes, the electronic excited state created by single-photon absorption can be regarded as a discrete level embedded in a quasicontinuum consisting of vibrational levels of the lower electronic states. The coupling between the discrete levels and the quasicontinuum (*i.e.*, IC and ISC) is very strong because of this large density of states. Thus, multiple-photon absorption might eventually produce a superheated C_{60}^{**} at an energy sufficient to "boil off" an electron.

We now examine whether the thermionic emission model is compatible with the observed laser power dependence. A set of rate equations describing the excited state population can be introduced to describe the incoherent photoexcitation process.[7] Because of the rapid ISC and the low fluxes used here, a constant photoabsorption cross-section can be assumed,[8] resulting in a Poisson distribution of the population of the C_{60} in the different states with the energy $nh\nu$,

$$P(\bar{n},n) = \frac{\bar{n}^n}{n!}exp(-\bar{n}) \qquad (1)$$

where $\bar{n}$ is the average number of photons absorbed, estimated simply by $\bar{n} = \sigma I \tau$, where σ is the photoabsorption cross section, I is the number of photons/cm^2/sec, and τ is the molecule-photon exposure time. This exposure time is given by the C_{60} transit time through the laser beam. The predicted number of ions, produced from thermionic emission of C_{60}^{**}, can then be estimated from a summation of weighted probabilities from eq(1)

$$S(\bar{n}) = \sum_{n=0}^{\infty} P(\bar{n},n)[1 - exp(-k(E)\tau_0)] \qquad (2)$$

where $k(E)$ is the rate constant for thermionic emission from the superheated C_{60}^{**} with an internal energy $E = nh\nu + E_{th}$ (E_{th} is the contribution from thermal heating and is equal to 3.3 eV at T=400 °C) and τ_0 is the observation time for the electron emission, typically of the order of 10^{-4} sec.

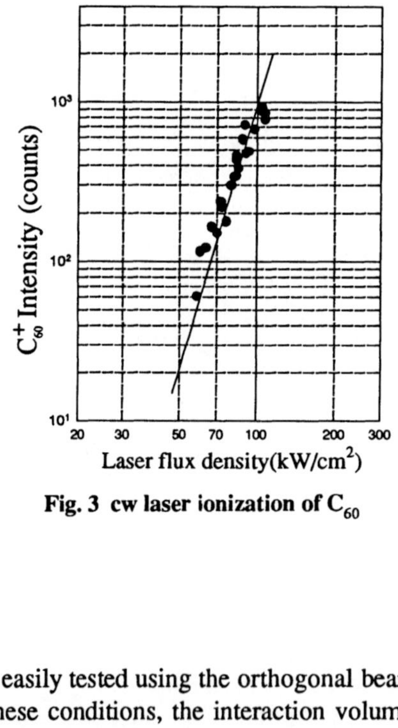

Fig. 3 cw laser ionization of C_{60}

This model is most easily tested using the orthogonal beam set-up and a long (35cm) focal length lens. Under these conditions, the interaction volume can be readily estimated. At the low power densities used in these cw laser experiments, a value of $\sigma \sim (1.3 \pm 0.7) \times 10^{-17}$ cm^2 may be adopted from the absorption spectra of the triplet and singlet states.[8] For photon fluxes in the range covered by the data of Fig.3 a slope of 5.6 ± 1 for the log-log dependence of count rate with laser flux is then calculated. That slope is much less than the total number of photons actually needed to induce observable thermionic emission, yet is clearly compatible with the data. An alternative mechanism invoking a small number of triplet excitons, while not excluded[9], is thus not needed.

Electron attachment to C_{60}

Figure 4 shows the C_{60}^- ion signal versus the electron energy in the energy range from 0 to 12eV. Remarkably, electrons with a wide range(up to 14eV) of energies are observed to attach to C_{60} as previously reported by Märk et al.[3,4] On top of this broad signal, two peaks are observed; one at 1eV and the other at 5.5eV, as well as a shoulder at 8eV. Figure 5 shows an expanded view of the low energy region. The low energy attachment cross section shows a near zero(0.15eV) energy resonance. This peak is clearly seen between the SF_6^- peak (~.05eV) and SF_5^- peak (0.35eV). It also exhibits an attachment threshold of ~0.15eV which is close to the activation energy of 0.26eV reported by Smith et al.[5] This ultra-low energy behavior was not reported in the studies of Märk et al.[3,4]

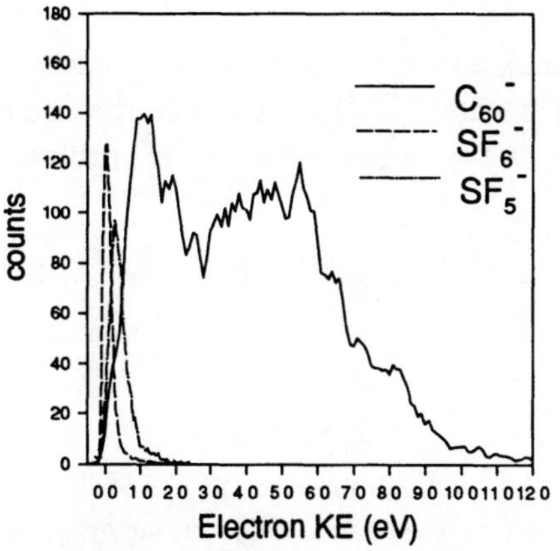

Fig. 4 Relative electron attachment cross section for C_{60}

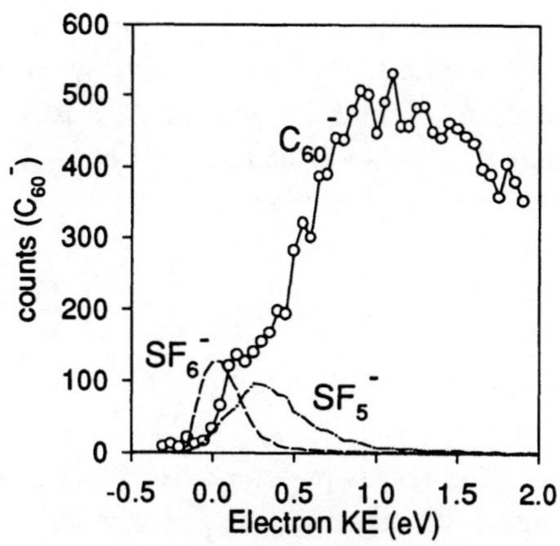

Fig. 5 Low energy electron attachment cross section for C_{60}

The resonance features observed in the low energy attachment cross section for C_{60} can be related to the unoccupied electronic states of C_{60} as predicted by theory and observed in experimental studies of inverse photoemission[10]. In figure 6 we show the lower unfilled molecular orbitals of C_{60} along with the approximate effective potential (polarization plus angular momentum barrier) corresponding to the angular momentum multiplet(ℓ) to which each state belongs. The effective potential is calculated from

$$V(r) = -\frac{e^2 \cdot (k-1) \cdot R_0^3}{2 \cdot (k+2) \cdot r^2 \cdot (r^2 - R_0^2)} + \frac{\ell \cdot (\ell+1) \cdot \hbar^2}{2 \cdot \mu \cdot r^2} \qquad (3)$$

where e is the electron charge, k is the dielectric constant of C_{60}, R_0 is the van der Waals radius of C_{60}, r is the distance from the center of C_{60} cage, and μ is the reduced mass which is the mass of the electron in this case. We adopt 3.8 for the dielectric constant.

As an incident electron penetrates the potential barrier(quantum tunneling), it can be initially trapped into one of the unoccupied states of C_{60} (Fig.6), then subsequently release its kinetic and potential energy to ro-vibrational modes of the anion. This would correspond to the formation of an initial shape resonance whose lifetime is greatly extended by the electronic to ro-vibrational coupling in the anion. Under certain conditions, the excited anion could also emit a photon and relax into the ground electronic state. The unoccupied electronic states of C_{60} have been calculated by several groups[11,12,13]. As shown in Fig. 6, the h_g or t_{2u} states are possible candidates for the 0.2eV resonance. The next peak at 1eV is close to the g_g or h_u state. Above these two resonance states, the broad band between 3 and 10eV could also be attributed to attachment into states of higher angular momentum(ℓ=8, etc). Another interesting possible explanation for the ~5.5eV peak involves excitation of the known plasmon resonance at ~6.3eV[14] followed by(or with simultaneous) attachment to C_{60}^-. Thus, we suggest that the broad peak at 5.5eV is a result of a Multi-Electron-Excited Feshbach Resonance(MEEFR) or alternatively a potential plasmon excitation attachment. This proposed mechanism involves simultaneous excitation of and attachment to the C_{60} plasmon. Since the plasmon energy is located at 6.3eV(the anti-symmetric state)[15] and 8eV(the symmetric state), an electron with energy greater than 3.6eV can attach to C_{60} and form a plasmon excited state(C_{60}^{-**}) if we assume the electron affinity of C_{60}^{**} is ~2.7eV(i.e., that of C_{60}). The onset and peak of the ~5.5eV feature are compatible with this criterion.

POTENTIAL APPLICATIONS

The results presented herein suggest potential applications to the field of dielectric strengths of materials. C_{60} is known to be very robust and able to accommodate considerable internal energy before dissociation and ionization occurs. Thus despite its low ionization potential (7.65 eV), ionization may not readily occur. Secondly, since C_{60} is a spherical molecule, its coverage on a surface may reduce the effect of surface irregularities giving rise to corona discharges. The very large electron capture cross section over a large range of energies suggests that corona discharges might be self-extinguished. In applications towards dielectric strengths of materials where the attachment of low energy electrons is a contributing factor, fullerenes could be especially useful. Due to their low vapor pressure, it is unlikely that fullerenes will be applicable as gaseous dielectrics unless a sufficiently volatile derivative could be prepared. However, C_{60} as an additive may affect the gas dielectric properties by influencing the behavior of surfaces and other important processes such as electron detachment producing "seed" electrons. The addition of fullerene to oils, plastics or other media may also prove useful.

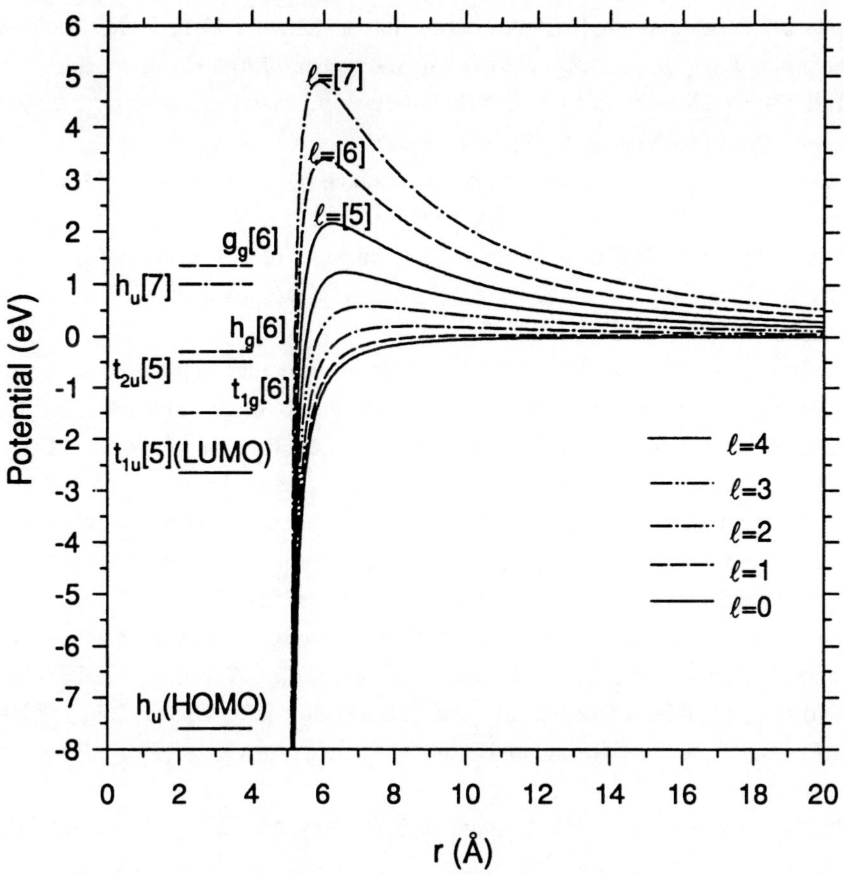

Fig. 6 Effective Potential

REFERENCES

1 W. Krätschmer, L.D.Lamb, K. Fostiropoulos, and D.R. Huffman, Solid C_{60}: a new form of carbon, *Nature* **347**, 354(1990).

2 H.W. Kroto, J.R. Heath, S.C. O'Brien, R.F. Curl, and R.E. Smalley, C_{60} : Buckminsterfullerene, *Nature* **318**, 162(1985).

3 M. Lezius, P. Scheier, M. Foltin, B. Dunser, T. Rauth, V.M. Akimov, W. Krätschmer, T.D.Märk, Internation of free electrons with C_{60}: ionization an attachment reactions, *Int. J. Mass Spectrom. Ion Processes.* **129**, 49(1992).

4 M. Lezius, P. Scheier, and T.D. Märk, Free electron attachment to C_{60} and C_{70}, *Chem.Phys.Lett.* **203**, 232(1993).

5 D. Smith, P. Spanel, and T.D. Märk, Electron attachment to C_{60} at low energies, *Chem.Phys.Lett.* **213**, 202(1993).

[6] J.G.Black, E.Yablonovitch, N.Bloembergen, and S.Mukamel, Collisionless multiphoton dissociation of SF_6: A statistical thermodynamic process, *Phys. Rev. Lett.* **38**, 1131(1977); N.Bloembergen and E.Yablonovitch, in *Laser-Induced Processes in Molecules* (Springer-Verlag, 1979).

[7] The rate of a coherent excitation process, determined by the laser-induced Rabi frequency, $\omega_r=\hbar^{-1}|\mu|\cdot|E|$, is much less in the region of the laser flux used than that of the damping process (ISC or IC process). See the paper of J.R.Ackerhalt and J.H.Eberly, Coherence versus incoherence in stepwise laser excitation of atoms and molecules, *Phys. Rev.* A **14**, 1705(1976).

[8] Use of an average cross section, σ, is reasonable because successive photoabsorption may take place from the same electronic state. The approximate value of 1.3±(0.7cm^2) is estimated from: Y.Kajii et al.,Transient absorption, lifetime and relaxation of C_{60} in the triplet state,*Chem.Phys.Lett.* **181**, 100(1991); J.R.Heath, R.F.Curl, and R.E.Smalley, The UV absorption spectrum of C_{60} (buckminsterfullerene): A narrow band at 3860Å, *J.Chem.Phys.* **87**, 4236(1987); B. BBrady and E.J. Beiting, Absolute absorption cross sections of gas phase C_{60} at 600°C, *J.Chem.Phys.* **97**, 3855(1992).

[9] Y.Zhang and M.Stuke, Direct observation of transition from delayed ionization to direct ionization for free C_{60} and C_{70}: Thermionic emission?, *Phys.Rev. Lett.* **70**,3231(1993).

[10] M.B. Jost, N. Troullier, D.M. Poirier, José Luís Martins, J.H. Weaver, LP.F. Chibante, and R.E. Smalley, Band dispersin and empty electronic states in solid C_{60}, *Phys. Rev.* **B** **44**, 1966(1911).

[11] B.I.Dunlap, D.W.Brenner, J.W.Mintmire, R.C.Mowrey, and C.T.White, Local density funtional electronic structures of three stable icosahedral fullerenes, *J.Chem.Phys.* **95**, 8737(1991).

[12] P.D.Hale, Discrete-variational-Xα electronic structure studies of the spherical C_{60} cluster: Prediction of ionization potential and electronic transition energy, *J.Am.Chem.Soc.* **108**,6087(1986).

[13] S. Satpathy, Electronic structures of the truncated-icosahedral C_{60} cluster, *Chem.Phys.Lett.* **130**, 545(1986).

[14] G.Gensterblum, J.J.Pireaux, P.A. Thiry, R.Caudano, J.P.Vigneron, Ph.Lambin, A.A.Lucas, and W.Krätschmer, High-resolution electron-energy loss spectroscopy of thin films of C_{60} on Si(100), *Phys.Rev.Lett.* **67**, 2171(1991).

[15] D. Östling, P. Apell, and A. Rosén,Surface plasmons of C_{60}, Supplement to Z. *Phys.* **D** **26**, S 282(1993).

DRIFT OF POSITIVE IONS IN METHANE

J. de Urquijo, C. Cisneros, H. Martínez,
and I. Alvarez

Instituto de Física, UNAM
P.O. Box 139-B, 62191, Cuernavaca, Mor.
México

ABSTRACT

Measurements of the mobility of CH_5^+, $C_2H_5^+$, $C_3H_5^+$, and H_3O^+ in methane gas with a drift tube-mass spectrometer are presented. The density-normalised electric field strength, E/N, was varied over the range 15-520x10^{-17} V cm^2. The low-field mobilities of $C_2H_5^+$ and $C_3H_5^+$ scale very well according to the mass dependence of the polarisation-limit theory, but are 15-23% lower than their predicted absolute values. For CH_5^+ and H_3O^+, an additional lowering of the mobility with respect to that of $C_2H_5^+$ is observed for E/N between 100-200x10^{-17} V cm^2, due probably to a proton transfer process between these two ions with methane and 200 ppm water vapour.

INTRODUCTION

In the past several years, there has been a growing interest in the understanding of charge transport and ion-molecule reactions in methane discharges, mainly because of its present and potential applications in the fields of plasma enhanced chemical vapour deposition, and diffuse electrical switching. However, measurements of the mobility of methane daughter ions in parent gas are scarce. In addition to its basic applications, these transport data are useful for calculating mean ion energies, recombination coefficients, and reaction rates.

This paper deals with the measurement of the mobility of the above four more abundant ion species in methane over a wide range of the density-reduced electric field strength, E/N, from 15 to 520 Td (1 Td=10^{-17} V cm^2).

Gaseous Dielectrics VII, Edited by L.G. Christophorou
and D.R. James, Plenum Press, New York, 1994

EXPERIMENTAL

The drift tube-mass spectrometer (DTMS) has been described in detail previously[1]. Hence, only a brief description will be given here. The DTMS consists of three vacuum enclosures: a movable-ion source chamber, a drift chamber, and detection chamber. A schematic of the apparatus is shown in Fig. 1. Ions are produced in the source by electron bombardment· (hot filament) and focused by an array of electrostatic lenses through an orifice of 0.2 mm diameter into the drift region, which is started by a 3.1 cm long drift space where the ions relax the translational energy acquired in the ion source, before entering the drift region. The ion source chamber can be moved over a distance from 0-40 cm. Due to the discrete arrangement of the 14 guard rings that produce the uniform electric field, 13 fixed positions, separated by 3.1 cm are available for measurement. The ion shutter mesh is normally biased to a potential of such polarity that it repells the ions. The experiment is operated by pulsing the ion swarm into the drfit region by a 10-20 μs rectangular pulse superposed to the repulsive potential, at intervals in the range 0.5-20 ms, for 1 to 10 min, depending on the E/N and source position conditions. A sample of the drifting ions passes through a second orifice at the other end of the drift space and enters a quadrupole mass spectrometer. The selected ions are detected by a continuous dynode multiplier, and the pulses are passed on to a multichannel scaler. The MCS was operated to a dwell time of 2 us, so that each arrival time spectrum contains at least 50-100 points useful for analysis. The start-sweep output signal of the MCS is used to trigger the whole experiment.

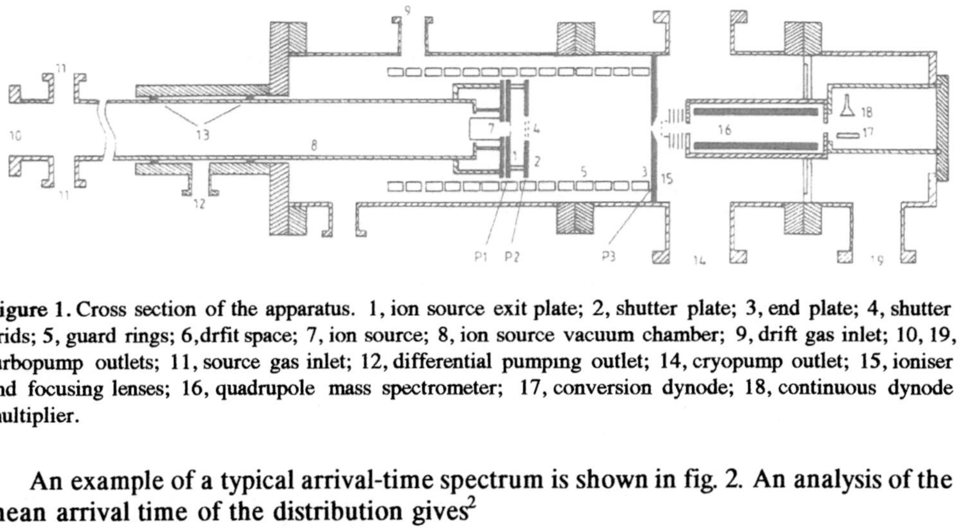

Figure 1. Cross section of the apparatus. 1, ion source exit plate; 2, shutter plate; 3, end plate; 4, shutter grids; 5, guard rings; 6,drfit space; 7, ion source; 8, ion source vacuum chamber; 9, drift gas inlet; 10, 19, turbopump outlets; 11, source gas inlet; 12, differential pumping outlet; 14, cryopump outlet; 15, ioniser and focusing lenses; 16, quadrupole mass spectrometer; 17, conversion dynode; 18, continuous dynode multiplier.

An example of a typical arrival-time spectrum is shown in fig. 2. An analysis of the mean arrival time of the distribution gives[2]

$$\langle t \rangle = z/v_d + D_L/v_d^2, \tag{1}$$

where z is the drift distance, v_d is the ion drift velocity, and D_L is the longitudinal diffusion coefficient. A means of eliminating the so-called end effects, and because the second term of (1) is usually small, the evaluation of v_d, consists of measuring the arrival time spectra at several source positions. Thus, from Eq. 1, a plot of $\langle t \rangle$ as a function of z would give a straight line, the slope of which is the inverse of v_d. It is customary to report the reduced mobility, K_o, instead of the drift velocities, K_o being defined as

$$K_o = (N/N_o)v_d/E, \tag{2}$$

where N_o=2.69x10^{19} cm^{-3} is gas density at STP.

The UHP methane gas sample supplied by Matheson was quoted to be 99.97% pure. Further analysis of the gas in the drift chamber indicated a 200 ppm content of water. The measurements herein reported were taken over the ambient temperature range 293-310 K.

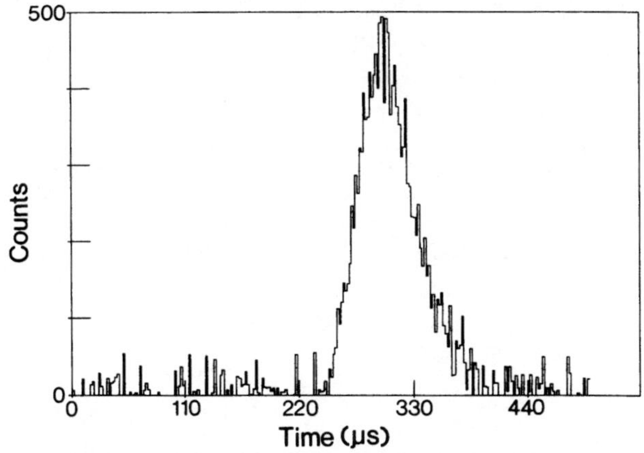

Figure 2. An example of arrival time spectrum of $C_2H_5^+$ in CH_4. E/N=190 Td

RESULTS AND DISCUSSION

Among the most abundant ion species found to be formed by secondary and tertiary reactions of primary positive ions in methane were CH_5^+, $C_2H_5^+$, and $C_3H_5^+$. Additionally, H_3O^+ was found to be abundant up to E/N=200 Td. The uncertainty in the mobility measurement was typically 1-2% up to E/N=300 Td, and 2-4% above. The mobilities of the above ion species are shown plotted in Fig. 3 as a function of E/N, over the range 15 to 520 Td.

Here, we shall briefly discuss some of the relevant features of these data:

a) At low E/N, the trend of the data suggest a tendency toward a rather constant mobility. An average of these data gives values for the low-field mobilities, K_{oo}, which are given in Table 1. It is customary to compare the K_{oo} value with that given by the polarisation limit formula[3]

$$K_{pol} = 13.853(m_r \alpha)^{-1/2} \quad [cm^2V^{-1}s^{-1}], \tag{3}$$

where m_r is the reduced mass of the ion-neutral system in amu, and α is the dipole polarisability of the neutral in Å^3. This formula is used here for qualitative comparisons only, since it is strictly applicable to a simple point charge-induced dipole interaction potential varying as r^{-4}, and for $T \rightarrow 0$ K. Studies in other systems have demonstrated that this approach is indeed valid at very low temperatures[4,5]. In the present case, the low field mobilities of all ions are 15-23% lower than their predicted polarisation values, also given in Table 1. Nevertheless, it is interesting to test the reduced mass dependence of the experimental K_{oo} data. From Eq.2, the product $m_r^{1/2}K_{oo}$ would be expected to be a constant for any ion drifting through the same gas.

This is indeed the case for $C_2H_5^+$ and $C_3H_5^+$, as it can be seen from Table 1, but it is not the case for CH_5^+ and H_3O^+. It is worth noting that for the SF_6 system studied previously, all the low-field mobilities of the ions (SF^+, SF_2^+, SF_3^+, SF_5^+, and SF_5^-, scaled very well according to the above law, although their absolute values were about 15% lower than predicted[1,6,7], as is the present case.

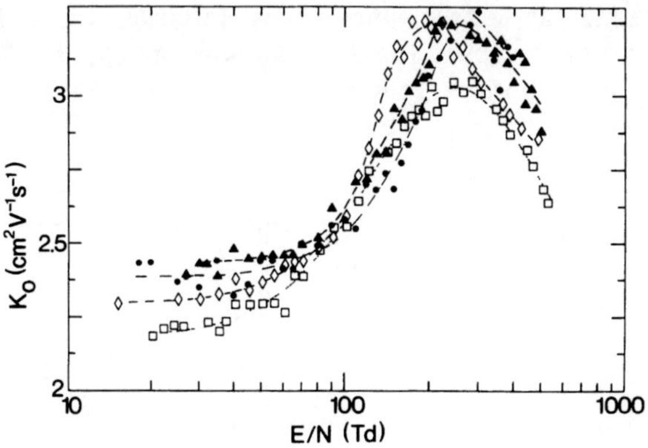

Figure 3. Reduced mobilities of CH_5^+ (▲), $C_2H_5^+$ (◇), $C_3H_5^+$ (□), and H_3O^+ (●) in CH_4 as a function of E/N.

Table 1. Summary of low-field mobility data[a] and [b]$K_{oo}m_r^{1/2}$

Author	CH_5^+	$C_2H_5^+$	$C_3H_5^+$	H_3O^+
Ridge[13]	2.32±0.02	2.28±0.06		
Theory[3]	3.00	2.68	2.54	2.92
This work	2.43±0.03	2.31±0.05	2.20±0.05	2.40±0.04
$K_{oo}m_r^{1/2}$	6.98	7.42	7.46	7.07

a) Mobility in units of $cm^2V^{-1}s^{-1}$
b) $K_{oo}m_r^{1/2}$ in units of $cm^2V^{-1}s^{-1}amu^{1/2}$

b) From the same plot in Fig.3, it is seen that the reduced mobilities of CH_5^+ and H_3O^+, the two lighter ions in the system, are lower than that of $C_2H_5^+$, which is heavier. This fact may be explained in terms of an additional mechanism to that of scattering. More specifically, proton transfer reactions may be the cause of the lowering of the mobility[3]. Thus, it would be very likely that the following proton transfer reactions are present[8-12]:

$$CH_4^+ + CH_4 \rightarrow CH_5^+ + CH_3$$
$$CH_4^+ + H_2O \rightarrow H_3O^+ + CH_3$$
$$CH_5^+ + H_2O \rightarrow H_3O^+ + CH_4,$$

and occur with fast rate constants in the range $1\text{-}3.7\times10^{-9}$ cm^3s^{-1} at thermal energies. Field and Munson have shown that the secondary ions CH_5^+, $C_2H_5^+$ are unreactive toward methane. To our knowledge, no evidence for a symmetrical proton transfer reaction such as $CH_5^+ + CH_4 \rightarrow CH_4+ + CH_5^+$ exists. The only studies[9] that have been carried out are those of deuteron transfer to CH_4

$$CD_5^+ + CH_4 \rightarrow CH_4D^+ + CD_4$$

with a small rate constant of 3.3×10^{-11} cm^3s^{-1}.

c) For the purposes of comparison with other data, those of Ridge and Beauchamp[13] for CH_5^+ and $C_2H_5^+$ are in excellent agreement with our low field data. These authors used and ion-cyclotron technique, and worked at low E/N<60 Td. To our knowledge, no other work on these ions has covered an E/N range as wide as the present one.

CONCLUSIONS

We have presented the mobilities of four of the predominant ion species drifting in methane over a wide E/N range. Ocurrence of proton transfer processes is suggested by the lowering of the mobility of CH_5^+ and H_3O^+. Further evidence is being presently gained by measurements of the relative abundance of these species as a function of E/N, gas density N, and drift distance.

ACKNOWLEDGEMENTS

This work was partially supported by CONACYT and DGAPA, Project IN104391. The authors are thankful to Mr. S. Pérez for his technical assistance.

REFERENCES

1. J. de Urquijo, I. Alvarez, C. Cisneros and H. Martínez, Mobility and longitudinal diffusion of SF_3^+ and SF_5^+ in SF_6, *J. Phys. D: Appl. Phys* 23:778 (1990)

2. J. de Urquijo, C. Cisneros, H. Martínez and I. Alvarez, *To be published.*

3. E.A. Mason and E.W. McDaniel. *"Transport Properties of Ions in Gases,"* Wiley, New York (1988)

4. R. Johnsen, R. Tosh, and L.A. Viehland, Mobility of helium ions in nenon: comparison of theory and experiment, *J. Chem. Phys.* 92:7264 (1990)

5. A.A. Viggiano, R.A. Morris, and E.A. Mason, Mobilities and interaction potentials for the O^+-He and O^--He system, *J. Chem. Phys.* 98:6483 (1993)

6. J. de Urquijo, I. Alvarez, H. Martínez, and C. cisneros, Mobility and longitudinal coefficients of SF_5^- and SF_6^- in SF_6, *J. Phys. D: Appl. Phys.* 24:664 (1991)

7. J. de Urquijo, C. Cisneros, H. Martínez and I. Alvarez, Transport coefficients of SF^+ and SF_2^+ in SF_6, *J. Phys. D: Appl. Phys.* 25:1277 (1992)

8. M.S.B. Munson and F.H. Field, Chemical ionization mass spectrometry. I. General considerations, *J. Am. Chem. Soc.* 88:2621 (1966)

9. F.P. Abramson and J.H. Futrell, Ion-molecule reactions of methane, *J. Chem. Phys.* 45:1925 (1966)

10. W.T. Huntress and R.F. Pinnizzotto, Product distributions and rate constants for ion-molecule reactions in water, hydrogen sulfide, ammonia and methane, *J. Chem. Phys.* 59:4742 (1973)

11. N.G. Adams and D. Smith, Reactions of hydrocarbon ions with hydrogen and methane at 300 K, *Chem. Phys. Lett.* 47:383 (1977)

12. D. Betowski, J.D. Payzant, G.I. Mackay and D.K. Bohme, Rate coefficient at 297 K for proton transfer reactions with H_2O. Comparisons with classical theories and exothermicity, *Chem. Phys. Lett.* 31:321 (1975)

13. D.P. Ridge and J.L. Beuachamp, The interation of ions with nonpolar neutrals: The collision broadening of ion cyclotron resonance lines of ions in hydrogen and methane, *J. Chem. Phys.* 64:2735 (1976)

DISCUSSION

R. J. VAN BRUNT: How were the measurements made at E/N >600Td and what was the upper limit on E/N in your drift tube?

J. DE URQUIJO: The ion abundance measurements at E/N >600Td were made at low pressures in the range 6.3 to 25 mTorr, drift voltages of less than 600V (to avoid breakdown of the guard ring structure with the vacuum vessel), and drift distances larger than 20 cm, to allow for at least a few ion collisions with the neutrals in the drift/reaction space. Under such conditions, the upper limit on E/N was 7.5 kTd.

A. GARSCADDEN: In your experiments varying E/N, did you keep the pressure constant? If N was varied, how did you distinguish between field and heavy particle collisional effects?

J. DE URQUIJO: The present ion abundance measurements are aimed to cover the widest possible variations of E/N, N and drift distance z. Because of breakdown limitations, it is not possible to keep a constant pressure over the whole E/N range from 50 to 7500 Td. Preliminary observations indicate a decay of the ion abundance as the drift distance is increased, but the gas density dependence has not been assessed.

ELECTRON MOBILITY IN HIGH-DENSITY ARGON GAS

A.F. Borghesani[1], M. Santini[1], P. Lamp[2], and G. Buschhorn[2]

[1]Dipartimento di Fisica "G.Galilei", Università di Padova
and Consorzio Interuniversitario Fisica della Materia
Via F.Marzolo 8, I-35131 Padova, Italy

[2] Max-Planck-Institut für Physik, W.Heisenberg-Institut
Föhringer Ring 6, W-80805 München, Germany

INTRODUCTION

The transport properties of excess electrons in non polar gases or liquids give important information on the electronic states in a disordered medium and on the relationship among the electron-atom interactions and the properties of the fluid. Anomalous density effects on electron mobilities were discovered in a number of compressed gases[1]. A negative density effect, i.e., the density-normalized zero-field mobility $\mu_0 N$ of electrons decreases with increasing the gas density N, is shown by gases such as He and Ne, whose interaction with electrons is dominated by short-range repulsive exchange forces and whose scattering length a is positive. On the other hand, a positive density effect, i.e., $\mu_0 N$ increases with N, is shown by Ar, Kr, and Xe, whose scattering length is negative owing to the overwhelming contribution of the long-range attractive polarization forces to the atom-electron interaction. The classical single-scattering picture of kinetic theory predicts $\mu_0 N$ independent of N for thermal electrons[2]

$$(\mu_0 N)_c = \frac{4e}{3(2\pi m)^{1/2}(k_{\rm B}T)^{5/2}} \int_0^\infty {\rm d}\,\epsilon\, \frac{\epsilon}{\sigma_m(\epsilon)} \exp\left(-\epsilon/k_{\rm B}T\right) \tag{1}$$

where σ_m is the momentum transfer cross section. At high densities, however, this picture breaks down because the electron mean free path ℓ becomes comparable to its thermal wavelength $\lambda_T = h(2\pi m k_{\rm B}T)^{-1/2}$ and multiple-scattering effects act so as to make $\mu_0 N$ density dependent. Different multiple-scattering theories were developed in the limit of small reduced field,i.e., $E/N \to 0$, in order to explain the two types of effects. For the negative density effect, it is assumed that the main physical effect is an enhancement of the collision rate when $\ell \to \lambda_T$, thus yielding

$$\mu_0 N = (\mu_0 N)_c \exp\left(-A\lambda_T N \bar{\sigma}_t\right) \tag{2}$$

where A is a constant of order unity[3] and $\bar{\sigma}_t$ is the value of the total scattering cross section at thermal energies. Eq.(2) describes He[4] quite well at low to medium density

Gaseous Dielectrics VII, Edited by L.G. Christophorou
and D.R. James, Plenum Press, New York, 1994

because the cross section σ is large and nearly energy independent, but is largely inadequate for Ne[5] where both the momentum transfer as well as the total cross sections σ_m and σ_t are small and strongly energy dependent.

On the other hand, the positive density effect in Ar is explained in terms of a quantum shift of the electron ground state energy in the dense medium. So, for negative scattering length gases multiple-scattering theories yield the following formula[6]

$$\mu_0 N = \frac{4}{3} \frac{eN}{m[\pi(k_B T)^5]^{1/2}} \int_0^\infty \mathrm{d}\epsilon \frac{\epsilon^{3/2} \exp(-\epsilon/k_B T)}{\nu(\epsilon - V_0)} \tag{3}$$

where $\nu(\epsilon) = (2\epsilon/m)^{1/2} N\sigma_m(\epsilon)$ is the collision frequency and $V_0 \approx 2\pi\hbar^2 Na/m$ is the real part of the total energy shift[7]. Eq.(3) is in disagreement up to 50 % with the experimental data in Ar[8] at $T = 297\,K$ for densities up $N = 25 \times 10^{20}\,\mathrm{cm}^{-3}$ and is moreover in contrast with a different formulation of the same equation due to O'Malley[3]. Even neglecting the disagreement of the present multiple scattering theories with experimental data, we believe that the description of the well-defined physical process of scattering of (epi)thermal electrons off noble gas atoms should be described in an unified way independently of the sign of the atom scattering length. In fact, it is not clear why the quantum shift of the ground state energy of the electrons in the conduction band of the medium should be effective only for the negative scattering length gases and why the quantum interference effect, which depends on the ratio ℓ/λ_T, should be effective only in positive scattering length gases.

The results of the electron mobility measurements in Neon[5,9] led to the conclusion that all multiple-scattering effects, i.e., a quantum shift of the ground state energy of the electrons in the medium, the mean free path reduction due to quantum interference, and eventually correlations among the scatterers, can be accounted for in a simple and general way and we believe that the different behavior shown by different gases is a consequence of the relative weight of these multiple scattering effects. When the scattering cross section is large and nearly energy independent like in He the main effect is the mean free path reduction due to quantum interference and depends on the ratio ℓ/λ_T. When the scattering cross section is small and strongly energy dependent like in Ne, the main effect is due to the energy quantum shift. In He and Ne, this approach has proven successful, but we decided to test it by carrying out measurements in Ar, whose total-energy shift is surely negative[10] and whose scattering cross section is strongly energy dependent, showing a Ramsauer minimum. Owing to the scarcity of accurate measurements covering a large density range at constant temperature, we carried out a complete set of mobility measurements in Ar at $T = 162.7\,K$ in an extended density range $(4 \leq N \leq 70) \times 10^{20}\,\mathrm{cm}^{-3}$.

EXPERIMENTAL RESULTS AND DISCUSSION

Details of the experimental technique can be found elsewhere[11,12]. In Figure 1 we show the experimental values of the density-normalized mobility μN as a function of the reduced field E/N ($1\,\mathrm{mTd} = 10^{-20}\,\mathrm{Vcm}^2$). We observe that for each N μN is independent for E/N for $E/N \leq 1\,\mathrm{mTd}$. The threshold E/N value below which μN approaches the value pertaining to thermal electrons decreases with increasing density. All curves show a very pronounced maximum in the range $(2 \leq E/N \leq 4)\,\mathrm{mTd}$. Since this maximum can be followed from the low density gas, it has to be attributed to the well-known Ramsauer-Townsend (RT) minimum of the scattering cross section[13]. It is interesting to note that the reduced field value at the mobility maximum, $(E/N)_m$, decreases as N increases, as shown in Figure 2. This fact was observed also at room temperature[14]. Finally, for $E/N \leq 8 - 10\,\mathrm{mTd}$ $\mu N(E/N)$ converges to a single curve for each densities.

The existence of such a mobility maximum as a function of E/N at constant N is a clue that the gas-phase scattering cross section can still be used even at the large densities of our experiment. A similar conclusion was drawn in the case of Ne.

At the mobility maximum, the mean electron energy $\bar{\epsilon}$ is essentially that of the RT minimum, $\bar{\epsilon} \approx \epsilon_{RT}$. Part of this energy is gained from the field and is an increasing function of E/N. Since we have found that $(E/N)_m$ decreases with N, it therefore seems that the electrons require a smaller energy gain from the field at increasing densities in order to reach ϵ_{RT}.

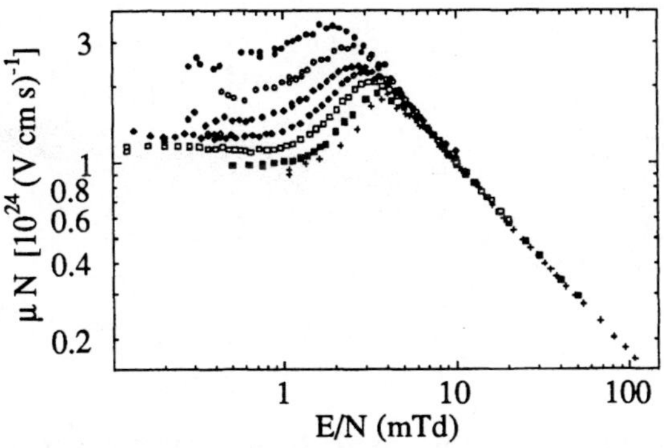

Fig. 1. Density-normalized mobility μN as a function of reduced electric field E/N (1mTd=10^{-20}Vcm^2) in Ar at T=162.7 K. N=(60.8, 50, 39.8, 30.2, 19.9, 7.91, 3.7)$\times 10^{20}$ cm^{-3}(from top).

The conclusion is that the missing energy is supplied to the electrons by a quantum shift. This shift, yet to be determined, must however be positive and increase with N. This picture is supported by the results of our mobility measurements in Ne.

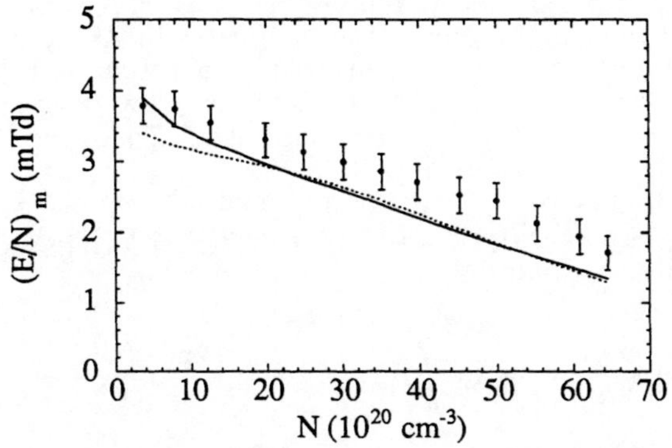

Fig. 2. Reduced electric field at the mobility maximum, $(E/N)_m$ for T=162.7 K. Open points: experimental data, solid line: calculated with cross section of Weyhreter et al.[13], dashed line: calculated with cross section of Haddad et al.[15]

In Fig. 4 we plot the thermal value $\mu_0 N$, i.e., μN extrapolated to $E/N \rightarrow 0$. Also its behavior can be understood with similar considerations, by recalling that for $E/N \rightarrow 0$ electrons do not practically gain energy from the field. Their mean energy $\bar{\epsilon}$ is therefore well below ϵ_{RT}. In this energy range the e-Ar scattering cross section σ_m is large but decreases strongly with increasing energy. With the aid of Eq. (1) we can think of the density-normalized zero-field mobility $\mu_0 N$ as a suitable weighed average of the inverse of the momentum transfer scattering cross section. If the mean electron energy had not any density-dependent contribution in addition to the thermal one, the average inverse cross section would not vary with N. On the contrary, if the mean electron energy depends on N, increasing with N itself at constant T, owing to the quantum energy shift, the average inverse cross section and, hence, $\mu_0 N$, increase because the scattering cross section decreases with increasing energy. The increase of $\mu_0 N$ with N is therefore a combined result of a positive quantum shift of the electron ground state energy and the decrease of σ_m with increasing energy.

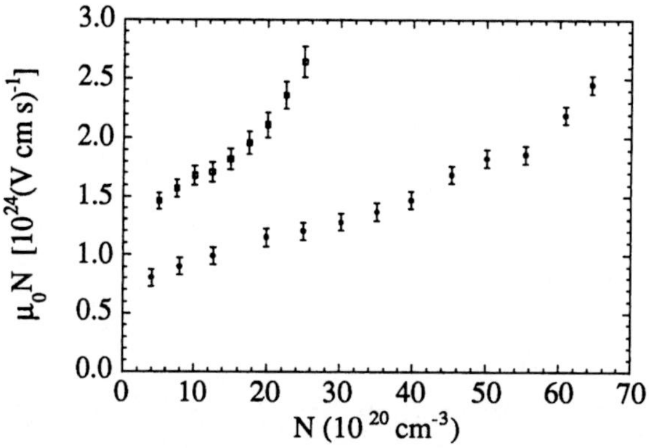

Fig. 3. $\mu_0 N$ as a function of N at room temperature[8] (squares) and at T=162.7 K (closed points).

As a consequence, the behavior of μN at low fields and that of $(E/N)_m$ suggests that the density-dependent quantum shift contribution to the electron energy must be positive. For Ar, the quantum shift of the total energy, V_0, is negative[10] because the scattering length is negative, $a \approx -0.8$ Å, and therefore it cannot be the correct one. Generally speaking, the electron energy shift is made up by two contributions, $V_0 = E_k + U_p$, where E_k is the positive zero-point kinetic energy essentially arising from excluding the electrons from the hard-core volume of the atoms, and U_p is a negative polarization potential-energy contribution. In He and Ne E_k dominates owing to the small atomic polarizabilities. In Ar U_p is overwhelming. We obtained a good agreement with the measured mobility in Ne in a wide density and temperature range by shifting the electron energy only by the kinetic contribution E_k calculated according to the Wigner-Seitz method[16]. Therefore we use also here for Ar the same heuristic model as in Ne.

In this model the electron kinetic energy is shifted to $\epsilon' = \epsilon + E_k = \hbar^2 k^2/2m + E_k$ and the resulting group velocity is $v = \hbar k/m = [2(\epsilon' - E_k)/m]^{1/2}$. The classical formula is modified into the following

$$\mu_0 N = \frac{4}{3}\frac{e}{h(k_B T)^2}\lambda_T \int_{E_k}^{\infty} \mathrm{d}\,\epsilon' \frac{(\epsilon' - E_k)\exp\left[-(\epsilon' - E_k)/k_B T\right]}{\sigma_m(\epsilon')} \tag{4}$$

Upon a change of variables and multiplying the r.h.s. by $\exp\left[-(\pi^{-1/2}\lambda_T N\bar{\sigma}_t)\right]$ in order to take into account the quantum interference effect due to multiple scattering, we obtain

$$\mu_0 N = \frac{4}{3}\frac{e}{h(k_B T)^2}\lambda_T \exp\left[-(\pi^{-1/2}\lambda_T N\bar{\sigma}_t)\right] \int_0^{\infty} \mathrm{d}\,\epsilon \frac{\epsilon \exp(-\epsilon/k_B T)}{\sigma_m(\epsilon + E_k)} \tag{5}$$

where it appears that the cross section has to be evaluated at the total shifted kinetic energy $\epsilon' = \epsilon + E_k$. This formula, valid at small electric field, is in very good agreement with th results obtained in Ne far from the critical point. The exponential factor in (5) for Ne is close to unity because the cross sections are very small and it is a good approximation to evaluate σ_t at thermal energy yielding $\bar{\sigma}_t$. In Ar, however, the cross sections are much larger than in Ne and strongly energy dependent. Therefore the quantum interference effect has to be treated more carefully. Moreover, since the critical temperature of Ar ($T_c = 150.89\,K$) is close to the temperature of our experiment ($T = 162.7\,K$), we cannot neglect correlations among scatterers as a result of density fluctuations. According to the Lekner model[17] σ_m has to be replaced by $\mathcal{F}(k)\sigma_m$, where

$$\mathcal{F}(k) = \frac{1}{4k^4}\int_0^{2k} \mathrm{d}q\, q^3 S(q) \tag{6}$$

$S(q)$ is the static structure factor of the gas. By inserting the experimental results[18] for $S(q)$ and recalling that[19] $S(q) = [S(0)+(qL)^2]/[1+(qL)^2]$ with $L = 0.1l^2[S(0)-1]$, where $l \approx 10\,\text{Å}$ is the so called " short range correlation length", we obtain

$$\mathcal{F}(\epsilon) = 1 + [S(0)-1]\left\{\frac{(8mL^2\epsilon)/\hbar^2 - \ln[1+(8mL^2\epsilon)/\hbar^2]}{8(2mL^2\epsilon/\hbar^2)}\right\} \tag{7}$$

It is easy to note that $\mathcal{F}$ decreases sharply towards 1 as energy increases, so that the multiple scattering effects weaken as energy increases. This is seen by observing that all μN curves as function of E/N converge to the low density curve for large E/N. As the electron energy increases, for instance by increasing the electric field, there is a gradual transition from a multiple-scattering situation to one where single scattering precesses dominate. In fact, as energy increases the electron wavelength decreases and the size of the electron wavepacket decreases. Thus, the effects of multiple scattering decrease.

As a consequence of the experimental results, we have put into evidence three simultaneously present multiple scattering effects: (i) a positive, density-dependent quantum shift of the electron kinetic energy forcing the evaluation of the cross section at the shifted kinetic energy, (ii) an enhancement of the scattering rate due to quantum self-interference of the electron when its wavelength becomes comparable to its mean free path, and (iii) correlations among scatterers to be accounted for by the complete wavevector dependence of the static structure factor. In order to finally determine E_k we have used the usual classical theory for electron mobility not restricted to thermal electrons[20]

$$\mu N = -\frac{2}{3}\left(\frac{e}{m}\right)^{1/2}\int_0^{\infty} \frac{\epsilon}{\sigma_m(\epsilon)}\frac{\mathrm{d}g(\epsilon)}{\mathrm{d}\epsilon}\mathrm{d}\epsilon \tag{8}$$

We used the Davydov-Pidduck distribution function

$$g(\epsilon) = B\exp\left\{-\int_0^{\epsilon}\frac{\mathrm{d}z}{k_{\mathrm{B}}T + \frac{1}{6}\frac{M}{m}\left(\frac{eE}{N}\right)^2\frac{1}{z\sigma^2{}_m(z)}}\right\} \tag{9}$$

where B is the normalization constant, m, and M are the electron and atom masses. In order to take into account the previous multiple scattering effects, in Eqns. (8) and (9) we used the linearized form for the mean free path reduction factor since in our case $\lambda/\ell < 1$, yielding an enhancement of the collision rate by a factor $(1 + N\lambda\sigma_m/\pi)$ as predicted by Atrazhev and Iakubov[21]. Furthermore, the scattering cross section is multiplied by the Lekner factor $\mathcal{F}$ and evaluated at the shifted energy $\epsilon + E_k$. We used E_k as a free parameter to be adjusted so as to fit the $\mu_0 N$ values at low electric field. The values of E_k determined by this procedure obtained by using two different cross sections available in literature are shown in Fig. 4. They are in good agreement with the zero-point kinetic energy values calculated with the Wigner-Seitz model using the same cross sections. With these values of E_k we also calculated the values of $(E/N)_m$ shown in Fig. 3. In Fig. 5 we report a sample calculation of $\mu N(E/N)$ to show the good agreement of our model in the whole range of electric fields. This fact shows the good degree of consistency of the model, in the sense that the E_k values determined from the fit of the mobility data at low fields is a good value over more than two orders of magnitude in E/N.

We can conclude that multiple scattering effects give origin to a shift of the electron kinetic energy, to an increase of the collision rate depending on the ratio λ/ℓ, and to an increase of the cross section due to correlations among scatterers. The main point is that the electron energy shift that enters in the scattering process and determines the scattering properties is the positive kinetic contribution to the total energy shift and that it can be calculated with the Wigner-Seitz model to a good approximation. This appears to be true for both positive as well as negative scattering length gases. All these multiple scattering effects are simultaneously present and only their relative weight determines if the mobility shows a negative or a positive density effect. Correlations are effective close to the critical point. If the cross section is large and nearly constant the main effect is that of the increase of the collision rate and is ruled by the ratio λ/ℓ. If the cross section is small and strongly energy dependent the energy shift effect is overwhelming.

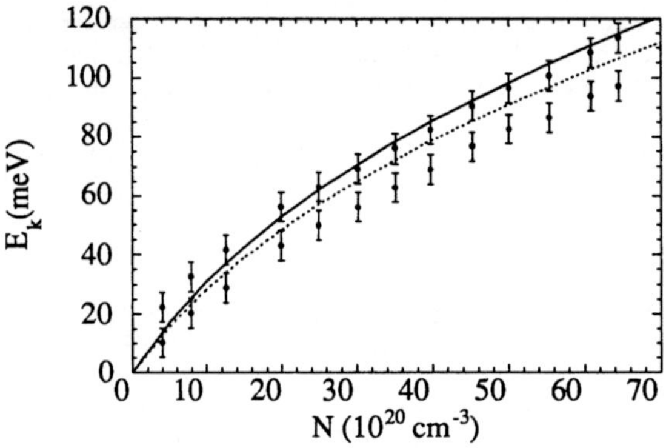

Fig. 4. Density dependence of the kinetic-energy shift determined from mobility data (points) and calculated according to the WS model (lines). Closed points and solid line: cross section of Weyhreter et al.[13]. Open points and dashed line: cross section of Haddad al.[15].

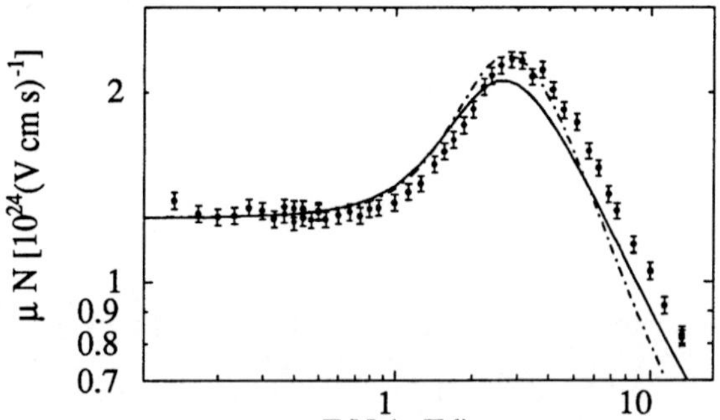

Fig. 5. Calculation of $\mu N(E/N)$ for N=30.2×10^{20} cm^{-3}. Points: experimental data. Solid line: cross section of Weyhreter et al[13]. Dashed line: cross section of Haddad et al[15].

In Ar the cross section is strongly energy dependent, is very large at small energies, and yields large energy shifts. These are the main features determining the behavior of μN. Moreover, T_c is only 12 K below our working temperature so that intrinsic density fluctuations yielding correlations among scatterers cannot be neglected.

As a final remark, we believe that measurements in Ar helped unify the physical picture of the low-energy excess electron scattering processes in dense non polar gases.

REFERENCES

1. S.R.Hunter and L.G.Christophorou, Interphase Physics: Linking Knowledge on Electron-Molecule Interactions in Gases to Knowledge on Such Processes in Condensed Matter, *in:* " Electron-Molecule Interactions And Their Applications", vol.**2**:221, and Electron Motion in Low- and High-Pressure Gases, *ibid.* vol.**2**:89, L.G.Christophorou and K.Siomos Eds., Academic Press, Orlando (1984).
2. L.G.Huxley and R.W.Crompton, "The Diffusion and Drift of Electrons In Gases", Wiley, New York (1974).
3. T.F.O'Malley, Multiple Scattering Effects on Electron Mobilities in Dense Gases, *J.Phys.* **B 13**:1491 (1980).
4. A.K.Bartels, Density Dependence of Electron Drift Velocities in Helium and Hydrogen at 77.6 K, *Appl.Phys.* **8**:59 (1975).
5. A.F.Borghesani, L.Bruschi, M.Santini, and G.Torzo, Electron Mobility in Neon at High Densities, *Phys. Rev.* **A 37**:4828 (1988)
6. G.L.Braglia and V.Dallacasa, Theory of Electron Mobility in Dense Gases, *Phys.Rev.* **A 26**:902 (1982).
7. E.Fermi, Sopra lo Spostamento per Pressione delle Righe Elevate delle Serie Spettrali, *Il Nuovo Cimento* **11**:157 (1934).
8. A.K.Bartels, Density Dependence of the Electron Drift Velocity in Argon, *Phys.Lett.* **44 A**:403 (1973).
9. A.F.Borghesani and M.Santini, Electron Mobility and Localization Effects in High-Density Ne Gas, *Phys. Rev.* **A 42**:7377 (1990).
10. R.Reininger, U.Asaf, I.T.Steinberger, and S.Basak, Relationship between the Energy V_0 of the Quasi-free-Electron and its Mobility in Fluid Argon, Krypton, and Xenon, *Phys.Rev.* **B 28**:4426 (1983).
11. A.F.Borghesani, M.Santini, and P.Lamp, Excess Electron Mobility in High-Density Argon Gas, *Phys.Rev.* **A 46**:7902 (1992).
12. R.Eibl, P.Lamp, and G.Buschhorn, Measurement of Electron Mobility in Liquid and Critical Argon at Low Electric-Field Strengths, *Phys.Rev.* **B 42**:4356 (1990).
13. M.Weyhreter, B.Barzick, A.Mann, and F.Linder, Measurements of Differential Cross Sections for e-Ar, Kr, Xe Scattering at E=0.05-2 eV, *Z.Physik* **D 7**:333 (1988).
14. L.Christophorou and D.L.McCorkle, Experimental Evidence for the Existence of a Ramsauer-Townsend Minimum in Liquid CH_4 and Liquid Ar (Kr and Xe), *Chem.Phys.Lett.* **42**:533 (1976).
15. G.N.Haddad and T.F.O'Malley, Scattering Cross Sections in Argon from Electron Transport Parameters, *Aust.J.Phys.* **35**:35 (1982).
16. E.Wigner and F.Seitz, On the Costitution of Metallic Sodium, *Phys.Rev.* **43**:804 (1933).
17. J.Lekner and A.R.Bishop, Electron Mobility in Simple Fluids Near the Critical Point, *Philos.Mag.* **27**:297 (1973) *and* J.Lekner, Scattering of Waves by an Ensemble of Fluctuating Potentials, *Philos.Mag.* **18**:1281 (1968).
18. G.E.Thomas and P.W.Schmidt, X-Ray Study of Critical Opalescence in Argon, *J.Chem.Phys* **39**:2506 (1963).
19. M.Fixman, Density Correlations, Critical Opalescence, and the Free Energy of Nonuniform Fluids, *J.Chem.Phys.* **33**:1357 (1960).
20. G.H.Wannier, "Statistical Physics", Dover Publications, New York (1987).
21. V.M.Atrazhev and I.T.Iakubov, The Electron Drift Velocity in Dense Gases, *J.Phys.* **D 10**:2155 (1977) *and* A.G.Khrapak and I.T.Iakubov, Electrons and Positrons in Denses Gases, *Sov.Phys.Usp.* **22**:703 (1979).

LOW TEMPERATURE O_2^- MOBILITY IN HIGH-DENSITY Ne GAS

A.F. Borghesani, D. Neri, and M. Santini

Dipartimento di Fisica "G.Galilei", Università di Padova
and Consorzio Interuniversitario Fisica della Materia
Via F.Marzolo 8, I-35131 Padova, Italy

INTRODUCTION

The motion of ions in gases or liquids under the influence of an electric field is a major source of information on the interaction among charged and neutral species as well as on the microscopic structure of the host medium[1]. In low-density gases, where the kinetic theory holds true, the atom-ion interaction potential can be extracted from mobility data[2,3]. In dense liquids, where ordinary hydrodynamics applies, ions have been used as a probe in order to investigate the microscopic behavior and structure of the host medium. However, there are very few systematic studies of the ionic mobilities in the intermediate density region of compressed gases between the low-density gas and the high-density liquid, although the transition from the kinetic regime to the hydrodynamic one is a very important physical topic.

In the experiments aimed at the determination of the ionic mobilities in dense fluids ions in the parent gas are usually observed because they are generated by irradiating the sample with X-ray pulses. As a consequence, the probe changes according to the gas used. On the other hand, in experiments based on the pulsed photoinjection technique, where a bunch of electrons is released by a photocathode irradiated with a short pulse of UV light, O_2^- ions are generated by the resonant attachment process of electrons to O_2 impurities[4] that are normally present even in very pure gases. It is evident that this ion producing technique has the useful feature that the ionic species is the same in any gas, so that specific characteristics of the particular gas under investigation can be put into evidence. Moreover, by reducing the O_2 content to very low levels and suitably introducing a different electron attaching species (such as SF_6) in ppm amounts, the behavior of different ionic species can be studied. On the other hand, this technique has the drawback that at too small gas densities the number of produced ions can be so low as to be not easily detected.

O_2^- mobility measurements in dense He gas at $T = 77.4\,K$ were already carried out in the past in an unsystematic way[5]. The results were unexpected, because the density-normalized zero-field mobility $\mu_0 N$ increases linearly with the gas density in contrast to the prediction of the kinetic theory and because the reduced mobility extrapolated to $N = 0$ is in strong disagreement with the prediction of the Langevin theory based on the scattering of thermal ions by polarization potentials[6].

We therefore decided to carry out systematic studies of O_2^- mobilities in dense gases exploiting the same apparatus and technique developed in our laboratory in order to measure the electron mobility in dense non polar gases[7]. Here we report the results obtained in Neon gas at $T = 45\,K$ in an extended density range[8].

Gaseous Dielectrics VII, Edited by L.G. Christophorou
and D.R. James, Plenum Press, New York, 1994

EXPERIMENTAL RESULTS

The pulsed photoemission technique is very well-known and our experimental apparatus has been extensively described elsewhere[7]. We recall here only its main features. A massive brass cell containing the electrode assembly can withstand pressures up to $\approx$ 14 MPa and can be cooled by means of a commercial cryocooler down to $T \approx 25\,K$. Electrons are released from the gold-coated brass cathode by a UV light pulse of a Xe flashlamp, whose lenght ($\Delta t \approx 4\,\mu$s) is very short compared to the ionic transit time in the drift space. Approximately 2×10^4 to 2×10^6 electrons are released at each light pulse. The gas (Ne "45", 99.995 % vol.) can be purified in a recirculation loop including an active-charcoal liquid N_2 cooled trap and an Oxisorb filter. At $T = 45\,K$ the amount of impurities was estimated to be less than 10 ppm.

Owing to the small number of released electrons, the ion number is also small and direct ion-ion interactions can be neglected. The current induced in the anode circuit by the motion of the ions in the drift space is integrated by means of a passive *RC* circuit and linearly amplified. The signals are recorded by a storage oscilloscope and stored in a personal computer for the analysis. The analysis of the waveform is normally very simple, since the signals are usually straight lines because the ions are formed very close to the anode as a consequence of the strong attachment of electrons to O_2 molecules at high densities. At lower densities, the efficiency of the attachment process is smaller and the signal waveform is more complicated. The shape of the signals and the analysis algorithms are described elsewhere[8,9]. As a result, the overall accuracy of the experimental data is $\approx \pm 5\,\%$.

In Figure 1 we report the mobility μ of O_2^- ions in Neon gas at $T = 45\,K$ as a function of the reduced field E/N for some of the explored densities.

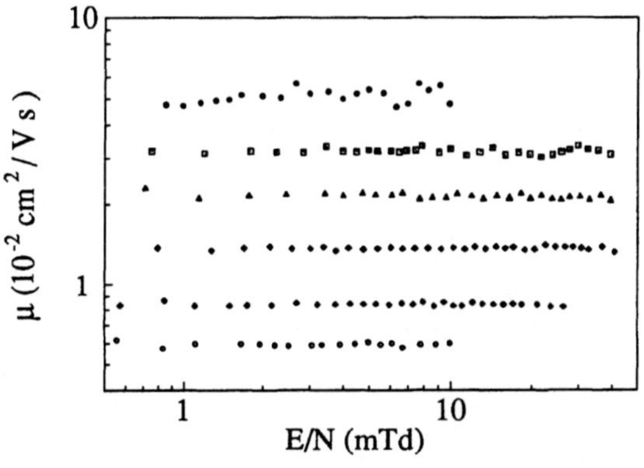

Fig. 1. O_2^- mobility as a function of reduced electric field E/N (1mTd=10^{-20}V cm^2) in Ne at T=45 K for some of the explored densities. From top: N=(15.7, 26.3, 39.4, 63.5, 98.8, 234.5)$\times 10^{20}$ cm^{-3}.

For nearly all densities μ turns out to be field independent over more than one order of magnitude in E/N. Only at the lowest density ($N = 15.7 \times 10^{20}$ cm^{-3}) μ shows a very small field dependence that is probably an instrumental effect due to the difficulty of analyzing the signal wave form when the ionic contribution is small. The independence of μ on E/N means that the mean electron energy is essentially thermal and that the measurements have been carried out in the low electric-field limit.

In Figure 2 we show the zero-field mobility values μ_0 (open points) and the zero-field density-normalized mobility values $\mu_0 N$ as a function of the gas density N. The behavior of $\mu_0 N$ is rather complex, but is very similar to the behavior of

Ar^+, N_2^+, and CH_4^+ in their parent gases[10] and can be divided in three branches as function of N. There is a first region from $N = 0$ up to $N \approx 120 \times 10^{20}\,\mathrm{cm}^{-3}$ where $\mu_0 N = (8.2 \pm 0.4) \times 10^{19}\,(\mathrm{V\ cm\ s})^{-1}$ is nearly density independent. Then, we observe a transition region for $(130 \leq N \leq 200) \times 10^{20}\,\mathrm{cm}^{-3}$, where $\mu_0 N$ increases linearly with N and, finally, $\mu_0 N$ saturates to a constant value $\mu_0 N \approx 14 \times 10^{19}\,(\mathrm{V\ cm\ s})^{-1}$.

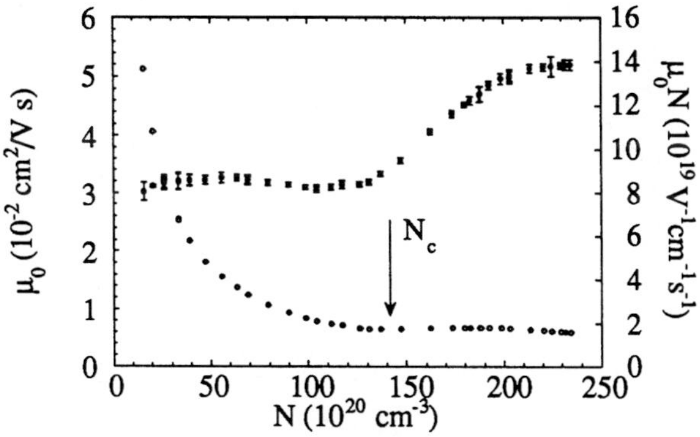

Fig. 2. μ_0 (open points) and $\mu_0 N$ (closed points) of O_2^- at T=45 K as a function of the gas density N. N_c is the critical density.

Classical kinetic theory, in the regime where binary scattering controls the ionic transport, predicts a constant density-independent value for the reduced mobility $\mu_0 N$. Therefore, the first density region would seem that of validity of kinetic theory. Unfortunately, this is not the case. In fact, since the temperature and, hence, the mean ionic energy are fairly low and scattering is governed by the polarization potential due to the interaction of the ionic charge and the dipole moment induced in the Ne atom, the prediction of the classical Boltzmann transport theory is then given by $(\mu_0 N)_P = 4.81 \times 10^{-4}/(\alpha M_r)^{1/2}$, where α is the atomic polarizability of the host gas, and M_r is the reduced mass of the ion-atom system. For Ne, $\alpha = 0.3946 \times 10^{-24}\,\mathrm{cm}^{-3}$ and $M_r = 2.055 \times 10^{-23}\,\mathrm{g}$, and we obtain $(\mu_0 N)_P = 16.9 \times 10^{-19}\,(\mathrm{V\ cm\ s})^{-1}$. This is nearly twice as large as experimentally observed. Even assuming completely clustered ions, i.e., $M_r = M_{Ne}$, we obtain $(\mu_0 N)_P = 13.2 \times 10^{-19}\,(\mathrm{V\ cm\ s})^{-1}$, again much greater than the experimental result.

The problem is that the classical kinetic theory can be used if the ionic mean free path ℓ is larger than the atomic diameter. In our case, a rough estimate of ℓ can be obtained as $\ell = M_r \bar{v}_r \mu / e$, where $\bar{v}_r = (3k_B T/M_r)^{1/2}$. For instance, for $N = 15.7 \times 10^{20}\,\mathrm{cm}^{-3}$ and $T = 45\,K$ we obtain $\ell \approx 2\,\text{Å}$. This value has to be compared with the Lennard-Jones diameter of Ne, $\sigma_{LJ} = 2.76\,\text{Å}$ or with the O_2^-–Ne hard-sphere diameter, which can be estimated to be $\sigma_{O_2^-} \approx 3.2\,\text{Å}$, and with the average interatomic spacing $d \approx N^{-1/3} = 8.6\,\text{Å}$. It is evident that never in our experiment, even at the lowest densities, the conditions for the use of classical kinetic theory have been met. Also the Enskog theory for transport phenomena in dense gases[11], which takes into account the finite atomic size, is in disagreement with our experiment since it predicts a monotonic, though small decrease of $\mu_0 N$ with increasing N.

At the other end of the density range, the gas density is comparable to that of the liquid (at the critical point, $T_c = 44.38\,K$, the critical density is $N_c = 144.3 \times 10^{20}\,\mathrm{cm}^{-3}$) and it is assumed that hydrodynamics can be used to describe the ionic motion. In fact, the mean free path of the Ne atoms for $N = 240 \times 10^{20}\,\mathrm{cm}^{-3}$ is $\ell_{Ne} \approx 1.7\,\text{Å}$, and is smaller than the dimensions of the ion. Moreover, at such densities it is plausible that the ion is surrounded by a solvation shell of atoms of the gas so that its equivalent radius is much larger than that of the bare ion. The mobility in the hydrodynamic regime is given by the Stokes formula

$$\mu_0 = \frac{e}{6\pi\eta R} \tag{1}$$

where η is the gas viscosity. In Figure 3 we plot the $\mu\eta$ product as a function of N.

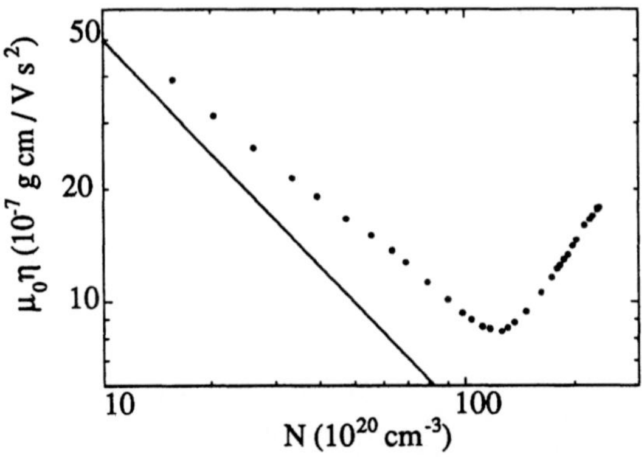

Fig. 3. $\mu_0\eta$ as a function of N. The line is the prediction of kinetic theory.

At large densities, from Eq.(1) $\mu_0\eta$ should be constant if the hydrodynamic radius R were constant and at low densities in the kinetic regime $\mu_0\eta$ should be inversely proportional to N. It is clear that at the lowest density the kinetic regime has not been attained yet and that in the high density range the assumption of a constant hydrodynamic radius is incompatible with the experimental results. On the other hand, the validity of Eq.(1) can be retained by assuming that the hydrodynamic radius of the ion depends on N and thus the mobility data can be used to determine it. In order to account for the fact that our data cover a transition region from the kinetic to the hydrodynamic region we use an interpolation fromula due to Cunningham connecting the ideal gas and ideal liquid limits of ion transport [12]:

$$\mu_0 = \frac{e}{6\pi\eta R}\left(1 + \frac{3\ell}{2R}\right) \tag{2}$$

where ℓ is the ion mean free path. In Figure 4 we show the hydrodynamic radius calculated according to the pure Stokes law Eq.(1) and to Eq.(2) as a function of the gas density N. In this last case we used both the kinetic expression $\ell = \ell_\eta = 3\eta/(M_r\bar{v}_r N)$ for the mean free path, which is not very realistic at high densities because it increases with N, or a more realistic, though still kinetic expression $\ell = \ell_Q = 1/(NQ)$, where we used the hard-sphere cross section for the binary collision of a clustered ion off a Ne atom $Q = \pi\sigma^2 \approx 110\,\text{Å}^2$, with $\sigma = (\sigma_{LJ} + \sigma_{O_2^-})/2 + \sigma_{LJ} \approx 5.74\,\text{Å}$. All the three resulting curves share a similar density dependence. They increase rapidly with N up to a maximum at a density $130 \times 10^{20}\,\text{cm}^{-3}$ close to, but lower than N_c. They then decrease for larger N, approximately saturating to a constant values for large N. At very high N, where a pure hydrodynamic regime is expected, Stokes law Eq.(1) gives a radius R close to the value 4.36 Åcorresponding to a complete solvation shell of Ne atoms clustered around the O_2^-. Also Eq.(2) with $\ell = \ell_Q$ gives nearly the same value, as expected because the correction term ℓ/R is small. On the other hand, the use of the kinetic value ℓ_η fails to predict a reasonable value for R at large N because of the unrealistic behavior of ℓ_η. Moreover, we observe that the hydrodynamic formula Eq.(1) is inadequate at low density, since it predicts too small a radius. Also Eq.(2) with the kinetic mean free path calculated from the viscosity is inadequate since it gives an unreasonably large R value. On the contrary, the semihydrodynamic formula Eq.(2) with $\ell = \ell_Q$ predicts a low-density radius self-consistently close to the value $\sigma = 5.74\,\text{Å}$ used to compute the hard-sphere cross section for the collision of the clustered ion off a Ne atom.

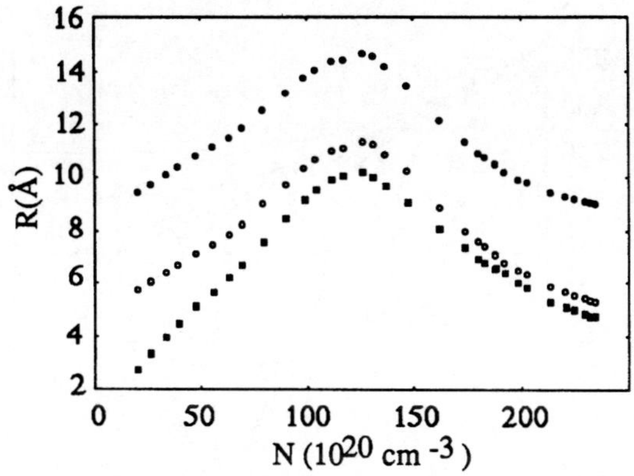

Fig. 4. Semihydrodynamic radius of the ion as a function of N calculated according to Eqns. (1),(2) with $\ell=\ell_Q$, and (2) with $\ell=\ell_\eta$ from bottom.

In any case, the computed radii show the general feature of a maximum at a density close but lower than N_c. In order to explain this fact we adopted the electrostriction model developed by Atkins for ions in liquid He[13]. The fluid is assumed to be a classical continuum described by a given dielectric constant $K(N)$ related to the local density by the Clausius-Mossotti equation. The ion in the gas produces a very strong, non-uniform electric field polarizing the fluid, whose density increases therefore in proximity of the ion with respect to the value in the unperturbed fluid. Since the gas viscosity is an increasing function of the density also the viscosity is larger close to the ion than anywhere else. To determine the drag force on the ion (assumed spheric) the Navier-Stokes equations have thus to be solved in presence of nonuniform density and viscosity profiles. These density profiles are given by the equation

$$[K(N)]^2 \int_{N_\infty}^{N} \frac{1}{N}\left(\frac{\partial P}{\partial N}\right)_T \mathrm{d}N = \frac{1}{2}\alpha\frac{e^2}{4\pi\epsilon_0 r^4} \tag{3}$$

where P is the gas pressure and N_∞ the density of the unperturbed fluid. Some density profiles are shown in Figure 5. The effect of electrostriction is to enhance the local gas density around the ion. For densities $N_\infty \le N_c$ there is a spatial region around the ion where the local density passes through the critical value and here the density profiles are very steep owing to the large compressibility of the gas. The mobility of a sphere with such N and η profiles can be cast in a Stokes-like formula[14]

$$\mu = \frac{e}{6\pi\eta_\infty RF} \tag{4}$$

where R is the radius of the bare ion, η_∞ is the viscosity of the unperturbed fluid, and F is numerically calculated as a function of T and N. The product RF can be thus considered as the hydrodynamic ionic radius.

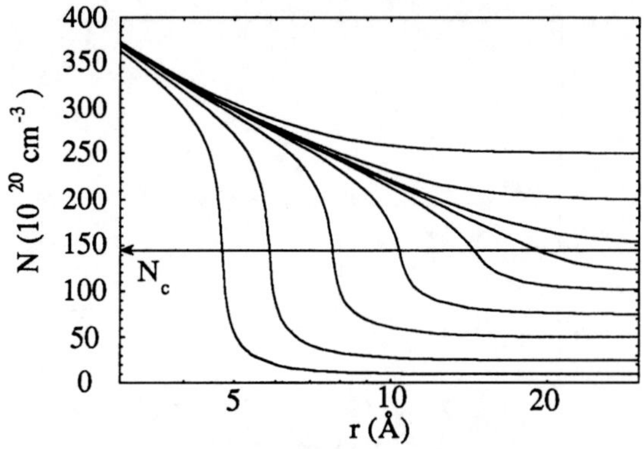

Fig. 5. Density profiles calculated according to the electrostriction model for several unperturbed gas densities. From top: N_∞=(250, 200, 150, 120, 100, 75, 50, 25, 10)×10^{20} cm^{-3}.

In Figure 6 we plot F as a function of N for three different values of the radius of the bare ion. We see that F has a maximum at a density close to, but lower than the N_c, like the radii calculated from the mobility data.

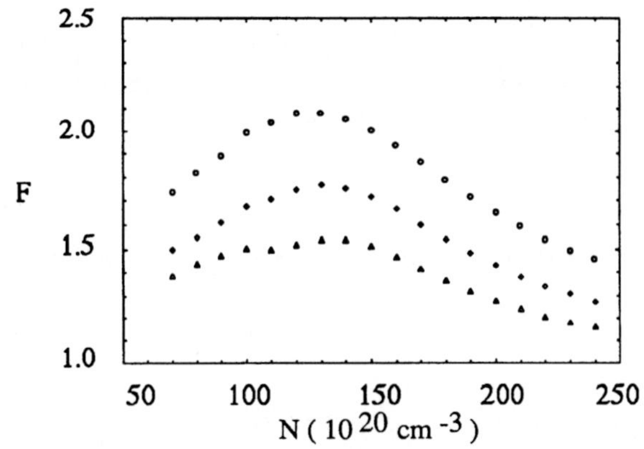

Fig. 6. Correction factor F of the Stokes formula calculated according the electrostriction model for several values of the bare ion radius. From top: R=3.2, 4.0, 5.0 Å.

Finally, we compare in Figure 7 the experimental $\mu_0 N$ data with the prediction of the electrostriction model Eq.(3), as well with the pure Stokes formula results obtained with the same bare ion radius. We see that Eq.(1) does not reproduce the experimental data, while the agreement obtained with the electrostriction model is much better, showing deviations not larger than 40% . We point out that the residual discrepancies shown by the electrostriction model might be attributed to the fact that (i) a continuum model has been assumed throughout, (ii) the steepness of the density profiles for $N_\infty \leq N_c$ is not very realistic and is in contrast with the assumption of local thermodynamic equilibrium. Moreover, we point out that the largest discrepancies are close to the N_c, where the sound velocity in the gas becomes smaller than the thermal velocity of the ions, so that the fluid cannot react readily to the passage of the charged particle and the density profiles cannot fully establish.

The conclusions drawn from this experiment are that in a dense gas the ionic transport mechanism cannot be simply explained in terms of either the kinetic or of the pure hydrodynamic theory. The measurements indicate that the experimental conditions correspond to a transition regime. Extensive measurements have to be further carried out in order to better clarify this problem.

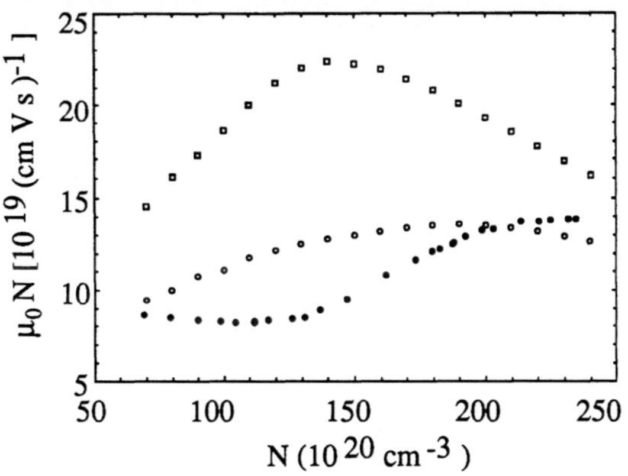

Fig. 7. $\mu_0 N$ as a function of N. Closed points: experimental data. Open squares: Eq. (1). Open circles: Eq.(4). R=4 Å has been used throughout.

REFERENCES

1. G.R.Freeman and D.A.Armstrong, Electron and Ion Mobilities, *in*: "Advances of Molecular Physics", D.Bates and B.Bederson Eds., vol.**20**:267, Academic Press, Orlando (1985).
2. E.A.Mason and E.W.McDaniel. "Transport Properties of Ions in Gases", Wiley, New York (1988).
3. E.W.McDaniel and E.A.Mason. "The Mobility and Diffusion of Ions in Gases", Wiley, New York (1973).
4. S.R.Hunter and L.G.Christophorou, Interphase Physics: Linking Knowledge on Electron-Molecule Interactions in Gases to Knowledge on Such Processes in Condensed Matter, *in*: "Electron-Molecule Interactions And Their Applications", L.G.Christophorou and K.Siomos Eds., vol.**2**:221, Academic Press, Orlando (1984).
5. A.K.Bartels, Messungen der Elektronendriftgeschwindigkeit in He, Ar, H_2 Und N_2 bei sehr hohen Gasdichten, Ph.D. Thesis, University of Hamburg, Germany (1971), unpublished.
6. G.H.Wannier, "Statistical Physics", Dover, New York (1987).
7. A.F.Borghesani, L.Bruschi, M.Santini, and G.Torzo, Electron Mobility in Neon at High densities, *Phys. Rev.* **A 37**:4828 (1988) .
8. A.F.Borghesani, D.Neri, and M.Santini, Low-Temperature O_2^- Mobility in High-Density Neon Gas, *Phys. Rev.* **E 48**:1379 (1993).
9. A.F.Borghesani and M.Santini, Electron Swarm Experiments in Fluids - Signal Waveform Analysis, *Meas. Sci. Technol.* **1**:939 (1990).
10. N.Gee, S.S.-S. Huang, T.Wada, and G.R.Freeman, Comparison of Transition from Low to High Density Transport Behavior for Ions and Neutral Molecules in Simple Fluids, *J. Chem. Phys.* **77**:1411 (1982).
11. J.O.Hirschfelder, C.F.Curtiss, and R.B.Bird, "Molecular Theory of Gases and Liquids", Wiley, New York (1954).
12. A.M.Tyndall,"The Mobility of Positive Ions in Gases", Cambridge University Press, Cambridge (1938)
13. K.R.Atkins, Ions in Liquid Helium, *Phys. Rev.* **116**:1339 (1959).
14. R.M.Ostermeier and K.W.Schwarz, Motion of Charge Carriers in Normal He^4, *Phys. Rev.* **A 5**:2510 (1972).

MEASUREMENT OF ELECTRON TRANSPORT IN HIGH DENSITY GASEOUS ARGON IN CROSSED ELECTRIC AND MAGNETIC FIELDS

P. Lamp, G. Buschhorn

Max-Planck-Institut für Physik
Föhringer Ring 6
D-80805 München, Germany

INTRODUCTION

In low density gases, where only classical single–scattering has to be regarded, it is found that the quantity $\mu_D N$ which is the drift mobility μ_D – given by the electron drift velocity v_D in an electric field E by the relation $\mu_D = E/v_D$ – multiplied by the gas density N is a universal function of the reduced field E/N, i.e. independent of the gas density. At higher densities in a number of gases anomalous density effects have been observed [1, 2, 3, 4, 5, 6, 7, 8, 9], demonstrating the need for multiple scattering corrections. Several attempts [10, 11, 12, 13, 14], have been made to explain this density effect within Boltzmann transport theory by modifying the single electron–atom scattering cross section in a density dependent way. These models were only partly successful and they in particular treat gases with positive and negative scattering lengths showing opposite density effects in a different way. For example for He, Ne (positive scattering length) $\mu_D N$ decreases with increasing N while for Ar, Kr, Xe (negative scattering lenght) $\mu_D N$ increases. Recently a phenomenological model has been put forward which explains both the negative density effect in Ne [15] and the positive density effect in Ar [16] in the same way. The main features of the model are a positive shift of the kinetic energy of the electrons in the dense medium and the consideration of correlations among scatterers by taking into account the static structure function. This model is able to explain the electric field dependence of the drift mobility data in Ar for densities up to $N \leq 70 \cdot 10^{20}\,cm^{-3}$ [16].

The measurement of the electron transport parameters in a combined electric and magnetic field is an important test of this model. The solution of the Boltzmann equation for the electron drift in an electric and magnetic field at low densities is available [17]. The density dependent changes of the scattering cross section introduced in [16] for the special case $B = 0$ can be applied also for the full formulas for $B \neq 0$. In contrast to measurements at very low densities with crossed electric and magnetic

Gaseous Dielectrics VII, Edited by L.G. Christophorou
and D.R. James, Plenum Press, New York, 1994

fields [19, 20, 21], to our knowledge no similar measurement exists in high density gas were the above mentioned density effects are observed. Therefore we have set up an experiment which allows to measure both components of the drift velocity vector (along $\vec{E}$ and $\vec{E} \times \vec{B}$) of an electron moving in a crossed electric and magnetic field in dielectric gases of high densities up to the liquid. By means of a superconducting solenoid magnetic fields of $5\,T$ are reached. We have done measurements in high density Ar gas and compared the experimental results with calculations performed using the model of [16].

EXPERIMENT

The measuring principle is to detect the deflection angle θ_H (Hall-angle) between the electron drift velocity $\vec{v}_D$ and the electric field vector $\vec{E}$ which appears if a magnetic field $\vec{B}$ perpendicular to the electric field is present. With the convention $\vec{E} = (E, 0, 0)$ and $\vec{B} = (0, 0, B)$ it follows for the components of the drift velocity $v_{D,x}, v_{D,y} \neq 0$, $v_{D,z} = 0$ and $tan\theta_H = v_{D,y}/v_{D,x}$. The velocity component $v_{D,x}$ is directly obtained by the time-of-flight (TOF) method. Therefore, measuring $tan\theta_H$ is equivalent to determining $v_{D,y}$. The drift mobility μ_D is calculated by $\mu_D(E) = v_{D,x}(B = 0)/E$.

The excess electrons are generated using the pulsed photoemission method as in our previous measurements [16, 22], i.e. by irradiating a gold photocathode with 1μs UV-pulses from a Xe-flashlamp (Hamamatsu model L2435). The area illuminated by the UV-light has a diameter of $6\,mm$. The electrodes are of circular shape with a radius of $12.5\,mm$ and are separated by a gap of $2.3\,mm$. The photocathode is held on a negative potential to provide the electric drift field $\vec{E} = (E, 0, 0)$. The anode is separated into two halves, with the deviding cut parallel to the direction of the magnetic field. The current induced by the drifting electrons in each part of the anode is integrated by means of a charge–sensitive amplifier (time constant $\approx 1ms$). The output of each amplifier is fed into a two–channel digital storage oscilloscope which is connected via GPIB–bus to a Personal Computer. With no magnetic field applied the measured current is the same for both parts of the anode. Increasing the magnetic field leads to an increasing deflection of the drifting electrons perpendicular to the magnetic field and therefore to a difference between the signals from the two parts of the anode. A schematic drawing of the electrode configuration is shown in figure 1 .

The analysis of the time dependence of the sum of both signals yields the drift time t_D from which the velocity component $v_{D,x} = d/t_D$ is calculated. With $B = 0$ this technique is identical to the usual TOF method to measure the drift mobility [16, 22]. If a magnetic field is applied the number of electrons collected on each part of the anode differs due to the Lorentz–force acting on the drifting electrons. The charges Q_l and Q_r on the two halves of the anode which are deduced from the amplitude of the signals allow to determine Δx, i.e. the deflection of the electrons perpendicular to the direction of the electric field. Then $tan\theta_H$ is given by the relation $tan\theta_H = \Delta x/d$. An example of the charge measurement is shown in figure 2 . The behaviour of the charge ratio $(Q_{l,r}(B) - Q_o)/Q_o$ (Q_o is the value measured for $B = 0$) is symmetric with respect to the reversal of the magnetic field. Any asymmetry would indicate a misalignment of the electrode configuration or a systematic error of the measuring circuits. Nevertheless, a possible misalignment of the electrodes is taken care of by calculating the deflection Δx from the combination of the measurements at $\pm B$.

The Ar drift cell itself is mounted in a variable temperature insert which fits into the central bore of a superconducting solenoid. Within the drift volume of about $1\,cm^3$

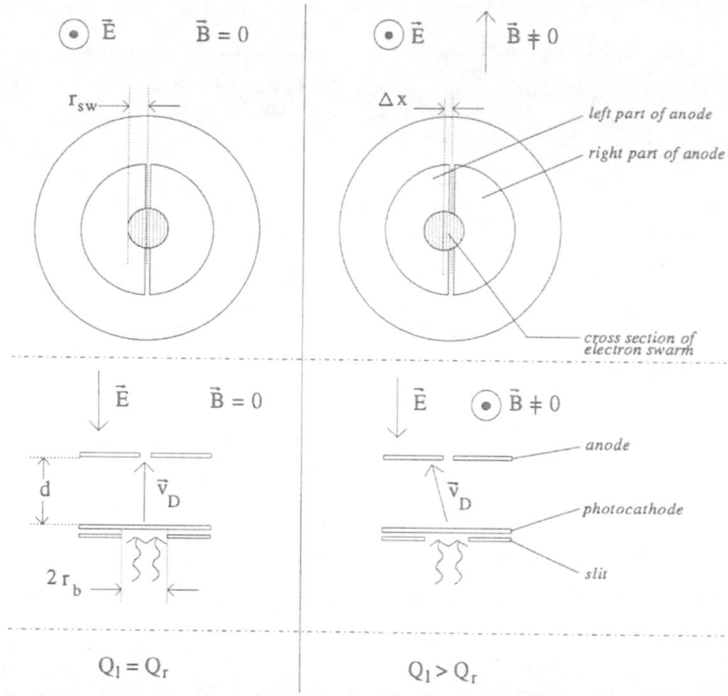

Figure 1. Schematics of the electrode configuration. Q_l, Q_r: charge collected on the left and right part of the anode; Δx: deflection of the electron swarm; d: distance between the electrodes, r_{sw}: radius of the electron swarm; r_b: radius of the diaphragm defining the UV–light beam.

the magnetic field variation is $< 0.1\,\%$. The usual precautions and cleaning procedures have been applied to obtain Ar gas with an impurity content in the *ppb*–range.

The temperature the argon cell is regulated with a resolution of about $20mK$ and a stability of approximately $50mK$ over several hours. The pressure is determined to a accuracy of $\pm 0.05\,bar$ and the density calculated from the temperature and pressure following [23].

The measuring error is of the order of 10 %, being slightly higher for $v_{D,y}$ compared to $v_{D,x}$.

RESULTS AND DISCUSSION

In figure 3 we show $v_{D,x}$ and $v_{D,y}$ as function of the magnetic field strength B for a density of $N = 10.0 \cdot 10^{20}\,cm^{-3}$. The measurements have been performed at electrical field strengths of 11.0, 22.0, 52.9 and 110.1 V/cm .The open symbols are the experimental data while the dashed lines are results of the calculation which will be explained below. The decrease of the velocity component $v_{D,x}$ with increasing

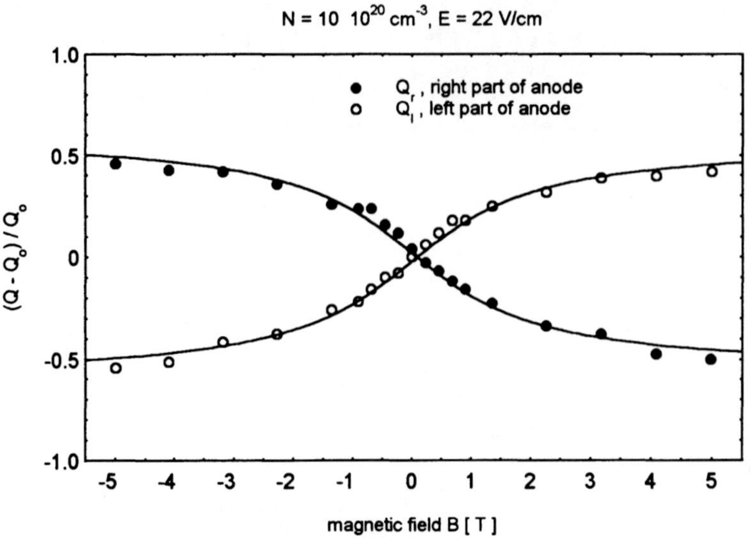

Figure 2. Normalized charge collected on each half of the anode as function of the magnetic field. Q_o is the value for $B = 0$. Note the symmetric behaviour of the normalized charge with respect to a reversal of the magnetic field and an exchange of left and right. The lines are to guide the eye.

magnetic field reflects the fact that a strong magnetic field leads to curved trajectories between two consecutive scattering events thus leading to a shorter 'effective' mean free path. The velocity component $v_{D,y}$ increases with increasing magnetic field due to the Lorentz–force ($\propto B$) acting on the electrons but saturates (or even decreases again) if the drift mobility is high. The relative strength of the influence of the magnetic field depends on the drift mobility and not on the absolute value of the drift velocity as can be understood qualitatively from the usual description of the Hall effect in solid state physics [24]. The important factor governing this effect is $\omega_c\tau$, i.e. the cyclotron frequency $\omega_c = eB/m$ (e, m: electron charge and mass) times the mean scattering time τ . For example $v_{D,x}$ is given by

$$v_{D,x}(B) = \frac{eE}{m}\langle\frac{\tau}{1+\omega_c^2\tau^2}\rangle \tag{1}$$

where the brackets indicate the average over the electron distribution function. If $\omega_c\tau \ll 1$ to a first order approximation the drift velocity $v_{D,x}$ is independent of B while $v_{D,y}$ increases linearily with B. Only for $\omega_c\tau \approx 1$ the strong influence observed in this experiment follows. At $N = 10 \cdot 10^{20}\,cm^{-3}$ and $E = 52.9\,V/cm$ the drift mobility is $\mu_D = 1370\,cm^2/Vs$. Estimating τ from the approximate relation $\mu_D = e\tau/m$ leads to $\omega_c\tau = 0.68$ at $B = 5\,T$ resulting in a strong dependence of the drift velocity on the magnetic field as actually observed.

For more quantitative calculations the specific form of the electron distribution function and the energy dependence of the mean scattering time has to be taken into account. Starting with the drift velocity of an electron swarm in crossed electric and magnetic fields and in a gas of very low density, i.e. only single electron–atom scattering

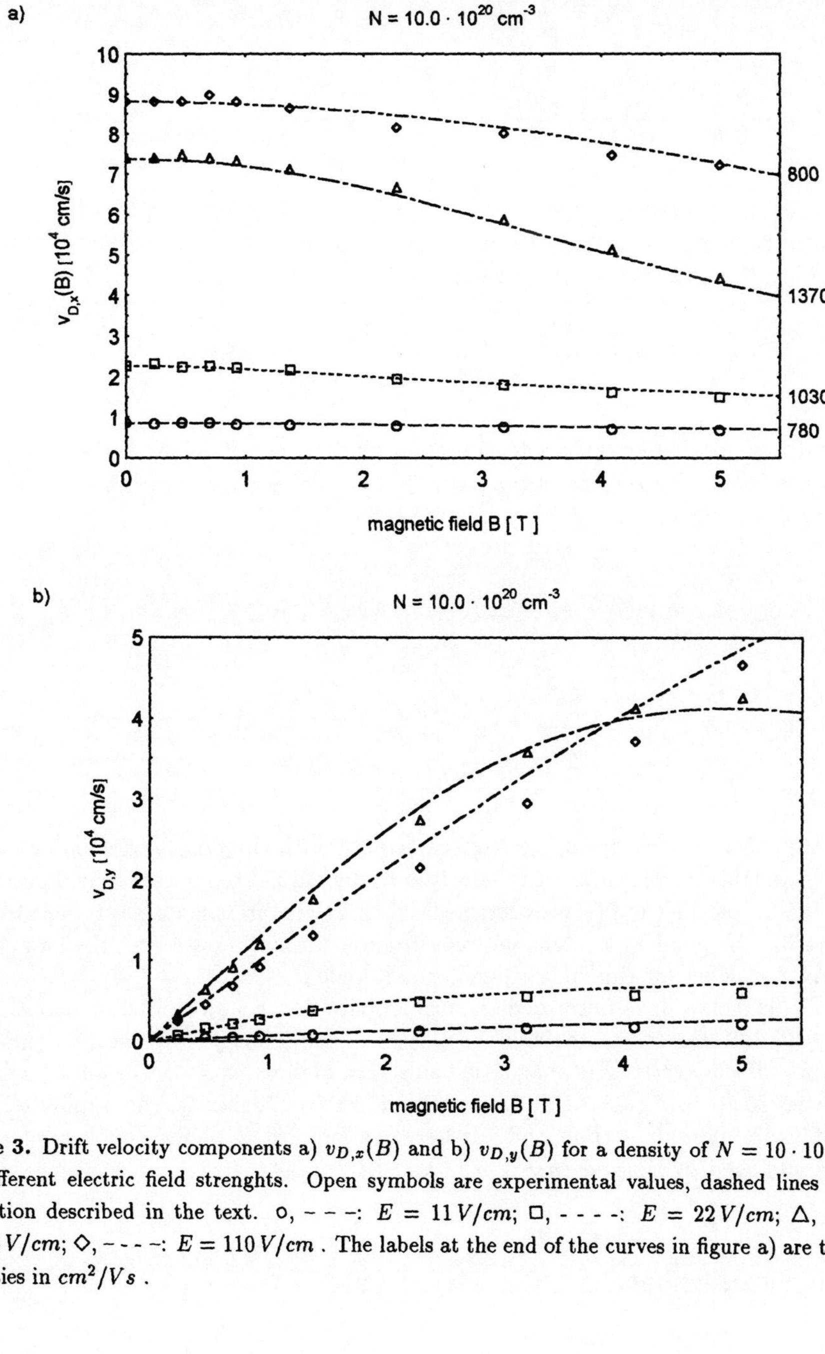

Figure 3. Drift velocity components a) $v_{D,x}(B)$ and b) $v_{D,y}(B)$ for a density of $N = 10 \cdot 10^{20}\,cm^{-3}$ and different electric field strenghts. Open symbols are experimental values, dashed lines are the calculation described in the text. o, – – –: $E = 11\,V/cm$; □, - - - -: $E = 22\,V/cm$; Δ, – - – -: $E = 53\,V/cm$; ◇, - - - –: $E = 110\,V/cm$. The labels at the end of the curves in figure a) are the drift mobilities in cm^2/Vs .

prevails, one has [17]:

$$v_x = \frac{e}{m} E I_1 \quad ; \quad v_y = \frac{e}{m} E \omega_c I_2 \tag{2}$$

$$I_1 = -\frac{2}{3}\int_0^\infty \frac{\nu(\epsilon)\epsilon^{3/2}}{\nu^2(\epsilon)+\omega_c^2}\frac{dg(\epsilon)}{d\epsilon}d\epsilon \quad ; \quad I_2 = -\frac{2}{3}\int_0^\infty \frac{\epsilon^{3/2}}{\nu^2(\epsilon)+\omega_c^2}\frac{dg(\epsilon)}{d\epsilon}d\epsilon \tag{3}$$

with ϵ = electron energy, $\nu(\epsilon) = 1/\tau(\epsilon)$ = mean scattering rate, $\sigma_M(\epsilon)$ = scattering cross section for momentum transfer.

The distribution function $g(\epsilon)$ with magnetic field is given by

$$g(\epsilon) = A \cdot exp\{-\int_0^\epsilon [\frac{e^2 M E^2}{6mN^2 z \sigma_M^2(z) + 3m^2\omega_c^2} + k_B T]^{-1} dz\} \quad . \tag{4}$$

In order to adapt this equations to the high density gas we follow the model of [16] and replace in eq. 2 to 4 the atomic scattering cross section $\sigma_M(\epsilon)$ by the effective scattering cross section

$$\sigma_M^*(\epsilon') = F(\epsilon')\sigma_M(\epsilon')[1 + N\lambda_B F(\epsilon')\sigma_M(\epsilon')/\pi] \tag{5}$$

with $\epsilon' = \epsilon + \epsilon_o$ and

$$F(k) = \frac{1}{4k}\int_0^{2k} q^3 S(q) dq \quad , \tag{6}$$

taking into account correlations among scatterers. With the density dependent values of ϵ_o given in [16] this provides a model without any adjustable parameter. The results of the calculation for the given experimental conditions are shown as dashed lines in the figure 3. The good agreement between the experimental data and the calculation gives strong support for this phenomenological model.

In conclusion we have described in this contribution a new and direct method to measure the different components of the electron drift velocity in dielectric gases (and liquids) in crossed electric and magnetic fields. For high density Ar gas and0 magnetic field strengths up to $5\,T$ the results presented here demonstrate the applicability of the used experimental method and furthermore support a phenomenological model recently published by Borghesani, Santini and Lamp [16].

ACKNOWLEDGEMENT

We want to thank Mr. W. Erbe for the technical support by setting up the experiment and Prof. A.F. Borghesani for encouraging discussions and the leaving of his computer codes.

REFERENCES

[1] J.C.Bowe, Phys. Rev., **117**, 1411 (1960).

[2] J.A.Jahnke, L.Meyer, S.A.Rice, Phys. Rev., **A 3**, 734 (1971).

[3] A.K.Bartels, Phys. Lett., **44 A**, 403 (1973).

[4] S.S.-S. Huang, G.R.Freeman, Phys. Rev., **A 24**, 714 (1981).

[5] R.Grünberg, Z. Naturforsch., **23 a**, 1994 (1968).

[6] J.L.Levine, T.M.Sanders, Phys. Rev., **154**, 138 (1967).

[7] K.W.Schwarz, Phys. Rev., **B 21**, 5125 (1980).

[8] A.F.Borghesani, L.Bruschi, M.Santini, G.Torzo, Phys. Lett., **108 A**, 255 (1985).

[9] N.L.Allen, B.A.Prew, J. Phys., **B 3**, 1113 (1970).

[10] E. Fermi, Nuovo Cimento **11**, 157 (1937).

[11] M. Lax, Rev. Mod. Phys. **23**, 287 (1951).

[12] L.L. Foldy, Phys. Rev. **67**, 107 (1945).

[13] T.F. O'Malley, J. Phys. **B 13**, 1491 (1980).

[14] G.L. Braglia, V. Dallacasa, Phys. Rev. **A 26**, 902 (1982).

[15] A.F. Borghesani, L. Bruschi, M. Santini, G. Torzo, Phys. Rev., **A 37**, 4828 (1988).

[16] A.F. Borghesani, M. Santini, P. Lamp, Phys. Rev. **A 46**, 7902 (1992).

[17] L. G. H. Huxley, R. W. Crompton, 'The Diffusion and Drift of Electrons in Gases', Wiley, New York 1974.

[18] K.F. Ness, Phys. Rev., **E 47**, 327 (1993).

[19] B. Schmidt, J. Phys., **B 24**, 4809 (1991).

[20] T. Kunst, B. Götz, B. Schmidt, Nucl. Instr. & Methods, **A 324**, 127 (1993).

[21] B. Schmidt, Comments At. Mol. Phys., **28**, 379 (1993).

[22] R. Eibl, P. Lamp, G. Buschhorn, Phys. Rev., **B 42**, 4356 (1990).

[23] D.G. Friend, NIST Thermophysical Properties of Pure Fluids Database, Version 3.0, National Institut of Standards and Technology, Gaithersburg 1992.

[24] R.A. Smith, Semiconductors, Cambridge University Press, Cambridge, 1978.

PRESSURE SHIFTS AND ELECTRON SCATTERING IN ATOMIC AND MOLECULAR GASES

K. Rupnik[1], U. Asaf[2], and S. P. McGlynn[1]

[1]Chemistry Department, Louisiana State University, Baton Rouge, LA 70803
[2]Racah Institute of Physics, The Hebrew University, Jerusalem, Israel 91904

INTRODUCTION

In this work, we focus on one aspect of Rydberg electron scattering, namely number density effects in molecular gases. Our recent study of Rydberg states of CH_3I and C_6H_6 perturbed by H_2 is the first attempt to investigate number density effects of a molecular perturber on Rydberg electrons[1,2]. Highly excited Rydberg states, because of their "large orbital" nature, are very sensitive to the surrounding medium. Photoabsorption or photoionization spectra of CH_3I have also been measured[3] as a function of perturber pressure in 11 different binary gas mixtures consisting of CH_3I and each one of eleven different gaseous perturbers. Five of the perturbers were rare gases {He, Ne, Ar, Kr, Xe}and six were non-dipolar molecules {H_2 , CH_4, N_2, C_2H_6, C_3H_8}. The goal of this work is to underline similarities and differences between atomic and molecular perturbers. We first list some results of the molecular study:

(1) The energy shifts of CH_3I Rydberg states become independent of n, the principal quantum number, for $n \geq 10$.

(2) The energy shifts for $n \geq 10$ vary in a linear fashion with perturber number density.

(3) A particular advantage of molecular absorbers (relative, that is, to alkali metal absorbers) is that they may be studied at room temperature. Indeed, the lower temperature limit, being dependent only on the freezing point of the particular absorber/perturber system, can lie considerably below room temperature. That is, molecular perturbers can be studied well below their vibrational thermal excitation thresholds in a molecular absorber system whereas such is not possible with the common alkali metal absorber systems.

(4) Since molecules can rotate and vibrate, the density of states in them is usually much larger than in atoms. Consequently, the band widths of molecular absorbers are always considerably broader than those of atoms. There is considerable disadvantage in this because the minimal shift size which can be measured for a molecule is larger than that for an atom, the minimal shift sizes being roughly in the ratio of the Rydberg excitation bandwidths. There is also some advantage in this: asymptoticity of energy may occur in molecular absorbers at principal quantum number values as low as n=10; and large perturber densities, certainly as large as $\rho = 10^{21} cm^{-3}$, may be investigated before the

Gaseous Dielectrics VII, Edited by L.G. Christophorou
and D.R. James, Plenum Press, New York, 1994

effects of pressure on band width and shape begin to confuse the measurements of band shift . The "signatures" of molecular rovibronic spectra are not totally insensitive to number density effects, and investigations on this matter are in progress in our Laboratory.

(5) The measurement of the pressure shifts of absorption spectra has been restricted to perturbers that do not absorb in the region of Rydberg excitation of the absorber. This restriction may be removed by doing photoionization experiments[3] in which synchrotron radiation excites the autoionization region of the designated absorber in a narrow sheath in the immediate vicinity of the surface of optical incidence. These radiations, however, must be insufficiently energetic to ionize the designated perturber. Since the vast majority of potential perturber systems absorb in the VUV region, this technique broadens the scope of the pressure shift technique. (LSU has a new synchrotron facility and the equipment for such investigations.)

Spectrographic studies of density dependent shifts of atomic Rydbergs have been known since the pioneering 1934 work by Amaldi and Serge[4], which used gaseous perturbers to modify the excited states of Na and K atoms. Fermi presented a simple model[5] of the observed energy shifts. According to Fermi, the interaction between a highly excited absorber and a perturbing medium can be decomposed into two independent, additive terms. These are the scattering term, $\Delta_{sc}(\rho)$, describing the interaction between the optical electron and perturber, and a polarization term, $\Delta_p(\rho)$, describing the interaction between the cationic core and perturber. The sum of these two terms is the total energy shift, $\Delta_t(\rho)$, of the high-n Rydberg state. The density dependence of the scattering shift $\Delta_{sc}(\rho)$ speaks to the evolution of band structure in insulators. The $\Delta_{sc}(\rho) \equiv V_0(\rho)$ equivalence, where V_0 is the electron energy at the bottom of the conduction band, is the crucial connector. V_0 is known to be polarization dependent.

COMPARISON OF PRESSURE SHIFT, TIME-OF-FLIGHT AND SWARM SCATTERING LENGTHS

Since the total energy shift of all high-n Rydberg states of CH_3I becomes constant for all $n \geq 10$, the average asymptotic energy shift for CH_3I can be measured and can be presented as a function of perturber number density. These density dependencies are consistently linear and are given as "averaged total shift rates" in the first column of Table 1. The typical experimental slope error lies in the range 0.1-0.3x10^{-23} eVcm^3 when ρ is of the order of 10^{20}cm^{-3}. The precise individual slope errors are cited in Table 1. The accuracy of the shift measurements was approximately $\pm$ 0.1Å, corresponding to ~0.7 meV in the spectroscopic range of interest. These data agree well with both very old[8] and quite recent[9] pressure shift data, the latter of which used "Doppler-free" laser techniques to investigate the Rydbergs of alkali metal absorbers. (For pressure shift studies at high number densities, deviations from ideal gas behavior will occur. Consequently, number densities in this work were calculated using the Beattie-Bridgeman equation with the parameter sets for the various perturber gases.[10]) Perturber polarizabilities are also collected in Table 1. The polarizability of radon has been cited as 39.5a.u. by Tuck[11] and 42.5a.u. by Miller[12].

The scattering term, which describes the electron scattering contribution, can be compared with results from other electron scattering experiments. Using a modified Fermi approach[13], electron scattering lengths, A, can be extracted directly from measured shifts. Thus, we compare three sets of scattering length data in this work: one from density effects on high-n Rydberg states of CH_3I, a second from TOF (time-of-flight) data, and a third based on the drift of electron swarms. The latter two data sets yield A only after extrapolation to zero energy using MERT theory. MERT (modified effective range theory)[14] pertains to the phase shifts of the partial waves of a scattering system consisting of an electron and a neutral polarizable atom or molecule, and can be used for extrapolation to otherwise experimentally inaccessible low electron energies. The primary adjustable parameter in MERT is the electron scattering length. Comparisons of scattering lengths from pressure shift data for the CH_3I Rydberg states n=10-16 with those from TOF and swarm measurements are given in Table 1. The values of A from both the TOF and swarm sources were extrapolated by MERT to zero energy; those from shift data were not. The agreement of the three data sets is good.

Table 1. The averaged total shift rates for the high-n Rydbergs of CH_3I (n=10-16), cited in units of 10^{-23} eV cm^3, followed by polarizabilities, cited in units of 10^{-24} cm^3, followed by electron scattering lengths (in units of Bohr radius, a_0) from pressure shift (A), and from MERT extrapolation of both TOF (A_{tof}) cross-sections and swarm (A_{sw}) cross-sections. Average polarizabilities are taken from Landolt-Börnstein, *Zahlenwerte und Funktionen*, I Bd., Springer, Berlin, 1951; those for Rn are taken from from refs. 11 and 12. References for all other data sources are given in Table 1 [in brackets]. The scattering lengths for Rn, shown in parentheses, are calculated from the atomic correlation line.

gas	Δ_t/ρ	α	(A)	(A_{tof})	(A_{sw})
He	2.45±0.06[16]	0.20	1.08±0.03	1.95[22]	1.19[30]
				1.16[23]	
Ne	0.09±0.03[16]	0.40	0.17±0.01	--	0.21[30]
Ar	-4.86±0.4[16,1,6]	1.65	-1.63±0.09	-1.593[23]	-1.48[30]
			-1.55±0.09	-1.492[24]	
				-1.449[25]	
Kr	-8.6±0.2[16,17]	2.48	-3.02±0.11	-3.478[24]	-3.39[30]
				-3.19[26]	
Xe	-16.8+2.0[18]	4.04	-6.12±0.49	-6.527[24]	-6.09[30]
Rn	- - -	**5.86**	**(-9.05)**	--	--
	- - -	**6.3**	**(-9.8)**	--	--
H_2	1.62±0.10[1]	0.81	0.93±0.04	1.27[27,28]	1.27[27,28]
N_2	0±0.15[19]	1.76	0.35±0.06	--	0.44[22]
CH_4	-7.86±0.15[20]	2.60	-2.61±0.04	-2.475[29]	-2.755[31]
C_2H_6	-10.05±0.2[20]	4.47	-3.33±0.08	--	
C_3H_8	-12.93±0.26[20]	6.29	-4.31±0.06	--	--
CO_2	-11.8±0.5[21]	2.65	-4.23±0.17	--	--

THE DEPENDENCE OF SCATTERING LENGTH ON POLARIZABILITY

All available models for the electron scattering length, A, of atomic and molecular systems, as well as the expression for Vo, contain some kind of polarizability dependence. Certainly, scattering cross-sections appear to increase monotonically with polarizability. Consequently, we have investigated the effect of polarizabilities on the number density shifts and on A. We find, as previously,[2] that a linear correlation exists between the electron scattering length, A, and the polarizability α for He, Ne, Ar, Kr, and Xe gases. This correlation is shown in Fig. 1. A calculated value for Rn is also shown. This simple correlation is considerably different in intent and complexity from the modified effective-range models which have been developed for rare gases. In specific, it makes use of only one parameter: the polarizability; and it presumes to yield scattering lengths at the low-energy limit. In addition, the nature of the scattering potential (whether repulsive or attractive) and the shift (whether blue or red) are contained in the correlative behavior, which predicts both the size and sign of the electron scattering length. As our work is primarily aimed at the study of molecular scatterers, we also present data for H_2, N_2, CO_2, CH_4, C_2H_6 and C_3H_8. If we exclude CO_2 from the molecular set because of its very large (absolute value) quadrupole moment of $-14.8 \times 10^{-14}Cm^2$, we find that these two sets, atoms and molecules, define two reasonable linearities: $A(a_o) = -0.265\alpha + 1.23$ for atoms, with $d = 0.1a_o$, d being the standard deviation in atomic units; and $A(a_o) = -0.144\bar{\alpha}+1.304$ for molecules, with $d = 0.4a_o$, $\bar{\alpha}$ being the average polarizability. The slopes of these two lines are quite different, but the intercepts are very similar. The line for molecules is also shown in Fig. 1. These linearities represent a very novel finding. Yet, this linearity is not too surprizing since, at least for large atoms, it is well known that the scattering cross-section increases monotonically with increasing polarizability. However, the observed linearity is very precise and it indicates, in agreement with Omont[15], that the monotonicity referred to in the previous sentence does not extend to low atomic masses. The cross-section ($\sigma \propto A^2 \propto \alpha^2$) is parabolic in the polarizability, with the left-hand leg of the parabola being truncated. These results indicate the following: 1), the maximum value of the scattering length (i.e., that for $\alpha=0$) appears to be a natural constant; and 2), scattering cross-sections are not monotonic in the polarizability when low atomic masses are considered, whereas the scattering length is.

CONCLUSION

The spectroscopic study of perturber number density effects is a very efficient approach to mechanisms of momentum and energy transfer and deposition in highly-energetic electronic processes. It is found that the electron scattering lengths and cross sections of rare gases obtained in this work agree well with the existing data for pressure shift studies using atomic Rydbergs. A recent empirical model which provides a linear correlation of the electron scattering lengths of rare gases (He to Xe) and atomic polarizabilities, has been extended to molecular scatterers. The results confirm the system representation as the polarization of perturber molecules and/or atoms by an external (Rydberg) electron. This mechanism is based on "soft"/"hard" perturber deformations which result in attraction/repulsion.

The new findings contained in this work are:

--- It has been shown for both atoms and molecules that the A vs α correlation lines are not even approximately the same. Further data, particularly for non-polar, low-quadrupolar molecules are required to clarify this point.

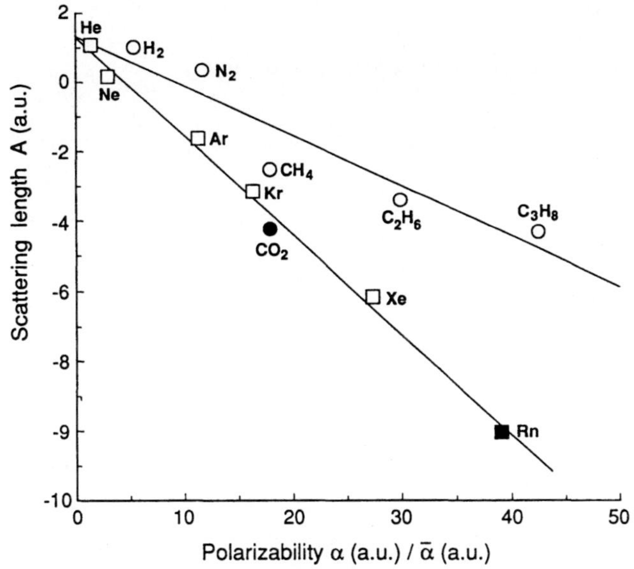

Figure 1. Scattering lengths of rare gases (denoted by open squares) and molecules (denoted by open circles) are shown plotted against perturber polarizabilities. The best straight line fits are the solid lines which have a slope = -0.265 for the rare gases and a slope = -0.144 for the molecules. The point for CO_2 is shown as a solid circle; it is not included in either correlation for reasons cited in the text. The point for Rn, shown as a solid square, corresponds to the polarizabilty α=39.5 a.u. The equations for the lines are given in the text.

---The scattering lengths for molecules are smaller in absolute value than for rare gases of the same polarizability. That is, molecules attract electrons less than do atoms.

--- Molecular structure plays an important role: the methane molecule, CH_4, lies closer to the atom line than to the molecular line. Whether or not this is a consequence of the tetrahedral (i.e., near-spherical) symmetry of CH_4 we cannot say. The investigation of other tetrahedral (e.g., SiH_4) and octahedral (e.g., SF_6) molecules would be of considerable interest.

--- We have also investigated the role of polarizability anisotropy which is responsible for many molecular properties[2]. Polarizability is a tensor. The importrance of polarizability components for use in the correlation of scattering length and polarizability is not known to us. The average value, $\bar{\alpha}$, seems to be more physically-realistic quantity.

--- The linearity of the α dependence and the existence of both positive and negative scattering lengths suggests a parabolic dependency of scattering cross-section on α (or α) , the left branch of each parabola being truncated at α (or $\bar{\alpha}$) =0. This parabolic dependency has been observed, the minima occurring at $\alpha \sim 5a_0^3$ for atoms and $\bar{\alpha} \sim 10a_0^3$, for molecules.

It is clear that much more experimental work is neccessary. Such efforts are in progress.

ACKNOWLEDGMENT

This work was supported by the U.S. Department of Energy.

REFERENCES

1. U. Asaf, W. S. Felps and K. Rupnik, S. P. McGlynn and G. Ascarelli, Density effects on high-n molecular Rydberg states: CH_3I and C_6H_6 in H_2 and Ar, *J. Chem. Phys.* **91**:5170 (1989).
2. K. Rupnik, U. Asaf and S. P. McGlynn, Electron scattering in dense atomic and molecular gases: An empirical correlation of polarizability and electron scattering length, *J. Chem. Phys.* **92**:2303 (1990).
3. U. Asaf, K. Rupnik, G. Reisfeld, and S. P. McGlynn, Pressure shifts and electron scattering lengths in atomic and molecular gases, *J. Chem. Phys.* **99**:2560 (1993).
4. E. Amaldi and E. Segré, Effetto della pressione sui termini elevati degli alcalini, *Nuovo Cimento* **11**:145 (1934).
5. E. Fermi, Sopra lo spostamento per pressione delle righe elevate delle serie spettrali, *Nuovo Cimento* **11**:157 (1934).
6. R. Reininger, U. Asaf and I. T. Steinberger, The density dependence of the quasi-free electron state in fluid xenon and krypton, *Chem. Phys. Lett*, **90**:287 (1982).
7. U. Asaf, R. Reininger and I. T. Steinberger, The energy V_0 of the quasi-free electron in gaseous, liquid and solid methane, *Chem. Phys. Lett.* **100**:363 (1983).
8. S. Y. Ch'en and M. Takeo, Broadening and shift of spectral lines due to the pressure of foreign gases, *Rev. Mod. Phys.* **29**:20 (1957)
9. K. H. Weber and K. Niemax, Impact broadening and shift of Rb nS and nD levels by nobel gases, *Z. Phys. A* **307**:13 (1982).
10. Parameter sets are available in Table 4.2-2 of J. O. Hirschfelder, C. F. Curtiss and R. B. Bird, *Molecular Theory of Gases and Liquids,* J. Wiley & Sons, New York, (1954).
11. D. G. Tuck, The polarizability of radon, *J. Phys. Chem.* **64**: 1775 (1960).
12. G. A. Miller, The intermolecular force constants of radon, *J. Phys. Chem.* **64**: 163 (1960).
13. V. A. Alekseev and I. I. Sobel'man, A spectroscopic method for the investigation of elastic scattering of slow electrons, *Sov. Phys. JETP* **22**:882 (1966).
14. T. F. O'Malley, Extrapolation of electron-rare gas atom cross sections to zero energy, *Phys. Rev.* **130**:1020 (1963).
15. A. Omont, On the theory of collisions of atoms in Rydberg states with neutral particles, *J. Physique* **38**:1343 (1977).
16. A. M. Köhler, R. Reininger, V. Saile and G. L. Findley, Density effects on high-n molecular Rydberg states: CH_3I in He, Ne, Ar and Kr, *Phys. Rev. A* **35**:79 (1987).

17. U. Asaf, M. Sielberg, J. Meyer, R. Reininger, G. Reisfeld and I. T. Steinberger, (unpublished)

18. I. T. Steinberger, U. Asaf, G. Ascarelli, R. Reininger, G. Reisfeld and M. Reshotko, Extrinsic photoconduction and Rydberg states due to a methyl iodine impurity in xenon, *Phys. Rev. A* **42**:3135 (1990).

19. U. Asaf, J. Meyer, R. Reininger and I. T. Steinberger, High Rydberg states of methyl iodide perturbed by nitrogen: a mutual cancellation of shift terms, *J. Chem. Phys.*, **96:**7885 (1992).

20. J. Meyer, R. Reininger, U. Asaf and I. T. Steinberger, Autoionization spectra of CH_3I in dense gaseous methane, ethane, and propane observed by photoconduction, *J. Chem. Phys.* **94**:1820 (1991).

21. U. Asaf, I. T. Steinberger, J. Meyer and R. Reininger, Electron scattering in dense CO_2 gas: Photoionization spectra of CH_3I perturbed by CO_2, *J. Chem. Phys.* **95**:4070 (1991).

22. J. Ferch, W. Raith and K. Schröder, Total cross section measurements for electron scattering from molecular hydrogen at very low energies, *J. Phys. B: At. Mol. Phys.* **13**:1481 (1980).

23. S. J. Buckman and B. Lohmann, Low-energy total cross section measurements for electron scattering from helium and argon, *J. Phys. B: At. Mol. Phys.* **19**:2547 (1986).

24. M. Weyhreter, B. Barczik, A. Mann, and F. Linder, Measurements of differential cross sections for e-Ar, Kr, Xe scattering at E=0.05-2 eV, *Z. Phys. D.* **7**:333 (1988).

25. J. Ferch, B. Granitza, B. Masche, and W. Raith, Electron-argon total cross section measurements at low energies by time-of-flight spectroscopy, *J., Phys. B: At. Mol. Phys.*, **18:** 967 (1985).

26. S. J. Buckman and B. Lohmann, The total cross section for low-energy electron scattering from krypton, *J. Phys. B: At. Mol. Phys.* **20**: 5807 (1987).

27. E. S. Chang, Modified effective-range theory for electron scattering from molecules, *J. Phys. B: At. Mol. Phys,* **14**: 893 (1981).

28. R. W. Crompton, D. K. Gibson and A. I. McIntosh, The cross section for J=0->2 rotational excitation of hydrogen by slow electrons, *Aust. J. Phys.* **22**: 715 (1969).

29. J. Ferch, C. Masche, and W. Raith, Total cross section measurement for e-CO_2 scattering down to 0.07eV, *J. Phys. B: At. Mol. Phys.* **14**: L97 (1981).

30. R. W. Crompton (Private communications, 1989).

31. B. Schmidt, Anisotropic low energy electron collision cross section for methane derived from transport coefficients, *J. Phys. B: At. Mol. Phys.* **24**:4809 (1991).

SECTION 2: BASIC MECHANISMS/LEADER BREAKDOWN

A MACROSCOPIC-FIELD MODEL OF LEADER BREAKDOWN IN NONUNIFORM FIELD GAPS

N. Giao Trinh

Cables et Isolants, Technologie et IREQ
Varennes, Québec, Canada J3X 1S1

INTRODUCTION

The main features of the leader discharge, a leader channel terminated by a streamer discharge, are illustrated in Fig. 1a. Free electrons produced in the streamer discharge maintain the ionization activity in the leader channel, thus sustaining the leader discharge. Neither the leader nor the streamer is fully conducting, which results in voltage drops along the leader (ΔUl) and streamer (ΔUs) channels. Under positive polarity, if the applied voltage is sufficient to allow an increase in the leader voltage drop, the leader discharge will develop further into the gap with the tip potential remaining fairly constant. Breakdown occurs when the continuous leader reaches a critical length and manages to break down the remaining gap through a streamer discharge in the so-called final jump.[1] Under these conditions, the breakdown voltage Ub of the gap can be expressed as:

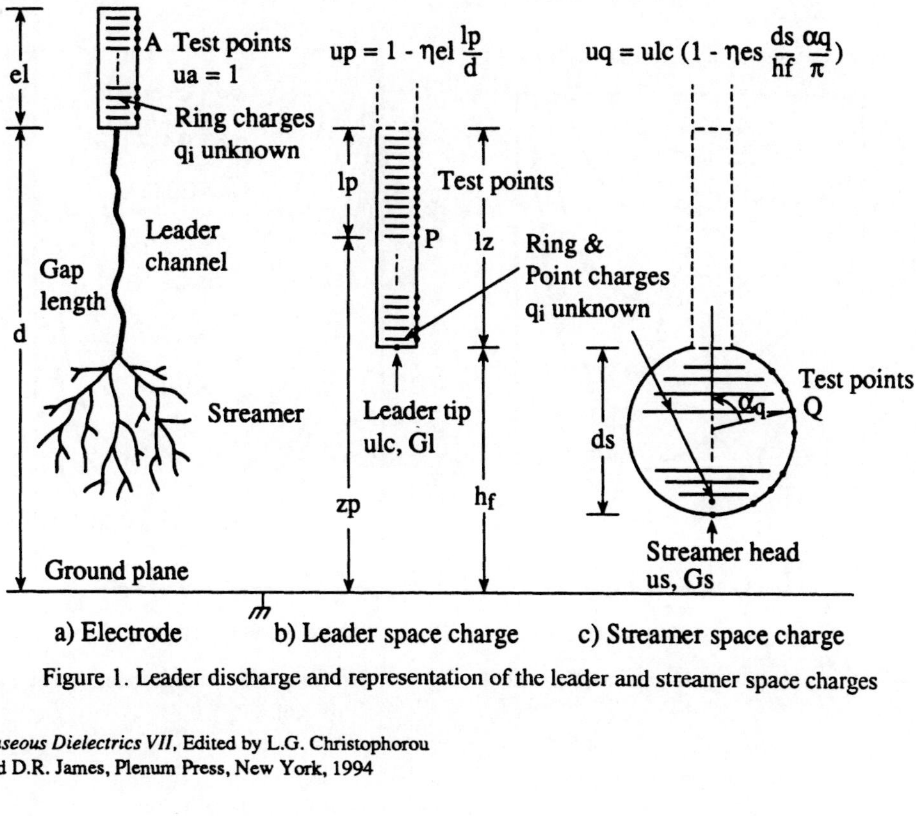

Figure 1. Leader discharge and representation of the leader and streamer space charges

Gaseous Dielectrics VII, Edited by L.G. Christophorou
and D.R. James, Plenum Press, New York, 1994

$$Ub = \Delta Ul + Ulc \tag{1}$$

where ΔUl *is* the leader voltage drop and Ulc the streamer breakdown voltage of the remaining gap at the final jump (also known as the continuous leader inception voltage).

The macroscopic-field model of leader breakdown of a nonuniform field gap relies on the specific properties of the macroscopic fields produced by the leader and streamer space charges to determine the characteristics of the leader discharge at the final jump and the breakdown voltage of the gap respectively. Numerical methods of field calculation require that the applied voltage be known. To avoid this problem, the normalized parameters, i.e. the potential up and the field intensity (also called the gradient factor) gp, are systematically evaluated assuming a nominal applied voltage of 1 V at the anode. To obtain the actual field intensity E (in V/m), the gradient factor must be multiplied by the applied voltage U (in V). Inversely, the required applied voltage U for a given field intensity in the gap is obtained by dividing E by the corresponding gradient factor.

MACROSCOPIC FIELDS OF LEADER AND STREAMER SPACE CHARGES

Critical leader length at the final jump

The representation of the leader space charge: a cylindrical channel of length lz and a voltage drop of ΔUl, is shown in Fig. 1b. During development, the leader channel extends into the gap along the gap axis. The normalized potential up at a point P along the leader channel can be expressed as:

$$up = \frac{Up}{Ub} = 1 - \frac{\Delta Ul}{Ub}\left(\frac{d - zp}{lz}\right) = 1 - \frac{\Delta Ul / lz}{Ub / d}\left(\frac{d - zp}{d}\right) \tag{2}$$

where Ub is the applied voltage, which becomes the breakdown voltage of the gap when the leader discharge is at the final jump, d the gap length and zp the height of P above ground.

By defining the average leader field ratio ηel as the ratio of the average leader field to the mean breakdown field:

$$\eta el = \frac{\Delta Ul / lz}{Ub / d} \quad \text{with } 1 \leq \eta el \leq 0, \tag{3}$$

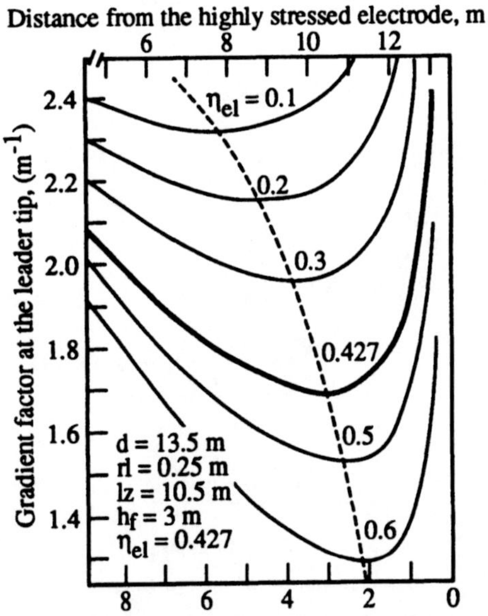

Figure 2.Leader space charge macroscopic field

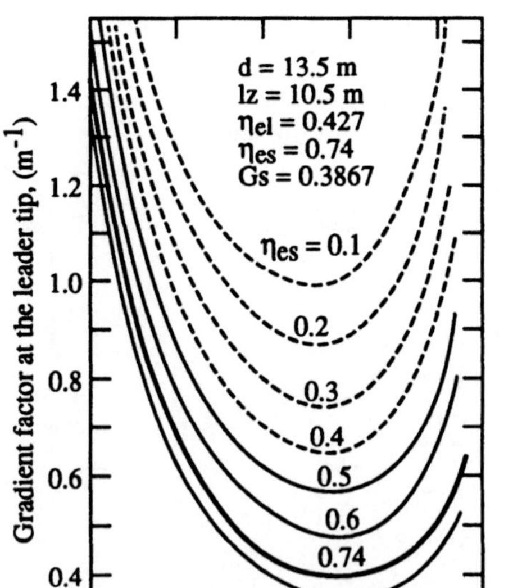

Figure 3.Streamer space charge macroscopic field

the normalized potential at the point P on the leader channel can be rewritten as

$$up = 1 - \eta el \frac{d - zp}{d} \qquad (4)$$

and the normalized potential at the leader tip is

$$ulc = 1 - \frac{\Delta Ul}{Ub} = 1 - \eta el \frac{lz}{d} \qquad (5)$$

Figure 2 shows variations in the gradient factor at the tip of the leader channel as a function of the leader length in a 13.5-m air-insulated gap for different values of ηel. The radius of the leader space charge channel was taken as 0.25 m. It can be seen that, as the leader extends into the gap, the field intensity at its tip decreases at first, passes through a minimum *Gl*, and increases again when the leader approaches the ground plane. Such behavior implies that, once the leader discharge is past the point of minimum gradient factor, it will cross the remaining gap. Since breakdown of the remaining gap occurs at the final jump through a streamer discharge, the leader does not cross the remaining gap; instead, **the position of the minimum field intensity corresponds effectively to the final jump while its distance from the ground plane corresponds to the height of the final jump, *hf*.**

Breakdown voltage

The representation of the streamer space charge, a spherical volume of diameter ds equal to the streamer length and a streamer voltage drop ΔUs, is shown in Fig. 1c. Assuming a uniform potential distribution, the normalized potential at a point Q along the streamer space charge is given by

$$uq = \frac{Uq}{Ub} = \left[1 - \frac{\Delta Ul}{Ub}\right]\left[1 - \frac{\Delta Us / ds}{Ulc / hf}\left(\frac{ds}{hf}\right)\left(\frac{\alpha q}{\pi}\right)\right] \qquad (6)$$

By defining the average streamer field ratio ηes as the ratio of the average streamer space charge field to the mean breakdown field across the height of the final jump:

$$\eta es = \frac{\Delta Us / \mathrm{ds}}{Ulc / \mathrm{hf}} \qquad \text{with } 1 \leq \eta es \leq 0, \qquad (7)$$

the earlier expression for the normalized potential at point Q can be rewritten as

$$uq = \left[1 - \eta el \frac{lz}{d}\right]\left[1 - \eta es \frac{ds}{hf}\left(\frac{\alpha q}{\pi}\right)\right] \qquad (8)$$

and the normalized potential at the streamer head is

$$us = \left[1 - \eta el \frac{lz}{d}\right]\left[1 - \eta es \frac{ds}{hf}\right] \qquad (9)$$

Figure 3 illustrates variations in the gradient factor at the streamer head as a function of the streamer length along the gap axis. The calculation was made for the leader at the position of the final jump in a 13.5-m air-insulated gap, a leader length of 10.5 m corresponding to a final jump height of 2.2 m, an average leader field ratio of 0.427, and different values of the average streamer field ratio. It can be seen that, here also, the field intensity at the streamer head decreases at first, passes through a minimum and then increases sharply as the streamer approaches the ground plane. This behavior again implies that, once the streamer has developed beyond the point of minimum field intensity, it will effectively cross the gap and break it down. **By equating the minimum field at the streamer head with the critical field *Ecr* for stable propagation of the streamer discharge across the remaining gap,** the breakdown voltage of the gap is readily determined from

$$Ub = \frac{Ecr}{Gs} \tag{10}$$

MACROSCOPIC-FIELD MODEL FOR LEADER BREAKDOWN

It follows from the above considerations on the macroscopic fields of leader and streamer space charges that for a given gap, if the average leader and streamer field ratios ηel and ηes are known, the various characteristics of the leader discharge at the final jump can be determined.

- Knowing ηel, the critical leader length lz and the corresponding final jump height, $hf = d - lz$, are determined from the position of the minimum gradient factor at the leader tip.
- With ηes known and lz defined, the critical streamer length ds and corresponding minimum gradient factor Gs at the head of the streamer space charge are determined.
- By equating the minimum field at the streamer head with the critical field Ecr for stable propagation of the streamer discharge across the remaining gap, the breakdown voltage is readily determined according to Eq. (10).
- The leader voltage drop ΔUl and the streamer-breakdown voltage at the final jump Ulc are given respectively by

$$\Delta Ul = Ub\left(\eta el\frac{lz}{d}\right) \text{ and} \tag{11}$$

$$Ulc = Ub - \Delta Ul = Ub\left(1 - \eta el\frac{lz}{d}\right) \tag{12}$$

The model calls for a number of parameters to be determined, namely: the average leader and streamer field ratios and the critical field for stable streamer propagation. The following hypotheses were used:

- A one-to-one correspondence between the average leader and streamer field ratios;
- A one-to-one correspondence between the average leader field ratio and the gas pressure;
- A constant value, independent of the gas pressure, of the critical field for stable propagation of the streamer discharge.

Air-insulated gaps

Average leader field ratio: The average leader field ratio ηel is an intrinsic characteristic of the leader discharge and therefore dependent essentially on the leader length and the insulating medium (gas type and pressure) although in air at atmospheric pressure, it is a function of the leader length only. From Eq. (3), it can be seen that ηel has a value less than 1.0, varies with the leader length, and approaches unity in very long gaps where the average leader field approaches the mean breakdown field. The variations in ηel with the leader length can be approximated by an exponential function:

$$\eta el = 1 - [1 - \eta e0]\,\mathbf{exp}\left(-\frac{lz}{LC}\right), \tag{13}$$

where $\eta e0$ is the initial value of ηel when lz is zero, and LC the critical leader length for which the term $(1 - \eta el)$ equals 37% of its initial value.

In the present study, the average leader field ratio was evaluated according to Eq. (3) for a range of gap lengths using the formulas developed by Rizk for the height of the final jump, the leader voltage drop and the breakdown voltage.[2] Least-mean-squares fitting of the results yielded $\eta e0$ = 0.315 and LC = 58.5 m. The above approximation of ηel was verified and found to be in good agreement with Rizk's results for gap lengths up to 40 m, which is adequate for most practical situations.

Average streamer field ratio: The average streamer field ratio η_{es} is an intrinsic characteristic of the streamer discharge. Its value can be determined from the measured breakdown voltage of the gap. In this study, η_{es} was evaluated from the breakdown voltage estimated with Rizk's formula[2] and was found to remain more or less constant, between 0.71 and 0.74, for gap lengths ranging from 4 to 13.5 m, with final-jump heights of 1.97 to 3.02 m.

The reason for this stability may be attributed to the fact that streamer discharges develop along a multitude of filamentary branches extending into a virgin insulating medium of relatively low field intensities, i.e. under constant conditions. By contrast, the leader channel is formed by the convergence of streamer branches at its tip and the incoming electrons produce intensive ionization in the leader channel, giving the leader discharge an arc-like behavior[2]. Its conductivity and, hence, the average leader field ratio are a function of the leader length.

Critical field for stable streamer propagation. The critical field for stable streamer propagation established by Gallimberti[3] is 7 10^5 V/m. Figure 4 compares the breakdown voltage and other leader characteristics, i.e. the leader voltage drop ΔUl and the leader inception voltage Ulc, predicted by the macroscopic-field model with those obtained by Rizk for air gaps under switching impulses with critical time-to-crest[2] of lengths from 4 to 13.5 m. As expected, since the values of the average leader and streamer field ratios were derived from Rizk's formulas for the breakdown voltage and leader characteristics, excellent agreement was achieved. It should be pointed out here that by neglecting the effect of the ellectrode, the radius rl of the leader space must be "fined tuned" to obtain full concordance with Rizk's results.

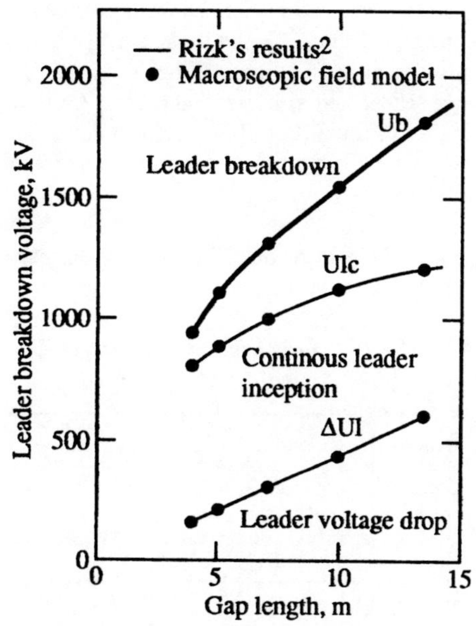

Figure 4. Leader-breakdown of long air gaps

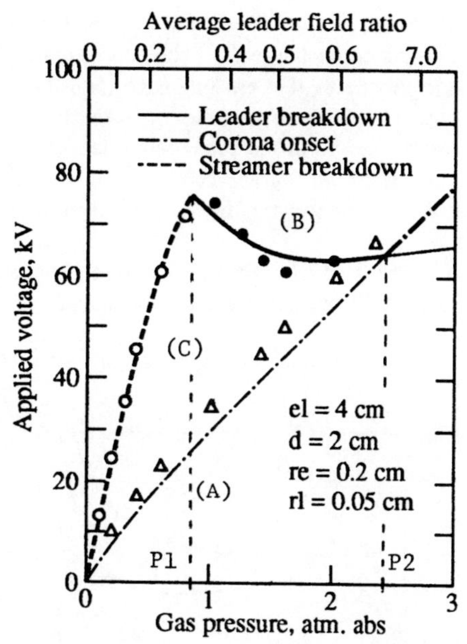

Figure 5. Ub-p curve of a SF6-insulated gap[4]

SF6-insulated gaps

Figure 5 illustrates the breakdown characteristics of a rod-plane SF6-insulated gap as a function of the gas pressure.[4] The gap spacing is 20 mm and the hemispherical rod is 4 cm long with a diameter of 4 mm. The *Ub-p* curve shows a characteristic Z-shape with three distinct regions defined by the pressures *P1* and *P2*. It results from two distinct breakdown processes, namely the streamer and leader breakdown, due to the relative predominance of one breakdown process over the other in a particular condition defined by the gap configuration and the gas pressure. At low gas pressures, below P1, streamer-initiated breakdown is the most likely process, with breakdown occurring whenever a streamer succeeds in bridging the gap.

For gas pressures higher than P1, the predominant process is leader initiated-breakdown, whose *Ub-p* characteristic generally shows a level or negative slope. P2 then corresponds to the intersection point of two curves representing the leader breakdown and streamer onset conditions as a function of the gas pressure. At gas pressures below P2, the leader breakdown voltage is higher than the streamer onset. Corona discharges can develop over a finite range of applied voltages. At gas pressures higher than P2, the leader-breakdown voltage is lower than the streamer onset. Since breakdown cannot occur before streamer onset, this condition determines the breakdown of the gap. The corona-onset condition can be determined from the familiar streamer criterion, curve A. The leader-breakdown characteristic, curve B, was determined from the macroscopic-field model as discussed below.

Critical leader lengths. For a constant gap length, the average leader field ratio ηel is a function of the gas pressure p only. The critical leader length was evaluated for the whole range of variations of ηel. Figure 6 shows variations in the gradient factor at the leader tip as a function of the leader length and for different values of ηel. For small values of ηel, the gradient factor at the leader tip increases continuously with length, which implies that the leader discharge, once initiated, will continue to develop across the gap and cause it to break down. Since, at these values of ηel, the onset field is lower for streamer than for leader discharge, breakdown of the gap will occur when the streamer successfully bridges it. Leader breakdown can occur only when the normalized leader voltage drop exceeds a critical value of $\eta e1$, characterized by the existence of a critical leader length for which the field intensity at the leader tip is minimum. This is in conformity with the experimental fact that breakdown is governed by a streamer process in gas pressures below $P1$, and by a leader process in higher gas pressures. A correspondence between the critical value of $\eta e1$ and $P1$ can thus be established.

Average streamer field ratio. The average streamer field ratio ηes is an intrinsic characteristic of the streamer discharge. Its value is less than 1.0 and higher than or equal to the average leader field ratio ηel owing to the less intensive ionization activities in the streamer channels. In this study, a single relationship was assumed between the average leader and streamer field ratios (Fig. 7a) and was established from the breakdown data, following an essentially trial and error process aimed at reproducing the appropriate shape of the *Ub-p* curve. Figure 7b represents variations in the minimum gradient factor at the streamer head *Gs* as a function of ηel.

Critical Field for Stable Streamer Propagation. The critical field for stable propagation *Ecr* was evaluated from the breakdown voltage at *P1* and Eq (10).

$$Ecr = Ub * Gs = Ub(P0) * Gs(\eta e0) = 76\ 8\ 10^3 * 156.25 = 12 \text{ MV/m, or } 12 \text{ kV/mm}$$

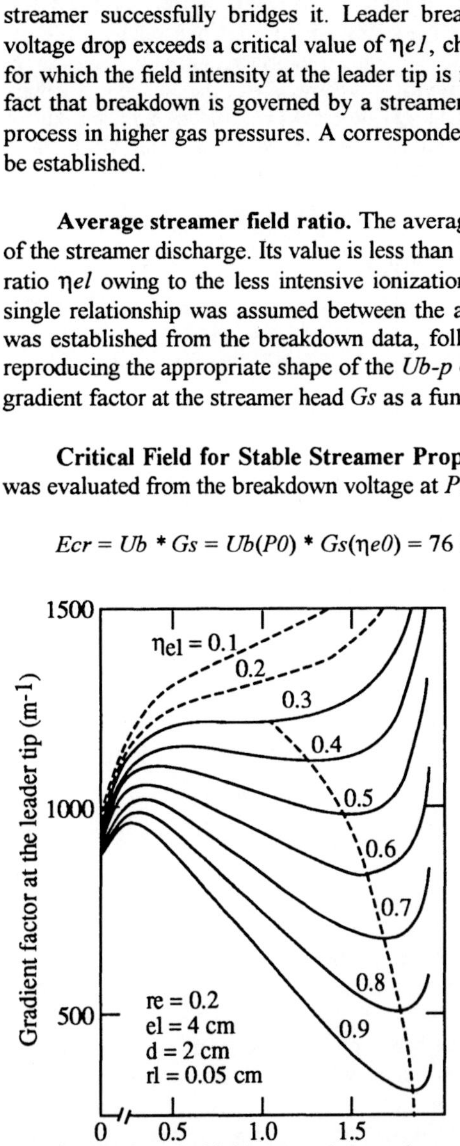

Figure 6. Leader space charge's field in a SF6 gap

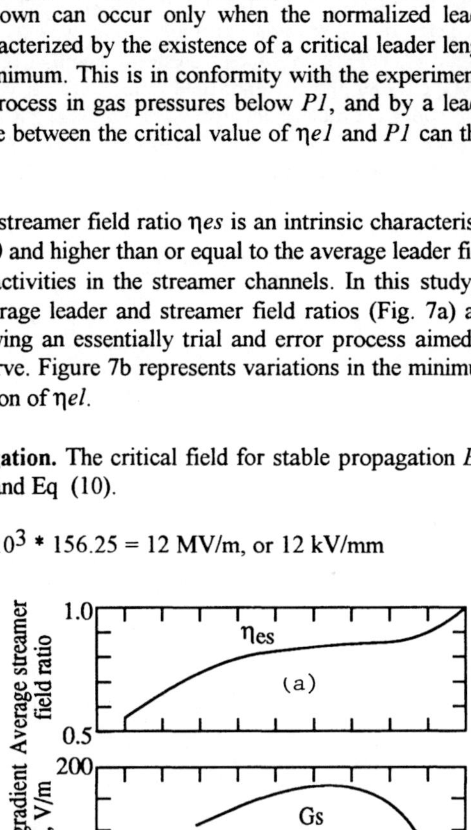

Figure 7. Ub-ηel curve of a SF6-insulated gap

With the value of Ecr defined, the Ub-ηel curve is readily determined (see Fig. 7c). Translation of the Ub-ηel to the Ub-p curve for leader breakdown in Fig. 5 was made by first determining the relationship between ηel and p. Noting that the average leader field ratio approaches unity for high gas pressures, its variations with the gas pressure can be approximated by an exponential function:

$$\eta el = 1 - [1 - \eta e0]\exp\left(-\frac{p}{PC}\right) \tag{14}$$

where $\eta e0$ is the initial value of ηel when p is zero, and PC critical gas pressure for which the term (1 - ηel) equals 37% its initial value. For the Ub-p curve of Fig. 5, values of $\eta e0 = 0$ and PC = 2.243 were used.

Streamer-initiated breakdown. The streamer-initiated breakdown characteristic (Fig. 5 curve C) was calculated for gas pressures below $P1$ by assuming a constant ratio between the streamer-initiated breakdown voltage and the streamer onset voltage $Uc(p)$ over this range of gas pressures:

$$Ub(p) = \frac{p}{P1} \frac{Ub(P1)}{Uc(P1)} \quad Uc(p) \tag{15}$$

The results presented in Fig. 5 compare favorably with the experimental data from Ref. 4 and validate the macrospic-field model for leader-breakdown of nonuniform gas-insulated gaps.

CONCLUSION

From the present study of breakdown in nonuniform gas-insulated gaps, a new model has been developed for leader breakdown of nonuniform gas-insulated gaps based on the macroscopic fields produced by leader and streamer space charges. The model relies on a number of parameters determined from experimental breakdown data, namely the average leader and streamer field ratios, ηel and ηes, and the critical field for stable streamer propagation, Ecr. It has been validated for long air gaps subjected to switching impulses of critical time-to-crest and can be used to determine the Ub-p characteristic of SF6-insulated gaps.

ACKNOWLEDGEMENT

The author would like to thank Mrs. Lesley Kelley-Régnier for the linguistic revision of the text.

REFERENCES

1. Les Renardières Group, "Research on Long Air Gap Discharges at Les Renardières," ELECTRA 23:53 (1972), ELECTRA 35:49 (1974), ELECTRA 53:31 (1977).
2. F.A.M. Rizk, "A Model for Switching Impulse Leader Inception and Breakdown of Long Air Gaps," IEEE Trans., Vol. PWRD-4:596 (1989).
3. I. Gallimberti, "The Mechanism of the Long Spark Formation," Journal de Physique - Colloque C6, Vol. 40:C7.193 (1979).
4. O. Farish, O.E. Ibrahim, A. Kurimoto, "Prebreakdown Corona Processes in SF6 and SF6/N2 Mixtures", Proc. 3rd International Symposium on High Voltage Engineering, Milan (1979), Paper No. 31.15.

DISCUSSION

D. SCHWEICKART: The model you developed for long gaps (air), Fig. 5, refers to an application with a 20 mm gap in SF_6. Is there a minimum critical spacing, or similar criterion, for which the model is not valid?

N. G. TRINH: The model was developed for leader breakdown and has been shown to apply for long air gaps as well as short nonuniform SF_6 gaps. Typical range of application: several tens of meters in air and several centimeters in SF_6 for which experimental data were available.

E. MARODE: Is there any relationship between the voltage gradient at the leader head and the amount of charge Q in the streamer head?

N. G. TRINH: The filed intensity at the leader tip varies considerably with the radius of the leader space charge used in the calculation. As such, this parameter is not reliable for modelling. On the other hand, the position of minimum field intensity at the leader tip was relatively unaffected by the leader space charge radius. This parameter was used to determine the critical leader length and the height of the final jump. The field intensity at the streamer head is a function of the streamer length, i.e., the diameter of the streamer space charge. Its value is uniquely defined and when made equal to the critical field for stable streamer propagation, will yield the breakdown voltage of the gap.

O. FARISH: The peaked V-P characteristic in SF_6 results from the effect of the corona stabilization which occurs under DC or AC conditions. Under impulse conditions the characteristic is "flat" between P_1 and P_2 in Fig. 4. How does your model incorporate the effect of corona space charge on the macroscopic field distribution?

N. G. TRINH: The model was developed from SF_6 breakdown data under DC and step impulse voltages. As it is, the model may not apply to the case of impulse voltage.

L. A. PINNADUWAGE: Have breakdown calculations been completed for both positive and negative corona?

N. G. TRINH: The leader breakdown model was developed for positive breakdown under switching impulses in air, DC and step impulses in SF_6, due to the availability of breakdown data.

CHARACTERIZATION OF THE ACTIVE ZONE FOR LEADER PROPAGATION IN SF_6

N. Femia,[1] V. Tucci,[2] M. Vitelli[1]

[1]Dipartimento di Ingegneria dell'Informazione e Ingegneria Elettrica
Università di Salerno
[2]Dipartimento di Ingegneria Elettrica
Università di Napoli "Federico II"
Italy

INTRODUCTION

The propagation of step-leader discharges in SF_6 depends on the inception of electronic avalanches in the streamer phase that occur in the region where the electrical field strenght exceeds a critical threshold E_{cr}. Such region surrounds the propagating discharge leader tips and may be considered as an *active zone* since it is the locus where new streamers and leaders take place. The active zone is crucial for the propagation of leader discharges since it determines the main directions and the branching of the discharge channels. Its shape and local properties are strictly related to those of the propagating leader path, the *pattern*. In the context of physical and chemical sciences it has been shown[1] that the characteristic scaling laws of frozen structures generated by diffusion-limited aggregation or chemical reactions may be related to the scaling exponents of the growing active zone which, in turn, depend on the frozen pattern. In that case some conditional probabilistic laws are used to find out the characteristics of the active zone given those of the pattern. In the leader discharge the new leader tips come from the active zone but their location may not be simply deduced from the preceding leader pattern because of the underlying physical processes. In fact, the streamers propagate within the active zone following statistically the background electrical field distorted by the space charge. Some streamers only change into leaders after a sufficiently intensive charge injection, usually occurring if a critical tip charge amount Q_{cr} is excedeed. Moreover, during the discharge propagation some leaders may not propagate and become inactive because either the local electrical field does not exceed the threshold E_{cr} or they are shielded by faster neighbouring leaders. When this happens the recombination phenomena yield the *cutting* of such leader branches. The active zone and the discharge pattern are therefore desordered sets that come in turn one from the other and evolve in time following the random add and cut mechanism mentioned above. Such mechanism makes almost impossible to find out a simple relation between active zone and pattern except in terms of fractal scaling laws, which is a rather common practice today to characterize both streamer and leader discharge patterns.

Gaseous Dielectrics VII, Edited by L.G. Christophorou
and D.R. James, Plenum Press, New York, 1994

The purpose of this work is the characterization of the active zone for surface step-leader discharges in SF_6 and the correlation of its scaling properties with those of the discharge pattern. A specific simulation algorithm has been developed by adopting an *add and cut* model that accounts for the elimination of dead branches from the leader pattern and two probabilistic laws for the modeling of the discharges' stastistical features. The first law reflects the spread of characteristic times for streamer formation in the surroundings of leader tips throughout the discharge, while the second one expresses the electrical field dependence of streamers directions. The Hurst's Rescaled Range Analysis has been used to compare the growth rates of active zone and pattern.

ADD AND CUT NUMERICAL MODEL FOR LEADER PROPAGATION

In order to study the relationships between active zone and pattern, a special model for the numerical simulation of leader discharge propagation has been adopted, accounting for the relevant physical mechanisms mentioned in the previous section. A brief description of the model and its numerical implementation follow.

At any instant of the leader discharge propagation we have two disordered sets, the *pattern* and the *active zone*, embedded for simulation on a 3D square lattice whose bond size equals the average step leader length (~1mm for SF_6[2]). Even if the instants of avalanches inception around the pattern tips are uncorrelated, we assume that the avalanches start all together at each iteration. Since the characteristic times for streamers formation depend inversely on the local electrical field[3], a list of streamer inception probability may be made:

$$p(P_l) = \frac{\max\limits_{P_a}\left(|\mathbf{E}(P_l, P_a)|\right)}{\max\limits_{P_a, P_l}\left(|\mathbf{E}(P_l, P_a)|\right)} \tag{1}$$

where $p(P_l)$ is the probability that streamer incepts from a leader tip P_l towards the surrounding points P_a of the active zone. The law (1) derives from two facts: firstly, the streamers develop more fastly in the regions where the electrical field is stronger; secondly, faster streamers may shield and kill neighbouring slower streamers. After the selection of the leader tips from which new streamers incept, the directions of propagation are determined. It has been shown[4] that the probabilistic law:

$$q(P_l, P_a) = \frac{|\mathbf{E}(P_l, P_a)| - E_{cr}}{|\mathbf{E}(P_l, P_a)|}, \quad |\mathbf{E}(P_l, P_a)| > E_{cr} \tag{2}$$

correctly reproduces the field dependence of the streamers directions and allows to select, among the active zone points P_a, the points P_s reached by streamers.

A space charge $Q(P_s)$ accumulates at the tips of new streamers due to the slow ionic drift which takes place along the streamer channels (precursor mechanism). It causes a strong electrical field increase ahead and a strong electrical field decrease along the streamers. The voltage drop across them may be roughly neglected and this boundary condition adopted for the calculation of the distorting space charge. Wherever $Q(P_s)$ exceeds a critical threshold Q_{cr} the corresponding streamer becomes leader and the lattice node P_s is transferred from the active zone to the pattern.

The failure of leader propagation may be generally related to a not much high electrical field and denotes a reduced activity around some tips with respect to others. It is reasonable to assume that, after a certain time, the activity around a non propagating leader tip definitively ceases and the corresponding leader branch vanishes. On the basis of this consideration, in the simulation model the leader branches from whose tips no propagation occurs for a certain number of iterations are removed and the corresponding lattice nodes reassigned to the neutral gas. This mechanism of dead branches cutting yields two relevant consequences in terms of discharge figures and growth rates. In fact, unlike photographs which are *static* records affected by an integral effect, the simulation carried out by means of the ***add and cut*** model gives the possibility of observing at each step the actual leader pattern and reproducing the discharge through a more realistic ***movie***. Moreover, the trend of the growth rate of both pattern and active zone during the discharge propagation can be evaluated and correlated with the applied voltage.

As an application a surface discharge in SF_6 has been simulated by means of the model illustrated above for a point-to-plane configuration under a positive applied voltage α times the inception voltage V_{in}, with $\alpha=1.4, 1.6, 1.8$. The experimental values of the average electrical field E_{ch} along the leader pattern (~2 kV/cm for SF_6[2]) and critical field E_{cr} (~90 kV/cm at 100kPa for SF_6[2]) have been considered. The value of Q_{cr} has been evaluated following the way indicated by the authors in a previous paper[5]: shortly, Q_{cr} is the average charge at the tips of the earliest streamer corona when the inception voltage V_{in} is applied to the electrodes. With regard to the life-time τ of leaders, since no specific data were found in literature, we roughly assumed that τ equals the average time for a leader step formation, so that the leader branches added at the n-1-th iteration which do not propagate at the n-th iteration are removed. A further study has been undertaken to fully clarify this topic.

In fig.1 some sequences of intermediate frames of the active zone and the pattern obtained by simulation are shown. Note that the active zone tends to crumble under low applied voltage whereas under high voltage it looks like a compact cloud surrounding the pattern, which in turn, shows an increasing branching. The discharges show a lower surface filling with respect to those ones descending from previous[5] *add only* models. These last were post-processed with a filtering procedure, that accounted also for the limited sensitivity of photographic emulsions, consisting in the elimination of minor (dead) leader branches from the final pattern. With the add and cut model proposed in this paper such operation is no more necessary. In order to correlate the properties of active zone and pattern the technique described in the following section has been adopted.

CHARACTERIZATION OF ACTIVE ZONE AND PATTERN

The problem of characterizing irregular or disordered sets arises in many physical sciences. In most cases a concise analytical description is found out by considering geometrical scaling and conservation laws whose exponents are correlated with the fractal dimension D. Such approach has been often adopted in the field of electrical breakdown to describe the branching of discharge patterns and usually consists in the *static analysis* of the Lichtenberg figures obtained either by experiments or by numerical simulations. Several methods for fractal dimension calculation may be found in literature. Barclay et al.[7] made an overview on this topic with specific reference to the fractal characterization of electrical trees. In the context of the present paper it is worth remarking the fundamental difference between spatial and temporal methods. In both cases the aim is to concisely express the relation existing between the microscopic mechanisms which are responsible for the growth and the complex macroscopic shapes and behaviours shown by the phenomena.

Spatial methods consider the geometrical scaling of a measure of the fractal set (e.g. the number N(l) of elements of the set lying within a square box of side l). They are oriented to find out geometrical self-similarities and analyze the final record of the breakdown process, the discharge figure, which usually includes even minor branches having a life-time smaller than the total time taken by the discharge propagation. In a previous paper[6] the authors have shown that the application of one of these methods to intermediate records of the discharge during the propagation yields a time-varying fractal dimension with an asymptote at *1.73*. It follows that the static fractal analysis allows a concise description of electrical discharge scaling properties only if the branching rate is invariant (and usually high) during the propagation; in this case the calculation of D may be performed on the final discharge pattern since few dead branches are removed during the propagation.

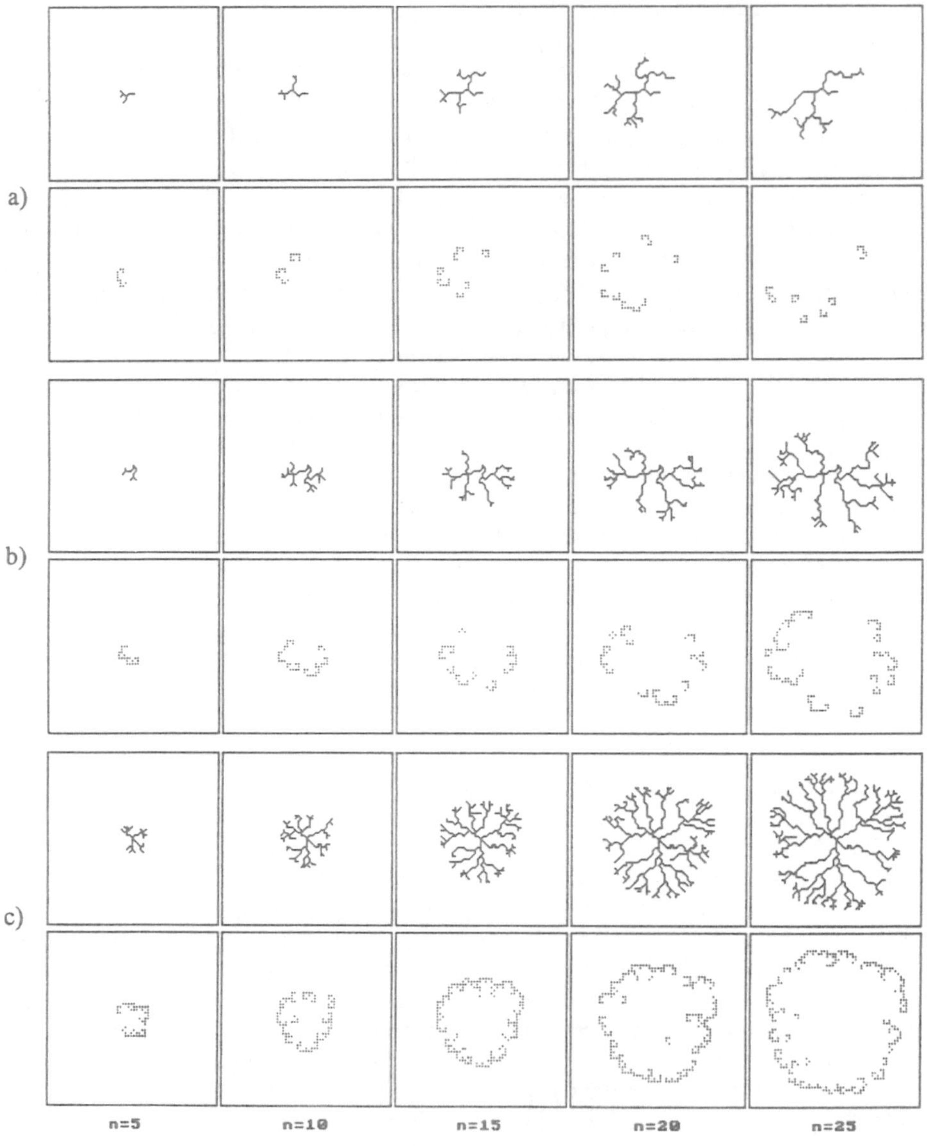

Figure 1. Pattern and active zone during the propagation: a) α=1.4; b) α=1.6; c) α=1.8.

More in general alternative methods must be adopted to analyze the scaling properties of discharges affected by sharper time fluctuations of the growing rate. Temporal methods, indeed, consider the time scaling of a measure of the fractal set (e.g. the gyration radius of the set with respect to a given origin). They allow to evaluate the growth rate of the set. For our purposes, temporal methods seem more convenient. In fact, the add and cut process described in the previous section may produce relevant time fluctuations both in the active zone and in the pattern, so that the final shapes of such sets may not resemble the familiar visual image of fractals. However, their effective growth rates, namely the net number of points added at each step, show rather erratic time records which may exhibit fractal characteristics. For the analysis of such records the authors have adopted the *Hurst's Rescaled Range Analysis* (shortly *R/S analysis*).

The purpose of R/S analysis is to study the time scaling properties of a function $\xi(t)$ expressing the effective time variation of a set. It is widely used for the study of time records of temperature, rain fall, rivers discharge, and other natural phenomena. For a full explanation of R/S analysis the reader is addressed to the literature[8]; in the following the relevant elements for pattern and active zone characterization only will be considered.

For the study of electrical discharges obtained by numerical simulation we consider the function $\xi(n)$ which gives the net number of points added to the set (active zone or pattern) at the n-th iteration. Let $X(n,N)$ be defined as:

$$X(n,N) = \sum_{k=1}^{n} \left\{ \xi(k) - \langle \xi \rangle_N \right\} \tag{3}$$

where $\langle \xi \rangle_N$ is the average of $\xi(n)$ over N iterations. The *range* $R(N)$ of $X(n,N)$ over N iterations:

$$R(N) = \max_{n=1,\dots,N} X(n,N) - \min_{n=1,\dots,N} X(n,N) \tag{4}$$

expresses a measure of the growth rate fluctuations. $R(N)$ may be normalized by means of the standard deviation $S(N)$ of $\xi(n)$:

$$S(N) = \sqrt{\frac{1}{N} \sum_{k=1}^{N} \left\{ \xi(k) - \langle \xi \rangle_N \right\}^2} \tag{5}$$

to get the dimensionless parameter R/S. Hurst found[8] that for long-term time records of many natural phenomena the dimensionless ratio R/S fulfils the empirical relation

$$R/S = (N/2)^H \tag{6}$$

The ratio R/S can be used to synthesize the growth rate trend of active zone and pattern over N iterations. In fact, it is easy to verify that a smooth $\xi(n)$ exhibits a high ratio R/S whereas a jagged $\xi(n)$ exhibits a small ratio R/S. Moreover, for self-affine records the number H, called *Hurst exponent*, fulfils the relation $D=2-H$ where D is the fractal dimension.

The plots of R/S shown in fig.2 were obtained by averaging the results of *10* simulated discharges for each value of the applied voltage. Firstly, we observe that the Hurst exponent may be defined for both active zone and pattern, namely they scale in a fractal manner in time. Moreover, the Hurst exponent of the active zone is always smaller than that one of the pattern: this corresponds to a pattern fractal dimension D_p smaller than the active region fractal dimension D_a. We remark that the fractal dimension of erratic records increases with their indentation.

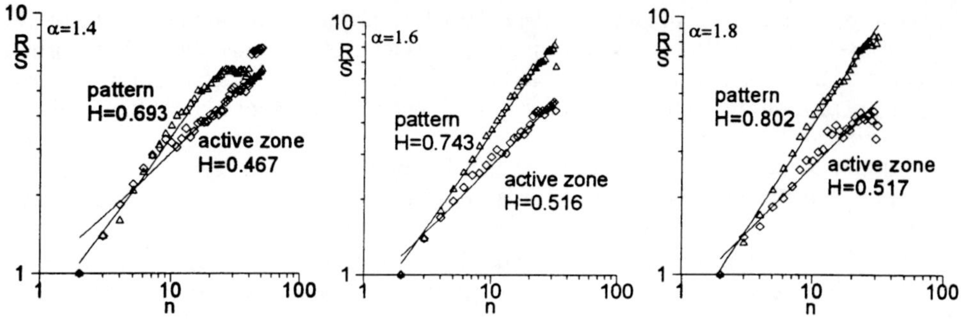

Figure 2. Log-log plots of pattern and active zone *R/S* ratio vs the number *n* of iteration for three different values of the applied voltage.

Therefore, the relation $D_p < D_a$ means that the active region is characterized by a more irregular growth rate with respect to the pattern. Such difference decreases under low applied voltage. The higher pattern fractal dimension at low voltages indicates a more disordered growth: the low electrical field leads to more frequent branches death, which contributes to the formation of protruding tips with a higher local field and then higher branching. After branching the field decreases and involves once again branches cutting. Such alternation of branching and cutting is particularly evident in the pattern shown in fig.1.a). At higher voltages, instead, the pattern exhibits a more regular growth rate trend since new branches are added ahead while dead branches are removed in the inner part of the pattern, like in fig.1.b,c). Note that, although under lower voltage the discharges are less branched and slower, they also cross the electrodic gap in some cases.

The plots of fig.2 show that the Hurst exponent of the active zone is a lower bound for the Hurst exponent of the pattern; the joining of the two Hurst exponents corresponds to the discharge stop. Moreover, for the active zone the Hurst exponent seems to be only weakly dependent on the applied voltage, in spite of the evident crumbling occurring at low voltages. The records of $\xi(n)$ plotted in fig.3 allow to clarify this apparent discrepancy.

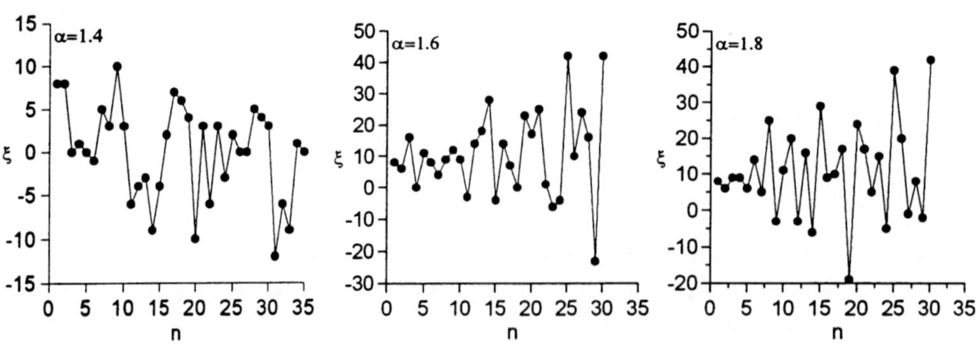

Figure 3. Effective number of points added to the active zone at each iteration.

They show a relevant indentation of the active zone growth rate at high as well as at low voltages even if with different peak-to-peak amplitude. This means that under high applied voltage the growth rate trend of the active zone is not uniform as one could expect, probably because it is conditioned mainly by the mechanism of propagation. The cutting of a leader branch may be not much relevant for a bush-like pattern such the one of fig.1.c), whereas the active zone may be severely modified. Nevertheless, the weak dependence of the active zone Hurst exponent on the applied voltage seems to be in contrast with the shape of active zone which, instead, is much influenced by the applied voltage. At present a work is in progress concerning the application of density-density correlation function to characterize the active zone in addition to the *R/S* ratio. Such additional tool allows to correlate the spatial distribution of the active zone with the applied voltage. In particular some results have been already obtained that indicate the conditions for the crumbling of the active zone into a number of *active centres* located about the tips of protruding discharge branches. The informations provided by *R/S* ratio and correlation function may be combined for getting a unitary interpretation of spatial and temporal scaling of the active zone. Such argument will be treated in a forthcoming paper.

CONCLUSIONS

In this paper the characterization of the active zone for leader propagation has been carried out by means of the Hurst's Rescaled Range Analysis. A surface discharge in SF_6 under positive polarity has been simulated by means of a numerical model accounting for the life-time of leader channels, thus obtaining more realistic patterns with respect to static photographic records. The use of R/S analysis allowed a correlation between the characteristics of the active zone with those of the propagating leader pattern. It has been found that growth rates of active zone and pattern scale with two different fractal dimensions; these latter converge for decreasing applied voltages and the occurrence of this eventuality corresponds to non propagating discharges. Moreover, it has been shown that the fractal dimension of the active zone for high applied voltages is greater than for low applied voltages, in spite of the more regular shape exhibited.

ACKNOWLEDGEMENTS

The work has been carried out with the financial grants M.U.R.S.T. 40%-60%.

REFERENCES

1. Z.Racz: Continuum Description of Active Zones, Fractals' Physical Origin and Properties, Plenum Press, New York (1989).
2. L.Niemeyer: A Fractal Lichtenberg Figure, Proc. of XX ICPIG, Pisa (1991), pp.937-938.
3. L.Pietronero, H.J.Wiesmann· From Physical Dielectric Breakdown to the Stochastic Fractal Model, Z.Phys.B-Condensed Matter 70, pp. 87-93, (1988).
4. L.Egiziano, N.Femia, G Lupò, V.Tucci: Effective numerical procedures for the calculation of electrical discharge parameters, Proc. of 10th Int. Conf. on Gas Disch. and their Appl., Swansea (1992), Vol.2, pp. 856-859.
5. N.Femia, V.Tucci, M.Vitelli: Generalized Models for the Simulation of Step-wise Propagating Gas Discharges, Proc. of XXI Int. Conf. on Phen. in Ion. Gases, Ruhr University, Bochum (1993), Vol.3, pp.411-412.
6. N.Femia, L.Niemeyer, V.Tucci: Fractal Characteristics of Electrical Discharges: Experiments and Simulation, J. Phys D Appl. Phys., 26, pp. 619-627, (1993)
7. A.L.Barclay, P.J.Sweeney, L.A.Dissado, G.C.Stevens: Stochastic Modeling of Electrical Treeing: Fractal and Statistical Characteristics, J. Phys. D: Appl. Phys., 23, pp. 1536-1545, (1990).
8 J.Feder: Fractals, Plenum Press, New York (1988)

DISCUSSION

W. KINSNER: The authors should be commended for addressing a very difficult problem of the dynamics of the active zone by Hurst's analysis.

E. MARODE: Perhaps we can point out the difference of approach of the streamer to leader transition, in this paper and the previous one by G. Trinh. Here one individual streamer is moving into a leader stage, while in the former, a bunch of streamers merge together to build a leader channel.

LAPLACIAN FRACTALS IN DIELECTRIC DISCHARGE PHENOMENA

W. Kinsner, G. Stacey and E. Kuffel

Department of Electrical & Computer Engineering
University of Manitoba
Winnipeg, Manitoba R3T 2N2, Canada
eMail: kinsner@ee.umanitoba.ca

ABSTRACT

Irreversible random-growth phenomena have been of great interest to physics and engineering, but difficult to model and solve in a closed form. This paper discusses dynamical stochastic modelling of dielectric discharges and breakdown in gaseous, liquid, solid and polymer insulators, with emphasis on the computation of their fractal dimensions.

1. INTRODUCTION

1.1 Irreversible Random-Growth Phenomena and Fractal Measures

The dielectric discharge problem, fracture propagation, viscous fingers in liquids, and electrochemical deposition are examples of *irreversible random-growth phenomena*[1]. Such kinetic critical phenomena have been of great interest to physics and engineering, but difficult to model and solve in a closed form. However, a recent discovery that the phenomena exhibit *scale invariance* (self-similarity) and fractal properties has lead to a new stochastic modelling and numerical solutions of such nonlinear problems. The models can be used to *synthesize* such growth phenomena and measure their fractal (non-integer) or multifractal dimensions. As an objective measure of the complex growth structure, the fractal dimension(s) can be correlated to the environmental conditions affecting the growth, thus providing a basis for *analysis* of such phenomena, and for prediction of their consequences to the dielectric environment.

Mandelbrot has defined a fractal as "a set for which the Hausdorff-Besicovitch dimension strictly exceeds the topological dimension"[2]. The importance of such mathematical fractals appears to be in their potential to represent complex shapes of natural objects compactly. Coastlines, mountain ranges, clouds, snowflakes, the leaves of a fern, trees, music, the human vascular system, and the retina are just a few examples of the many fractal objects found in the every day physical world[3,2]. What do they all have in common? They are self-similar or self-affine; that is, they are either exact or similar reduced copies of themselves on smaller scales. Can such objects be measured with the classical topological dimensions, i.e., the customary integer dimensions, ranging from 0 (a point) to 3 or more (a volumetric object)? Unfortunately, the topological dimensions lead to fundamental inconsistencies when applied to fractal objects (e.g., while the length is infinite, the surface is

Gaseous Dielectrics VII, Edited by L.G. Christophorou
and D.R. James, Plenum Press, New York, 1994

zero). Consequently, the dimension must be defined as fractal (non-integer). The fractal dimension is a good objective measure of the *complexity* (roughness) of such objects.

A corona is defined as a *partial* (i.e., no punch-through breakdown) discharge in either gaseous, or liquid or solid dielectrics. The corona develops in a short time period of the order of microseconds. As a first measure, the entire corona can be viewed as an object with the same complexity from its interior to the outer shell, thus having a single fractal dimension. Consequently, many different coronas may have the same fractal dimension. By using fractals varying with distance (multifractals), the differences may be discerned. Furthermore, since the spatial complexity in the interior and the exterior corona regions differs, multifractals should be considered to distinguish between the regions. In this sense, multifractal features can be used as *virtual sensors* of the corona and its segmentation.

1.2 Models of the Discharge Process

A stochastic model of the discharge process was first introduced by Sawada *et al.*[4] who assigned an arbitrary priority factor (R=1 to 200) to the growth of tips in the expanding pattern. Their model produces unrealistic fractals whose dimension is 2 (Euclidean) because the tips are allowed to grow with the same rate not only outside the enclosed fjords but also inside them (i.e., there is no screening effect). The first plausible stochastic dielectric breakdown model (DBM) was introduced by Niemeyer, Pietronero and Wiesmann who linked the growth of the pattern to the local electric field in 2D from a point electrode[5]. The DBM simulations were extended to 3D Laplacian fields and Debye screening (due to mobile carriers in the discharges)[6]. Satpathy[7] calculated the 3D Laplacian with the growth starting from a point electrode into either full 3D space or limited to 2D surface. His calculations were done on a cubic lattice within a sphere. Murat[8] calculated 2D discharge between two parallel lines. This was extended to a cylinder by Evertsz[9] who also discovered a self-affine nature of the growth transition phase in the cylinder. In general, the DBM model produces open tree-like patterns that closely resemble diffusion-limited aggregates (DLA) with fractal dimension of 1.7 on a square and triangular lattices[10]. These patterns are called Laplacian fractals because the dynamic growth probability is governed by a Laplacian field. They grow preferentially at the tips (extremities), while leaving the interior of the fjords (Faraday cages) relatively stagnant due to screening. Such aggregation phenomena have also been modelled by self-avoiding random walks (SAW) (i.e., every site can be visited only once), kinetic SAW in which one-step probabilities depend on the local environment[11], as well as percolation[12], and possibly cellular automata.

2. MODELLING OF DIELECTRIC DISCHARGES

2.1 Dynamics of the Growth Process

Although the microscopic process leading to the breakdown depends on the dielectric used, the structural properties of the discharge patterns are similar for all of them. The breakdown process goes through at least four phases: a nonconducting (insulating) phase, P_n, the growth transition phase, P_t, the growth steady-state phase, P_s, and/or the catastrophic punch-through phase, P_p. Following the nonconducting phase, P_n, a local conducting region may occur within the dielectric at a seed location p_0, if the material is subjected to an electric field whose value exceeds a critical level, E_c. A set of channels develops, each containing a non-equilibrium plasma with mobile electrons due to the critical field $\boldsymbol{E}_c$ and electron impact ionization. This region expands with time in a stochastic manner in at least two stages: the first growth transition phase, P_t, is characterized by a frontal interface with all the uninhibited discharge branches (containing large cumulative charge), as well as the inner screened discharge branches (with a lower charge). The transition phase continues until either a steady-state equilibrium (P_s) bounded in space or a catastrophic punch-through is reached (P_p). An understanding of the physical phases is important in dealing with their mathematical model.

The relationship between the electrostatic potential ϕ, charge density Q, and electrostatic field $\boldsymbol{E}$ (without photoionization in dark discharges) is described by Poisson's equation[6]

$$\nabla^2 \phi = -Q \tag{1}$$

where the total charge Q is given by

$$Q = e \int_{m,\sigma} \sigma n_m(\sigma) \tag{2}$$

Here, e is the electron charge, $n_m(\sigma)$ is the density of charge carriers, each with charge σe, σ is the infinitesimal volume, and m is an index to distinguish between different materials. During the discharge process (stages P_t, P_s, P_p), charge is created, annihilated, and transported (both field and diffusion currents) according to the following transport equation

$$\frac{\partial n_m(\sigma)}{\partial t} = D_m(\sigma) n_m(\sigma) \tag{3}$$

where $D_m(\sigma)$ is a local (but nonlinear) operator dependent on both the electric field and $n_m(\sigma)$. It should be remembered that during P_n, the right-hand-side of Eq. 3 becomes zero. The field $\boldsymbol{E}$ at each active corona site is calculated from the potential gradient

$$\mathbf{E} = -\nabla \phi \tag{4}$$

The boundary of the corona is moved by one bond from the most favourable point, according to a probability obtained from the field. But how do we solve the nonlinear coupled system of Poisson and transport equations to obtain the field? Is the dynamics equally complex in all the phases? Could and analytical solution describe the stochastic nature of the growth process? And finally, how do we determine the displacement of the corona boundary?

Since the charge density is zero during the P_n phase, Eq. 1 reduces to the Laplace equation with Dirichlet and/or Neumann boundary conditions. However, during the P_t, P_s, and P_p phases, an analytical solution is very difficult because the coupled system of Poisson and transport equations must now be solved. Furthermore, even if the equations could be solved analytically, the formulation does not describe the stochastic nature of the breakdown process.

2.2 Dielectric Breakdown Model (DBM)

The dielectric breakdown model (DBM) approximates the continuous material by a lattice on which the field is computed. The essential question is how to model the stochastic nature of the discharge growth. In their DBM, Sawada *et al.*[4] allowed the corona to expand according to a *tip growth priority factor*

$$R = \frac{p_t}{p_s} \tag{5}$$

where p_t is the probability of the tip to grow, and p_s is the probability of a side branch to grow. Since R was selected *a priori*, the corona was not shaped by the local field effect which produces caging, and consequently the patterns would fill the space completely.

Niemeyer *et al.*[5] improved the model by substituting the priority factor with a *tip growth probability* at a given time defined as

$$p_j \equiv \frac{\nabla_j^{\eta}}{\sum_j \nabla_j^{\eta}} \tag{6}$$

where

$$\nabla_j \equiv |(\nabla\phi)_j| = |(\hat{\mathbf{n}}\bullet\nabla\phi)_j| \tag{7}$$

is the potential gradient at a point j, normal to the branch. The j point is defined as being one lattice spacing away from the discharge pattern, thus all the points j define the set sites where growth may occur. The exponent η defines the growth complexity. The probability is normalized to all the possible growth bonds. This approach mimics the tip effects and the Faraday screening. For a small η, the discharge pattern is dense, spherical and strongly ramified, but becomes more elongated with fewer branches and scarce ramifications for a larger η (as found in liquids, solids and polymers). For η=0, we have the Eden (cancer) model, while for η=1, we have the standard DLA model with fractal dimension of 1.7[1].

When a bond breaks, the potential changes everywhere, and should be recalculated on the entire lattice. This iterative process ends when the discharge reaches either the steady state or a punch-through condition. From a practical point of view, the potential could be recalculated in the neighbourhood of the breakdown only, with less frequent updates on the entire lattice.

2.3 Fractal Dimension

The fractal dimension D is obtained from the "mass-length" relation in the power-law form

$$L(r) \propto r^{\mathrm{D}} \tag{8}$$

where $L(r)$ is the total number of discharged bonds enclosed in a sphere of radius r. If the $\log L(r)$ is plotted against $\log r$, then the linear part of the plot provides an estimate of D. There are also other measuring techniques such as the box dimension algorithm (see next section), and the density correlation function, also called density-density or pair correlation function[10].

3. EXPERIMENTAL RESULTS

3.1 Example 1: Parallel Electrodes with Gauss-Seidel Iteration

Let us consider two long parallel electrodes separated by a homogeneous dielectric (Example 1). The electrodes and the dielectric region occupy a square region which is subdivided into a square lattice with dimension is M=512 by N=512 . The electrodes are treated as fixed Dirichlet boundary conditions, while the other sides are treated as Neumann boundary conditions. Using unsigned integer arithmetic, the bottom electrode located at n=1 was set to C_1=0, and top electrode located at $n=N$ was set to a value of $C_2=10^9$. The gradient across the Neumann boundaries was set to zero. The initial value of the potential in the lattice, $\phi^{(0)}_{\mathrm{m,n}}$, was set as a linear distribution given by

$$\phi^{(0)}_{\mathrm{m,n}} = C_2\,(n\text{-}(\mathrm{N}\text{-}1)) \qquad \text{for } m=1 \text{ to } M \text{ and } n=2 \text{ to } N\text{-}1 \tag{9}$$

where the superscript in parenthesis denotes the iteration count. The exponent η in Eq. 6 was set to 1, thus resulting in the DLA-type discharges.

The discharge process can now begin. The seed point p_0 of the discharge is set midway on the bottom electrode. The Gauss-Seidel (GS) technique is used to solve

Laplace's equation on the lattice. In this technique, a potential calculated at a new point is used in the calculation of the potential at the next point. After each bond is broken on the lattice, local region of 50x50 pixels was iterated 50 times before doing a global recalculation of 8 iterations. This local-field iteration technique was also used by Fukai and Sano[13]. Once the simulated discharge reaches from the bottom electrode to the top, the simulation stops. (Actually, the simulation was stopped when the sum of all the gradients around a pixel in the fractal exceeded 10^9. This stops the simulation about two steps before the complete punch through.) Figures 1a and 1b show the negative of the fractal tree image and the corresponding field distribution obtained by Gauss-Seidel iteration, respectively. The field distribution is plotted with the equipotential lines in a linear scale.

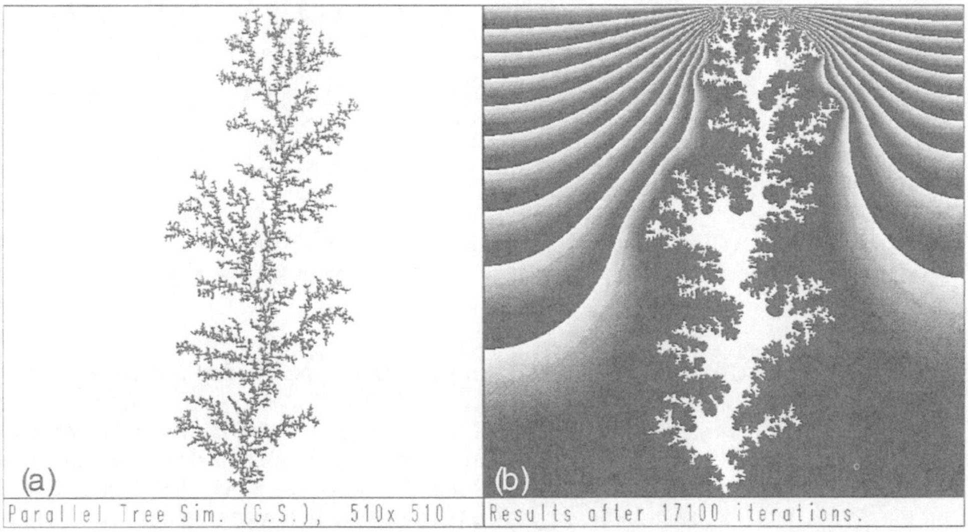

Figure 1. Parallel configuration of electrodes with Dirichlet and Neumann boundary conditions. (a) Laplacian fractal and (b) corresponding field distribution obtained by Gauss-Seidel iteration.

It is seen that the fractal tree in Fig. 1a resembles a DLA tree with a main discharge path and moderate ramifications, which is expected because η=1. To provide more information about how the simulated discharge develops with time (i.e., the intensity of the pixels added to the tree depends upon when the breaking of the bond occurred), initial pixels are given the intensity "white" while progressively later pixels are given darker and darker shades of gray. This field profile is displayed on the computer screen during the simulation process, so that one knows how far the simulation has progressed.

Notice that the field image in Fig. 1b has a large equipotential (zero) area surrounding the fractal tree; this signifies the effects of Faraday caging.

3.2 Example 2: Parallel Electrodes with Jacobi Iteration

The same parallel configuration of electrodes, with the same boundary conditions, has been recomputed using the parallel-update algorithm of Jacobi in order to test the sensitivity of a numerical technique on the corona formation (Example 2). Figures 2a and 2b show the negative of the fractal image and the corresponding field distribution obtained by Jacobi iteration, respectively.

Again, the tree has the characteristic DLA form, but appears to be wider at its tip. The spreading can be attributed to the difference in the propagation speed of the boundary conditions during the iteration process. Since the entire potential vector is displaced in one step during each iteration, the convergence is slower (21,100 versus 17,000 iterations for GS), and there is a greater uniformity of the potential everywhere. This contributes to the spread of the main branches of the tree.

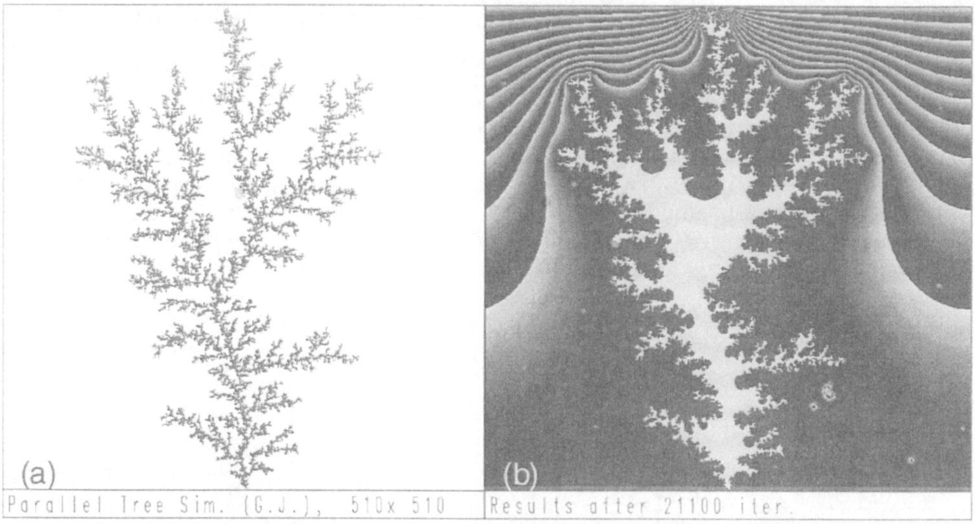

Figure 2. Parallel configuration of electrodes with Dirichlet and Neumann boundary conditions. (a) Laplacian fractal and (b) corresponding field distribution obtained by Jacobi iteration.

3.3 Example 3: Circular Point Electrodes with Gauss-Seidel Iteration

Let us consider the same square region with a 512 by 512 square lattice, and place at its centre a point electrode with a zero potential (C_1=0) and a concentrical circular outer electrode with a high potential (C_2=10^9). As before, the Gauss-Seidel (GS) technique is used to solve Laplace's equation on the lattice between the electrodes. After each bond is broken on the lattice, local region of 50x50 pixels is iterated 50 times before doing a global recalculation of 8 iterations. Figures 3a and 3b show the negative of the fractal tree image and the corresponding field distribution, respectively.

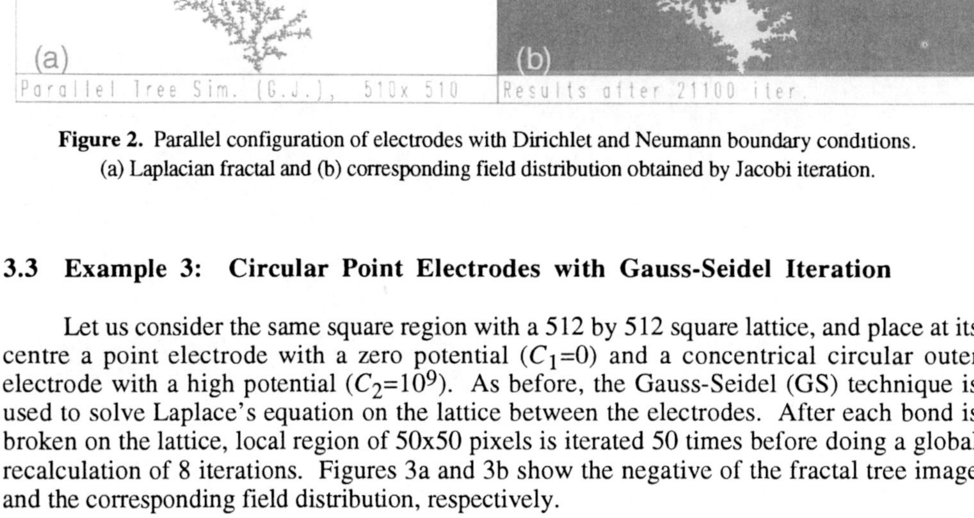

Figure 3. Circular point configuration of electrodes with Dirichlet boundary conditions only. (a) Laplacian fractal and (b) corresponding field distribution obtained by Gauss-Seidel iteration.

3.4 Example 4: Circular Disc Electrodes with Gauss-Seidel Iteration

The last example is similar to Example 3, but with a disc electrode. Figures 4a and 4b show the negative of the fractal tree image and the corresponding field distribution, respectively.

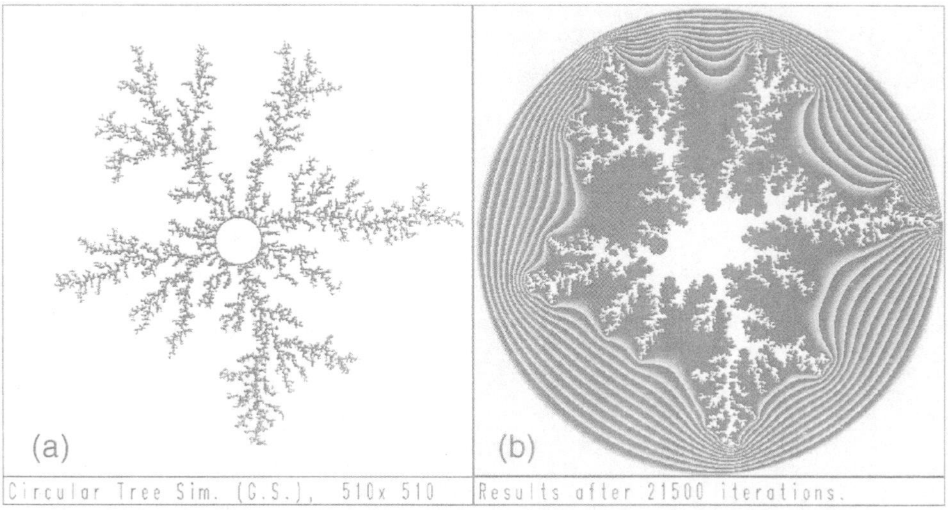

Figure 4. Circular disc configuration of electrodes with Dirichlet boundary conditions only. (a) Laplacian fractal and (b) corresponding field distribution obtained by Gauss-Seidel iteration.

It is seen that both the point and disc electrodes produce discharges of the same nature, after almost the same number of iterations.

3.5 Estimating Fractal Dimension

Once an image of the fractal tree has been obtained, estimating the dimension was performed using the box dimension (BD) algorithm. First, the above gray-scale image of a fractal tree is converted to a white on black image. Then the image is broken down into boxes of 1x1, 2x2, 3x3, 4x4, 5x5, 6x6, 10x10, and 15x15 pixels in size, thus giving a range of 15 to 1. The number of boxes containing at least one part of the image are counted at each of these sizes. The logarithm of both the number of boxes counted and the scale at which they were counted is calculated. The results are then fitted using a least squares fit to a line within the log count against log scale coordinates. The slope of this line gives the estimate of the fractal dimension of the image.

The image size (M and N) must also be rounded down to the nearest multiple of 30 pixels. Note that all box sizes will divide into this resultant evenly. This ensures that the BD calculation uses the same area for calculation at each box size. Figure 5 shows a result of the BD calculation (D=1.500432) for Example 1.

The simulated discharges corresponding to Examples 2 to 4 have fractal dimensions D=1.490368, D=1.496302, and D=1.492069, respectively. As expected, the dimensions are very close numerically. This means that although the fractal trees have a very different appearance, their character is similar as long as they are governed by the same underlying process (η=1).

The line in Fig. 5 also illustrates another feature of dielectric discharges in that they are multifractals[9] rather than single fractals. A single fractal would have a *straight* line over the entire range of box counting. The line in Fig. 5 is not straight, but curved, with at least one visible change in slope in the middle of its range.

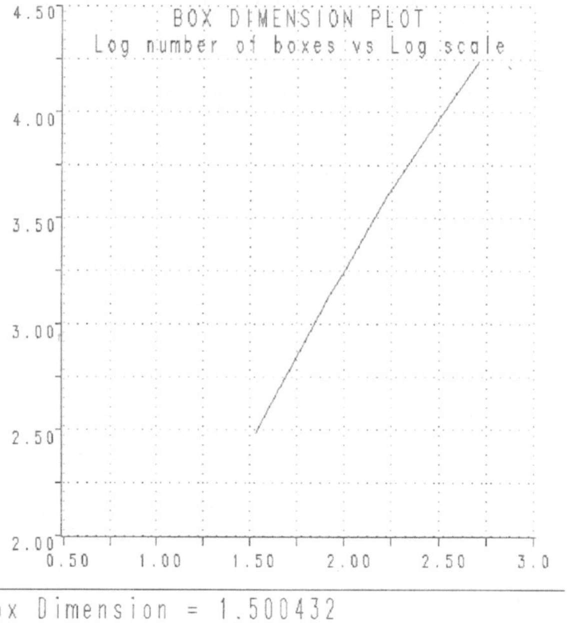

Figure 5. Box dimension calculation of the fractal dimension corresponding to Example 1.

3.6 Verification of Fractal Dimension Calculations

The reliability of a specific fractal dimension calculation can be verified by generating fractals with known dimensions and comparing them to those calculated. A good technique for generating fractals with known dimensions is the parallel rewriting system (L-systems) as introduced by Lindenmeyer[14]. Three examples of such fractals are shown in Fig. 6. Their dimensions are D_a =1.5000, D_b =1.4649, and D_c =1.5849, respectively, and the corresponding dimensions obtained from box counting are D_{Ba} =1.4906 (with error of –0.63%), D_{Bb} =1.4551 (–0.67%), and D_{Bc} =1.5749 (–0.63%). It is seen that the box counting technique underestimates the fractal dimension.

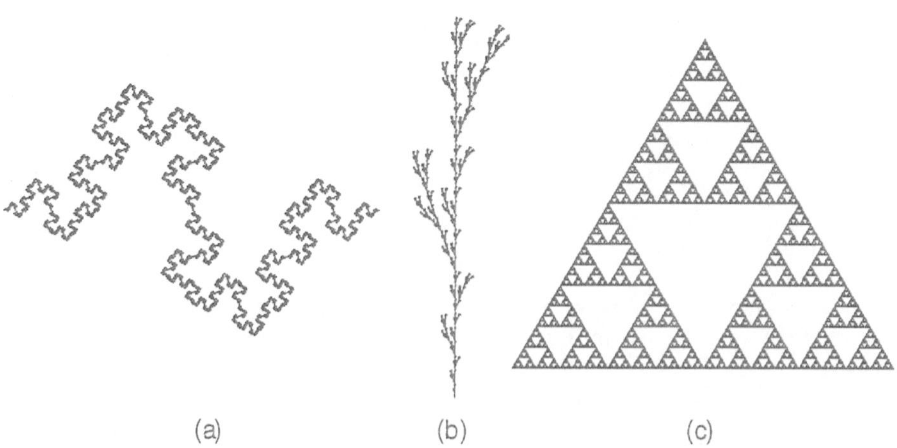

Figure 6. Fractals generated with L-systems have known dimensions.

In order to prepare the system for measuring fractal dimensions of real discharge photographs, the L-system fractals were also printed and scanned back into the computer where their fractal dimensions was calculated. There are numerous problems with accurate determination of fractal dimension from photographs as they are always blurred due to small ramifications that appear and disappear during the aperture time. We have been studying edge detection and thinning to improve the estimation, and will report on the findings later.

4. CONCLUSIONS

We have developed a tool for simulating discharges in dielectrics, as well as for estimating their fractal dimensions. The ability to simulate the dynamics of such discharges and their field distributions helps in the visualization of the irreversible random-growth phenomena in scaled time. Computations were carried out on large 2D lattices of size 512×512 for discharges, as well as on 1024×1024 for verification. Several experiments have shown that the fractal dimensions computed by the box dimension algorithm are lower than expected. The simulations were also tested with two fundamental techniques for solving Laplace's equation, the Gauss-Seidel and Jacobi iterations, and produced discharges having the same character. As expected, both techniques exhibited monotonic convergence. The local-field recalculation approach speeded up the computations considerably. The work continues to improve: (i) the computational speed of the corona simulator in 3D, and (ii) the accuracy of computing the fractal dimension for multifractals.

ACKNOWLEDGEMENTS

This work was supported in part by the Natural Sciences and Engineering Research Council (NSERC) of Canada

REFERENCES

1. T. Vicsek. "Fractal Growth Phenomena," World Scientific (2nd ed.), Singapore (1992).
2. B.B. Mandelbrot. "The Fractal Geometry of Nature," W.H. Freeman and Co., New York, NY (1983).
3. H.O. Peitgen, H. Jurgens and D. Saupe. "Chaos and Fractals," Springer Verlag, New York, NY (1992).
4. Y. Sawada, S. Ohta, M. Yamazaki, and H. Honjo, Self-similarity and a phase-transition-like behavior of a random growing structure governed by a nonequilibrium parameter, *Phys. Rev*, A26(6): 3557-3563 (December 1982).
5. L. Niemeyer, L. Pietronero, and H.J. Wiesmann, Fractal dimension of dielectric breakdown, *Phys. Rev. Lett.*, 52(12): 1033-1036 (19 March 1984).
6. H.J. Wiesmann and L. Pietronero, Properties of Laplacian fractals for dielectric breakdown in 2 and 3 dimensions, *in:* "Fractal in Physics," L. Pietronero and E. Tosatti (eds.). North Holland, New York, NY, 151-157 (1986).
7. S. Satpathy, Dielectric breakdown in three dimensions, *in:* "Fractal in Physics," L. Pietronero and E. Tosatti (eds.). North Holland, New York, NY, 173-176 (1986).
8. M. Murat, 2D dielectric breakdown between parallel lines, *in:* "Fractal in Physics", L. Pietronero and E. Tosatti (eds.). North Holland,New York, NY, 169-171 (1986).
9. C. Evertsz, Self-affine nature of dielectric-breakdown model clusters in a cylinder, *Phys. Rev*, A41(4):1830-1841 (15 February 1990).
10. T.A. Witten and L.M. Sander, Diffusion-limited aggregation, a kinetic critical phenomenon, *Phys. Rev. Lett.*, 47(19): 1400-1403 (9 November 1981).
11. J.W. Lyklema and C. Evertsz, The Laplacian random walk, *in:* "Fractal in Physics," L. Pietronero and E. Tosatti (eds.). North Holland, New York, NY 87-92 (1986).
12. L. Pietronero, W.R. Schneider and A.L. Stella, Percolation and invasion percolation as fractal growth problems, *Phys. Rev*, A41(12): 7496-7499 (15 December 1990).
13. J. Fukai and N. Sano, Tree growth simulation associated with dielectric breakdown, *IEEE Proc. Conf. Electrical Insulation & Dielectric Phenomena*, 432-439 (1987).
14. A.Lindenmeyer, Mathematical models for cellular interaction in development, Parts 1 and 2, *J. Theoretical Biology*, 18:280-315 (1968).

DISCUSSION

E. MARODE: What is the relation between the size of an individual computational step and the physics associated with the studied object?

W. KINSNER: The step sizes used in our numerical experiments were 512 x 512 and 1024 x 1024. The minimum step size for physical fractals should not be smaller than their bond size or another characteristic length.

E. MARODE: Is it possible to derive an approximate value of the fractal dimension of an object from the visual appearance of its compacted image?

W. KINSNER: The visual appearance of simulated fractals gives a key to their fractal dimension estimation. This is based on the fundamental procedure of Hansdorff covering of an object with volume elements, with the size of the volume elements diminishing to zero. The covering sequence provides the basis for the estimation of box-counting dimension. The mass dimension and gyration dimensions are also easily estimated. Less evident, though more accurate, are the correlation dimension, information dimension, and variance dimension.

A POSSIBLE NEW MECHANISM INVOLVED IN NON-UNIFORM FIELD BREAKDOWN IN GASEOUS DIELECTRICS[1]

Lal A. Pinnaduwage and Loucas G. Christophorou

Atomic, Molecular, and High Voltage Physics Group, Health and Safety Research Division, Oak Ridge National Laboratory, Post Office Box 2008, Oak Ridge, Tennessee 37831-6122, and Department of Physics, The University of Tennessee, Knoxville, Tennessee, 37996

I. INTRODUCTION

The electrical breakdown of gases under uniform field conditions is fairly well understood [1-3] in terms of the Townsend's breakdown theory. In most cases involving uniform fields, the breakdown voltage can be estimated via this theory using basic electron impact parameters (including electron attachment cross sections) for molecules in their ground electronic states [1-3].

In contrast, a consistent model of gaseous breakdown under non-uniform fields is not available at present although substantial progress has been made recently [1-5]. We point out the possibility that electron impact processes (in particular electron attachment processes) involving high-lying electronically-excited states [6,7] may play a significant role under non-uniform field conditions. Thus, such processes may need to be included in order to obtain a better understanding of non-uniform field breakdown phenomena.

The general, breakdown characteristics of highly non-uniform field gaps can be illustrated by that for a point-plane geometry, shown in Fig. 1 [4]. It has been found that the breakdown voltage for such a gap can be calculated by a simple streamer criterion [8] if the pressure, P, is above a critical value, P_c; for $P < P_c$, the estimated breakdown voltage is found [8] to coincide with the corona inception voltage, with the actual breakdown occurring at a higher voltage; corona discharges occur only for $P < P_c$. In other words, the presence of corona in the pressure region below P_c seems to prevent the breakdown from occurring at the predicted value.

[1]Research supported in part by the U.S. Air Force Wright Laboratory under Contract No. F33625-92-C-2221 and the National Science Foundation under Contract No. CHE-9022903 with the University of Tennessee and by the Office of Health and Environmental Research, U.S. Department of Energy, under Contract No. DE-AC05-84OR21400 with Martin Marietta Energy Systems, Inc.

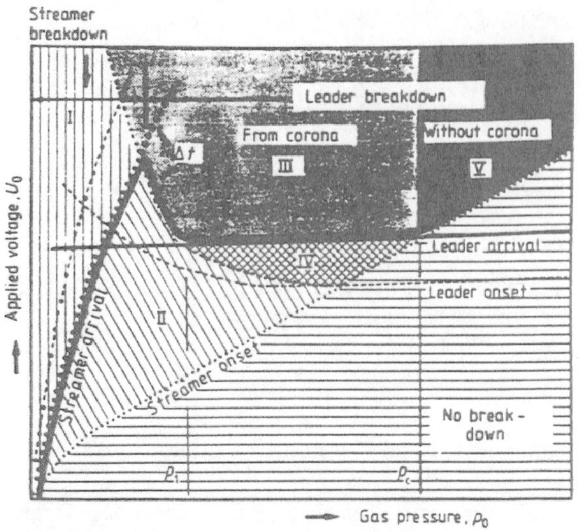

Figure 1. General breakdown characteristics for a point-to-plane gap [4].

This has led to the term "corona stabilization" to describe the enhancement in the breakdown voltage for pressures below P_c. Non-uniform field breakdown measurements in gases will be discussed in Section III.

In Section III, we will discuss the possibility that the "corona stabilization" (under both negative and positive corona) is due to the prevention of avalanche progression by attachment of free electrons to molecules in their high-lying electronically-excited states. Information on electron attachment to electronically-excited states of molecules was not available up until the late 1980's, see below.

II. RECENT DEVELOPMENTS IN ELECTRON ATTACHMENT PROCESSES

Recent basic studies on electron attachment to low-lying [9,10] and high-lying [7,11] excited electronic states of molecules indicate that the cross sections for these states can be orders of magnitude larger compared to those for the corresponding ground electronic states. In particular, electron attachment to superexcited states, SES, (electronically-excited states lying above the ionization potential, I.P., of a molecule) is an extremely efficient process. For example, it was shown [11] that while the electron attachment rate constant, k_a, is $< 10^{-12}$ $cm^3\ s^{-1}$ for the ground electronic state of triethylamine, $(C_2H_5)_3N$, (for electron energies < 1 eV), k_a is $\geq 10^{-3}$ $cm^3\ s^{-1}$ for the SES; this electron attachment to the SES of triethylamine (TEA) is more than 4 orders of magnitude larger compared to thermal electron attachment to SF_6.

Furthermore, electron attachment to SES appears to be of common occurrence in molecules. We have observed efficient electron attachment to SES in saturated tertiary amines (including TEA), nitric oxide, hydrogen, deuterium, methyl iodide, and benzaldehyde. The general requirements for this process to occur are: (i) molecules must be excited to an energy within a few eV of the I.P. where the efficiency for formation of SES is high, and (ii) at least one of the possible molecular fragments has a positive electron affinity. Of the molecules we have studied so far, only N_2 did not attach molecules in the SES; this is understandable since the nitrogen atom does not have a positive electron affinity, i.e., it does not form a stable negative ion.

In our basic studies, we populated SES of molecules via laser multiphoton excitation [11]. Resonant two-photon excitation of a molecule (via a low-lying electronic state) to an energy above its ionization threshold is schematically shown in Fig. 2. As a result of the excitation the molecule may undergo direct ionization (the excited electron coming off within ~ 10^{-15} s) or formation of a SES. A molecule in a SES may decay via preionization/predissociation (the SES lifetime is presumed to be < 10^{-9} s) or it can attach a closeby electron dissociatively [11]. In our experiments, the same laser pulse provided molecules in SES and attaching electrons (via direct ionization and preionization), see Fig. 2. Since the attaching electrons and SES were produced concomitantly and in close proximity, electron attachment can occurr prior to the decay of the SES [11].

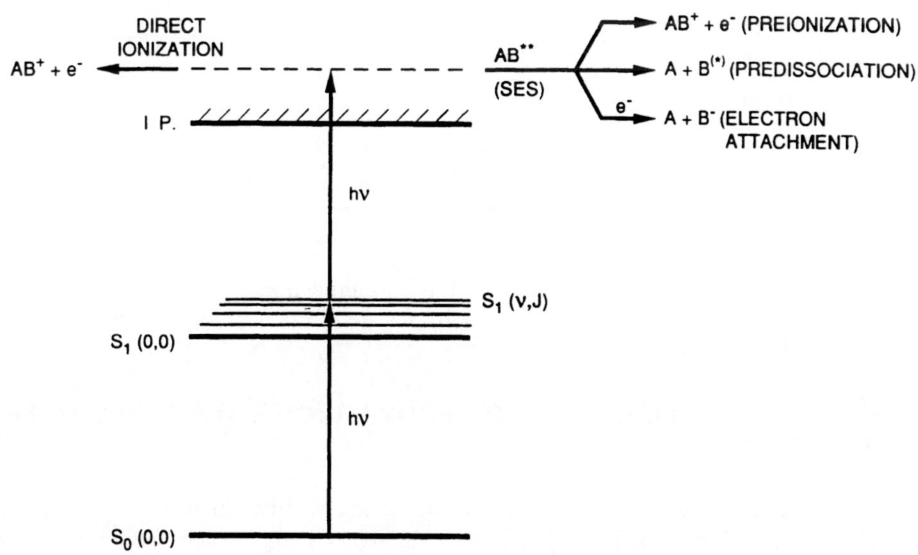

Figure 2. A typical two-photon excitation of a large molecule to an energy above its ionization potential (I.P.). This two-photon excitation will result in direct ionization or population of superexcited states (SES) of the molecule. A molecule in a SES (populated via photon absorption, electron impact or by other means) can decay via preionization, predissociation, or by dissociatively attaching a close-by electron.

Figure 3 shows our data [11] for TEA excited via the KrF (248 nm) excimer laser line; 15 microns of TEA were mixed in with 200 Torr of N_2 for the data of Fig. 3. Nitrogen was used as a buffer gas to lower the effective E/P (E = applied electric field and P = total pressure) value and hence to separate the electronic and the ionic components in the signal waveform [11]. The laser light did not affect N_2 (I.P. ~ 15.8 eV), but excited TEA (I.P. ~ 7.5 eV) to an energy ~ 2.5 eV above its I.P. via two-photon excitation. The positive ions produced via photoionization were rejected using a 3-electrode arrangement and only the negative ions and unattached electrons were detected [11]. The total signal, V_T, of Fig. 3 is the sum of the detected signals due to negative ions, V_I, and unattached electrons, V_E. Thus, V_T is proportional to the number of electrons initially produced via laser photoionization; some of these were converted to negative ions via attachment to SES. As seen in Fig. 3, at higher laser intensities almost all the electrons were converted to negative ions.

The basic electron attachment studies [9-11] have so far concentrated on electronically-excited states produced via laser excitation; however, in a corona discharge these excited states may be produced also via electron impact [6,7], excitation transfer via collisions with metastables or by light absorption.

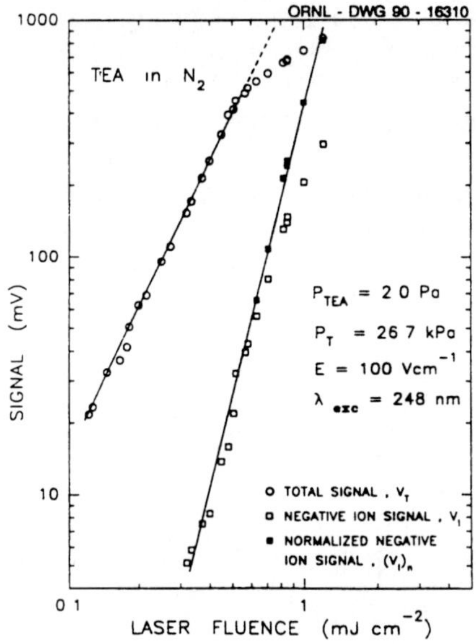

Figure 3. Laser fluence dependence of the measured total, V_T, and negative ion, V_I, signal and the normalized negative ion signal, $(V_I)_n$ [11] for TEA under the experimental parameters indicated in the figure.

III. QUALITATIVE MODEL FOR BREAKDOWN UNDER NON-UNIFORM FIELD CONDITIONS

In this section we point out that the enhancement in breakdown voltage under non-uniform field conditions (see Fig. 1) can be qualitatively explained by taking into account electron attachment to highly-excited molecules and in particular to SES, produced in the corona discharges.

1. Breakdown of Unitary, "Non-Electron Attaching" Gases

The streamer theories first proposed, independently, by Meek, Raether, and Loeb deal primarily with electron multiplication via electron impact and photoionization [12]. Implicit in these theories are two main assumptions: (i) electron multiplication must occur rapidly to produce large space charge fields, and (ii) these localized discharges produce radiation (light) which induces photoionization in front of the corona, especially for positive corona [12]. Once a critical field is reached to satisfy these conditions, breakdown ensues.

The above picture works well with unitary gases that do not attach electrons in their ground or excited electronic states: breakdown occurs almost simultaneously with the appearance of first corona. The so-called "corona stabilization region" (see Fig. 1) is absent or minimal in such gases. In both positive and negative point-to-plane gaps the absence of a significant stabilization region has been reported for He [13], Ar [13], and N_2 [13,14]. To quote Pollock and Cooper [13], "In these gases the individual streamers are able to propagate entirely across a short gap before a sufficient number of streamers form simultaneously to give burst corona, ..." . Any evidence for stabilization in these gases observed by some experimenters is probably due to the presence impurities [12]. It must be noted that for a gas at atmospheric pressure, contamination of 1 part per million corresponds to ~ 10^{13} impurity species per cm^{-3}.

Therefore, in the absence of electron attachment to excited states the first streamers formed can lead to breakdown. The above discussed gases do not form stable negative ions via electron attachment to either ground or electronically-excited states.

2. Breakdown of Electronegative Gases

When electron attachment to the ground electronic state is taken into account, the streamer criterion can predict [8] the breakdown voltage for electronegative gases under non-uniform fields for pressures above P_c, see Fig. 1. Below P_c, the estimated voltage coincides with the corona inception voltage, but the actual breakdown voltage lies at a higher value, see Fig. 1; i.e., for pressures below P_c, an additional mechanism for inducing an enhancement in breakdown voltage is present, which comes in to play with the inception of corona. We propose that the enhancement in breakdown voltage for $P < P_c$ is due to the attachment of electrons to high-lying excited electronic states of the molecules produced in corona discharges.

At the onset of the corona, localized "mini discharges" occur in the gas where electron impact induces ionization and excitation of the gas. In such mini discharges, as many excited species compared to ionization events can be produced [15]. Thus, at a total pressure of 300 Torr, the electron number density [16], and hence the excited species number densities, are in the range of ~ 10^{11} to 10^{16} cm^{-3} in various stages of discharge development. Since the electron attachment cross sections for the high-lying excited electronic states can be much more efficient (with $k_a > 10^{-3}$ cm^{-3} s^{-1} [11]) compared to ground state electron attachment, under such conditions electron attachment to these excited states can completely dominate even for a strongly electronegative gas such as SF_6 ($k_a \leq 10^{-7}$ cm^3 s^{-1} for the ground electronic state). Therefore, the stabilization of breakdown voltage under both positive and negative corona conditions can be predominantly due to the prevention of electron avalanches by electron attachment to high-lying excited electronic states.

Any electronegative molecule (i.e., a molecule which attaches electrons in the ground electronic state) should display electron attachment to high-lying excited states, provided that such states do not undergo rapid dissociation, see Section II.3. Thus, with a possible few exceptions, any electronegative gas should display the stabilization effect shown in Fig. 1. At pressures higher than P_c the loss of stabilization could be due to the shortening of the lifetime of the excited species via collisional quenching compared to the electron attachment time. It must be noted that, high non-uniformity of the electric field (e.g., sharper point) will lead to higher excited species number densities and thus small attachment times, i.e., the stabilization effect is more pronounced under higher field non-uniformity.

We also like to point that the terms electronegative and non-electronegative may have to be redefined since up to now they refer only to electron attachment to the ground electronic state of a species. A molecule which does not attach electrons in the ground state could attach electrons in the electronically-excited states, in particular the SES. For example, while TEA does not significantly attach electrons in the ground electronic state, it attaches electrons efficiently when excited to SES, see Section II.

3. Enhanced Stabilization by Additive Gases of Low Ionization Potentials

It has been shown [2, 17-23] that the stabilization effect in gases like SF_6 can be enhanced by the addition of a few percent of a gas with low I.P. for both positive and negative points in point-to-plane gaps. In these cases the highly-excited states of the additive lying both above and below its I.P. are populated by electrons with energies well below the I.P. of the main constituent (buffer gas) of the mixture; also they may be populated via energy transfer from comparatively low-lying electronic states of the main constituent, and by radiation emitted by the excited states of the main constituent. Even though the additive concentration is only a few percent, efficient excitation together with the ultra efficient electron capture can lead to efficient removal of accelerating electrons from the discharge. This is confirmed by the observed reduction in corona activity [17,18, 22] in the discharge with additives.

One of the additives used [18, 19] with SF_6 is triethylamine (TEA) where enhancement in stabilization was observed for both positive and negative points. The I.P. of TEA is ~ 7.5 eV compared to ~ 16.0 eV for SF_6. Electron attachment to ground state TEA is extremely weak (rate constant < 10^{-12} cm^3 s^{-1}) for electron energies of ≤ 1 eV [11]; electron attachment data for electron energies above 1 eV are not available. However, we observed that electron attachment to laser excited superexcited states of TEA is extremely large, rate constant > 10^{-3} cm^3 s^{-1} [11].

In a very recent study reported in these proceedings, Chalmers et al. [23] observed enhanced stabilization in SF_6 with different freon additives. Of the four freons studied, they did not observe enhanced breakdown only in R116 (C_2F_6). This molecule seems to be dissociating rapidly when in a SES: in both photoionization [24] and electron-impact ionization [25] studies the parent ion $C_2F_6^+$ was not observed and only fragment positive ions were observed; upon excitation to energies above the I.P., the superexcited molecule dissociated rapidly via repulsive potential energy surfaces. Evidence for the inability of C_2F_6 to capture electrons via superexcited states formed in a corona discharge comes also from a recent corona ion source mass spectrometric study [26]. In this study, ions formed in a corona discharge were extracted into a quadrupole mass spectrometer; the mass spectra of a pure SF_6 discharge and a mixture of 1% C_2F_6 and SF_6 were similar, i.e., no negative ions characteristic of C_2F_6 appeared in the spectrum of the mixture. On the other hand, a mixture of 0.5% R113 (CCl_2F - $CClF_2$) and SF_6 clearly showed Cl^- and ClF^- peaks in addition to the peaks characteristic of SF_6; this is consistent with the observed enhancement in breakdown characteristics of R113/SF_6 mixtures observed by Chalmers et al. [23].

From the discussion so far, we can make some predictions that can be tested in breakdown experiments: (i) C_2F_6 should not display a significant breakdown stabilization region (see Fig. 1) in pure form or as an additive in a mixture with other gases, (ii) even though TEA does not attach electrons in the ground electronic state, ions characteristic of TEA (possibly $C_2H_5^-$ and/or $C_2H_5N^-$) should be observable in a corona source mass spectrometric study on a mixture of TEA and SF_6, and (iii) since we have observed [11] enhanced electron attachment to the superexcited states of nitric oxide, the stabilization effect should be present in nitric oxide even though the electron attachment to the ground electronic state is extremely weak [11]. Furthermore, used as an additive, NO (I.P. = 9.6 eV) should display enhanced stabilization in SF_6 or in other gases with I.P. > 10 eV.

IV. CONCLUSIONS

We have introduced a novel concept that may lead to a better understanding of electrical breakdown in gases under non-uniform field conditions. Evidence was presented that the loss of electrons via attachment to high-lying excited electronic states of molecules could play a significant role in corona discharges, leading to the observed enhancement in breakdown voltage under non-uniform field conditions. Further basic and breakdown studies are needed to test this concept.

REFERENCES

1. Meek, J. M. and Craggs, J.D. (Eds.), Electrical Breakdown of Gases, (John Wiley, New York, 1978).

2. Christophorou, L. G., Sauers, I., James, D. R., Rodrigo, H., Pace, M. O., Carter, J. G., and Hunter, S. R., Recent Advances in Gaseous Dielectrics at Oak Ridge National Laboratory, *IEEE Trans. Electr. Ins.* **EI-19**, 550-566 (1984).

3. Christophorou, L. G. and Pinnaduwage, L. A., Basic Physics of Gaseous Dielectrics, *IEEE Trans. Electr. Ins.* **EI-25**, 55-74, (1990).

4. Pinnekamp, F and Niemeyer, L., Qualitative Model of Breakdown in SF_6 in Inhomogeneous Gaps, *J. Phys. D* **16**, 1293-1302 (1983).

5. Gallimberti, I., Breakdown Mechanisms in Electronegative Gases, in Gaseous Dielectrics V, Ed., by L. G. Christophorou and D. W. Bouldin, (Pergamon, N. Y., 1987, pp. 61-80).

6. Christophorou, L. G., Electron-Excited Molecule Interactions, Proc. of the XXth Intern. Conf. on Phenomena in Ionized Gases, (Ed., V. Palleschi, D. P. Singh, and M. Vaselli). pp. 3-13.

7. Pinnaduwage, L. A. and Christophorou, L. G., Verification of H^- Formation in UV Laser Irradiated Hydrogen; Implications for Negative Ion and Neutral Beam Technologies, *J. Appl. Phys.* (March, 1994 issue).

8. Nitta, T. and Shibuya, Y. Electrical Breakdown of Long Gaps in Sulfur Hexafluoride, *IEEE Trans. Power App. Systems* **PAS-90**, 1065-1071 (1971).

9. Christophorou, L. G., Hunter, S. R., Pinnaduwage, L. A., Carter, J. G., and Christodoulides, A. A., Optically Enhanced Electron Attachment, *Phys. Rev. Lett.* **58**, 1316-1319 (1987); Pinnaduwage, L. A., Christophorou, L. G., and Hunter, S. R., Optically Enhanced Electron Attachment to Thiophenol, *J. Chem. Phys.* **90**, 6275-6289 (1989).

10. Kuo, C. T., Ono, Y., Hardwick, J. L., and Moseley, J. T., Dissociative Attachment of Electrons to the A $^2\Sigma^+$ State of Nitric Oxide, *J. Phys. Chem.* **92**, 5072-5074; Mock, R. S. and Grimsrud, E. P., Optically Enhanced Electron Capture by p-Benzoquinone and Its Methylated Derivatives, *J. Phys. Chem.* **94**, 3550-3553 (1990); Jaffke, T., Hashemi, R., Christophorou, L. G., Illenberger, E., Baumgärtel, H., and Pinnaduwage, L.A., Photoenhanced Dissociative Electron Attachment to SO_2, *Chem. Phys. Lett.* **203**, 21-27 (1993).

11. Pinnaduwage, L. A., Christophorou, L. G., and Bitouni, A. P., Enhanced Electron Attachment to Superexcited States of Saturated Teritary Amines, *J. Chem. Phys.* **95**, 274-287 (1991); Pinnaduwage, L. A. and Christophorou, L. G., Enhanced Electron Attachment to Superexcited States of Nitric Oxide, *Chem. Phys. Lett.* **186**, 4-10 (1991); Pinnaduwage, L. A. and Christophorou, L. G., H^- Formation in Laser-Excited Molecular Hydrogen, *Phys. Rev. Lett.* **70**, 754-757 (1993).

12. Loeb, L. B., Electrical Coronas, (Univ. of California Press, Berkeley, 1965).

13. Pollock, H. C. and Cooper, F. S., The Effect of Pressure on the Positive Point-to-Plane Discharge in N_2, O_2, CO_2, SO_2, SF_6, CCl_2F_2, Ar, He, and H_2, *Phys. Rev.* **56**, 170-175 (1939).

14. Ford, R. T., Positive Point-to-Plane Spark Breakdown of Compressed Gases, *Nature* **166**, 688-689 (1950).

15. Loeb, L. B. and Meek, J. M., The Mechanism of Spark Discharge in Air at Atmospheric Pressure I, *J. Appl. Phys.* **11**, 438-447 (1940).

16. Stritzke, P., Sander, I., and Raether, H., Spatial and Temporal Spectroscopy of a Streamer Discharge in Nitrogen, *J. Phys. D* **10**, 2285-2300 (1977).

17. Weissler, G. L. and Mohr, E. I., Negative Corona in Freon-Air Mixtures, *Phys. Rev.* **72**, 249-297 (1947).

18. Farish, O., Chalmers, I. D.,and Vidaud, P., SF_6 Mixtures with Enhanced Corona Stabilization, in Proc. of the 7th Intern. Conf. on Gas Discharges and Their Applications (London, 1982), pp. 223-226.

19. Lakdawala, V.K., James, D.R., and Christophorou, L.G., Effect of Ionization Processes on the Corona Stabilization Breakdown in SF_6 and SF_6-Mixtures, Proc. of the 4th Intern. Symposium on High Voltage Engineering (Athens, Greece, 1983), pp. 1-3.

20. Nelson, J. K., Modification of Corona Stabilization in Electronegative Gases by Low Ionization Potential Additives, in Gaseous Dielectrics IV, (Ed. by L. G. Christophorou and M. O. Pace, *Pergamon*, New York, 1984) pp. 175-183.

21. Chalmers, I.D., Qiu, Y., Sun, A., and Xu, B., Extended Corona Stabilization in SF_6/Freon Mixtures, *J. Phys. D* **L107-112** (1985).

22. Qiu, Y. Sun, A., and Kuffel, E., Improved Dielectric Strength of SF_6 Gas With a Trichlorotrifluoroethane Vapor Additive, *IEEE Trans. Electr. Ins.* **EI-22**, 763-768 (1987).

23. Chalmers, I.D., Farish, O., and Shelton, R. V., Insulation Performance of SF_6 Mixtures, in Gaseous Dielectrics VII, (Ed., by L. G. Christophorou and D. R. James, Plenum, New York, to be published, 1994).

24. Noutary, C. J. , Mass Spectrometric Study of the Photoionization of Some Fluorocarbons and Trifluoromethyl Halides, *J. Res. NBS* **72A**, 479-485 (1968).

25. Lifshitz, C. and Long, F. A., Some Observations Concerning the Positive Ion Decomposition of C_2F_6 and C_3F_8 in the Mass Spectrometer, *J. Phys. Chem.* **69**, 3746-3751 (1965).

26. Sigmond, R. S. and Sigmond, T., Corona Ion Source Mass Spectrometry of SF_6 Gas and Mixtures, Proc. of the IXth Intern. Conf. on Gas Discharges and Their Applications (Venezia, 1988), pp. 212-214.

DISCUSSION

M. GOLDMAN: How do polar molecules which must be orientated in the high field present near the point react?

L. A. PINNADUWAGE: In the case of polar molecules, the attachment to superexcited states will be enhanced due to the dipole moment of the molecule.

E. ILLENBERGER: Can you comment on the mechanism of dissociation when an electron is captured by a superexcited molecule?

L. A. PINNADUWAGE: The approaching electron easily polarizes the superexcited molecule and is captured with high efficiency. The next step of dissociation into a negative ion and a neutral fragment is not clear. There must be some efficient way to transfer electronic energy to the nuclear motion.

R. NAGPAL: For stabilization in gas mixtures it is apparently desirable that the superexcited state of the additive should be stable. Is there an adverse effect if the attachment to the superexcited state is dissociative?

L. A. PINNADUWAGE: The superexcited states do not have to be stable; they have short lifetimes ($\leq 10^{-9}$ s). The role of the superexcited states in this particular case is to attach electrons dissociatively to form stable negative ions thus effectively removing free electrons from the discharge.

I. SAUERS: I would like to point out that gas components added to SF_6 can result in negative ions by other processes such as charge transfer, which could explain the Cl^- observed in the freon/SF_6 mixtures reported by Sigmond et al. In our work on corona ion chemistry in SF_6 the presence of small amounts of water also results in the formation of water ions [OH^- and $OH^-(H_2O)_n$].

L. A. PINNADUWAGE: In the study of Sigmond et al., while they did see Cl^- and ClF^- from $C_2F_3Cl_3/SF_6$ mixtures they did not see any ions characteristic of C_2F_6 from C_2F_6/SF_6 discharges.

SECTION 3: GAS BREAKDOWN

NUMERICAL STUDY OF THE AVALANCHE TO STREAMER TRANSITION IN N_2 AND DRY AIR

J.T. Kennedy and J.M. Wetzer

High Voltage and EMC Group
Department of Electrical Engineering
Eindhoven University of Technology
Eindhoven, The Netherlands

INTRODUCTION

In this paper, we will present the results of a numerical study of the avalanche to streamer transition in an overvolted uniform field gap in atmospheric N_2 and dry air. The model used is fully two-dimensional in the solution of the representative set of continuity equations, as well as in the determination of the space-charge field, gas-phase photoionization and cathode photoelectron emission. The spatio-temporal development of the electron density as well as the temporal development of the cathode and anode directed streamer velocities are presented and will be discussed.

Regarding the streamer formation process in N_2 gas phase photoionization plays the dominant role. In air, the process of electron detachment increases the propagation velocity of both streamer fronts. At the voltages studied photoelectron emission from the cathode is observed to be of secondary importance in the avalanche to streamer transition for both air and N_2. During the transition from avalanche to streamer, the maximum electron density in the anode directed streamer is located off the axis of symmetry in both gasses, confirming the importance of a fully two-dimensional simulation.

THEORETICAL MODEL

The following set of continuity equations describing the spatio-temporal development of electrons, positive ions, and negative ions (if applicable) apply:

$$\frac{\partial N_e}{\partial t}+\nabla\cdot\left(V_e N_e\right)=\nabla\cdot\left(D_e \nabla N_e\right)+\left(\alpha-\eta_u-\eta_s\right)\left|V_e\right|N_e+N_{un}k_{ud}+N_{sn}k_{sd}-\varepsilon_{pe}N_e N_p+S_{ph} \quad \textbf{(1)}$$

$$\frac{\partial N_{un}}{\partial t}+\nabla\cdot\left(V_{un}N_{un}\right)=\eta_u\left|V_e\right|N_e-N_{un}\left(k_{ud}+k_c\right)-\varepsilon_{pn}N_{un}N_p \quad \textbf{(2)}$$

Gaseous Dielectrics VII, Edited by L.G. Christophorou
and D.R. James, Plenum Press, New York, 1994

$$\frac{\partial N_{sn}}{\partial t}+\nabla\cdot\left(V_{sn}N_{sn}\right)=\eta_s\left|V_e\right|N_e+N_{un}k_c-N_{sn}k_{sd}-\varepsilon_{pn}N_{sn}N_p \tag{3}$$

$$\frac{\partial N_p}{\partial t}-\nabla\cdot(V_pN_p)=\alpha\left|V_e\right|N_e-N_p(\varepsilon_{pe}N_e+\varepsilon_{pn}(N_{un}+N_{sn}))+S_{ph} \tag{4}$$

In the above equations, $N_{e,un,sn,p}$ (x,ρ,t) are the number densities (cm^{-3}) of electrons, unstable (O^-) and stable (O_2^-, O_3^-) negative ions, and positive ions respectively, and $V_{e,un,sn,p}$ (x,ρ,t) are the respective drift velocities. Here t is time, and ρ and x are the radial and axial coordinates in the cylindrical coordinate system. The cathode is located at x = 0 (x_o on the computational mesh) and ρ = 0 (ρ_o on the computational mesh) is the axis of rotational symmetry. The coefficients α, η_u , η_s , k_{ud} , k_{sd} , k_c , ·ε_{pe} , ε_{pn} , S_{ph} , represent the ionization, unstable and stable electron attachment coefficients, unstable and stable electron detachment frequencies, charge exchange (ion conversion) frequency, pos. ion-electron and pos. ion-neg. ion recombination coefficients, and gas phase photoionization. Because N_2 has no known electron attachment in the time scale of interest, only Eqs. (1) and (4) are applicable with the terms η_u , η_s , ε_{pn} , k_{sd} , and k_{ud} set to zero. Electron diffusion is included in Eq. (1) where D_e is the electron diffusion tensor consisting of a longitudinal and a transverse component. Swarm data for air and N_2 from Badaloni, Wen, Novak, Davies, and Dutton are used in this study. For a complete listing of the data sources see references 1 and 2 and the references therein.

Photon Processes

Cathode Photoelectron Emission. A method similar to that used by Wu and Davies has been applied to this study [2,3,4]. The emission of secondary photoelectrons is taken proportional to the normal component of the photon flux density per second at the cathode surface due to excited species in the discharge volume. The number density of secondary photoelectrons generated at the cathode is:

$$N_e(0,\rho,t)=\Gamma\frac{F(\rho,t)}{\left|V_e(0,\rho,t)\right|} \tag{5}$$

Here, F(ρ,t) is the photon flux density (photons/cm^2-sec) at the cathode surface and Γ is the cathode quantum efficiency. For both N_2 and air, the excitation coefficient is assumed to be that representing the 2nd Positive System of N_2 [5]. A time constant of 36 nsec is used with a quenching pressure of 60 Torr for N_2 and 10 Torr for air [5,6]. Unless otherwise stated the values of Γ used are $4x10^{-4}$ and $3x10^{-4}$ for N_2 and air respectively [2].

Gas Phase Photoionization. The formation of photoelectron-ion pairs in the background gas due to the electron avalanche is handled as follows. A method similar to that proposed by Wu [3] in accordance with the experimental geometry and data of Penney and Hummert [7] is used in this study. The photoelectron-ion density N_{ph} generated at time t and position (ρ,x) is:

$$N_{ph}(\rho,x,t)=\frac{p_q}{p+p_q}\int_0^d\int_0^\infty pN_e(\rho',x',t)\,\alpha(\rho',x',t)\left(\int_0^{2\pi}\frac{\Psi(r'p)\,D(\rho,\rho',|x-x'|,\varphi)}{r'^2}d\varphi\right)\rho'd\rho'dx' \tag{6}$$

In Eq. (6) $r'=\left\{(x-x')^2+\rho^2+\rho'^2-2\rho\rho'\cos\varphi\right\}^{1/2}$ is the total distance between the photon source (ρ',x') and absorption locations (ρ,x), p is the gas pressure, p_q is the quenching pressure (10 Torr for both N_2 and air)[5], α is the ionization coefficient, and Ψ is the photoionization coefficient [7]. The function D in Eq. (6) is the photon absorption distance of the cylindrical volume element designated by the coordinates $x_j\rho_i$, $x_{j+1}\rho_i$, $x_j\rho_{i+1}$, and $x_{j+1}\rho_{i+1}$ on the computational mesh.

Space-Charge Potential Calculation

The space-charge potential throughout the discharge volume has been determined analytically using the following equations [8],

$R \leq \rho_\imath$:

$$\Phi(R,\theta)=\frac{eN\Delta x}{2\varepsilon_o}\left[\left[\rho_{\imath+1}\sum_{n=0}^{\infty}\binom{\frac{1}{2}}{n}\left(\frac{R}{\rho_{\imath+1}}\right)^{2n}P_{2n}(\cos\theta)\right]-\left[\rho_\imath\sum_{n=0}^{\infty}\binom{\frac{1}{2}}{n}\left(\frac{R}{\rho_\imath}\right)^{2n}P_{2n}(\cos\theta)\right]\right] \quad (7)$$

$\rho_\imath < R < \rho_{\imath+1}$:

$$\Phi(R,\theta)=\frac{eN\Delta x}{2\varepsilon_o}\left[\left[\rho_{\imath+1}\sum_{n=0}^{\infty}\binom{\frac{1}{2}}{n}\left(\frac{R}{\rho_{\imath+1}}\right)^{2n}P_{2n}(\cos\theta)\right]-\left[\rho_\imath\sum_{n=1}^{\infty}\binom{\frac{1}{2}}{n}\left(\frac{\rho_\imath}{R}\right)^{2n-1}P_{2n-2}(\cos\theta)\right]-RP_1(\cos\theta)\right] \quad (8)$$

$R \geq \rho_{\imath+1}$:

$$\Phi(R,\theta)=\frac{eN\Delta x}{2\varepsilon_o}\left[\left[\rho_{\imath+1}\sum_{n=1}^{\infty}\binom{\frac{1}{2}}{n}\left(\frac{\rho_{\imath+1}}{R}\right)^{2n-1}P_{2n-2}(\cos\theta)\right]-\left[\rho_\imath\sum_{n=1}^{\infty}\binom{\frac{1}{2}}{n}\left(\frac{\rho_\imath}{R}\right)^{2n-1}P_{2n-2}(\cos\theta)\right]\right] \quad (9)$$

Here $\Phi(R,\theta)$ is the potential in the spherical coordinate system due to the total charge density N in the cylindrical volume element designated by $x_j\rho_\imath$, $x_{j+1}\rho_\imath$, $x_j\rho_{\imath+1}$, and $x_{j+1}\rho_{\imath+1}$ on the computational mesh. R is the spherical radius measured from (x_j,ρ_o) to the location where Φ is to be determined (x_k,ρ_m), and θ is the angle R sweeps out between (x_k,ρ_m) and the axis ρ_o. Δx represents the axial distance given by $x_{j+1}-x_j$, e is the electronic charge, ε_o is the permittivity of free space, $P_m(\cos\theta)$ are the Legendre polynomials of the first kind in $\cos\theta$, and $\binom{\frac{1}{2}}{n}\equiv\left(\frac{\frac{1}{2}!}{n!\left(\frac{1}{2}-n\right)!}\right)$ is the binomial coefficient. By taking the gradient of the above equations an analytical expression for the space charge field is obtained. Because of the small radius of the discharge body (< .1 cm) only one image of N is required to adequately represent the electrode boundaries. The total space-charge field distribution is obtained by superimposing the contributions from all charges located in the discharge volume.

Simulation Conditions and Numerical Method

The pressure and gap width are kept constant at 760 Torr and .5 cm respectively, resulting in a breakdown voltage of approximately 17.5 kV for both N_2 and air [13]. The maximum simulation radius is .1 cm. The discharge volume region is represented by a 200 point uniform mesh in the axial direction and a 22 point logarithmic mesh radially. For both N_2 and air the simulation voltage is varied from 26 kV to 32 kV resulting in a overvoltage range of (47% - 81%). The simulation starts from an initial Gaussian electron density distribution with a total number of 10^4 electrons and a variance of $2D_et'$. Here, t' is the time required for the avalanche to grow from one to 10^4 electrons in a space charge free field. The peak electron density at $x'=v_et'$ appears to be almost independent on the applied voltage and approximately equal to 10^{10} cm^{-3}.

The convective contribution to the species density derivative in Eqs. (1-4) is solved using the generalized flux corrected transport routine [9]. An eighth order modified Euler difference approximation [10,11] and a donor cell algorithm [9] are used to implement the high and low order fluxes respectively. To ensure monotonicity in the solution only the one- dimensional flux limiter [9] is used requiring a time-splitting scheme to incorporate the axial and radial flows. In this time-splitting scheme the radial direction is integrated first followed by the axial direction after which the order of integration is reversed and the process repeated [12]. The two results are averaged to obtain the final solution. The diffusive contribution to the electron density in Eq. (1) is solved using an explicit first order finite difference approximation [11]. The integration of both the space charge field and gas phase photoionization (Eqs. 6-9) introduces four nested loops. To decrease computation time only charge densities within four orders of magnitude from the maximum charge density are used in these calculations. This approximation introduces a small error only in the regions distant from the main discharge body.

RESULTS AND DISCUSSION

In the following discussion electron density contours in r-x space are presented for both N_2 and air outlining the spatio-temporal development of the avalanche to streamer transition. In addition, the temporal development of the axial electron velocity for both the anode and cathode discharge fronts is presented. The front velocity is determined graphically by recording the displacement that occurs over a fixed time increment of .2 ns.

Nitrogen

The simulated electron density contours are shown in Fig. 1. After 5.586 ns the initial phases of the avalanche to streamer transition begin to occur. Most apparent is the spatial broadening of the gaussian pulse along the axial direction due to local space-charge distortion of the Laplacian field. In addition, the initial stage of the development of the cathode directed streamer is apparent by the step-like discontinuity at the cathode side of the electron distribution. The formation of the cathode streamer is attributed to the "delayed" electrons undergoing enhanced ionization and inward radial drift resulting from the positive ion space charge created by the main electron body. Both gas-phase photoionization of the background gas and cathode photoelectron emission produce these "delayed"

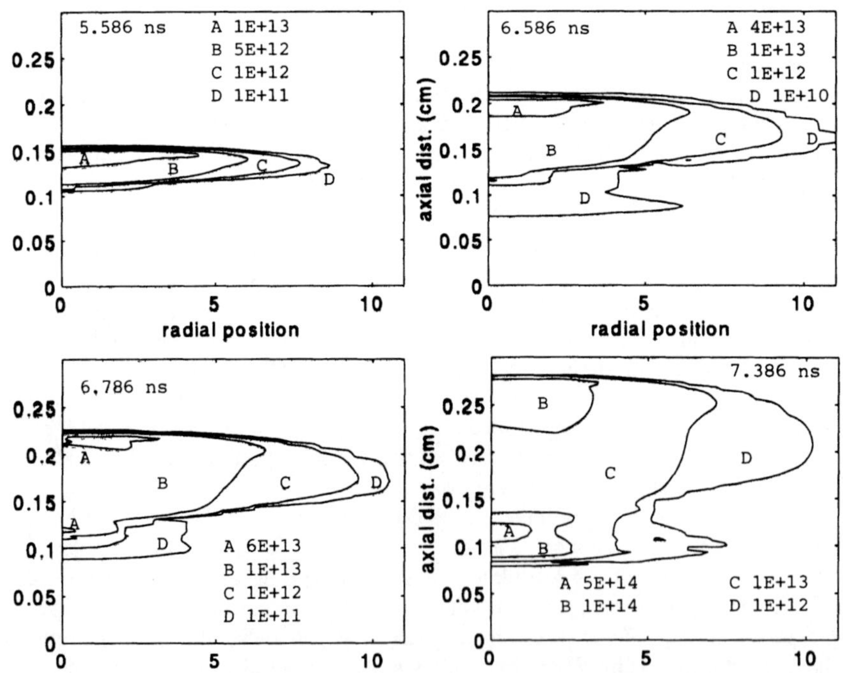

Figure 1. Constant electron density contours describing the spatio-temporal evolution of the avalanche to streamer transition for N_2 at 30 kV. The cathode and anode are located at an axial position of 0 and .5 cm respectively. The following function relates the radial position to a radial width: radial width (cm) $= 3.493 \times 10^{-4}$ (radial position)$^{1.8263}$

electrons. In the next contour plot (6.586 ns) these space-charge effects become more apparent. At 6.786 ns two regions of maximum electron density exist, one located on the anode side off the axis of symmetry, and the second located on the axis of symmetry at the cathode side of the discharge. The fact that at the anode side the density is larger off axis indicates a strong radial field. This has also been observed by Wu in simulations in N_2 at a 68% overvoltage [3]. In the final plot both anode and cathode streamers are fully developed with the maximum electron density now located in the cathode streamer.

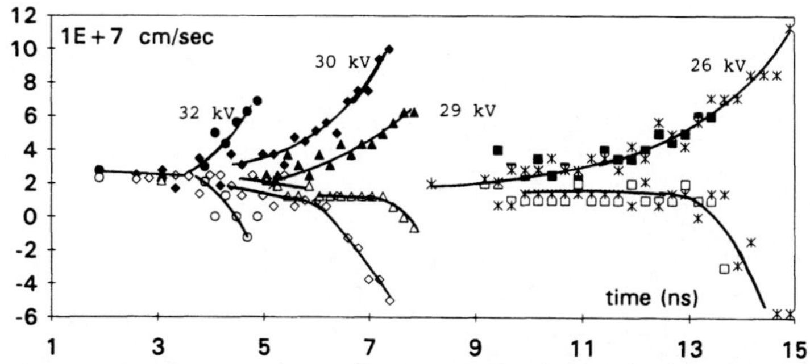

Figure 2. Temporal development of the axial electron velocity for both the anode (dark) and cathode (light) discharge fronts in N_2. The curves labeled by stars (26 kV, Γ=3x10^{-3}).

The anode and cathode electron front velocites are plotted in Fig. 2. These results agree with those presented by Wu (see Ref. 3 and references therein). From these results there is a definite temporal dependence in the development of the anode and cathode streamer fronts. At 26 kV the anode front velocity gradually increases away from the space-charge free electron drift velocity (v_{dr}= 2.05x10^7) while that of the cathode front shows two decreasing step-like transitions seperated by a period of relatively uniform velocity at approximately 75% of v_{dr}. The first transition occurs at the time when the anode front velocity begins to deviate from v_{dr} while the more pronounced, transition occurs when the cathode streamer is fully developed and starts its propagation toward the cathode. As the voltage is increased the deviation of the anode front velocity away from v_{dr} occurs at a higher rate, and the period seperating the two cathode velocity transistions decreases. On the other hand, the rate of change of the fully developed cathode streamer velocity shows no dependence on the applied voltage.

When increasing the cathode quantum efficiency to 3x10^{-3} at a voltage of 26 kV no change in the observed velocity trends are observed. This suggests cathode photoelectron emission plays a secondary role in the avalanche-streamer transition process. At lower voltage levels (15 kV) it has been shown that cathode photoelectron emission is of critical imporatance with regards to the formation of the cathode directed streamer [1].

Dry Air

The simulated electron density contours are shown in Fig. 3. Similar to N_2 the initiatory phase of the cathode directed streamer can be seen at 3.565 ns. At 4.365 ns the maximum electron density is now located at the cathode side of the charge body, and the anode side density maximum is located off the axis of symmetry. As the discharge proceeds in time the effect of the space-charge field on the electron attachment process becomes apparent by the pinching and eventual seperation

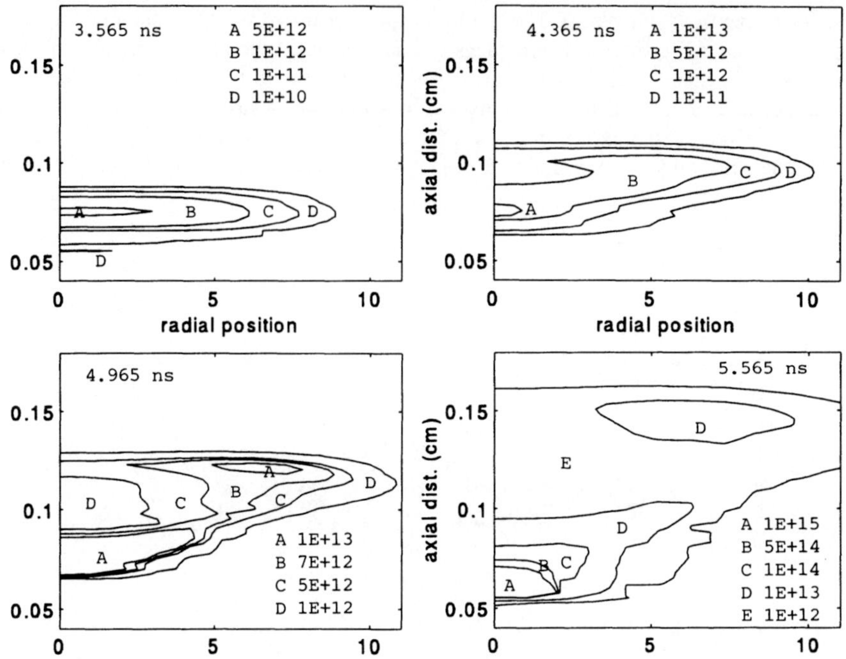

Figure 3. Constant electron density contours describing the spatio-temporal evolution of the avalanche to streamer transition for air at 28 kV. The cathode and anode are located at an axial position of 0 and .5 cm respectively. The following function relates the radial position to a radial width: radial width (cm) = 9.281×10^{-5} (radial position)$^{2.227}$

into two regions of high electron density. The simulations also reveal that the electron density in the cathode directed streamer is two orders of magnitude larger then that in the anode streamer. When no electron detachment is incorporated, similar spatio-temporal distributions are observed but with a cathode streamer electron density that is one order of magnitude smaller. As of present a more

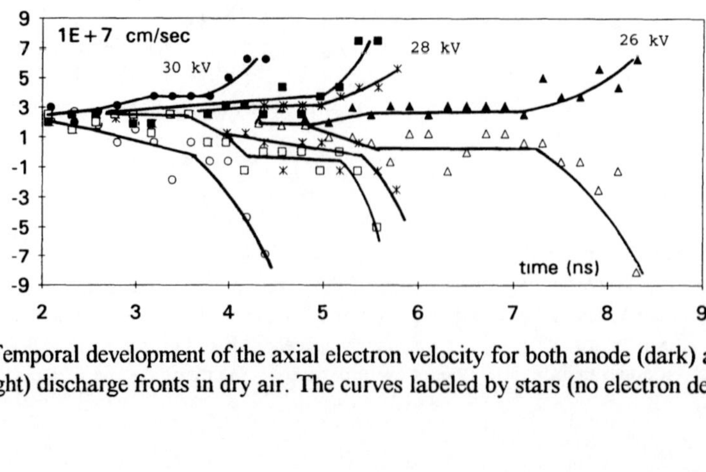

Figure 4. Temporal development of the axial electron velocity for both anode (dark) and cathode (light) discharge fronts in dry air. The curves labeled by stars (no electron detachment).

complete simulation study is needed to fully ascertain the inter-relationship of the parameters involved in an air discharge.

The anode and cathode electron front velocites are plotted in Fig. 4. Currently we are not aware of any experimental velocity data for a comparison of our results. Similar to N_2 we observe two transistions in the cathode front velocity. The first coincides with a slight increase in the anode front velocity and the second occurs when the cathode streamer is fully developed. The cathode streamer transition is not as abrupt as in N_2. Also the rate of change of the streamer velocity is smaller. This may be due to the net reduction of the positive charge density, essential for cathode streamer propagation, by the negative ion density in the discharge bulk. Simultaneously with the cathode streamer transition the anode front velocity also undergoes a transition into the anode streamer phase with a more gradual slope then that observed in N_2. Increasing the voltage appears to cause an increase in the rate of change of both cathode and anode streamer velocites. When electron detachment is not included only the slope of the cathode and anode streamer velocites are reduced, all other trends appear to be unaffected. At lower voltage values (15 kV) it has been shown that electron detachment plays the dominant role in the development of the cathode streamer [1].

CONCLUSIONS

Initial results of fully two-dimensional simulations of the avalanche to streamer transition in N_2 and dry air in overvolted gaps at atmospheric pressure have been presented. In N_2 and air gas phase photoionization plays the dominant role in the cathode and anode streamer development. In both N_2 and air cathode photoelectron emission is of secondary importance at the overvoltage levels studied. In air, electron detachment appears to increase the propagation speed of both streamer fronts. The simulated behavior in overvolted gaps is substantially different from that observed in undervolted gaps. A more detailed study is required to ascertain the dependencies of the observed trends on the discharge parameters.

REFERENCES

1. J.T. Kennedy, J.M Wetzer, and P.C.T. van der Laan, Space charge induced breakdown in under-volted gaps in N_2 and air, *8th Int. Conf. High Volt. Engr - Yokohama* 1:383 (1993).
2. J.T. Kennedy, M.G.M. Megens, and J.M. Wetzer, Cathode photoelectron emission during a gas discharge in N_2 and dry air, *IEEE Int. Symp. El. Ins. - Pittsburgh.* (1994).
3. C. Wu and E.E. Kunhardt, Formation and propagation of streamers in N_2 and N_2-SF_6 mixtures, *Physical Review A.* 37:4396 (1988).
4. A.J. Davies, C.J. Evans, P. Townsend, and P.M. Woodison, Computation of axial and radial development of discharges between plane parallel electrodes, *Proc. IEE.* 124:179 (1977).
5. S. Badaloni and I. Gallimberti, Basic data on air discharges, *Universita di Padova Report. Upee-72/05.* (1972).
6. T.H. Teich and R. Braunlich, UV and other radiation from discharges in artificial air and its constituents, *Proc. VIII Int. Cnf. Gas. Dschrgs. and Appl.* 1:441 (1985).
7. G.W. Penney and G.T. Hummert, Photoionization measurements in air, oxygen, and nitrogen, *J. Appl. Phys.* 41:572 (1970).
8. J.D. Jackson. Classical Electrodynamics, Wiley & Sons, New York (1962).
9. S.T. Zalesak, Fully multidimensional flux-corrected transport algorithms for fluids, *J.Comp. Phys.* 31:335 (1979).
10. S.K. Dhali and P.F. Williams, Two-dimensional studies of streamers in gases, *J. Appl.Phys.* 62:4696 (1987).
11. D. Potter. Computational Physics, Wiley & Sons, New York (1977).
12. J.P. Boris, LCPFCT - A flux corrected transport algorithm for solving generalized continuity equations, *Naval Research Lab., NRL/MR/6410-93-7192* (1993).
13. M. Aguet and M. Ianovici. Traité d' Électricité Vol. XXII, Presses polytechniques romandes, Lausanne (1982).

DISCUSSION

E. MARODE: We did some similar work in nitrogen which we presented at the Gas Discharges Conference in Swansea in 1992. We found out that a propagating cathode directed streamer depends on the photon absorption length for photoionizing photons. If this absorption length is too short or too large no streamer is launched. Did you find a similar behavior? What are your initial conditions concerning electrons? Have you information on the net charge distribution?

J. T. KENNEDY: The experimental photoionization data used are from Penney and Hummert whereby the absorption distance is included in the photoionization coefficient data. Yes, I would expect that if the absorption coefficient is too large the newly formed electrons would not have ample distance to undergo several ionizing collisions before entering the "lower" field region of the avalanche body, thus retarding the launching of the cathode directed streamer. On the other hand, I would expect gas phase photoelectrons created far from the avalanche body to result in the launching of the cathode directed streamer, although at a later time. A good upper limit to this line of thought is cathode photoelectron emission which has been shown to be adequate to launch the cathode directed streamer. Finally, I would like to add that in the case of air the presence of electron detachment decreases the time whereby the cathode directed streamer is launched, and also results in an enhancement of its subsequent propagation velocity.

As stated in the paper, the initial conditions are as follows: the simulation starts from an initial Gaussian electron density distribution with a total number of 10^4 electrons and a variance of $2D_e t'$. Here, t' is the time required for the avalanche to grow from one to 10^4 electrons in a space-charge-free field. The peak electron density is approximately $10^{10}cm^{-3}$.

With regards to the net charge density distribution, this information is available and was used for the calculation of the field distribution. I only presented the electron density data as I felt it was quite representative of the spatio-temporal variation of the species densities involved.

R. J. VAN BRUNT: Could you comment about how your method differs from that of Wu and Kunhardt?

J. T. KENNEDY: I am aware of the work of Wu and Kunhardt and have referenced it in my paper [Ref. (3)]. Our model differs from the one by Wu and Kunhardt with respect to the method of determining the space charge field as well as the high order flux algorithm used. Primarily though, our work in nitrogen agrees well with their results. For dry air on the other hand our work provides, to our knowledge, the first fully two dimensional simulation. The results support the necessity of using such a fully 2-D model.

TIME-RESOLVED AVALANCHES IN SF_6

UNDER GEOMETRICALLY-PERTURBED UNIFORM FIELD

M.F. Fréchette,[1] N. Bouchelouh,[2] and R.Y. Larocque[1]

[1]Câbles et Isolants, Technologie et IREQ, Hydro-Québec
[2]Engineering Physics, Ecole Polytechnique
Québec, Canada, J3X 1S1

INTRODUCTION

There is technical interest for the ability of experimentally producing rapid space-charge growth in a stressed gaseous dielectric. For instance, to simulate residual-charge effects during the recovery phase of an interrupting medium. Or, to help in the assessment of a mixed-voltage situation, e.g. when both DC and time-varying voltages superposed to stress a gas-insulated gap.

Fast and reproducible control of the space-charge development should be achievable over the first transit time in an attempt to avoid breakdown. On this time-scale, there are essentially two basic approaches to concentrate a sufficient high carrier number inside the diffusing avalanche head: an initial large amount of photo-released electrons at the cathode, or a considerable carrier amplification over the transit time.[1] There are several instances where sufficient carrier amplification is unattainable, offering a rather limited spectrum of operating conditions. In other respects, using strong illumination allows functionality and versatility in a variety of situations, the demand on amplification being less drastic and the dependence on secondary-emission processes less pronounced. This last context is referred to as an undervoltage-breakdown condition. Typically, a gas discharge is initiated in a DC pre-stressed gap following the UV-release of photoelectrons at the cathode, and resulting into the occurrence of a voltage withstand lower than normal. Pending on the selection of experimental parameters, space-charge growth may lead to a cathode-directed streamer right passed the first electron transit time.

It is usually contended that the initial stage of avalanche growth occurs under a space-charge free environment. Using an experimental artifice, namely a decreasing geometrical field particularly depressed towards the anode, Fréchette *et al.*[2] have shown that the avalanche growth could be affected at an early stage of its development, exhibiting an enhanced rate of growth. These earlier measurements were performed in atmospheric N_2 and under a non-zero probability of an undervoltage breakdown. Such an experiment based on strong illumination (at the cathode surface) and low amplification (geometrically decreas-

Gaseous Dielectrics VII, Edited by L.G. Christophorou
and D.R. James, Plenum Press, New York, 1994

ing field) is not readily adaptable to the investigation and control of space-charge formation in atmospheric SF_6. Under DC stress, avalanche growth is totally quenched.

However, at lower values of the product pressure times gap length, conditions for an undervoltage breakdown are accessible. Thus, the same approach can be used to examine the early development of the discharge. The present paper reports on avalanche growth observed in low-pressure SF_6 for undervoltage-breakdown conditions.

EXPERIMENT AND CONDITIONS

The experiment consisted of measuring time-resolved avalanches under conditions preceding undervoltage breakdown. Measurements were taken as a function of E/p, where E and p are respectively the uniform electric field and pressure. This was realized via an increase of the applied DC voltage pre-stressing the gap.

Figure 1 offers a schematic view of the experimental layout. A pair of electrodes forming a parallel-plane arrangement was used. The anode is Bruce-profiled with a diameter of 12.2 cm. The flat portion of the anode has a 0.5-cm hole in its center: it allowed for the laser light to pervade through, and produces a geometrical field-nonuniformity. The earthed electrode is fragmented;[3] it is formed of two concentric plates separated by 0.2 mm. The central plate has a diameter of 5.2 cm, insuring that the current is fully induced at this location for gap lengths smaller than 1.3 cm. The electrodes are confined in a high-vacuum stainless-steel cylindrical chamber. The interelectrode distance could be adjusted to a precision of 0.03 mm, by moving the earthed plane. Typically, the system was pumped down to 10^{-6} Torr, then flushed and filled with the gas under study and finally, the gap length adjusted. Nitrogen (N_2 UHP grade) and sulfur hexafluoride (SF_6 CP grade) were used, the gas pressure being referenced to the room temperature 22 °C.

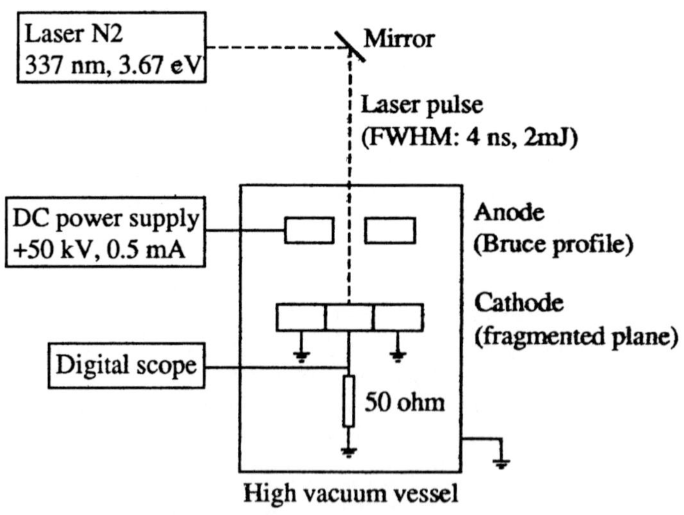

Figure 1. Diagram illustrating the experimental setup.

Positive DC was applied to the anode, and varied in order to establish the conditions for which the laser pulse triggers an induced breakdown. The UV pulse (FWHM = 4 ns) produced by a high-power N_2 laser was focussed (spot = 0.5 x 0.2 cm^2) on the central part of the fragmented plane. An approximate energy amounting to 2 mJ was injected in the gap.

In vacuum, the photoelectrons initially released at the cathode ranged in the order of 10^9, when Aluminum was used as an emissive material. Recorded in vacuum and at low field (4 kV/cm), the signal was detected using a fast digital scope (Single-shot bandwidth = 400 MHz). The measured signal showed a FWHM comparable to that of the UV pulse. Many waveforms were recorded in low-pressure SF_6 under nonuniform-field conditions, and fed to a computer for further processing. Gap length (g) and pressure were kept constant throughout: g = 1 cm, and p = 5 Torr.

The field distributions characterizing the present gap geometry are shown in Fig. 2. The gap length was set at 1 cm, as to geometrically produce quasi-uniform field-conditions over a large portion of the gap. The axial Laplacian distribution (Fig. 2, the cathode being located at zero on the discharge axis) stays within 3% of the uniform field (1 kV/cm) for at least 60% of the interelectrode distance. Farther on, the field is seen to steeply decrease when approaching the anode. The presence of the hole at the anode will also affect the off-axis field distribution, the field being enhanced at its periphery.[4] This effect is illustrated in Fig. 2, where a radial field distribution is displayed. It features the field distribution at the coordinate x = 1 cm, the distribution extending radially from the hole center.

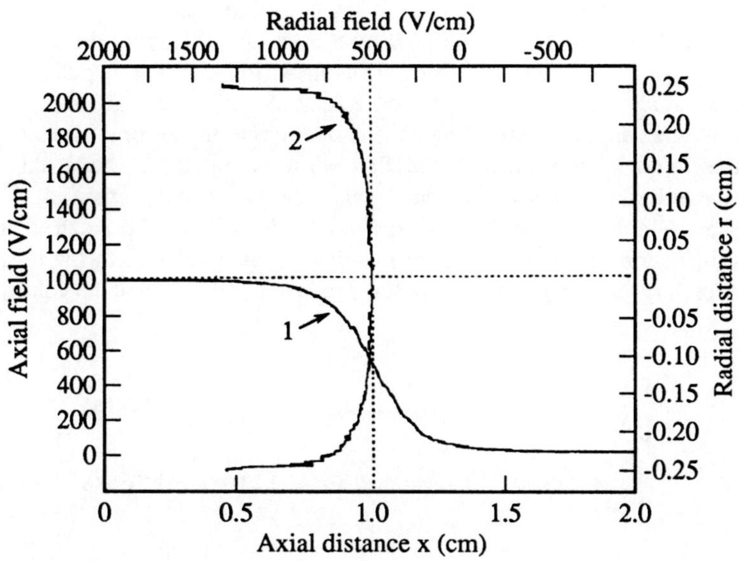

Figure 2. Field distribution corresponding to the present gap geometry; 1 - Axial field on the discharge axis (r = 0); 2 - Radial field calculated at x = 1 cm.

RESULTS AND DISCUSSION

Under low pd-values in SF_6 and uniform-field conditions, it is recognized that the final stage leading to breakdown is controlled by a Townsend mechanism. Spontaneous (and unirradiated) breakdown behavior of our gas-insulated arrangement was assessed. Present results compare favorably with data reported in the literature.[5,6] On average, the present breakdown voltage (1.21 kV) was found to be slightly lower than that corresponding to the Paschen-value, by approximately 6%. This withstand drop can be explained by the slight geometrical field-nonuniformity. When a large amount of photoelectrons (released at the cathode) initiated the discharge, the insulation properties could be modified. Laser-induced breakdown was possible at a lower applied voltage, viz. at 90% of the spontaneous breakdown-value. The minimum conditions for which breakdown could be induced, corresponded to an E/p-value of 220 $Vcm^{-1}Torr^{-1}$. Reduction in voltage withstand provides supporting evidence in favor of an active role of space-charge during discharge development.

In order to investigate potential space-charge effects in the early development of the discharge, i.e. for times smaller than an electron transit time, a particular gap geometry was selected (see Fig. 2). Considering the low-pressure conditions (5 Torr), it was contended that space-charge effects could be counteracted by local diffusion processes. Consequently, the interelectrode distance was set as to produce geometrical uniform field-conditions near the cathode. To help in the waveform interpretation, time-resolved signals were recorded in N_2, where electron attachment and detachment are inoperative. Figure 3 shows a typical waveform, registered at low reduced field-value, E/p = 112 $Vcm^{-1}Torr^{-1}$. The initially photo-released electrons are detectable, the curve exhibiting a shoulder shape near the starting time. Not all the photoelectrons contribute to the discharge, since some are backscattered due to collisions with the gas molecules. Furthermore, diffusion and electrostatic repulsion result into the enlargement of the initial photoelectron-pulse. The initial number of contributing electrons was estimated to reach about 3×10^7. As long as the electron growth exceeds the counteractions resulting from the decreasing field, the signal slope will remain positive. The induced electron current (or voltage) being proportional to $n_e(x).E(x).v_e(x)$, where $n_e(x)$, $E(x)$ and $v_e(x)$ are respectively the electron density, the electric field and the electron drift velocity, the signal peak may not correspond to the arrival time of the majority of carriers. Having considered this, the electron transit time could be estimated to occur around 27 ns, where an abrupt change in the slope is observed (Fig. 3). Dividing the gap length by this time estimate gives a drift velocity that was found to exceed substantially (14%) that reported by Aschwanden.[7] Considering that the field is geometrically decreasing over almost half the gap, a longer transit time should be expected. Following the photoelectrons release, the field surrounding the cathode, thus during the early development of the discharge, could also be somewhat enhanced. Finally, the waveform shown in Fig. 3 displays an enlarged tail beyond the electron transit time. Longitudinal diffusion of the swarm, path lengthening associated with the presence of the hole, and secondary emission due to photon feedback are believed to contribute to this enlargement.

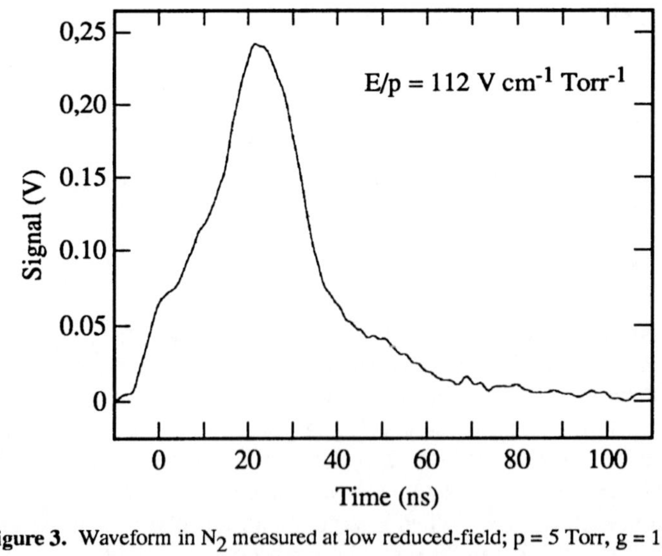

Figure 3. Waveform in N_2 measured at low reduced-field; p = 5 Torr, g = 1 cm.

Figure 4 exemplifies a typical waveform, observed in SF_6 at a high value of reduced field, E/p = 180 $Vcm^{-1}Torr^{-1}$. If the shoulder exhibited on the curve is taken as being representative of the initial value of the current, the thus estimated number of free electrons is found to be reduced considerably (by one order of magnitude) as compared with the

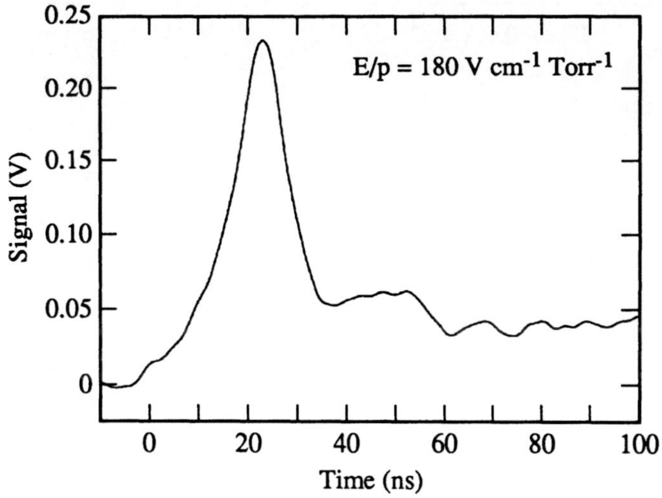

Figure 4. Time-resolved avalanche in SF_6; p = 5 Torr, g = 1 cm.

amount evaluated in the case of low-pressure N_2. Collisional processes in SF_6 differ. Some photoelectrons will be captured due to attachment, while some negative ions will detach. This will diminish the intensity of the electron source as well as expanding it in time. The initial swarm will leave behind a negative space-charge. For the following few nanoseconds (about 10 ns), the swarm growth is observed to be rapid (Fig. 4). At larger times, the swarm enters the region where the geometrical field becomes depressed. This is also a region where the charge carrier number is in excess of 10^6. The electron density will eventually pass through a maximum, that corresponding to the time where the limit field is attained. The field depression intensifies as the swarm approaches the anode: at first, the electron growth will be slowed down, and further on, a large number of electrons will suffer collisional attachment. Therefore, in accordance with this argument, there will be a negative-ion swarm of substantial density reaching the anode.

The features exhibited by the waveform presented in Fig. 4 tend to concur with the discussed scenario. The signal peak is seen to occur too early (around 23 ns) to represent the transit of the majority of carriers, particularly considering that the geometrical field is decreasing. Furthermore, the FWHM of the waveform is rather narrow, the signal magnitude falling quite rapidly with time, to join with the ionic component. This rapid fall is not consistent with the tail enlargement as observed and discussed in the case of N_2. However, it would be with a minimum delayed contribution from detached electrons, whereas a high density of negative ions is submitted to rather depressed field-conditions in the anode surroundings. If the intersection between the total ion-current curve (constant signal starting around 60 ns) and the extrapolated line from the falling part of the peaked waveform is considered as giving the total electron transit time, an estimated value of 37 ns is obtained. This value is just slightly higher than that corresponding to uniform-field conditions.[7] As seen on Fig. 4, the main peak is followed by a secondary avalanche. The underlying mechanism leading to this second peak can be ascribed to photon feedback at the cathode. The occurrence of this peak so close to the estimated transit time is consistent with an active building up of the positive space-charge in the early development of the discharge.

Figure 5 offers a comparison of two waveforms taken at different E/p-values. The time-resolved signals recorded at 200 $Vcm^{-1}Torr^{-1}$ displayed similar features. When compared to waveforms registered at 180 $Vcm^{-1}Torr^{-1}$, the measured signals exhibited a larger peak-amplitude, a narrower width, multi-avalanche features, and a larger ionic component. In general, the growth rate in the early stage of the avalanche was found to vary exponentially, as demonstrated in Fig. 6. In the next graph (Fig. 7), the experimental values of the

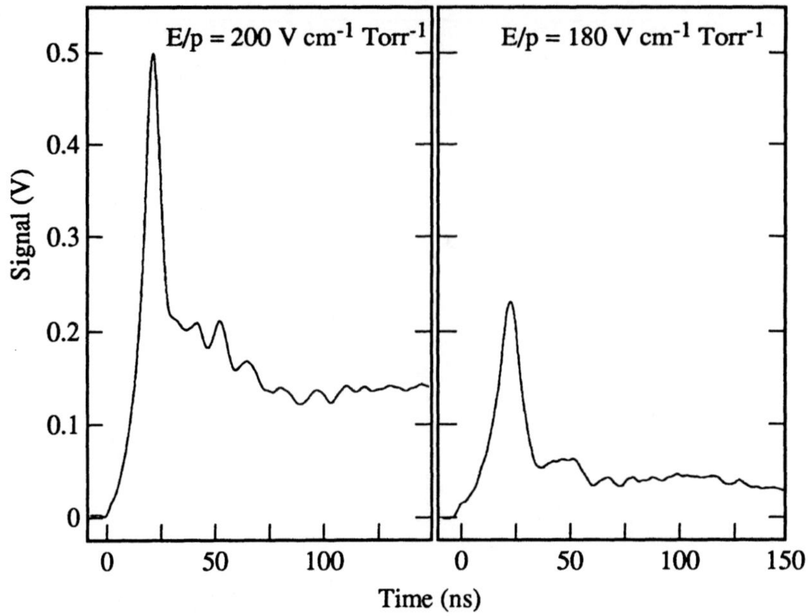

Figure 5. Waveforms in SF_6 at high reduced-fields; p = 5 Torr, g = 1 cm.

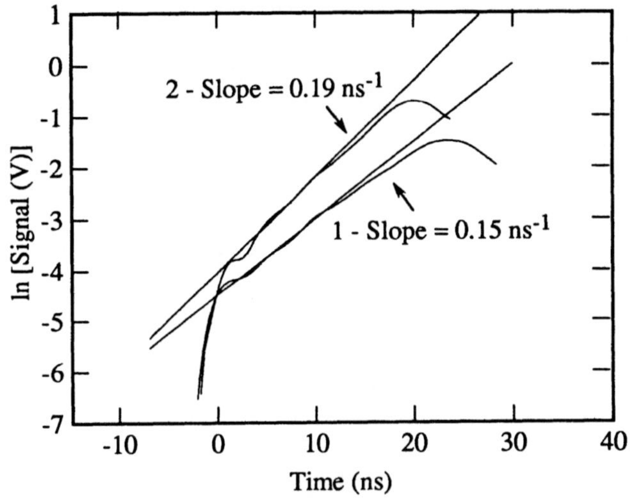

Figure 6. Avalanche growth in SF_6; 1 - E/p = 180 $Vcm^{-1}Torr^{-1}$, 2 - E/p = 200 $Vcm^{-1}Torr^{-1}$.

growth rate are compared with experimental data featuring the product of the effective ionization coefficient times the drift velocity. Aschwanden's data[7,8] were obtained in a Pulsed-Townsend experiment, performed under highly uniform field-conditions. Figure 7 presents the rate of growth as a function of the applied E/p. The trend exhibited by the present experimental data shows an increasing rate as a function of the applied reduced field, the observed variation being linear. The present results (squares) lie considerably lower than those corresponding to uniform field-conditions (circles). This difference was observed to grow with E/p increasing. The discrepancy can not be explained solely on the basis of the slight geometrical field-nonuniformity in the first portion of the gap. Since the electron amplification increases with the applied field, it is reasonable to assume that the buildup of the positive ion-density will be enhanced as well, affecting the local field early in the development of the discharge. This would also imply that the initial negative ion-density is formed mostly of stable ions.

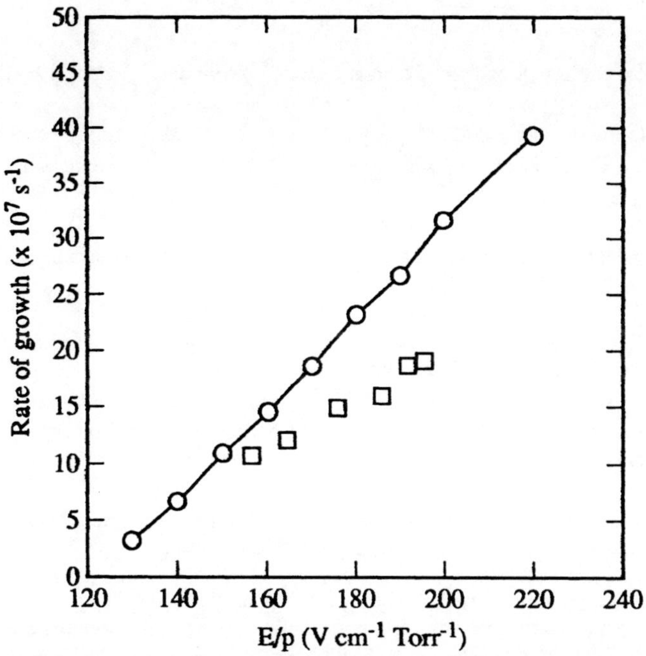

Figure 7. Growth rate in SF_6 as a function of reduced field; circles - Ref. 7, squares - present results.

CONCLUSION

Time-resolved avalanches in low-pressure SF_6 have been investigated. The investigation was performed for conditions associated with undervoltage breakdown, using a particular gap-geometry. The resulting geometrical field-distribution has helped to evidence the avalanche growth in its early stage of development. In this context, it is worth underlying that the geometrical characteristics of the gap may serve in controlling both space-charge magnitude and location, illustrating the concept of space-charge structural design.

It was demonstrated that the breakdown voltage could be reduced under static fields, upon the photo-release of a large amount of initial electrons. Waveforms were recorded at high E/p-values, almost up to the reduced field-value at which an undervoltage breakdown could occur. The time-resolved signals have shown particular features over the first transit time. In the early stage of development, the avalanche growth was found to vary exponentially with time, exhibiting a lower rate than that observed under uniform field-conditions. This observation, plus the particular time-location of the delayed secondaries, concur in the supporting of an early buildup of space-charge. Finally, the presence of depressed field-conditions towards the anode affected the registered waveform: collisional attachment dominates in this region, resulting into a decrease of electron density. The rapid fall of the time-resolved signal might not be fully associated with neutralization at the electrode, the electron growth being quenched.

ACKNOWLEDGMENTS

This work was sponsored by *Direction Distribution*, Hydro-Québec.

REFERENCES

1. H. Raether. "Electron avalanche and breakdown in gases," Butterworths, London, (1964).

2. M.F. Fréchette, N. Bouchelouh, and R.Y. Larocque, Laser-induced undervoltage breakdown in atmospheric N_2 correlated with time-resolved avalanches, *Proc. 1994 Int. Symp. on Electrical Insulation*, Pittsburgh, Penn. USA (1994).

3. H.F.A. Verhaart and P.C.T. van der Laan, Fast Current Measurements for Avalanche Studies, *J. Appl. Phys.*, 53:1430 (1982).

4. M.F. Fréchette, SF_6 breakdown characteristics with a specially-designed cathode, *Proc. 1992 IEEE Int. Symp. on Electrical Insulation*, Baltimore MD USA (1992).

5. T. Hosokawa, Y. Sekiya, and Y. Miyoshi, Pre-breakdown phenomena in uniform field gap in SF_6, *Electrical Engineering in Japan*, 105:9 (1985).

6. T.W. Dakin, G. Luxa, G. Oppermann, J. Vigreux, G. Wind, H. Winkelnkemper, Breakdown of Gases in Uniform Fields - Paschen Curves for Air, N_2 and SF_6, *Electra*, 32:61 (1974).

7. Th. Aschwanden. "Die ermittlung physikalischer entladungsparameter in isoliergasen und isoliergasgemischen mit einer verbesserten swarm-methode," Doctorate dissertation, ETH, Zurich, Switzerland, unpublished (1985).

8. Th. Aschwanden, Swarm parameters in SF_6 and SF_6/N_2-mixtures determined from a time resolved discharge study, *Proc. Fourth Int. Symp. on Gaseous Dielectrics*, Knoxville TN USA (1984).

DISCUSSION

E. MARODE: The current pulse seems to show several avalanches due, you said, to photoionization effect. This is similar to a Townsend ionization growth. Why is the second avalanche smaller compared to the first one? Is it due to a space-charge effect? Does the radial electric field, which should be maximum near the anode hole not lead to a maximum in the current value?

M. F. FRÉCHETTE: Over the first transit time, the shape and height of the signal are mainly linked to the electron population. Therefore, the first peak exhibited by the recorded signal is strongly proportional to the large amount of initially photo-released electrons. The overall yield leading to secondaries is less efficient. Contribution to the signal associated with these secondaries appears past the first transit time. The secondary avalanches grow under the influence of a space-charge reduced local field. In SF_6 and the present conditions, secondary avalanches are efficiently quenched: occurrence of undervoltage breakdown shows a low probability over the applied voltage resulting into a spontaneous breakdown slightly reduced (irradiation intensity kept constant).

The anode hole perturbs the radial field. However, it does so in a region where the axial geometrical field is decreasing. In this region, the electron population is quenched. This would concur in explaining why the radial component is assumed to have a minor role. Assuming low-intensity UV pulse, a one-dimensional simulation of the growth has shown that the signal shape could be reproduced without reference to the radial component.

J. M. WETZER: You use a nonuniform field for the purpose of rapid space-charge growth. Considering the amount of initial electrons released from the cathode ($\sim 10^9$), such growth can also be achieved in a uniform field. We have measured and modeled "space-charge induced breakdown in undervolted gaps in nitrogen and air" in uniform fields (see paper by Kennedy and Wetzer presented at ISH, 1993). Why do you use a nonuniform field geometry in your work?

F. M. FRÉCHETTE: The ultimate goal of the ongoing research program concerns discharges occurring in real gas-insulated systems, where geometrical field-nonuniformity is common. Moreover, considering the status of knowledge accumulated over the years and the understanding of the uniform-field case, the time is right to transfer this body of knowledge to the nonuniform-field case. In that sense, the present experiment is an extrapolated situation: the anode hole is also present in the uniform-field time-resolved studies. Here, its enlargement allows to produce an enhanced non-uniformity. In fact, the design (or variety) of the gap, and other particularities of the experiment, play an important role in the control of both space-charge magnitude and location, introducing the concept of space-charge structural design. Finally, the selection of a decreasing field (in the direction of the axial growth of the avalanche) allows to have a better measurement sensitivity in the early stage of discharge development.

S. MAHAJAN: Was 200 V cm^{-1} $Torr^{-1}$ the highest value of E/p used in these measurements? If so, can you comment on the absence of secondary avalanches in your records? What kind of scheme was used for the protection of the oscilloscope against accidental breakdown in the SF_6 gap?

M. F. FRÉCHETTE: The minimum conditions for which breakdown could occur with a non-negligible probability corresponded to an E/p value of 220 Vcm^{-1} $Torr^{-1}$ (laser irradiation kept constant). The highest value of E/p used for the time-resolved measurements was 200 Vcm^{-1} $Torr^{-1}$. Due to the geometrical features of the gap, avalanche growth approaching the anode is somewhat quenched. Space-charge effects play a role as well. This also applies to secondaries. In SF_6 using the present experimental arrangement, it was observed that the border conditions for having an undervoltage breakdown were well defined. Therefore, no specific protection relative to the scope was used.

J. CASTANGUAY: Can you explain a little more why a relatively small change in field strength causes such a large change in both signal peak intensity and peak width, and why the decay is so rapid?

F. M. FRÉCHETTE: In both examples, the experimental conditions were the same, except for the applied voltages which defined the given E/p-values. Notwithstanding the space-charge effects, the variations brought upon the time-resolved signals due to the increase in reduced fields are linked mainly to the buildup processes. The effective ionization coefficient increases almost linearly with E/p, whereas the induced current will be an exponential function of this parameter. In the present geometry, the signal width is a rather complex function. However, a larger E/p-value means an increase in drift velocity, resulting in a more rapid crossing of the majority of carriers. The sharp decay of the signal (past the first peak) results from the quenching effect due to the decreasing field towards the anode. Reduction in electron detachment is believed to contribute to the sharpness of the descent.

APPLICATION OF STREAMER BREAKDOWN CRITERION FOR INHOMOGENEOUS FIELDS IN DRY AIR AND SF_6

W.S. Zaengl and K. Petcharaks

Swiss Federal Institute of Technology
8092 Zürich, Switzerland

INTRODUCTION

It is widely accepted that the simplified streamer breakdown criterion

$$\int_0^{x_c} \bar{\alpha}(x) \cdot dx \quad = \quad K \tag{1}$$

where $\bar{\alpha}$ is the effective coefficient of ionization, x is a coordinate along an electric field line and K is a dimensionless number representing various secondary ionization process, can well be used to calculate breakdown voltages in many gases subjected to homogeneous or weakly inhomogeneous field distributions. For field distributions of this kind, x_c is either identical with the gap distance (g) or a critical electron avalanche length, which is defined by a position within the gap, where $\bar{\alpha} \leq 0$. Though K is often taken to be 18 for calculating onset of breakdown (thus taken to be the natural logarithm of a critical number N_{crit} of about 10^8, assumed as a number of electrons produced by ionization in a critical avalanche), it has been shown by A. Pedersen et. al.[1] and W.S. Zaengl et. al.[2] that K for SF_6 and synthetic air is much lower.

The application of eq. (1) to inhomogeneous field gaps for a wide range of gas pressures is the aim of this investigations, as at least some uncertainty exists, whether K remains constant for a given gas under such conditions. Investigations in nonuniform asymmetrical fields will also provide informations about polarity effects which are often encountered within gas discharge investigations. Furtheron, it is well known that for highly nonuniform fields the application of eq. (1) is restricted to the calculation of (partial) discharge inception voltage, U_i, which is of very high interest for engineering applications.

INCEPTION VOLTAGE CALCULATION

As the effective ionization coefficient, $\bar{\alpha}$, is strongly dependent on the electric field distribution E(x) in the gas and electrode system used for the experiment, the first prerequisite for a numerical calculation of the inception voltage U_i is to accurately calculate E(x) along field lines providing largest values of the left hand side of eq. (1). The HSSSM program (High Speed Surface Charge Simulation Method), developed by S. Sato[3], was used for numerical field calculation. Very probably, this program provides the highest accuracy today[4] at least for problems as in question for the electrode configuration used.

Gaseous Dielectrics VII, Edited by L.G. Christophorou
and D.R. James, Plenum Press, New York, 1994

For the electrode system used, the most effective field line acts along the axis of symmetry, and E(x) was calculated along this axis at 900 up to 1700 points, where the largest number of points was used for largest gap spacing. The calculated values were then fitted to a polynomial of order 14 to provide an analytically integrable equation, i.e.

$$E(x) \quad = \quad U \cdot \sum_{n=0}^{14} a_n x^n \tag{2}$$

where a_n are the polynomial coefficients, U the voltage across the gap, and x the distance along the axis of symmetry. Though the absolute accuracy of field calculation programs can only be checked by comparing them with analytical solutions for similar field distributions (such a comparison was made for sphere - to - infinite plane configurations resulting in error of less than 0.04 %), the accuracy of E(x) based on numerical calculations for the actual *sphere-pot* arrangement and the fitting procedure was certainly less than 1 %.

The second prerequisite for the calculations is the knowledge of the relationship between $\bar{\alpha}/p$ and $f(E/p)$ for a given temperature and gas. For synthetic air (80 % N_2 : 20 % O_2) and $E/p <$ 7.94 kV/mm·bar, we used the relation

$$\frac{\bar{\alpha}}{p} \quad = \quad 1.6053 \cdot \left[\frac{E}{p} - 2.165\right]^2 \quad - \quad 0.2873 \qquad \text{in } (\text{mm} \cdot \text{bar})^{-1} \tag{3}$$

where pressure p is quantified for 20°C. The above relation is based on swarm parameter measurements and breakdown experiments published by Zaengl et al.[2]. As the nonlinear increase for $\bar{\alpha}/p$ with E/p will gradually change to a linear one for $E/p >$ 7.94 kV/mm·bar, this region was accounted for by the relation given below, provided by the data from Rao as published by Dutton[5] :

$$\frac{\bar{\alpha}}{p} \quad = \quad 16.7766 \cdot \frac{E}{p} - 80.006 \qquad \text{in } (\text{mm} \cdot \text{bar})^{-1} \tag{4}$$

For SF_6, the data from Th. Aschwanden[6] were used. As in inhomogeneous gaps and at low pressures investigated, quite high E/p - values become active, Th. Aschwanden data had also to be subdivided into two E/p regimes. This field dependencies are

$$\text{for } E/p < 12.36\frac{\text{kV}}{\text{mm} \cdot \text{bar}} \quad : \quad \frac{\bar{\alpha}}{p} \quad = \quad 27.9 \cdot \left[\frac{E}{p} - 8.9246\right] \quad \text{in } (\text{mm} \cdot \text{bar})^{-1} \tag{5}$$

$$\text{for } E/p > 12.36\frac{\text{kV}}{\text{mm} \cdot \text{bar}} \quad : \quad \frac{\bar{\alpha}}{p} \quad = \quad 22.3595 \cdot \frac{E}{p} - 180.1709 \quad \text{in } (\text{mm} \cdot \text{bar})^{-1} \tag{6}$$

Combining equations (2) with (3) and (4) or (5) and (6) respectively, the inception voltage U_i in dry air and SF_6 can be computed, if the right hand side (K) of eq. (1) is known. In this report, we display the results based on some knowledge about the K - values. Therefore, the critical avalanche length x_c in eq. (1) was iteratively varied so that the difference between the two sides of the equation was less than 0.01 %.

EXPERIMENTAL ARRANGEMENT AND PROCEDURE

To introduce different degrees of field inhomogeneities within the gap spacing, metal spheres fixed by cylindrical shafts were centered within a metallic pot (diameter 185 mm), the cylindrical part of which (200 mm in length) was large enough to avoid any influence of external potentials, as e.g. the walls of the vacuum - tight vessel in which this electrode arrangement was placed. Extremely smooth steel spheres of high geometrical precision, as used for bearings, provided high quality field stress control in the vicinity of these spheres up to the plane (bottom) of the pot. Together with variable gap spacing (g) between spheres (diameter 5, 10 and 20 mm) and plane the field factors could be varied between about 2.6 and 7.7. The gap spacing was measured with a micrometer providing an accuracy of ± 10 μm. The configuration was *upside down* to facilitate the use of ultraviolet (UV) irradiation and to avoid accumulation of particles at the bottom of the pot.

UV irradiation was provided by placing a *pencil UV lamp* (Oriel pen ray lamp, type 6035 - Hg/Ar) with main emission spectrum at about 250 nm. No attempt has been made to measure UV intensity or the effect of UV wavelength on U_i. Thus, we subdivide the results of measurements *with* and *without* UV irradiation.

The high voltage source consisted of a DC power supply (SPELLMAN type RHR100PN100, 0 - 100 kV, 1 mA), a 50 kΩ charging resistor, and a 25 nF filter capacitor. The voltage was measured with a 200 MΩ resistor voltage divider together with an electrometer (KEITHLEY 617). The voltage measurement error was less than 0.5 % for $U_i \leq 5$ kV and less than 0.3 % for $U_i > 5$ kV.

We used dry bottle air (from Carbagas) and commercial grade SF_6 (from Kalichemie) in this experiment. For each gap spacing, the chamber was evacuated down to less than 0.005 mbar for about one hour before introducing the gas. Gas was introduced slowly via a 2 μm dust filter without any further treatment. The pressure was measured with a vacuum or a pressure gauge with an uncertainty of ± 0.1 %. All experiments were made at room temperature, and small deviations from 20°C have been corrected by adjusting the gas pressure.

Inception voltage U_i was quantified with voltages of both polarities. First, the voltage was increased from zero up to about 97 % of the expected U_i - values with a rate of rise of 2kV/s. Then, the voltage was increased in steps of 4, 10 or 20 V, depending on measured U_i. After every step, the voltage was kept constant for 2 s, except in the case of dry air without UV irradiation where this period was 10 s. This procedure was repeated 15 - 20 times for each measuring point, and mean values of U_i are used for comparison with those obtained in computations.

INCEPTION VOLTAGE DETERMINATION

As is well known, the quantitative application of the streamer breakdown criteria, eq. (1), results in breakdown voltages for electric field distributions within gas gaps, which are homogeneous or inhomogeneous to a less extent. The rapid growth of electron avalanche within field regions, where $E(x) \geq p\cdot(E/p)_0$ and where $0 \leq x \leq x_c$ defines the path of avalanche growth, produces instability of the current growth due to well known secondary effects, e.g. the development of secondary streamers. The number K in eq. (1) is in general assumed to be $\ln(N_{crit})$, where N_{crit} is a certain critical number of electrons in the head of the initial avalanche.

If the field distributions become much more inhomogeneous and field intensities too low in regions where $E(x) < p\cdot(E/p)_0$, secondary effects are not any more able to produce self-propagation of secondary streamers and direct breakdown. For this situation, we define, as usual, a corona or partial discharge inception voltage U_i, the experimental detection criterion of which, however, is not yet well defined. Without thorough discussion of this criterion, U_i may be the voltage level at which the (discharge) current flowing into the gap is abruptly changed periodically or this current reaches a certain value, or at which light is emitted by the ionization process or even sound is generated etc. Our definition of U_i is based on measuring methods quantifying a sudden or high nonlinear current growth. Current signals were, therefore, be quantified by three methods, namely : A narrow band partial discharge (PD) detector (Bandwidth 30 kHz, center frequency 200 kHz, TETTEX instruments), an electrometer which can measure in fA range (KEITHLEY 617) and a digital storage oscilloscope, DSO (LeCroy 9140).

As quite different types of discharge currents appear for both gases investigated, the definition of U_i will further be explained by their shapes as shown in figure 1 :

Without UV radiation applied, figure 1 a) illustrates fairly long duration current pulses in dry air at the positive inception voltage and two different pressures as recorded by the DSO. The electrometer records average currents of less than 1 nA below U_i and a sudden increase to many nA at U_i. With UV radiation, the discharge currents are continuous but highly fluctuating for low pressures, and appear as current pulses for higher pressures, see figure 1 b). For low pressures (< 500 mbar) and voltages of only $1.005\cdot U_i$, the well know *silent discharges* produce only continuous currents.

If, for dry air, the sphere was cathode, the Trichel type of discharge pulses as shown in figure 1 c) appear with or without UV radiation. As the current pulse amplitudes are relatively high and charge magnitudes in the range of 100 pC are combined with those pulses, the PD detector can easily be used to quantify U_i.

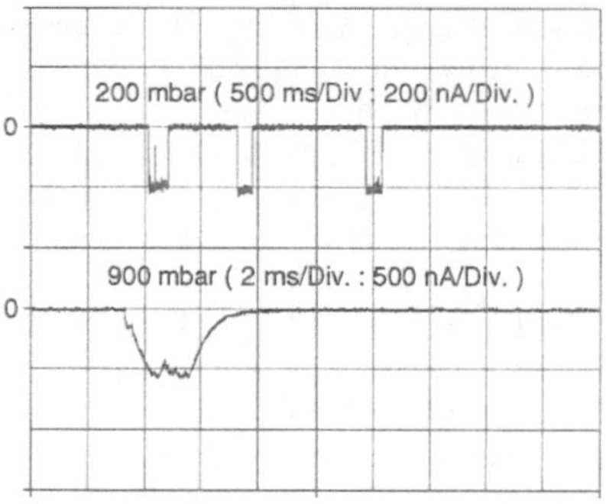

a) Sphere anode *without* UV in dry air at U_i, g = 30 mm

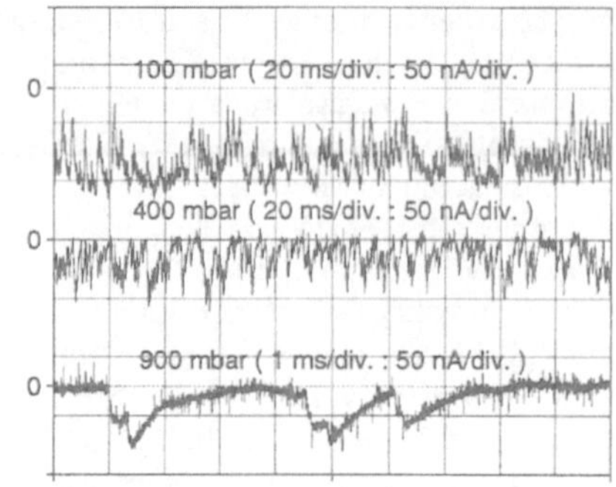

b) Sphere anode *with* UV in dry air slightly above U_i, g = 30 mm

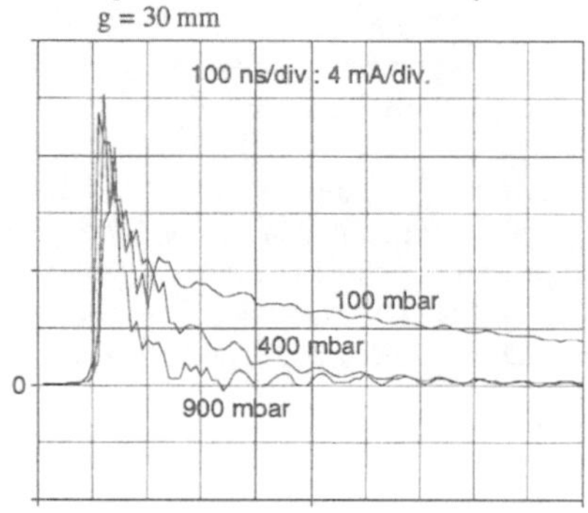

c) Sphere cathode *with* UV in dry air at U_i, g = 30 mm

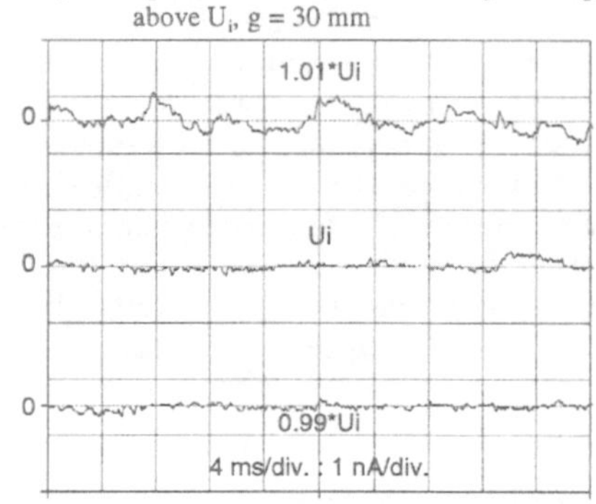

d) Sphere anode *with* UV in SF_6 at 600 mbar, g = 40 mm

Figure 1. Current waveform in the case of the sphere as anode and as cathode, with and without UV irradiation in dry air and SF_6. Sphere radius = 5 mm.

In the case of SF_6, without UV radiation and even negative potential at the highly stressed electrode, no inception voltage could be identified which would have been consistent with application of eq. (1). With UV radiation, however, for both polarities discharge currents appeared quite suddenly at a voltage level which may clearly be identified with U_i. In figure 1 d), current waveforms of this kind are shown for voltages at U_i, and 1 % lower and higher than U_i. This records could be made with a built-in current-to-voltage converter of the electrometer together with the DSO. As the mean value of the discharge currents is less than about 1 nA at U_i-level and only relatively slow fluctuations of the currents appear above U_i, neither the PD detector nor the DSO could be used to quantify U_i.

RESULTS

Since the inception voltage U_i as quantified by the experiments or calculated by eq. (1) will vary within a wide range due to the different gas pressures, we display the results in figure 2 as percentage deviation from the calculated values, defined as

$$\% U_i = \frac{(\text{Experiment} - \text{Calculation})}{\text{Calculation}} \cdot 100$$

The calculations of U_i are based on $K = 10.5$ in SF_6[1] and 9.15 in synthetic air[2]. The results are plotted against the product $p{\cdot}r$ for dry air without and with UV irradiation (figure 2 a and 2 b) and for SF_6 with UV irradiation when the sphere was anode (figure 2 c) and cathode (figure 2 d) respectively. The scatter in the results (based on standard deviation) are not shown in the figures for the sake of clarity. In brief, the maximum deviation of data from the average value (results from the experiment) are in general less than 1 % in SF_6, 0.5 % in dry air without UV and 0.3 % in dry air with UV.

In dry air, UV irradiation reduces the measured U_i-values by only about 1 % for both polarities of the sphere. The results compared to the calculation agree quite well when the sphere was cathode, but show large negative deviations at low pressure which decrease with increasing pressure when the sphere was anode.

In SF_6, the results show for both polarities the general decreasing trend with pressure, which may indicate that *K* in eq. (1) depends on gas pressure for this type of gas.

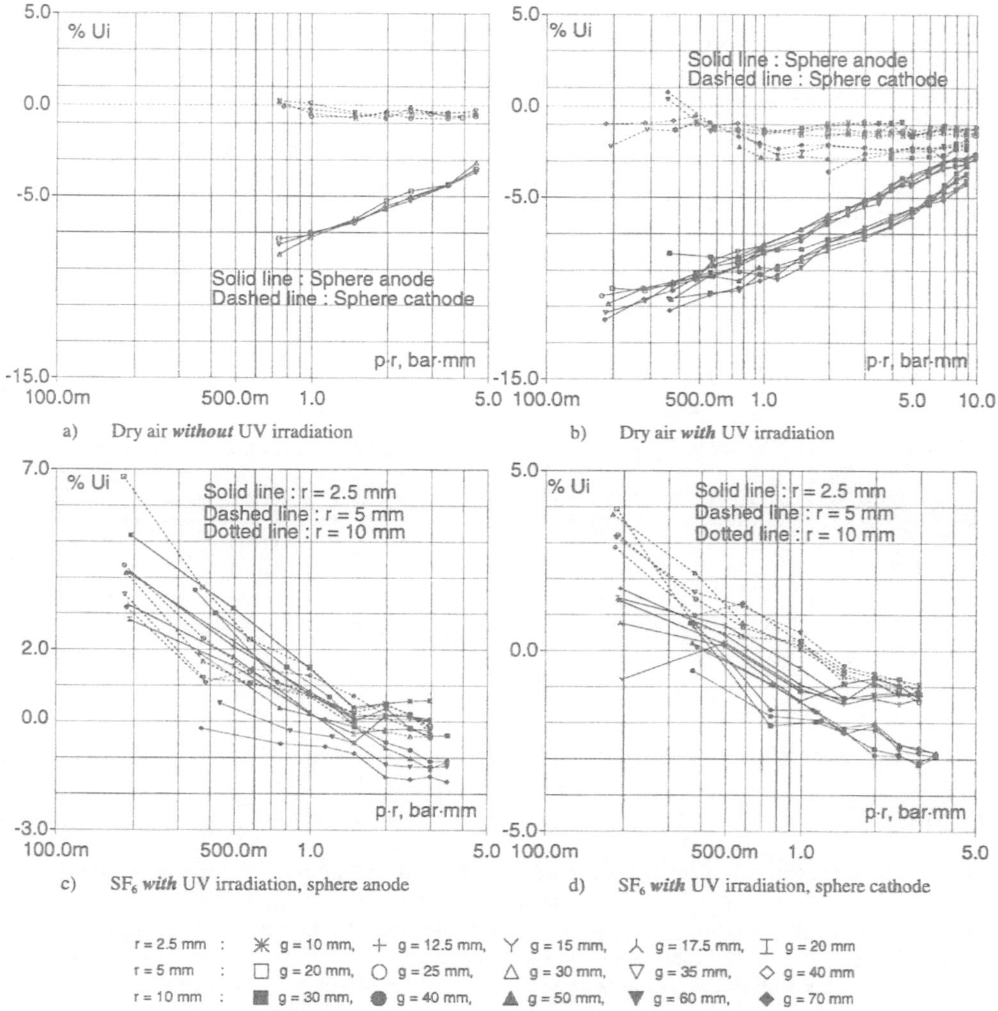

a) Dry air ***without*** UV irradiation
b) Dry air ***with*** UV irradiation
c) SF_6 ***with*** UV irradiation, sphere anode
d) SF_6 ***with*** UV irradiation, sphere cathode

Figure 2. Comparison between the experiment results and calculation U_i.

DISCUSSION

Before some details of our results are discussed, remember the object of this investigation : Insulation systems based on gaseous dielectrics contain quite often electrode configurations in which some parts of the electrodes are curved or bent. This yields local high field intensities decreasing strongly with distance from such electrodes. As numerical, computer aided field calculations can readily be performed nowadays with high accuracy, the application of the streamer criterion, eq. (1), provides an effective tool to calculate a (partial) discharge inception voltage. Additional prerequisites for this calculations are the knowledge about the effective ionization coefficients, see eq. (3) to (6) , and the value of *K*, see eq. (1), the magnitude of which is still under discussion. Numerical values of *K* which are independent of gas conditions or field distributions are most favorable for practical applications.

The results of our investigations as can be visualized in figure 2 do obviously not confirm a complete independence. Taking, however, the conditions of the experiments into account, it is easy to demonstrate that for small products of pressure p and curvature r the value of K has a large influence on the numerically computed inception voltages. This effect is well known from the similarity law of gas breakdown, which relates the pressure reduced breakdown fields to the products of pressure and gap distance for homogeneous fields or radii of cylindrical or spherical conductors, see e.g. 7. This influence of K can also be shown for our specific sphere-pot arrangement, see figure 3. In these figures, for air as well as SF_6 the percentage deviations of U_i in dependence of $p \cdot r$ are *computed* based on a field distribution for the sphere of $r = 5$ mm and different gap distances of g = 25 to 40 mm, which, however, have no influence on the results. If these results are compared with the experimental results of figure 2, one can easily recognize that only minor changes of K are necessary to bring the results of the experiments in close agreement with the results as computed with $K = 9.15$ or 10.5 respectively.

However, apart from minor measuring errors and the procedures applied to quantify U_i, the results of figure 2 display some basic trends and deviations, which only shortly can be discussed for each individual gas.

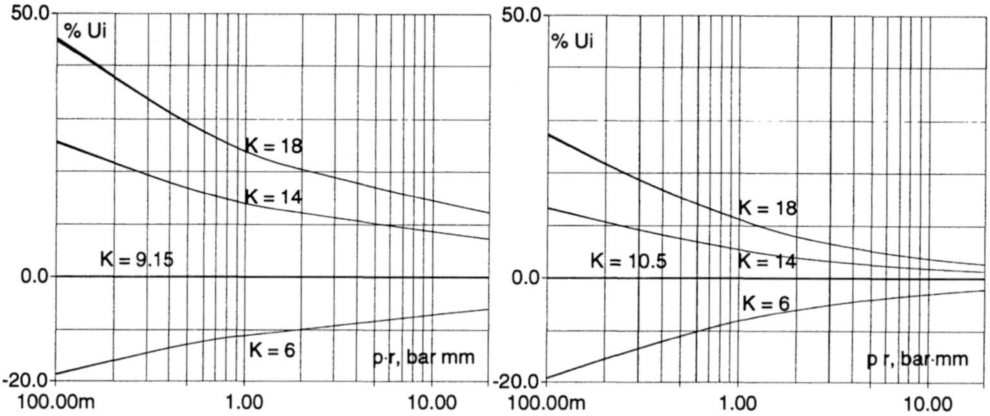

a) Synthetic air, $\bar{\alpha}/p = f(E/p)$ from eq. (3) and (4)   b) SF_6, $\bar{\alpha}/p = f(E/p)$from eq. (5) and (6)

Figure 3. Effect of K on inception voltage in synthetic air and SF_6, $r = 5$ mm, g = 25 to 40 mm in 5 mm step.

Dry Air

If the discharges start at cathode, the value of K is obviously quite independent from $p \cdot r$. UV radiation reduces U_i by an order of 1 %, which may be explained by the generation of a high number of primary electrons as starting electrons for critical avalanche. As the number of primary electrons reduces K due to an enlarged feedback - coefficient (see e.g. 8), this reduction of U_i is plausible.

If the highly stressed electrode is anode, U_i is only for high $p \cdot r$ - values about equivalent to negative polarity inception but becomes lower for decreasing pressures or $p \cdot r$ - products. This physical effect, that positive inception voltage (for e.g., thin wires) is lower than the negative one for pressures below atmospheric is known since a long time, see W.O. Schumann[9]. F. Llewellyn Jones[10] explains this effect by differences of a generalized secondary ionization (or feedback) coefficient which is based on different electron emission at the cathode of inhomogeneous gaps by γ - and δ - effects, i.e. by the incidence of positive ions and photons. As these explanations are, however, somewhat qualitative, also other effects not known up to now by the authors may cause this phenomena. A quantitative reduction of K for positive inception voltages, the magnitude of which can be judged by figure 3 a), must thus be assumed for low pressures.

SF_6

As already mentioned before, the lack of primary electrons in the strongly electron attaching gas, SF_6, increases the discharge inception voltage for the gas gaps as investigated, to levels which are far beyond the calculation values. Therefore, only UV radiation is able to produce stable conditions.

But then the measured values of U_i as visualized in figure 2 c) and d) have similar trend for both polarities of discharges, and in contrast to air, U_i - values for positive discharge inception are about 2 - 3 % higher than those for negative. The trend that U_i is some percent higher for small $p \cdot r$ - values could be related to wrong values of $\bar{\alpha}$, especially in the high regions of E/p, for which eq. (6) has been assumed. Though not quantitatively shown in this paper, the real E/p - values for $p \cdot r = 0.2$ bar·mm (and for $r = 2.5/5/10$ mm) are as high as 20 kV/mm·bar. To the best knowledge of the authors, there exists up to now no experimental confirmation that the effective ionization coefficients of eq. (6) are applicable for practical breakdown conditions. Again, for high $p \cdot r$ - values the inception voltages are very close to the computed ones, though the value of $K = 10.5$ was used together with eq. (5) and (6), i.e. effective ionization coefficients which are somewhat smaller than that used by A. Pedersen[1] for the establishment of this K - value.

Thus, there remains some uncertainty about the value of K in eq. (1).

SUMMARY

Based on experiments in dry air and SF_6 concerned with the measurement of discharge inception voltages for quite inhomogeneous gas gap, it was shown that the streamer breakdown criteria may well be applied in practice to calculate such voltages. Prerequisites for this calculations are accurate field distribution calculations, the availability of the effective ionization coefficients, and the knowledge about the parameter K of the particular gas.

The investigations showed, that for both gases investigated K is in the order of 10 and that this value can also be applied under inhomogeneous gas gap conditions.

ACKNOWLEDGEMENT

The authors would like to thank Mr. H. Wymann for providing all the supports in mechanical works and Mr. B. Waldesbühl, the late Mr. H. Hürlimann, and Mr. C. Sigrist for providing all the supports in electrical works.

REFERENCE

1. A. Pedersen, I.W. McAllister, G.C. Crichton, and S. Vibholm, *Arch. Elek.*, Vol. 67, pp. 395 - 402, 1984.
2. W.S. Zaengl, S. Yimvuthikul, and G. Friedrich, *IEEE Trans. Elect. Insul.*, Vol. 26, No. 3, pp. 380 - 390, 1991.
3. S. Sato and W.S. Zaengl, *Effective 3 - dimensional electric field calculation by surface charge simulation method*, Proc. IEE, Vol. 133, Pt. A, pp. 77 - 83, 1986.
4. WG 22.03 report, *Comparative electric field calculations and measurements on high voltage insulators*, Electra, No. 141, pp. 69 - 77, 1992.
5. J. Dutton, *A survey of electron swarm data*, J. Phys. Chem. Ref. Data, Vol. 4, No. 3, p.730, 1975.
6. T. Aschwanden, Ph. D. Thesis, ETH No.7931, Swiss Federal Institute of Technology, Zurich, pp. 186 - 195, 1985.
7. W.S. Zaengl, *Electronegative gases : present state of knowledge and applications, future prospects*, Nordic Symp. on Electrical Insulation, Trondheim, p. I.1 - I.39, 1988.
8. T.H. Teich and W.S. Zaengl, *The dielectric strength of an SF_6 gap*, In " Current interruption in high-voltage networks ", Plenum Press, pp. 269 - 297, 1978.
9. W.O. Schumann, *Elektrische Durchbruch-Feldstärke von Gasen : Theoretische Grundlagen und Anwendung*, Springer Verlag, 1923.
10. F. Llewellyn Jones, *Encyclopedia of Physics, Vol. XXII, Gas Discharges II*, Springer Verlag, 1956.

DISCUSSION

E. KUFFEL: I am surprised with the low value of K (for SF_6) of approximately 10. The very extensive studies conducted by T. Nitta in the 1970-80s gave good agreement with experiments using a value of approximately 18.

K. PETCHARAKS: Two plausible explanations for the apparently good agreement between the calculation and the experiment in the studies of Nitta are: (1) he used a different relationship of $\frac{\bar{\alpha}}{p} = f(\frac{E}{p})$ in the calculation. (2) If his study is performed at high pressure, the different value of K has only a small influence on the calculation results for the breakdown voltage.

CALCULATION OF DIELECTRIC STRENGTH OF SOME SF_6 GAS MIXTURES

Y. Qiu and D.M. Xiao

High Voltage Division
Xi'an Jiaotong University
Xi'an, China 710049

INTRODUCTION

Interest in the possible use of compressed electronegative gas mixtures in high-voltage apparatus has stimulated many investigators to try to develop simple methods for calculating the dielectric strength of gas mixtures from the basic parameters of the constituent gases. A simple method is to calculate the effective ionization coefficient($\bar{\alpha}$) in a gas mixture by the sum of the partial-pressure-weighted $\bar{\alpha}$ in the constituent gases, which is based on the assumption that no interaction takes place between constituent gas molecules. This simple method was shown to apply to SF_6/N_2, SF_6/CO_2 and SF_6/air mixtures[1]. And it was pointed out later by Chantry and Wootton[2] that the assumption of this method would be best satisfied by those gas components, which individually have similar electron energy distributions. This criterion for determining the applicability of the simple calculation method is certainly true, but calculation of electron energy distributions in the constituent gases is not an easy task. The present paper presents a simple criterion to determine whether the above-mentioned calculation method is applicable to a certain gas mixture.

THE PROPOSED CRITERION

Calculation of the ionization and attachment coefficients in a binary electronegative gas mixture by the sum of the partial-pressure-weighted coefficients in the constituent gases can be expressed as

$$\alpha_m = x\alpha_1 + (1-x)\alpha_2 \qquad (1)$$

$$\eta_m = x\eta_1 + (1-x)\eta_2 \qquad (2)$$

or

$$\bar{\alpha}_m = x\bar{\alpha}_1 + (1-x)\bar{\alpha}_2 \qquad (3)$$

Gaseous Dielectrics VII, Edited by L.G. Christophorou and D.R. James, Plenum Press, New York, 1994

where the subscripts 1, 2 and m denote the two constituent gases and the gas mixture, respectively; and x is the fractional concentration of the first gas in the mixture.

In Eq.(3) if we set a different E/p value each time and let $\bar{\alpha}_m=0$, then we can obtain the $(E/p)_0$ values as a function of x for the gas mixture.

It can be seen from Eq.(1) through Eq.(3) that α_m, η_m and $\bar{\alpha}_m$ are all a linear function of x. And it is quite obvious that Eqs.(1) and (2) are true only when the measured α_m and η_m are a linear function of x, which can therefore be a simple criterion for determining the validity of linear addition of the partial-pressure-weighted coefficients.

SF_6 AND NITROGEN GAS MIXTURE

Equations (1) and (2) can apply very well to the SF_6/N_2 gas mixture because the electron energy distribution in SF_6 is quite similar to that in N_2, and the drift velocity in these two gases is also about the same[2]. Therefore, it is expected that the measured α_m and η_m in SF_6/N_2 should be a linear function of x. Figure 1 shows the measured α_m/p and η_m/p as a function of x in the SF_6/N_2 mixture at an E/p value of 67.7kV/mm·MPa using the steady-state Townsend method. It can be seen in Fig. 1 that both $\alpha_m/p=f(x)$ and $\eta_m/p=g(x)$ are very close to a straight line, and it is interesting to note that the ionization coefficient increases with the increase of SF_6 content.

Figure 2 shows the good agreement between the measured relative electric strength (RES) of the SF_6/N_2 mixture and the calculated curve. In calculating $\bar{\alpha}_m$ in SF_6/N_2, the following expressions were used.

$$\bar{\alpha}_1/p=28[E/p-88] \quad (4)$$

$$\alpha_2/p=6600\exp[-215/(E/p)] \quad (5)$$

$$\eta_2=0 \quad (6)$$

In these expressions E/p is in kV/mm·MPa; α/p and α/p are in 1/mm·MPa.

SF_6 AND ARGON GAS MIXTURE

Figure 2 also shows the agreement between the measured RES of the SF_6/Ar mixture[3] and the calculated curve. In calculating $\bar{\alpha}_m$ for SF_6/Ar, a graphic method was used with the ionization coefficients in argon taken from Ref.[4]. It is not surprising that the agreement for SF_6/Ar is not as good as that for SF_6/N_2 because the electron energy distribution in SF_6 is slightly different from that in Ar[5]. The difference in electron energy distribution between the two constituent gases is demonstrated by Fig. 3, where it can be seen that both $\alpha_m/p=f(x)$ and $\eta_m/p=g(x)$ are quite away from a linear function.

SF_6 AND HELIUM GAS MIXTURE

It was reported that the above mentioned simple calculation method did not apply to the SF_6/He mixture because the electron energy distributions computed for SF_6 and helium were quite

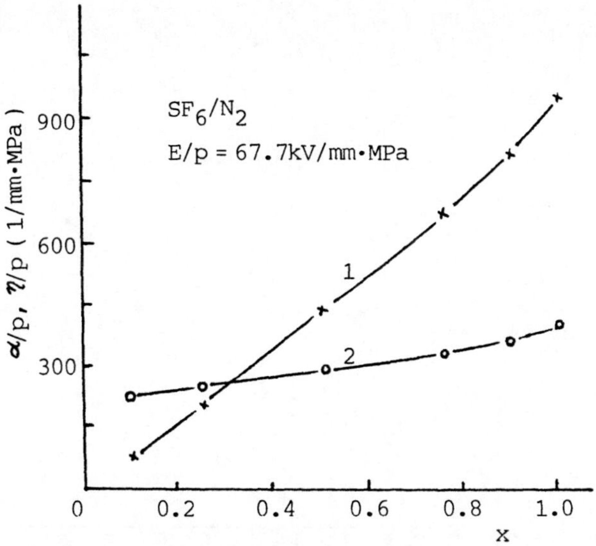

Figure 1. η/p(curve 1) and α/p(curve 2) as a function of x measured in SF_6/N_2

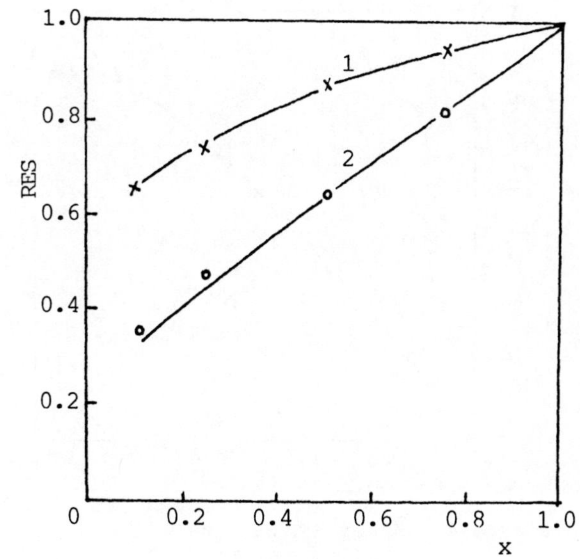

Figure 2. RES of SF_6/N_2(1) and SF_6/Ar(2)
x,o: measured
——: calculated

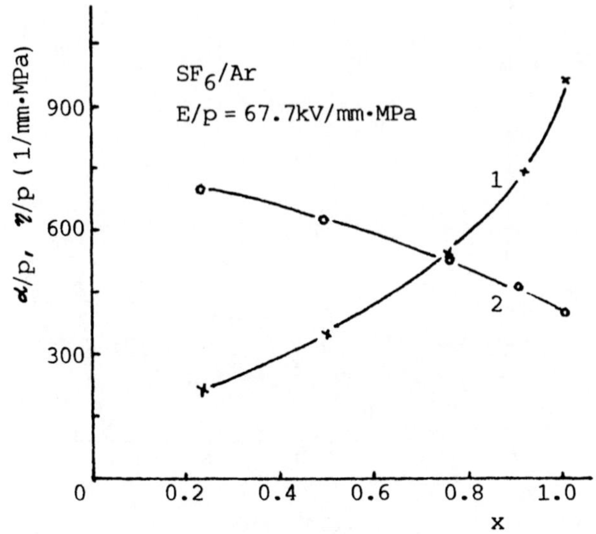

Figure 3. $\eta/p(1)$ and $\alpha/p(2)$ as a function of x measured in SF_6/Ar

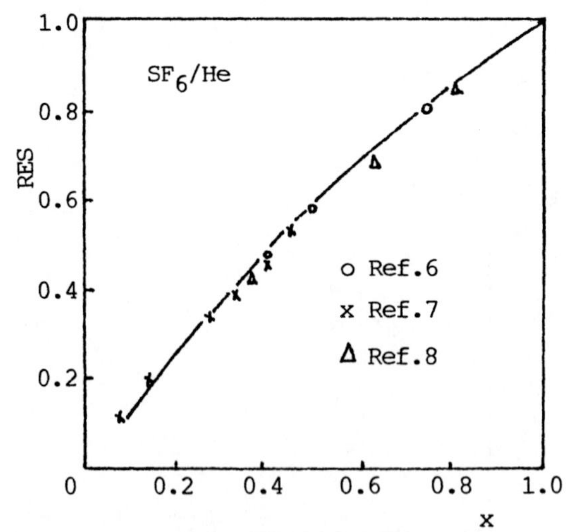

Figure 4. RES of SF_6/He
o,x,Δ: measured
——: calculated with α_2 taken from He with impurities

different from each other[2]. But our previous work showed that the simple calculation method could be used for SF_6/He if the Penning ionization process in helium was taken into account, which means the ionization coefficients in helium with impurities should be used in the calculation[6] instead of those measured in pure He as shown in Fig.4.

Figure 5 gives the measured α_m/p and η_m/p in SF_6/He as a function of x. It can be seen that although both $\alpha_m/p=f(x)$ and $\eta_m/p=g(x)$ are not quite close to a linear function, the calculated RES is still in reasonably good agreement with the experimental data.

DISCUSSION

It is quite interesting that the simple method of the calculation of RES can be used for SF_6/Ar and SF_6/He mixtures although in these two cases α_m, η_m and $\bar{\alpha}_m$ are not quite a linear function of x. The reason for that lies in the great c value in the following expression.

$$\bar{\alpha}_m/p=c[E/p-(E/p)_0] \qquad (7)$$

where $(E/p)_0$ is the intercept of the oblique line at the abscissa. It is quite apparent that even if there is a significant error in $\bar{\alpha}_m$ at a certain E/p value, the error in $(E/p)_0$ will be much smaller because of the great slope of the oblique line.

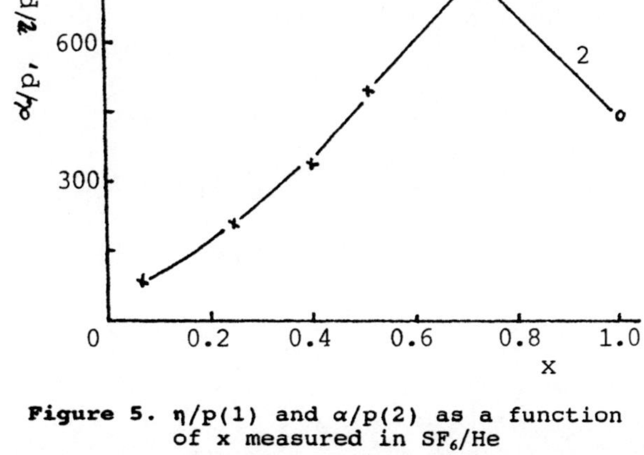

Figure 5. $\eta/p(1)$ and $\alpha/p(2)$ as a function of x measured in SF_6/He

CONCLUSIONS

A simple criterion is presented for determining whether the linear addition of the partial-pressure-weighted coefficients applies to a certain gas mixture. This method can be used for calculating the dielectric strength of some gas mixtures even if the proposed criterion is not strictly observed.

ACKNOWLEDGMENT

The financial support of the National Natural Science Foundation of China is gratefully acknowledged.

REFERENCES

1. N.H.Malik and A.H.Qureshi, Breakdown gradients in SF_6-N_2, SF_6-air and SF_6-CO_2 mixtures, IEEE Trans. Elec. Ins., 15:413(1980).
2. P.J.Chantry and R.E.Wootton, A critique of methods for calculating the dielectric strength of gas mixtures, J. Appl. Phys., 52:2731(1981).
3. Y.Qiu, M.C.Zhang and E.Kuffel, Ionization and attachment coefficients measured in SF_6 and argon mixtures, Proc. 20th Int. Conf. on Phenomena in Ionized Gases, 1029(1991).
4. D.E.Davies and G.C.Milne, First ionization coefficients in hydrogen, neon, and argon, British J. Appl. Phys., 10:301(1959).
5. S.Okabe and T.Kouno, Flashover properties and calculated values of binary mixtures of SF_6, N_2, O_2 and Ar, Electrical Engineering in Japan, 106:29(1986).
6. Y.Qiu and X.Weng, Predication of dielectric strength of SF_6/He gas mixtures, J. Phys.D: Appl. Phys., 23:363(1990).
7. J.L.Moruzzi and J.D.Craggs, Ionization and attachment and breakdown measurements in mixtures of SF_6 with helium, argon and hydrogen, Proc. 12th Int. Conf. on Phenomena in Ionized Gases, 225(1975).
8. M.Shimozuma and H.Tagashira, Measurement of the ionisation and attachment coefficients in SF_6 and helium mixtures, J. Phys. D: Appl. Phys., 16:1283(1983).

DISCUSSION

R. J. VAN BRUNT: I would like to comment that in the SF_6/He mixture, Penning ionization can occur with significant probability. It can, therefore, be argued that equality of the electron energy distributions in He and SF_6 required for validity of the Wieland approximation does not apply unless proper account is taken of Penning ionization. Moreover, the possible existence of "impurities" in helium could have a major influence on the electron energy distribution in that gas.

Y. QIU: Penning ionization is likely to occur in SF_6/He, and this should be taken into account when the Wieland approximation is applied to SF_6/He as was pointed out in the present paper and in Ref. [6]. It is also my understanding that the impurity greatly modifies the electron energy distribution in the helium gas.

A. GARSCADDEN: I endorse Van Brunt's comments. The nature and amount of impurity in your helium (or argon) must be defined. It can be shown from a Boltzmann analysis, that a few hundred ppm of nitrogen in helium will "cool" the electron energy distribution function and greatly change the alpha coefficient. The changes to the alpha coefficient and to the drift velocity are very sensitive to the impurity level at low concentrations.

Y. QIU: The nature of the impurity was defined in a previous publication (Ref. [6]), i.e., it is either N_2 or SF_6. The amount of the impurities was a few hundred ppm. It has been found that both impurities have nearly the same effect on the alpha coefficient in helium. This is probably because the electron energy distributions in N_2 and SF_6 are about the same and both impurities have almost the same ionization energy level.

BARRIER EFFECTS IN A ROD/ROD AIR-GAP UNDER DC VOLTAGE

Li Ming*, Mats Leijon*, Tord Bengtsson* and Mat Darveniza†

* ABB Corporate Research
  Sweden
† The University of Queensland
  Australia

ABSTRACT

The effects of dielectric barriers on partial discharge and breakdown characteristics in a 50 mm rod/rod air-gap have been investigated under DC voltage. The results show that a barrier with large diameter and situated in a favourable position leads to a significant improvement in PD characteristics and a remarkable increase in the breakdown voltages. For example, the breakdown voltage of the air gap with a barrier of diameter 400 mm is 3 times as large as that without barrier. The effects of dielectric barriers depend on DC voltage polarity, size of barrier and position in the gap.

The favourable position of a barrier for improvement in PD characteristics is near the HV rod-tip. For effects on the breakdown voltage, the favourable positions are near both rod-tips. Barrier effects in breakdown voltage are due to elongation of the flashover channel, and the breakdown voltage increases proportionally to the barrier diameter up to a critical diameter which depends on the dielectric strength of the barrier.

INTRODUCTION

It is known that PD characteristics and breakdown voltages of a non uniform gas-gap can be strongly affected by inserting a dielectric barrier in the gap. This is called the barrier effect.

The study of barrier effect on the breakdown voltages dates back to the 1920´s and 1930´s [1-3]. It is well known that the positive breakdown voltage of a point/plane air gap can be increased considerably by a barrier, depending upon the barrier position. The negative breakdown voltage is much less affected by the barrier. The increase of positive breakdown voltage is due to the creation of a nearly homogeneous field between barrier and plane.

Since then, the investigation on barrier effects has been continually carried out in air and SF_6 gas [4-11]. The results show that in some cases the dielectric barrier is effective under impulse, AC and DC voltages, depending on the polarity of the applied voltage, the barrier size, shape and location in a gap.

Gaseous Dielectrics VII, Edited by L.G. Christophorou
and D.R. James, Plenum Press, New York, 1994

The possible mechanisms for the barrier effect are:

-The accumulated charge on a barrier surface make the field more uniform;

-The flashover channel is elongated due to the presence of a barrier.

In the previous works, point/plane or very small sphere/plane electrode systems with low field utilisation factor were commonly used. These were very different from actual technical arrangements. The studies also did not cover the barrier effect on PD characteristics.

Therefore, the barrier effects on both PD and breakdown characteristics in a rod/plane and rod/rod air-gaps under DC voltage were studied. The results for the rod/plane system were presented in paper [12]. This paper presents the results for the rod/rod system.

EXPERIMENTAL SET-UP

The experiments were performed in the HV laboratory with an ambient temperature of about 20±2 °C. All the measured values of breakdown voltages were corrected to the standard conditions.

Figure 1 shows the 50 mm rod/rod electrode system used in the experiments. The hemispherical rods were made of copper and the diameter of the rod-tips was 10 mm.

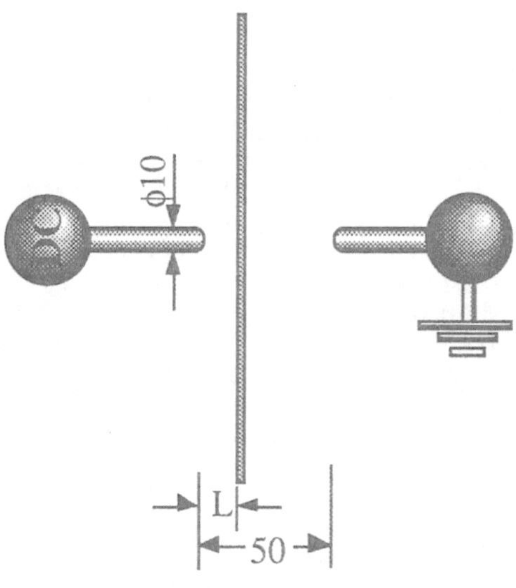

Figure 1. 50 mm rod/rod electrode system and dielectric barrier

In the tests, all the dielectric barriers were round sheets and made of 1 mm thick thermoplastic polyester (Valox) film. The barrier position in the gap is described by the distance L between high voltage electrode and barrier as shown in Fig. 1.

In the experiments, two DC power supply systems with maximum output voltage of 150 kV and 600 kV were used as test sources. For the PD measurements under DC stress, the applied voltage was started at 10 kV, then increased by steps of 5 kV. The voltage remained for 2 minutes at each step. PD levels were recorded continuously and average values are given in the results. For the DC breakdown tests, the applied voltages were raised at a rate of 2 kV/s from 0 until breakdown. In the figures, each value of breakdown voltages is based on at least 3 independent measurements.

EXPERIMENTAL RESULTS

The test results show that barriers with various diameters in the rod/rod air gap can lead to a significant improvement in PD characteristics and a considerable increase in breakdown voltages under DC voltage. Figure 2 and 3 show the barrier effects on the PD characteristics and breakdown voltages of the air gap.

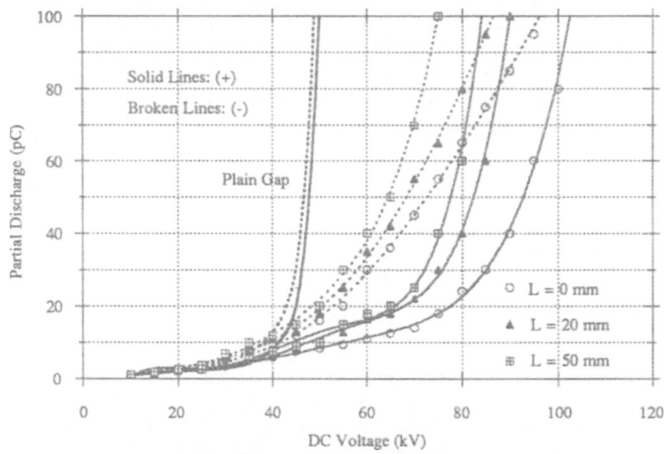

Figure 2. The effects of a barrier with diameter 180 mm on the partial discharge of a 50 mm rod/rod air gap with the barrier placed at different positions

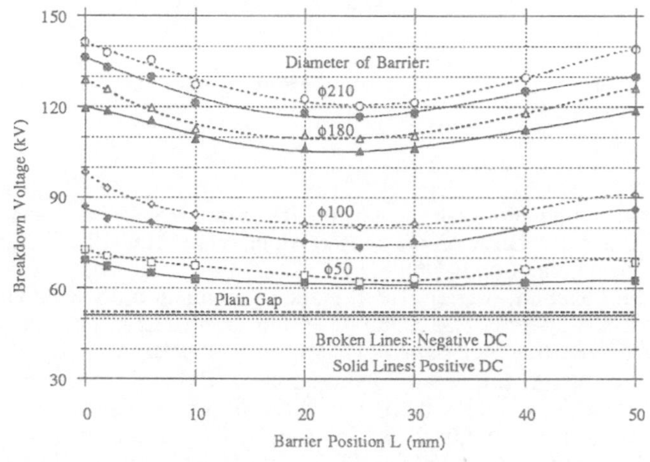

Figure 3. The effects of barrier with various diameters on the breakdown voltages of a 50 mm rod/rod air gap as a function of barrier position

From the figures, it can be seen that the barrier effects depend on DC voltage polarity and the size and position of barrier. For PD characteristics, the barrier has a more beneficial effect under positive DC voltage than that under negative DC voltage and the favourable position of a barrier is near the HV rod-tip. When the barriers are placed in the middle of the gap, the breakdown voltages are lowest. As the barriers are moved towards either of the rod-tips, the breakdown voltages increase. The barrier diameter shows strong influence on the breakdown voltages. Increase of the barrier diameter leads to an increase of the breakdown voltages.

DISCUSSION

The partial discharge phenomena of a gap with barrier are found to be distinctly different from the plain air gap. The transition from a single pulse discharge to sustained partial discharges take place at a much higher voltage. During the period, the PD level increases gradually with the applied voltage.

The charges are generated by the discharge activity and drift in air along the field line to the barrier surface. Because the conductivity of the barrier is very low, space charges build up on the surface. This charging process takes place very rapidly. The initial field distribution in the rod-barrier-rod gap is altered by the accumulated charges as shown in figure 4. The field strength near the high voltage rod electrode is more or less reduced, depending on the barrier position. As the barrier is more close to HV rod electrode, the field near rod-tip becomes lower and PD characteristics become better.

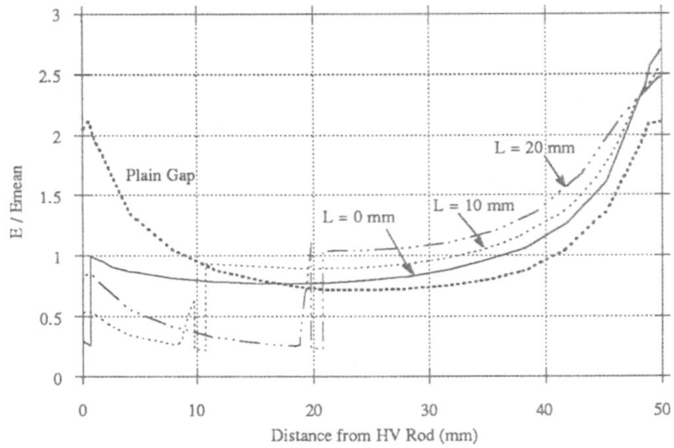

Figure 4. Field distribution along the 50 mm rod/rod gap axis for a barrier of diameter 210 mm situated at different positions

There are two explanations for the barrier effects on breakdown voltages. First, barrier effect is due to the surface charge on the barrier, which makes field distribution in the air gap more uniform. This explanation is suitable for non-symmetrical electrode systems, such as rod/plane electrode. Second, barrier effect is due to elongation of flashover channel. The barrier effects in a non-symmetrical electrode system as well as symmetrical electrode system, such as rod/rod electrode in the present case, can be explained by this mechanism.

For a better understanding, many photographs of flashovers were taken. The results show that when the barrier diameter is smaller than 400 mm, the flashover channel is always around the edge of barrier. A predischarge initiated by the surface and / or space charges is usually observed prior to total breakdown. The predischarge can last till whole gap breakdown occurs as shown in figure 5. When the barrier diameter is larger than 400 mm, the barrier is usually punctured. The critical diameter is dependent on the dielectric strength of the barrier. This critical diameter can be increased, for example, by increasing the barrier thickness. The test results of flashover voltage as a function of barrier diameter shown in figure 6.

The increase of flashover channel length in the rod-barrier-rod air gap is shown in figure 7. From the figure, it can be seen that as a barrier is placed near the rod-tip, the flashover channel is longest and in the middle is shortest. The tendency is in a good agreement with the breakdown voltage shown in figure 3. However, the breakdown voltages were not directly proportional to the increase of flashover channel length because the flashover propagated partly along the barrier surface.

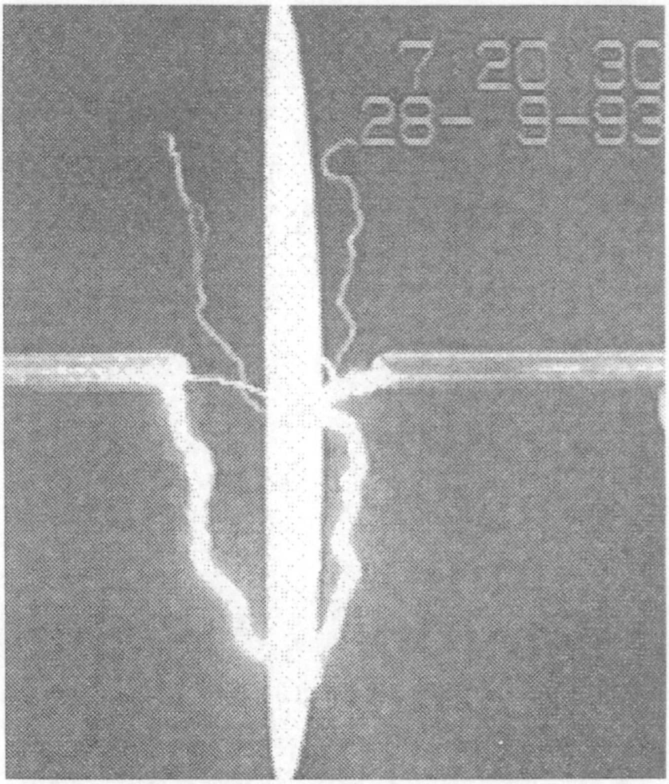

Figure 5 Flashover in 50 mm rod/rod air-gap with a barrier

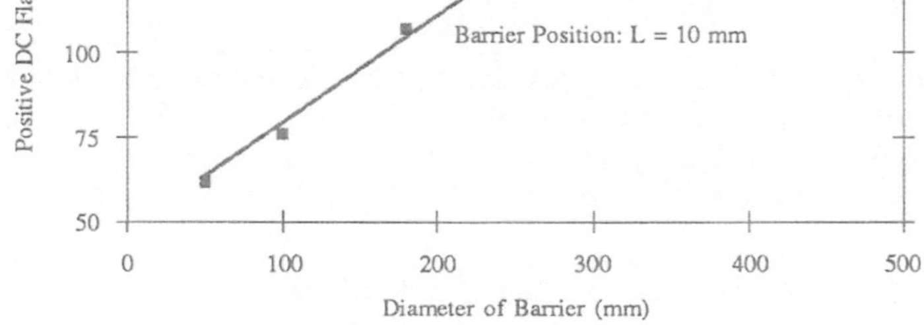

Figure 6. Flashover voltage as a function of diameter of the barrier placed at L = 10 mm under positive DC voltage

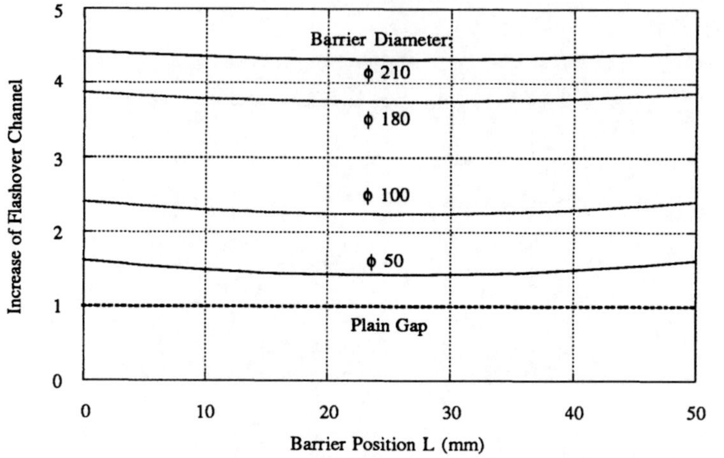

Figure 7. Increase of flashover channel in a 50 mm rod/rod electrode system

CONCLUSIONS

The effects of dielectric barriers on partial discharge and breakdown characteristics in a 50 mm rod/rod air-gap have been investigated under DC voltage. The results obtained in this study are summarised as follows.

(1) A barrier with large diameter and situated in a favourable position leads to a significant improvement in PD characteristics and a remarkable increase in the breakdown voltages. For example, the breakdown voltage of the air gap with a barrier of diameter 400 mm is 3 times as large as that without barriers. The effects of dielectric barrier depend on DC voltage polarity, size of barrier and barrier position in the gap.

(2) The barrier effects on PD characteristics for positive DC voltage are larger than for negative DC voltage. However, the negative DC breakdown voltages of the gap with a barrier are little higher than for positive voltages.

(3) For improvement in PD characteristics, the favourable position of a barrier is near the HV rod-tip. For best effects in the breakdown voltage, the favourable positions are near both rod-tips.

(4) Barrier effects on breakdown voltage are due to elongation of the flashover channel. Therefore, the breakdown voltages are directly proportional to the barrier diameter. If the diameter of a barrier is larger than a critical diameter which depends on the dielectric strength of barrier, the barrier is usually punctured and the breakdown voltages vary with the puncture position.

ACKNOWLEDGEMENTS

The authors wish to thank Mr. Lars-Gunnar Dahlberg and Mr. Jan Karlsson for their valuable help during the experimental work.

REFERENCES

1. A.Roth, "Hochspannungstechinik", Berlin Julius Springer, 177-180 (1928)
2. E.Marx, Der durchschlag der luft imum-homogenen elektrischen felde bei verschiedenen spannungsarten, E.T.Z., Vol.A33, pp.1161-1165 (1930)

3. H.Roser, Schrime zur erhörung der durchschran-spannung in luft, E.T.Z., Vol.B17, pp.411-412 (1932)
4. H.E.Remde and H.Boecker, Voltage-current characteristics during propagation of a surge breakdown of a point-to-plate gap with insulating barrier, IEEE, TP-616-PWR, pp.271 276 (1971)
5. T.Kouno, Breakdown of composite dielectrics: the barrier effect, IEEE, EI-15, pp.259-263 (1980)
6. D.König and H.G.Lautenschläger, The barrier effect in air and SF_6-insulated nonuniform electrode configurations, VII International Conference on Gas Discharges and Their Applications, London, pp.292-295 (1982)
7. P.A.Bruyn and R.G.Van Heeswijk, Influence of a dielectric barrier on the DC and AC breakdown strength of a point-to-plane gap in air, Conference Record of 1984 IEEE Inter. Symp. on Electrical Insulation, Montreal, June 11-13 (1984)
8. K.Hidaka, M.Chiba and T.Kouno, Barrier effect on electrical breakdown in air, IX International Conference on Gas Discharges and Their Applications, Venezia, pp.479-487 (1988)
9. M.Miyagawa, T.Yoshida, N.Masaki, S.Ohtake and I.Ohshima, Barrier effect under nonuniform field in atmospheric pressure SF_6 gas, Electrical Engineering in Japan, Vol.111, No.7, pp.60-67 (1991)
10. F.V.Topalis and I.A.Stathopulos, Barrier effect in small and medium air gaps, 7th International Symposium on High Voltage Engineering, Paper 42.23, Dresden, Aug. 26-30 (1991)
11. A.Beroual and A.Boubakeur, Influence of barrier on the lightning and switching impulse strength of mean air gaps in point/plane arrangements, IEEE, EI-26, No.6, pp.1130-1139 (1991)
12. Li Ming, Mats Leijon and Tord Bengtsson, Barrier effects in air-insulated systems under DC voltage, 8th International Symposium on High Voltage Engineering, Paper 43.03 Yokohama, Aug 23-27 (1993)

DISCUSSION

I. D. CHALMERS: When you measure PD's do you find that they are stable with time or is it not the case that as PD's take place, charge is deposited on the dielectric surface or gradually reduce the residual field in the gas gap to the inception field at which discharges should cease?

L. MING: This is exactly why the barriers show beneficial effects on PD characteristics. The charges produced due to high field at rod-tip deposited on the barrier surface and reduce the field so that discharges cease. In the present case, the time constant is small. Therefore, the results show that the PD inception voltage with a barrier takes place at higher voltage than in plain gap.

S. W. ROWE: This paper proposes to improve PD and flashover performance for DC conditions. How would the performance be affected by transient voltage reversal due to a lightning inpulse? Would this type of barrier be applicable in full scale GIS?

L. MING: The impulse flashover voltage of a gap with a charged barrier is dependent on the voltage polarity. If the polarity is opposite to the surface charge, the flashover voltage will decrease. If the barrier surface is well-treated, it can be used in gas insulated transformer and GIS with low pressure.

E. KUFFEL: Did you attempt measuring the charge distribution on the barrier using, for example, a probe? In typical GIS spacers with charges under DC conditions the typical reduction in breakdown voltage is down to 40% of the uncharged spacer.

L. MING: I have measured the spacer surface potential distribution in HVDC GIS (for example, see author's paper at the 10th International Conference on Gas Discharges, Swansea, 1992). The surface field distribution can be disturbed by the surface charge and the flashover voltage can be reduced. Therefore, the surface charge density and distribution must be well controlled as the barrier is used in electrical equipment.

M. GOLDMAN: Is your Fig. 5 a general phenomenon? Does it mean that you have no surface discharges? Are your results stable with time?

L. MING: The surface discharge before the whole gap breakdown was observed, but it was not shown in Fig. 5. The PD characteristics in the gap with barrier is dependent on time under DC voltage. In the present case, the stable condition can be reached in a very short time.

MEASUREMENTS OF SYNERGISMS OF STRONGLY ELECTRONEGATIVE GASEOUS MIXTURES

Li Zheng–ying and C. Zhang

Deptarment of Electrical Engineering
Huazhong University of Science and Technology
Wuhan 430074, China

INTRODUCTION

Of many strongly electronegative gases, SF_6 has found a wide range of applications such as in circuit breakers and gas insulated systems (GIS) as an insulation medium due to its higher breakdown stength and arc quenching characteristic than other gases (for instance, air). The use of SF_6 is, however, limited owing to the higher liquefyimg temperature when it is operated at higher pressures, and the sensitivity to the non–uniformity of the electric field, as well as the high cost. Therefore, in recent years, many studies on the strongly electronegative gases and gas mixtures have been conducted, aiming at the finding of new gases or gas mixtures with high dielectic strength and of low cost, to be used instead of SF_6[1–3].

Discussed in this paper are some aspects of electronegative gases, including whether some strongly electronegative gases or mixtures can be used instead of SF_6 gas, whether the gases with higher limiting breakdown strength than that of SF_6 may be chosen as insulation mediums, and whether the figure of merit of the electronegative gas is dependent upon the limiting breakdown strength of the gases, as well as how to assess and select the electronegative gases as insulation mediums.

To assess the electronegative gases, there are two parameters which have to be considered. One is the ratio of the limiting breakdown strength of the gas to the gas pressure, $(E/P)_{lim}$. From the point of view of application in power apparatus, the value of $(E/P)_{lim}$ of the gas should be as high as possible, and then the apparatus can be made as compact as possible. The other is the figure of merit M of the electronegative gas, which is a measure of quantifying the sensitivity of the gas to the effects of the electrode surface roughness; the smaller the M, the more sensitive the gas to non–uniformity of the electric fields[4] . The electrode surface roughness and conductive particles in the power apparatus are unavoidable due to manufacturing,

Gaseous Dielectrics VII, Edited by L.G. Christophorou
and D.R. James, Plenum Press, New York, 1994

installng and maintenance. Therefore the figure of merit of the gas has been expected to be as high as possible.

For the determination of $(E/P)_{lim}$ and M, there are many methods which are rather complicated and troublesome. A simple and precise method, however, for determination of $(E/P)_{lim}$ and M from the paschen's curve of the gas has been proposed[5]. For any gas, what has to be carried out to determine $(E/P)_{lim}$ and M is only the measurement of the right part of the minimum of the Paschen's curve.

RESULTS AND DISCUSSIONS

Firstly, the values of $(E/P)_{lim}$ and M for various electronegative gases and binary gas mixtures were measured using the Paschen's curve approach as mentioned above. The results obtained show that the values of $(E/P)_{lim}$ would vary with the concentrations of the gas in the mixtures in which at least one of the constituent component gases is an electronegative gas. The behaviour of the variation of $(E/P)_{lim}$ with the relative concentration can be classified into four types; the linear relation, the synergistic relation, the positive synergistic relation and the negative synergistic relation.

The results as plotted in Fig. 1 show the linear relation. The variation of $(E/P)_{lim}$ with gas concentrations for eight N_2–containing gas mixtures, and for SF_6 –or C_2ClF_5–containg gas mixtures are shown in Figs, 1(a) and(b) respectively. When an electronegative gas is mixed with another gas (electonegative or not), $(E/P)_{lim}$ of the resultant mixture may vary considerably, depending on the electron–molecule reaction processes occuring within the mixture. The values of the limiting breakdown strength for the linear relation gas mixtures can be obtained by the sum of the partial pressure weighted limiting breakdown strengths of the constituent component gases. This linear relation, where each gas acts independently of the other, has been observed in many binary gas mixtures, including most of the N_2–containing mixtures.

For the gas mixtures in which the two constituent component gases are electronegative, the linear relation also exists. In Fig. 1(b). for example, $(E/P)_{lim}$ of $C–C_4F_8$ is higher than that of SF_6 in the mixture of $C–C_4F_8/SF_6$, and $(E/P)_{lim}$ of C_2ClF_5 is higher than that of SF_6, but $(E/P)_{lim}$ of $CB_r F_3$ is lower than that of SF_6 in the mixture of $C_2 ClF_5/CB_r F_3$.

For comparison, the values of $(E/P)_{lim}$ in Fig. 1 have been normalized to that of SF_6, and the same has been done in the other Figures in this paper.

The results, as shown in Fig. 2, demonstrate the synergistic relation of binary gas mixtures, When a highly attaching electrongative gas whose electron attaching rate is pressure–independent is admixed into a weakly ettaching gas whose electrin attaching rate is also pressure–independent, or a non–attaching gas, a marked increase in $(E/P)_{lim}$ of the resultant gas mixture will appear. The increase in $(E/P)_{lim}$ is much larger than that predicted by summing the partial pressure weighted $(E/P)_{lim}$ of the individual gas components. This effect is termed a synergistic relation. To determine $(E/P)_{lim}$ of the gas mixture with synergistic relation, the most reliable approach is by the Boltzmann equation analysis. This approach requires, however, a detailed knowledge of elastic, inelastic collision coefficients and electron attachment and ionization cross sections of the individual gases. Such knowledge is at least very difficult to obtain, or even nonexistant, for most mixtures of interest. Therefore, in practice, a synergistic technique is used whichinvolves summing the partial pressure

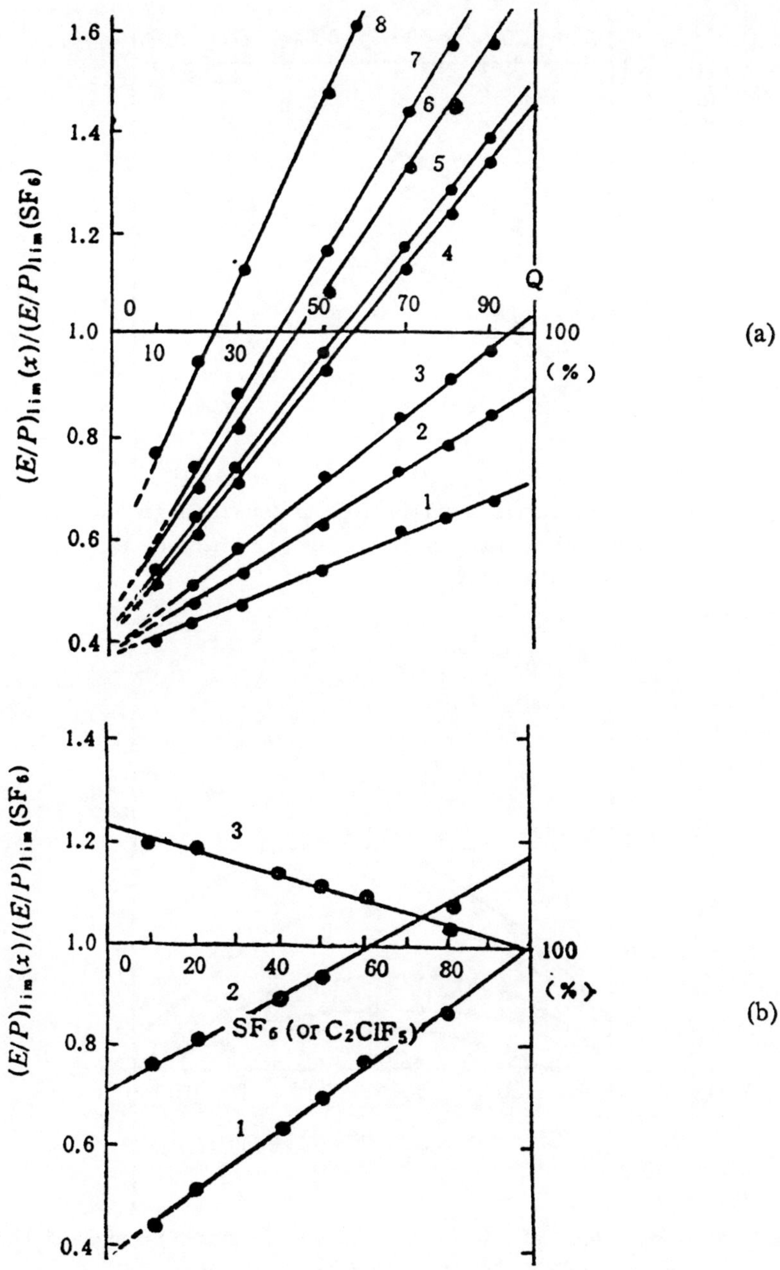

Fig.1 Variation of the relative limiting breakdown strengths of gas mixtures with percentages of electronegative gases (the linear relation).
(a) N_2–containing mixtures (Q is the percentages of electronegative gases in the mixtures, %). 1, $CBrF_3/N_2$; 2, $CHCl_2F/N_2$; 3, CC_2F_2/N_2; 4, CBr_2F_2/N_2; 5, $CBrClF_2/N_2$; 6, 1, 2–$C_2Cl_2F_4/N_2$; 7, CCl_3F/N_2; 8, 1, 1, 2–$C_2Cl_3F_3/N_2$.
(b) SF_6–or C_2ClF_5–containing mixtures. 1, CF_4/SF_6; 2, $CBrF_3/C_2ClF_5$ 3, C–C_4F_8/SF_6.

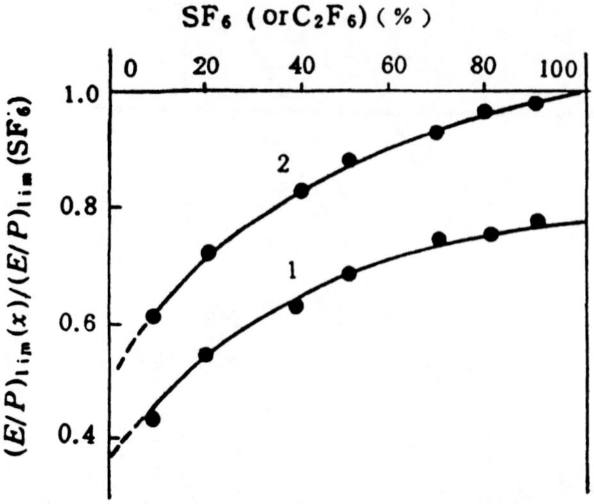

Fig.2 Variation of the relative limiting breakdown strengths of gas mixtures with percentages of SF_6(or C_2F_6) in the mixtutes (the synergistic relation).
1, C_2F_6/N_2; 2, SF_6/N_2.

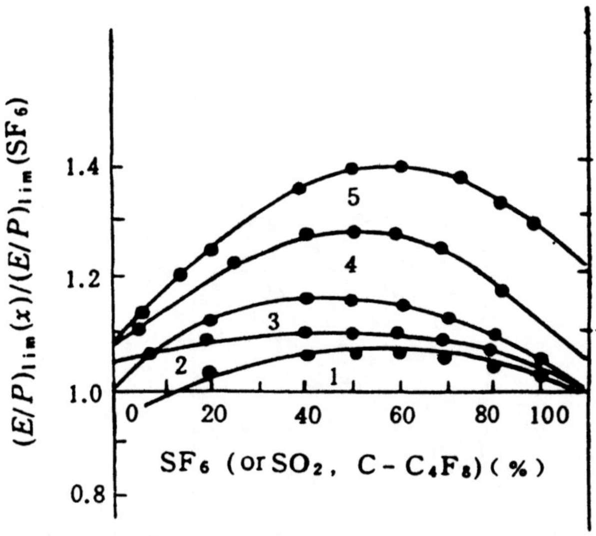

Fig.3 Variation of the relative limiting breakdown strengths of gas mixtures with percentages of SF_6(or SO_2, or C–C_4F_8) in the mixtures (the positive synergistic relation).
1, C_3F_8/SF_6; 2, CCl_2F_2/SF_6; 3, SO_2/SF_6 ; 4, 1–C_3F_6/SO_2 ;
5, 1–C_3F_6/C–C_4F_8.

weighted effective ionization coefficients to determine the $(E/P)_{lim}$ values. In Fig. 2, both mixtures with synergistic relation are N_2 –containing mixtures. It is note that SF_6/N_2 mixtures are very attractive in practical applications.

The results in Fig. 3 exhibit the positive synergistic relation which is quite different from the relations described above. For this type of mixtures, the values of $(E/P)_{lim}$ are higher than those of either constituent component gases in a certain range of gas concentrations, The prediction of $(E/P)_{lim}$ of this type of mixtures is impossible using the Boltzmann or the synergistic approach. The appearance of the positive synergistic relation may rely upon the variation of the electron attaching rate with pressure for one or even borh constituent component gases. This can be explained as resulting from an increase in stable negative ion production in the gas mixture as compared with the constituent component gases. The positive synergistic relation has been interpreted elsewhere[6] . An important judgement from the discussion is that the positive synergistic relation requires that at least one of the constituent component gases in the mixture possesses pressure–dependent electron attachment process and that the additon of the second component produces more stable negative ions.

In Fig. 4 there plotted is the result, indicating the negative synergistic relation, In this situation, the electron attachment rate constant, and hence the limiting brekdown strength of the mixture, can be smaller than those of both pure component gases. It means that the stable negative ion production in the mixture is less than that in the gases. This behaviur has been observed during this study only in $CBrClF_2/SF_6$ mixture.

From the point of view of application of insulation medium in high voltage equipment, the gas mixtures with syuergistis or positive synergistic relations may be adopted, and the gas mixtures with negative synergistic relation will be rejected.

The emphasis of the above discussions is only on the limiting breakdown strengths of the gas mixtures. It is necessary to consider the figure of merit M of the gas mixtures to assess the dielectric characteristics of the gas mixrures. For this purpose, measurements of M of pure strongly electronegative gases and gas mixtures have been carried out. The results obtained show that if a gas or gas mixture possesses smaller of M, i. e. more sensitve to the non– umiformity of the electric field, then it would possess higher value of $(E/P)_{lim}$ and vice versa. Obviously, this is useful for the assessment and selection of strongly gases and gas mixtures.

Acknowledgments

The authors are very much indebted to A. Pedersen and hiscolleagues, physics Lab. II, Technical University of Denmark, for thelaboratory facilities and many helpful discussions, without which thisstudy would not have been possible.

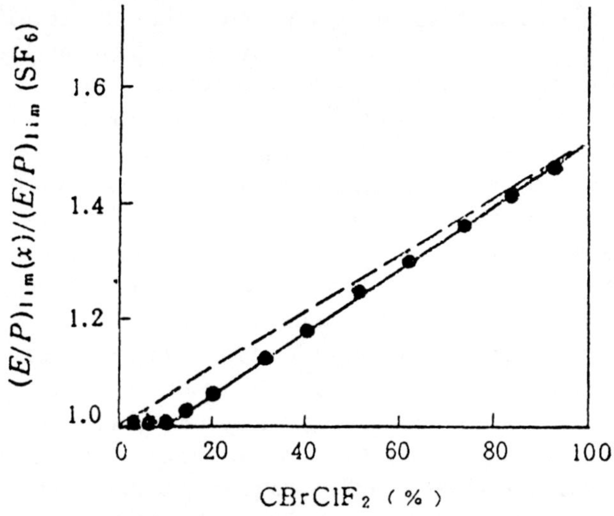

Fig.4 Variation of the relative limiting breakdown strengths of $CBrClF_2$ / SF_6 with percentages of $CBrClF_2$ in the mixtures (the negative synergistic relation). ---, Straight line.

References

1. S. Yanabu, Y Murayama and S. Matsumoto, SF_6 Insulation and its Application to HV Equipment, IEEE Trans on Electrical Insulation, Vol. 26, NO 3, 358–366 (1991)
2. Z. Y. Lee, Measurements of Breakdown Potential and Ionization and Attactment in SF_6–CO_2 Mixtures, IEEE Trans on Electrical Insulation, Vol. 18, No. 6, 637–641 (1983)
3. A. Pedersen, On the Electrical Breakdown of Gaseous Dielectrics, IEEE Trans. on Electrical Insulation Vol. 24, No. 5, 721–739 (1989)
4. A. Pedrsen, The Effect of Surface Roughness on Breakdown in SF_6, IEEE Trans. Vol. PAS–94, No. 5, 1749–1754 (1975)
5. A. Pedersen, I. W. McAllister, G. C. Crichton and S. Vibholm, Formulation of the Streamer Breakdown Criterion and its application to Strongly Electronegative gases and gas Mixtures, Archiv fur Elektrotechnik, 67, 395–402 (1984)
6. S. R. Hunter and L. G. Christophorou, Pressure–dependent Electron Attachment and Breakdown Strengths of Unary Gases and Synergism of Binary Gas Mixtures: A Relationship, J Appl. Phys. 57(9), 4377–4385 (1985)

SF_6 VOLTAGE/PRESSURE CHARACTERISTICS MODIFICATION WITH ADDED N_2 AND OTHER GASES MIXTURES

Anwar H. Mufti and Abdul-Aziz O. Arafa

High Voltage Laboratory
Electrical & Computer Eng. Dept.
King Abdulaziz University
Jeddah, Saudi Arabia

INTRODUCTION

The fast development of gas insulated switchgear (GIS) equipment using pressursied SF_6 as an insulant, and the recognition that insulation failures are often encountered in practical systems due to electrode surface imperfections, presence of free particles and sharp edges, there is a great concern to substitute SF_6 for promising candidates gas. In this regard, nonuniform field breakdown data of SF_6/gas mixtures are of considerable interest as they provide valuable information on the fundamental physical processes.

SF_6/N_2 and SF_6/air gas mixtures have breakdown properties superior to pure SF_6 at nonuniform field[1,2,3]. The SF_6/N_2 mixtures have been used with perfluorocarbons of large electron attachment rates and low ionization onsets to modify the space charge behaviour of the gas[4,5]. More experimental investigations have been reported on the breakdown behavior of SF_6/CO mixed with CO_2, or N_2. Both gas mixtures exhibit higher dielectric strength[6,7]. In another work the behaviour of N_2/freon mixtures and SF_6/freon mixtures have been studied under different pressure ranges with different concentrations[8,9]. Considerable theoretical work on the effective ionization coefficient and attachment coefficient in SF_6/freon mixtures was made [10]. The mixtures of SF_6 and N_2 with tetrachloroethylene are also reported to increase the flash over voltage in a nearly uniform field[11].

In this paper the authors extended their work[12] to contribute to a better understanding of the effect of N_2 as synergism gas with electron attaching gases which can increase corona stabilization peak and modify the corona inception. In this work mixturs of SF_6 with small concentrations (1.5% and 2.5%) of perfluorocarbon ($C_8F_{16}O$), freon R113 ($C_2CL_3F_3$) and tetrachloroethylene (C_2CL_4) are tested. For each SF_6/gas mixtures different percentage of N_2 have been added.

EXPERIMENTAL SETUP

The investigations are carried out in plexiglass pressure vessel 52 cm high and 13 cm in diameter and fitted with a high voltage bushing. Details of the vessel and gap arrangements have been reported in an earlier paper[12]. The high voltage a.c. transformer used is capable of supplying up to 100 kV. A current limiting resistor is placed in series with the high voltage

Gaseous Dielectrics VII, Edited by L.G. Christophorou
and D.R. James, Plenum Press, New York, 1994

line. The 10 mm long rod electrode used in most experiments is shaped as a sharp point with tip diameter of 0.25 mm or paraboloidal point with a tip diameter of 1 mm. The plane electrode has a diameter of 100 mm and is machined to 90° Rogowski profile. The rod-to-plane spacing was varied in the range 5-10 mm.

The vessel is evacuated to a pressure of about 0.01 KPa and flushed a few times with the appropriate gas prior to pressurization. The chamber is first evacuated and the lower constituent is admitted first to a partial pressure corresponding to the desired mixture ratio. The second constituent is then added to the predetermined total pressure.

RESULTS

The effect of adding N_2 to $SF_6/C_8F_{16}O$ mixture is examined using different electrode gap separations. Figure 1 shows the voltage/pressure characteristics of SF_6 in 5 mm gap and 0.25 mm tip diameter. The breakdown voltage, increased as the pressure is increased. The breakdown voltage is significantly higher than the corona onset level. As the low percent (0.6%) of N_2 is added with SF_6/1.5% $C_8F_{16}O$ the breakdown voltage is reduced by 5.5% at 0.2 MPa and by 8.5% at 0.5 MPa. The corona onset in this mixture decreases slightly than that of pure SF_6. When the N_2 percentage is increased to 1.5% in SF_6/1.5% $C_8F_{16}O$ mixtures the breakdown voltage of SF_6 is reduced by 2.7% at 0.2 MPa and by 6.7% at 0.5 MPa. The corona onset by adding 1.5% N_2 in this mixture becomes of the same level as in pure SF_6. In a longer gap the breakdown voltage is reduced by an average of 3% with modification in the corona stabilization curve. The corona onset exhibits a higher level than that of pure SF_6 as shown in Figure 2.

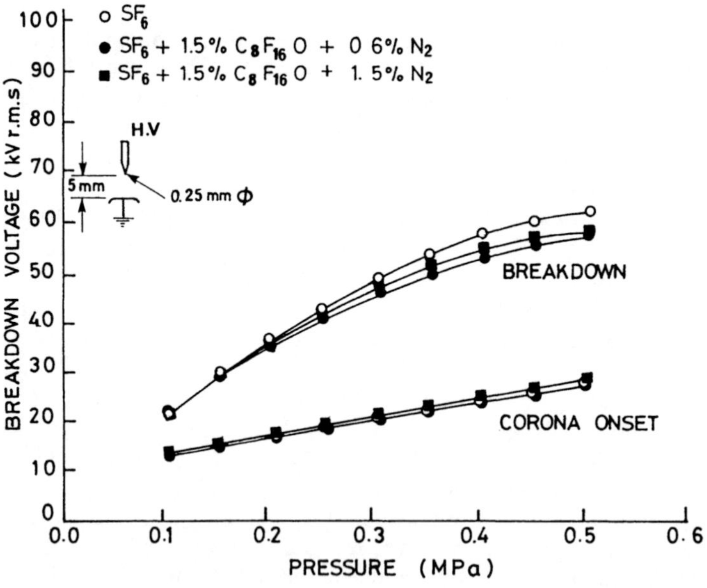

Figure 1. V/P characteristics for a 5 mm point-plane gap (ϕ=0.25mm) for SF_6, $SF_6/C_8F_{16}O + N_2$ mixtures.

More work is carried out to study the gap and rod diameter effect on using N_2 with $SF_6/C_8F_{16}O$ mixture. Figure 3 shows the typical corona stabilization curve of pure SF_6 in 10 mm gap and 1 mm tip diameter. The maximum breakdown voltage is recorded at a pressure of 0.35 MPa with 68 kV. When SF_6 gas is mixed with 1.5% $C_8F_{16}O$ and 1.5% N_2 the breakdown voltage is increased by (8% - 3%) up to a pressure of 0.35 MPa to pronounce 70 kV as a maximum breakdown voltage. At a higher pressure the breakdown voltage is decreased by 4%. The corona onset in this mixture has a higher level than that of pure SF_6. As the N_2 percent is increased to 2.5% with SF_6/1.5% $C_8F_{16}O$, the breakdown voltage is increased slighly by an average of (10% - 3%) up to a pressure of 0.35 MPa with a maximum breakdown voltage of 70 kV. At a higher pressure the breakdown voltage is decreased by 3%. The corona stabilization curve is modified without any change in the corona onset level. For a percentage of 5% N_2 in the mixture of SF_6/1.5% $C_8F_{16}O$ the breakdown voltage records an increase in the range of (12% - 4%) up to a pressure of 0.35 MPa to pronounce a maximum breakdown voltage of 71 kV. As the pressure is increased the breakdown voltage decreases slightly and the corona stabilization curve of SF_6 is improved.

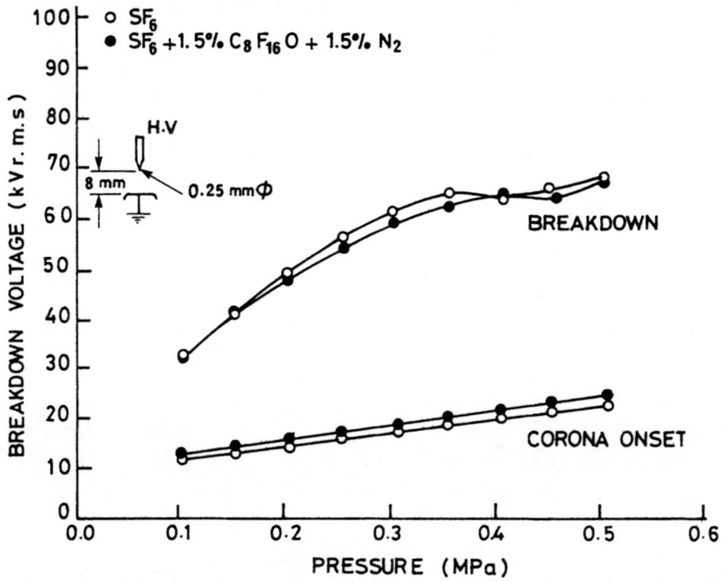

Figure 2. V/P characteristics for an 8 mm point-plane gap (ϕ=0.25mm) for SF_6, $SF_6/C_8F_{16}O + N_2$ mixtures.

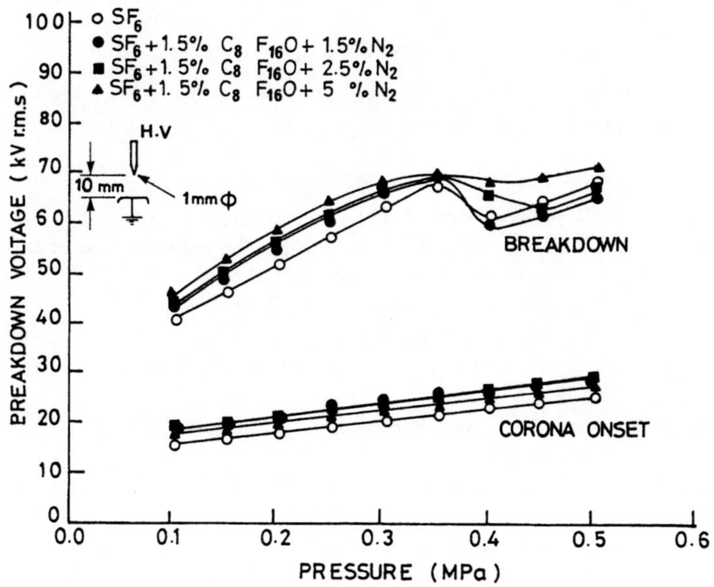

Figure 3. V/P characteristics for a 10 mm point-plane gap (ϕ=1mm) for SF_6, $SF_6/C_8F_{16}O + N_2$ mixtures. The N_2 percentages are 1.5%, 2.5% and 5%.

An extensive investigation of a higher N_2 percent added to $SF_6/C_8F_{16}O$ mixture is carried out. Figure 4 shows SF_6/1.5% $C_8F_{16}O$ added to different N_2 percent. When the 10% N_2 is used the breakdown voltage is increased by an average of 15% up to a pressure of 0.35 MPa than that of pure SF_6. The maximum breakdown voltage is recorded at a pressure of 0.4 MPa with 81 kV. Whereas for 20%N_2 the breakdown voltage is decreased by (9% - 4%) up to a pressure of 0.35 MPa relative to that of SF_6/1.5% $C_8F_{16}O$ + 10% N_2 with a maximum breakdown voltage of 76 kV. As the N_2 percentage is increased to 50% the breakdown voltage is decreased by (13% - 6%) up to a pressure of 0.35 MPa relative to that of pure SF_6. In this mixture the maximum breakdown voltage is extended to a pressure of 0.4 MPa with 67 kV. The corona onset level is significantly reduced by an average of 60%. In another experiment as shown in Figure 5 the 50% of N_2 is used with an increase in the percentage of $C_8F_{16}O$ to 2.5%. The breakdown voltage is decreased by (50% - 14%) up to a pressure of 0.35 MPa than that of pure SF_6 where the corona onset is slightly increased.

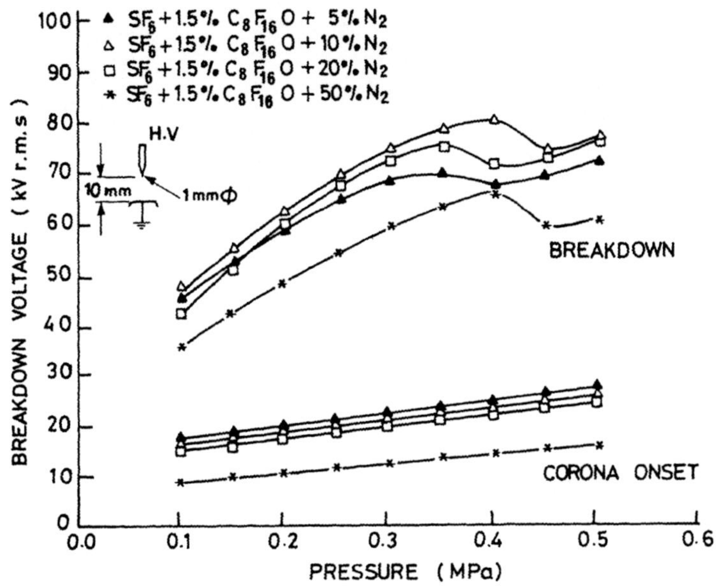

Figure 4. V/P characteristics for a 10 mm point-plane gap (ϕ=1mm) for SF_6, $SF_6/C_8F_{16}O + N_2$ mixtures. The N_2 percentages are 5%, 20% and 50%.

Freon R113 $C_2Cl_3F_3$ is used with 10%N_2 and compared with $C_8F_{16}O$ as shown in Figure 6. With SF_6/1.5% $C_2CL_3F_3$ + 10%N_2 mixture the corona stabilization curve is modified and exhibits a higher breakdown voltage than that of pure SF_6. The corona onset level increases slightly. Whereas in the mixture of SF_6/2.5% $C_2CL_3F_3$ + 10% N_2 the corona stabilization curve is modified at a higher pressure above 0.3 MPa with lowering corona onset level. These two mixtures with $C_2CL_3F_3$ have a lower breakdown voltage than that of SF_6/1.5% $C_8F_{16}O$ + 10% N_2.

The tetrachloroethylene C_2CL_4 is also used as an additive to SF_6 and 10% N_2. Figure 7 shows the corona stabilization curve of SF_6/2.5% C_2CL_4 + 10% N_2. In this mixture the breakdown voltage is significantly increased than that of pure SF_6. The corona onset level in this mixture is reduced significantly by 60% than that of pure SF_6. The above mixture shows higher breakdown voltage than that of SF_6/2.5% $C_2CL_3F_3$ + 10% N_2. Whereas the mixture of SF_6/2.5% $C_8F_{16}O$ + 10%N_2 exhibits a higher breakdown voltage than all the above mixtures. The corona stabilization curve is improved with a maximum breakdown voltage of 83 kV. Small portion of N_2 exhibits excellent synergism with $SF_6/C_8F_{16}O$ mixtures.

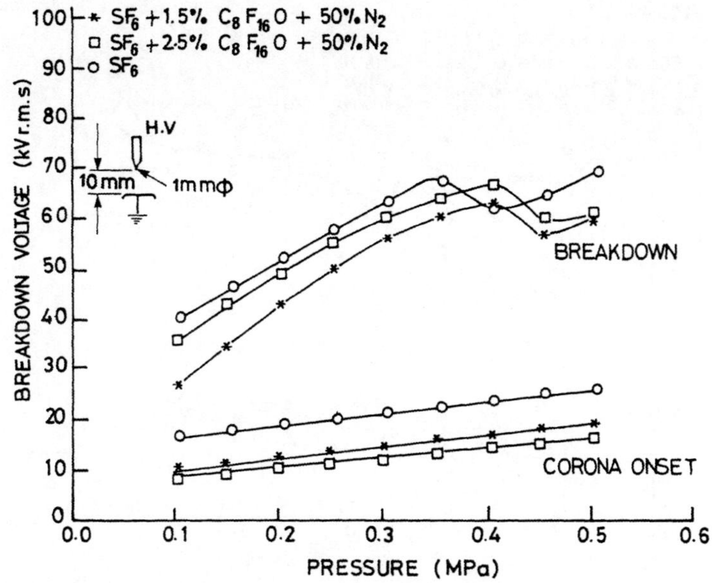

Figure 5. V/P characteristics for a 10 mm point-plane gap (ϕ=1mm) for $SF_6/C_8F_{16}O + N_2$ mixtures. The N_2 percentage is 50%.

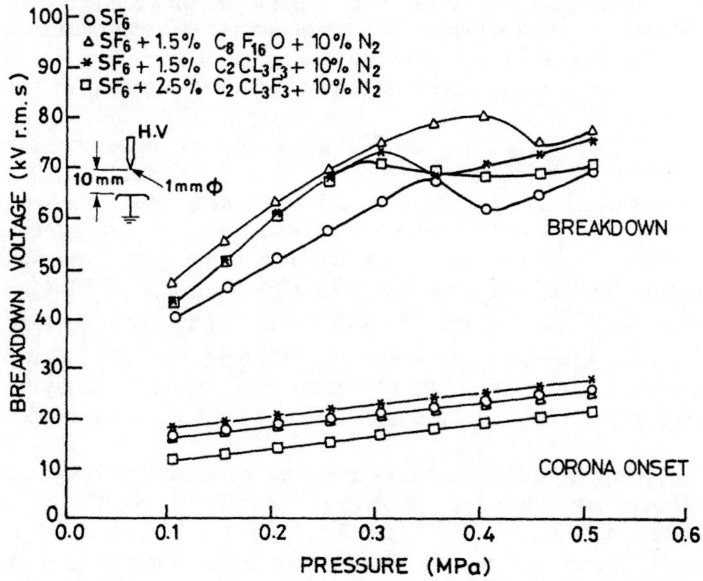

Figure 6. V/P characteristics for a 10 mm point-plane gap (ϕ=1mm) for $SF_6/C_8F_{16}O + N_2$ and $SF_6/C_2CL_3F_3 + N_2$. The N_2 percentage is 10%.

DISCUSSIONS

At present a great potential for new gaseous insulators appears to be in the development of synergistic multicomponent gas mixture[4,5,7,12]. In view of this some factor can be considered such as[13] the favourable of electron attachment cross-section with electron energy distribution function; electron slowing-down properties of the gas; electron affinity of the gas molecules; and the different degrees of fragmentation via dissociative attachment. For example, the freons capture slow electrons dissociatively producing mainly CL^-, whereas the

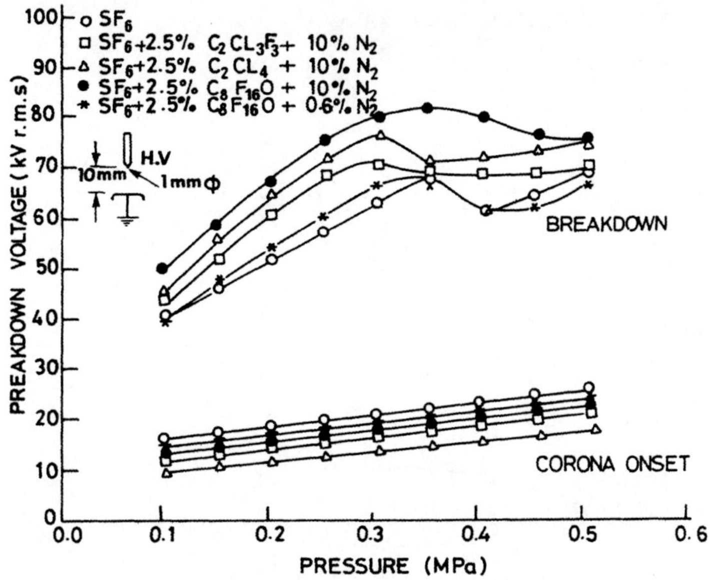

Figure 7. V/P characteristics for a 10 mm point-plane gap (ϕ=1mm) for $SF_6/C_8F_{16}O + N_2$, $SF_6/C_2CL_3F_3 + N_2$ and $SF_6/C_2CL_4 + N_2$. The N_2 percentage is 10%.

perfluorocarbons have positive electron affinities (i.e. the negative ion is often a little more stable than the neutral species) and capture electron at thermal and near thermal energies non dissociatively [13]. The perfluorocarbon molecules lead predominantly to parent negative ions and processes many negative ion states at an electron energy less than the ionization energy [14].

In figure 7 the $SF_6/2.5\%\ C_2CL_4 + 10\%\ N_2$ mixture shows a higher breakdown voltage than that of SF_6 and $SF_6/2.5\%\ C_2CL_3F_3 + 10\%\ N_2$. This may refer to the increase in the number of CL atoms which increase the magnitude of the attachment cross-section, and to the synergism of N_2 gas. The combination of freons with N_2 as a good slowing down gas due to its negative ion state is significantly beneficial in effecting a large value of breakdown voltage[13,14]. The mixtures of $SF_6/1.5\%\ C_8F_{16}O + 10\%\ N_2$ and $SF_6/2.5\%\ C_8F_{16}O + 10\%\ N_2$ exhibit a higher breakdown voltage. The perfluorocarbon has been reported[15] to have a high peak intensity of parent negative ion than that of fragment ions F^- which can increase the breakdown voltage. In addition, the synergism of N_2 works perfectly with electron attaching additives whose electron attachment cross-section are large at moderate thermal energies[5,13].

However, SF_6/perfluorocarbon mixtures increase the breakdown voltage and enhance the corona stabilization curve than that of SF_6/freon. The high electron affinities of CL atoms make the electron detachment less significant for the freons than that for the perfluorocarbon[13]. The careful combination in mixtures is an important factor to make stronger candidate to replace pure SF_6 in practical equipment.

CONCLUSION

In the present work voltage/pressure characteristics for $SF_6/C_8F_{16}O$, $SF_6/C_2CL_3F_3$ and SF_6/C_2CL_4 with different percentages of N_2 ($\leq$ 10%) shows a higher breakdown voltage with modification in the corona stabilization curve and corona onset level. Small portion of N_2 exhibits excellent synergism with $SF_6/C_8F_{16}O$ mixtures.

REFERENCES

1. N. Malik, A.Qureshi, and G. Theophilus, "Static field breakdown of SF_6-N_2 mixtures in rod-plane gaps", IEEE Trans. Electrical Insulation, Vol. EI-14, No. 2, pp. 61-69 (1979).
2. A. Kurimoto, "Prebreakdown phenomena in long nonuniform field gaps in SF_6/N_2 and SF_6/air mixtures", 7th International Conference on Gas Discharge and Their Applications, London, England, pp. 211-214 (1982).
3. K. Nakanishi,"New gaseous insulation", IEEE Trans. Electrical Insulation, Vol. EI-21, No. 6, pp. 933-937 (1986).
4. R. Pai, L. Christophorou, I. Sauers, and A. Fatheddin, "Measurement of properties of perfluorcarbon/SF_6 mixtures relevant to applications", 6th International Conference on Gas Discharge and Their Applications, Edinburgh, Scotland, Part 1, pp. 232-235 (1980).
5. L. Christophorou, I. Sauers, D. James, H. Radrigo, M. Pace, J. Carter, and S. Hunter, "Recent advances in gaseous dielectrics at oak ridge national laboratory", IEEE Trans. Electrical Insulation, Vol. EI-19, No. 6, pp. 550-566 (1984).
6. Y. Qiu and E. Kuffel, "Dielectric strength of gas mixtures comprising SF_6, CO, c-C_4-F_8 and SF_6, N_2, c-C_4F_8", IEEE Trans. Power Apparatus and Systems, Vol. PAS-102, No. 5, pp. 1445-1451 (1983).
7. Y. Qiu, Y. Feng, C. Lu, and Z.Wu., "Breakdown of SF_6 and some SF_6 mixtures in a point-plane gap", 7th International Symposium on High Voltage Engineering, Dresden, Germany, Paper No. 32.09, Vol. 3, pp. 83-86 (1991).
8. T. Rajendran, C. Lakshminarasimha, and M. Naidu, "Lightning and Switching impulse breakdown of rod/plane gaps in nitrogen and nitrogen/freon (CCL_2F_2) mixtures", IEE Proc., Vol. 130, Pt. A, No. 3, pp. 134-139 (1983).
9. Y. Qui, A. Sun, and E. Kuffel, "Improved dielectric strength of SF_6 gas with a trichlorotrifluoroethane vapor additive", IEEE Trans. Electrical Insulation, Vol. EI-22, No. 6, pp. 763-768 (1987).
10. Y. Qiu, X. Weng, and M. Zhang, "Measurement of ionisation and attachment coefficients in trichloro-trifluoroethane and SF_6 mixtures", Journal of Physics. D: Appl. Phys., Vol. 20, pp. 801-802 (1987).
11. M. Yashima, H. Fujinami, and T. Takuma, "Breakdown characteristics of gases mixed with tetrachloroethylene mist under nearly uniform fields", IEEE Trans. Electrical Insulation, Vol. 25, No. 2, pp.371-379 (1990).
12. A. Mufti and A.A.O. Arafa, "Non-uniform field breakdown characteristics of SF_6/gas mixtures containing small amount of N_2", 10th International Conference on Gas Discharge and their Applications, Swansea, Wales, Vol. I, pp. 394-397 (1992).
13. L. Christophorou, R. Mathis, D. James, and D. McCorkle, "On the role of electron attachment in the breakdown strength of gaseous dielectrics", Journal of Physics. D: Appl. Phys., Vol. 14, pp. 1889-1901 (1981).
14. L. Christophorou, D. James, and R. Mathis, "Dielectric gas mixtures with polar components", Journal of Physics. D: Apl. Phys., Vol. 14, pp. 675-692 (1981).
15. I. Sauers, L. Christophorou, and J. Carter, "Electron attachment to perfluorocarbon compounds, III. Fragmentation of aliphatic perfluorocarbons of interest to gaseous dielectrics", Journal of Chem. Phys., Vol. 71, No. 7, pp. 3016-3024 (1979).

EVIDENCE FOR A GAS PHASE IN LIQUID PREBREAKDOWN

M. O. Pace, T. V. Blalock, I. Alexeff

Electrical & Computer Engineering Department
University of Tennessee
Knoxville, TN 37996-2100 USA

I. INTRODUCTION

Experimental studies are reviewed to understand the very earliest prebreakdown events in liquids. Emphasis is on cathode events that begin at lower voltages than anode events[1]. Other events such as anode events or those with pulsed voltages much greater than the DC or 60 Hz breakdown voltages are found in the literature as well. Photographs of stressed plane cathodes have shown events similar to those at points.[2]

Previously prebreakdown current pulses in stressed liquids were considered random processes; e.g., in 1969 their autocorrelation function was measured[3]. In 1985 we undertook a study of these pulses as a random process, but reported in 1986[4] that each event previously thought to be one "random pulse" was actually a well-structured burst of pulses, each burst starting at a random time. In 1987 we suggested a possible model[5]. In 1988[6] we discussed resolution techniques, and in 1989 others confirmed the individual pulses[7].

Kelley at the National Institute of Standards and Technology (NIST) developed an optical delay for fast photography of randomly initiated events from their very beginning, or before. In 1987, 1988, and 1989 we combined our systems with those at NIST for synchronized photographs and resolved current pulse records[8,9]. The region accompanying the pulses is called the "low density region" (LDR) here.

II. EXPERIMENTAL APPARATUS

A. Liquid Test Cell

A closed system was used to purify hexane and clean and evacuate the cell[4], producing a clean liquid in a cell otherwise at vacuum, and thus at the vapor pressure of the liquid. In later tests the pressure was varied and some contamination was deliberately added.

A needle cathode (1 mm diameter, 3 micron tip radius) and a high voltage plane (typically 3.8 cm. nickel disk with rounded edges) were separated by a 4.5 to 5 mm gap. The needle was mounted on a micrometer drive and electrically isolated by teflon, with connection to a vacuum feedthrough.

B. Current Pulse Detection

Pulses had not been well resolved previously because the bandwidths of the earlier detectors were lowered to reduce detector noise. The individual pulses were too fast for these bandwidths[6].

Gaseous Dielectrics VII, Edited by L.G. Christophorou
and D.R. James, Plenum Press, New York, 1994

Here one detection system[10] sensed the voltage on a 50Ω resistor from the needle to ground with a bandwidth of 29 MHz, and a minimum detectable current of 1.1μA. The second[11] operated in a preamplifier transimpedance mode (current to ground sensed by a feedback resistor), with 50 MHz bandwidth, minimum detectable signal of 60 ηA, and minimum detectable charge in a fast pulse of 2.8 fC. The amplified current pulses were digitized for a computer by a Tektronix 7612D digitizer, using its pretrigger mode to view current and light in separate channels before, during, and after the burst.

C. Light Pulse Detection and Spectrum

Light accompanying the current pulses was focused on a cooled photomultiplier. The spectrum of the light was observed by introducing a monochromator[12].

D. Photography at NIST

Our apparatus was combined with that at the National Institute of Standards and Technology (NIST) for photography. An image converter camera photographed the shadow of the low density region in the beam of a 5W cw argon laser. Current pulses were recorded with framing pulses from the camera for synchronization. An optical delay was developed by Kelley[8] to photograph the beginning of the prebreakdown event even though that stage had passed when the camera could be triggered. A 1μm tip needle with gap of 3.2 mm was connected to the detection electronics in **IIB** Also, a synchronizing circuit detected a burst of current pulses and controlled the triggers on the digitizer and camera. NIST later increased magnification with a microscope objective in the test cell[9].

III. PREBREAKDOWN CURRENT PULSES

Using the first system described in **IIB** above, pulse bursts such as that shown in Fig. 1 were recorded[13] in hexane at its vapor pressure. The charge per pulse is shown in Fig. 2 for three typical bursts. Adding particles to the purified liquid tended to increase the burst rate, but not to affect the pulses within a burst.

Using the second (and faster) of the two systems in **IIB**, pulse bursts such as that in Fig. 3 were recorded[14]. The lower bandwidth in Fig. 1 caused a pile-up of pulses. Note in Fig. 3 the first pulse is actually two pulses.

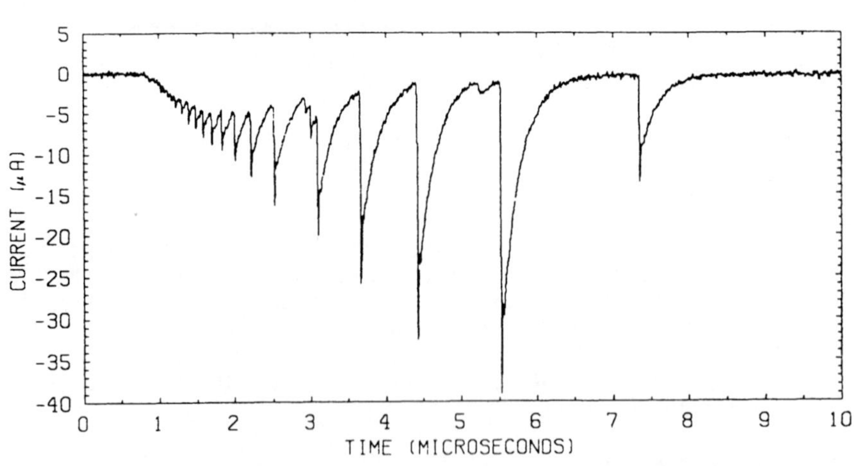

Figure 1. A typical complete current pulse burst with the lower (29 MHz) bandwidth.

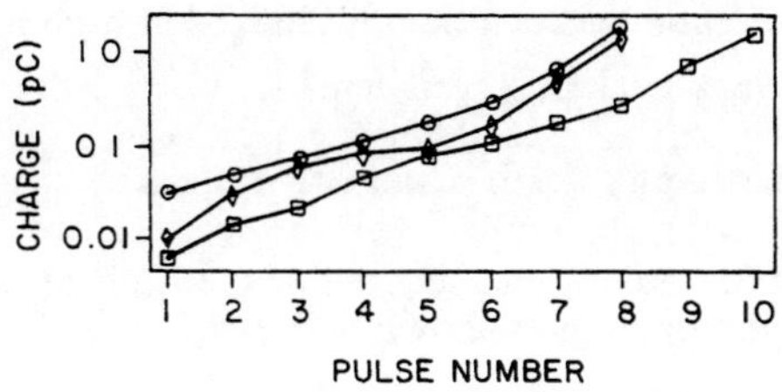

Figure 2. Pulse charge vs. number of the pulse in the burst, for three typical bursts.

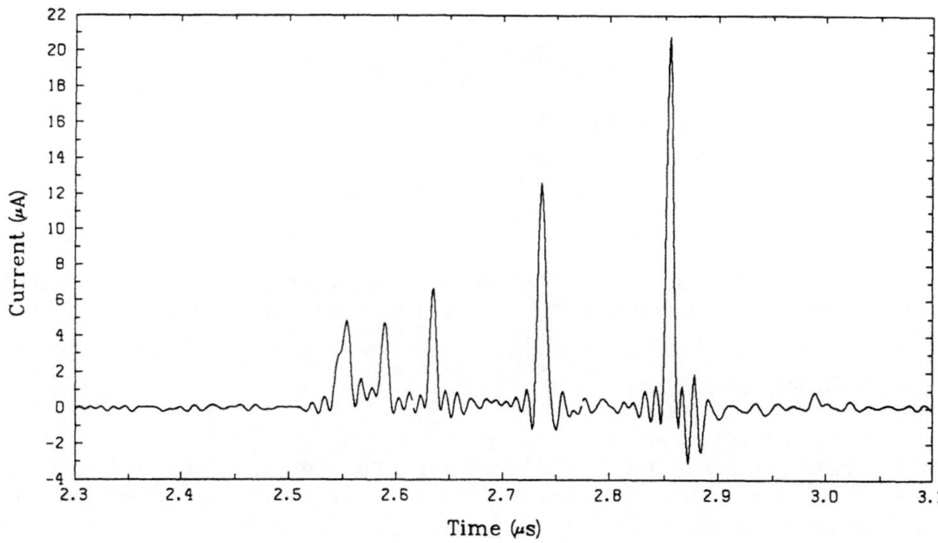

Figure 3. First part of a typical current pulse burst using the higher bandwidth (50 MHz) system.

IV. PHOTOGRAPHS AND SPECTRA

A. Emitted Light: Timing, Photographs and Spectrum

Time between each current and light pulses was measured[12]. They were found coincident within 5 ηs.

The LDR events were photographed by their emitted light with an intensified video camera[14]. The light came from a region about the maximum size (100 μm) attained by the LDR. Low density regions occurred on different sides of the needle tip.

Means of observing the spectrum of the emitted light were described in **IIC.** A passband of 5.5 ηm was centered in 30 steps of 10 ηm each from 330 to 620 ηm. The total area of all light pulses at each wavelength was normalized by the total area of corresponding light pulses to minimize effects of amplitude variation. Strong peaks near 430 and 480 ηm and several weaker peaks resulted. Hydrocarbon reactions in the sun's atmosphere produce a strong peak near 430 ηm, and hydrocarbon discharges at high pressure produce a peak at 590 ηm, which was one weak peak observed here[15].

B. Fast Shadow Photography with Current Pulses

The means of photographing the shadow of the LDR at NIST[8,9] was described in IID. A typical series of photographs is shown in Fig. 4.

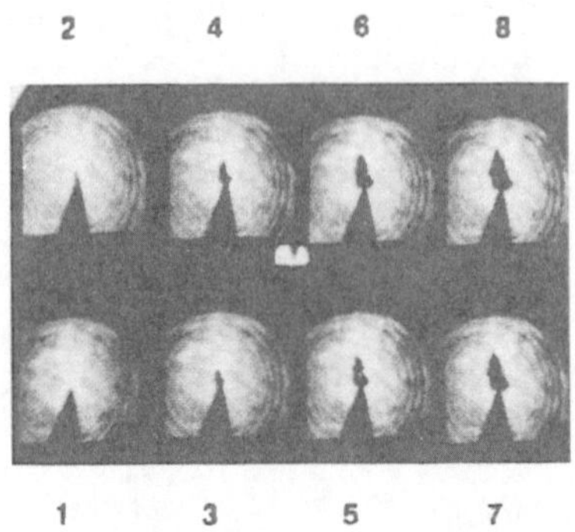

Figure 4. A typical sequence of LDR shadow photographs, framing period 200 ηs. Frames are numbered, starting at lower left. The needle tip points upward.

V. PRESSURE EFFECTS

The absolute pressure was usually 150 torr, the vapor pressure of hexane. For pressure effects an auxiliary cell with controlled temperature was connected to the main cell.[16]. Contrary to reports, strong pressure effects existed down to below one atm. Fig. 5a and 5b show respectively the considerable pressure effects [17].

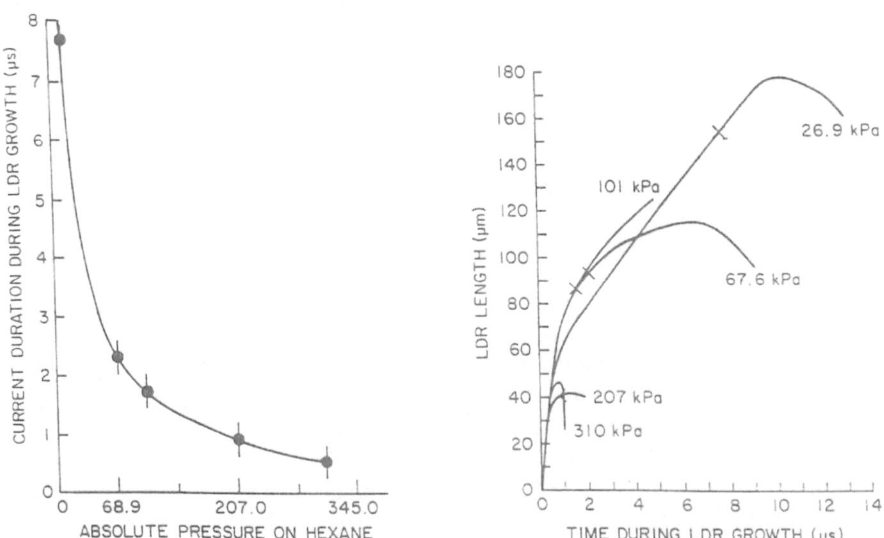

(a) (b)

Figure 5. (a) Current burst duration vs. pressure; (b) LDR size vs. time with pressure as a parameter.

VI. MODELS

The LDR began at random times (more frequent with contamination), grew during the current pulse burst, and dissipated in an instability. An explanation first given by Watson[18] for the mechanical aspects of the growth was used here. The models of Alexeff for the structured current pulse burst[5], as well as for the initiation and instability[19] are used.

A. Initiation

The LDR initiated at apparently random times, but more often with higher voltage (in the range of 10 to 20 kV in these experiments) and with increased added contamination. Alexeff[19] estimated the voltage required for initiation to be approximately 18 kV by equating the electrostatic repulsion of liquid, charged to the potential of the needle tip, to its adhesion to the needle as calculated from surface tension. The random timing could be attributed to contaminants drifting.

B. Growth and Current Pulses

A computer simulation was made for the coupled mechanical and electrical events in the growth of the LDR[20]. The LDR expanded by the electrostatic force on the charged liquid surface increasing the liquid's kinetic energy, as suggested by Watson[18]. The moving liquid boundary carried trapped charge while expanding, thus increasing the potential between it and the cathode. until it reached a critical value and a discharge occurred. This process repeated cyclically, with each discharge being larger than the one before and thus creating the structured burst of pulses. It was found the field required for these discharges was similar to that for field emission.

The simulated electrostatic pressure on the liquid surface is on the order of 100 atm. initially, but it drops quickly to about 1 atmosphere (zero net pressure) in about 0.4 μs, when the LDR radius is on the order of 30 μm. The momentum carries the LDR to about 100 μm. in 5 or 6 μs before it begins to shrink under the ambient pressure, but at some time during this phase the wall instability usually causes the LDR to dissipate. The energy in the current pulses was much less than that of work from electrostatic energy, so the latter accounted for the kinetic energy of the moving liquid outside the LDR.

The simulation results were compared with experiment in five ways[20]: (1) the charge in pulses (Fig. 2), (2) the burst duration vs. pressure (Fig. 5), (3) the LDR size vs. time at various pressures, (4) the potential in the LDR, and (5) the damping of the LDR, due to viscosity[21]. In these ways the simulation agreed reasonably well with the experimental results, within the limits allowed by the random nature of the LDR wall instability. The simulated potential was as much as 6 kV, while others[21] had estimated it to be only hundreds of volts from the Kerr effect in nitrobenzene; however, the Kerr effect observations were affected by the macroscopic field, and the breakdowns could be affected by higher localized fields, as pointed out by Forster[22].

C. Instability

In photographs it was clear that the LDR wall instability led to its dissipation. The instability was modeled as the electrostatic force mentioned earlier working against surface tension[19]. The equations had solution only for wavelengths (of the instability shape) of about one μm. and larger. The photographed instability was found to have wavelengths on this order.

VII. CONCLUSIONS

Major present issues in prebreakdown in liquids are (1) whether the current pulses cause the LDR growth or vice-versa, and (2) whether liquid breakdown really begins as gas breakdown. The results here suggest that the LDR growth causes potential changes that lead to the current pulses, and the energy for LDR expansion comes mainly from electrostatic force and not heat.

The hypothesis, that the LDR causes the current pulses, is supported by three main results: photographs synchronized with current pulse records, modeling of the LDR growth and current pulses, and the first pulse in a burst being double. It is assumed that this double pulse consists of a displacement current when the liquid breaks away from the cathode surface, followed very quickly by the first breakdown pulse as the LDR expands. Simulation indicated that the pulse heat energy was not sufficient to provide the kinetic energy of the liquid outside the LDR, but that the electrostatic force did provide sufficient energy.

This provides evidence on the medium in which breakdown begins. The earliest prebreakdown currents in liquids are occurring in the low density region, which is likely to be gaseous in some sense but which is not yet well identified. It seems to consist of a void left by liquid departing the region at tens of meters per second, under electrostatic pressures up to about 100 atmospheres on the liquid surface. (It must be noted that this pressure is not that of any gas in the LDR). The critical fields for the simulated breakdown were between 10^9 and 10^{10} V/m, suggesting field emission into the LDR. The light spectra seemed to correspond to gas dischage spectra.

Several groups[23,24] have studied liquid prebreakdown in detail for a wider range of cases than the case reported here. A common observation is one current pulse accompanying one "bubble", with the conclusion that the current caused the bubble. One possible explanation is that their pressure or voltage was much higher than those here; another possible explanation is that the burst of current pulses was not resolved into individual pulses. For example, in one experiment[25] the pressure was on the order of 50 atm., which would cause the LDR episode to occur so quickly there would seem to be only one pulse. In other papers[26,27] there are hints of several pulses in the current records, or a stepped nature to the charge waveform, both suggesting several separate pulses.

REFERENCES

1. O. Lesaint, R. Tobazeon, Streamer Generation and Propagation in Transformer Oil Under AC Divergent Field Conditions, *IEEE Trans. Elec. Insulation* EI23:941 (1988).
2. W. R. L. Thomas, "An Ultra-High Speed Laser Schlieren Technique for Studying Electrical Breakdown in Dielectric Liquids", *Proc. Conf. on Elec. Ins. and Diel. Phen.,* 293 (1973).
3. I. Megahed and A. Nosseir, "Measurement of Autocorrelation Function and Power Spectral Density of Prebreakdown Conduction Currents in Liquid Dielectrics", *J. Phys. D* 2: 880 (1969).
4. A.L. Wintenberg, M. O. Pace, T. V. Blalock, J.V. Foust, "Statistics of Prebreakdown Current Pulses in n-Hexane", *Proc. Conf. on Elec. Ins. and Diel. Phen.,* 293 (1986).
5. I. Alexeff, T.V. Blalock, A.L. Wintenberg, J.V. Foust, M.O. Pace, "Multipulse Prebreakdown Phenomena In Liquid Dielectrics", *Proc. 14th IEEE International Conf. on Plasma Sci.,* 55 (1987).
6. A.L. Wintenberg, M.O. Pace, T.V. Blalock, "Low Noise, High Bandwidth Amplification of Prebreakdown Current Pulses in Liquid Dielectrics", *Rec. IEEE Int. Symp. on Elec. Ins.,* 383 (1988).
7. P. B. McGrath, H.I. Marsden, "Prebreakdown Events in AC and DC Stressed n-Hexane", *Proc. Conf. on Elec. Ins. and Diel. Phen.,* 93 (1989).
8. E.F. Kelley, M. Nehmadi, R.E. Hebner, M.O. Pace, A.L. Wintenberg, T.V. Blalock, J.V. Foust, "Measurement of Partial Discharges in Hexane Under DC Voltage", *IEEE Trans. on Elec. Ins.* EI-24: 1109 (1989).
9. K.L. Stricklett, C. Fenimore, E.F. Kelley, H. Yamashita, M.O. Pace, T.V. Blalock, A.L. Wintenberg, I. Alexeff, "Observation of Partial Discharge in Hexane Under High Magnification", *IEEE Trans. on Elec. Ins.* 26:692 (1991).
10. T.V. Blalock, A.L. Wintenberg, M.O. Pace, "Low Noise Wideband Amplification System for Acquiring Prebreakdown Current Pulses in Liquid Dielectrics", *IEEE Trans. on Elec. Ins.* 24:641 (1989).
11. A.L. Wintenberg, T.V. Blalock, M.O. Pace, "High Bandwidth Measurement of Low Level Prebreakdown Currents in Liquid Dielectrics", *Conf. Rec. 1990 IEEE Int. Symp. on Elec. Ins.,* 422 (1990).

12. D.N. Pittman, "Spectrum of Prebreakdown Light and Temporal Correlation of Prebreakdown Current with Light Pulses at a Point Cathode in Hexane", *M.S. Thesis,* University of Tennessee (1988).
13. A.L. Wintenberg, M.O. Pace, T.V. Blalock, J.V. Foust, "Prebreakdown Current Pulses in n-Hexane and Other Dielectric Fluids", *IEEE Trans. on Elec. Ins.* 24:67 (1989).
14. I. Alexeff, A.L. Wintenberg, D.N. Pittman, M.O. Pace, F.F. Dyer, M.S. Rader, "Self-Luminous Prebreakdown Phenomena in Liquid Hexane", *Proc. Conf. on Elec. Ins. and Diel. Phen.,* 340 (1988).
15. W.G. Driscoll, *Handbook of Optics,* McGraw Hill, NY, section 11 (1978).
16. M.O. Pace, D.N. Pittman, A.L. Wintenberg, I. Alexeff, T.V. Blalock, "Effects of pressure Near One Atmosphere on Prebreakdown Current Pulses at a Needle Cathode in Hexane", *Conf. Rec. 1988 IEEE Int. Symp. on Elec. Ins.,* 23 (1988).
17. M.O. Pace, A.L. Wintenberg, T.V. Blalock, E.F. Kelley, G.J. FitzPatrick, C. Fenimore, H. Yamashita, "Pressure Effects on Partial Discharges in Hexane Under DC Voltage", *Proc. Conf. on Elec. Ins. and Diel. Phen.,* 87 (1987).
18. P.K. Watson, "Electrostatic and Hydrodynamic Effects in the Electrical Breakdown of Liquid Dielectrics", *Proc. Conf. on Elec. Ins. and Diel. Phen.,* 370 (1981).
19. I. Alexeff, M.O. Pace, T.V. Blalock, A.L. Wintenberg, "Possible Models for the Earliest Prebreakdown Events in DC Stressed Hexane", *Proc. Tenth Int. Conf. on Conduction and Breakdown in Dielectric Liquids,* 387 (1990).
20. M.O. Pace, I. Alexeff, T.V. Blalock, "Spherical Geometry, Electromechanical Modeling of Cathode Liquid Prebreakdown", *Proc. Conf. on Elec. Ins. and Diel. Phen.,* 387 (1993).
21. S. Oliveri, R. Kattan, A. Denat, "Numerical Study of Single Vapor Bubble Dynamics in Insulating Liquids Initiated by Electrical Current Pulses", *J. Appl. Physics* 71:108 (1992).
22. E.O. Forster, "The Metal/Liquid Interface: the Charge Injection Process", *IEEE Trans. on Elec. Ins.* EI19: 524 (1984).
23. W.G. Chadband, "From Bubbles to Breakdown, or Vice-Versa", *Proc. 11th IEEE International Conference on Conduction and Breakdown in Dielectric Liquids,* 184 (1993).
24. R. Tobazeon, "Prebreakdown Phenomena in Dielectric Liquids", *Proc. 11th IEEE International Conference on Conduction and Breakdown in Dielectric Liquids,* 172 (1993).
25. O. Lesaint, R. Kattan, A. Denat, "Generation and Growth of Gaseous Bubbles in Hydrocarbon Liquids Under High Divergent Fields", *Proc. Conf. on Elec. Ins. and Diel. Phen.,* 269 (1988).
26. A. Beroual, R. Tobazeon, "Prebreakdown Phenomena in Liquid Dielectrics", *IEEE Trans. on Elec. Ins.* EI21: 613 (1986).
27. A. Beroual, C Marteau, R. Tobazeon, "Behavior of Streamers in Liquids Under Step Voltages in Point-Plane Geometry", *IEEE Trans. on Elec. Ins.* EI23: 427 (1988).

AUTOMATION OF A 400*KV* IMPULSE GENERATOR

H. Rodrigo and R.J. Hellner III

Department of Electrical Engineering
Michigan Technological University
Houghton, MI 49931

Abstract

The statistical nature of high voltage testing requires the application of multiple impulses to the test objects. The repetitive action of firing the impulse generator and recording the resulting data lends itself to the automation of the test procedure. A $400KV$ impulse generator that produces standard $1.2/50\mu s$ waveforms has been brought under computer control. The control system consists of the existing manual controls, constructed TTL and CMOS based interfacing hardware, and an A/D & D/A data acquisiton board,and controlling FORTRAN software. The entire system is contained in a Faraday cage, yielding noise free accurate waveforms. The system uses a novel electronic breakdown detection circuit that responds to the acquired voltage waveform.

1. **Introduction** The area of high voltage testing and research normally entails large amounts of data. Much of the experimental work is of a repetitive nature that lends itself to effective automation. In this paper we present the software and hardware necessary to accomplish data gathering, that is normally required in insulation research. A novel technique has been developed to sense the occurence or otherwise of a breakdown of a test gap. This technique is based on the actual waveform of the output signal at the bottom of the capacitor divider. The waveform is captured and compared using two comparators; one for fast events that occur at time scales of the order of submicroseconds to $5\mu s$ and the other for slow events that occur at time scales of the μs to several tens of μs. This is an improvement on previously published material[1] [2] [3] which relied for all measurements on the transient ground currents that generally accompany breakdown events. Reliance on ground currents alone lead to spurious and erroneous results.

2. **Experimental Setup** The experimental setup consists of the impulse generator, whose output goes directly onto the capacitor divider, and the test object is connected to the impulse generator via the capacitor divider. The measuring system is housed in a fully shielded Faraday cage. The Faraday enclosure is made of $2mm$ thick Copper sheet. The Faraday enclosure is extended using Copper tubing to the bottom end of the capacitor divider. All the manual controls of the "Hypotronics" control console were converted to electronic switches and relays, and these were brought into the Faraday enclosure. The switches and relays were interfaced onto an IBM PC and Tektronix $100MHz$ storage oscilloscope. The PC

monitors the captured waveforms, and the software enables it to make decisions as to whether a breakdown has occured or not and it records each such event accordingly. A manually operated toggle switch enables full manual control and overrides computer control. Figure 1. shows the experimental setup.

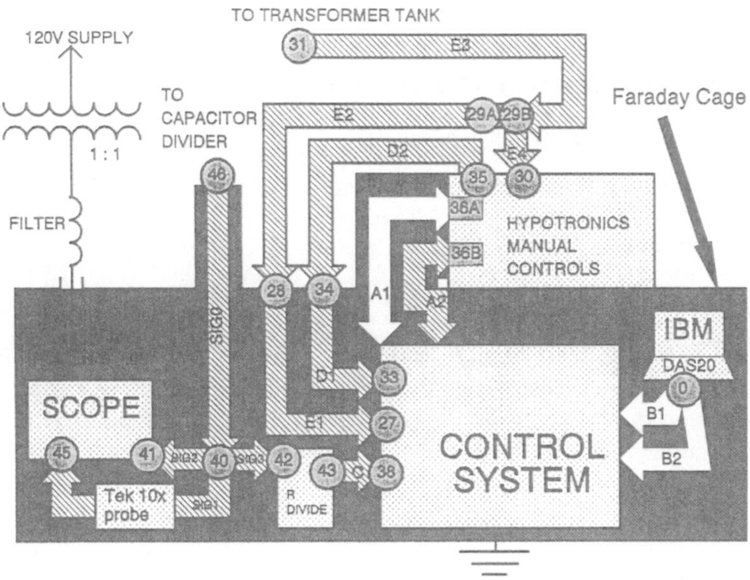

Figure 1. The Overall System Diagram.

3. **Results and Discussion** The output voltage signal from the main capacitor divider is between $0V$ and $400V$. Since this level of voltage is much too high for the logic circuitry, it is divided by a factor of 50, yielding a voltage range between $0V$ and $8V$. Thus the peak voltage output from the impulse generator is divided in the ratio $1 : 50,000$. Figure 2 shows the A/D system. The $DAS20$ box is a junction for all the *input/output* signals of the system. The analogue inputs have a minimum overvoltage of $\pm 35V$ and a minimum input current of $1nA$ (hence the need for rectifier protection circuits). The A/D converter has 12 bit resolution, and 7 software selectable instrument amplifier gain ranges. In FORTRAN, the analogue channels are read by a call to the function $DAS20$, which performs a single A/D conversion on the required channel.

(a) **The Digital Output System** The digital output byte (DOUT) from the $DAS20$, is fed by cables B1 and B2 into the $DAS20$ box, and then onwards to the demultiplexer board through cable V, Figure 3. The DOUT byte is manupulated in the same manner as any $DAS20$ operation. A call is made to the function $DAS20$ in the FORTRAN language. The digital output bits are individually referred to as bits DOUT 0 to DOUT 7 and denoted by O_0 to O_7. On the demultiplexer board, the DOUT bits O_0 to O_7 are sent through buffer chips.

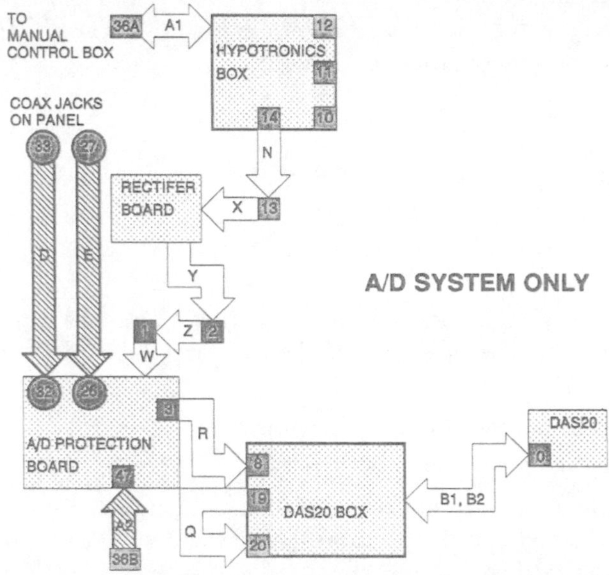

Figure 2. A/D Control System Diagram.

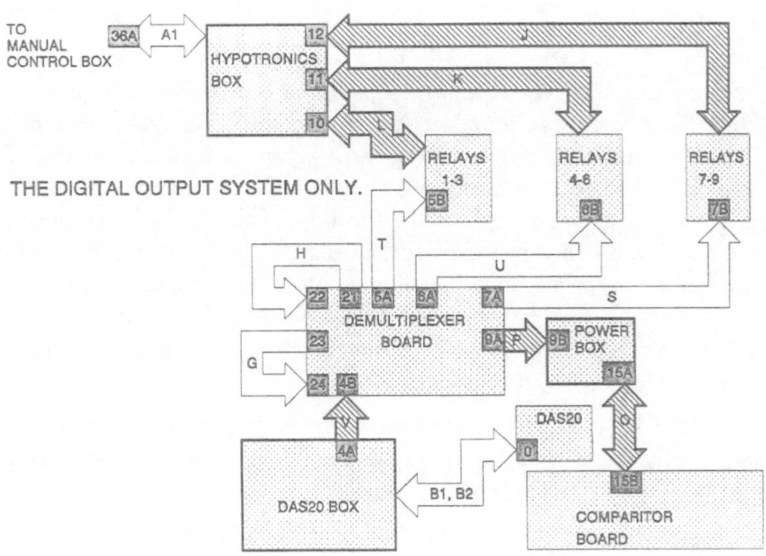

Figure 3. Digital Ouput System Diagram.

One buffer chip sends its output to the LED diplay of the control box through connector 25. The other chip feeds the bits to two $HC373D$ type transparent latches. These serve to demultiplex the DOUT bits. The controlling output bits of the $HC373$s are then sent to connectors 5A, 6A, 7A, and 9A. An overall diagram of this action is shown in Figure 4. The $HC373$ latches have a general output equation of the form:

$$\delta = \delta\overline{\alpha} + \alpha\gamma \tag{1}$$

where; α is the enable or pass bit
γ is the data bit
and δ is the control bit

(b) **Comparator Board** The comparator board utilizes both the digital input and output system. The operation of the comparator board is two fold:

i. differentiate between a breakdown and a withstand.

ii. determine the voltage range the impulse reached.

The comparator board is complex, but the need to understand its operation is critical because it differentiates between breakdowns and withstands, the whole basis for impulse testing.

(c) **Comparator operation** The comparator compares two voltages and outputs a logical level on the basis of comparision. One of the input signals can be a set DC reference voltage V_{ref}. The other can be an analogue waveform V_{in} that one wishes to measure. Figure 5 shows the outputs of three comparators in idealised form. By changing the response times of the comparators we have obtained two sets namely; fast comparator circuit and slow comparator circuit.

(d) **Fast comparator** The comparator will start to go from a high state to a low state like the output signal of V_{03a} in Figure 5. Since the peak of the pulse is a transient, the output voltage will again go to a high state. The result is a dip in the output voltage. If the dip is deep enough and lasts long enough, it will trigger the clock of the flip-flop. Therefore, the comparator creates a clock pulse that switches the flip-flop output Q from low to high when an impulse reaches above a reference setting. Figure 6 shows the actual result of a comparator response to a $0.220mV$, $284ns$ pulse. The slight $0.36V$, $200ns$ drop from $2.6V$ in the comparator output being enough to trigger the flip-flop.

(e) **The Slow Comparator** The slow circuit output starts at low and goes high similar to output V_{01} to V_{03} in Figure 5. An oscilloscope output showing this transition is shown in Figure 7.

(f) **Full Wave Generation and Detection** Identical waveforms are produced for a $1.2/50\mu s$ output peak voltages of up to $400KV$. Figure 8 shows the breakdown waveform of a 1cm gap in atmospheric air.

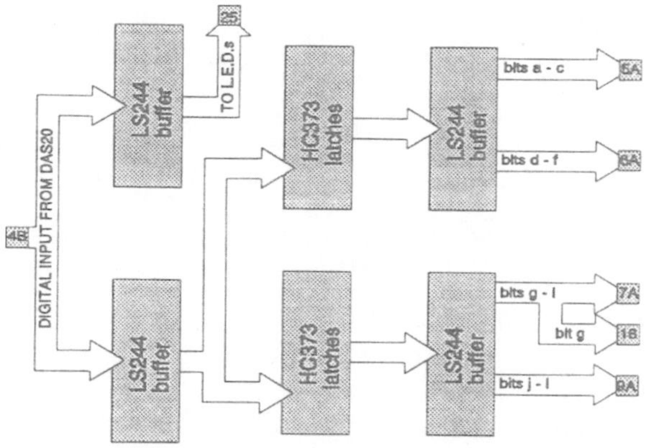

Figure 4. Demultiplexer Data Flow Chart.

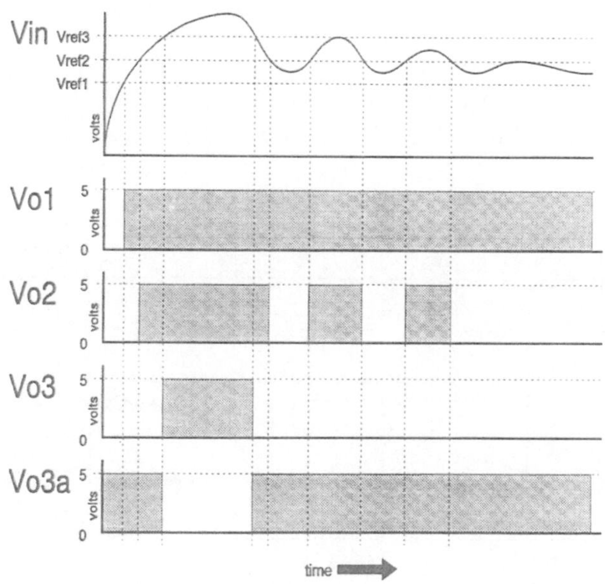

Figure 5. Comparator responses for varying references.

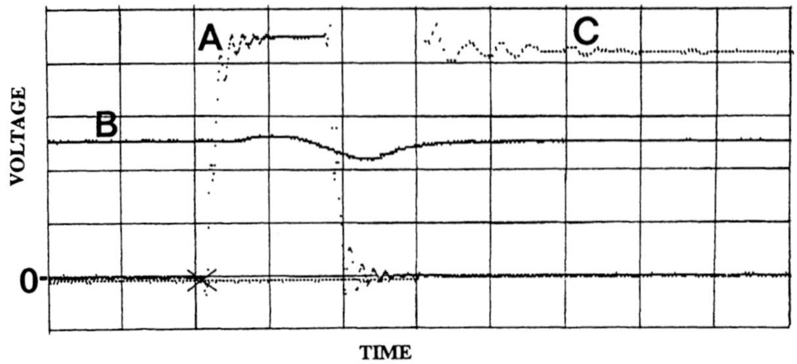

Figure 6. Examples of a fast comparator response.
(A) Input pulse to the comparator. Average peak: $220mV$, Duration: $280ns$ ($50mV/div$); (B) Comparator output showing a $0.36V$ dip from $2.6V$ ($1V/div$); (C) Flip-Flop output, achieving steady state at $4.32V$ ($1V/div$). There is a $590ns$ delay from start of input pulse to reaction of the Flip-Flop. Time Scale is $200ns/div$.

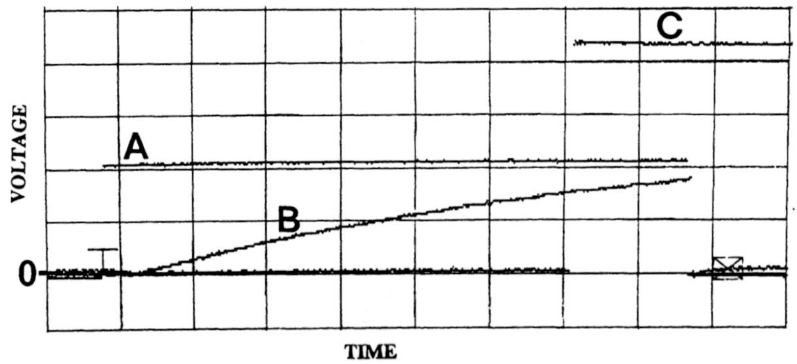

Figure 7. Example of the response of a slow comparator circuit.
(A) Input pulse to the comparator. Peak: $220mV$, duration: $39\mu s$. ($0.1V/div$). (B) Comparator output. ($0.5V/div$). (C) Flip-Flop output, achieving steady state at $4.28V$. ($1V/div$). The Flip-Flop responds, when The comparator output reaches about $0.8V, 31.5\mu s$ after the start of the pulse. Time scale is $5\mu s/div$.

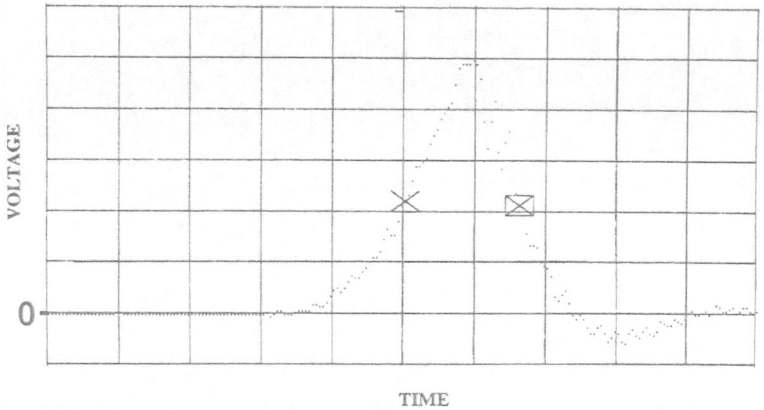

Figure 8. Breakdown of a $1cm$ gap from a $150KV$ applied impulse. This breakdown was detected by the control system. Note that the voltage waveform is above $11KV$ for $320ns$. Scale: $5KV/div., 200ns/div.$

References

[1] W.Eangauer, A.Johannsen, H.Lennartz, and H.G.Kranz. Erahrungen bei der automatisierung betreibs und der dalenerfassung von hochspannungs-versuchsanlagen. In *Int. Symp. on H.V. Zürich*, pages 311–315, 1975.

[2] K.Feser and R.Niederhauser. Trends towards automation of impulse voltage tests. In *Int. Symp. on H.V. Zürich*, pages 301–306, 1975.

[3] A.Fatheddin, V.K.Lakdawala, R.Y.Pai, and L.G.Christophorou. Automatic control of high voltage impulse generator. In *IEEE Southeastcon, Orlando, Florida*, pages 366–371, 1983.

SECTION 4: PARTIAL DISCHARGES/DIAGNOSTICS

PARTIAL DISCHARGE DIAGNOSTICS : A UTILITY PERSPECTIVE

R. Bartnikas

Institut de recherche d'Hydro-Québec (IREQ)
Varennes, Québec, Canada J3X 1S1

INTRODUCTION

It is now more than fifty years that partial discharge measurements have maintained a pre-eminent role as a non-destructive diagnostic test procedure for much of the electrical power apparatus acquired and placed in service by electrical utilities. The utilities have been particularly predisposed to the use of the partial discharge test procedures, which are capable of yielding data on the state of quality and aging process or deterioration rate of the electrical apparatus under service conditions. Since it has not been possible to relate either the apparent charge value of the discharge or the total dissipated discharge energy to the degradation rate or life expectancy of insulating systems exposed to discharges, it has been common practice to insist on the complete absence of partial discharges in electrical apparatus above some preset sensitivity level of the partial discharge pulse detectors in common use. For example with solid dielectric extruded cables, which are highly susceptible to discharge induced degradation, no discharges with apparent charge transfers greater than 5 pC are tolerated, whilst with oil-filled transformers the specifications are much more lenient by allowing an apparent charge level up to 500 pC at the test voltage.

It is desirable to monitor, in at least the larger and more critical apparatus, the partial discharge behavior in order to determine whether discharges appear and subsequently vanish or whether, if continually present, their intensity diminishes or increases with time; and, finally, whether their intensity is inordinately high so as to forebode imminent failure, so that the equipment can then be removed from service and either replaced or, if it is possible to locate the discharge sites, repaired. The form of discharge monitoring carried out depends on the sophistication of the partial discharge measuring equipment disponible[1,2]. The basic conventional RCL-type partial discharge detection circuit will indicate at least the presence or absence of discharge at some preset sensitivity level and will measure the peak discharge amplitude in apparent charge units[3]. More advanced partial discharge measurement systems with suitable pulse shaping circuits in conjunction with multichannel analyzers will be capable of providing partial discharge pulse-height, pulse interval and pulse phase distributions and an estimate of the total energy dissipated in the discharge process[4]. Use of computers will permit manipulation of the data, whereby three dimensional displays, of the aforegoing distributions with testing time as a parameter may be readily obtained.[5] Similarly suitable computer programs will allow storage of test data with time, such as variation of the average apparent pulse charge with time, thereby permitting to receive large amounts of

Gaseous Dielectrics VII, Edited by L.G. Christophorou
and D.R. James, Plenum Press, New York, 1994

discharge data for storage and subsequent plotting and interpretational purposes[6]. Since partial discharge patterns of pulse amplitude, shape, time interval separation and discharge epoch or phase bear some relationship to the type of discharge source and its location within the electrical apparatus structure of the specimen under test, a considerable amount of effort has gone into the application of intelligent computer systems such as neural networks[7,8] for discharge pattern recognition. The results are encouraging, but much work remains to be done to render the neural network techniques applicable with some degree of reliability on complex discharge patterns involving numerous and different discharge sources. However as concerns equipment operating in the field, one can make the general observation with some certainty that the majority of tests are intrinsically of a very simple nature and are confined to a measurement of the partial discharge inception and extinction voltages, the peak apparent charge and some form of intensity measure proportional to discharge magnitude and pulse density (as in the case of radio frequency and acoustical measurements). A notable exception, are tests on rotating machines, where pulse amplitude distribution constitutes the basis of the measurement.

Discharge source location techniques tend to center on acoustical methods[9] and rf probes[10], with some exception for cable specimens where time domain reflectometry correlation methods[11] and X-ray induced ionization procedures[12] have yielded some success. With unshielded cables the old Gooding train method[13] is still the most reliable cavity-defect locating procedure. In general, it can be said however, that the acoustical and radio frequency (rf) procedures have been the most successful approaches for locating cavities. But they tend to suffer due to their inherently low sensitivity and the difficulties associated with relating the acoustical or rf signal values to the apparent charge levels of the discharge pulses detected electrically.

The intent of this paper is to present a compendium of the various partial discharge test methods utilized on different types of specimens installed in the field. In this regard we shall consider each category of specimens separately, namely : SF_6 gas insulated systems, cables, capacitors, transformers and rotating machines.

SF_6 GAS INSULATED SYSTEMS

SF_6 gas insulated lines with solid epoxy spacers constitute perhaps one of the simplest form of test specimens in that they approach in their behaviour lossless transmission lines, thereby greatly facilitating the interpretation of the partial discharge data obtained. Very high frequency detection methods (350 MHz to 1 GHz), developed principally by Boggs and co-workers[14,15], are employed where the detection impedance Z_d and the coupling capacitor, C_c, are incorporated as parts of a transmission line to yield the required frequency response (cf. Fig. 1). Location of discharge sites in either SF_6 bus ducts or lines may be accomplished within 1 m by measuring the arrival times of the discharge pulses at two displaced coupling capacitors. However, in actual SF_6 gas insulated substations where there are a number of T-type joints, multiple reflections occur from the various joints and the method tends to become impractical. But it can be used to determine the partial discharge inception and extinction voltage readily even in the presence of excessive extraneous noise and interference from radio stations which have frequencies well below the lower limit of 350 MHz of the high frequency detection circuit. The rapid circuit detects only the apparent charge due to the electron transfer, so that only half of the charge is measured as the ionic contribution to the overall current is omitted; however, the total charge transfer, ΔQ, may be obtained from the relation,[14]

$$\Delta Q / 2 = (\pi)^{1/2} \Delta V_d . \Delta t / Z_d \qquad (1)$$

where Δt is the width of the pulse. The highest sensitivity obtained with this circuit is of the order of 0.1 pC, which is that approximately achievable with RCL-type conventional pulse detectors with small lumped capacitance specimens in the absence of extraneous noise. The sensitivity is appreciably reduced if the coaxial coupler, which is more convenient to install and use, is employed in lieu of the solid dielectric coupler (cf. Fig. 1).

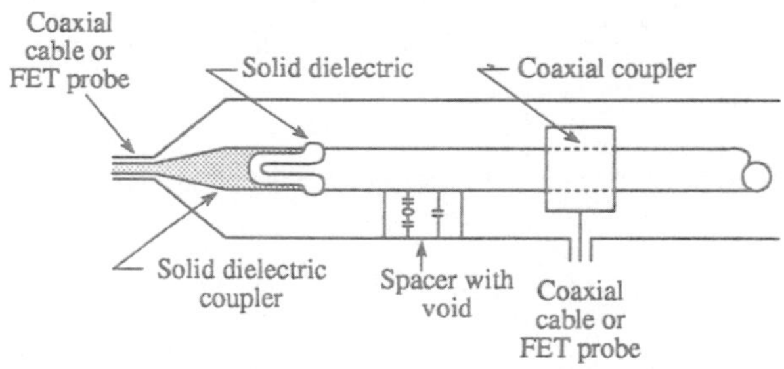

Figure 1. Couplers for partial discharge measurements on an SF_6 line specimen (after Boggs and Stone[14]).

In situ location of discharge sites in SF_6 gas insulated equipment can be also accomplished using acoustical detection techniques[9] (cf. Fig. 2)[16]. The acoustical probe is simply moved slowly along the surface of the SF_6 line or duct to establish points of maxima in the signal intensity. Compressed gas lines using SF_6 gas at 5 atm transmit ultrasonic signals much more efficiently than at 1 atm, though considerable attenuation occurs at the flanges and spacers; thus it is more expedient to scan the line in parts between flanges. The attenuation, α, between two spacers along an SF_6 line may be approximated by Kirchhoff's formula, which is valid for a tube geometry[9],

$$\alpha = 27.6 \times 10^{-5} \left(f^{1/2} / r \right) \tag{2}$$

where f is the frequency in Hz, r the radius of the tube in cm and the attenuation factor, α, in dB/cm. Although the attenuation is much lower at audio frequencies (af) (≤ 20 kHz), higher frequencies must be used because of the high ambient interference over the af band; hence measurements are carried out in the ultrasound region over which acoustic emission probe-sensors are ideally suited. In general acoustic methods are less sensitive than electrical methods, yielding sensitivity values *ca.* 10 to 25 pC.[9] However recent claims of sensitivity as high as 2 pC have been made in conjunction with an ultrasonic detector operating at a frequency of 50 kHz.[17] The ultrasound intensity is proportional to both the discharge amplitude as well as the discharge pulse density with the consequence that an appreciable response of the ultrasonic detector to very low intensity discharges may imply a very dense discharge pattern, consisting of numerous minute pulses. In contradistinction an electrical partial discharge detector could quite readily detect a single very rapid pulse due to a 1 or even 0.1 pC apparent charge transfer.

The presence of ultrasound in SF_6 lines is not necessarily indicative of partial discharges. Often a variation in voltage is employed to modulate the discharge intensity for ascertaining whether the acoustic response is due to a discharge source; however, such variation in the applied voltage level is not feasible under operating conditions. Perhaps one of the main difficulties that afflicts acoustic detection is that it is not always possible to distinguish between the acoustical signals caused by partial discharges and those due to the movement of metallic particles; also, the acoustical method is relatively insensitive to discharges emanating from cavities occluded within the epoxy spacer insulators themselves.

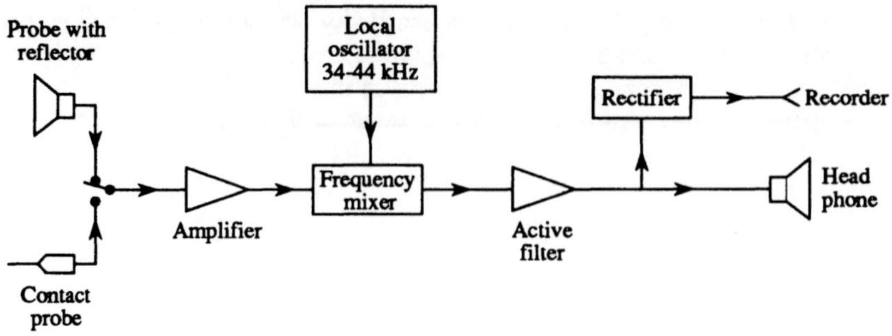

Figure 2. Schematic circuit diagram of a typical commercial ultrasonic detector (after Morin *et al*[16]).

CABLES

Most of the partial discharges measurements in the field are performed primarily on solid dielectric extruded distribution cables, which operate within the voltage range of 5 to 35 kV.[18] In the early days, the tests were carried out at 60 Hz, using an RCL-type balanced partial detection circuit in order to reduce the effects of extraneous noise. A balanced system is limited to tests on polyphase circuits, whereby one phase of the cable is balanced against one of the other two such that the partial discharge signals are always measured from the cable phase having the lower discharge inception voltage and the higher discharge pulse magnitude. Thus, by a comparison of all three phases, the cable phase with the highest discharge intensity is readily determined. An appreciable portion of the interference originating in the remainder of the power system is transmitted along the cable specimens themselves, which are connected to it. In order to eliminate the conducted interference, it is desirable to disconnect the cables from the system and test them with an independent voltage supply. To circumvent the requirement of a high power capacity source required for the relatively large cable specimen capacitance, it is common practice to employ in conjunction with either a balanced or straight RCL-type detector a portable low frequency high voltage supply operating typically at 0.1 Hz as delineated in Fig. 3.[19,20] Here the inductances L_1 and L_2 isolate the cables C_x under test and filter the power supply noise; the high voltage discharge-free capacitors C_1 and C_2 isolate the low voltage portion of bridge from the cables; R_1, R_2 and C_3 comprise bridge arms for balancing. The discrete discharge pulses are indicated on the oscilloscope and the peak apparent charge on the meter. Long lengths of cable up to 1000 ft have been measured using the 0.1 Hz power source and the determined values of the partial discharge inception voltage and peak discharge pulse amplitudes were found to be essentially the same to those obtained with a 60 Hz power source. This non-destructive test procedure has been found to be effective in ascertaining deterioration of cables installed in the field due to shipment and faulty installation, separation of the semiconducting shields from the insulation and development of cavities due to load cycling and faulty joints and terminals.[19]

When a cable fails a partial discharge test, it is desirable to locate and establish the source that gives rise to partial discharges. It is possible to remove the cable and measure the discharge intensity variation as the cable is passed along an X-ray source, which enhances discharge activity in the existing voids by a provision of copious free electrons necessary to initiate discharge formation[12]. Alternatively the insulation shield of the cable may be removed and the cable scanned in a Gooding train system, that consists of an insulating pipe filled with dionized water with a hollow HV scanning electrode at the center and grounded tanks at each end. But the latter, although perhaps constituting the most reliable cavity locating system , has two inherent disadvantages, namely; firstly, it is only

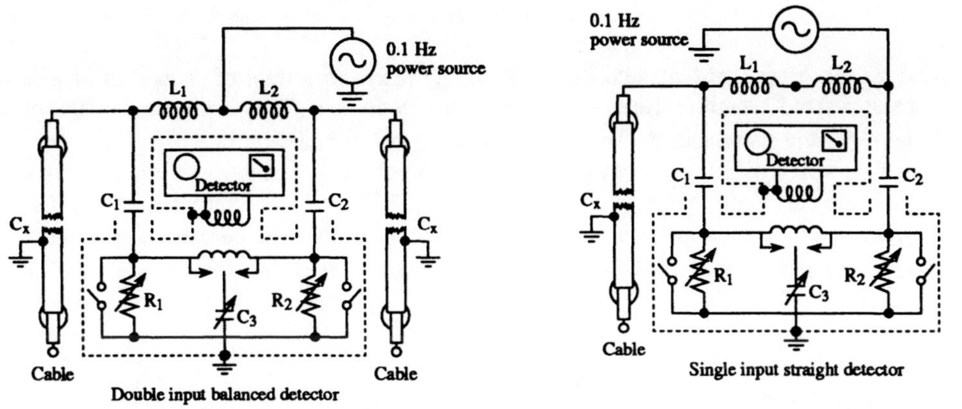

Figure 3. Schematic circuit diagrams of a low frequency energized double input balanced and single input straight-type partial discharge bridge-detectors.[19,20]

suitable for locating cavities within the insulation and between the insulation and the semiconducting conductor shield, and, secondly, the scanned cable is not suitable for re-use.

Various time delay pulse reflectometry techniques have been proposed, and some of these utilizing digital circuits and various correlation procedures have shown promise[11,21], but only one of these[21] was successfully applied on cables in the field. The latter method, using a digitized and computerized time domain technique, was found to function well with a small number of concentrated discharge sites involving cavities and electrical trees. However, at the moment it appears that the most prevalent and effective test procedure for locating discharge faults on cables in the field is based on the rf probe technique. Acoustical methods have been less successful because of the high attenuation of ultrasound through the outer layer of the cable formed by the semiconducting shield, outer conductor and sheath and, finally, armor if used. Perhaps the most successful method for locating cavities in polymeric cables, splices and joints is the radio frequency (rf) technique developed by Morin *et al*[16]. It operates in the frequency range from 5 to 10 MHz, well above the low frequency moving contact interference spectrum, the radio broadcast band and other normal communication frequency bands; its basic circuit is depicted in Fig. 4. Two types of probes are utilized : one is a capacitively coupled - bidirectional contact probe consisting of a small metallic copper plate of rectangular configuration covered with a dielectric layer to avert direct contact with the concentric neutral of the cable under test as well as other grounded metallic conductors in its vicinity. The overall area and thickness of the dielectric-covered-copper contact probe is adjusted to suit the sensitivity required in the measurements. The second probe is a vertical omnidirectional probe and consists of a rod with a copper conductor. The latter is a general test probe, which is employed to determine whether there is any discharge activity present in a cable specimen -- *i.e.* it is passed through the openings of manhole covers to sense the presence of partial discharges over the entire cable. If discharge activity is detected, the contact probe is then employed to scan the surface of the cable. The output from the probes is applied to the input of the rf detection circuit which comprises an attenuator in tandem with an amplifier, filter, frequency mixer and detector with meter. As the contact probe, mounted on a 1 m long insulating stick, is moved along the cable, maxima in the response of the meter indicate partial discharge sites along the cable, splices and joints.

The output of the rf partial discharge detector in decibels (dB) is related to the apparent charge in the discharge pulse, ΔQ, in pC by the empirical relation,[16]

$$dB = K \ln(\Delta Q) \pm b \quad (3)$$

where K and b are constants and the value of ΔQ refers to a train of calibration discharge pulses at a fixed repetition rate -- *i.e.*, the meter readout in dB is a function of the pulse charge, ΔQ, and its repetition rate. At a calibration pulse repetition rate of 120 per second, the rf detector sensitivity was established to be 5 pC. It is palpably evident that since in practice, the partial discharge pulse patterns consist of pulses of different charge transfers, ΔQ, with their own intrinsic repetition rates, a direct calibration between the rf detector output in dB and the average detected discharge charge transfer ΔQ is not possible.

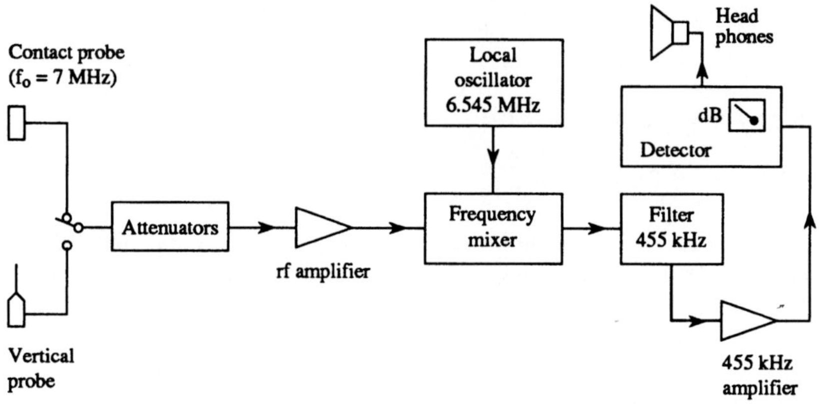

Figure 4. Schematic diagram of an rf partial discharge detector (after Morin *et al*[16]).

TRANSFORMERS

It is common to carry out partial discharge tests on some power transformers by simply applying the voltage across the low potential winding, so that essentially the voltage stress is developed between the individual turns and sections of the windings as in service. Such tests are referred to as induced voltage tests, whereby the need of a discharge free coupling capacitor is obviated by utilizing the capacitor bushing tap in service[10]. With larger transformers, however, it is common practice to use the third harmonic (180 Hz) of the power frequency voltage in order permit an overvoltage test on the transformer without saturating the magnetic core and thereby causing damage to the transformer (cf. Fig. 5). Thus partial discharge tests on high power transformers in service necessarily involve the transport of mobile generators for 180 Hz or sometimes 400 Hz operation. For small transformers, a separate 60 Hz power supply must also be transported, in order to isolate the transformer from the rather high conducted interference originating from the remainder of the system, which generally occurs well within the resonant frequency regimes of the RCL-type discharge detectors used for transformer tests (*ca.* 40 to 200 kHz). In this test the insulation between the high voltage winding and the grounded low voltage winding, core and tank is stressed (cf. Fig. 6). Transport of mobile power sources constitutes a costly proposition and, consequently, electrical *in situ* tests on power transformers do not represent a common practice with most utilities; they are generally only carried out after repair work or to verify whether a given transformer has been damaged during transport or installation. The transformer is considered to have failed if the detected peak partial discharge amplitude exceeds 500 pC. Some utilities may also require a second type of discharge test, namely the radio influence voltage test (RIV). The RIV detection circuit differs from the usual type of RCL partial discharge detection circuit in that the RCL detection impedance is replaced with a 600 Ω resistor and a radio noise meter, tunned at a frequency anywhere within the range of 0.85 to 1.15 MHz, is employed. Suitable weighting circuits are employed and the detected signal is expressed in μV quasi-peak values. The transformer is considered to have failed the RIV test if the reading exceeds 100 μV.[22] The RIV value in μV is difficult to interpret physically as its relationship to the apparent charge value in pC is complex, because it is dependent upon the discharge pulse amplitude and its repetition rate. Historically, the RIV measurement had its beginnings in the early concerns on the radio interference produced by electrical apparatus.

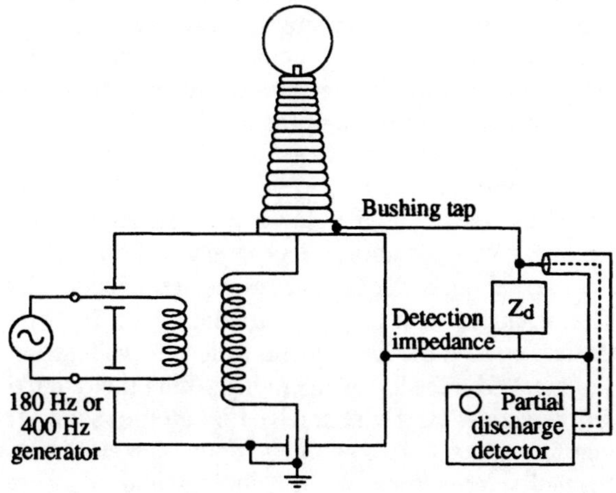

Figure 5. Schematic circuit arrangement of an induced voltage partial discharge test on a power transformer specimen.

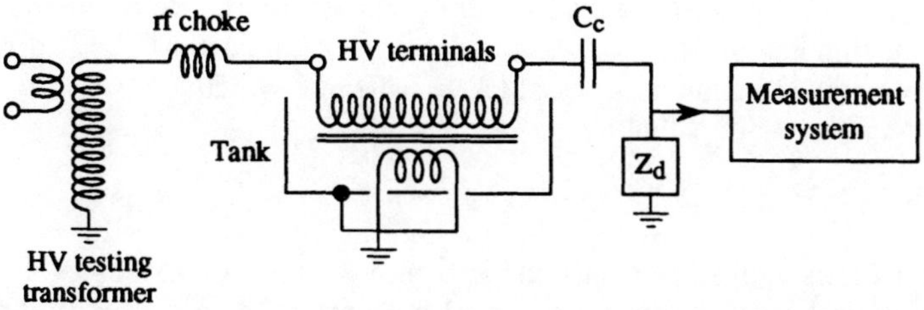

Figure 6. Schematic circuit diagram for a discharge test on a small transformer, using a separate 60 Hz high voltage test source.

Electrical partial discharge monitoring in transformers is not a common practice, though there has been some work reported on current transformers where the discharge behaviour was monitored using dissipation factor measurements[23]. In another study, a partial discharge monitoring technique was implemented on a bank of current and potential transformers in a substation[24]. Here, the extraneous noise pulses were distinguished from those occurring in the instrument transformers in terms of their polarity and thus eliminated. A central processor was employed to perform pulse polarity recognition and record the discharge repetition rate of the actual partial discharge pulses emanating from each unit.

Routine monitoring of partial discharges as well as the location of discharge sites in transformers under operating conditions is normally accomplished by means of acoustical techniques. Historically, acoustical methods for the detection and location of partial discharges were first developed and applied to transformer specimens by Anderson[25] in the 1950's. In a number of respects oil-filled transformers constitute ideal specimens on which acoustical techniques may be applied to extract information on the location of partial discharge sites. Little acoustical attenuation takes place in the liquid dielectric and, compared to a gas medium, an order of magnitude more of the discharge energy is converted to acoustical energy[9]. A triangulation method developed by Anderson (*loc. cit.*) involves the use of three acoustical sensors located at positions selected by trial and error on the face of the transformer tank. The arrival of the first acoustical pulse is employed to trigger an oscilloscope so that the pulse arrival times at the two remaining transducers can be compared and an estimate made of the position of the partial discharge fault. It can be appreciated that difficulties in discharge site location may arise with transformers that contain numerous discharge sites. An alternate method developed by Allan *et al*[26] is based on the positioning of two transducers on a locus of points along a straight line of a transformer tank and then recording the position of the acoustical sensor that receives the acoustical signal first, whereby the location of the discharge fault may be estimated.

The ultrasound techniques for in service transformers must be capable to operate in high ambient noise environments -- *e.g.*, in the vicinity of HV transmission lines and substations, which are characterized by high levels of Barkhausen (magnetostriction) noise[9]. Since the frequencies of Barkhausen noise are generally centered around 40 kHz, acoustical discharge detection is carried out at higher ultrasound frequencies; Train *et al*[27] have made use of an ultrasonic transducer with an upper frequency of 200 kHz, whose output was applied to a 10 kHz high pass filter in tandem with an oscilloscope, while Howells and Norton[28] employed ultrasound transducers operating within the range of 100 to 150 kHz.

The intensity of the acoustical signal received by an ultrasound transducer situated on the surface of a transformer tank is contingent upon the electrical discharge pulse and acoustical wave attenuation properties of the windings and structure of the transformer. Common sensitivity values are of the order of 70 pC/mV of the detected ultrasound signal. The sensitivity is augmented and the influence of external noise diminished if the transducers placed on the surface of transformer tanks are immersed in an oil medium; this reduces appreciably the reflections of the acoustical waves at the interface between the enclosure tank and the transducer, thereby enhancing transmission of the detected signal. However, there are still outstanding difficulties with the detection of acoustical signals from discharges occurring deep within the inner sections of the transformer winding.

CAPACITORS

Capacitors act as lumped elements and in this respect represent a category of specimens on which partial discharge tests may be carried out with relative ease with the proviso that they be of low capacitance. If C_p designates the major portion of the capacitance of the capacitor, which is much greater than the capacitance of the series combination formed by that of the cavity in series with the remaining dielectric, then in terms of the peak detected discharge pulse magnitude, ΔV_d, the associated apparent charge transfer is given by

$$\Delta Q = \Delta V_d \cdot C_p \tag{4}$$

Thus, the detected voltage pulse magnitude, ΔV_d, is a monotonically decreasing function of the specimen capacitance. As a consequence, discharge detection by means of electrical methods in capacitances in excess of 0.5 μF becomes exceedingly difficult. In this regard, acoustical measurements not only represent a viable approach in service areas with high extraneous noise environments, but constitute perhaps also one of the most effective techniques for partial discharge detection in high capacitance specimens.

It has been established that ultrasound methods are capable of detecting discharges in capacitances as high as 40 μF.[9] Coupling of acoustical sensors to the casing of capacitors may be readily achieved by the use of glycerin. Capacitances equal to or greater than 1 μF, in small case sizes, are most effectively tested using acoustic detectors with resonance at 20 kHz. Whereas resonance frequencies within the regime from 60 to 80 kHz are found to be more efficient on capacitors with large container sizes. Apparent charge transfer sensitivities range from 10 to 20 pC.

ROTATING MACHINES

Rotating machines, in particular hydrogenerators, are perhaps the most monitored of all apparatus in electrical utilities. This is a direct consequence of the successful application of the balanced partial discharge detection principle to the high frequency case by Stone and co-workers[29,30] and the rf monitor technique[31] based on the an earlier technique[32] for partial discharge pulse detection at the neutral point of a generator. In the method of Stone *et al* (*loc. cit.*) delay lines are utilized in conjunction with a balanced partial discharge measurement scheme as portrayed in Fig. 7 : two permanent couplers having a capacitance of 80 pF are installed at the ends of the ring buses that are connected to each phase of a waterwheel generator. If x and r represent the respective circuit bus lengths while y and s are the respective coaxial cable lengths, the balanced measurement condition stated in transmission line context requires that[33],

$$x+(y/0.65)=r+(s/0.65) \tag{5}$$

where the factor 0.65 is introduced to take into account the reduced velocity of propagation in the solid polyethene dielectric coaxial cables as compared to free space. The coupler pairs denoted by capacitances C_1 and C_2 in Fig. 7 with the respective coaxial lines and bus bar lengths are thus matched in their equivalent electrical lengths at the input of a differential amplifier, where the incident interference pulses (arriving from the machine with equal times of travel) are annihilated in the differential amplification mode. Whereas discharge pulses occurring in a phase close to one of the couplers are transmitted by the differential amplifier. A pulse height analyzer, in tandem with the differential amplifier, provides a discharge pulse-height (in mV) distribution characteristic of the discharge process within each phase, which when analyzed as a function of time yields some indication on the deterioration rate of the stator bars. A different approach is employed for partial discharge monitoring on turbine generators, where the ring buses are either insufficient in length or entirely absent[30]. Here electromagnetic couplers are installed beneath the stator wedges; these particular couplers respond only to the rapid discharge pulses that occur in their vicinity, but act as rejection filters to the slower pulses due to extraneous interference that are arriving with highly degredated rise times. All discharge pulse amplitudes are recorded in mV, since no suitable calibration procedure has yet been developed, thereby preventing a direct comparison of the discharge intensity between machines with different winding characteristics.

The radio frequency method, which is suitable for both waterwheel and turbine generators and which measures the rf signal at the neutral point across a high frequency current transformer is based on an entirely different principle[31]. Here the partial discharge spectrum is recorded over a wide frequency range with certain frequency bands being associated with different forms of discharge such as slot discharges and surface corona at the coil-ends. External ambient interference is rejected by the tunning mode of the measurement.

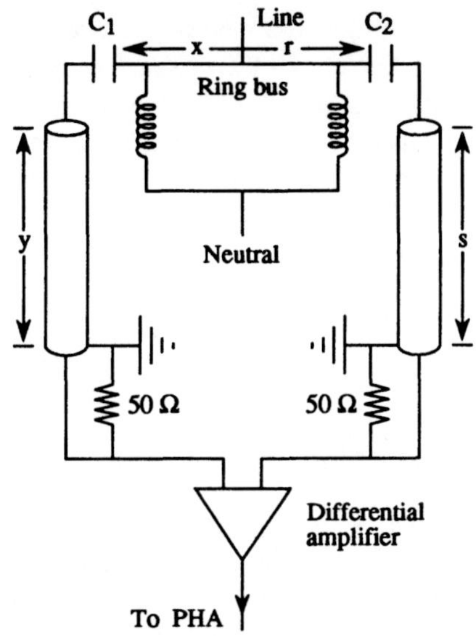

Figure 7. Balanced permanent coupler connections for a waterwheel generator[29,30,33].

Once the partial discharge characteristics of the machines are determined by either the balanced coupler technique or the rf method, it remains incumbent to locate the discharge sites responsible for the detected discharge activity. Location of the discharge sites requires the removal of the rotor so that access with an rf probe may be made to permit scanning of the individual bars in the stator; this is normally carried out using an inductive probe of the type described by Dakin *et al*[34]. The probe operates at a frequency of 5 MHz in conjunction with a quasi-peak voltmeter indicator and is designed to be more sensitive to repetitive discharge pulses rather than the sporadic bursts of interference. However this procedure may fail to locate all of the slot discharge sites, since some of the slot gaps may only exist during the electromagnetically induced bar vibrations whilst the machine is in operation.

REFERENCES

1. R. Bartnikas, "A commentary on partial discharge measurement and detection", *IEEE Trans. on Electrical Insulation*, EI-22 : 629 (1987).

2. R. Bartnikas, "Detection of partial discharges (corona) in electrical apparatus", *IEEE Trans. on Electrical Insulation*, 25 : 111 (1990).

3. E.B. Curdts, Chapter 3, Fundamentals of partial discharge detection : System sensitivity and calibration in : "Engineering Dielectrics, Vol. I, Corona Measurement and Interpretation", R. Bartnikas and E.J. McMahon, eds., STP 669, ASTM, Philadelphia (1979).

4. R. Bartnikas, "Use of multichannel analyzers for corona pulse-height distribution measurements in cables and other electrical apparatus", *IEEE Trans. on Instrumentation and Measurement*, IM-22 : 403 (1973).

5. G. Vailancourt and R. Malewski, "Digital acquisition and processing of partial discharges during acceptance tests of HV transformers", *IEEE Trans. on Power Delivery* 4 : 421 (1989).

6. C. Hudon, R. Bartnikas and M.R. Wertheimer, "Spark to glow discharge transition due to increased surface conductivity on epoxy resin specimens", *IEEE Trans. on Electrical Insulation*, 28 : 1 (1993).

7. H.-G. Kranz, "Diagnosis of partial discharge signals using neural networks and minimum distance classification", *Ibid.*, p. 1016.

8. Amira A. Mazroua, M.M.A. Salama and R. Bartnikas, "PD pattern recognition with neural networks using the multilayer perceptron technique", *Ibid.*, p. 1082.

9. R.T. Harrold, Chapter 10, Acoustical techniques for detecting and locating electrical discharges in "Engineering Dielectrics, Vol. I, Corona Measurement and Interpretation", R. Bartnikas and E.J. McMahon, eds., STP 669, ASTM, Philadelphia (1979).

10. T.W. Dakin, Chapter 6, Measurement of partial discharges in inductive apparatus : transformers and rotating machines, *Ibid.*

11. J.P. Steiner, P.H. Reynolds and W.L. Weeks, Estimating the location of partial discharges in cables, IEEE Trans. on Electrical Insulation, 27 : 44 (1992).

12. A.H. Baghurst, A new method for the location of partial discharge sites using modulated X-rays, *1985 Annual Report, Conf. on Electrical Insulation and Dielectric Phenomena*, IEEE Conf. Record 85CH 12165-9 : 471, Buffalo (Oct. 20-24, 1985).

13. F.H. Gooding and H.B. Slade, Corona level scanning of high voltage power cables, *AIEE Trans. on Power Apparatus and Systems*, 76 : 999 (1957).

14. S.A. Boggs and G.C. Stone, Fundamental limitations in the measurement of corona and partial discharge, *IEEE Trans. on Electrical Insulation*, EI-17 : 143 (1982).

15. S.A. Boggs, G.L. Ford and R.C. Madge, Coupling devices for the detection of partial discharges in gas insulated insulated switchgear, *IEEE Trans. on Power Apparatus and Systems*, PAS-100 : 3969 (1981).

16. R. Morin, A. Gonzales, P. Christophe and B. Poirier, Partial discharge detector for in service cables and joints under normal network operating conditions, *Proc. IEEE Transmission and Distribution Conference*, 23 : Dallas, (Sept. 23-28, 1991).

17. L.E. Lundgaerd, G. Tangen, B. Skyberg and K. Faugstad, Acoustic disgnosis of gas insulated substations : a theoretical and experimental basis, *IEEE Trans. on Power Delivery*, 5 : 1751 (1990).

18. N.B. Timpe and S.V. Heyer, Laboratory and field partial-discharge studies by a utility, *IEEE Trans. on Electrical Insulation*, EI-12 : 159 (1977).

19. G. Bahder, G.S. Eager, R. Suarez, S.M. Chalmers, W.H. Jones and W.H. Mangrum, In service evaluation of polyethylene and crosslinhed polyethylene insulated cables rated 15 to 35 kV, *IEEE Trans. of Power Apparatus and Systems*, PAS-96 : 1754 (1977).

20. N.B. Timpe, Chapter 4 Partial discharge measurements in distributed parameter systems : cables, in "Engineering Dielectrics, Vol. I, Corona Measurement and Interpretation", R. Bartnikas and E.J. McMahon, eds., STP 669, ASTM, Philadelphia (1979).

21. M.S. Mashikian, R. Luther, J.C. McIver, E.J. Jurcisin and P.W. Spencer, Evaluation of field aged crosslinked polyethylene cables by partial discharge location, Paper 93SM363-2-PWRD, IEEE PES Summer Power Meeting, Vancouver, July 18-22, 1993.

22. *Revised IEEE Standard 62*, "Guide for Diagnostic Field Testing of Power Apparatus".

23. H.B. Cummings, J.R. Boyle and B.W. Arp, Continuous on line monitoring of free standing oil-filled current transformers to predict imminent failure, *IEEE Trans. on Power Delivery*, 3 : 1776 (1988).

24. M. Kurtz, G.C. Stone, P. Daechsel and B.K. Gupta, Fault anticipator for substation equipment, *IEEE Trans. on Power Delivery*, 2 : 722 (1987).

25. J.G. Anderson, Ultrasonic detection and location of electric discharges in insulating structures, *AIEE Trans. on Power Apparatus and Systems*, 75 : 1193 (1956).

26. D.J. Allan, J.A.C. Forrest, E.L. Howitt and A.B. Petchell, Electrical and acoustical location of discharges in transformers, *IEE Conf. on Diagnostic Testing of High Voltage Power Apparatus in Service*, Publication No. 94, Part. I, p. 64 (March 6-8, 1973).

27. D. Train, A. Mercier and D. Thorne, Detection of partial discharges in high voltage potential transformers, *IEEE Trans. on Power Apparatus and Systems*, PAS-93 : 1909 (1974).

28. E. Howells and E.T. Norton, Detection of partial discharges in transformers using acoustic emission techniques, *IEEE Trans. on Power Apparatus and Systems*, PAS-97 : 1538 (1978).

29. G.C. Stone, H.G. Sedding, N. Fujimoto and J.M. Braun, Parctical implementation of ultrawideband partial discharge detectors, *IEEE Trans. on Electrical Insulation*, 27 : 70 (1992).

30. H.G. Sedding, S.R. Campbell, G.C. Stone, G.S. Klempner, A new sensor for detecting partial discharges in operating turbine generators, *IEEE Trans. on Energy Conversion*, 6 : 700 (1991).

31. J.E. Timperly, Incipient fault identification through neutral rf monitoring of large rotating machines, *IEEE Trans. on Power Apparatus and Systems*, PAS-102 : 693 (1983).

32. J.S. Johnson and M. Waren, Detection of slot discharge in high-voltage stator windings during operation, *AIEE Trans. on Power Apparatus and Systems*, 70 : 1998 (1951).

33. J.C. Bromley and W. McDermid, Application of the CEA partial discharge analyser, *Doble Conference*, Boston (April 20, 1982).

34. T.W. Dakin, C.N. Works and J.S. Johnson, An electromagnetic probe for detecting and locating discharges in large rotating machine stators, *IEEE Trans. on Power Apparatus and Systems*, PAS-88 : 251 (1969).

DISCUSSION

I. HERBST: In SF_6, with fixed protrusions, the slow signal component (ion contribution) can be 2 to 12 times more than that of the fast pulse (electronic) depending on the pressure and the geometry. The physics of the discharge is such that if you take the full spectrum down to the kilohertz region, you will pick up much more signal with the slower technique.

R. BARTNIKAS: Dr. I. Herbst has correctly pointed out that the actual apparent charge, ΔQ, measured by the EHF method is often very substantially less than that given by Eq. (1), which gives the theoretically expected charge transfer value due to the electronic component in the PD current pulse. This indeed constitutes one of the major disadvantages of the EHF technique as concerns both the accuracy and sensitivity, since it is only with the low frequency RCL-type circuits that generally operate within the resonant frequency range of 30 to 200 kHz (and that behave essentially as charge integrators) that the total apparent charge transfer due to a discharge pulse may be determined. During this measurement the GIS line may have to be disconnected from the remainder of the network to eliminate conducted interference effects.

R. J. VAN BRUNT: In your paper, you cite a recent paper by Mashikian et al. [21] in which a pulse time reflectometry method was used to locate partial discharge in a practical cable. Could you comment on the success and promise of this method?

R. BARTNIKAS: It is to be noted that the method by Mashikian et al. (see Ref. 21 in my paper) was found to perform well on distribution cables in the field that were found to contain only a few isolated discharge sources along their length. The effectiveness of the time reflectometry method still remains to be evaluated under more stringent conditions on cable specimens, which may contain greater numbers of distributed cavities along their length. The results should then be compared to data obtained with the rf probe method to establish reproducibility and accuracy criteria.

O. FARISH: You mentioned that the HF electrical method is not useful for locating discharges in practical GIS with "T" connections. Hampton and coworkers [B. F. Hampton et al. CIGRE 1992, report of WG 15/23-03 "Experience and progress with UHF diagnostics in GIS"; B. F. Hampton, R. J. Meats, "Diagnostic Measurements at UHF in GIS," Proc. IEE, 135, 137-144, 1988] have shown that it is possible to locate PD to within 20 cm even in practical GIS, using multiple UHF couplers.

A. H. MUFTI: Most of the electrical utilities and manufacturers try to use and to develop the UHF method in the GIS. Can you comment on this method? What is the promising method for future application to detect and to locate PD in the GIS system?

R. BARTNIKAS: It is correct that utilities and manufacturers are employing UHF methods in the detection of partial discharges in GIS systems. As pointed out also by Prof. O. Farish, PD site location even in GIS systems with many interconnections and "T" joints may be achieved within 20 cm -- though this requires the use of a large number of couplers situated at various strategic points along the GIS system. UHF methods are thus quite capable of locating PD sites even in highly interconnected GIS systems and the limitation is only practical in regard to the degree of complexity of the PD measurement system itself. In contradistinction, the acoustical procedure for the location of the PD discharge sites is much simpler, but suffers in the sense that the quantities measured are difficult to relate to the pulse apparent charge values in pC.

Ideally, the PD site location procedure should involve the use of both the electrical and the acoustical methods so that the position of the PD sites determined by both methods would be compared and the discharge magnitude at the site recorded in apparent charge units (pC).

K. FESER: Can you get 0.1 pC sensitivity with the electrical method on a complete SF_6 installation on site?

R. BARTNIKAS: That level would be fairly difficult to achieve because we would have some reflection effect. As far as the interference is concerned, once you are operating at these very high frequencies, you are above most of the interference. The other problem is calibration.

PD RELATED FIELD ENHANCEMENT IN THE BULK MEDIUM

A. Pedersen[1], G.C. Crichton[2] and I.W. McAllister[2]

[1]Physics Department, Building 309
[2]Electric Power Engineering Department, Building 325
The Technical University of Denmark
DK-2800 Lyngby, Denmark

INTRODUCTION

The occurrence of a partial discharge in a void embedded in a dielectric leads to the accummulation of charge on the wall of the void. This charge gives rise not only to the familiar PD transient through electricification of the detecting electrode by induction,[1] but also brings about a field enhancement within the bulk dielectric in the vicinity of the void.[2] In the present paper we discuss the evaluation of this field enhancement from measurement of the PD transients.

VARIATION WITH TIME OF THE PD-TRANSIENT

Space-Charge Fields and Induced Charge

Let us assume that the electric field within a small gas-filled void first attains the partial discharge inception level E_i at exactly the peak of the applied voltage. Prior to this no free charge is assumed to be present within the void. The field strength E_v within the void will therefore follow the field E_a associated with the applied voltage in the time interval from $0 \leq t < T/4$, where T is the duration of the period of the applied voltage; see Figure 1. The first discharge occurs at $t = T/4$ and E_v drops to E_ℓ because of the space charge field which now exists within the void. E_ℓ represents the limiting field strength for the void gas; i.e. the field value below which ionization growth is impossible for the gas pressure in question. The collapse of the field to E_ℓ is depicted in Figure 1 as an instantaneous event as the formative time lag of the discharge may be considered as negligible with respect to the period of the applied voltage.

The space charges now present within the void establish almost immediately electricification by induction on the detecting electrode of the system. This means that at $t = T/4$ a Poissonian induced-charge of magnitude q_1 is created at this electrode and that it subsequently remains at this value until the next discharge occurs; see Figure 1. As discussed previously,[1] there are two induced-charge components; *viz.* a Poissonian and a Laplacian component. The latter, which is equal in magnitude but opposite in polarity to the former, is the directly

Gaseous Dielectrics VII, Edited by L.G. Christophorou
and D.R. James, Plenum Press, New York, 1994

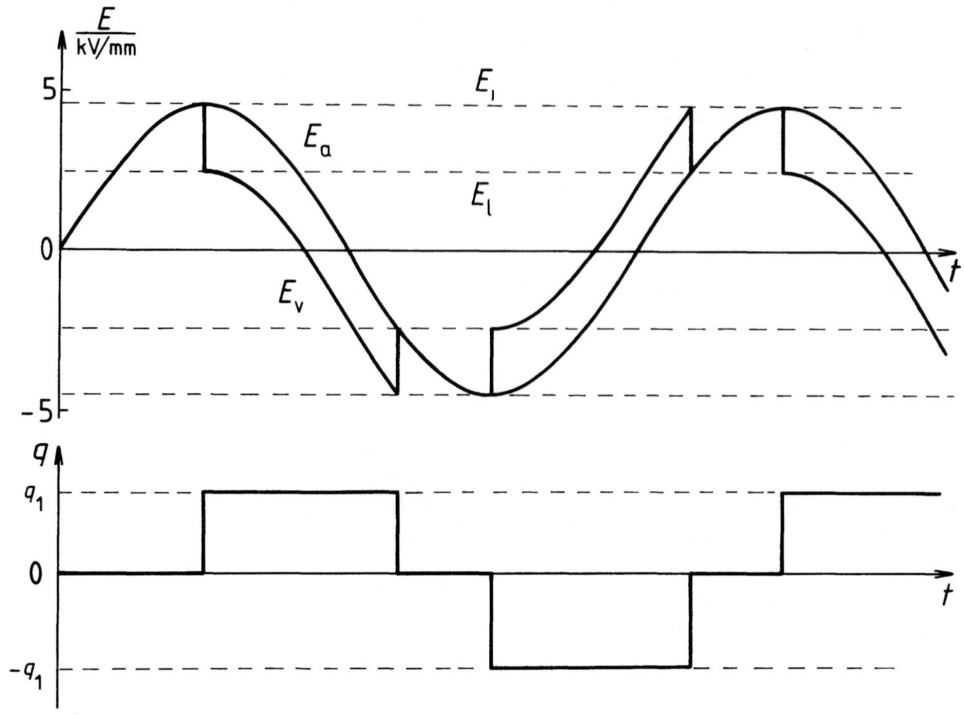

Figure 1. Electric field within the void and the corresponding induced-charge for N = 2.

measurable quantity. However, unless discrimination is essential, we will use only the common term induced-charge.

The space charge field of magnitude (E_i - E_ℓ) which now exists within the void will continue to oppose the applied field inside the void until the applied field becomes negative. Sometime before the first negative peak of the externally applied field is reached, the now space-charge-augmented internal-field E_v will again attain the inception value E_i. A second discharge now occurs within the void but in a direction opposite to that of the first. Assuming that these two discharges are of equal magnitude the space charges within the void will cancel, and the internal field is again synonymous with the applied field E_a. Since space charges are no longer present the induced-charge on the detecting electrode reduces to zero within the time span of the second dischage; see Figure 1.

At t = $3T/4$, i.e. at the negative peak of the applied voltage, E_v again attains the discharge inception value E_i, and a third discharge occurs leading again to the establishment of a space charge field within the void, and consequently to the establishment on the detecting electrode of an induced-charge of magnitude q_1 but now of negative polarity. The sequence of events from t = $T/4$ to t = $5T/4$ is thereafter repeated for as long as the externally applied field is unchanged in frequency and magnitude.

From the preceding temporal behaviour of the partial discharge process we see that a sequence of two positive and two negative discharges is contained in each period of the applied voltage. This results in an induced-charge which varies between q_1 and $-q_1$ per period as shown in Figure 1. Furthermore, it is seen that the magnitude of the induced-charge attains the maximum value already on the very first discharge event.

The field E_a created in a void by an externally applied AC voltage will normally reach the inception field strength E_i well before the peak of the applied voltage. This leads in time to a sequence of discharges during which the field E_v within the void will vary between E_i and E_ℓ as shown schematically in Figure 2. This pattern results from a superposition of the internal space charge field and the void related applied field. It is seen that E_v becomes negative at a time well before that of the applied field. This behaviour leads in an *idealized* system to a sequence of events per period consisting of N positive and N negative discharges as indicated in Figure 2. In a *practical* system, owing to differences in positive and negative partial discharge characteristics, we may see a difference between successive positive and negative half-cycle patterns, and N may vary with time.

Void Dipole Moment and Induced Charge

The first discharge creates space charges within the void of both +ve and -ve polarity. External to the void, these wall-deposited charges can be considered as an electric dipole configuration of dipole moment $\vec{\mu}_1$. This dipole moment $\vec{\mu}_1$ is strictly proportional to the induced-charge, and thus the variation with time of this dipole moment is the same as that of the induced-charge. Consequently, the field distortion created in the bulk of the dielectric by this dipole source will also vary with time in the same manner. From Figure 2 we see that the very first sequence of discharges in the time interval $0 < t < T/4$ consists of $N/2$ discharges in the same direction. This results in a stepwise buildup of a dipole moment, the maximum value of which is given by

$$\mu_m = \frac{1}{2} N \mu_1 \qquad (1)$$

Each step in this buildup process results in an increase in the field distortion outside of the void in the bulk of the dielectric.

Simultaneous with this buildup of the dipole, a similar stepwise buildup occurs of the induced-charge $q(t)$ on the electrode, see Figure 2. Each step results in an actual charge increase of q_1. The resulting maximum value q_m of this induced-charge is therefore given by

$$q_m = \frac{1}{2} N q_1 \qquad (2)$$

Such stepped induced-charge waveforms are reported in the literature.[3-5]

EQUIVALENT DIPOLE MOMENT $\vec{\mu}_o$

The induced-charge $q(t)$ which arises from a dipole configuration of space charge is linked to the dipole moment through the relationship[6]

$$q(t) = -\vec{\mu}(t) \cdot \vec{\nabla}\lambda \qquad (3)$$

λ is a dimensionless positive scalar function which relates the charge in the dielectric to the charge on the detecting electrode; it is simply a proportionality factor.[6]

If we determine $q(t)$, and know the location and shape of the void, the dipole moment of the charges within the void can be evaluated from (3). This requires a knowledge of λ. Owing to the presence of the void, a direct calculation of λ becomes cumbersome. We can simplify the procedure by considering the value which λ would attain at a location identical to that of the void, but with the entire dielectric system now assumed void-free. Let us denote this function by λ_o. The two functions are related by[7]

$$\vec{\nabla}\lambda = h\vec{\nabla}\lambda_o \,, \qquad \text{with } 1 < h < \varepsilon_r \tag{4}$$

where ε_r is the relative permittivity of the bulk dielectric in which the void is embedded. If we restrict this concept to voids of simple form,

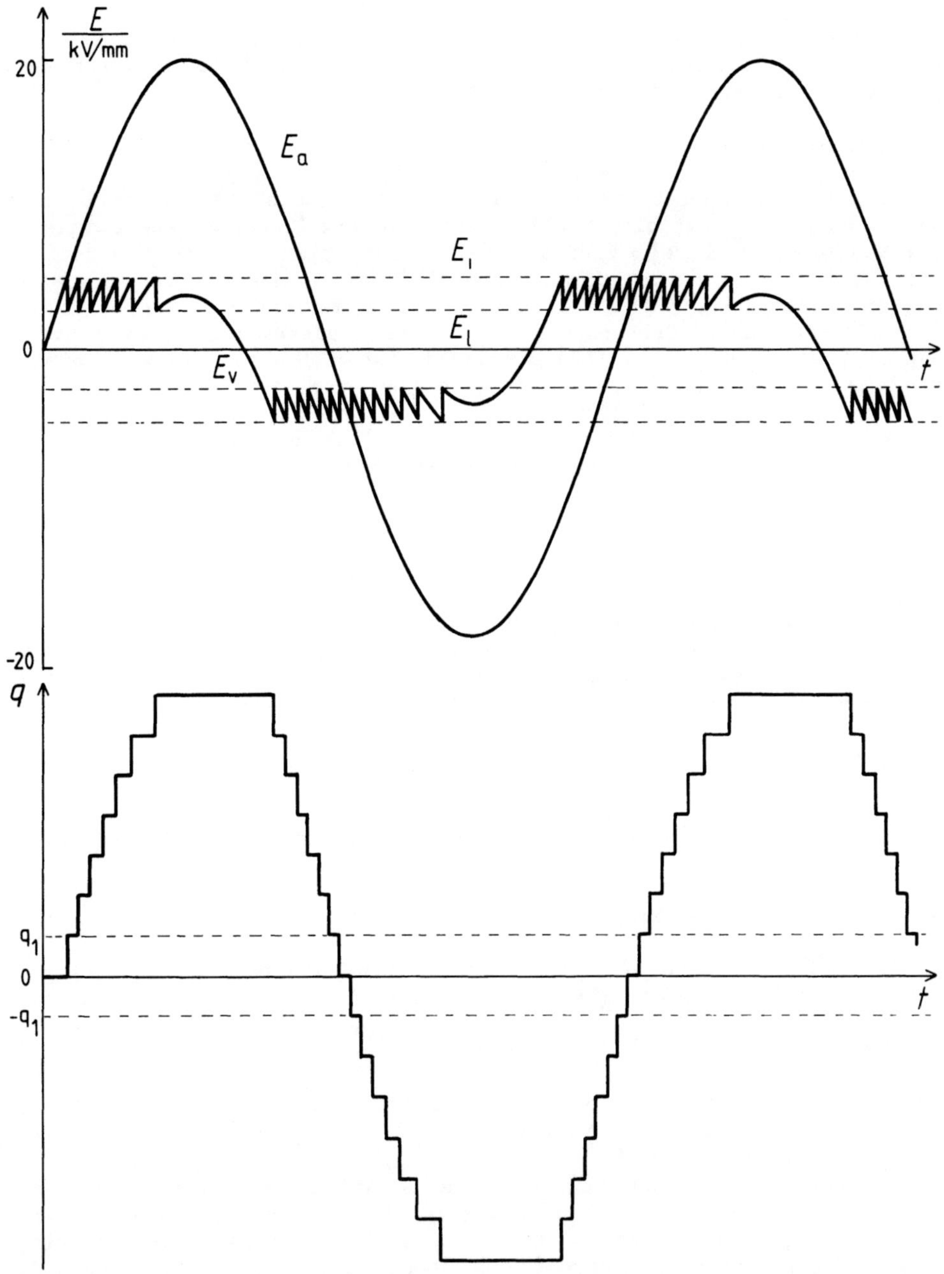

Figure 2. Electric field within the void and the corresponding induced-charge for N = 14.

such as spheroids, and to isotropic dielectrics, then h is a scalar. For a spherical void h is given by

$$h = \frac{3\varepsilon_r}{1 + 2\varepsilon_r} \tag{5}$$

The relationship between the Poissonian induced-charge and the space-charge dipole-moment may thus be written in the form

$$q(t) = -\vec{\mu}(t) \cdot h\vec{\nabla}\lambda_o \tag{6}$$

We note that h depends only on ε_r and the shape of the void, and thus mathematically it is permissible if we choose to connect h with the dipole moment $\vec{\mu}$, i.e.

$$q(t) = -[h\vec{\mu}(t)] \cdot \vec{\nabla}\lambda_o \tag{7}$$

or by introducing an equivalent dipole moment $\vec{\mu}_o$ defined by

$$\vec{\mu}_o \equiv h\vec{\mu} \tag{8}$$

$q(t)$ can be expressed in the final form

$$q(t) = -\vec{\mu}_o(t) \cdot \vec{\nabla}\lambda_o \tag{9}$$

The equivalent dipole moment $\vec{\mu}_o$ is the dipole which, if placed in a void-free system at the location of $\vec{\mu}$, would generate exactly the same induced-charge on the electrode as $\vec{\mu}$. The field perturbation associated with $\vec{\mu}$ will therefore, at some distance from the location of the void, be the same as that associated with $\vec{\mu}_o$. The advantage of the concepts of λ_o and the equivalent dipole $\vec{\mu}_o$ is that these enable us to deal with the field perturbations in the bulk of the dielectric even if in the actual system the shape of the void is unknown.

If we could obtain a knowledge of the function $q(t)$ from suitable measurements,[1] (9) would enable us to compute the variation with time of $\vec{\mu}_o$ for any location in the dielectric. However the problem can be reduced as all we require to know in order to dimension an insulation correctly is a knowledge of the maximum possible value μ_{om} of $\vec{\mu}_o(t)$.

FIELD ENHANCEMENT IN THE BULK DIELECTRIC

We now assume that we know from such $q(t)$ measurements the value of $\vec{\mu}_{om}$ in a given point in the dielectric system. The field perturbation in the bulk dielectric associated with $\vec{\mu}_{om}$ at some distance r from the dipole is then given by

$$\Delta V(r,\theta) = \frac{\mu_{om} \cos\theta}{4\pi\varepsilon_o\varepsilon_r r^2} \tag{10}$$

$\Delta V(r,\theta)$ is the scalar potential of the field perturbation, and θ is the angle between the direction of $\vec{\mu}_{om}$ and the line from the dipole to the point in question. To this potential should be added the perturbations from charges on the electrodes created by electricification by induction. These latter contributions can, however, be neglected as we are mainly considering the field in the vicinity of the void. Furthermore, it is required that r must be greater than a certain critical length which depends on the actual shape of the void. As a rough guide this critical distance can be taken as the greatest distance to the surface from that point within the void where we have chosen to locate the dipole. Let this minimum value of r be denoted by a.

The field strength derivable from the potential function can be considered as the sum of two components. A component $\Delta\vec{E}_1$ in the direction of $\vec{r}$, and $\Delta\vec{E}_2$ at right angles to $\vec{r}$ in the plane containing $\vec{r}$ and $\vec{\mu}_{om}$. These components are given by

$$\Delta E_1 = -\frac{\partial \Delta V}{\partial r} = \frac{2\,\mu_{om}\cos\theta}{4\pi\varepsilon_o\varepsilon_r r^3} \tag{11}$$

and

$$\Delta E_2 = -\frac{1}{r}\frac{\partial \Delta V}{\partial \theta} = \frac{\mu_{om}\sin\theta}{4\pi\varepsilon_o\varepsilon_r r^3} \tag{12}$$

This field attains its maximum value in the solid dielectric along the fields line $\theta = 0$ and $\theta = \pi$. Hence the maximum value of the field perturbation occurring in the dielectric at the void interface is given by

$$\Delta E_m = \frac{2\mu_{om}}{4\pi\varepsilon_o\varepsilon_r a^3} \tag{13}$$

ΔE_m has the same direction as the applied field. Lacking experimental values of q_m with which to quantify μ_{om}, and thereafter ΔE_m, we must resort to an indirect approach to determine μ_{om}.

APPLICATION TO A SPHERICAL VOID

Consider a small spherical void of radius a, in which the electric field remains effectively uniform following partial discharge activity. With this approach to the problem, the authors have shown[7] that the dipole moment of the surface charges deployed by a *single* discharge in such a spherical void may be expressed as

$$\vec{\mu}_1 = \frac{3}{h}\Omega\varepsilon_o\varepsilon_r(\vec{E}_i - \vec{E}_\ell) \tag{14}$$

where Ω is the volume of the void. Hence, from (8), we have

$$\vec{\mu}_{o1} = 3\Omega\varepsilon_o\varepsilon_r(\vec{E}_i - \vec{E}_\ell) \tag{15}$$

Upon substituting (15) in (13), we obtain the maximum value of $\Delta\vec{E}_m$ for a single discharge, *viz.*

$$\Delta\vec{E}_{m1} = 2(\vec{E}_i - \vec{E}_\ell) \tag{16}$$

Owing to the nature of spherical field solutions, (16) is identical with the expression derived previously.[2] In that study, the difference $(E_i - E_\ell)$ is shown to exhibit values of the order of several kV/mm.

The expression (16) represents the distortion per step in the $q(t)$-waveform. Consequently, with reference to Figure 1 and (16), the maximum value of the field distortion can therefore be expressed in the form

$$\Delta\vec{E}_m = N(\vec{E}_i - \vec{E}_\ell) \tag{17}$$

in which N is either the total number of positive or negative discharges within one period of the applied field. On the basis of (17), we arrive at the conclusion that field enhancement in the bulk dielectric may reach levels of the order of 10-20 kV/mm.

DISCUSSION

The dramatic cumulative field distortion in the dielectric, within the vicinity of the void, which results from a sequence of partial discharges, cannot be accounted for in terms of circuit analysis.

This field distortion does not disappear with the cessation of each discharge current pulse. If the surface conductivity of the void wall is taken as zero, then the distortion remains at a constant level until the occurrence of the next discharge at which point the distortion is further increased if the discharge occurs in the same direction. Accordingly, in the time interval between two successive discharges, the dielectric is subjected to an additional constant stress. With reference to Figure 2, it is seen that such enhanced stress levels in the dielectric attain their maximum value in the time interval between two discharge-pulse sequences of opposite polarity. The Poissonian induced-charge $q(t)$ on the electrode has its maximum value q_m during all of this time interval. Consequently, it is also in this time interval that the dielectric is subjected to both a maximum additional- and total-stress. The acceptable discharge level can therefore be linked to the value of q_m. Furthermore, it is seen from Figure 2 that, compared with the extremely short time duration of each partial-discharge current-pulse, the maximum value of the induced-charge persists virtually as a DC signal.

CONCLUSION

The space charges in a spherical void create a significant distortion of the electrostatic field in the bulk dielectric in the vicinity of the void. This field enhancement can be evaluated through the equivalent dipole moment $\vec{\mu}_0$ derived from measurements of the induced-charge $q(t)$. The magnitude of this external field distortion may easily exceed the maximum permissible field strength of the dielectric. These conclusions can be extended with confidence to voids of any shape.

REFERENCES

1 A. Pedersen, G.C. Crichton, and I.W. McAllister, Partial discharge detection: theoretical and practical aspects, International Conference on Partial Discharge, IEE Conference Proceedings No.378:21 (1993).

2 A. Pedersen, G.C. Crichton, and I.W. McAllister, PD-related stresses in the bulk dielectric and their evaluation, 1993 Annual Report - Conference on Electrical Insulation and Dielectric Phenomena, IEEE Publication 93CH3269-8:474 (1993).

3 W. Boeck, Verfahren zur Ladungs- und Leistungsmessung bei Korona-entladungen an Isolierstoffen, *ETZ-A* 84:752 (1963).

4 D. König, Alterung von Epoxydharz-Formstoffen durch innere Teilentladungen, *ETZ-A* 88:27 (1967).

5 D. Kind and D. König, AC breakdown of epoxy resins by partial discharges in voids, *IEEE Trans. Electrical Insulation*, EI-3:40 (1968).

6 A. Pedersen, G.C. Crichton, and I.W. McAllister, The theory and measurement of partial discharge transients, *IEEE Trans. Electrical Insulation*, 26:487 (1991).

7 G.C. Crichton, P.W. Karlsson, and A. Pedersen, Partial discharges in ellipsoidal and spheroidal voids, *IEEE Trans. Electrical Insulation*, 24:335 (1989).

DISCUSSION

I. D. CHALMERS: Does your analysis apply equally in the case of microvoids?

G. C. CRICHTON: Yes, if charge separation occurs within the void, e.g., if $\Gamma \neq 0$. In XLPE microvoids are known to be spherical in form. The relationship between ambient and internal field is independent of the size of the void. Consequently, an internal field reduction will cause an external field enhancement, the magnitude of which is independent of void size. Furthermore, if there is a line-up of the microvoids in the ambient field direction, then the field between such voids will be further enhanced.

L. NIEMEYER: For GIS insulators design fields are of the order of $E_o \simeq 5$ kV/mm typically leading to local field enhancements in the insulating material which should not exceed $E_o \simeq 30$ kV/mm. The inception fields for the electrofracture mechanism is typically in the range of 200 to 500 kV/mm. It is therefore not to be expected that this damage mechanism will play a role in GIS insulators.

J. KENNEDY: As the charge may be distributed on the surface of the void, and not point localized (ideal dipole) this would alter ΔE_m, and would probably result in a lower ΔE_m . Could you comment on this?

G. C. CRICHTON: What we had in mind was XLPE, where microvoids are known to cause concern. It should be remembered that a breakdown is not the only mechanism responsible for reduction of the internal field of the void. A finite surface conductivity would be sufficient. Any reduction of the internal field leads automatically to an increased field external to the void, and the field amplification factor is independent of void size.

J. KENNEDY: Concerning microvoids, ($x_c \approx 1\ \mu$m), is breakdown possible when the x_c value is extremely small? If a breakdown requires a secondary process for its occurrence (see paper by J. M. Wetzer et al.) some significant value of x_c is required for any significant charge (10 pC) to be transferred. As 1 μm is approaching the mean free path in air, is a breakdown possible?

G. C. CRICHTON: It should be noted that we are not discussing the field of an actual point dipole. Our dipole moment is the dipole moment associated with the real distribution of wall charges, and arises from a multipole expansion. In the case studied, we consider a cosine variation of wall charge-density, which yields an internal uniform-field distribution. It is only external to the void that the field distribution is dipolar. Our estimations of field enhancement are with respect to a point charge configuration extremely conservative.

R. J. VAN BRUNT: It is known that, for some types of materials such as cast epoxies, the occurrence of partial discharge in a cavity will cause the cavity to become conductive or partially conductive. Could you comment on the applicability of your model to a void that has a finite surface conductivity so that charge can migrate and redistribute during the time between discharge pulses?

G. C. CRICHTON: The effect of a possible neutralization of the PD wall charges due to a finite surface conductivity would be counteracted by the conductivity related current flow associated with the externally applied field.

A TIME-DEPENDENT MODEL OF PARTIAL DISCHARGES IN VOIDS

J.M. Wetzer, J.T. Kennedy and E.H.R. Gaxiola

High-Voltage and EMC Group
Department of Electrical Engineering
Eindhoven University of Technology
Eindhoven, The Netherlands

ABSTRACT

A time-dependent, hydrodynamic, model of partial discharges in voids in dielectrics is presented. The partial discharge is regarded as a space charge dominated electron avalanche. The model incorporates the relevant collisional and radiative processes in the void and at its boundary, and the field distortion due to the presence of space charge and void surface charges. Also the coupling with the external circuit is taken into account. Model simulations show that, as a result of field quenching, the associated charge versus applied field shows saturation above a certain threshold. The threshold field depends on geometry and initial electron number, but is higher than the inception values usually quoted. The saturation charge may deviate by a factor of 2-10 from values obtained with simplified models.

1. INTRODUCTION

Partial discharges in voids in dielectrics are a prime cause of insulation degradation. The damage is a result of the interaction between the discharge and the dielectric surface. The experimental study of the physical processes in naturally occurring voids is complicated because the void is not accessible, and because interpretation requires knowledge of the void shape and location. We have previously presented time-resolved experiments with artificially produced voids[1]. In this work we present a time-dependent model of the partial discharge for a cylindrical void and an electrode geometry as shown in Fig.1.

2. MEASURED PARTIAL DISCHARGE WAVEFORMS

The experimental setup used for the measurement of partial discharge waveforms with high time resolution (0.6 ns) was presented in detail in earlier publications[1,2]. A subdivided electrode arrangement is used, and the design is such that inductance, stray capacitance and traveling wave effects are minimized. The dielectric consists of polyethylene sheets of 0.2 mm thickness. An artificial void is produced by making a circular hole in one or more sheets. Typical waveforms measured at different void heights are shown in Fig.2.

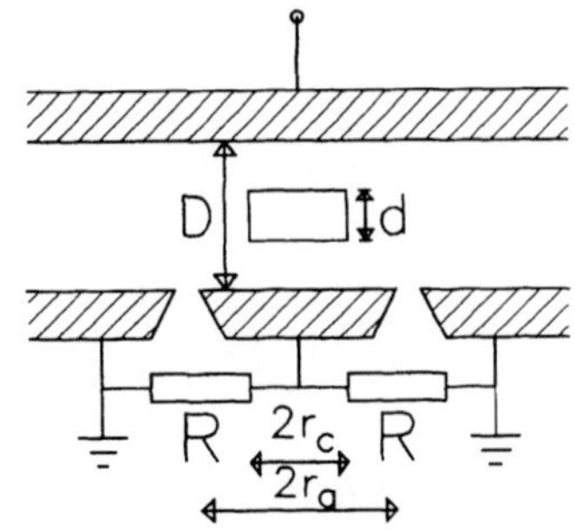

Figure 1. Electrode geometry with cylindrical void

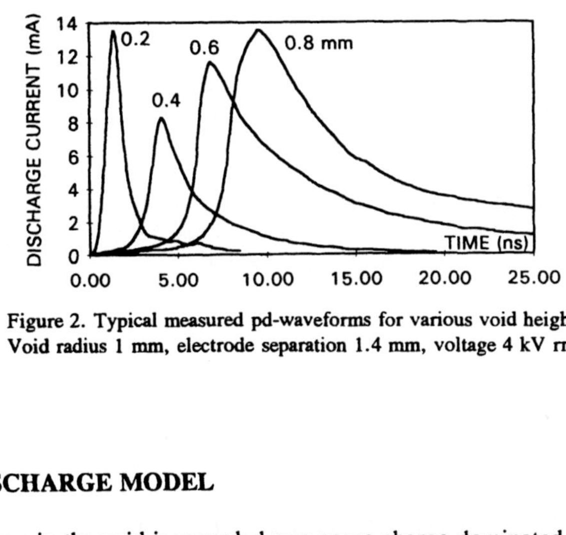

Figure 2. Typical measured pd-waveforms for various void heights. Void radius 1 mm, electrode separation 1.4 mm, voltage 4 kV rms.

3. PARTIAL DISCHARGE MODEL

The discharge in the void is regarded as a space charge dominated electron avalanche in atmospheric air. We represent the avalanche as a cylinder with a charge distribution that varies only in axial direction. The discharge model consists of a set of 1-D continuity equations for the production and transport of charged particles, in combination with a field calculation routine which includes the effects of space charge and surface charges. The processes considered, and the corresponding coefficients and time constants, are drift (v), ionization (α), recombination (ϵ_{pe}), attachment (η_u and η_s), detachment (τ_{ud} and τ_{sd}), charge exchange (τ_c), gas phase photo-ionization and photo-emission of secondary electrons from surfaces. The charged particles considered are electrons (index e), positive ions (index p) and negative ions. We distinguish stable and unstable negative ions (indices sn and un). The unstable ions are formed upon attachment, and may undergo charge exchange processes. Stable negative ions are formed upon attachment or charge exchange but are not subject to charge exchange collisions. Because of the short avalanche duration and the high pressure (1 bar), diffusion is neglected in the present model. The partial differential equations are given by:

$$\frac{\partial \rho_e}{\partial t} + \nabla\cdot(v_e\rho_e) = (\alpha-\eta_u-\eta_s)|v_e|\rho_e + \frac{\rho_{un}}{\tau_{ud}} + \frac{\rho_{sn}}{\tau_{sd}} - \epsilon_{pe}(\rho_e\rho_p) + S_{ph} \quad (1)$$

$$\frac{\partial \rho_{un}}{\partial t} + \nabla\cdot(v_{un}\rho_{un}) = \eta_u|v_e|\rho_e - \frac{\rho_{un}}{\tau_{ud}} - \frac{\rho_{un}}{\tau_c} \quad (2)$$

$$\frac{\partial \rho_{sn}}{\partial t} + \nabla\cdot(v_{sn}\rho_{sn}) = \eta_s|v_e|\rho_e - \frac{\rho_{sn}}{\tau_{sd}} + \frac{\rho_{un}}{\tau_c} \quad (3)$$

$$\frac{\partial \rho_p}{\partial t} - \nabla\cdot(v_p\rho_p) = \alpha|v_e|\rho_e - \epsilon_{pe}(\rho_e\rho_p) + S_{ph} \quad (4)$$

Here ρ_i is the number density and $\mathbf{v}_i$ is the drift velocity vector of species i, whereas $|v_e|$ is the magnitude of the electron drift velocity. The units are [cm^{-1}] for α, η_u and η_s, [s] for τ_{ud}, τ_{sd} and τ_c and [cm^3s^{-1}] for ϵ_{pe}. Values for coefficients, time constants and drift velocities are taken from Badaloni and Gallimberti[3] (BG), with the following exceptions: v_e-data are taken from Wen[4], v_p-data from Novak[5]; α-data in the E/p-range from 31 to 45 V/(cm.Torr) are taken from Davies[6], outside this range BG-values are used; η_u-data are taken from Wen[4] except for E/p-values below 25 V/(cm.Torr) where BG-values are used; the BG-values for τ_{ud} are adjusted by a factor of two[7], and are further adjusted to agree with Wen[4] in the E/p-range from 30 to 50 V/(cm.Torr). S_{ph} is the source term for gas phase photo-ionization:

$$S_{ph}(x) = \frac{p_q}{p+p_q}\cdot\int_0^d \psi(p|x-\xi|)\Omega(x-\xi)\alpha(\xi)\rho_e(\xi)v_e(\xi)p\, d\xi \quad (5)$$

Here x is the distance from the "cathode" side of the void, p is the pressure, and p_q is the quenching pressure at which collisional deexcitation starts dominating over radiative transitions. ψ is an overall photo-ionization coefficient for which experimental curves from Penney and Hummert[8] are used, and Ω is the solid angle:

$$\Omega(x-\xi) = 2\pi\left[1-\left(\frac{|x-\xi|}{\sqrt{r_p^2+(x-\xi)^2}}\right)\right] \quad (6)$$

The quenching pressure p_q is taken 10 Torr[3], and, as proposed by Kline[9], r_p is set at twice the discharge radius.

Secondary electron production from void surface

The production of secondary electrons by ion impact plays no role on the timescale considered. Photo-emission from dielectric surfaces (at the cathode side of the void) requires an energy exceeding 5 eV (far UV)[10,11,12]. Sources are the different N_2 and O_2 molecular bands. The secondary electron production at position x=0 is described by:

$$\rho_e^{sec}(0) = \sum_{i=1}^{k} \Delta_i \int_0^d \rho_e(x)\omega_i\alpha(x)g(x)e^{[-\mu_i x]}dx \quad (7)$$

Here Δ_i is the quantum efficiency for electron production, ω_i is the photon production efficiency upon electronic collisions, g(x) is a geometrical factor and μ_i is the absorption coefficient for photons of band i. The quantum efficiency values are taken from Tom[10], the other data are obtained from Badaloni[3].

Space charge field calculation

The avalanche is regarded as a cylinder of charge, its distribution varying only in axial direction. The space charge field is two-dimensional because of the finite radius of the cylinder. Without dielectric or metallic boundaries, the axial component of this field is[13]:

$$E_{sp} = \frac{1}{2\varepsilon_0}\left[\int_{-x}^{0} \varrho(x+\zeta)\left(-1-\frac{\zeta}{\sqrt{\zeta^2+r^2}}\right)d\zeta + \int_0^{d-x} \varrho(x+\zeta)\left(1-\frac{\zeta}{\sqrt{\zeta^2+r^2}}\right)d\zeta\right] \quad (8)$$

$\varrho(x)$ is the net charge density at position x, r is the cylinder radius and d is the cylinder (or void) height. The proximity of polarization charges at the dielectric boundaries and induced charges in the metal electrodes is incorporated in Eq.8 by adding appropriate image charges. For a charge q located at position x, the space charge field is obtained as follows: we place an image charge behind the dielectric surface at the same distance as the charge q is in front of it. Then we place the image of this first image charge in the opposite dielectric and so on. Each dielectric surface is represented by four image charges, their magnitudes given by[14]:

$$q_1 = q\frac{\varepsilon_r-1}{\varepsilon_r+1} \qquad q_n = q_{n-1}\frac{\varepsilon_r-1}{\varepsilon_r+1} = q\left[\frac{\varepsilon_r-1}{\varepsilon_r+1}\right]^n \quad (9)$$

Next, the dielectric image charges are replaced by two disks of uniform surface charge, one at each side of the void, that produce the same void field as do the image charges. The proximity of the metal boundaries is incorporated by using sets of four image charges per electrode for the original charge q and for the two surface charges. These 12 image charges represent the field of one single charge q in the void. The total space charge field is obtained by integration over the void volume. The field calculation procedure was checked to be accurate within a few percent by comparing it with FEM field calculations.

Boundary conditions

All charges reaching a dielectric boundary are considered permanently trapped. When a photo-electron is emitted from a dielectric surface, a permanent positive charge is left behind. The resulting surface charges are responsible for the quenching of the electric field and, consequently, the current. The experiment is designed such that the external circuit maintains a constant voltage across the electrodes. An avalanche simulation is started with a number of electrons in the first disk at the cathode side of the void.

Numerical procedure

The set of differential equations is solved with a flux-corrected transport (FCT) technique similar to that introduced by Boris and Book[15]. The source terms are incorporated using a two-step process[16], in which the number densities are determined twice, but the space charge field and the field dependent coefficients only once, in each time step Δt.

Induced current in external circuit

Several approaches are reported in literature for relating moving charges to induced currents. For a 1-D discharge in a cylindrical void these approaches all give the same result, provided that the external circuit is such that the induced charges are compensated without causing a voltage drop across the electrodes:

$$i(t) = \frac{\iiint \varrho v \, dxdydz}{d + \dfrac{D-d}{\varepsilon_r}} = \frac{\pi r^2}{d + \dfrac{D-d}{\varepsilon_r}} \int_0^d \varrho v \, dx \tag{10}$$

4. RESULTS

Figure 3 shows four simulated current waveforms. The initial Laplacian field applied to the void is varied whereas the void geometry is kept constant (radius 0.2 mm, height 0.2 mm, total insulation thickness 1.4 mm). The simulation starts with a number of 100 electrons at the cathode side of the void. Apart from the current simulated for an ideal measuring system, also the current for a practical system with a time resolution of 0.6 ns was calculated.

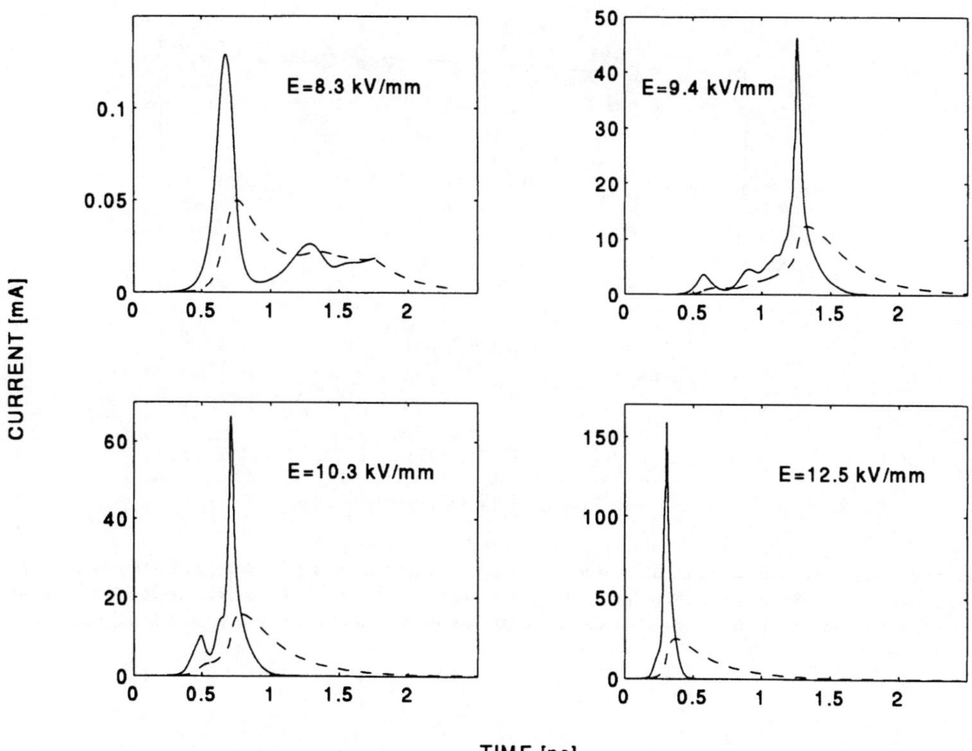

Figure 3

Simulated waveforms for different values of the Laplacian field prior to the discharge. Void: radius 0.2 mm, height 0.2 mm, insulation thickness 1.4 mm, 100 initial electrons. Drawn line: simulated current; dashed line: corrected for limited bandwidth (600 MHz).

For a low void field the discharge behaves like an avalanche with secondaries caused by photon emission. At higher fields these secondaries grow into a streamer discharge. The secondary emission of photons is of crucial importance to the development of discharges. The "measured" waveforms (dashed lines) appear to be dominated by the time-resolution of the system, they do however produce the correct charge. Waveforms such as shown in Fig.3 for E = 9.4 kV/mm have also been observed experimentally by Morshuis[17] for larger voids (1 mm height). In that case the discharge develops slower, and time-resolution is less of a problem.

Figure 4 gives the ("real" or "void") charge as a function of the field applied to the void prior to the discharge, for two void geometries. For one geometry also the number of initial electrons is varied. The charge is determined from the simulated currents using Eq.10 (after integration). It is observed that the charge strongly depends on the applied field until, as a result of field quenching, saturation is reached. The saturation field depends on the void geometry, as expected, but also on the number of initial electrons. Also when varying the initial number of electrons, a saturation threshold is observed, above which the charge produced remains constant.

The measured waveforms (Fig. 2) can only be reproduced with this model by assuming at least 100 initial electrons. This would indicate that discharge activity has taken place prior to the observed discharge. It may however also be due to the fact that the real space charge effect is not correctly described by a 1-D model with constant discharge radius.

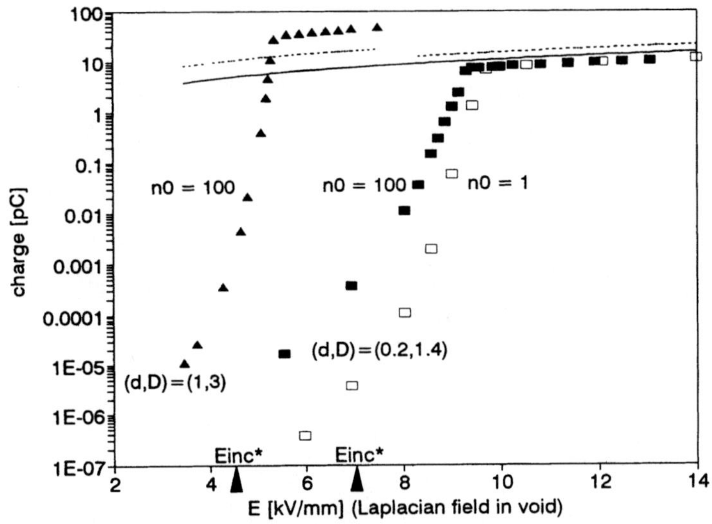

Figure 4

Real charge versus applied Laplacian field prior to the discharge, for two void geometries, and different numbers of initial electrons. Literature values for the inception field are shown[18]. The drawn line (1) represents the charge accumulated at the void boundary prior to the discharge, the dashed lines (2) follow from a simplified model presented earlier[1].

Saturation occurs at a rather well-defined field value, below which the charge drops sharply. For of comparison the inception field values according to Crichton[18] are shown in Fig.4. This is the field that, when applied between metal electrodes, would cause a breakdown. The simulations suggest that fields higher than this inception field are required for void breakdown. The saturation charge was first of all compared with the amount of charge accumulated at the void boundaries prior to the discharge (εAE, drawn curve). Also it is compared with a simple model presented earlier[1] (dashed curves) which calculates the amount of charge that is required to reduce the field from the inception value to the extinction value. For the curves in Fig.4 the extinction field is chosen zero, and the inception field is replaced by the actual field at the moment that the discharge starts.

Although this does not dramatically affect the observed trends, the present model is restricted to 1-D transport only. A recently developed 2-D version of the model, and experimental studies with a variety of void geometries and dielectric materials, will be employed to verify and optimize the discharge model.

5. CONCLUSIONS

Simulations with a 1-D model show that, as a result of field quenching, the associated charge versus applied field shows saturation above a certain threshold. The threshold field depends on geometry and initial electron number, but is higher than the inception values usually quoted. The saturation charge may deviate by a factor of 2-10 from values obtained with simplified models. The observed trends will have to be verified with a 2-D model and with experiments

REFERENCES

1. J.M. Wetzer, 7th.Int.Symposium on High-Voltage Engineering, Dresden, paper 71.02 (1991)
2. J.M. Wetzer, P.C.T. van der Laan, IEEE Transactions on Electrical Insulation, **24**, pp.297 (1989)
3. S. Badaloni, I. Gallimberti, "Basic Data of Air Discharges", Universita di Padova Report, UPee-72/05 (1972)
4. C. Wen, Ph.D. thesis, Eindhoven University of Technology (1989)
5. J.P. Novak, R. Bartnikas, Proc.Xth Int.Conf.Gas Discharges & Appl., **2**, p.864 (1992)
6. A.J. Davies, C.J. Evans, P.M. Woodison, Comp.Phys.Comm., **14**, p.278 (1978)
7. L.E. Kline, J.Appl.Phys., **46**, p.1994 (1975)
8. G.W. Penney, G.T. Hummert, J.Appl.Phys., **41**, p.572 (1970)
9. L.E. Kline, J.Appl.Phys., **45**, p.2046 (1974)
10. J. Tom, H.F.A. Verhaart, A.J.L. Verhage, C.S. Vos, 2nd.Int.Conf.on Conduction and Breakdown in Solid Dielectrics, Erlangen, p.301 (1986).
11. M. Yumoto, T. Sakai, 1980 Annual Report of the CEIDP, p.75 (1980)
12. S.M. Mahajan, T.S. Sudarshan, IEEE 6th. Pulsed Power Conference, Arlingto VA, p.191 (1987)
13. A.J. Davies, C.J. Evans, F. Llewellyn-Jones, Proc.Roy.Soc.London Ser.A, **281**, p.164 (1964)
14. P. Lorrain, D. Corson, Electromagnetic Fields and Waves, W.H. Freeman and Company, San Francisco (1970)
15. J.P. Boris, D.L. Book, Journ.of Comput.Phys., **20**, p.397 (1976)
16. R. Morrow, Journ.of Comput.Phys., **43**, p.1 (1981)
17. P.H.F. Morshuis, Ph.D. thesis, Delft University of Technology, The Netherlands (1993)
18. G.C. Crichton, P.W. Karlsson and A. Pedersen, IEEE International Symposium on Electrical Insulation, pp.238-241 (1988)

DISCUSSION

R. BARTNIKAS: We have used a two-dimensional model to examine the current waveforms between parallel plane electrode systems and find that the rise time and magnitude of the pulse is also very much affected by the overvoltage across the gap. At high overvoltages the rapid rise time electron component in the current tends to obscure the ion component of the current pulse. But the ion current component still exists in every current pulse [R. Bartnikas and J. Novak, *IEEE Trans. on Electrical Insulation*, Dec. 1993].

J. M. WETZER: It can be stated quite generally that the rise time and the magnitude of a discharge pulse strongly depend on the amount of overvoltage. In void discharges an additional effect is the change in waveshape due to field quenching. The ion component of the current is small compared to the electron component, but indeed it is present as long as we have a remanent void field, i.e. when the applied void field is below the saturation field. For higher applied fields the void breaks down and the field drops to virtually zero. Without a driving force, the ion current will be zero (see Fig. 3, 9.4 kV/mm and higher).

M. F. FRÉCHETTE: What are the initial conditions relative to the source? Can you change these conditions to take account of the finite size of an emitting site and of activities from multiple sites?

J. M. WETZER: The initial condition assumes a number of electrons between 1 and 100, spread out over the discharge area and located at the cathode side of the void. In our 1-D model the discharge radius is constant and the size of the emission site can only be taken into account by reducing the discharge radius. This would give an unrealistically high space charge field. With a 2-D model we will be able to correctly account for the finite emission site and for the space charge. The incorporation of multiple emission sites requires a considerable extension of the model (and an increase in computing time) because: (1) we lose radial symmetry and (2) the different discharges are not independent as a result of photon processes and space charge effects.

I. HERBST: Figure 4 puts constraints on measurement techniques for the apparent charge. Only at the highest possible fields will one be able to measure discharges. Monitoring of voids in real GIS will therefore be extremely difficult, if not impossible.

J. M. WETZER: First of all it should be noted that the applied field in Fig. 4 is the void field (prior to the discharge), which is enhanced as compared to the unperturbed field without void. Nevertheless, it is true that the limited real charge amplitude (even for real size insulators), combined with the small ratio of apparent and real charge, makes it difficult to monitor voids in real GIS. In previous work [Ref. (1) of this paper] we have given estimates of the observable charge for real size insulators with voids. The relatively high saturation field presented in this paper increases the difficulties in measuring void discharges.

S_2F_{10} IN SF_6 INSULATED EQUIPMENT

Lutz Niemeyer

ABB Corporate Research Center
CH 5405 Baden, Switzerland

ABSTRACT

The highly toxic SF_6 decomposition product S_2F_{10} is suspected to constitute a health risk associated with gas insulated equipment. In this paper a method is presented by which maximal S_2F_{10} concentrations in operating GIS can be estimated under worst case conditions. The method is based on simple integral discharge models and makes use of the results of the CRADA activity. It is found that neither partial discharge activity nor disconnector switching can lead to risk relevant S_2F_{10} concentrations.

INTRODUCTION

The highly toxic compound S_2F_{10} has been observed to be produced in small quantities by electrical discharges in SF_6 and is suspected to constitute a health hazard for SF_6 insulated equipment. Extensive background research on this issue has been carried out in the last years, mainly in the frame of the "CRADA" cooperative research activity. The results have been recently reviewed [1] and allow a preliminary assessment of the role of S_2F_{10} in operating GIS.

The scope of the present paper is to discuss a method for estimating the order of magnitude of S_2F_{10} concentrations that may occur in operating gas insulated equipment under worst case conditions. The discussion will be restricted to discharges of the corona and the spark type for which adequate background data have now become available. S_2F_{10} produced in switching arcs has been recently shown not to be health relevant [2].

The main elements of the estimation method are

- an S_2F_{10} production-destruction balance in the insulation compartment based on measured production and surface decomposition rates
- integral models relating the discharge parameters to the design of the equipment and to the characteristics of the defects causing partial discharges
- a cumulation of worst case assumptions for all involved mechanisms and quantities such that the S_2F_{10} concentration is definitely overestimated.

Gaseous Dielectrics VII, Edited by L.G. Christophorou
and D.R. James, Plenum Press, New York, 1994

S_2F_{10} PRODUCTION-DESTRUCTION BALANCE

Gas insulated equipment is subdivided in compartments which are hermetically sealed against each other and against the outside environment. The quantity M of S_2F_{10} that develops in such a compartment in the case of discharge activity is determined by a production-destruction balance of the general form

$$dM/dt = P - D \tag{1}$$

where P is a production rate and D a destruction rate. In this paper, M, P, and D will be measured in kmol and kmol/s . The solution of eq.(1) with specified values of P and D yields the temporal development M(t) of the S_2F_{10}-inventory and of the corresponding volume concentration

$$c(t) = M(t) / M_{SF6} = M / (\rho V/\mu) = RT\, M / (p_F \cdot V) \tag{2}$$

where μ = 146 kg/kmol is the molecular weight of SF_6, R = 8317 J $kmol^{-1}K^{-1}$ is the universal gas constant, and V, ρ, p_F, and T are the volume, density, filling pressure, and temperature of the SF_6, respectively.

S_2F_{10} Production

S_2F_{10} can be produced by two types of discharges, namely, "cold" non-equilibrium discharges like glow and streamer coronae and thermal discharges like sparks and arcs. For cold discharges it was shown both theoretically [3] and experimentally [4] that the production rate P scales with the average discharge current I

$$P_c = R_c \cdot I \quad \text{for cold discharges (glows and streamers)} \tag{3}$$

where R_c [kmol/As] is a corona yield coefficient which depends on various parameters such as humidity [3,4], polarity, and the surface state of the electrode [4]. From the data reviewed in reference[1] the average AC value of R_c can be derived to be in the range

$$2 \cdot 10^{-9} \le R_c \le 10 \cdot 10^{-9} \text{ kmol } A^{1}s^{-1} \tag{4}$$

For thermal spark discharges with core temperatures of the order of 20,000 K the kinetics of S_2F_{10} formation has not yet been understood. Based on thermodynamic and radiative transfer considerations the S_2F_{10} formation rate is expected to be related to the power $\dot{W}$ dissipated in the discharge:

$$P_s \approx R_s \cdot \dot{W} \quad \text{for hot discharges (sparks and arcs)} \tag{5}$$

where R_s [kmol/J] is a yield coefficient which for sparks was shown to depend on the discharge energy, the concentrations of oxygen and humidity present in the SF_6 [5], and on the presence of insulator surfaces [6]. The yields hitherto measured fall into the range

$$0.7 \cdot 10^{-14} \le R_s \le 36 \cdot 10^{-14} \text{ kmol } J^{-1}$$

the lower limit pertaining to high impurity levels and high discharge energies and the upper one to very dry and clean SF_6 and low discharge energy. S_2F_{10} measurements in an operating disconnector switch [7], i. e. under realistic GIS conditions, allow to estimate a spark production rate in the range (1....2)$\cdot 10^{-14}$ kmol J^{-1}. These relatively low values might be due to residual humidity and/or oxygen release from the sparking electrodes. For the estimates in the present paper the much higher value

$$R_s = 10 \cdot 10^{-14} \text{ kmol } J^{-1} \quad \text{(for sparks)} \tag{6}$$

will be used as a worst case upper limit.

Table 1. Experimental values of the surface decomposition rate coefficient δ_s together with major experimental parameters temperature T, water vapor concentration h, initial concentration c_o of S_2F_{10}, surface-to-volume ratio A/V, and measured exponential decay time τ. Humidity if not measured is specified "low" if water was not purposefully added.

	Surface	T [°C]	h [ppmv]	co [ppmv]	A/V [m^{-1}]	τ [s]	δ_s [m/s]	Reference
(a)	stainless	18	low	36	100	$1.8 \cdot 10^6$	$5.6 \cdot 10^{-9}$	Basile [12]
	steel	50	low	37	100	$8 \cdot 10^5$	$1.2 \cdot 10^{-8}$	Basile [12]
	"dry"	20	<1000	500	50	$>10^9$	$<10^{-11}$	Piemontesi [13]
		20	600	935	150	$2.5 \cdot 10^7$	$2.7 \cdot 10^{-10}$	Casanovas [14]
		20	low	1000	200	10^7	$5 \cdot 10^{-10}$	Sauers [15]
		20	low	5000	~70	10^9	$1.4 \cdot 10^{-11}$	Sauers [15]
(b)	idem "wet"	20	5000	95	50	$8 \cdot 10^5$	$2.5 \cdot 10^{-8}$	Piemontesi [13]
(c)	distribution GIS	12		0.19	25	$2.2 \cdot 10^5$	$1.8 \cdot 10^{-7}$	Basile [12]
	HV GIS type 1	20		0.11	8	$1.8 \cdot 10^5$	$7 \cdot 10^{-7}$	Basile [12]
	HV GIS type 2	20	<1500	37	19	$8 \cdot 10^5$	$7 \cdot 10^{-8}$	Piemontesi [13]
	HV GIS type 3		low	10	~7	$\sim 1.4 \cdot 10^6$	$\sim 10^{-7}$	Morrison[7]
(d)	molec. sieve	12		0.12		$2 \cdot 10^4$		Basile[12]
	silicagel	12		0.09		$5 \cdot 10^4$		Basile [12]

S_2F_{10} Destruction

S_2F_{10} may be removed from a GIS compartment by volume and surface reactions including unimolecular decomposition (pyrolysis) [8], catalysis [9], and sorption at surfaces. The hitherto published data indicate that volume decomposition is only important at temperatures above some 200 to 300 C [8,9]. It therefore is expected to play an important role in reducing the production rates in hot discharges of the spark and arc type. At normal GIS operating temperatures volume decomposition seems to be negligible and surface decomposition is the dominant destruction mechanism.

The surface decay of S_2F_{10} is controlled by the rate dN/dt at which S_2F_{10} molecules collide with the surface. This rate is proportional to the volume density n of the molecules and to the active surface area A:

$$dN/dt = \delta_s \cdot n \cdot A \tag{7}$$

where δ_s is a coefficient characterizing the decomposing activity of the surface. It depends on the temperature, the material, and the state of the surface. With $n = N/V$ one obtains

$$(dN/dt) / N = (dn/dt) / n = \tau_s^{-1} = \delta_s (A/V) \tag{8}$$

where τ_s is an exponential decay time constant.

Table 1 shows some of the presently available experimental surface decay data. They include published data and unpublished measurements both with SF_6 as background gas.

A first set of data (a) pertains to "dry" stainless steel surfaces when water was not purposefully added in the experiment. The obtained δ_s values are seen to spread over four orders of magnitude which is believed to be mainly due to uncontrolled surface conditions.

The measurement (b) was carried out with the method described in [10] and the humidity in the gas was monitored. It shows that purposefully added water results in δ_s values at the upper end of the "dry" data range (a). A similar result was obtained from decay measurements in argon as buffer gas [11]. This suggest that a sufficiently high level of surface adsorbed humidity enhances decomposition.

A third set of data (c) was directly measured in four different types of commercial gas insulated transmission and distribution equipment not fitted with adsorbents [12,13]. These data are in the range

$$\delta_s = (0.7 \ldots 7) \cdot 10^{-7} \text{ m/s} \tag{9}$$

i.e. an order of magnitude lower than the "wet" steel data (b). It is believed that sand blasted aluminium surfaces or the surfaces of polymer composite insulators might be the cause herefore.

A fourth set of data (d) pertains to adsorbents with which gas insulated compartments are usually equipped to remove humidity and toxic or corrosive decomposition products. The related decay time constants are in the range

$$\tau_f = (2 \ldots 5) \cdot 10^4 \text{ s} \sim 1 \text{ day} \tag{10}$$

They are of the same order of magnitude as for other corrosive decomposition products and are about one order of magnitude shorter than the ones pertaining to surface decomposition.

S_2F_{10} EQUILIBRIUM CONCENTRATIONS IN EQUIPMENT

From a health risk point of view, estimates of the S_2F_{10} concentration in gas insulated equipment should be cautiously based on worst case considerations. In this sense, the following assumptions will be made:

(1) The S_2F_{10} generating discharge source is permanently active.
(2) The upper limits of the yield coefficients R under GIS conditions are used (eqs.(4) and (6)).
(3) Adsorbents are assumed to be absent so that S_2F_{10} can only be destroyed by surface decomposition.
(4) For the surface decomposition coefficient δ_s the lowest value measured in equipment is used, namely, $\delta_s = 0.7 \cdot 10^{-7}$ m/s (eq.(9)).

It is clear that such a cumulation of worst case assumptions may drastically overestimate real S_2F_{10} concentrations.

The equilibrium concentration c_{eq} that will eventually be established in a compartment can be determined by equating the production and destruction terms P and D in eq.(1). With the above assumptions and using eqs.(2) and (8) one obtains

$$c_{eq} < [RT/(\delta_s A \cdot p_F)]\ P = C \cdot P \tag{11}$$

where the production term P is given by eqs (3) or (5) depending on the discharge type. Note that c_{eq} is inversely proportional to the SF_6 filling pressure p_F and to the active surface area A and that the compartment volume V does not enter.

EXAMPLES

As a reference we consider a 245 kV GIS (peak phase voltage $\hat{U}_o = 200$ kV) and assume a typical compartment volume $V \sim 0.2\ m^3$ with an internal surface area $A = 2\ m^2$ which is filled with SF_6 at a pressure of $p_F = 450$ kPa at room temperature. The factor C in eq.(11) then becomes $C = 4 \cdot 10^7$ s $kmol^{-1}$.

Corona at a mobile particle

We consider loose conducting particles which move in the vicinity of the GIS enclosure and do not cross the insulation gap. In the worst case, such particles move up and down in resonance with the AC voltage and exchange charge with the enclosure at every AC voltage peak. In a worst case sense, we assume that the discharge occurs in the form of a corona to obtain the highest possible S_2F_{10} production. For a particle of length ℓ the maximal charge transferred at electrode contact is

$$q = 2 \cdot [2\,\pi\varepsilon_0 / (\ln(2l/r) - 1)]\, \ell^{\,2}\, \hat{E}_0 \tag{12}$$

where $\hat{E}_0$ is the peak background field at the enclosure and l/r the particle aspect ratio. The factor 2 accounts for the fact that the particle returns to the enclosure after it has left it at the previous half cycle peak field where it was charged to the opposite polarity. This repetitive discharge activity corresponds to a time-averaged corona current

$$I_c = q/T = [4\,\omega_0\, \varepsilon_0 / (\ln(2l/r) - 1)] \cdot \ell^{\,2}\, \hat{E}_0 \tag{13}$$

where $T = \pi/\omega_0$ is the duration of the AC half cycle. Using eqs.(3) and (11) then gives the equilibrium concentration

$$c_{eq} < [RT / (\delta_s A \cdot p_F)] \cdot [4\,\omega_0\, \varepsilon_0 / (\ell n(2\ell/r) - 1)]\, R_c \cdot \ell^{\,2}\, \hat{E}_0 \tag{14}$$

Note that c_{eq} is proportional to the square of the particle length l and to $\hat{E}_0$. Assuming a slender particle with $l/r \approx 50$, a maximal length of $l = 10$ mm, and a peak electric background field of 20 kV/cm one obtains

$$c_{eq} = 8\; 10^{-7} < 10^{-6} = 1 \text{ ppmv} \tag{15}$$

Sparking at a floating shield

Floating shields are known to be the strongest PD sources in GIS. They can be represented by a series combination of two capacitances C_1, C_2 where C_1 is the capacitance of the shield to the opposite electrode and C_2 the capacitance of the loosened contact gap. The gap C_2 periodically breaks down at a voltage U_{S2}. The discharge is of the spark type because the capacitive energy dissipated is sufficient to create a hot plasma channel. The time averaged power dissipated by a continuous spark sequence is

$$\dot{W} < N \cdot (1/2)\, C_2\, U_{S2}^2 / (\pi/\omega_0)$$

where $(1/2)\, C_2 U_{S2}^2$ is an upper bound to the capacitive energy fed into one single spark and N is the number of sparks per AC half cycle $T = \pi/\omega_0$. N is equal to the number of times the spark gap can be recharged during a half cycle to its breakdown voltage U_{S2} and is therefore given by the ratio

$$N \approx 2\, \hat{U}_2 / U_{S2} = 2\, (\hat{U}_0 / U_{S2})\, (C_1/C_2)$$

where $\hat{U}_2 \approx \hat{U}_0\, C_1/C_2$ is the peak voltage that would appear over the gap C_2 when no breakdown would occur ($C_1 << C_2$). The true charge $q = C_2 U_{S2}$ transferred by the spark induces the apparent charge

$$q' = q\, (C_1 / C_2) = C_1 U_{S2}$$

in the external circuit. Combining the last three equations yields the time averaged power dissipated by the spark sequence as

$$\dot{W} = (\omega_0/\pi)\, \hat{U}_0\, q' \tag{16}$$

Inserting this value into eqs.(5) and (11) yields

$$c_{eq} < [RT / (\delta_s A \cdot p_F)]\, (\omega_0/\pi)\, R_s\, q'\, \hat{U}_0 \tag{17}$$

Note that c_{eq} is proportional to the apparent charge q' and to the peak of the applied voltage $\hat{U}_o$. If one assumes a PD magnitude of q' = 10 000 pC as an upper limit of what is beeing observed in reality, one obtains

$$c_{eq} < 8 \cdot 10^{-7} < 1 \text{ ppmv} \quad (18)$$

Disconnector Switching

We assume that a disconnector switches regularly k times a day a bus section of length L with a capacitance $C = C' \cdot L$ ($C' \sim 100$ pF/m). Each switching operation consists of a sequence of sparks the breakdown voltages of which vary during contact travel [16]. In the case of an opening operation the sparking voltage U_s increases with the contact separation x approximately as

$$U_s(t) = \hat{U}_o \, x / x_{sp} = \hat{U}_o \, t / t_{sp} \quad (19)$$

where t is the time after contact separation and t_{sp} the time interval during which sparking occurs. t_{sp} is determined by the contact travel velocity and is typically in the range of some tenths of a second. The average number of sparks n_s occuring during one AC period is inversely proportional to the ratio $\hat{U}_o/U_s$ and reaches a minimum of 2 at the end of the sparking period, i.e. at $t = t_{sp}$. Hence

$$n_s(t) = 2 \hat{U}_o / U_s(t) = 2 \, t_{sp} / t \quad (20)$$

The energy dissipated in a single spark is $w < (1/2) \, C \, U_s^2$. The total enrgy dissipated in the spark sequence then results by integration as

$$W = \quad (1/2) \, C \, U_s(t)^2 \, n_s(t) \, dt/T = (1/2) \, C \, \hat{U}_o^2 \, t_{sp} / T \quad (21)$$

With k disconnector operations per day the long term averaged power dissipation becomes

$$W < k \cdot C'L \, \hat{U}_o^2 / (1 \text{ day} \sim 10^5 \text{ sec})$$

Inserting this into eqs. (5) and (11) yields

$$c_{eq} < [\, RT / (\delta_s \, A \cdot p_F)] \; R_s \cdot C'L \, \hat{U}_o^2 \, (t_{sp}/T) / 10^5 \quad (22)$$

Typically, for L = 20 m, $\hat{U}_o$ = 200 kV, t_{sp} = 0.5 s, and k = 2 one obtains

$$c_{eq} < 4 \cdot 10^{-8} < 0.1 \text{ ppmv} \quad (23)$$

CONCLUSIONS

(1) A simple method is proposed to estimate the order of magnitude of S_2F_{10} concentrations to be expected in operating gas insulated equipment due to PD activity and disconnector switching.

(2) Worst case assumptions are systematically cumulated so that real S_2F_{10} concentrations are overestimated.

(3) Partial discharge activity caused by mobile particles or loose shields is expected to produce S_2F_{10} concentrations below 1 ppmv.

(4) S_2F_{10} concentrations generated by disconnector switching are expected to remain far below the ppmv level even if no adsorbers are provided.

(5) The presence of adsorbers further reduces the S_2F_{10} concentrations by at least one order of magnitude.

(6) If a conservative leakage scenario is assumed[1] health risk may only have to be taken into consideration when the enclosed S_2F_{10} concentration exceeds 10 ppmv. Hence, the small quantity of S_2F_{10} that may be produced in GIS under worst case conditions does not seem to comport a health risk.

ACKNOWLEDGEMENTS

The author gratefully acknowledges the unpublished data supplied by G. Basile and M. Piemontesi. He also appreciates many valuable discussions with G. Mauthe, F. Perdoncin, M. Piemontesi, and R. Pietsch.

REFERENCES

1. D.R. James, I. Sauers et al., "Investigation of S_2F_{10} Production and Mitigation in Compressed SF_6 -Insulated Power Systems", IEEE EI Insul. Magazine, 9, 29, (1993)
2. H. D. Morrison, F. Y. Chu et al., "Decomposition of SF_6 and production of S_2F_{10} in power arcs", 7th Internat. Sympos. on Gaseous Dielectrics, Knoxville TE, 1994
3. J.T. Herron, "S_2F_{10} Formation in Computer Simulation Studies of the Breakdown of SF_6", IEEE Trans. Elec. Insul., Vol. EI-22, No. 4, pp. 523-525, (1987).
4. R.J. Van Brunt, J.K. Olthoff, and Manish Shah, "Rate of S_2F_{10} Production from Negative Corona in Compressed SF_6", Conference Record of the 1992 IEEE Int. Symp. Elec. Insul., June 7-10, 1992, Baltimore, IEEE Pub. No. 92CH3150-0, pp. 328-331, 1992
5. I. Sauers and S.M. Mahajan, and R. Cacheiro, "Production of S_2F_{10}, S_2OF_{10}, and $S_2O_2F_{10}$ by spark discharges in SF_6", 7th Internat. Sympos. on Gaseous Dielectrics, Knoxville TE, 1994
6. I. Sauers, S. M. Mahajan, et al., "Effect of solid insulator on the spark yield of S_2F_{10} in SF_6", 1994 IEEE Intern. Sympos. on Electr. Insul., Pittsburgh PA, 1994
7. H. D. Morrison, V. P. Cronin et al., "Production and decay of S_2F_{10} in a disconnect switch", 7th Internat. Sympos. on Gaseous Dielectrics, Knoxville TE, 1994
8. J.T. Herron and Wing Tsang, "Formation and Stability of SF_5 and S_2F_{10}", 5th Intern. Sympos. on Gaseous Dielectrics, Knoxville, 199 (1987)
9. J.T.Herron, "A Critical Review of the Chemical Kinetics of SF_4, SF_5, and S_2F_{10} in the Gas Phase", Int. J. Chem. Kinet., Vol. 19, pp. 129-142, 1987.
10. M. Piemontesi, R. Pietsch, and W. Zaengl, "Analysis of Decomposition Products of Sulphur Hexafluoride in Negative DC Corona with Special Emphasis on Content of H_2O and O_2", IEEE Internat. Sympos. on Electr. Insulation. Pittsburgh 1994
11. J.K. Olthoff, R.J. Van Brunt, J.T. Herron, I. Sauers, and G. Harman, "Catalytic Decomposition of S_2F_{10} and Its Implication on Sampling and Detection from SF_6-Insulated Equipment", Conference Record of the 1990 IEEE International Symposium on Electrical Insulation, Toronto, June 3-6. 1990, IEEE Pub. No. 90-CH2727-6, pp. 248-252
12. G. Basile, private communication
13. M. Piemontesi, private communication
14. A.M.Casanovas, J. Casanovas et al., "Study of the decomposition of SF_6 under negative polarity corona discharge (point to plane geometry): influence of the metal constituting the plane electrode", J. Appl. Phys.72, 3344 (1992)
15. I. Sauers, M. Siddanagupta et al., "Production and stability of S_2F_{10} in SF_6 corona discharges", 6th ISH, New Orleans (1989) 23.08
16. CIGRE WG 33/13-09, "Very fast transient phenomena associated with gas insulated substations", CIGRE (1988) paper 33-13.

DISCUSSION

R. J. VAN BRUNT: Could you comment on the different S_2F_{10} decay rates that you have observed with different surface materials?

L. NIEMEYER: Our main data refer to various types of GIS compartments for which δ_s was found to be in the range $10^{-7} \leq \delta_s \leq 10^{-6}$ m/s, and are in accordance with the data of Morrison et al. (these proceedings, page 433). However, we are not in a position to explain these results consistently. Preliminary experiments by Piemontesi with relatively well defined surfaces seem to indicate that (a) surface adsorbed humidity promotes decomposition at stainless steel surfaces and (b) sand blasting promotes decomposition at aluminum surfaces.

I. SAUERS: How did you choose this destruction rate for S_2F_{10}? Since it can depend significantly on the particular equipment, wouldn't a lower destruction rate be more useful for a "worst" case scenario?

L. NIEMEYER: The destruction rate was chosen according to your single spark measurements which are in the range 0.01 to 0.02 nmol/J. In the meantime it has become clear that these values may increase by an order of magnitude in the absence of humidity. This would increase the worst case estimated concentrations by one order of magnitude which would still be noncritical from a health risk point-of-view. I have tried to use the lowest destruction rate we have measured. The value taken in the paper was the lowest we knew at that time. We have not gotten lower ones in real equipment. I can of course find well cultivated surfaces pretreated in a special way that have a much lower destruction rate. But in the real device, and we checked three different designs, we always got the value that I used and it is also about the same as in the paper by Morrison et al.

E. KUFFEL: Your modeling focuses on the worst scenarios specifically on the cases of disconnect switches and particle initiated discharge. There have been many studies conducted on the particle initiated discharges. Has anyone reported values of S_2F_{10} of concentration values of 40 to 50 ppmv in practical or laboratory studied systems?

L. NIEMEYER: To the author's knowledge S_2F_{10} concentrations in the ppmv range have not been reported in operating GIS. A survey of field data is to be published at the CRADA workshop to be held in Pittsburgh in June 1994.

M. F. FRÉCHETTE: What is a surface decomposition coefficient and how is it measured?

L. NIEMEYER: The surface decomposition coefficient δ_s is defined by Eq. (9) in my paper and characterizes the specific decomposition efficiency of the surface irrespective of the absolute values of the surface area A and gas volume V. δ_s can be determined from the measured exponential concentration decay time constant τ, the decomposing area A, and the gas volume V [Eq. (10)].

J. CASTONGUAY: It seems that you used the last CRADA value for the formation rates of S_2F_{10} in sparks ($\approx$ 0.02 nmole/J), but these are 10 times less than the older value of Sauers and $\approx$ 50 times less than mine. What will be the effect of this higher value?

L. NIEMEYER: The spark formation rates by Sauers in fact refer to situations in which humidity is present. In the absence of humidity they are expected to increase by about one order of magnitude as indicated by Dr. Sauers. As your rates are still much higher it would be important to know what the humidity was in your experiment and if the evaluation of the energy dissipated in the spark is comparable with the one used by Sauers.

C. X. WANG: (1) What are the surface decomposition products of S_2F_{10}? Are they no longer toxic? (2) Can you offer some estimates of S_2F_{10} concentration after a power fault?

L. NIEMEYER: (1) The decomposition products are supposed to be SF_4 and SF_6. I do not know if this is really the case since we do not at all understand the catalytics at the surface. The sequence of steps is probably that the S_2F_{10} is decomposed into SF_5 radicals which undergo reactions at the surface to produce eventually SF_4 and SF_6. Of course SF_4 can hydrolyze when there is humidity. But these processes are not well understood. (2) There is not enough information to allow estimates. We do know that S_2F_{10} is decomposed rapidly above temperatures on the order of 300°C so what must create S_2F_{10} in a hot discharge must be some equilibrium mechanism probably at the boundary of the discharge but we do not know the details. [Editors' note: See the paper by Morrison et al. on power arcs in Chapter 8.]

R. J. VAN BRUNT: When S_2F_{10} decomposes on a surface what we see is SOF_2. We are not sure why it is SOF_2 but it appears that water adsorbed on the surface is the key ingredient to the destruction. The surface destruction rate depends very much on the surface conditions, the surface temperature, and the diffusion rate to the surface. There can be a lot of variability for the destruction rate in different kinds of equipment so one has to be cautious in deciding on what is a reasonable destruction rate for this compound.

L. NIEMEYER: We have some preliminary data on surface decay rates since the paper was written. We have looked at various pretreatments of the surface: stainless steel containers like those used for transporting samples and sand-blasted aluminum because that is the material in equipment. Water concentration plays an important role and should be monitored during the experiment.

SIMPLE CHEMICAL ASSAYS FOR PRESENCE OF S_2F_{10} IN DECOMPOSED SF_6

Guy D. Griffin, Maureen R. Baker and Isidor Sauers

Health Sciences Research Division
Oak Ridge National Laboratory
P.O. Box 2008, MS-6101
Oak Ridge, TN 37831-6101

INTRODUCTION

Disulfur decafluoride (S_2F_{10}; also referred to as sulfur pentafluoride) is a highly toxic gas which can be produced by electrical discharges in SF_6.[1,2] It would be very useful, both for safety in handling decomposed SF_6 and for diagnosis of fault conditions, to have a convenient, rapid and sensitive field assay for the presence of S_2F_{10}. A number of different assays for S_2F_{10} with varying levels of sensitivity have been developed in the course of a CRADA initiative "Investigation of S_2F_{10} Production and Mitigation in Compressed SF_6-Insulated Power Systems."[3] These techniques include GC-MS[4], cryogenic GC enrichment[5] and IR absorption.[6] Although all of these procedures have been shown to have utility in detecting S_2F_{10}, and some have great sensitivity (GC-MS and cryogenic enrichment), they do not, in general, lend themselves to rapid field analyses. Ideally, one might wish to have an assay modeled after the "Draeger Tubes," where a gas sample is drawn through some chemical compound, and the resulting color change can be used to quantify the gas of interest. Disulfur decafluoride had been rather extensively studied in World War II as a candidate chemical warfare agent[7], and along with biological studies of its toxicity, chemical detection techniques had also been developed.[8] Of the chemical assays described, two were particularly promising, one involving reaction of S_2F_{10} with NaI to produce free I_2 which could be titrated with sodium thiosulfate, and the other involving reaction of S_2F_{10} with p-phenylenediamine to produce a pink/red color. This second assay appeared to hold some promise, and accordingly, experiments were carried out to see if this assay could be useful for detecting S_2F_{10} in electrically decomposed SF_6 samples.

METHODS AND MATERIALS

The standard procedure by which various gas samples were analyzed by colorimetric assay was as follows. Filter paper discs (Whatman 3MM) were wet with 100 μl of the test chemical in appropriate solvent, and the filter papers were then

Gaseous Dielectrics VII, Edited by L.G. Christophorou
and D.R. James, Plenum Press, New York, 1994

immediately inserted into a 10 ml glass vacutainer tube (Becton-Dickinson 6441) by removing the rubber septum sealing the tube. The rubber septum was immediately replaced (to avoid evaporation of solvent from the filter paper), and the pressure inside the sealed vacutainer was reduced to below atmospheric, by using a plastic syringe and needle (the syringe was used to sequentially remove three 10 ml aliquots of the gas from inside the vacutainer). The gas sample of interest was then introduced into the tube using a needle and syringe. Usually, 8 ml of the gas was drawn into the tube by the reduced pressure in the tube. For dilute gas samples, an attempt was made to introduce the needle directly over the filter paper, so that as the gas entered the tube, it would impinge directly onto the moist paper. The color change on the filter paper as a whole or in a linear region of the paper over which the gas flowed, was observed. When p-phenylenediamine was used for the assays reported in Table 2, the assay solution was 5% (w/v) of p-phenylenediamine in acetone.

For spectrophotometric assays of absorption bands due to reactions of gases with the assay chemical, the colored material was either eluted from the paper using the same solvent the chemical was originally dissolved in, or a 1.0 ml aliquot of the chemical solution was placed in a 10 ml vacutainer tube, and the gas sample introduced as explained above. In this case, the gas sample was mixed with the chemical in solution by vigorous mixing and inversion. Absorption spectra were determined using an IBM Spectrophotometer, model 9420. Fluorescence measurements were made using a Perkin-Elmer LS50B Luminescence Spectrometer.

Analysis of individual gas mixtures was by mass spectrometry and/or gas chromatography[9]. Individual gases were from the following sources: SiF_4, HF, SF_4, and SO_2, Matheson Gas Products; SO_2F_2, Union Carbide, Linde Division; SOF_2, Columbia Organic Chemicals Co.; SOF_4, S_2F_{10} and $S_2O_2F_{10}$ were synthesized and provided by Dr. D. DesMarteau, Clemson University.

All solvents used were reagent grade and from standard laboratory supply companies. The p-phenylenediamine (1, 4-phenylenediamine) used was from Aldrich (catalog #27, 515-8), 99+% purity, sublimed, and zone-refined. It was important to use this high quality p-phenylenediamine in the assay, as experiments with lower purity grades of this chemical were not as satisfactory. Suppliers for other chemicals used were Aldrich (melamine, phenylhydrazine, diphenylamine, 2,4,6-triaminopyridine, 4-nitroaniline, and 2,4-dinitrophenylhydrazine), Fluka (benzidine), Eastman Kodak (3-aminoquinoline) and Fisher Scientific (catechol).

RESULTS AND DISCUSSION

World War II era attempts[8] to develop a chemical assay included development of an analytical method based upon oxidation of NaI to I_2 by S_2F_{10}, and subsequent titration of the free iodine using thiosulfate. This assay, however, was found to be not nearly as sensitive as the reaction of S_2F_{10} with p-phenylenediamine, which was recommended as a qualitative test for the presence of S_2F_{10}. About 30 easily oxidizable organic compounds had been tested as possible detectors for S_2F_{10}. Based upon the range of compounds used in this old report, we selected a variety of organic compounds which we felt might serve as suitable detectors for S_2F_{10}, and tested them for visual evidence (i.e., a colored product or a color change) of a reaction with S_2F_{10}, when the compound, dissolved in suitable solvent and spotted onto white filter paper, was exposed to a stream of gas containing S_2F_{10}. The concentration of S_2F_{10} used in these studies varied from 100 to 500 ppm. It was felt that no evidence of reaction with these concentration ranges of S_2F_{10} precluded the compound from being a useful indicator of the presence of S_2F_{10}. Table 1 shows the results of these studies. It is clear that of the compounds tested, only p-phenylenediamine showed any real promise as an S_2F_{10} indicator compound. The color change seen upon reaction of S_2F_{10} with p-phenylenediamine was observed to have the following characteristics. The solution of the high purity compound in acetone was a very light tan color when freshly prepared (upon standing, air oxidation produced progressive darkening until the solution became black after ~24 hrs). Upon spotting on filter paper, the fresh solution did not noticeably color the filter paper at all. When a stream of gas containing S_2F_{10} impinged on the paper, a red color developed as a line along the paper, indicative of the direction of gas flow. The intensity (and hue) of this colored product varied with the concentration of S_2F_{10} in the gas stream (see below for further discussion). Certain of the compounds in Table 1 (i.e. p-phenylenediamine and diphenylamine) were investigated as the basis for a fluorimetric assay for S_2F_{10}. In this case, the fluorescence excitation and emission spectra of the compounds was determined in various solvents, and after reaction of a solution with S_2F_{10}, the spectra were re-examined for any changes which might be indicative of a reaction between the S_2F_{10} and the indicator chemical. No consistent change in fluorescence spectra could be discerned following treatment with S_2F_{10}.

Table 1. List of Chemicals Tested for Suitability as a Detector of S_2F_{10}

Chemical	Solvent*	Reaction with S_2F_{10}**
p-Phenylenediamine	Acetone	+
Catechol	Ethanol	-
Benzidine	Ethyl Acetate	-
Diphenylamine	Ethyl Acetate	-
3-Aminoquinoline	Ethyl Acetate	-
4-Nitroaniline	Ethyl Acetate	-
Melamine	Ethyl Acetate	-
2,4,6-Triaminopyrimidine	Dimethyl Sulfoxide	-
Phenylhydrazine	Ethanol	-
2,4-Dinitrophenylhydrazine	Methanol	-

*A solution of 4 or 5% (w/v) of compound in appropriate solvent was made up in each case.

**+ = visible colored product on filter paper; - = no evidence of a color change

Using the p-phenylenediamine reagent, a number of studies were made to see if other decomposition products of SF_6 might react with this reagent, and to determine what the approximate limit of detection (LOD) of various decomposition products might be. Table 2 gives the results of these experiments. The p-phenylenediamine reagent proved to be a quite sensitive indicator of S_2F_{10}. Concentrations of S_2F_{10} in the range of 20-50 ppm produced a definite pink to pink-red line on the filter paper. The limit of detection for an 8 ml sample of gas containing S_2F_{10} was approximately 6 ppm of S_2F_{10} in SF_6. A rather faint pink band appeared at this concentration which required about 30 to 60 sec. to reach its final intensity. Nevertheless, the band was very discernible against the white background of the paper.

This chemical detection technique for S_2F_{10} had originally been developed for detecting the use of S_2F_{10} in air as a chemical warfare agent, and it was therefore not of any particular importance to detect S_2F_{10} in the presence of numerous other sulfur or sulfoxy-fluor compounds. The chemical detection of S_2F_{10} in decomposed SF_6, however, demands a considerable degree of specificity, since a variety of other closely related compounds are produced by electrical decomposition of SF_6.

As can be seen from Table 2, neither SF_6 nor many of the common decomposition products of SF_6 (e.g. HF, SO_2, SO_2F_2, SOF_4 and SiF_4) produced any visible reaction with the color reagent, even at relatively high concentrations (thousands of ppm). Thionyl fluoride, SOF_2, produces a yellow color, totally different from the color produced by S_2F_{10}. Furthermore, the reaction of SOF_2 with p-phenylenediamine is different from that with S_2F_{10} in two respects. First, the reaction of the reagent with SOF_2 is considerably less sensitive than that with S_2F_{10}(LOD$\cong$50ppm for SOF_2). Second, the yellow color produced by the reaction of SOF_2 with the color reagent takes considerable time (minutes) to develop, and this development time lengthens as the concentration of SOF_2 decreases. In contrast, the color of the S_2F_{10} reaction product develops almost instantaneously, even at very low (near LOD) concentrations of S_2F_{10}. In fact when a mixture of S_2F_{10}/SOF_2 (250/475 ppm) was tested with the reagent, the red color due to S_2F_{10} was seen immediately, and eventually the filter paper turned an orange-yellow color, as the SOF_2 reacted. We therefore believe that SOF_2, even when present in fairly high concentrations, may not be a hinderance to detecting S_2F_{10} by this method. Of course many more trials using mixtures would need to be carried out before one could validate the limit of detection of S_2F_{10} in the presence of SOF_2.

Sulfur tetrafluoride (SF_4), on the other hand, produces a color with p-phenylenediamine which is indistinguishable from the color produced by S_2F_{10}. Spectrophotometric analysis of the colored solutions eluted from filter papers reacted with SF_4 and S_2F_{10} revealed the absorption spectra to be indistinguishable. Thus SF_4 could certainly interfere with the detection of S_2F_{10}. Since SF_4 is one of the primary breakdown products produced by electrical discharge in SF_6[9], its presence is certainly a possibility in various SF_6 samples. It should be noted that the LOD of SF_4 is much higher than S_2F_{10}.

Sulfur tetrafluoride is rapidly hydrolyzed by trace moisture to SOF_2 and HF[10], so the amount of SF_4 remaining in a particular sample is very dependant upon conditions within the electrical apparatus, sampling procedures, etc. Whether the presence of SF_4 would preclude detection of S_2F_{10} would depend in part, upon absolute concentrations of both components. Possible strategies to eliminate SF_4 (e.g. exposure of samples to moist atmospheres) might also be successful.

Table 2. Reaction of p-Phenylenediamine with Various Gases

Compound	Description of Reaction- Approximate Limit of Detection (LOD) or Highest Concentration Tested Giving No Reaction
S_2F_{10}	Red to Pink (higher to lower concentrations) LOD $\cong$ 6 ppm (faint pink) 20 ppm - immediately discernible pink
SOF_2	Yellow to Pale Yellow (higher to lower concentrations) LOD $\cong$ 50 ppm (pale yellow which develops only after a delay of several minutes)
SF_4	Red to Pink (higher to lower concentrations) LOD $\cong$ 100 ppm (faint pink)
$S_2O_2F_{10}$	Very faint reddish tan at 333 ppm*
HF	No Reaction-5,000 ppm
SO_2F_2	No Reaction-2,000 ppm
SOF_4	No Reaction-2,000 ppm
SO_2	No Reaction-2,000 ppm
SiF_4	No Reaction-2,000 ppm
SF_6	No Reaction-100%

*Preliminary Results

The compound $S_2O_2F_{10}$ also produced a reddish color in the color assay. This is a very preliminary result, however, as only one assay (3 different concentrations) was performed, and we are unsure of the purity of the compound, which had to be custom-synthesized for us.

The preliminary results to date suggest that this colorimetric assay may have some utility in assaying for the presence of S_2F_{10} in decomposed SF_6 samples. This colorimetric assay is unlikely to be able to detect S_2F_{10} at its TLV (0.01 ppm), approximately three orders of magnitude below our current LOD. On the other hand, the LOD listed in Table 2 of 6 ppm is based upon a gas sample of ~8-9 mls total volume. Preliminary experiments have indicated that concentrations of S_2F_{10} lower than 6 ppm (i.e. ~1 ppm) can be detected, if the sample volume of gas passed over the filter is increased. The color reaction appears to be accumulative; that is, color develops as a function of total amount of S_2F_{10} reacting on the filter paper surface. If S_2F_{10} is present in the gas in lower concentrations, then more gas volume must flow over the reactive surface, than if it is present at higher concentrations. We therefore feel that the LOD of the colorimetric reaction for S_2F_{10} could be lower than indicated in Table 2. The assay procedure would have to be modified, however, to incorporate larger gas volumes. Even though this colorimetric assay may not be sensitive enough to detect S_2F_{10} at TLV levels, it may serve a useful purpose as a screening assay for evidence of potentially hazardous levels of certain highly toxic SF_6 decomposition products (e.g. SF_4 and S_2F_{10}, with TLV's of 0.1 and 0.01 ppm respectively).[11,12] The speed and simplicity of the assay both recommend it for this purpose. Use of this colorimetric assay in the field will require further development, however. One potential problem is the instability of the p-phenylenediamine in acetone. The organic compound oxidizes rapidly (overnight) to produce a black solution, of no use for assay purposes. The reagent solution would need to be stabilized, or perhaps made freshly before an assay.

ACKNOWLEDGEMENTS

The helpful assistance of Beverly Norton in all phases of manuscript preparation is gratefully acknowledged. This work was conducted under a Cooperative Research and Development Agreement supported by the Office of Energy Management, U.S. Department of Energy under contract DE-AC05-84OR21400 with Martin Marietta Energy Systems, Inc., and the Bonneville Power Administration, Canadian Electrical Association, Electric Power Research Institute, Empire State Electric Energy Research Corporation, Hydro-Quebec, National Institute of Standards and Technology, Ontario Hydro, and Tennessee Valley Authority.

REFERENCES

1. I. Sauers, P.C. Votaw, and G.D. Griffin, Production of S_2F_{10} by SF_6 Spark Discharges. *Proceedings of the IX International Conference on Gas Discharges and Their Applications*, Venice, Sept. 19-22, 1988, pp. 592-594.

2. I. Sauers, G. Harman, J.K. Olthoff, and R.J. VanBrunt, S_2F_{10} Formation by electrical discharges in SF_6: Comparison of spark and corona, *in* "Gaseous Dielectrics VI," L.G. Christophorou and I. Sauers, eds., Plenum Press, New York, pp. 553-562, 1991.

3. D.R. James (ed), I. Sauers, G.D. Griffin, R.J. VanBrunt, J.K. Olthoff, K.L. Stricklett, F.Y. Chu, J.R. Robbins and H.D. Morrison, Investigation of S_2F_{10} Production and Mitigation in Compressed SF_6-Insulated Power Systems, *IEEE Electrical Insulation Magazine* 9(3):29-40, 1993.

4. J.K. Olthoff, R.J. Van Brunt, J.T. Herron, and I. Sauers, Detection of Trace Disulfur Decafluoride in Sulfur Hexafluoride by Gas Chromatography/Mass Spectrometry, *Anal. Chem.* 63(7):726-723, 1991.

5. I. Sauers and R.A. Cacheiro, A Cryogenic Enrichment Technique for Gas Chromatographic Detection of S_2F_{10} in SF_6 Discharges, *Conference Record of the 1992 IEEE Int. Symp. Elec. Insul.*, June 7-10, 1992 Baltimore, MD, IEEE Pub. No. 92CH3150-0, pp. 340-344, 1992.

6. C.S. Viera, J.R. Robins, H.D. Morrison, and F.Y. Chu, The application of infrared absorption spectroscopy in gas-insulated equipment diagnostics, *in* "Gaseous Dielectrics VI," L.G. Christophorou and I. Sauers, eds., Plenum Press, New York, pp. 539-544, 1991.

7. B. Renshaw and M. Gates, Disulfur Decafluoride, *Chemical Warfare Agents and Related Chemical Problems Parts I-II*, Office of Scientific Research and Development, National Defense Research Committee, Washington, D.C. NTIS no. PB-158508, pp. 24-29, 1946.

8. D.S. Tarbell, The Chemical Detection of 1120, Office of Scientific Research and Development, National Defense Research Committee, Washington, D.C., OSRD 879, pp. 1-21, 1942.

9. I. Sauers, H.W. Ellis, and L.G. Christophorou, Neutral Decomposition Products in Spark Breakdown of SF_6, *IEEE Trans. Elect. Insul.* EI-21(2):111-120, 1986.

10. I. Sauers, J.L. Adcock, L.G. Christophorou, and H.W. Ellis, Gas Phase Hydrolysis of Sulfur Tetrafluoride: a Comparison of the Gaseous and Liquid Phase Rate Constants, *J. Chem. Phys.* 83(5):2618-2619, 1985.

11. G.D. Griffin, I. Sauers, K. Kurka, and C.E. Easterly, Spark Decomposition of SF_6:Chemical and Biological Studies, *IEEE Trans. Power Delivery* 4(3):1541-1551, 1989.

12. G.D. Griffin, M.G. Nolan, C.E. Easterly, I. Sauers, and P.C. Votaw, Concerning Biological Effects of Spark-Decomposed SF_6, *IEE Proceed.* 137(Part A, No. 4):221-227, 1990.

DISCUSSION

J. CASTONGUAY: Did you check the response of your new S_2F_{10} assay with mixtures (synthetic) of various SF_6 by-products?

G. D. GRIFFIN: Only one mixture was tested, i.e., 250 ppm of S_2F_{10}, 475 ppm of SOF_2 both in SF_6. This mixture produced an immediate red color, indicative of S_2F_{10}. We concluded that it was possible to detect S_2F_{10} in mixtures with relatively large amounts of SOF_2, because the reaction with SOF_2 is not nearly as sensitive as the reaction with S_2F_{10}, and also because the color produced by SOF_2 does not immediately develop, while the S_2F_{10} reaction is almost instantaneous.

J. CASTONGUAY: Did you check it with corona degraded SF_6 since it is "free" of SF_4 that could interfere with it?

G. D. GRIFFIN: The chemical detection assay for S_2F_{10} has not been carried out on corona-degraded SF_6 samples. Using the assay on arc-decomposed samples, we have observed an intense reddish-orange color, which might be presumed to be the result of reaction with SOF_2 (yellow), SF_4 (red), and perhaps S_2F_{10} (red). The contribution of S_2F_{10} to the final color would depend on the absolute amount of S_2F_{10} present.

APPEARANCE POTENTIALS OF IONS PRODUCED BY ELECTRON-IMPACT INDUCED DISSOCIATIVE IONIZATION OF SF_6, SF_4, SF_5Cl, S_2F_{10}, SO_2, SO_2F_2, SOF_2, AND SOF_4

Kenneth L. Stricklett, Jason M. Kassoff,* James K. Olthoff, and Richard J. Van Brunt

Electricity Division
National Institute of Standards and Technology†
Gaithersburg, Maryland 20899

INTRODUCTION

The identification of S_2F_{10}, a highly toxic compound,[1] in SF_6 that has been subjected to electrical discharges, including negative corona discharges,[2,3] sparks,[4] and arcs,[5] has raised issues of safety and proper handling of SF_6 removed from electrical equipment, and motivates the need for trace detection of S_2F_{10} in SF_6.[6] A reliable measurement protocol for trace detection of S_2F_{10} may be applied to: 1) develop safe handling procedures for SF_6 gas removed from operating machinery, 2) ensure compliance with regulatory agencies, 3) monitor the purity of SF_6 supplied both by manufacture and reprocessing, and 4) replace animal toxicity tests of commercial SF_6.

Conventional mass spectrometry is of limited value in detection of trace levels of S_2F_{10} in SF_6 due to the similarity in their mass spectra.[7] Two methods have shown detection sensitivity down to the peak exposure limit of 10 parts in 10^9 by volume (ppb): a technique employing cryogenic-enrichment[4] of S_2F_{10} and an indirect method that relies on gas-chromatographic separation followed by chemical conversion of S_2F_{10} to SOF_2.[8] Both methods have their advantages and limitations. The first is time consuming and laborious, while the second has a reduced sensitivity to S_2F_{10} when SOF_2 is present at high concentrations.[9] Disulfur decafluoride (S_2F_{10}) normally occurs in concert with other discharge by-products, e.g., SOF_2, SOF_4, and SO_2.

The present study was undertaken to evaluate non-conventional mass spectroscopy, i.e., mass spectroscopy at electron-impact energies well below the standard of 70 eV, as a possible method to increase sensitivity of S_2F_{10} detection. This method has been successfully used to detect SF_4 in SF_6.[10] An additional purpose of this work is to obtain appearance potential data for other compounds that may be present in SF_6 that has been exposed to electrical discharge.

*Summer student, Princeton University

†Electronics and Electrical Engineering Laboratory, Technology Administration, U. S. Department of Commerce

Gaseous Dielectrics VII, Edited by L.G. Christophorou and D.R. James, Plenum Press, New York, 1994

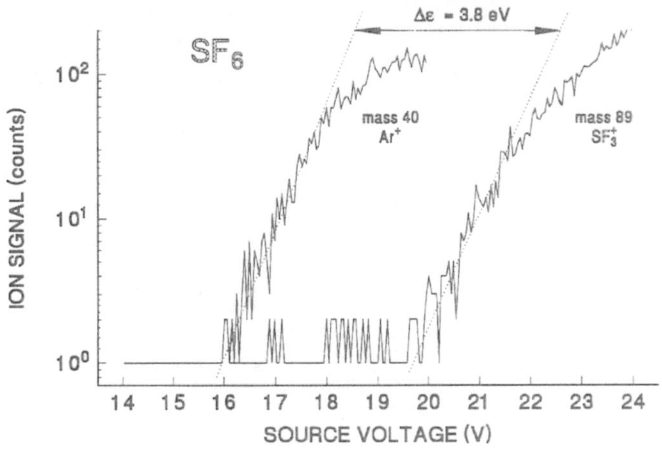

Figure 1. The observed ionization efficiency curves for Ar^+ and SF_3^+. The semi-log plots of the curves are fit by linear functions and the difference in their x-axis intercepts, $\Delta\varepsilon$. The appearance potential of SF_3^+ determined by this method is 15.76 + 3.8 eV or 19.6 eV, in good agreement with earlier reported values (see Table 1 below).

EXPERIMENTAL METHOD

Near the ionization threshold, the ion signal observed using a conventional quadrupole mass spectrometer has an exponential dependence on the accelerating potential. This is due primarily to the energy distribution of the electrons emitted from the filament.[11] Thus a semi-log plot of the ion signal near threshold may be fit by a straight line whose intercept with the baseline noise level is, to a good approximation, the appearance potential. This technique provides a means of determining ion appearance potentials using a commercially available quadrupole mass spectrometer. The electron impact energy may be varied by changing the accelerating potential between the filament and first lens element in the quadrupole ionizer while the ion of interest is detected using single-ion monitoring.

For the results presented here, the electron energy is determined by way of reference to the spectroscopic ionization potential of argon (15.76 eV).[12] The sample gas is admixed with argon gas and the ionization efficiency curves are recorded sequentially. The results of such a measurement for SF_3^+ produced by dissociative ionization of SF_6 are presented in Figure 1.

RESULTS

Appearance potential measurements for the compounds included in this study are summarized and compared with other similar measurements in Tables 1, 2, and 3 below. Overall, agreement with earlier published results is good, i.e., in all cases, the measured appearance potential lies within the combined measurement uncertainties. The stated measurement uncertainties are influenced by the quality of the linear fit, the difference in slopes between the reference and measured ions, and the baseline noise level. For minority ions, where the ion signal is comparable to the background noise, the measurement uncertainties for this method can be large (greater than ±1 eV). Additionally, as noted in the tables, many of the measured ionization efficiency curves

Table 1. Appearance potential measurements for SF_6, SF_4, and S_5Cl.

		Appearance Potential (eV)	
Compound	Ion	This Work	Previous Results
SF_6	SF_5^+	15.6±0.2	15.9±0.2[a], 16.2±0.2[b], 15.50±0.10[c]
	SF_4^+	18.7±0.5	18.9±0.2[a], 19.6±1.0[b], 18.44±0.10[c]
	SF_3^+	19.6±0.5	20.1±0.3[a], 19.6±0.5[b], 20.0±0.5[c]
	SF_2^+	26±3	26.8±0.3[a], 27.0±0.3[b], 27.5±0.5[c]
	SF^+	30±2	31.3±0.3[a], 37.6±3.0[b], 30.5±0.5[c]
	S^+	39±5†	37.3±1.0[a], 37.5±1.0[b]
	F^+	36±5†	35.8±1.0[a]
	SF_4^{++}	39±4	40.6±0.5[a]
SF_4	SF_4^+	12.2±0.5	12.08±0.10[c], 11.90[d]
	SF_3^+	12±1	12.63±0.10[c], 12.4[d]
	SF_2^+	18±2†	17.4±0.5[c], 16.90[d]
	SF^+	22±2†	
	S^+	30±5†	
	F^+	33±5†	
	SF_2^{++}	39±4†	
SF_5Cl	SF_5^+	12.5±0.7	12.32[e]
	SF_4Cl^+	14.5±0.5	14.76[e]
	SF_4^+	16±1	15.87[e]
	SF_3^+	18±4	16.2[e]
	SF_2^+	24±3	
	SF^+	29±2	
	S^+	38±3	

† Observed ionization efficiency curve showed an apparent onset below the value indicated.
[a] Appearance potentials determined by electron impact, Reference 14.
[b] Appearance potentials determined by electron impact, Reference 15.
[c] Appearance potentials determined by electron impact , Reference 16.
[d] Appearance potentials from photoeletron-photoion-coincidence measurements, Reference 13.
[e] Appearance potentials from photoion spectrum, Reference 17.

showed an apparent onset for ion production at low energies, which tended to obscure the threshold. Two possible explanations are offered for this behavior. First, the hot filament in the ionizer may introduce unwanted thermochemical reactions that produce species with ionization potentials lower than that of the sample gas. The mass spectra of SOF_2 shown in Figure 2, for example, show mass peaks corresponding to the presence of SO_2 and provide clear evidence of chemical conversion in the ionizer. Second, polar dissociation may produce positive ions at impact energies below dissociative ionization. This process is know to produce SF_3^+ from SF_4 at an excitation energy of 9.00 eV,[13] approximately 3.4 eV below the corresponding ionization appearance potential.

The potential for improved sensitivity to S_2F_{10} is demonstrated by data presented in Figure 3. With even modest electron energy resolution (the energy resolution for these measurements was approximately 1 eV), SF_3^+ from SF_6 and S_2F_{10} may be resolved.

Table 2. Appearance potential measurements for SOF_2, SO_2F_2, SOF_4, and SO_2.

		Appearance Potential (eV)	
Compound	Ion	This Work	Previous Results
SO_2	SO_2^+	12.0±0.5	12.4±0.1[a], 12.3[b], 12.50[c]
	SO^+	15.0±0.5†	16.2±0.2[a], 15.930±0.005[d]
	S^+	17±2†	17.5±0.2[a], 16.334[d]
	O^+	21±2†	20.6[a]
SO_2F_2	$SO_2F_2^+$	13.3±0.5	13.3±0.1[a], 13.0[c], 13.04±0.01[e]
	SO_2F^+	14.9±0.5	15.1±0.2[a]
	SOF^+	17.9±1.0	18.6±0.1[a]
	SO_2^+	21±2†	19.9±0.3[a]
	SO^+	25±3†	24.3±0.3[a]
	S^+	33±3†	
SOF_2	SOF_2^+	12.5±0.5	12.25[c], 12.19[f]
	SOF^+	13.8±1.0	
	SO^+	16.0±1.0†	
	S^+	25±4†	
SOF_4	SOF_4^+	12.7±0.5	
	SOF_3^+	13.0±0.2	
	SF_3^+	17.4±0.5	
	SOF_2^+	18±1†	
	SF_2^+	24±2†	
	SOF^+	20±2†	
	SF^+	29±2†	
	SO^+	28±3†	
	S^+	36±3†	

† Observed ionization efficiency curve showed an apparent onset below the value indicated.
[a] Electron impact, Reference 18.
[b] Vertical ionization potential from photoelectron spectrum, Reference 19.
[c] Adiabatic ionization potential from photoelectron spectrum, Reference 20.
[d] Appearance potentials from photoeletron-photoion-coincidence measurements, Reference 21.
[e] Adiabatic ionization potential from photoelectron spectrum, Reference 22.
[f] Adiabatic ionization potential from photoelectron spectrum, Reference 23

These data indicate that use of an electron energy of about 16 eV in the ion source will completely suppress SF_3^+ production from SF_6.

DISCUSSION AND CONCLUSIONS

The principle ions produced by electron-impact ionization of S_2F_{10} are clearly distinguished from those produced in SF_6 by their appearance potentials. The lower appearance potentials of the larger fragment ions such as SF_5^+ and SF_4^+ from S_2F_{10} is expected from the relatively weak S–S bond strength compared to the S–F bond strength

Table 3. Appearance potential measurements for S_2F_{10}.

		Appearance Potential (eV)	
Compound	Ion	This Work	Previous Results
S_2F_{10}	SF_5^+	13.0±0.2	13.2±0.3[a]
	SF_4^+	12.8±0.7, 18±1[†]	
	SF_3^+	12.9±0.2	13.3±0.3[a]
	SF_2^+	20±2[†]	
	SF^+	33±3[†]	

[†] Observed ionization efficiency curve showed an apparent onset below the value indicated.
[a] Electron impact Reference 7.

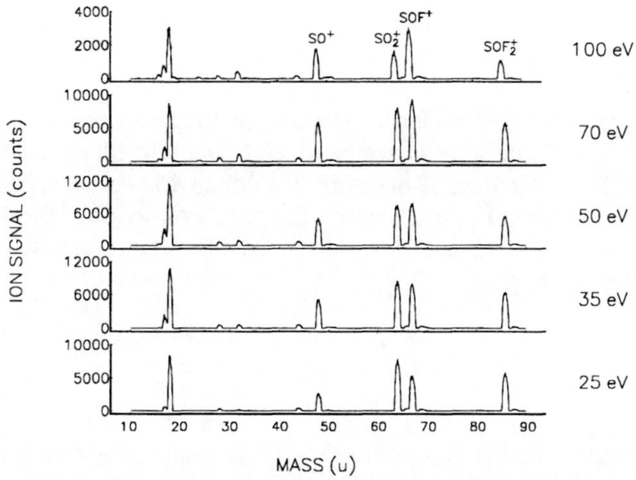

Figure 2. Mass spectra for SOF_2. The electron impact energy for each spectrum is indicated on the right. Note the presence of a mass peak corresponding to SO_2^+, a clear indication of the presence of SO_2. The presence of SO_2 is an artifact that is most likely due to the exposure of SOF_2 to the hot filament in the quadrupole ionizer.

in SF_6.[24,25] At the present time there is no explanation for why the appearance potentials of the fragments SF_5^+, SF_4^+, and SF_3^+ from S_2F_{10} all have nearly the same values. Although this could be due to contributions from polar dissociation near threshold, there is no evidence from other measurements[26] of significant negative-ion formation from S_2F_{10} at electron energies above 13 eV. The results presented here suggest that a judicious choice of electron-impact energy holds promise for improved sensitivity to S_2F_{10} in SF_6 when using a conventional quadrupole mass spectrometer and ionizer with or without a gas chromatograph.

The presence of SO_2 in the mass spectrum for SOF_2 shown above in Figure 2 suggests that thermochemical reactions may be catalyzed on the hot filament in the ionizer. Such reactions may also occur for many of the compounds examined here and may contribute to the low-energy onset for production of positive ions noted above. Polar-dissociation, i.e., electron impact followed by unimolecular dissociation to form

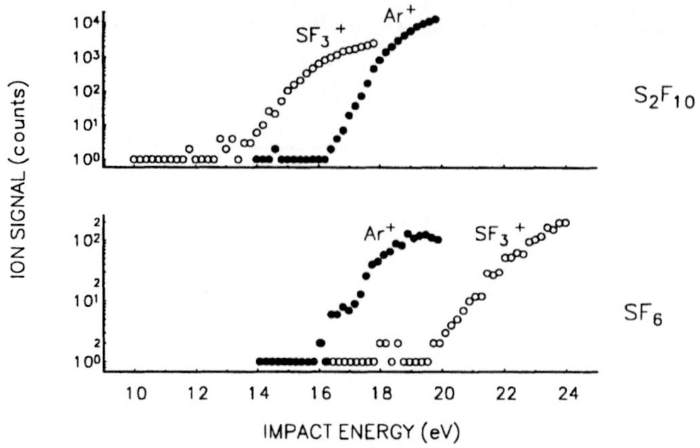

Figure 3. Comparison of the observed ionization efficiency curves for SF_3^+ from S_2F_{10} and SF_6.

stable positive and negative ions, may also contribute to low-energy production of positive ions.

ACKNOWLEDGMENTS

This work was conducted under a Cooperative Research and Development Agreemer supported by the Office of Energy Management, U.S. Department of Energy, Martin Mariett Energy Systems/Oak Ridge National Laboratory, the Bonneville Power Administration, Canadia Electrical Association, Electric Power Research Institute, Empire State Electric Energy Researc Corporation, Hydro-Québec, National Institute of Standards and Technology, Ontario Hydrc and Tennessee Valley Authority.

REFERENCES

1. L.A. Greenberg and D. Lester, The toxicity of sulfur pentafluoride, *Arch. Indust. Hygiene and Occupat. Medicine* 2:350 (1950).
2. R.J. Van Brunt, J.K. Olthoff, and M. Shah, Rate of S_2F_{10} production from negative corona in compressed SF_6, in: "Conference Record of the 1992 IEEE Symposium on Electrical Insulation," IEEE, Piscataway, NJ (1992).
3. A.M. Casanovas, J. Casanovas, F. Lagarde, and A. Belarbi, Study of the decomposition of SF_6 under dc negative polarity corona discharge (point-to-plane geometry): Influence of the metal constituting the plane electrode, *J. Appl. Phys.* 72:3344 (1992).
4. I. Sauers and S.M. Mahajan, Detection of S_2F_{10} produced in by a single-spark in SF_6, *J. Appl. Phys.* 74:2103 (1993).
5. F.J.G. Janssen, Measurements of sub-ppm level of sulfur-fluoride compounds resulting from the decomposition of SF_6 by arc discharge, *Kema Sci. Tech. Rep.* 2:9 (1994).
6. D.R. James, I. Sauers, G.D. Griffin, R.J. Van Brunt, J.K. Olthoff, K.L. Stricklett, F.Y. Chu, J.R. Robins, and H.D. Morrison, Investigation of S_2F_{10} production and mitigation in compressed SF_6-insulated power systems, *IEEE Elec. Ins. Mag.* 9:29 (1993).
7. J.K. Olthoff, R.J. Van Brunt, and I. Sauers, Electron-energy dependence of the S_2F_{10} mass spectrum, *J. Phys. D: Appl. Phys.* 22:1399 (1989).

8. J.K. Olthoff, R.J. Van Brunt, J.T. Herron, and I. Sauers, Detection of trace disulfur decafluoride in sulfur hexafluoride by gas chromatography/mass spectrometry, *Anal. Chem.* 63:726 (1991).

9. R.J. Van Brunt, J.K. Olthoff, K.L. Stricklett, and D.J. Wheeler, Procedure for measuring trace quantities of S_2F_{10}, S_2OF_{10}, and $S_2O_2F_{10}$, in SF_6 using a gas chromatograph-mass spectrometer, these proceedings.

10. I. Sauers, H.W. Ellis, L.C. Frees, and L.G. Christophorou, Detection of SF_4 in sparked SF_6, *IEEE Trans. Elec. Ins.* 17:284 (1982).

11. R.E. Honig, Ionization potentials of some hydrocarbon series, *J. Chem. Phys.* 16:105 (1948).

12. L. Minnhagen, Spectrum and the energy levels of neutral argon, Ar I, *J. Opt. Soc. Am.* 63:1185 (1973).

13. H.-W. Jochims, E. Rühl, and H. Baumgärtel, Photoionization and ion fragmentation of SF_4, *Z. Naturforsch.* 44b:13 (1989).

14. V.H. Dibeler and F.L. Mohler, Dissociation of SF_6, CF_4, and SiF_4 by electron impact, *J. Res. Natl. Bureau Standards* 40:25 (1948).

15. B.P. Pullen and J.A.D. Stockdale, Dissociative ionization of SF_6 by electron impact, *Int. J. Mass Spec. Ion Phys.* 19:35 (1976).

16. D.L. Hildenbrand, Mass spectrometric studies of some gaseous sulfur fluorides, *J. Phys. Chem.* 77:897 (1973).

17. H. Baumgärtel, H.-W. Jochims, and E. Rühl, Photoion spectroscopy of sulfurchloridepentafluoride SF_5Cl, the ionization potential of sulfurpentafluoride SF_5. *Z. Naturforsch.* 44b:21 (1989).

18. R.M. Reese, V.H. Dibeler, and J.L. Franklin, Electron impact studies of sulfur dioxide and sulfuryl fluoride, *J. Chem. Phys.* 29:880 (1958).

19. D.R. Lloyd and P.J. Roberts, The assignment of the photoelectron spectrum of sulfur dioxide, *Mol. Phys.* 26:225 (1973).

20. D. Chadwick, D.C. Frost, F.G. Herring, A. Katrib, C.A. McDowell, and R.A.N. McLean, Photoelectron spectra of sulfuryl and thionyl halides, *Can. J. Chem.* 51:1893 (1973).

21. M.J. Weiss, T.-C. Hsieh, and G.G. Meisels, Fragmentation of SO_2^+ prepared in state selected vibrational levels, *J. Chem. Phys.* 71:567 (1979).

22. R.L. DeKock, D.R. Lloyd, L.H. Hillier, and V.R. Saunders, Experimental and theoretical study of the electronic structures of sulphuryl fluoride and perchloryl fluoride, *Proc. R. Soc. Lond.* A328:401 (1972).

23. G.W. Mines, R.K. Thomas,and H. Thompson, Photoelectron spectra of compounds containing thionyl and sulphuryl groups, *Proc. R. Soc. Lond.* A329:275 (1972).

24. G.L. Gustev, The structure and stability of S_2F_{10} and S_2F_{11} and their anions, *Chem. Phys.* 158:33 (1991).

25. W. Tsang and J.T. Herron, Kinetics and thermodynamics of the reaction $SF_6 \leftrightarrow SF_5 + F$, *J. Chem. Phys.* 96:4272 (1992).

26. J.K. Olthoff, K.L. Stricklett, R.J. Van Brunt, J.H. Moore, J.A. Tossell, and I. Sauers, Dissociative electron attachment to S_2F_{10}, S_2OF_{10}, and $S_2O_2F_{10}$, *J. Chem. Phys.* 98:9466 (1993).

DISCUSSION

S. R. HUNTER: Why are there such large uncertainties stated for the appearance potentials of some ion fragments?

J. K. OLTHOFF: As can be seen in Fig. 2, there is evidence that the hot filament in the electron impact source affects the gas in the source region. For the ion fragments for which we have quoted large uncertainties, structure is observed in the ionization efficiency curves, which we attribute to excitation or dissociation occurring on the filament. This structure makes it impossible to get a unique linear fit to the efficiency curve (see Fig. 1) and therefore increases the uncertainties.

J. CASTONGUAY: (1) What detector absolute sensitivity reduction did you observe by lowering the electron energy from 70 to 16 eV? (2) How has lowering the electron energy affected the intensities of the SOF_2 ion fragments present in the mass spectrum of S_2F_{10}?

J. K. OLTHOFF: Lowering of the electron energy from 70 eV to 16 eV reduces the ion signal by nearly 2 orders of magnitude for S_2F_{10}. We have not yet attempted to adjust the electron impact energies for improved detection of S_2F_{10} by GC/MS. Obviously we must investigate the trade-off between the increased selectivity and the reduced sensitivity inherent in reducing the electron energy.

SECTION 5: BREAKDOWN CHARACTERISTICS UNDER SHORT TIME FRONTS AND FAST TRANSIENTS

DIELECTRIC STRENGTH OF SF_6 IN NONUNIFORM GAPS STRESSED BY VERY FAST TRANSIENT VOLTAGES

Th. Flohr, W. Pfeiffer, C. Zender, V. Zimmer

Technische Hochschule Darmstadt
Landgraf-Georg-Str.4
64283 Darmstadt
Germany

ABSTRACT

This paper deals with the dielectric strength of SF_6 in strongly nonuniform gaps stressed by very fast transient voltages (VFT-voltages) of both polarities. In these investigations the shape of the applied voltage simulates the voltages that may occur in gas insulated substations (GIS) due to switching operations. Therefore sinusoidal voltages of different frequencies up to 15 MHz were superimposed to a voltage step. The maximum voltage amplitude was 250 kV peak. The parameters that have been investigated, are the frequency of the superimposed sinusoidal voltage, the polarity, the gap width, and the gas pressure. The gap width varied from 15 mm to 25 mm and the gas pressure from 0.05 MPa to 0.4 MPa. For diagnostics both optical and electrical measuring techniques were used to analyse the prebreakdown and breakdown phenomena. Optical investigations were performed with ultraviolet- and infrared-sensitive photomultiplier tubes. For spatial resolved optical measurements of the predischarge development a high speed camera system with an exposure time of 1 ns was used.

The results presented here show that the dielectric strength slightly increases with the VFT-frequency for both, homogenous and inhomogenous fields. In comparison with homogenous field arrangements - where the dielectric strength rises linearly with the gas pressure independent of the polarity of the VFT - inhomogenous fields provide a significantly smaller increase of the dielectric strength as well as an area of constant breakdown voltage. This plateau of the dielectric strength moves with increasing frequency of the VFT and increasing gap width towards smaller gas pressures. Furthermore positve polarity of the VFT results in a smaller dielectric strength of inhomogenous fields than negative polarity.

1 INTRODUCTION

Steep fronted transients are generated in GIS during operation of disconnectors[1]. At each irregularity of the characteristic impedance inside the GIS and at the interfaces of the

Gaseous Dielectrics VII, Edited by L.G. Christophorou
and D.R. James, Plenum Press, New York, 1994

GIS with other equipment these waves are reflected and superimposed to the steady state voltage. The characteristic frequency of such very fast transient voltages (VFT) is determined by the distance between the disconnector contact and the irregularity next to the disconnector[2]. These overvoltages can cause insulation failure inside[1] and outside GIS[3].

The influence of the oscillation frequency of the VFT-voltage on the breakdown voltage is not clear until now, as different information has been published. Although there is a frequency mixture existing in real GIS, a monofrequent voltage shape has been used in order to evaluate the effect of the fundamental frequency upon the insulating behaviour of the test gap. With a multifrequent voltage shape the performance of an individual GIS-arrangement can be tested but such results do not provide sufficient basic information about the problem. Furthermore in many real GIS-arrangements a single frequency, which is dominant, can be observed[5].

2 EXPERIMENTAL SETUP

The experimental setup is shown in figure 1. All investigations were performed using a high frequency sinusoidal voltage superimposed to a voltage step. This voltage shape is generated by excitation of a damped series resonant circuit including a variable capacitor (C_x) and a variable inductor (L_x). The connection of a charged low-inductive high voltage capacitor (C_L) to the resonant circuit by a switching gap (SG) induces a damped oscillation. The damping mainly results from the resistive losses of the inductor and its electro-magnetic radiation. The ratio of the first peak of the damped VFT to the amplitude of the voltage step is in the order of 1.8 to 2.0 depending on the frequency of the VFT.

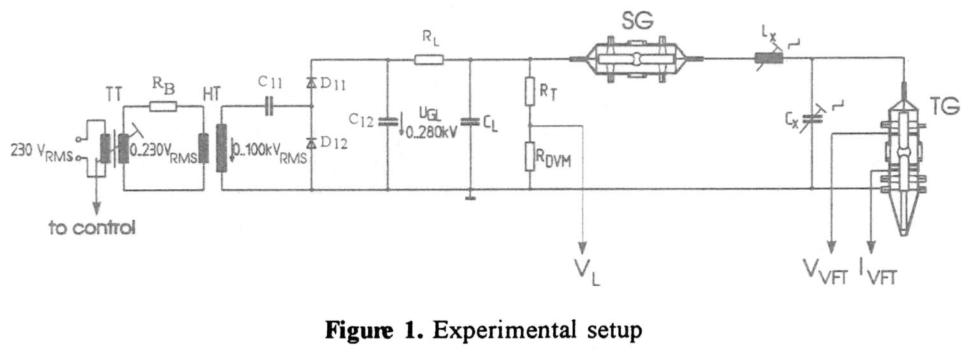

Figure 1. Experimental setup

The experimental setup was optimized to achieve a voltage shape across the test gap (see figure 2) oscillating at high frequencies with low distortion. The four frequencies regarded here are 2.6 MHz, 5.3 MHz, 10.6 MHz and 14 MHz. The maximum peak voltage is 250kV.

2.1 TEST GAP

A cross section of the test gap is shown in figure 2. The VFT-voltage is coupled to the test gap. To avoid reflections an exponential load resistor (3) which resistance equals to the characteristic impedance of the system (i.e. 50Ω) is used. The discharge phenomena are observed through two quartz glass windows (2) for optical analysis (i.e. streamer/leader detection with photomultipliers). The waveform of the applied voltage V_{VFT} and the current I_{VFT} are measured with capacitive probes (4,5,6). The two hemispherical electrodes (7) with

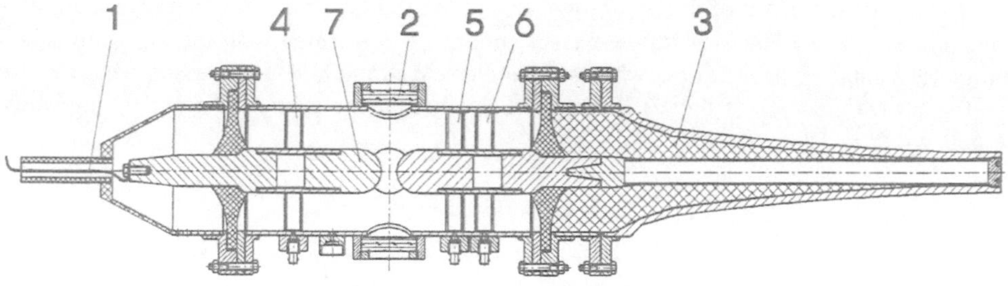

Figure 2. Test gap inside test vessel

a diameter of 60.8 mm are made of brass. The hemispheric top of the high voltage electrode contains a steel needle with 3mm length and a diameter of 100μm.

Measurements at a gap with a Rogowski profile with a nearly homogenous field distribution are performed for comparison.

2.2 MEASURING SETUP

Both signals V_{VFT} and I_{VFT} were recorded by a two channel digital storage oscilloscope. Additional to the electrical probes two photomultipliers were used to record the early predischarge phenomena; one solar-blind photomultiplier with high ultraviolet-sensitivity and an infrared photomultiplier in order to detect a streamer inception or the leader development, respectively.

The whole experimental setup can be controlled by a personal computer (PC) which also allows to perform series measurements. This is required due to the statistical scatter of some phenomena.

3 RESULTS

All investigations of the influence of the frequency of the VFT-voltage on the dielectric strength of SF_6 are performed in SF_6 with additional preionisation by the ultraviolet radiation of a mercury arc lamp. This reduces the scatter of the measured values. For gas pressures up to 0.2 MPa the preionization of the test gap reduces the maximum dielectric strength by about 10 kV. For greater gas pressures the influence of preionization on the dielectric strength can be neglected. According to IEC42[7] the dielectric strength is regarded as the smallest amplitude of V_{VFT} of the VFT first causing breakdown. Hence, this value was determined from a sample of 30 measurements.

The investigated pressure range is from 0.05 MPa to 0.4 MPa whereas the gap width is varied from 15mm to 25mm.

3.1 INFLUENCE OF THE FREQUENCY OF THE VFT-VOLTAGE ON THE DIELECTRIC STRENGTH

It is well known that the dielectric strength - as well as the increase of the dielectric strength with the gas pressure - is far less in an inhomogenous field than in a homogenous field.

To evaluate the effect of the inhomogeneity of the field on the insulating characteristics of SF_6 stressed by VFT, comparative measurements are performed with inhomogenous and homogenous gaps. Hence, figure 3 shows the dielectric strength versus the SF_6 gas pressure for different frequencies of the VFT and for both, homogenous and inhomogenous fields. The gap width is 15mm.

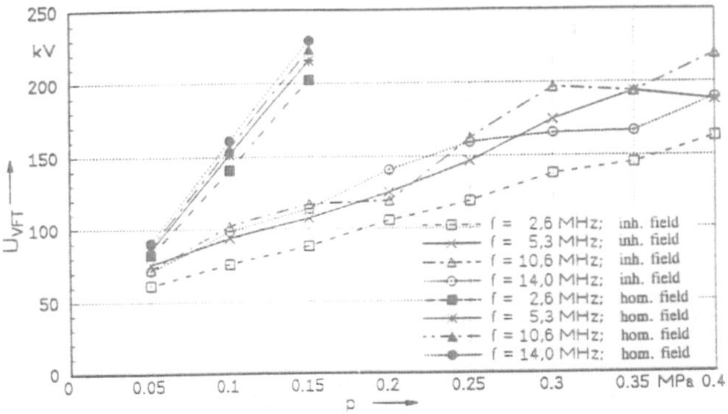

Figure 3. Dielectric strength of SF_6 for positive VFT-voltage for both, homogenous and inhomogenous field distribution (gap width d=15mm)

Breakdown in the homogenous field at low pressures (0.05 MPa) and for all frequencies is caused by a pure streamer mechanism. An increase of the gas pressure or an inhomogenous field distribution provides a streamer leader transition as detected by the optical analysis with the photomultipliers[6].

A pure streamer discharge needs a higher field intensity. This explains why in a homogenous field at a pressure of 0.05 MPa the maximum of the first period has to exceed the critical voltage by more than 20%.

Moreover, it is obvious that the dielectric strength increases with the frequency of the oscillating voltage within both, homogenous and inhomogenous field arrangements. With an increase in frequency, i.e. a decrease in time period, the time while the critical field intensity is exceeded decreases, provided that the amplitude is constant. According to the volume time law[4] the likelihood for the existance of an avalange causing electron in the critical gas volume decreases as well. For that reason the amplitude has to increase with higher frequencies to create a critical electron with the same likelihood. A streamer development is interrupted if the VFT-voltage becomes smaller than the critical voltage at any time. Hence, the VFT has to exceed the critical voltage for at least the time the streamer needs to cross the gap. For the homogenous field with a streamer velocity of v_s=1mm/ns [4] it would take about 15ns to cross a gap width of 15mm.

This leads to the conclusion that with increasing frequency the amplitude of the VFT-voltage has to increase to make sure that the critical voltage is exceeded for the time the streamer needs to cross the gap. For a voltage shape according to figure 4 the ratio V_{VFT}/V_{crit} can be estimated by the following equation:

$$\frac{V_{VFT}}{V_{crit}} = \frac{2}{1+\cos\left(\pi\cdot\frac{d}{v_s}\cdot f_{VFT}\right)}$$

According to this equation the amplitude of a 14MHz VFT exceeds the amplitude of a 2.6 MHz VFT by 12%. For the measured values shown in figure 3 the dielectric strengh at 14MHz is by (160kV-140kV)/140kV = 14% higher than at 2.6MHz. Therfore, the equation is well suitable to estimate the dielectric strength of a homogenous field at a given frequency.

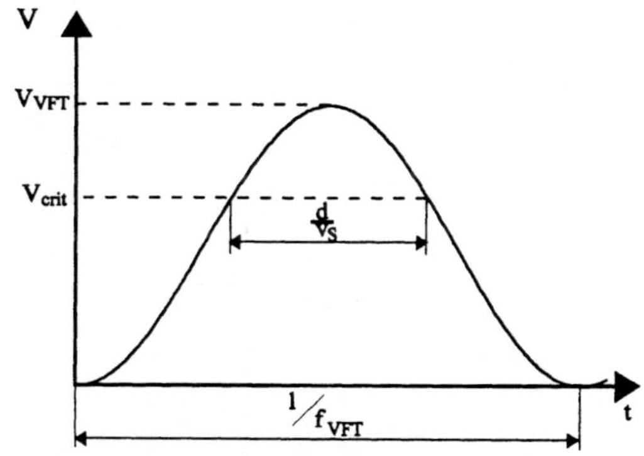

Figure 4. Calculation of V_{VFT}

Moreover, figure 3 shows that in inhomogenous field arrangements the VFT-voltage with the lowest frequency provides the smallest dielectric strength as well. However, the highest dielectric strength is not obtained for the highest frequency. The breakdown voltage of a 10.6MHz VFT is higher than that of a 14MHz VFT-voltage for gas pressures of more than 0.25MPa. Up to a pressure of 0.25MPa the dielectric strength always rises with the frequency.

Figure 3 shows plateaus with constant or even decreasing dielectric strength for inhomogenous fields. This plateau moves either with a dublication in frequency or with an increase in gap width by 65% by nearly 0.05MPa towards smaller gas pressures.

For inhomogenous field distribution the field strength in the vicinity of the inhomogeneity exceeds the critical field intensity for smaller VFT-amplitudes than for homogenous fields Provided that the duration of the VFT-voltage is long enough, a transition from a streamer to a leader can be observed[6]. After several leader steps a breakdown is caused even at small VFT-amplitudes.

3.2 INFLUENCE OF THE POLARITY OF THE VFT-VOLTAGE

For homogenous fields the influence of the polarity of the VFT-voltage on the dielectric strength is not significant. For inhomogenous fields, there is a strong influence of the polarity of the VFT-voltage on the breakdown voltage of SF_6 as shown in figure 5.

In part the measured dielectric strength for negative VFT-voltages is 60% higher. The reason for this behaviour are field distortions caused by the slow movement of the ions compared to the fast electrons. Provided, the HV-electrode is of positive polarity, in the vicinity of the inhomogeneity, i.e. the critical volume, avalanches of electrons are moving towards increasing field intensity to the anode. Therefore the amount of ionisation processses increases with the vicinity to the inhomogeneity. The electrons are collected by the anode whereas the slow positive ions nearly remain where they were created. On the one hand the negative space charge cloud in the vicinity of the anode reduces the field strength at the inhomogeneity, but on the other hand the positive space charges caused by the slow ions virtually reduce the gap width and hence, the dielectric strength. If the inhomogeneity has negative potential, electrons are mainly generated in the vicinity of the inhomogeneity. In that case, electrons move towards decreasing field strength to the anode. The electrons are attached by the SF_6 molecules whereas the slow positive ions are left were they were created. On the one hand the space charge increases the field strength at the electrodes but on the other hand the field distribution becomes more uniform. This provides a higher dielectric strength.

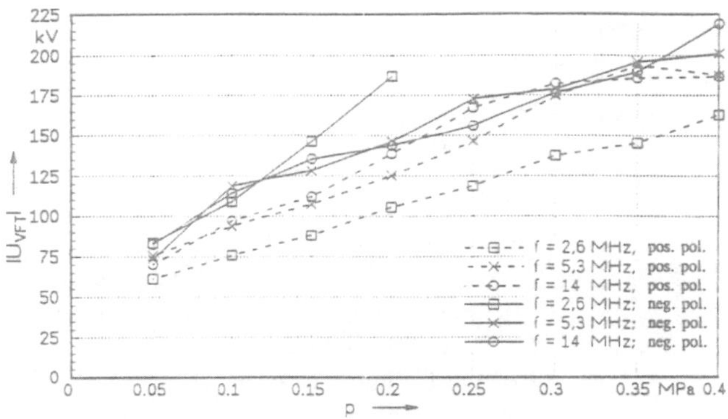

Figure 5. Breakdown voltage of SF_6 versus the pressure for positive and negative VFT-voltage (inhomogenous field, gap width=15mm)

Futhermore, figure 5 shows no area of constant dielectric strength for VFT-voltages of negative polarity.

3.3 INFLUENCE OF THE GAP WIDTH

Figure 6 shows the dielectric strength versus the pressure for different frequencies and for two different gap widths of 15mm and 25mm.

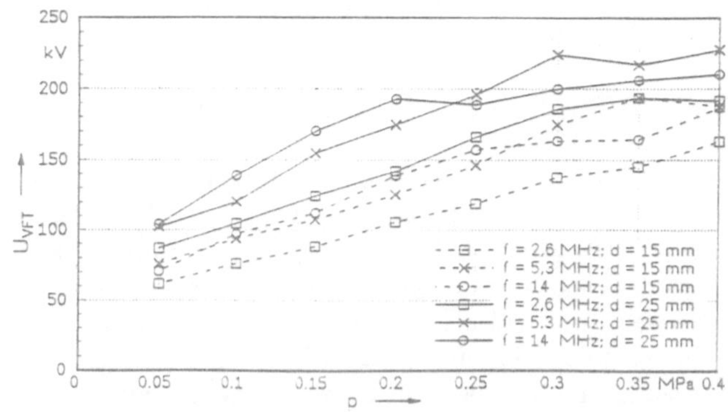

Figure 6. Dielectric strength of SF_6 for positive VFT-voltage for different gap width (inhomogenous field)

For the larger gap width the influence of the VFT-frequency becomes more significant. Still, the VFT with the smallest frequency provides the smallest dielectric strength. By an increase of the gap width of 66% an increase of the dielectric strength of 30%-50% is obtained. The plateaus of constant dielectric strength mentioned in 3.1 move with increasing gap width towards smaller gas pressures.

4 CONCLUSIONS

The following results about the insulating characteristics of SF_6 at VFT-voltage stress can be pointed out:

According to the Paschen law the dielectric strength rises in homogenous fields linearly with the gas pressure independent of the polarity of the VFT. The dielectric strength increases with the frequency of the VFT independent of the its polarity.

Inhomogenous fields lead to a much smaller dielectric strength as well as to a much smaller increase of the dielectric strength with the SF_6-pressure. Up to a pressure of 0.25MPa the dielectric strength increases with the frequency of the VFT as well.

A positive polarity of the VFT-voltage results in a smaller dielectric strength than a negative polarity.

The course of the breakdown voltage versus the gas pressure shows an area of a constant dielectric strength, but only for positive polarity of the VFT-voltage. This plateau moves either with a dublication in frequency or with an increase of the gap width by 65% by nearly 0.05MPa towards smaller gas pressures.

5 REFERENCES

1. Witzmann, R.; "Schnelle transiente Spannungen in metallgekapselten SF_6-isolierten Schaltanlagen", VDI Forschritt Berichte, Reihe 21, Nr. 55, Düsseldorf, VDI Verlag 1989

2. Hiesinger, H.; Witzmann, R.; "Very fast transient breakdown at a needle shaped protrusion", 9th Int. Conf. on Gaseous Diel. and their Appl., Venezia 88, pp. 323-326

3. Remde, H.; Meppelink, J.; Brand, K.-P.; "Elektromagnetische Verträglichkeit in Hochspannungsschaltanlagen", Elektrotechnik und Informationstechnik, Heft 9, 105. Jahrgang 1988, pp. 357-370

4. Beyer, M.; Boeck, W.; Möller, K.; Zaengl, W.; "Hochspannungstechnik", Springer Verlag Berlin, 1986, pp 125-126

5. Hiesinger, H.; "Der Hochfrequenz-Durchschlagsmechanismus in SF_6 bei schwingenden transienten Überspannungen", Dissertation, Universität München, 1991

6. Flohr, Th.; Pfeiffer, W.; Zender, C.; Zimmer, V.; "Diagnostics of prebreakdown phenomena in SF_6 in non-uniform gaps for very fast transient voltage stress"; 1994 IEEE International Symposium on Electrical Insulation; Pittsburgh; to be published

7. IEC42 (Sec)93 / January 1993; High-voltage testing techniques; High voltage testing with very fast transients

DISCUSSION

I. HERBST: How do breakdown values with lightning surge compare to VFT?

W. PFEIFFER: The breakdown voltages for VFT stress may be significantly lower than those for lightning impulse voltage stress. This is especially true in that area where the VFT-breakdown voltages versus pressure have a plateau. Therefore a general comparison figure cannot be given.

K. FESER: (1) When you compare the breakdown on the VFT to the breakdown on a lightning impulse, what waveform is more critical for the breakdown at protrusions? (2) What is the jitter in breakdown values in the measurements?

W. PFEIFFER: (1) Comparing breakdown for VFT-voltages and lightning impulse voltages the first waveform may be much more critical in the case of inhomogeneous fields (protrusions). However the outcome of this comparison depends greatly on the gap width and especially on the gas pressure. (2) There was a significant jitter of the measured breakdown voltages. Therefore, at least a series of 30 measurements have been performed and the lowest measured value is being reported.

E. KUFFEL: (1) Some years ago we measured V-t characteristics for SF_6 using a coaxial system and we included VFT down to 10 MHz - the breakdown values for VFT fell consistently below lightning impulse values. (2) Did the breakdowns always occur on the first half of the cycle?

W. PFEIFFER: The breakdown only exceptionally occurred in the first half cycle of the VFT-voltage for inhomogeneous field distribution. It may even happen that breakdown occurs after some ten microseconds, where the oscillation of the VFT has been completely damped.

O. FARISH: Have you measured time lags to breakdown and is there a difference for the inhomogeneous-field case in the mean time lag and scatter observed for positive and negative polarity conditions? Does this give any insight into the mechanism of breakdown in each case?

W. PFEIFFER: Measurements of the time lags to breakdown have been performed to a large extent. However, there is a large scatter in such curves which requires such a large number of measurements, which may be unrealistic. It seems to be preferable to investigate the breakdown mechanism by use of UV- and IR-sensitive photomultipliers, which have been reported elsewhere.

I. HERBST: What is the relation of U and the field E on the electrode for 15/25 mm gaps?

W. PFEIFFER: By increase of the gap width from 15 mm to 25 mm, that means by 66%, the increase of dielectric strength is only 30% to 50%. From that ratio the relation of the field strengths on the electrode can be estimated.

I. HERBST: What is the length of the needle used to get an inhomogeneous field?

W. PFEIFFER: The needle length was 3 mm in all experiments being presented here (diameter 0.1 mm).

BREAKDOWN TIME DELAY MEASUREMENTS IN SF_6 WITH A RADIOACTIVE GAS ADDITIVE: CF_3T

V.P. Cronin, H.D. Morrison and F.Y. Chu

Ontario Hydro Technologies
800 Kipling Ave, Toronto, Canada

INTRODUCTION

The concept of statistical time delay has been given much attention in gas breakdown studies, especially when dealing with fast-rising impulse stresses.[1-5] In some studies, researchers have used external radiation or light sources, mostly UV, to substantially increase the electron production rate and decrease statistical time delays. A similar effect can be achieved by seeding a dielectric gaseous media with a trace amount of radioactive gas, such as tritium. This concept was proposed for improving the effectiveness of tests of gas-insulated equipment.[6] Some other practical application of this concept could also be found in improving the time characteristics of gas switches by reducing their "statistical" jitter. In addition, the concept provides other experimenters with a useful tool in gaseous dielectrics research.

This paper describes some basic investigations into the dielectric behaviour of SF_6 mixed with trace amounts of a tritium doped gas, CF_3T. CF_3T is the tritiated form of trifluoromethane, CF_3H. CF_3H possesses dielectric properties similar to that of nitrogen. Both CF_3T and CF_3H are chemically stable. SF_6 was chosen as the primary dielectric gas because of its practical and commercial importance. Tritium atoms, as in CF_3T, decay through the emission of an electron with a mean energy of 5.7 keV and a half-life of 12.3 years. The addition of sub-ppm (parts per million) levels of CF_3T in SF_6 would raise the rate of ion production by many orders of magnitude compared to that produced by cosmic rays and natural background radiation. This would result in much higher breakdown probabilities (lower breakdown voltages), especially for positive polarity high voltage (HV) impulses. For high voltage testing procedures, this means that the effectiveness of the test in finding stress enhancements could be substantially improved in comparison with normal test conditions. A related technique using external x-ray irradiation to enhance the detection of partial discharge sites in a solid insulation has been well studied and has proved to be very effective.[7] We hope that the latter in a combination with the use of radioactive gas additives in SF_6 will result in a new class of test methodology superior to the conventional technique.

As a major producer of tritium, Ontario Hydro has been engaged in laser isotope

Gaseous Dielectrics VII, Edited by L.G. Christophorou
and D.R. James, Plenum Press, New York, 1994

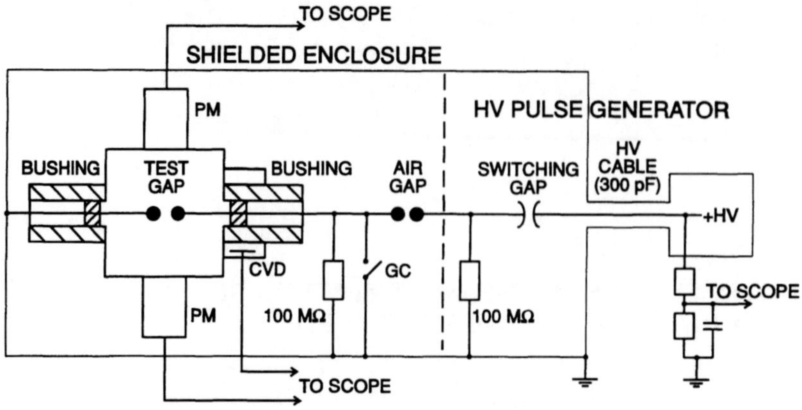

Figure 1. Experimental set up for measuring time delays with fast rising HV impulses. CVD is a capacitive voltage divider, GC is a ground connector, and PM is a photomultiplier.

separation experiments with the CF_3T/CF_3H system for many years. One experiment studied laser-induced electrical breakdowns in this gas mixture with pulsed, infrared, CO_2-laser radiation. For 0.8 ppm of CF_3T in CF_3H, the breakdown threshold was about 14% lower than in the non-tritiated system in terms of power of the laser beam.[8] Our experimental investigation was performed to study the breakdown characteristics of $SF_6/CF_3H/CF_3T$ mixtures under high voltage impulse conditions. Time delays to breakdown were measured in tritiated and non-tritiated mixtures. In this paper we present experimental results for positive polarity impulse breakdown.

EXPERIMENTAL SET UP

Our tests were made with sphere-to-sphere gaps with spacings of 8 mm and 6 mm. Stainless-steel spheres, 10 mm in diameter, were polished and installed horizontally in a stainless-steel vessel. The volume of the cell was about 2000 cm^3. The entire test vessel was placed in a large shielded enclosure to avoid electrical interference in the measuring instrumentation. Two photomultipliers mounted on two windows in the vessel were used to monitor visible light emissions. The cell could be evacuated to a pressure of 0.1 Pa. The SF_6 was of commercial purity and its absolute pressure, accurate to 0.1%, was 0.1 MPa in most of the tests. The temperature in the shielded enclosure was controlled to 23.5±0.5°C. The electrical diagram of the test arrangement is given in Fig. 1.

The steel test vessel and one of the electrodes were permanently grounded to the copper shielded enclosure. The high voltage DC supply was connected to a 3-m long PVC cable that served as the charging capacitance. The cable was discharged by a switching gap through a small (~1 mm) air gap. Immediately before initiating the impulse, the ground connector, GC, was opened. The small air gap was important since small leakage voltages from the charging circuit could change the ion distribution inside the cell.[3] The ground connector was also found to be very effective from the same point of view.[4] During the charging of the pulse generator, the induced stray voltages across the air gap were about 50 to 200 mV at frequencies of about 100 Hz. After charging the pulse generator and opening the GC, but before the impulse, the measured stray voltages between the electrodes were limited to about 20 to 50 mV. The voltage impulse that appeared at the tested gap

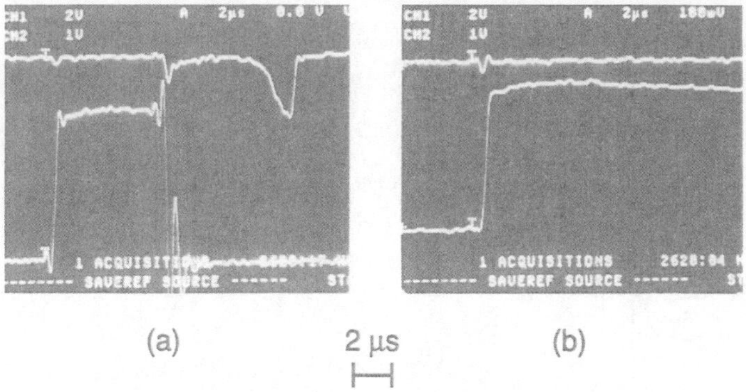

Figure 2. Signals from the capacitive voltage divider (bottom trace) and from a photomultiplier (upper trace) for two cases: (a) breakdown of the test gap, and (b) no breakdown.

had a rise time of 0.5 to 1.0 µs and the FWHM of the impulse was approximately 100 µs (Fig. 2). During experiments, the fast rising HV impulses were applied to the test cell at an interval of about 4 minutes.

A simple but effective device was used to pick up the time-dependent voltage signal. A 20 µm thick, 2 cm^2 square copper foil on a 0.2 mm PE tape was glued to the inner surface of the test cell enclosure. This capacitive voltage divider had an attenuation factor of about 10^5, and its step function response was less than 100 ps.[9] The voltage waveforms and signals from the photomultipliers (PM) were recorded by a Tektronix 2430A digitizer. Time delays were measured as the period between the beginning of the HV front rise and of the light signal of a breakdown from the PM. The latter signal was usually followed within 50 ns by the collapse of the high voltage.

RESULTS FOR POSITIVE POLARITY HV IMPULSES

Preliminary measurements using tritiated gas mixtures, adding sub-ppm quantities of CF_3T to SF_6, indicated a dramatic effect. The 50% probability breakdown voltages using fast-rising HV impulses were substantially lower for tritiated mixtures (Fig. 3) than for non-tritiated mixtures.

A model developed for positive polarity HV impulses[2,3] relates the probability, P, that a critical avalanche will be created during the time interval between t_{si}, the time at which the streamer inception level is exceeded, and a time $t_{si}+t_s$, i.e. the statistical time delay, for slightly nonuniform experimental arrangements. The model can be manipulated to the following form:

$$<n>V_W(t_s) = \ln(1 - P(t_s)). \tag{1}$$

The average ion concentration $<n>$ is a function of radiation activity in the gas volume.[2,10] The function $V_W(t_s)$ is a cumulative weighted volume of the SF_6 gas and is a function of pressure and electrical field. The probability $P(t_s)$ or cumulative distribution

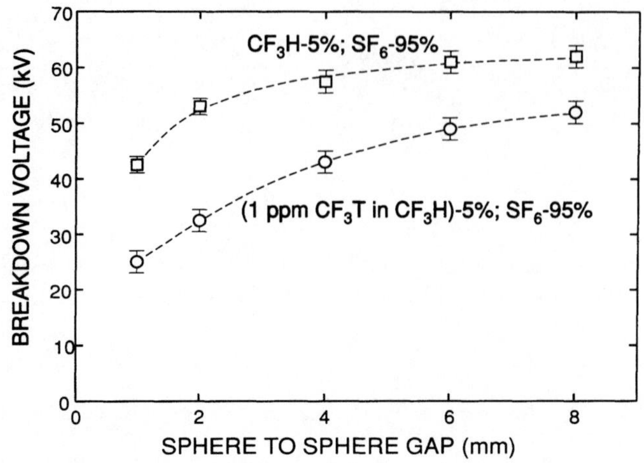

Figure 3. Breakdown voltage, defined at the level of 50% probability, for fast rising HV impulses in two gas mixtures: one with and one without a trace of the radioactive gas, CF_3T.

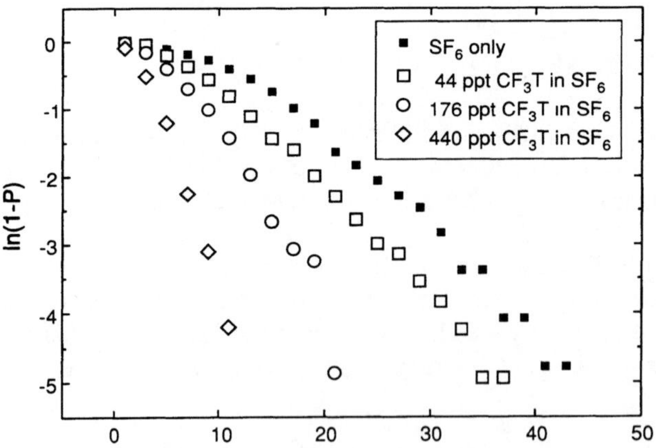

Figure 4. Laue plots of breakdown time delays for different gas mixtures of CF_3T in SF_6 for an 8 mm gap at 71 kV.

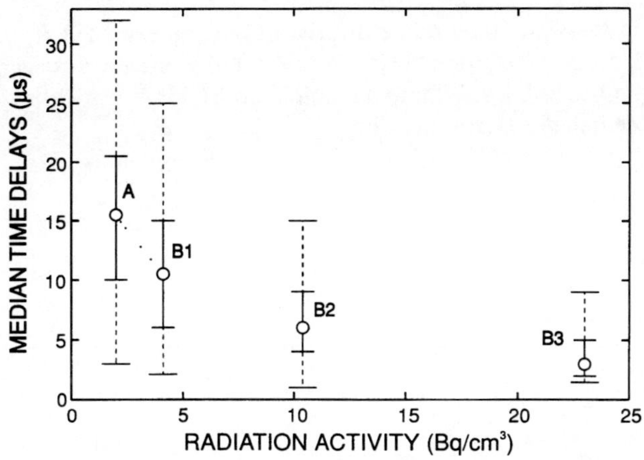

Figure 5. Measured median time delay distributions as a function of radiation activity in SF_6 for an 8 mm gap at 71 kV: circles represent the measured medians, t_{50}, solid lines show the 50% confidence interval, and dashed lines show the 90% confidence interval. Point A corresponds to background radiation activity level in the vessel. Points B1, B2 and B3 correspond to background plus 44, 176 and 440 ppt CF_3T in SF_6, respectively.

function of the statistical time lag, t_s, is the integral of the probability density over the interval (t_{si}, t_s+t_{si}). The cumulative distribution function is best represented in a Laue plot of $ln(1-P(t_s))$ as a function of t_s. The Laue plot has the advantage that it yields a straight line whenever the applied voltage and the electron production rate are constant.[2] The Laue plots in Fig. 4 show results for mixtures of 44, 176, and 440 ppt (parts per trillion) CF_3T in SF_6. From analysis of these distributions it is clear that statistical time delays are shorter for mixtures with higher tritium concentration.

The breakdown probability as a function of the time delay, t_s, is characterized by a median value, t_{50}, and the associated 90% and 50% confidence time intervals obtained from analysis of the probability density. Values for t_{50} and the confidence intervals for different gas mixtures, i.e. different radiation activities, are shown in Fig. 5. Clearly, there is a trend of substantial reduction in both t_{50} and the confidence intervals with the increase in radiation activity of the gas mixtures.

After statistical analysis of our experimental data, we introduced an additional parameter: the relative average ion concentration $<n_{rel}>$. This parameter can be estimated by manipulation of Equation 1:

$$<n_{rel}> = <n>/<n_0> = \ln(1 - P(t_s)) / \ln(1 - P_0(t_s)). \quad (2)$$

where $P(t_s)$, and $P_0(t_s)$ are the probabilities measured for two different gas mixtures, e.g. with some fixed amount of CF_3T in SF_6, and in pure SF_6, respectively. We observed that the relative average ion concentration $<n_{rel}>$ for each particular tritiated mixture is not very dependent on t_s in the 4 to 20 μs range of time delays, as expected from the basic model. The plot of $<n_{rel}>$ as a function of radiation activity yields a straight line for CF_3T concentrations of 44 to 440 ppt. Some deviations from this linear behaviour, however, are evident for greater radiation activities (Fig. 6).

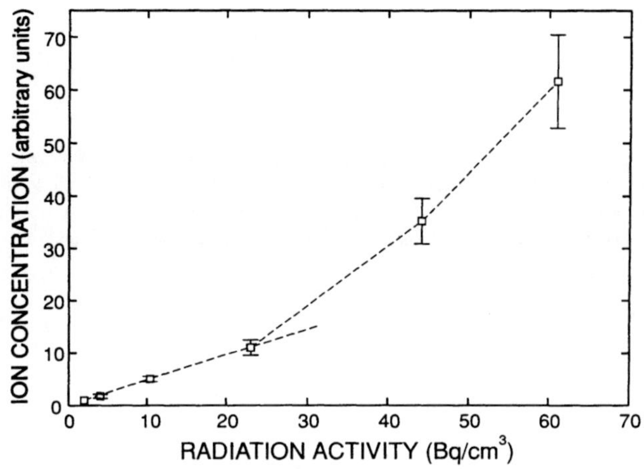

Figure 6. Variation in the relative average ion concentration with radiation activity in SF_6.

CONCLUSIONS

In our experiments, we studied statistical time delays to breakdown for gas mixtures of varying tritium concentration in SF_6, from 44 ppt to 1232 ppt. As a result of much greater radiation activity, breakdown time delays for all tritiated mixtures were significantly reduced in comparison with time delays in pure SF_6. The most interesting result is that the addition of relatively small amounts of radioactive gas in a dielectric gaseous media can substantially increase the probability of a breakdown. From a testing perspective, this means that the testing of HV equipment could be performed at lower voltages and in much shorter time with better overall effectiveness. These potential benefits follow from the much higher electron production rate at field enhancements, such as in the presence of defects. The overall reduction in statistical time delays and the reduced scatter could also have benefit in reducing the statistical jitter in gas switches. In addition, the technique provides another means of dealing with statistical time delay problems in dielectric studies, without the complication of external radiation sources.

ACKNOWLEDGMENTS

The authors would like to thank N. Fujimoto and S.A. Boggs for helpful discussions, and E. Alonso, J. O'Neil and L. Pateopol for practical assistance.

REFERENCES

1. L.G. Christophorou and L.A. Pinnaduwage, Basic Physics of Gaseous Dielectrics, *IEEE Trans. on Elect. Insulation*, EI-25:55 (1990).
2. J. Kindersberger, The Statistical Time-Lag to Discharge Inception in SF_6 , Ph.D. Thesis for Technical University, Munich (1986).
3. N. Wiegart, L. Niemeyer, F. Pinnekamp, W. Boeck, J. Kindersberger, R. Morrow, W. Zaengl, M. Zwicky, I. Gallimberti, and S.A. Boggs, Inhomogeneous Field Breakdown in GIS-the Prediction of Breakdown Probabilities and Voltages, Part 1-3, *IEEE Trans. on Power Delivery*, 3:923 (1988).
4. N. Wiegart, A Model for the Production of Initial Electrons by Detachment of SF_6^--ions, *IEEE Trans. on Elect. Insulation*, EI-20:587 (1985).
5. L. Niemeyer, private Communication with F. Y. Chu, 1992.
6. S.A. Boggs, F.Y. Chu, G.L. Ford, N. Fujimoto, and S.Harvey, "Gas-Insulated Substations: Reliability Research Program," EPRI Final Report, EL-5551, *Vol. 1* (1988).
7. S. Rizzetto, N. Fujimoto, and G.C. Stone, A System for the Detection and Location of Partial Discharges Using X-rays, *in*: "Proceedings of IEEE International Symposium on Electrical Insulation," Boston (1988).
8. E.M. Alonso, J.A. O'Neil, and L. Pateopol, Dielectric Breakdown of CF_3H and a CF_3H/CF_3T Mixture by Pulsed Infrared Radiation, *J. Phys. D. Appl. Phys.* 26:1650 (1993).
9. N. Machlitt, W. Assmann, and S. J. Skorka, Response Time and Maximum Overvoltage of Spark Gaps in Compressed SF_6 for Steeply Rising Pulses, *Nucl. Instrum. Methods* A244:93 (1986).
10. W. Knorr, K. Moller, and K.J. Diederich, Voltage-Time Characteristics of Slightly Nonuniform Arrangement in SF_6 Using Linearly Rising and Oscillating Lightning Voltage, *CIGRE*, 15-05 (1980).

DISCUSSION

C. X. WANG: (1) Why is it that you get a burst of light emission sometime after voltage collapse as seen from Fig. 2? (2) Is the purpose of measuring time delay of breakdown to measure ion concentration?

H. D. MORRISON: (1) We do not have an explanation for that at the present time. (2) The purpose was to demonstrate that CF_3T would be a useful tool for increasing the ion concentration and finding out what level would be needed to make it effective.

O. FARISH: Your paper presents information on the relative ion concentration in CF_3T-doped SF_6. Have you used the volume-time concept to analyze the time-lag data to yield estimates of the actual ion concentration as a function of CF_3T concentration? Also, what are the breakdown products resulting from sparking in CF_3T-SF_6 mixtures and would they contaminate the spacers in practical HV apparatus?

H. D. MORRISON: We have only calculated relative ion concentration as a function of CF_3T concentration. We have estimated, however, the ion production rate: ~8 ion pairs/(cm^3-sec) for 1 ppt CF_3T in 0.1 MPa SF_6. As for the breakdown products, we have not studied them but we would expect at least HF and TF to appear. At this time, we do not recommend this technique for practical application, only as a research tool.

M. F. FRÉCHETTE: Have you treated the gas in a particular way following testing? Does the contaminated SF_6 need cleaning treatment upon removal?

H. D. MORRISON: We have not treated the gas in any special way. The quantities of both SF_6 and CF_3T we used were small enough to be treated by dilution and venting.

K. FESER: You said you can improve GIS testing because you can run the test at a lower voltage! How should we understand this proposal? Do you propose to do the on-site test of GIS at low voltage, when the gas is tritiated? Is this a proposal from a utility point of view?

H. D. MORRISON: The use of CF_3T should be considered strictly as a research tool in a controlled research laboratory, not as a tool for on-site testing. The work shows that the breakdown would occur at a lower voltage with the tritiated mixture which might reduce the damage caused by breakdown because there is less energy discharged.

MEASUREMENT AND SIMULATION OF VFTO AFTER A BREAKDOWN IN GAS INSULATED SUBSTATIONS

Rainer Weber

Institute of High Voltage Engineering
Technical University of Munich
Germany

ABSTRACT

Very Fast Transient Overvoltages (VFTO) caused by line to enclosure faults can lead to secondary breakdowns within Gas Insulated Substations (GIS). Thus it is important to develop a suitable simulation model for the breakdown and the electrical behaviour of the burning arc.

Therefore high voltage measurements in a test setup with various defects were made. Based on these investigations Toepler's Spark Law has been modified and the results with this new computer model are compared with the measured voltages.

INTRODUCTION

Oscillating transient overvoltages in GIS are caused by switching operations or line to enclosure faults within the SF_6-installation. Defects inside the GIS like particles or fixed protrusions cause an inhomogeneous electrical field distribution and insulation breakdown can occur[1]. In this case the development of the internal VFTO is of special interest. Because of the behaviour of the travelling waves the resulting overvoltages can lead to secondary breakdowns within the GIS[2]. Moreover the transient potential rise of the encapsulation due to the VFTO causes problems for the Electromagnetic Compatibility (EMC) of electronic equipment. These problems are of special concern during tests on site[1].

According to the travelling surge phenomena the VFTO shape and amplitude is mainly influenced by the breakdown at the fault location itself, the configuration and the design of the GIS. Therefore it is important to have accurate electrical simulation models for the GIS components and the voltage collapse at the fault location.

The development of this simulation model is based on high voltage measurements made in a test arrangement to get detailed information about the arc. Inside the GIS different defects are implemented, which are stressed with impulse voltages.

Gaseous Dielectrics VII, Edited by L.G. Christophorou
and D.R. James, Plenum Press, New York, 1994

The time varying processes during voltage breakdown and the resulting VFTO are measured in these experiments.

Moreover a computer simulation model for this important first part of the breakdown is developed. Therefore calculations done by the Electromagnetic Transients Program (EMTP) are presented and compared with equivalent measurements.

EXPERIMENTS

The development of the computer model for a breakdown within a GIS is based on the analysis of high voltage experiments. During the measurement of the travelling surges, caused by a fault from the live conductor to the outer shielding, the time dependent processes at the beginning of the breakdown can be observed. Because of the comparatively long collapse time of the voltage for an insulation fault within a GIS[1] it is necessary to install a long setup to observe the voltage breakdown accurately. The experimental setup used for the high voltage investigations, shown in Fig. 1, consists of industrial 420 kV GIS components.

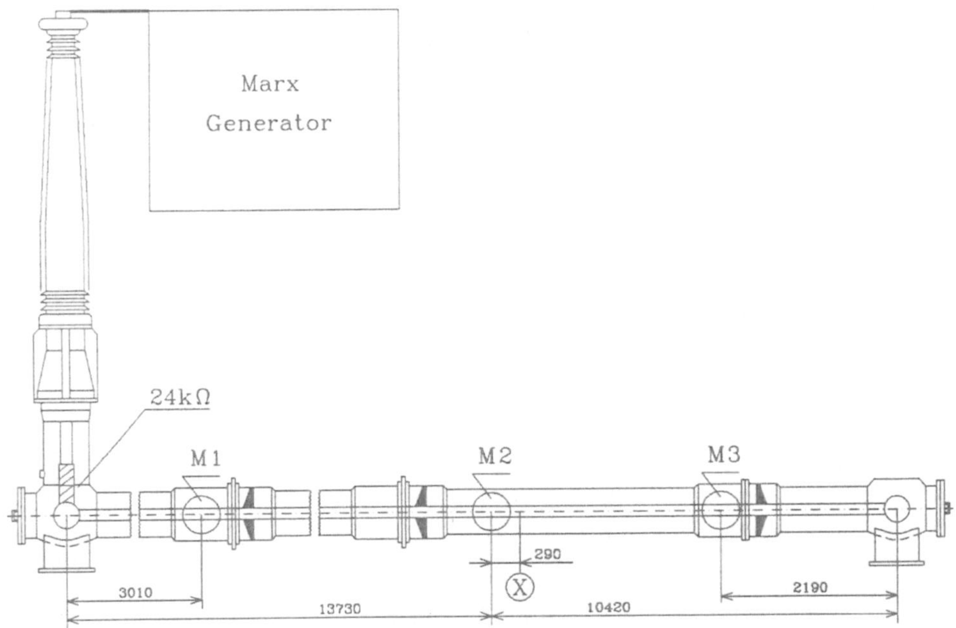

Figure 1. Test setup

A Marx generator is connected to a capacitive graded bushing by a short overhead line. The horizontal part of the bus duct with a length of nearly 25 m is separated from the bushing by a 24 kΩ resistor. For the travelling waves inside the GIS this resistor acts like an open end. Furthermore reflections generated by the impulse generator and the grounding system are effectively isolated from the GIS structure. The inner conductor with a diameter of 150 mm is fixed by three spacers within the outer shielding, which is 520 mm in diameter. As a result there are no big distortions along the whole length of the coaxial bus duct.

Within the experimental setup three measuring points M1, M2 and M3 are available. By means of capacitive dividers[3] and probes providing a bandwidth of about 500 MHz the voltage can be acquired simultaneously at two points. Therefore an optical link with a bandwidth of 600 MHz is used. The measurements are acquired by a digitizing oscilloscope with a maximum sampling rate of 1 GSa/s and transferred to a personal computer, where they are evaluated and prepared for further investigations. Special care has to be taken of the shielding of the whole measuring equipment referring to the EMC of the electronic systems.

To cause a breakdown at point X (Fig. 1) different electrodes can be installed on the inner conductor. The voltage collapse within the setup has been investigated for homogeneous and inhomogeneous field conditions. Therefore needle shaped protrusions with a length between 10 mm and 50 mm, a diameter of 2 mm, and a tip radius of 0.5 mm or a ring electrode 460 mm in diameter have been used. In all the experiments the distance to the measuring point M2 is 290 mm. The ring electrode generates a breakdown in a nearly homogeneous field, whereas the field enhancement factor of the needles is in the range of 20 to 60 compared to the coaxial bus arrangement.

The live conductor is loaded with a positive impulse voltage of different shape, supplied by the Marx generator. All measurements have been made at a SF_6 gas pressure of 0.3 MPa. With the fault location near the measuring point M2 it is possible to observe the voltage collapse over a comparatively long time without distortions. After a breakdown two travelling surges are sent to both ends of the setup with the speed of light. For the generated negative waves the distance to the 24 kΩ resistor on one side is about 14 m and to the open end on the other side only 10 m. Subsequent to its total reflection at the open end near M3 the surge is heading backward to the arc, where it is partly reflected and distorted. This latter part arrives 70 ns after the start at M2, corresponding to a covered distance of 21 m. This is very important for the development of a computer model, simulating the time varying processes at the beginning of the breakdown. On the other side the reflected travelling wave comes back from the resistance after 94 ns equivalent to a length of 28 m. An example for the voltage at the measuring point M2 is given in Fig. 3a) over 2 μs for a breakdown at a needle of 50 mm in length.

Because of the small amplitude the investigation of the damping after a longer time is impossible. Therefore it is important to have another measuring point near an open end like M1 or M3. The higher voltage oscillations occurring there after a breakdown are better for the evaluation of the burning arc long time behaviour. As an example for the voltage development near an open end Fig. 3c) and Fig. 4c) show the measured voltages at point M3 for a 50 mm needle and M1 for a 10 mm needle over a time period of 3 μs.

COMPUTER MODEL

The simulation models used for the components of the test setup have been developed out of measurements and calculations based on the EMTP[4].

The computer model for the breakdown within a GIS has to take into account two different periods of time. First the time varying process during the building up of the spark and second the behaviour of the burning arc with respect to the oscillation frequency and the damping of the resulting overvoltages.

According to the simplification, that the arc can be regarded as a short circuit between the live conductor and the encapsulation a simulation model consisting of a concentrated inductance to ground has already been developed[5]. Moreover spark laws are available, modelling the breakdown by means of a time varying resistance. Starting at a very high value, it decreases according to the particular spark law and finally remains at a small resistance. Based on the investigations for small SF_6 gaps with homogeneous electrical field the validity of Toepler's Spark Law in SF_6 has been proven[6]. The simulation model, used in this paper is shown in Fig. 2.

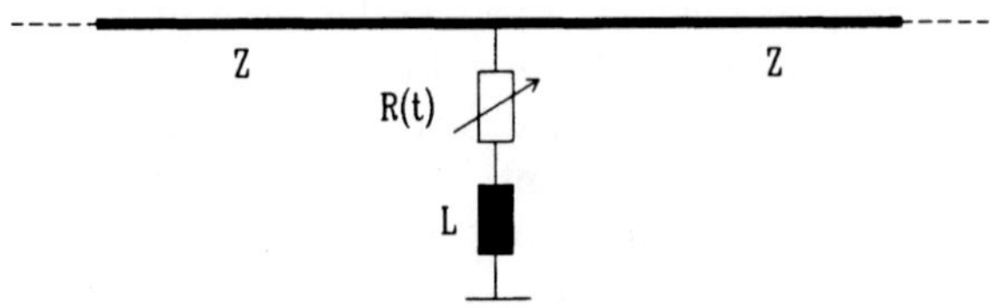

Figure 2. Simulation model

It consists of a series connection of a modified time varying resistance $R(t)$ and an inductance L, placed between the fault location and ground. The surge impedance Z in Fig. 2 represents the bus duct of the test setup, where the breakdown happens. According to the formula

$$R(t) = \frac{k_T \cdot s}{q_o + \int_o^t |i(t)|\, dt} + R_r$$

the value of the time varying arc resistance $R(t)$ is calculated until the rest value R_r in [Ω] is reached. The parameter k_T represents the Toepler constant in [Vs/cm] and s the gap width in [cm]. Furthermore the integral sums up the absolute value of the current i through the resistance $R(t)$ over the time, beginning at the breakdown inception. Thus it corresponds to the charge flown through the spark channel up to that time.

Moreover the initial charge q_o in [As] is a necessary parameter for the simulation, especially in case of inhomogeneous fields, which results out of the comparison of a large number of measurements with the equivalent simulations. For a needle shaped protrusion under impulse voltage stress the breakdown happens according to the leader mechanism. Therefore q_o has to be regarded as the charge flow within the conducting leader channel before the final jump, when the spark bridges the gap completely.

The inductance L is calculated out of the geometry of the electrode and an assumed radius of the initial spark channel. For the remaining resistance R_r a value between 0 Ω and 3 Ω can be taken.

MEASUREMENTS AND SIMULATIONS

The accuracy of the computer model is tested by comparing measured and calculated voltage shapes. A series of EMTP calculations are made to determine the computation parameters like the Toepler constant k_T, the initial charge q_o and the remaining resistance R_r. According to the geometrical dimensions of the gap and the electrodes the inductance L is calculated with an assumed initial spark radius of about 100 μm. Although the voltage is normalized to 1 pu in this paper, the absolute value of the breakdown voltage U_B is necessary for the calculation of $R(t)$.

The results for a needle with a length of 50 mm are presented as a first example for an inhomogeneous field breakdown in Fig. 3. It shows the measured voltage collapse

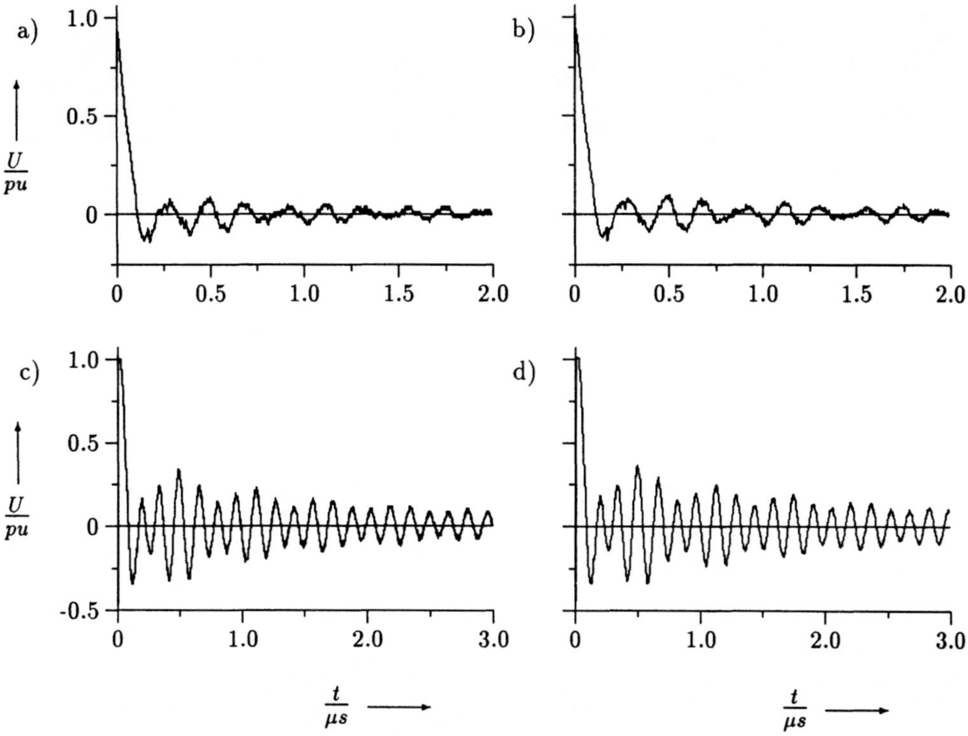

Figure 3. Needle 50 mm long
a) measured (M2) b) calculated (M2)
c) measured (M3) d) calculated (M3)

at point M2 (Fig. 3a)) with a damped oscillation within 2 μs. Higher amplitudes after the breakdown occur at measuring point M3 (Fig. 3c)) even at the end of the 3 μs time period. Furthermore Fig. 3b) and d) show the corresponding computed voltages for the gap width $s = 135$ mm and a breakdown value $U_B = 300$ kV. The parameters determined for the simulations are: $k_T = 2.5 \cdot 10^{-4}$ Vs/cm, $q_o = 5.7\,\mu$C, $R_r = 2.0\,\Omega$ and $L = 230$ nH.

Another example for a breakdown voltage $U_B = 402$ kV at a needle shaped protrusion of 10 mm length is presented in Fig. 4 showing the development of the voltage at point M2 in Fig. 4a) with a big damping during the time of 2 μs. Apparently higher amplitudes and a smaller oscillation frequency are given at point M1, as shown in Fig. 4c) over 3 μs. Reasons for that are the higher breakdown voltage and the bigger length of the bus duct between the fault location and the open end near M1. Within Fig. 4b) and d) the equivalent simulation results for the gap width $s = 175$ mm and the further parameters are shown. In detail the determined values are: $k_T = 2.5 \cdot 10^{-4}$ Vs/cm, $q_o = 0.11\,\mu$C, $R_r = 1.5\,\Omega$ and $L = 260$ nH.

A good agreement between measured and calculated data can be observed for all three measuring points M1, M2 and M3.

The new parameter q_o is caused by the leader breakdown at the fault location, which is characteristic of inhomogeneous fields. Moreover this seems to be the reason for the big k_T, which is bigger than the measured value of about $0.5 \cdot 10^{-4}$ Vs/cm for almost homogeneous fields[6].

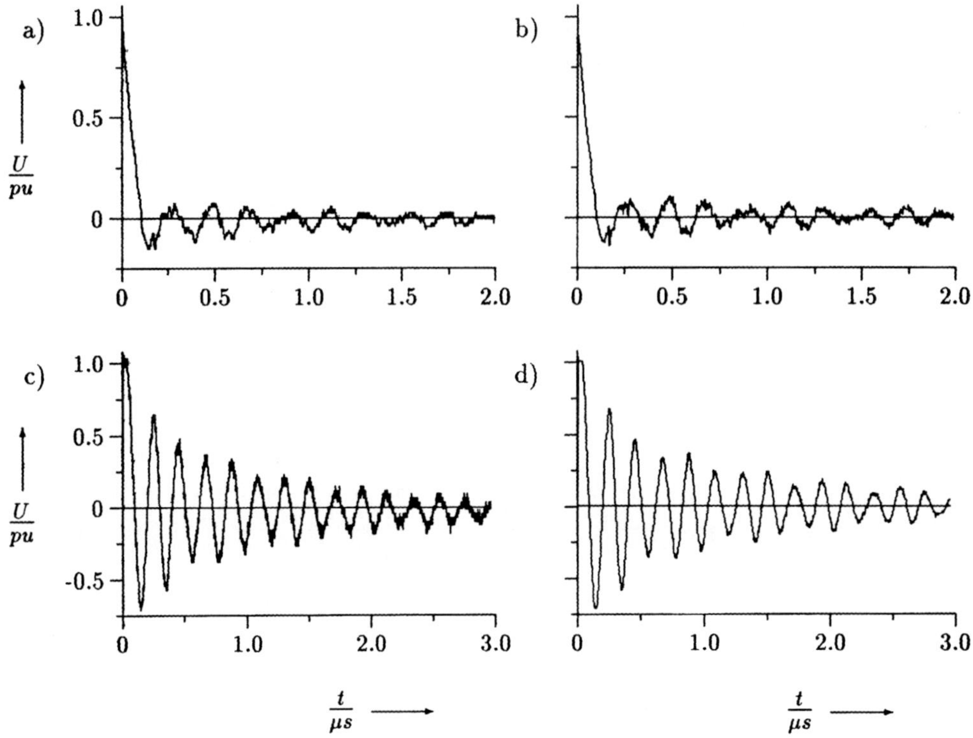

Figure 4. Needle 10 mm long
a) measured (M2) b) calculated (M2)
c) measured (M1) d) calculated (M1)

To approve this assumption a breakdown in a field caused by the ring electrode is presented in Fig. 5. In this third example the breakdown voltage is about 405 kV for a gap width of only 30 mm. Fig. 5a) shows the measured voltage at point M1 with comparative high amplitudes reaching from −1.4 pu to +1.3 pu at the beginning and only a small damping after 3 µs.

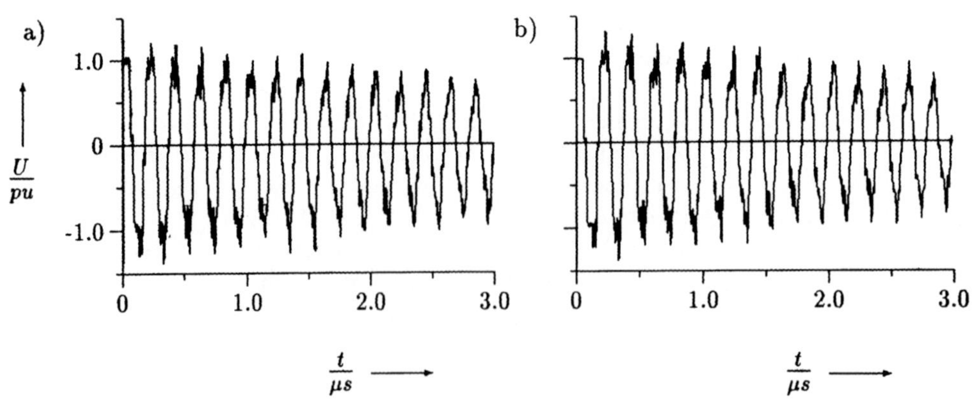

Figure 5. Ring electrode
a) measured (M1) b) calculated (M1)

The corresponding EMTP result in Fig. 5b), which shows a very good agreement, has been achieved with the following parameter values: $k_T = 0.5 \cdot 10^{-4}$ Vs/cm, $q_o = 0$ C, $R_r = 0.5\,\Omega$ and $L = 70$ nH.

Not only the Toepler constant but also the initial charge is completely different from the examples before. Because of the changed breakdown mechanism in the quasi-homogeneous field the usual value for k_T can be taken and no initial charge is necessary. Due to the small gap distance of approximately 30 mm a streamer breakdown has to be assumed, where no thermally heated leader channel exists before the final jump ($q_o = 0$). Moreover the remaining resistance and the inductance are smaller compared to the long spark at the needles.

CONCLUSION

The behaviour of the spark after a breakdown of the SF_6 insulation was investigated. Therefore measurements of the voltage development after a fault at different electrodes with various electrical field distributions were made.

Furthermore the model for the time varying processes at the beginning of the breakdown has been improved by the addition of a new parameter. This enables the simulation of the voltage collapse not only for a breakdown at the homogeneous, but also for the inhomogeneous field. Although some of the parameters used can be estimated out of the geometrical dimensions of the electrode and the gap, others have to be determined by comparing measured and calculated voltages. Nevertheless all the examples presented show a very good correspondence between the acquired and the computed waveforms, even after a few microseconds.

Thus it is possible to simulate the breakdown of any gap by the series connection of the time varying resistance and an inductance to ground, if the equivalent parameters are used.

REFERENCES

1. CIGRE WG 33/13-09 "Very fast transient phenomena associated with gas insulated substations," CIGRE-report, Paris, No. 13-33 (1988).
2. K. Feser, P. Kulik, L. Menten. "Dielectric field testing of GIS", CIGRE-Report , Paris, No. 33-09 (1986).
3. R. Witzmann. "Meßsystem zur Erfassung schneller transienter Vorgänge in metallgekapselten SF_6-isolierten Schaltanlagen", ETZ-Archiv, VDE-Verlag, Berlin/Offenbach, Bd. 9, H 6, pp. 189-194 (1987).
4. R. Witzmann. "Fast transients in gas insulated substations (GIS) - modelling of different GIS components", 5th ISH, Braunschweig, paper 12.06 (1987).
5. R. Weber. "An arc-model for the calculation of VFTO in case of breakdown in GIS," 8th ISH, Yokohama, pp. 319-322 (1993).
6. W. Pfeiffer. "Fortschritte bei der Untersuchung von Funkenentladungen in komprimierten SF_6," 2nd ISH, Zürich, pp. 365-369 (1975).

DISCUSSION

K. FESER: How do you choose the remaining resistance of 0.5 to 5Ω?

R. WEBER: First these values can be found in the literature. Moreover, the resistance depends roughly on the length of the spark. But for an exact agreement of the damping between the measured and the simulated voltages the remaining resistance has to be determined for every single measurement.

O. FARISH: Do the values of inductance used involve realistic assumptions about the spark channel diameter and was a constant value of channel diameter used throughout?

R. WEBER: The inductance of the spark has been calculated from the gap width, which represents the length, and an assumed radius of 100 μm for the initial spark channel radius.

DISCHARGE DEVELOPMENT IN SF_6 IN CASE OF COMPOSITE VOLTAGE STRESS

Dietmar Buchner

Institute of High Voltage Engineering
Technical University of Munich
Germany

ABSTRACT

The discharge development in SF_6 in case of strong inhomogeneous fields was investigated for different transient overvoltages of positive polarity. The evaluation of measured V-t curves up to 0.7 MPa shows the importance of Lightning Impulse (LI) and Very Fast Transient Overvoltages (VFTO) for the estimation of the insulation properties of SF_6. Therefore it is necessary to find a universal model, which describes the leader propagation in SF_6 under different voltage stress. Based on experimental data and an accurate field calculation a qualified model was developed. The corresponding simulation program allows to compute time and length of each leader step from the inception to the final breakdown in a pressure range from 0.1 to 0.3 MPa.

INTRODUCTION

The estimation of insulation properties is a fundamental point for the application of SF_6 in GIS. The main source for insulation faults are imperfections inside the GIS like moving particles or fixed protrusions.

The discharge development in case of such strong inhomogeneous fields is strongly influenced by the applied voltage. Depending on the voltage shape different breakdown mechanisms are active. In case of positive standard impulse conditions (LI) the precursor mechanism[1, 2] describes the streamer-leader transition. Different programs exist, which allow the calculation of the leader propagation for such voltages[3, 4]. Under VFTO stress with a high oscillation frequency and a small damping the high-frequency mechanism[5] has to be considered. The corresponding simulation model[6] permits the computation of the complete discharge development.

In addition a third type of transients can occur inside GIS. These overvoltages caused by switching operations consist of at least two components with different oscillation frequencies (FTO, VFTO). Up to now the relevant breakdown mechanism for such composite voltage shapes was unknown. Furthermore a universal discharge model for all transients is required.

Therefore experimental investigations were done under LI, VFTO and composite voltage stress. Based on these results a new mechanism was found, which describes the discharge development under such different voltage stress. The reliability of the corresponding simulation model is presented in this paper.

EXPERIMENTS

Two test setups were used for the experimental investigations. The first setup consists of an 1 MV–impulse generator, an external R–L–C circuit and usual 420 kV GIS components with realistic dimensions. Special details about this setup are described in a former paper of the author[7]. With this configuration LI, VFTO and composite voltage shapes with different oscillation frequencies can be generated. The advantages of this setup are the installed diagnostic tools, like voltage divider, photomultipliers with different spectral sensitivities and a high sensitive image converter camera including an image processing system. The investigations of the discharge development can be done at different defects up to 0.3 MPa.

Higher gas pressures are more and more in use for GIS. Therefore a second test setup was installed based on the geometry of the first setup. It consists of an 3 MV–impulse generator, an external R–C circuit and new GIS components designed for a higher gas pressure.

For an overview two different voltage shapes were applied to this second setup. In addition to former results up to 0.3 MPa the influence of LI and VFTO stress on the breakdown behavior was studied by means of measured voltage-time curves for different defects and gas pressures up to 0.7 MPa.

The results of all these experiments are presented in Fig. 1. The 50 %–breakdown voltages over pressure are compared for a needle / plane gap at the end of the bus duct with the needle on positive potential (needle length: l = 15, 4, 2 mm, tip radius: 0.25 mm and remaining gap width: 85, 96, 98 mm).

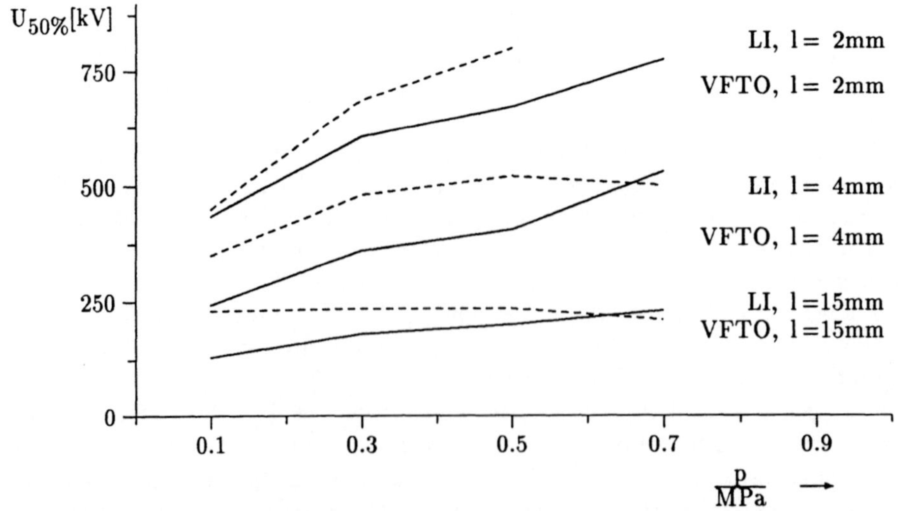

Figure 1. 50 %–breakdown voltages for LI (- - -) and VFTO (—) stress.

For pressures up to 0.65 MPa VFTO stress effects the lower breakdown levels, for higher pressures LI stress is more critical. As conclusion both voltage shapes have to be considered. Consequently an extended computer model must be developed, which allows the calculation of the breakdown levels for different types of transient overvoltages.

The development of the new model required extensive experiments by using the first of the above mentioned setups. Based on measured voltage, ultraviolet (UV), infrared (IR) signals and streak records typical characteristics of the voltage-dependent leader expansion can be recognized (The IR signals are not shown in this paper). Representative examples for 0.1 MPa are given in Fig. 2 for LI and VFTO stress, and in Fig. 3 for two types of composite voltage stress. The breakdown was investigated at a needle / plane gap with the needle on positive potential (needle length: 15 mm, tip radius: 0.25 mm and remaining gap width: 85 mm). For higher gas pressures similar results can be measured.

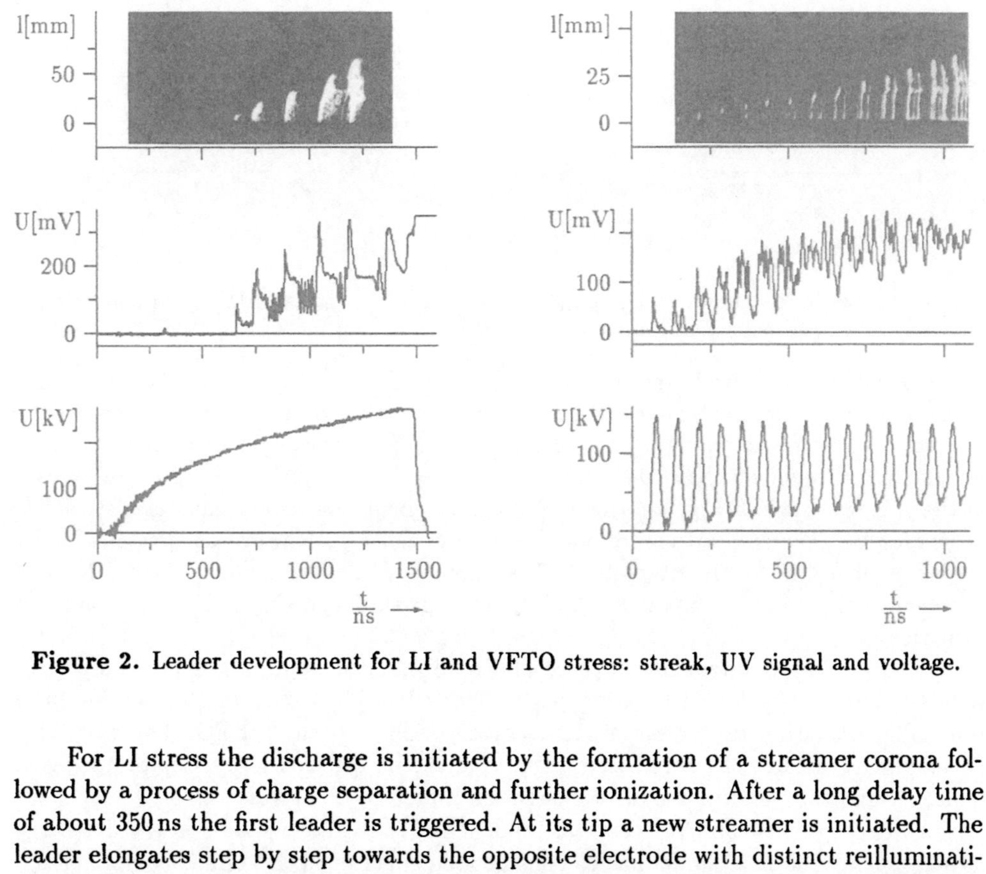

Figure 2. Leader development for LI and VFTO stress: streak, UV signal and voltage.

For LI stress the discharge is initiated by the formation of a streamer corona followed by a process of charge separation and further ionization. After a long delay time of about 350 ns the first leader is triggered. At its tip a new streamer is initiated. The leader elongates step by step towards the opposite electrode with distinct reilluminations in between. This development is detected by photomultipliers corresponding with the streak record. The diameter of the leader channel expands continuously between the steps and the gas density is reduced. As consequence the internal field strength rises and causes multiple ionization processes of the dissociated gas.

The leader development under VFTO conditions occurs periodically. At each positive slope of the VFTO beginning with the third one a leader step is initiated with a bright streamer head at its tip. Comparing the streak record and the UV–signal the essential process can be described. As result of intensive ionization inside the streamer region immediately the gas is heated up to its dissociation temperature. The streamer region becomes a part of the leader. Reilluminations between the steps can be detected in the negative slope of the VFTO.

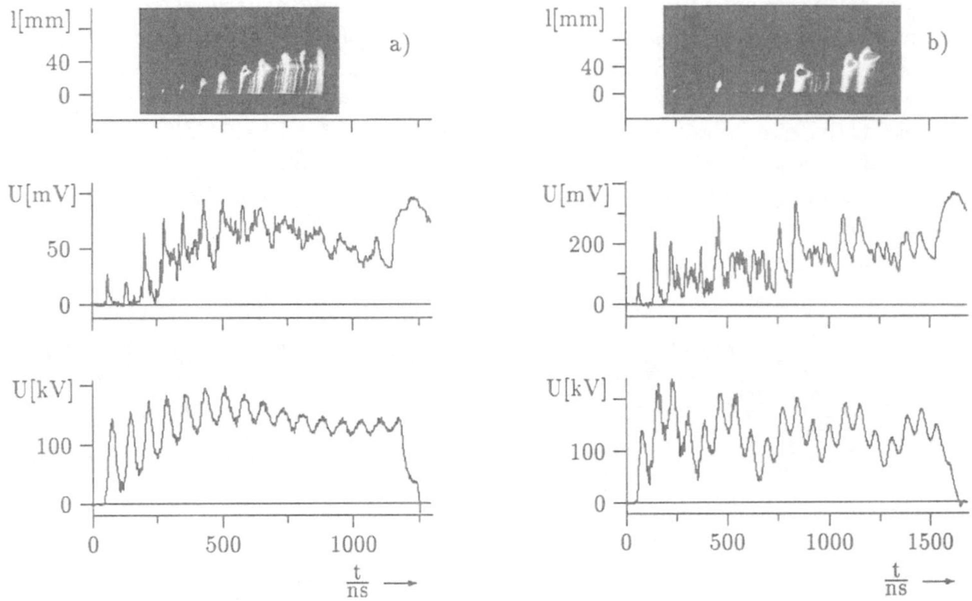

Figure 3. Leader development for composite voltage stress: streak, UV signal and voltage.

Considering the discharge process under composite voltage stress the characteristics for LI and VFTO stress can be observed depending on the relation between the two voltage components, FTO and VFTO. FTO (Fast Transient Overvoltage) describes the lower frequency, VFTO the superimposed higher frequency component. In case of the voltage shape of Fig. 3a) the leader development follows the characteristics described for VFTO stress with a periodic leader propagation. However, sometimes leader steps are "missing" for times greater 600 ns. Because of the falling part of the FTO and the decreasing VFTO component the ionization is slowed down and the propagation time rises.

This process is intensified in case of Fig. 3b). The dominant FTO component reduces the ionization in its falling parts evidently. Thus only in the maximum of the FTO oscillation leader steps occur corresponding to the VFTO. Therefore this development can be considered as a mixture of the irregular step by step propagation under LI and the periodic stepwise elongation under VFTO stress.

All experiments confirm the great influence of the voltage shape on the discharge development. The new model has to consider the effect of both voltage components.

MODELLING

The new model for the calculation of leader propagation in SF_6 is based on an energy balance in the streamer region. The discharge development starts, when the critical value $(E/p)_0 = 89\,kV/(mm\,MPa)$ is exceeded. The diameter of the corona $D_S(t)$ (Fig. 4a) is defined by the different field distribution inside and outside of the streamer

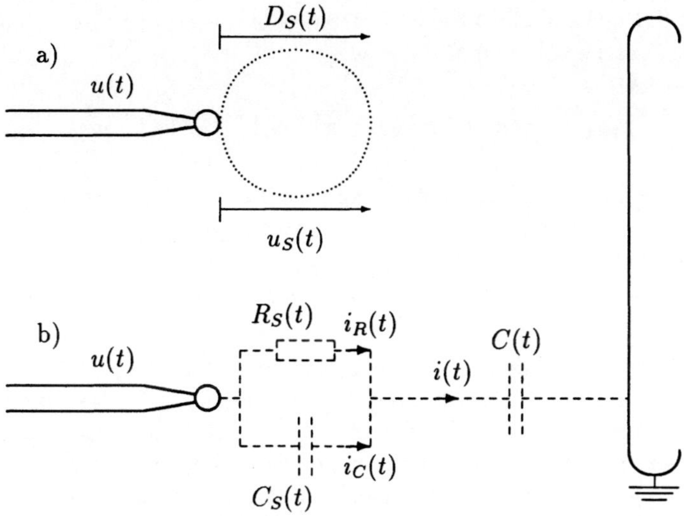

Figure 4. Modelling of the discharge development in SF_6.

corona. Within the streamer the field is balanced to $(E/p)_0$, outside the geometry of the electrodes determines the field distribution. $D_S(t)$ can be estimated by calculating the intersection of the internal and external potential curve for each time step. The capacitance $C(t)$ of this streamer region towards the grounded plane is computed by accurate field calculations using the Charge Simulation Method. The transient overvoltage $u^*(t) = u(t) - u_S(t)$ and the capacitance $C(t)$ cause a displacement current $i(t)$, which is responsible for the streamer to leader transition.

$$i(t) = C(t) \cdot \frac{du^*(t)}{dt} + u^*(t) \cdot \frac{dC(t)}{dt} \tag{1}$$

The physical behavior in the streamer can be described by a simple network (Fig. 4b). Due to the capacitance $C_S(t)$ of the streamer itself only a part $i_R(t)$ of the displacement current $i(t)$ results in thermal loss. This ionization current $i_R(t)$ is determined by the relation between the resistance $R_S(t)$ and the capacitance $C_S(t)$ of the streamer depending on the voltage shape. $i_R(t)$ can be calculated as follows:

$$i_R(t) + i_C(t) = i(t) \tag{2}$$

$i_C(t)$ can be replaced by:

$$i_C(t) = \frac{d}{dt}[C_S(t) \cdot u_S(t)] = \frac{d}{dt}[C_S(t) \cdot R_S(t) \cdot i_R(t)] \tag{3}$$

An infinitesimal consideration of $C_S(t)$ and $R_S(t)$ gives the following relation:

$$C_S(t) \cdot R_S(t) = \varepsilon_o \cdot \varrho_o \tag{4}$$

with ε_o as the dielectric constant and ϱ_o as the relative ohmic resistor, which describes the pressure-dependent ionization process in the streamer region. ϱ_o can be evaluated by comparison of a set of calculations with the corresponding measurements and has values of some kΩm for 0.1 MPa and some 100 Ωm for 0.3 MPa.

With Eq. 2, Eq. 3 and Eq. 4 the ionization current $i_R(t)$ can be computed according to Eq. 5, which can be solved numerically.

$$i_R(t) + \varepsilon_o \cdot \varrho_o \cdot \frac{d}{dt}[i_R(t)] = i(t) \tag{5}$$

The ionization current $i_R(t)$ and the voltage drop $u_S(t)$ along the streamer length effects a significant energy input into the streamer region:

$$W(t) = \int | u_S(t) \cdot i_R(t) | \, dt \tag{6}$$

If the enthalpy rise Δh exceeds 10^7 J/kg the gas is completely dissociated and heated up to about 2500 K[3]. The former streamer region is bridged by a leader channel with a high conductivity. The dynamic processes inside the leader channel are considered by using a time-dependent average leader field strength[3, 5].

At the tip of the leader a new streamer is initiated. Thus the next leader steps can be calculated in the same way as described before. However to take into account the changed field distribution caused by the leader growth new field calculations have to be done for each time step. As result a step by step elongation of the leader channel can be observed.

APPLICATION OF THE MODEL

The described simulation program allows to calculate the leader development from the inception to the final jump for pressures up to 0.3 MPa. The program requires the applied LI, VFTO or composite voltage shape, the gas pressure and the electrode geometry for the field calculation as input data.

As an example Fig. 5 shows the calculated leader steps under LI, VFTO and composite voltage stress for 0.1 MPa. The results coincide with the corresponding

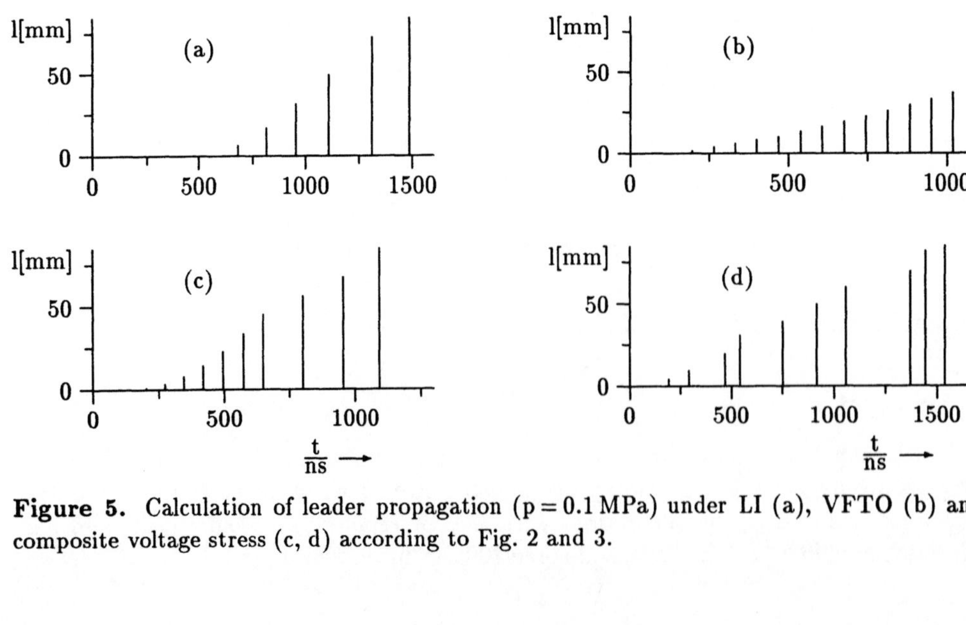

Figure 5. Calculation of leader propagation (p = 0.1 MPa) under LI (a), VFTO (b) and composite voltage stress (c, d) according to Fig. 2 and 3.

measured signals and streak records (Fig. 2, 3) concerning the number of steps, the delay time between the steps and the time to breakdown. Because of the statistical scatter of the discharge development under one voltage shape, especially for composite voltage stress, it is not possible to compute time and length of each leader step without any deviation. However the calculated leader elongation describes all characteristics of the corresponding real discharge development. Furthermore complete voltage-time curves can be calculated.

CONCLUSION

The leader development in case of strong inhomogeneous fields under LI, VFTO and composite voltage stress can be described by a new physical model based on an energy balance in the streamer region. The corresponding simulation program allows a very accurate calculation of the leader development and thus the prediction of the breakdown levels up to 0.3 MPa. Furthermore complete voltage-time curves can be calculated for all these transient overvoltages. Contrary to other models [3, 4, 6] the program is able to compute the discharge development for different types of transient overvoltages, LI, VFTO and composite voltage stress. Fundamental part of the program is a precise field calculation, which takes into account the geometry and the varying field conditions during the discharge development.

REFERENCES

1. N. Wiegart. "A semi-empirical leader inception model for SF_6," 8th Int. Conf. on Gas Discharges and their Applications, Oxford, pp.227-230 (1985).
2. I. Gallimberti, N. Wiegart. "Streamer and leader formation in SF_6 and SF_6 mixtures under positive impulse conditions, pts. I and II," J.Phys.D: Appl.Phys. 12, pp.2351-2379 (1986).
3. L. Niemeyer. "Leader breakdown in compressed SF_6: Recent concepts and understanding," 6th Int. Conf. on Gas. Dielectrics, Knoxville, Vol.VI, pp.49-59 (1990).
4. D. Buchner. "The calculation of leader propagation in point / plane gaps under LI and FTO stress," 8. ISH, Yokohama, No.30.01, Vol.1, pp.255-258 (1993).
5. H. Hiesinger. "The calculation of leader propagation in point/plain gaps under very fast transient stress," 6th Int. Conf. on Gas. Diel., Knoxville, Vol.VI, pp.129-135 (1990).
6. H. Hiesinger. "Leader breakdown for inhomogeneous fields in case of VFT conditions," 7. ISH, Dresden, No.32.05, Vol.3, pp.67-70 (1991).
7. D. Buchner. "Breakdown behavior of SF_6 insulation under composite VFT voltage stress," 10th Int. Conf. on Gas Discharges and their Applications, Swansea, Vol.1, pp.422-425 (1992).

DISCUSSION

I. HERBST: What is your exact electrode geometry?

D. BUCHNER: For the investigations of the physical behavior of SF_6 we used a needle/plane gap. A needle of 15 mm in length was fixed in a first plane. The second plane was installed in parallel to the first one at a distance of 100 mm. So we had a remaining gap width of 85 mm.

EXPERIMENTS ON THE DIELECTRIC BEHAVIOUR OF SF6 GAS IN NONUNIFORM FIELDS UNDER LI AND VFT

S. M. El-Makkawy and S. S. El- Dessouky

Electrical Engineering Department
Faculty of Engineering
Suez Canal University
Port Said, Egypt

ABSTRACT

Experimental results on the discharge development of SF6 dielectric caused by nonoscillating impulse voltage (LI) and very fast transient of oscillating impulse voltage (VFT) have been carried out.

In the experiments further breakdown characteristics for LI and VFT have been obtained in voltage-time (V-t) curves at different discharge parameters, such as, gap distance, needle radius, gas pressure, polarity and electrode roughness. The electrode roughness was fixed as hemispherical particle on the center of the plate electrode. For the nonoscillating voltage, the standard lightning impulse of 1/50 μs is used, while the oscillating impulse voltage have been generated with two higher frequencies of 1.12 MHz and 2.77 MHz.

Both statistical and formative time lags decreases with increasing the voltage stress and they are higher in the case of nonoscillating voltage than in case of oscillating voltage.

INTRODUCTION

Gas-insulated switchgear (GIS) has been making a great contribution to reducing the size and enhancing the reliability of transmission and substation equipment, owing to excellent dielectric properties of SF6 gas [1].

Gaseous Dielectrics VII, Edited by L.G. Christophorou and D.R. James, Plenum Press, New York, 1994

During the switch operation of disconnectors in Gas-insulated system (GIS) a great number of pre - or restrikes occur because of the slow speed of the disconnector moving contact. Due to the very fast travelling waves, fast transient overvoltages are generated. These overvoltages may cause a failure inside the GIS, transient potential rise of the sheath and transient overvoltages in low voltage control circuits. The shape and amplitude of these transients depend on the GIS design and on the overall system configuration. Today, the main source for insulation faults are imperfections inside the gas- insulated switchgear like moving particles or fixed protrusions. These imperfections become especially critical in case of VFT. Voltage- Time characteristics of electrical breakdown in compressed SF6 are of practical importance for the insulation co-ordination and overvoltage protection of gas insulated substations [2-5].

EXPERIMENTAL SETUP

The experiments were performed in a glass- transparent plastic pressure vessel 45 cm high and 12 cm diameter, which is capable of withstanding pressure up to 2.5 bar. The test chamber was evacuated before filling SF6 gas. The test arrangement, Fig.1 is a needle- plate gap configuration. The gap distance of 0.5 to 2.5 cm, needle with different tip radii of 3,5,10 µm and an ordinary profile plate with 50 mm diameter, 10 mm thickness have been used. The electrodes were throughly polished and cleaned before mounting them inside the chamber.

In order to simulate a distortion of the slightly inhomogeneous field between needle plate electrode a hemispherical protrusion with radius (1,2 mm) was fixed to the plate electrode as shown in Fig.2. The applied testing voltage was varied between very fast transient with oscillating frequencies 1.12 MHz, 2.77 MHz and lightning impulse voltage of 1/50 µs.

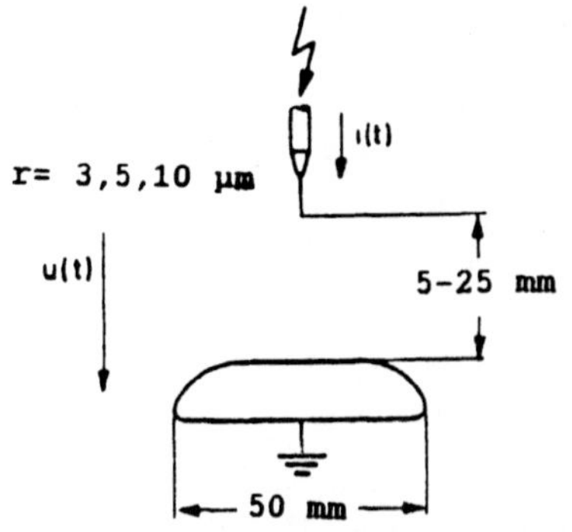

Figure 1. Needle Plate gap

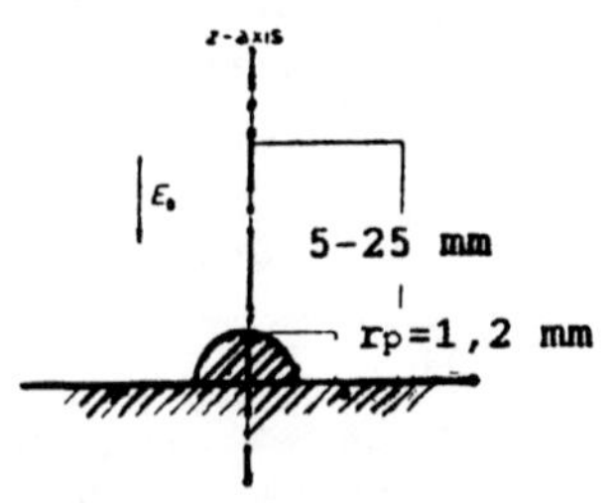

Figure 2. Model of a rough electrode surface.

For the evaluation the measured values of breakdown discharge probabilities for different voltages are suitably plotted in a probability net and one obtains the result shown in figure 3. If the measured points lie approximately on a straight line as indicated in the figure, it may be assumed that the breakdown discharge voltage of the investigated sample obeys a gaussian or normal distribution law [6].

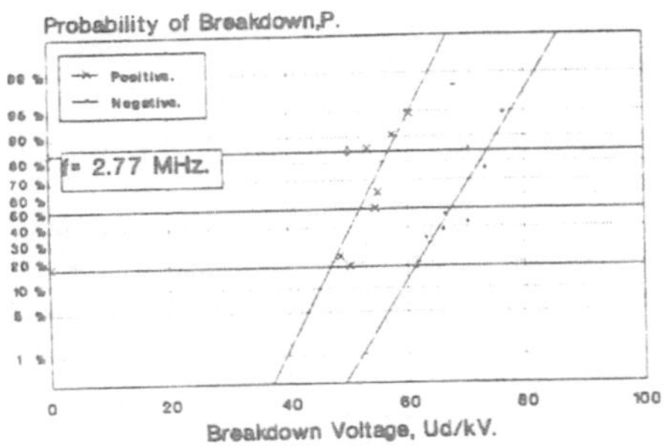

Figure 3. Breakdown probabilities as a function of breakdown voltage and polarities. d=1 cm, p=1.2 bar and r=10 µm.

EXPERIMENTAL RESULTS

Discharge Development

Fig.4 shows the instants of predischarge steps for different voltage shapes (impulse voltage and VFT at 2.77).

Concerning oscillating impulses the predischarges only appear in specified time intervals with increasing voltage (dU/dt>0). The discharge development stops as soon as the voltage and thus the driving electrical field is decreasing. The consequence is a large time lag between the individual discharge steps. In case of non-oscillating impulse the discharge development is much faster and leads to a breakdown shortly after the voltage crest.

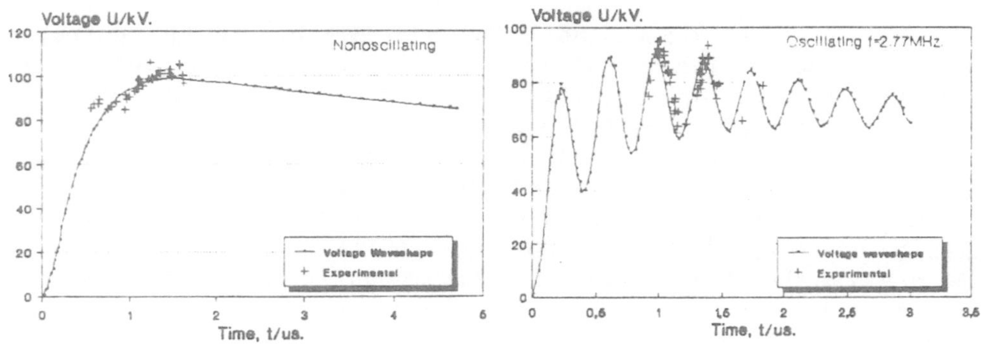

Figure 4. Instants of discharge for needle- plate at different voltage waveshape. p=1.6 bar, d=1 cm, r=10µm, positive polarity.

Voltage-Time Characteristics

V-t curves represent the breakdown behaviour of electrode arrangements under impulse voltage stress and serve as an essential basis for statistical insulation coordination. Consequently the V-t characteristic is represented either by a band which usually covers the 5% to 95% breakdown probability or by a single V-t curve which usually refers to a 5% breakdown probability (lower limit curve) [2].

Fig. 5 shows voltage time curves at positive polarity. The curves obtained for nonoscillating voltage (LI) and oscillating impulse with oscillating frequency f=2.77 MHz under the same conditions are compared. The results of the curves have been illustrated, that the oscillating breakdown occurs below the measured values for lightning impulse. These results demonstrate the strong reduction of the insulation strength in the case of oscillating stress than that of nonoscillating stress, i.e. the oscillating voltage (VFT) waveshape strongly affects the absolute breakdown values. Generally, it is concluded, that the influence of the voltage shape on the insulation strength can be explained considering the leader formation process and corona stabilization.

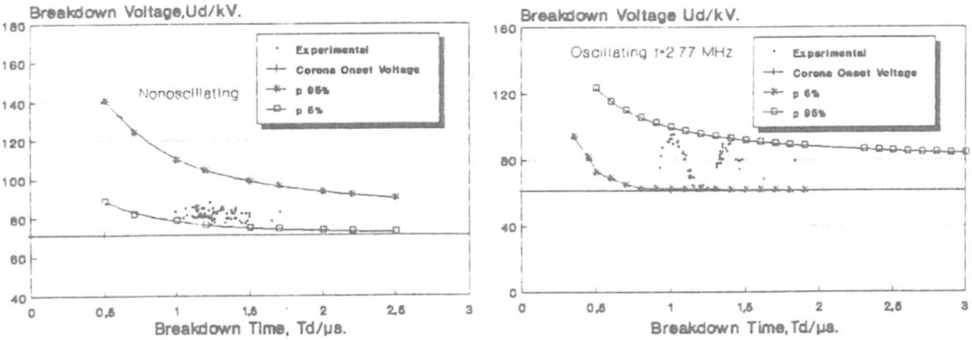

Figure 5. V-T characteristic of needle-plate at different voltage waveshape. p=1.2 bar, d=2 cm, r=10 µm, positive polarity.

Fig. 6 shows, that the voltage time curves at positive polarity under different gas pressure 2 and 1.4 bar and fixed metallic hemispherical particle of radius 2 mm was presented. From the curves, its clear that at higher pressure the breakdown value is greater than at lower pressure and also minimum breakdown. This means that increasing the gas pressure, increasing the breakdown voltage and increasing minimum breakdown value. This fact can be attributed to by the field divergence and gap irradiation. It is shown also, that the values of breakdown voltage and minimum breakdown with metallic particle is lower than in the case of without metallic particle.

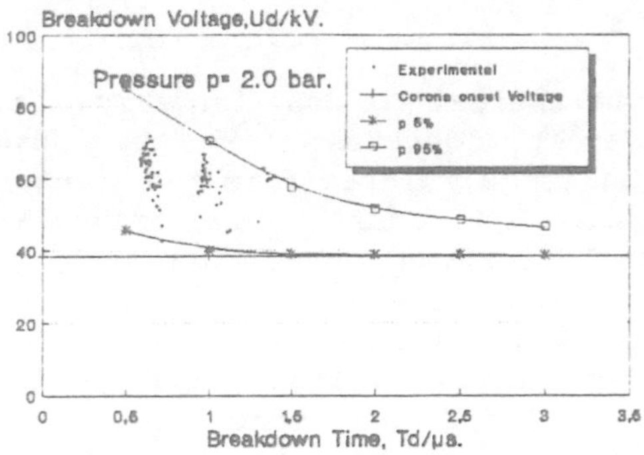

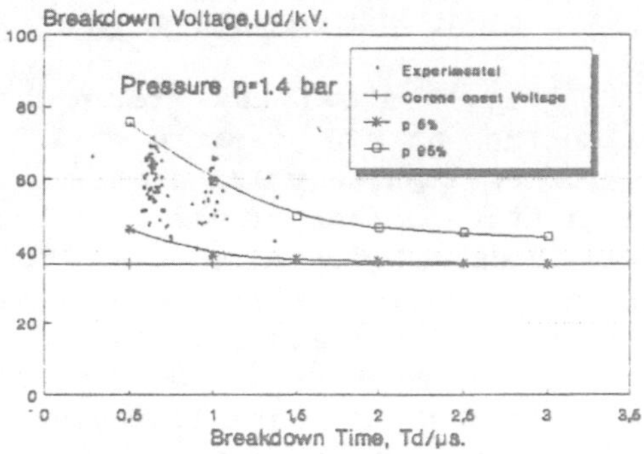

Figure 6. V-T characteristic of needle-plate at different gas pressure. d=1 cm,r=5 µm, positive, r_p=2 mm, oscillating,f=2.77 MHz

Fig. 7 gives respectively the measured positive and negative lower limit breakdown voltages for investigated point -to- plate geometry with smooth and rough plates (the roughness was simulated by a single protrusion attached to the grounded plate). It is shown from figures, that the lower limit curve in the case of without protrusion is greater than with protrusion for different polarities. It is interesting to observe, that the surface-state of the ground plate reduces significantly the breakdown voltage with negative polarity and appears to have little, if any, influence upon the breakdown voltage with positive polarity. Consequently, unlike what is observed in a homogeneous field gap, the positive polarity must be regarded and most critical one whatever the voltage stress. The reason of this behaviour is the faster discharge velocity for positive polarity with smaller formative time lag compared with negative polarity. Nearly all breakdowns for positive polarity occured in the steep front of the fast transient voltage, but in case of negative polarity much later in the oscillating tail [7].

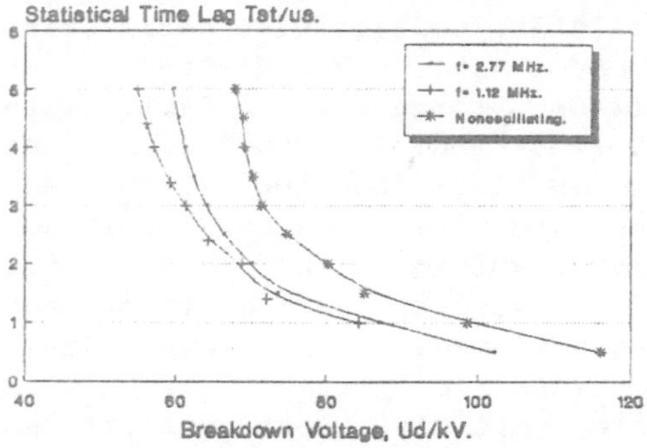

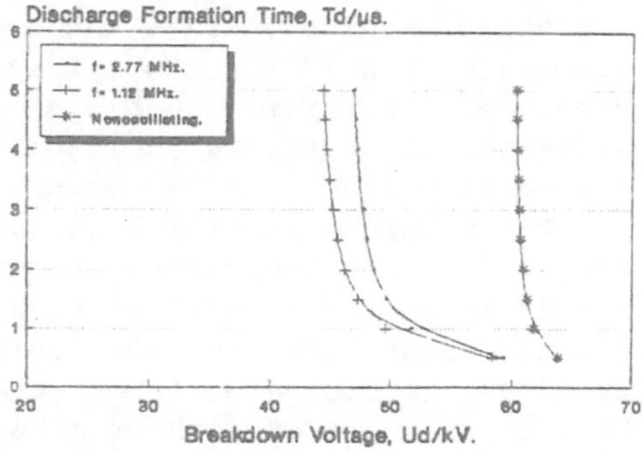

Figure 7. Relation between lower limit breakdown and time. d=1 cm, p=1.2 bar, r=5 µm, r_p=2 mm, ossillating, f=2.77 MHz.

Statistical and Formative Time Lags

Fig. 8 shows the computed statistical t_{st} and formative t_d time lags at SF6 pressure of 1.2 bar in dependence on the peak value of impulse voltage stress. t_{st} is calculated from the upper limitation curves of the corresponding V-t characteristics(U=95%) and t_d calculated from the lower limitation curves (U= 5%) [8]. From the curves it is shown, that both time-lags decrease with increase of voltage stress, and they are of the same order of magnitude. Both time lags for nonoscillating voltage are always greater than for oscillating voltage (VFT). It is shown also, that both time lags for oscillating voltage with oscillating frequency f=2.77 MHz is always greater than for oscillating voltage with oscillating frequency f=1.12 MHz. This mean that both statistical and formative time lags increases with increasing oscillating frequency.

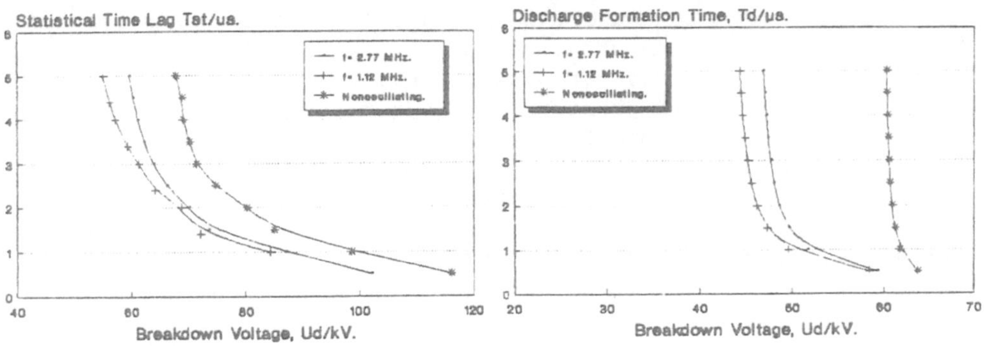

Figure 8. Statistical and formative time lages for needle-plate. p=1.2 bar, d=1.5 cm,r=10 μm, positive polarity.

CONCLUSION

V-t characteristics of SF6 gas is investigated in case of lightning impulse voltage and oscillating voltage with oscillating frequency of 2.77 MHz and 1.12MHz under various gap distance, gas pressure, polarity and electrode roughness effect with hemispherical particle with different radius. The oscillating breakdown voltage may be occur below the measured values for lightning impulse. For all waveshaps negative breakdown voltages are higher than the positive breakdown voltages. The lower limit breakdown increased with increasing pressure, gap distance and needle radius. The breakdown voltage values and the lower limit breakdown in the case of presented metallic particle are lower than in the case of without metallic particle. Both statistical and formative time lags decreases with increasing the voltage stress and they are in the same order of magnitude.

ACKNOWLEDGMENT

This work was carried out in the Institute für Hochspannungs-technik RWTH Aachen (Germany) under co-operative channel system with Institutsdirektor of RWTH Aachen Prof. Dr.-Ing. Klaus Möller. The authors wish to thank prof. Dr.-Ing. Klaus Möller of Institut für Hochspannungstechnik RWTH Aachen for his help ful suggestions and encouragements through this work.

REFERENCES

1. S. Matsumoto, H. Okubo,H. Aoyagi, S. Yanabu ' Nonuniform flash-over mechanism in SF6 gas under fast oscillating and non-oscillating impulse voltages' 6th ISH new orleans, 1989, paper No.32-02.
2. W. Knorr, K. Möller, K. diederich ' Voltage-time characteristic of slightly nonuniform arrangements in SF6 using linearly rising and oscillating lightning impulse voltages, Cigre, 1980, Report No. 15-05.
3. H. Hiesinger and W. Taschner ' V-t curves for different voltage shapes applied to inhomogeneous electrodes on fixed or free potential.' 7th ISH, Dresden, 1991, paper No. 32-04.
4. N. Izeki ' Voltage-time characteristics of SF6 Gas.' Electra, Vol. 101, pp. 41-59.
5. G. F. Luxa ' Recent research activity on dielectric performance of SF6 with special reference to fast transients.' Cigre, 1988, Report No. 15-06.
6. W. Mosch, W. Hauschild ' Statistik für Electrotechniker.' VEB Verlag Technik, Berlin, 1984.
7. M. Abdel-salam and W. Abul-Shohoud ' Effect on polarity on the breakdown voltage of gas gaps.' Proc. of the international symposium on Gaseous Dielectrics, Knoxville, 1980,pp 108-114
8. W. Knorr ' A model to describe the ignition time-lag of slightly nonuniform arrangements in SF6.' 3th ISH, Milan, Aug. 28-31, 1979, paper No. 31-11.

ALGEBRAIC APPROACH TO AN ANALYTICALLY INTEGRABLE FORM OF NONUNIFORM-FIELD DISTRIBUTIONS

M.F. Fréchette[1] and M. Côté[2]

[1]Câbles et Isolants, Technologie et IREQ, Hydro-Québec
[2]Engineering Physics, Ecole Polytechnique
Québec, Canada, J3X 1S1

INTRODUCTION

Most often, electrical-field calculations associated with non-uniform field geometries must resort to numerical solving techniques. The thus obtained numerical solutions may prove cumbersome for further treatment in the framework of a gaseous-dielectric problem. It is specially true for quasi-uniform field, where truncation of the field expression may not be warranted.

Indeed, under true-corona conditions, the critical length for which the discharge inception occurs is considerably smaller than the total gap length. Upon these circumstances, a simple field-expression in terms of the critical length can be assumed, and further used to predict the gap-insulated performance.[1,2] On the other hand, there is a whole category of geometrical field-conditions where this assumption does not hold. In particular, in quasi-uniform field conditions, where the critical length extends to a large portion of the gap. Recently, Ibrahim[3] addressed this problem for a point-to-plane geometry, the treatment including field enhancement as low as 2.5. The proposed field-expression and its numerical parameters were found not to be directly applicable to a similar geometry, viz. a plane/point/plane arrangement operating under quasiuniform field-conditions. Furthermore, the field integral over the gap length led to a series expression. Finally, the analytical form of the parameters entering the field expression required further numerical fitting.

The present work elaborates on a general algebraic method for converting the numerical solution into an analytical expression. The sought expression should correctly reproduce the field values, the algebraic form being preferably unique and applicable to a wide range of field nonuniformities. Furthermore, it should lead to an analytical integrable expression. This method was successfully developed and applied to a plane/point/plane geometry. The approach was put to further test: corona threshold and/or breakdown voltage in SF_6 according to the streamer criterion was estimated. Calculated values were compared with experimental data. The experiment was conducted for pressures up to 1.2 bar abs, and gap lengths varying from 3 to 15 mm. Finally, the use of the present approach in predicting breakdown fields when corona stabilization is non-negligible, is discussed.

Gaseous Dielectrics VII, Edited by L.G. Christophorou
and D.R. James, Plenum Press, New York, 1994

FORMULATION

Consider a plane/point/plane arrangement, the axial Laplacian-field distributions having the characteristics shown in Fig. 1a ($x = 0$ defining the tip location). The planes are Rogowski-profiled. For the sake of calculations, they can be modelled as parallel plates. The point-electrode is semi-hemispherically ended and has a 3.17-mm diameter. The rod extends into the inter-plane space, its length being 10 mm. The present geometry elicited nonuniformity factors ranging from 2.67 to 4.75, for gap lengths extending from 3 to 10 mm, respectively. The calculated field-distributions, using a finite-element method, were numerically fitted using a polynomial expression. It was found that the numerical distribution could be well rendered using the inverse of a fourth-degree polynomial with non-zero real coefficients:

$$E(x) = \frac{Emax}{[\, d.x^4 + c.x^3 + b.x^2 + a.x + 1]} = Emax.f(x), \qquad (1)$$

where a, b, c, d are fitting constants. Emax is the product of the uniform field times the nonuniformity factor. In order to get an adequate numerical fit, it may be necessary to use a summation of several of these functions [Eq. (1)]. It basically depends on the field divergence, and the extent of interelectrode distance overwhich the fitting is sought. Such polynomial function [Eq. (1)] can always be expressed as a sum of partial rational fractions with real coefficients, yet with the presence of irreducible quadratics.[4] Hence, it can be integrated analytically at all times, using the form:

$$\frac{Emax/d}{F_1.F_2} = \frac{A.x + B}{F_1} + \frac{C.x + D}{F_2}, \qquad (2)$$

where the functions F_j are:

$$F_j = [x^2 + r_j.x + del_j];\ j=1,2.$$

Simple algebraic manipulations give access to the several constants. The field integral over the axial field coordinate x, $\varphi(x)$, is therefore given by:

$$\varphi(x) = A/2.\ln[F_1] + C/2.\ln[F_2] + \qquad (3)$$

$$(2.B-A.r_1).E_1^{-0.5}.tg^{-1}[(2.x+r_1).E_1^{-.05}] +$$

$$(2.D-C.r_2).E_2^{-0.5}.tg^{-1}[(2.x+r_2).E_2^{-.05}].$$

where the functions E_j are:

$$E_j = [4.del_j - r_j^2];\ j=1,2.$$

The present technique was used to determine the corona threshold and/or breakdown voltage in SF_6. Applying the streamer criterion:[5]

$$K = \int_0^{x_c} (\alpha-\eta).dx \qquad (4)$$

with further considering that the net ionization coefficient $(\alpha\text{-}\eta)$ at constant pressure can be expressed as a linear function of the field $E(x)$, and that this coefficient becomes zero when the distance on the discharge axis reaches the critical length (x_c). The constant K associated with the critical charge was set at a value of $\ln[10^8]$.

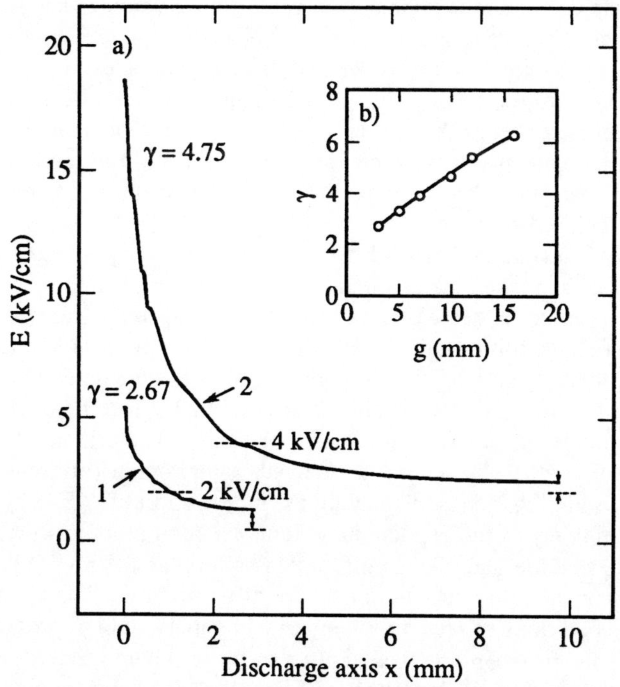

Figure 1. Field characteristics associated with the present electrode arrangement; a) Field distributions on the discharge axis: 1 - 3 mm, 2 - 10 mm; b) Nonuniformity factor (γ) as a function of gap length (g).

Let us define the pressure-reduced inception and/or breakdown field as:

$$(E/p)_{MAX} = \frac{V.\gamma}{g.p}, \tag{5}$$

where V, g, γ and p are respectively the applied voltage at threshold and/or breakdown, the gap length, the nonuniformity factor, and the gas pressure. Using Eqs. (3-4) and assumptions mentioned above, a general analytical expression linking $(E/p)_{MAX}$ to the critical length can be arrived at:

$$(E/p)_{MAX} = [K/(\beta.p) + (E/p)_c.x_c].[\mathrm{Int}(x_c)]^{-1}, \tag{6}$$

where $(E/p)_c$, 8.84 $kV^{-1}bar^{-1}$, is the critical field at which $(\alpha\text{-}\eta) = 0$; β, 27 kV^{-1}, is the slope of the linear function representing the variation of $(\alpha\text{-}\eta)/p$ in terms of E/p; and $\mathrm{Int}(x_c)$ is the integral of $f(x)$, as defined in Eq. (1), over the critical path.

EXPERIMENTAL COMPARISON AND DISCUSSION

The representative nature of the polynomial expression [Eq. (1)] was tested over a large range of nonuniformity factors. For the present geometry, applicability for fitting axial field-distributions was surveyed up to γ-values extending to 6.3, the longest gap being set to 16 mm. The variation of γ as a function of the gap length is displayed in Fig. 1b. By using a smaller rod with a conical tip in a similar arrangement,[6] true-corona conditions could be produced in atmospheric SF_6. In this instance, the calculated γ reached approximately 20. With both geometries, it was found that the present analytical form successfully represented the field distributions on the discharge axis. Experience has shown that a single fourth-degree polynomial expression may not be sufficient to give the best fit over an entire gap. The case being, a summation of several of these functions can be used, each featuring part of the gap. Or, only a fraction of the total gap length can be fitted, as long as the ending field coordinate exceeds the critical length. Under the present conditions, not more than two polynomial expressions were necessitated at one and the same time. Finally, it is worth mentioning that ounce the numerical fit is achieved, the remaining steps are fully analytical, giving a closed-form solution.

Figure 2a illustrates the variation of the relative critical length (x_c/g) as a function of the gas pressure, for the respective gap lengths 3 and 10 mm. Pronounced rise of this ratio is observed as the pressure is decreased, the variation being steeper for the shortest gap, i.e. for less enhanced-field conditions at the cathode. It is worth pointing out that as long as this ratio remains considerably smaller than one, the model will give an estimation of the inception field rather than the breakdown field. It is particularly true here for the 10-mm gap, although the magnitude of the ratio varies with pressure. The critical length as such does not vary very much as the gap length is changed, a maximum difference reaching about 15% as observed in Fig. 2b. Thus, displaying the inception fields $(E/p)_{MAX}$ as a function of pressure, irrespectively of the gap length, will give a good graphical representation of the insulating properties of the gap. This will remain as long as the normalized field distributions in the region surrounding the cathode do not differ greatly. This applies to breakdown fields as well, for conditions of relatively long critical paths and high pressures.

The present solving technique was applied to the problem of estimating negative DC corona-threshold and/or breakdown voltage in SF_6. A plane/point/plane arrangement hav-

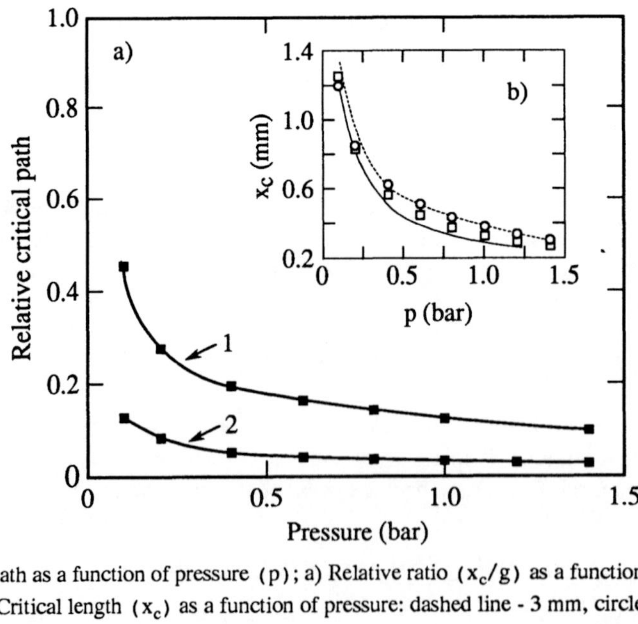

Figure 2. Critical path as a function of pressure (p); a) Relative ratio (x_c/g) as a function of pressure: 1 - 3 mm, 2 - 10 mm; b) Critical length (x_c) as a function of pressure: dashed line - 3 mm, circles - 5 mm, squares - 7 mm, full line - 10 mm.

ing the characteristics mentioned in the previous section was utilized. More details on the experimental setup and procedure were reported earlier.[7] The present experiment was conducted for pressures up to 1.2 bar abs, and gap lengths varying from 3 to 15 mm. Experimental results and theoretical values calculated according to Eq. (6) are compared in Figs. 3 and 4. Investigation of the characteristics corresponding to the 3-mm gap length (Fig. 3,

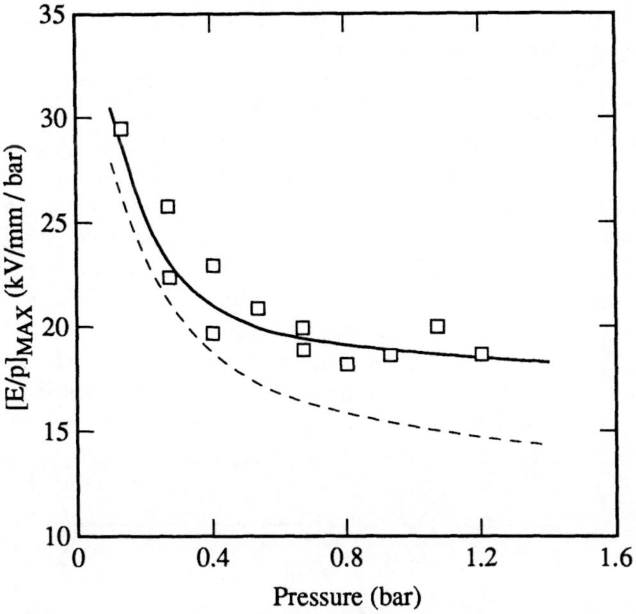

Figure 3. Reduced field $[E/p]_{MAX}$ as a function of pressure (p) for the 3-mm gap; squares - experimental data, dashed line - inception field, full line - breakdown field.

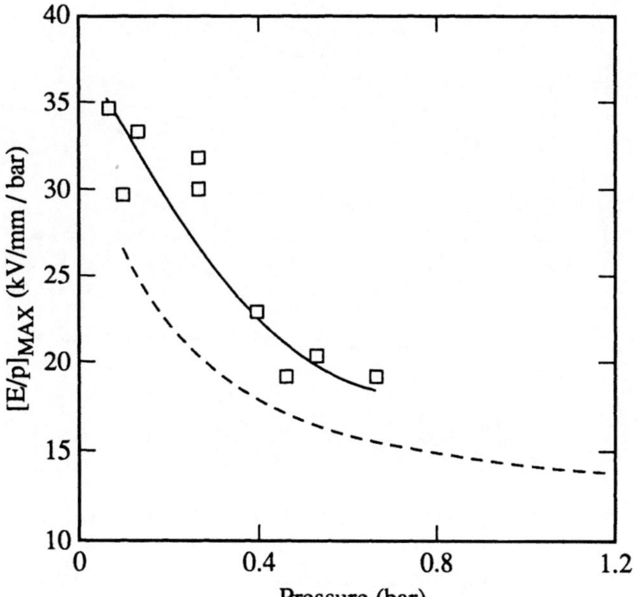

Figure 4. Reduced field $[E/p]_{MAX}$ as a function of pressure (p) for the 10-mm gap; squares - experimental data, dashed line - inception field, full line - breakdown field.

squares) showed no macroscopic signs of stable corona activity ($\gamma = 2.7$). The theoretical curve (dashed line) associated with the inception field compares favorably with its experimental counterpart. The best-fit curve (full line) representing the experimental data is slightly underestimated by the theoretical model, yet both curves display the same trend. Divergence between the two curves is seen to grow with the pressure increasing, the critical path becoming shorter: in this range, the model gives an estimate of the inception field. For a 10-mm gap ($\gamma = 4.75$), the model predicts the inception field over the whole pressure range, as illustrated in Fig. 4. Corona activity of low intensity was detected in this case even at atmospheric pressure.[7]

In an attempt to model the variation of the breakdown fields, Eq. (6) was used to fit the curves (Figs. 3-4, full lines) featuring the breakdown data. This approach basically implies that the critical path would be affected via corona shielding, when corona stabilization is present. Results are presented in Fig. 5. Fitting of the experimental fields required the use of longer critical lengths (x_f), the ratio x_f/x_c being at least larger than 2 over the whole pressure range. This relative ratio was found to vary almost linearly at low γ-values (Fig.5, curve 1). In this particular case, it is worth noticing that the same fit could be achieved using a constant critical path (x_u). This constant value (x_u) was determined by the inter-electrode distance at which the calculated field-distribution intersected that corresponding to the constant uniform field V/g. It is contended that this approximation can be used successfully as long as both ratios x_f/x_c and x_u/x_c display similar slopes in their variation with pressure, viz. at low γ-values or, at slightly larger γ-values and higher pressures. The restrictive implication is that the breakdown voltage must vary linearly with the pressure.

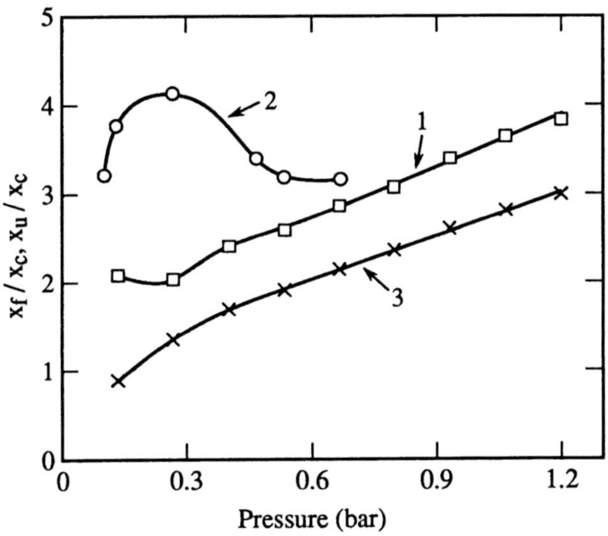

Figure 5. Ratio of (x_f/x_c) or (x_u/x_c) displayed as a function of pressure (p); (x_f/x_c), 1 - 3 mm and 2 - 10 mm; (x_u/x_c), 3 - 3 mm.

CONCLUSION

The problem of expressing numerical field-solutions as closed-form integrable functions was addressed. The present context was mainly concerned with electrode geometries eliciting quasiuniform field-conditions, i.e. when the critical path extends significantly into the gap. A plane/point/plane arrangement was considered, specifications eliciting non-uniformity factors that did not exceed approximately 6.

Calculated field-distributions were found to be well rendered by using the inverse of a fourth-degree polynomial. While only two particular arrangements were tested, the proposed polynomial function appears to be of broader use, whereas several of them can be utilized to represent the field distribution over the entire gap. Ounce the fit was done, the remaining steps were fully analytical.

In way of illustration, the present solving technique was applied to the estimation of corona-threshold and/or breakdown voltage in SF_6. Theoretical values compared favorably with experimental data. The analytical formulation elicited some insight comments on the relevance of analyzing $(E/p)_{MAX}$ as a function of pressure. Breakdown fields could be modeled, if corona shielding was assumed to lengthen the critical path. An analytical approximation was put forward, resulting into the prediction of breakdown fields when the γ-values are rather small.

ACKNOWLEDGMENTS

This work was sponsored in part by *Direction Distribution*, Hydro-Québec.

REFERENCES

1. W.L. Lama and C.F. Gallo, Systematical study of the electrical characteristics of the Trichel current pulses from negative needle-to-plane coronas, *J. Appl. Phys.* 45:103 (1974).

2. T. Nitta and Y. Shibuya, Electrical breakdown of long gaps in sulfur hexafluoride, *IEEE Trans. on Power App. and Sys.*, PAS-90:1065 (1971).

3. O.E. Ibrahim, An expression for the electric field distribution in rod-plane gaps, *IEEE Trans. on Electr. Insul.*, 23:493 (1988).

4. R. Godement, Algebra, Hermann, Paris, (1968).

5. A. Pedersen, I.W. McAllister, G.C. Crichton, and S. Vibholm, Formulation of the streamer breakdown criterion and its application to strongly electronegative gases and gas mixtures, *Archiv. fur Elektrotechnik*, 67:395 (1984).

6. M.F. Fréchette, M. Côté, N.G. Trinh, and R.Y. Larocque, Partial-discharge activity for a negative DC corona in atmospheric SF_6, *Proc. Int. Symp. on Electr. Insul.*, Pittsburgh, Penn. USA, June 5-8 (1994).

7. M.F. Fréchette, M. Côté, and R.Y. Larocque, Discharge characteristics in an asymmetrically perturbed SF_6 gap, *Proc. Conf. on Electr. Insul. and Dielec. Phen.*, Pocono Manor, Penn. USA, Oct. 17-20 (1993).

SECTION 6: SPECIAL APPLICATIONS OF GASEOUS DIELECTRICS

REDUCTION OF NO_x AND SO_x FROM COMBUSTION FLUE GASES BY A CORONA DISCHARGE-ELECTRON BEAM HYBRID SYSTEM

J.S. Chang[1], P.C. Looy[1], T. Yoshioka[2] and K. Nagai[2]

[1]Department of Engineering Physics
McMaster University
Hamilton, Ontario, Canada L8S 4M1
[2]Ebara Research Co.
4-2-1 Honfujisawa
Fujisawa 251, Japan

INTRODUCTION

The NO_x and SO_x emissions are the major cause of acid rain. The CO_2 and CH_4 emissions cause a greenhouse effect which leads to abnormal global heating of the atmosphere and can create a temperature inversion layer that traps pollutants. There is a vital interest in controlling these emissions -- in the near term for acid rain and in the long term for the greenhouse effect.

In a power plant gas cleaning system, a substantial modification has been proposed in recent years due to the tighter acid gas emission requirements. The options being considered to meet the stringent gas regulations include: burning low-sulphur coals, installing wet flue gas desulphurizers (FGD), catalytic ammonia reduction, in-furnace sorbent injection and energetic electron induced plasma processes.

In a gas cleaning of combustion flue gases, normally three approaches are proposed[2] as follows:

1. Conversion of acid and greenhouse gases to unharmful gases such as H_2O, N_2, O_2, etc.
2. Conversion of acid and greenhouse gases to a reusable products such as agricultural fertilizers, dry ices, etc.
3. Trapping a reformed acid and greenhouse gases byproducts such as ammonium sulfate aerosols, ammonium sulfate nitrate aerosols, carboxylic acid aerosols, carbon dusts, etc. for later disposals.

The energetic electron induced plasma process is one of the most effective methods of reducing SO_2, NO_x and CO_2 from industrial flue gases. This flue gas treatment consists of adding a small amount of ammonia, water vapour and/or argon gases, and irradiating the gas by means of an electron shower generated by an electron beam or corona discharge, thereby causing reactions which convert the acid and greenhouse gases to harmless gases, aerosols and other

Gaseous Dielectrics VII, Edited by L.G. Christophorou
and D.R. James, Plenum Press, New York, 1994

products. These products may then be collected from the flue gas by means of such conventional collectors as an electrostatic precipitator or baghouse.[1-12]

The mixture of the NH_3 is significantly reduces NO formations in a corona reactor due to the dissociation of NH_3 to NH and NH_2. Chang[1-4] proposed following reactions may play an important role in the system.

$$NO + NH_2 \rightarrow H_2O + N_2 \quad ; \; k = 2.1 \times 10^{-11}\,[cm^3/s] \tag{1}$$

$$NO + NH \rightarrow N_2 + OH \quad ; \; k = 4.7 \times 10^{-11}\,[cm^3/s] \tag{2}$$

$$NO + OH + N_2 \rightarrow HNO_2 + N_2 \quad ; \; k = 6.7 \times 10^{-31}\,[cm^6/s] \tag{3}$$

$$NO + H + H \rightarrow HNO + H \quad ; \; k = 3.5 \times 10^{-32}\,[cm^6/s] \tag{4}$$

$$NO_2 + H \rightarrow OH + NO \quad ; \; k = 1.1 \times 10^{-10}\,[cm^3/s] \tag{5}$$

$$NO_2 + OH + N_2 \rightarrow HNO_3 + N_2 \quad ; \; k = 2.6 \times 10^{-30}\,[cm^6/s] \tag{6}$$

Chang[1,2] suggested that NO_x can be reduced by injecting corona discharge generated NH_x and OH radicals in a flue gas based on the above reactions. In this case, corona discharge power consumptions may be reduced, since discharge power is not loss to activate other gas components in flue gases. Ohkubo et al.[12] replaced corona wire to pipe with nozzle gas injector in a wire-plate type electrostatic precipitator, and injected mixture of air and ammonia to the flue gas through corona discharge zone in nozzle gas injectors. In spite of more than one equivalent of ammonia is injected, no significant leak ammonia ($<$0.2 ppm) with 100% reduction of NO_x in optimum conditions was observed in air-NO mixture gases. Similar approaches are taken in the present investigation except we can install present corona radical injector in any flange locations in flue gas flow ducts.

In this work, an experimental investigation has been conducted to reduce NO_x and SO_x from combustion flue gases by a corona discharge-electron beam hybrid system. Experiments are conducted at Ebara Co. Ltd., Fujisawa pilot plant which combust heavy oils with flue gas flow rate of 1200 Nm^3/h.

FUJISAWA PILOT PLANT

A detailed description of the Fujisawa Pilot Plant has already been introduced by Kawamura et al.[8, 9] and is shown schematically in Figure 1. Gas flow rate of the pilot plant is 1500 m^3N/h. Flue gas is generated from a heavy oil-fired boiler. Flue gas enters the process with an initial concentration of 800-1000 ppm of SO_2 and about 200 ppm of NO_x and a temperature of 280°C. For initial cooling, to approximate large size boiler temperatures, the flue gas enters the shell and tube heat exchanger. A temperature of flue gas is cooled from 280°C to 150°C. The flue gas then enters the spray cooler. In the spray cooler, the flue gas is cooled from 150°C to 65°C by atomized water injection.

The flue gas enters the process vessel where it is irradiated by the energetic electron induced plasma system. The energetic electron irradiation produces reactive radicals which convert the SO_2 and NO_x to sulfuric and nitric acids. These acids immediately react with ammonia, and form the by-product particulate.

Particulate of ammonia sulfate and ammonia nitrate are collected by an electrostatic precipitator and a baghouse filter. The cleaned flue gas is then released through the stack by means of an induced-draft fan.

Energetic electron induced plasma system consists of corona discharge radical injection system (60 kV×5 mA) and an electron beam generator (500 kV×30 mA) as shown in Figure 2. In the present experiment, the corona radical injection system is located in the upstream of the electron beam generators at mixing chambers, and downstream position at flow tube before electrostatic precipitators ESP.

Infrared type SO_2, CO and CO_2 detectors, chemical luminescence type NO, NO_x detectors, and magnetic type O_2 detectors are used to analyze flue gas at upstream and downstream of an energetic electron induced plasma process. In addition, twin ion cathode type NH_3 detector and ozone monitor are installed downstream of the ESP system.

The core of the system is the data recorder which is capable of sixty data recordings. Data recorded include temperatures, pressures, concentrations, and flow rates. This data is recorded and plotted every 10 seconds. The recorder is connected to a personal computer which receives the recorded data and displays it on a monitor flow diagram.

ENERGETIC ELECTRON INDUCED PLASMA SYSTEM

Electron Beam

Electron beams generated by an electron gun are passing through a Ti made window and uniformly distributed by magnetic fields inside a reaction chamber, where combustion flow gases pass through. Unlike other pilot plant tests, only the single stage irradiation system was used.

Corona Discharge Radical Injection

Retro-fit type corona radical injector consisted of a ceramic tube (o.d. 3 cm and 10 cm long) with a hollow electrode (o.d. 0.5 cm and 5 cm long) in the front of tube, where Ar-NH_3 (3%) mixture or NH_3 (99.999% pure) gas exists and flow stabilized corona discharge is initiated. Grounded ring type electrode (4 cm wide) is located on outside surface of the ceramic tube.

EXPERIMENTAL RESULTS

Current-Voltage Characteristics

The time averaged current increases with increasing applied voltage. In contrast to the ordinary corona discharge, the time averaged current increases with increasing gas flow rate due to the flow stabilization of corona discharges[2] until initiation of a spark discharge. Optical emission from the corona discharge shows a flame like discharge pattern in the edge of the hollow electrode, and this flame length or highly excited plasma zone decreases with increasing NH_3 gas flow rate, and increases with increasing Ar-NH_3 gas flow rate until initiation of

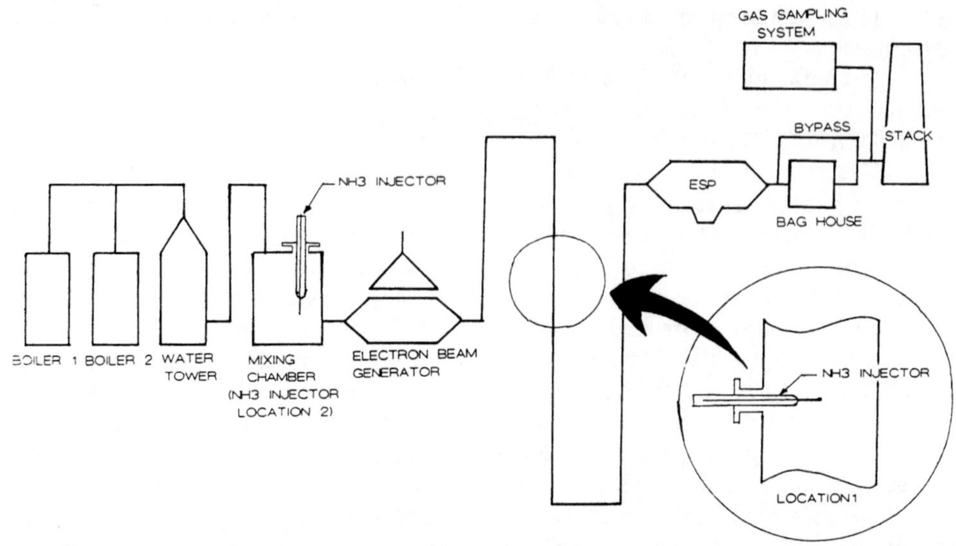

Fig. 1 Schematics of the Fujisawa pilot plant flue gas cleaning test loop.

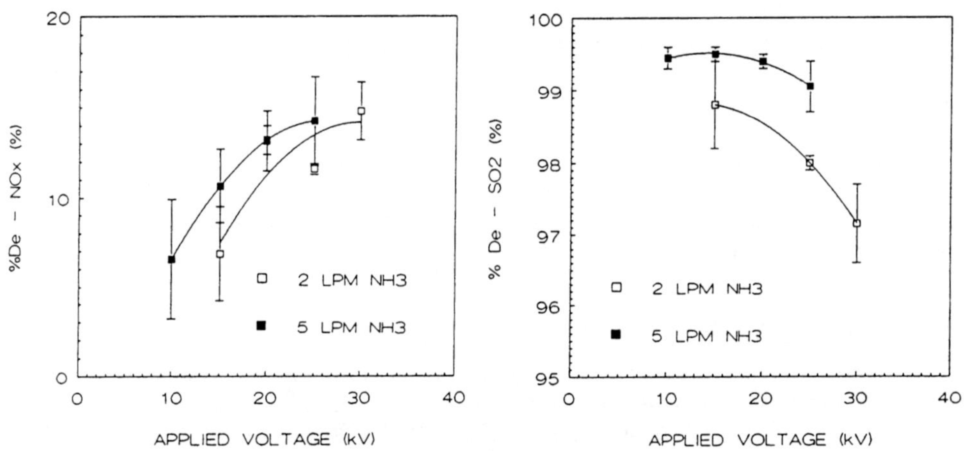

Fig. 2 SO_2 and NO_x reduction rates as a function of corona radical injector applied voltage without electron beams.

a spark discharge. Both discharge currents and flame lengths observed to be nonmonotonically dependent on the ammonia percentage mixture.

SO_2 and NO_x Reduction Rate With Corona Radical Injection Systems

SO_2 and NO_x reduction rates as a function of applied voltage is shown in Figure 2 for combustion flue gas flow rate $Q_{gf} = 1200$ m^{3}N/h. Figure 2 shows that NO_x reduction rate increases with increasing applied voltage or applied discharge power until initiation of spark discharges and SO_2 reduction rate slightly decreases with increasing applied voltage, where spark discharges not only reduce local electric field by space charges but also narrower plasma activated zone for ammonia radical particle generations. SO_2 and NO_x

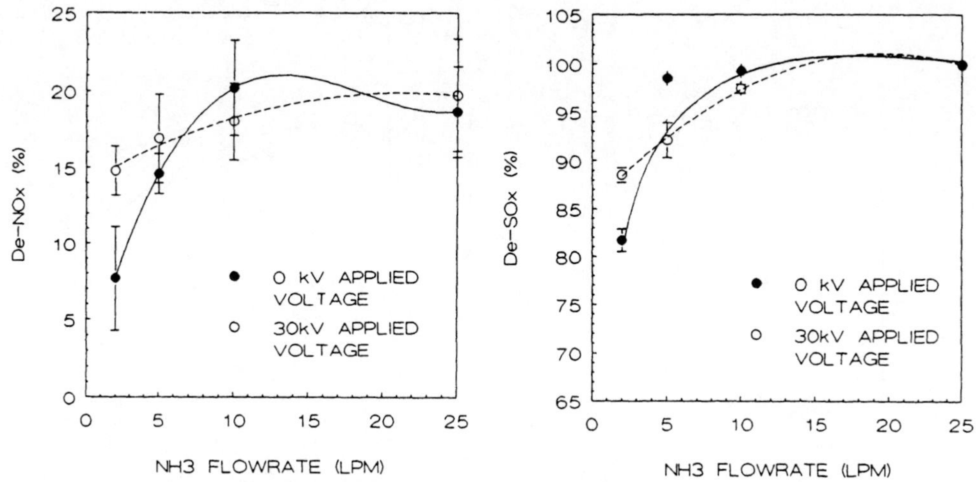

Fig. 3 **SO_x and NO_x reduction rates as a function of ammonia flow rates without electron beams.**

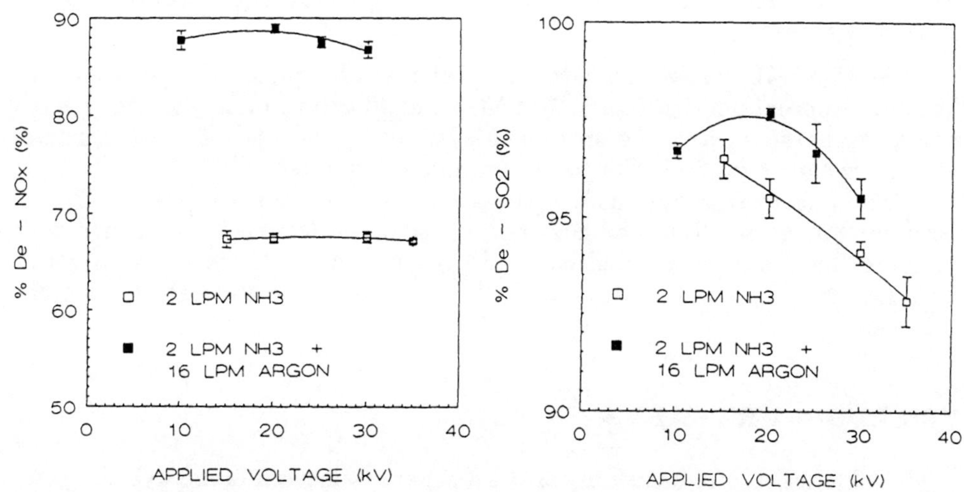

Fig. 4 **SO_2 and NO_x reduction rates as a function of corona radical injector applied voltage with electron beams.**

reduction rates as a function of NH_3 flow rate for various applied voltages is shown in Figure 3 for flue gas flow rate $Q_{gf} = 1200\ m^3N/h$. Figure 3 shows that the SO_2 and NO_x reduction rates increases with increasing NH_3 flow rate as one expected. Figures 2 and 3 show that approximately 5-20% of NO_x and 90 to 99% of SO_2 are reduced by single corona radical injection system with approximately 60 to 90 W of power, hence, multi-stage corona radical injections are required for much higher flue gas flow rates and NO_x concentrations or much higher NO_x reduction rates. Multi-stage injection tests are currently underway and will be reported elsewhere.

As we discussed in the Introduction, an injection of ammonia radicals will dissociate NO_x into nitrogen and water as well as forming aerosol particles. From the measurement, no significant O_3 or other N_xO_z gas was observed. However, small amount of unreacted ammonia also evident.

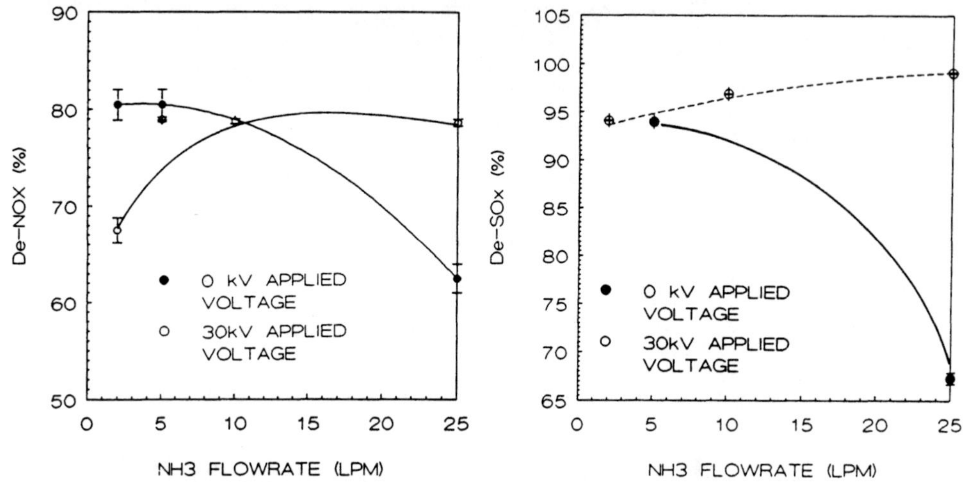

Fig. 5 SO_2 and NO_x reduction rates as a function of ammonia flow rates with electron beams.

SO_2 and NO_x Reduction Rate With Corona Discharge-Electron Beam Hybrid Systems

SO_2 and NO_x reduction rates as a function of applied voltage is shown in Figure 4. Approximately 65 to 85% of NO_x and 92 to 98% of SO_2 are removed by corona discharge-electron beam hybrid systems. The effect of corona radical injection is more significant for lower ammonia flow rates.

The effect of ammonia flow rates on the reduction of SO_2 and NO_x is shown in Figure 5. NO_x and SO_2 reduction rates decreases with increasing ammonia flow rate without corona radical generation, however, with corona discharge, NO_x and SO_2 reduction rates increases with increasing ammonia flow rates.

CONCLUDING REMARKS

Noble technique based on corona discharge-electron beam hybrid system for the reduction of SO_2 and NO_x from combustion flue gas is developed and following concluding remarks are obtained:

The results show that: (1) corona discharge was significantly enhanced when the electron beam system is operated; (2) up to 16% of NO_x and 95% of SO_x can be removed by operating the corona discharge radical injection system alone; (2) up to 98% of NO_x and 97% of SO_x are removed in flue gases under corona discharge-electron beam hybrid operations; (4) NO_x and SO_x removal rates are significantly influenced by flue gas temperatures, where the removal rate decreases with increasing gas temperature; (5) Leak ammonia and ozone generation are significantly reduced by the corona discharge ammonia radical injection method under corona discharge-electron beam hybrid operations; (6) Corona discharge current increases with increasing applied voltage and Ar-NH_3 flow rates; (7) Corona discharge zone increases with increasing applied voltage; (8) NO_x and SO_2 concentrations decrease with increasing applied voltage, hence corona current or discharge input power, and NH_3 flow rates;

(9) No significant N_xO_y and O_3 formations were observed; and (10) Formation of aerosol particles were observed.

ACKNOWLEDGMENT

The authors express their thanks to S. Aoki, A. Maezawa, M. Izutsu, R. Suzuki, J. Fujita, and T. Saito for variable discussions and comments. This work was supported by Ebara Co. Ltd.

REFERENCES

1.J.S. Chang, The role of H_2O on the formation of NH_4NO_3 aerosol particles and De-NO_x under the corona discharge treatment of combustion flue gases, *J. Aerosol Sci.*, 20:1087-1090 (1989).

2.J.S. Chang, Energetic electron induced plasma processes for a reduction of acid and greenhouse gases in a combustion flue gas, in "Non-Thermal Plasma Techniques for Pollution Control", B. Penetronte, Ed., Plenum Press, New York (1994).

3.J.S. Chang and S. Masuda, Mechanism of pulse corona induced plasma chemical process for removal of NO_x and SO_2 from combustion flue gases, in *IEEE Industry Appl. Soc. 1988 Meeting*, pp. 1645-1653 (1988).

4.J.S. Chang, P.A. Lawless and T. Yamamoto, Corona discharge processes, *IEEE Trans. Plasma Sci.* 19:1102-1166 (1991).

5.G. Dinelli, L. Civitano and M. Rea, Industrial experiments on pulse simultaneous removal of NO_x and SO_x from flue gases, *IEEE Trans. Industry Application,* 25:535-541 (1990).

6.M. Higashi, S. Uchida, N. Suzuki, and Fujii, K., Simultaneous reduction of soot and NO_x in a diesel engine exhaust by discharge plasma, *Trans. IEE Japan*, 111A:457-473 (1991).

7.S. Jordan, H.R. Pour, W. Cherdron and W. Lindner, Physical and chemical properties of the aerosol produced by the electron beam dry scrubbing of the flue gas, *J. Aerosol. Sci.*, 17:669-675 (1988).

8.K. Kawamura, S. Aoki, H. Kimura, K. Adachi, T. Katayama, K. Kengaku and Y. Sawada, Electron beam dry flue gas treatment process, *Environ. Sci. & Tech.*, 14:288 (1980).

9.K. Kawamura, S. Hitano, Y. Hirano, A. Maezawa, S. Aoki, M. Kaneko and R. Suzuki, Development of electron beam dry flue gas treatment process (EBA process) -- Process demonstration operation in USA coal-fired power station, *Ebara Jiho*, 141:2-10 (1988).

10.S. Masuda and Y. Wu, Removal of sulfur from stack gases by an electric discharge, *Electrostatics 87, Inst. Phys. Conf. Ser. 85*, IOP Press, Oxford, pp. 249-254 (1987).

11.F.J. Palumbo and F. Fraas, The removal of sulfur from stack gases by an electric discharge, *J. Air Pollution Contr. Assoc.*, 21:143-144 (1971).

12.T. Ohkubo, S. Kanazawa, Y. Nomoto, J.S. Chang and T. Adachi, NO_x removal by a pipe with nozzle-plate electrode corona discharge system, *Trans. IEEE Industry Applications* (In Press, 1993).

DISCUSSION

L. L. HATFIELD: What is the lifetime of the electron beam generator and the corona generator?

J. S. CHANG: The Ti window in the electron beam generator is the weakest point. With a suitable coating, the lifetime of 3000 h is the target at this moment. In the corona radical injection system, no major lifetime problem is expected except contamination of the electrodes. The frequency of cleaning will be determined in future tests.

E. MARODE: On figure 4, at what level do you expect the points to be for zero applied voltage, that is to say, without any corona? Would the total necessary amount of ammonia be lowered using this technique? Does the flow in the injection pipe decrease the probability of sparking effects?

J. S. CHANG: The De-NO_x efficiency with NH_3 flow only (0 kV applied) is approximately 60%. The major De-NO_x is carried out by electron beam processes. The effect of the corona radical injector is only a few percent. However, ammonia consumptions and ammonia leak will be substantially reduced by hybrid operations. Due to the flow effects, the corona discharge is stabilized and the sparking problem is reduced. However, too high gas flow rates enhance initiation of sparks.

S. W. ROWE: (1) The electron beam generator will produce strong X-ray emission. How is this question treated in industrial installations? (2) In the corona generator presented here, NH_3 is injected coaxially with respect to the corona. This has the advantage of "stabilizing" the corona discharge but has the disadvantage in that the injected flow will form a barrier which will tend to "screen" the corona from the flu gas to be treated. How much does this effect degrade the efficiency with respect to optimal performance? (3) In this work only DC corona has been used. In what manner would the use of a pulsed corona be expected to improve performance? (4) Would injection of the NH_3 "head-on" with the corona produce a higher overall efficiency?

J. S. CHANG: (1) In the Shim-Nagaya pilot plant and Research Cottsell Co. test conducted in the U.S., the electron beam reactor is installed underground to minimize X-ray shielding problems. (2) The effect of the gas flow is still not optimized in the present system. An experimental test is underway. (3) The pulse corona type radical injector is expected to generate a higher radical density without increasing power consumptions. Experiments are underway. (4) In the present hybrid system the major De-NO_x process was carried out by an electron beam. The role of corona radical NH_3 injections is to cut down NH_3 consumptions and NH_3 leaks. Electron beam power consumptions also are slightly cut down by hybrid operations.

E. ILLENBERGER: You use a window to inject the electrons to the reactor. Could you comment on the effect of the window on the electron energy?

J. S. CHANG: Currently the Ti window is only 36 μm thick. Hence, the effect of the window is minimized.

SOME EXCITED STATE REACTIONS OCCURRING IN DISCHARGES CONSIDERED FOR POLLUTION CONTROL

Timm H. Teich and Adriano Nasciuti

High Voltage Laboratory
Swiss Federal Institute of Technology, CH-8092 Zürich

INTRODUCTION

The work presented here is aimed at gaining some insight into *initial* processes occurring in a discharge intended to serve for gas treatment to reduce the content of noxious, toxic or otherwise unacceptable substances. The capabilities of various forms of discharges to serve this purpose has been amply demonstrated[1]; however, the energy requirement can be high and so efforts have to be made to find the most energy-efficient discharge configurations and ~conditions for a particular treatment task. Among the treatment problems receiving much attention is the at least partial removal of nitric oxide NO from gas flows originating from combustion.

It is here attempted to assess the possibilities to extract useful information from the optical emissions of *corona* discharges as considered for gas treatment. Such emissions might yield time histories of energetic electron populations and of certain excited state populations. The results of such observations should either give access to primary (or *fast* subsequent) processes in the discharge or at least be able to serve for a cross check on simulations which may contain such a plethora of possible processes that postulating reaction shares from final reaction results is fraught with uncertainties.

EMISSIONS FROM CORONA

Discharge configurations used here are negative and positive point-plane in self-pulsing modes with a DC high voltage source. For the observations reported here, the discharge is imaged *axially* on the input slit of a grating monochromator [Fig.1a]; the apparatus has now been modified also to permit the use of *radial* emissions in order to study the processes in dependence upon axial positions [Fig.1b]. The monochromator slit is then perpendicular to the discharge axis; the entire monochromator-photodetector arrangement is movable parallel to it. The advantages of this approach which has also been used by Ikuta and Kondo[2,3] will briefly be discussed in a later section.

Gaseous Dielectrics VII, Edited by L.G. Christophorou
and D.R. James, Plenum Press, New York, 1994

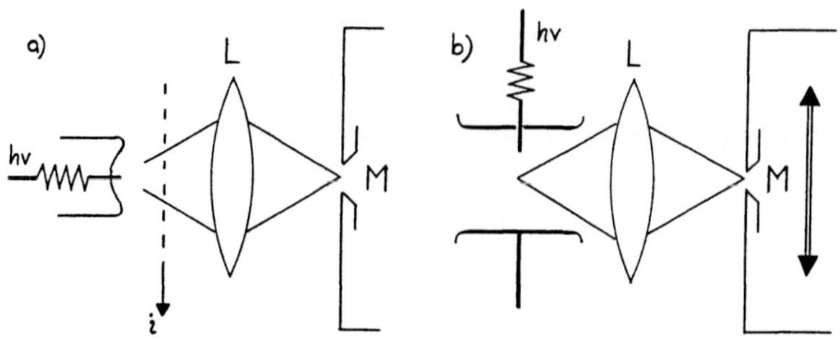

Fig.1. Configuration for recording a) axial and b) radial emissions from the point-plane discharge. L quartz optics; M monochromator entrance slit.

For the detection of monochromatic radiation, time-correlated single photon counting[2-5] is employed throughout. This permits the recording of *time-gated emission spectra* as well as the highly *time-resolved* measurement of particular emissions; it has the disadvantage that it will not deal correctly with large numbers of photons per discharge pulse; valid reproduction of the time history of emission demands that photons are registered only for a small fraction of the discharge cycles, so that one may have to operate with very small monochromator slits (making the measurement more susceptible to spatial fluctuations of the discharge) or with attenuating filters when recording the more intense emissions.

The first task is to find out which emissions are available for observation. While low-pressure discharges produce an extensive range of spectral bands and lines of reasonable intensity, spectra at higher gas density may show relatively few readily distinguished features (though the *high-resolution* study of emission from corona streamers in air of Hartmann[6] has nervertheless revealed a large number of features, some of which had to remain unidentified). As gases considered for discharge treatment are often associated with combustion processes, a standard power station flue gas composition has been taken here as a basis of investigations (13 % CO_2, 8 % H_2O, 6 % O_2, 73 % N_2, x ppm NO). In order to attain fairly realistic water contents (4 - 6 %) without heating the apparatus, many measurements were carried out for a total "target pressure" of 500 mb (hPa) at 23°C (partial pressures chosen to reflect the above percentages if the "complete" gas were made up to 500 mb at 23°C). The initial NO content - if any - had its origin in a mixture here referred to as BOX (British Oxygen "Spectron" calibration gas: 400 ppm NO in N_2).

Time-gated spectra

To facilitate the identification of features in the spectra of the gas mixtures, the emissions of individual constituents of the "target" mixture or of binary mixtures were recorded; then mixtures were varied towards the "target composition" in order to assess the effects of particular constituents. Some readily identified features are listed in the Table which also roughly indicates from which mixtures these features remain observable, as certain features clearly registered with the individual gases seem to disappear in the more complete mixtures due to quenching especially of the excited states with long radiative ("natural") lifetimes. A particularly conspicuous example is the suppression of the Fox, Duffendack, Backer bands (CO_2^+ $A\,^2\Pi$ - $X\,^2\Pi$) emitted from dry or moist CO_2. On the other hand, the emission of the Hopfield bands (O_2^+ $c\,^4\Sigma_u^-$ - $b\,^4\Sigma_g^-$), which is the only prominent feature of near-UV emission from O_2 at 1000 mb, is not affected by admixture of 3 % H_2O or 1 % N_2 (Fig.2).

Table. Readily observed emissions from a corona discharge

Emissions registered				inten-	quenching
abbrev.	states	degraded	gases	sity	by H_2O
N_2 (2+)	$N_2(C\,^3\Pi_u$ - $B\,^3\Pi_g)$	(V)	N_2 in flue N_2 in BOX trace in N_2O	H H L	M
N_2 (1-)	$N_2^+(B\,^2\Sigma_u^+$ - $X\,^2\Sigma_g^+)$	(V)	as above	M	M
N_2 (4+)	$N_2(D\,^3\Sigma_u^+$ - $B\,^3\Pi_g)$	(V)	as above	W/M	M
O_2^+ (c-b)	$O_2^+(c\,^4\Sigma_u^-$ - $b\,^4\Sigma_g^-)$	(V)	O_2 dry O_2 moist	W/M	W?
CO_2^+ (2883-2896)	$CO_2^+(A\,^2\Sigma^+$ - $X\,^2\Pi)$	(-)	CO_2, flue	M	S
CO_2^+(FDB)	$CO_2^+(A\,^2\Pi$ - $X\,^2\Pi)$	(R)	CO_2, (flue)	M/W	S
OH (3064)	$OH(A\,^2\Sigma^+$ - $X\,^2\Pi)$	(R)	H_2O, all moist gases	M	?
NO (A-X), NO_γ	$NO(A\,^2\Sigma^+$ - $X\,^2\Pi)$	(V)	NO in BOX flue, air impure Ar	H/M	M
NO (B-X), NO_β	$NO(B\,^2\Pi$ - $X\,^2\Pi)$	(R)	N_2O, (flue) **not** in BOX	M	VS
"3020"	not identified, very sharp band/line		flue dry flue wet	M/W	W?

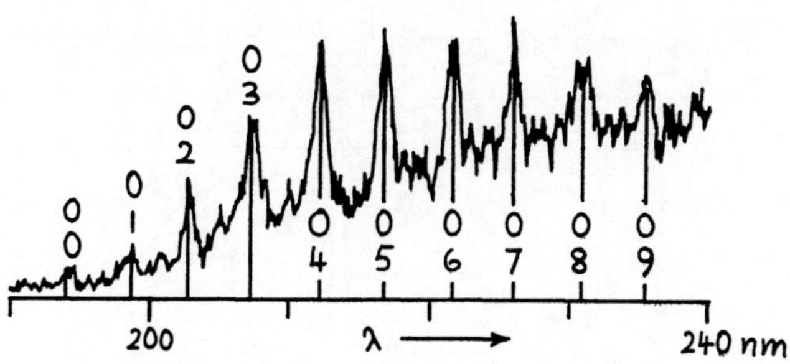

Fig.2. Hopfield bands emitted in atmospheric pressure O_2.

Time-gated spectra of a model flue gas have already been presented[4]. Besides the ubiquitous second positive group nitrogen radiation, N_2(C-B), the main features in the 200 - 330 nm range are the NO(A-X) and the OH(A-X) emissions. In Fig.3 the *early* emissions (approx. first 10 ns of the discharge) in the 306 - 319 nm range of a dry and a moist N_2/O_2 mixture are compared (365 mb BOX, 30 mb O_2, 0 and 20 mb H_2O); the *additional* emission of the moist gas is indicated by black infill. The partial identification of the OH(A-X, 0-0) emissions was carried out using the data of Dieke and Crosswhite[7]. For the spectrum of Fig.4 the early emissions were gated off and only *late* emissions in the time window 150 - 650 ns recorded; in this way the OH(A-X) features become more prominent and identification[7] is greatly facilitated. The details of the R_1 and R_2 heads can also be assigned in the original, but have not been entered in Fig.4 to avoid clutter. The two figures may illustrate one of the useful properties of time gating, which can in some cases serve to separate emissions the wavelengths of which coincide; in this way we identified e.g. a band head of NO_β not in the general list.

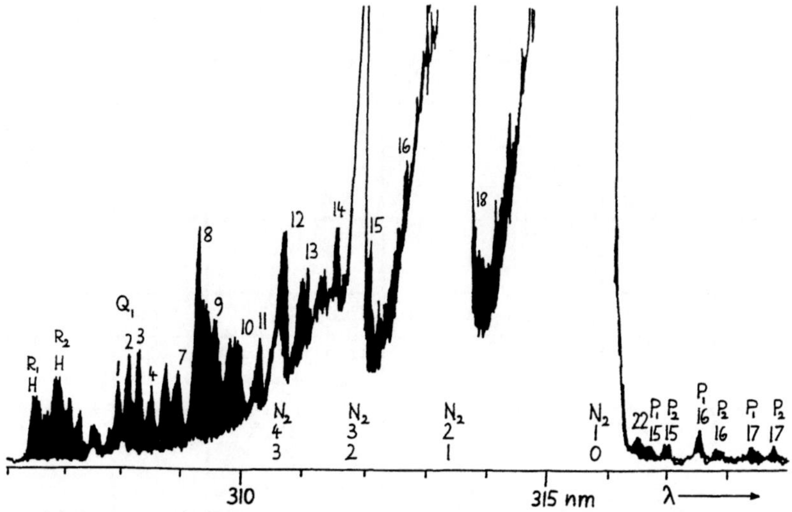

Fig.3. Short range time-gated spectrum of a positive streamer discharge in the mixture 365 mb BOX + 30 mb O_2, 0 (lower curve) and 20 mb H_2O (upper curve). The numbers serve partial identification of emissions belonging to branch Q_1 of OH(A-X, 0-0), first 10 ns of the discharge.

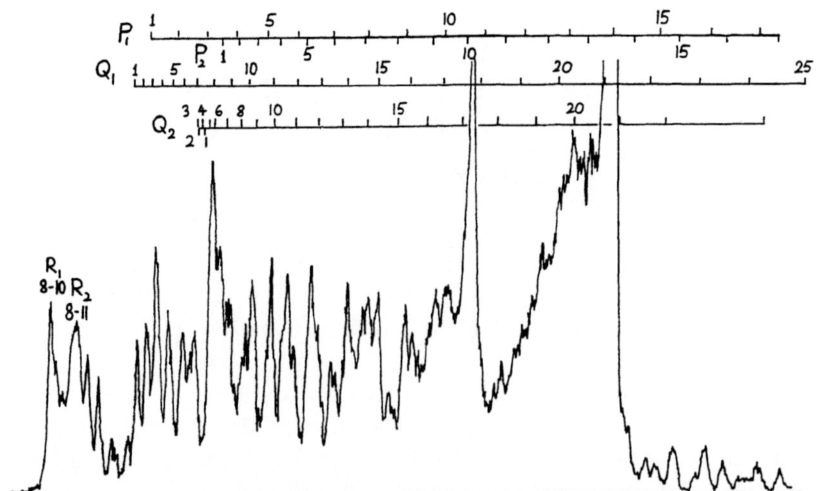

Fig.4. Late emissions (150 - 650 ns) for the moist gas of Fig.3, with more complete identification of the OH emissions.

Time history of emissions

The time dependence of four prominent emissions from flue gas components and their mixtures is presented here, in some cases as a sequence for several concentrations of one of the components. The specific emissions chosen are obviously those offering adequate intensity and - preferably - no overlap with other emissions. The latter aim cannot always be perfectly achieved when available spectral resolution is limited; for instance, the 391.44 nm $N_2^+(1-)$ emission will to a certain extent overlap emissions from the rotational tail of $N_2(2+)(2-5)$. As is obvious from Fig.3, *early* OH emissions in the "3064 Å band" can be affected in a similar way by the rotational tails of $N_2(2+)(4-3)$ and (3-2), except the band heads R_1 and R_2.

With the plots of time-dependent emission, the gas composition is given in terms of partial pressure at 23°C, units are mb = hPa. Operation of a repetitive streamer discharge is possible over the 0.5....100 μA range with transient peak currents from 1 mA (1000 mb O_2) to 40 mA (500 mb flue gas mix). Varying the current by applied total voltage change does not greatly affect the transient peak current but varies repetition frequency roughly proportional to mean current. Repetition frequencies encountered range from 0.5 to 100 kHz. - Time offset of the accurately calibrated time *scale* has not been corrected, the trigger from the first sharp current rise which coincides with a transient in light emission at the passage of the primary ionisation wave[2,3] is a suitable reference point.

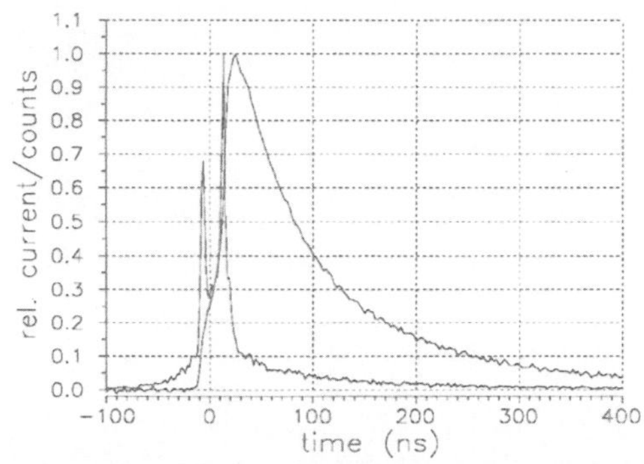

Fig.5. $N_2^+(1-)$ emission and simultaneous discharge current (upper curve from 30 ns; peak value 40 mA); 365 mb BOX, 30 mb O_2, 20 mb H_2O, 32.5 mb CO_2.

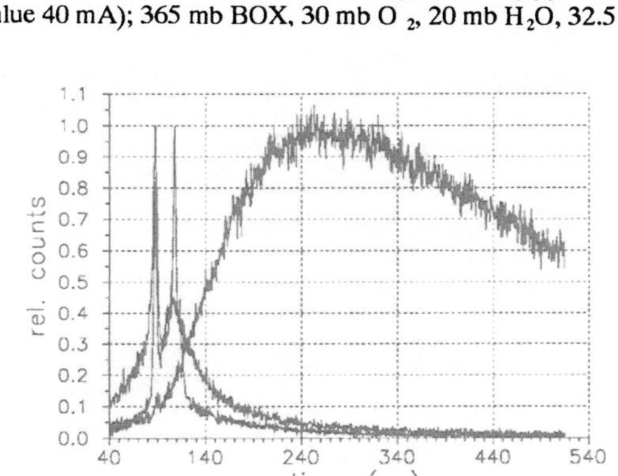

Fig.6. $N_2^+(1-)[0-0, 391.44 nm]$, $N_2(2+)[1-4, 399.84 nm]$ and $NO_\gamma[0-2, 247.1 nm]$ emissions; sequence from bottom to top at 140 ns (365 mb BOX + 30 mb O_2).

Figs.5 to 9 give examples of records of time-dependent emission: comparing current and energetic electron population (Fig.5); $N_2^+(1-)$, $N_2(2+)$ and NO_γ emission in a dry gas (Fig.6); $N_2^+(1-)$, OH(A-X) and NO_γ emission in a moist gas (Fig.7); NO_γ emission in dependence upon oxygen content in BOX (Figs.8a and 8b); NO_γ emission with different water content (Fig.9). These records will be commented upon in the next section.

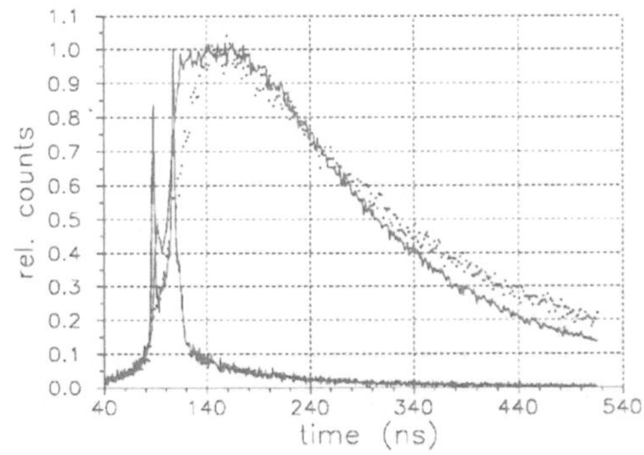

Fig.7. N_2^+, OH(R_2 band head) and NO_γ emissions (sequence from bottom at 440 ns; 365 mb BOX, 30 mb O_2, 20 mb H_2O).

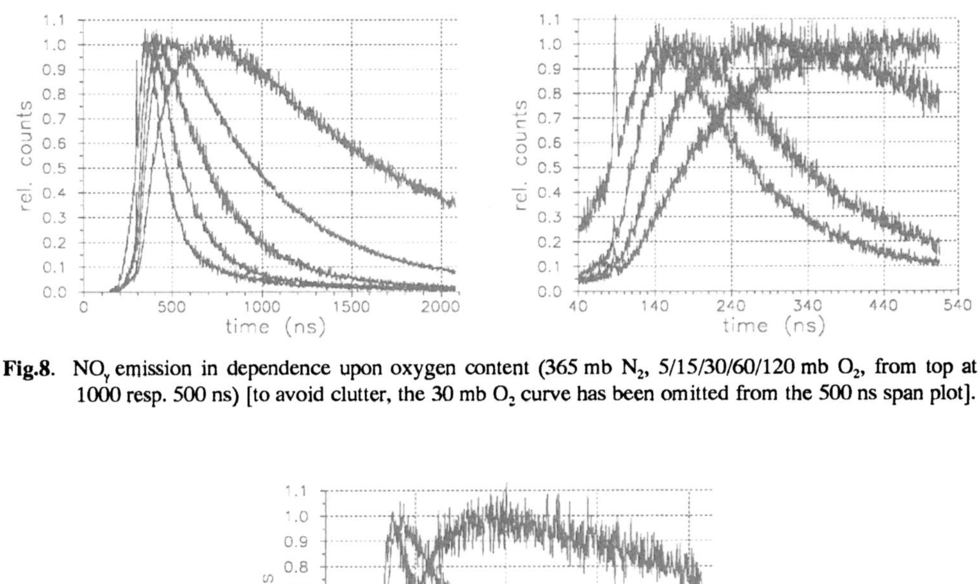

Fig.8. NO_γ emission in dependence upon oxygen content (365 mb N_2, 5/15/30/60/120 mb O_2, from top at 1000 resp. 500 ns) [to avoid clutter, the 30 mb O_2 curve has been omitted from the 500 ns span plot].

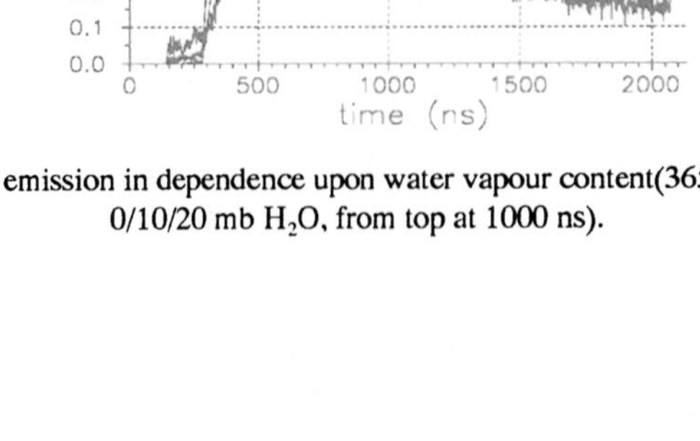

Fig.9. NO_γ emission in dependence upon water vapour content(365 mb BOX + 0/10/20 mb H_2O, from top at 1000 ns).

DISCUSSION

The short duration of N_2^+ first negative group (B-X) emission (Fig.5) which coincides with two sharp rises in current confirms that only the primary and secondary ionization waves of the streamer[2,3] contribute significantly to a population of very energetic electrons ($N_2^+(B\,^2\Sigma_u^+{}_{(v'=0)})$ lies 18.7 eV above the ground state of N_2). When we also consider $N_2(2+)(1\text{-}4)$ emission (the $N_2(C,\,_{v'=1})$ state is about 11.3 eV above the ground state), we recognize (Fig.6) a somewhat longer existence of electrons of the lower energy sufficient to bring about this excitation. - OH seems to be considered one of the *primary* products of the streamer (or barrier) discharge[8,9] . We find here emissions coincident with the streamer fronts (Fig.7), but always also a rise of emission *after* passage of these fronts and a persistent "plateau" of emission for some time after. This seems to indicate that there is still another process than direct excitation by electron collision generating a significant amount of excited OH. Subsequent reactions responsible for the decay of OH(A) have been catalogued by Schofield[10] .

Even when no NO is added to a gas mixture containing N_2 and O_2, there is a considerable intensity of NO_γ emission from the discharge. However, there is - even in the presence of a small amount of added NO - no or little *direct* excitation of NO(A); instead, there is a steady rise in emission (Figs.6,7) until well after the disappearance of the bulk of energetic electrons. This indicates the *indirect* formation of NO(A). As far as excitation of ground state NO is concerned, a likely precursor is $N_2(A)$:

$$N_2(A\,^3\Sigma_u^+) + NO(X\,^2\Pi_{(v=0)}) \rightarrow N_2(X\,^1\Sigma_g^+) + NO(A\,^2\Sigma^+_{(v=0\text{ or }1)})$$

For the mixtures not containing a significant amount of NO, the identity of the precursor(s) is not yet clear, but it seems that the ground state molecular partner is O_2. Formation of NO(A) from the abundantly formed $N_2(C)$ in reaction with O_2 is possible but this cannot be a dominant direct process here: under the prevailing conditions, the lifetime of $N_2(C)$ is of the order of 3 ns or less[2,11] , so that the excitation would have to be transferred during the existence of $N_2(C)$ as manifested by $N_2(2+)$ emission (Fig.6). The lifetime of NO(A) is also short due to various quenching processes[12] , so that the time-dependent NO_γ emission represents chiefly the time dependence of the concentration of the precursor(s). Fig.8 shows NO_γ emission from a N_2/O_2 mixture in dependence upon oxygen content. The model: δ-pulse generation of the precursor - subsequent excitation of NO(A) - leads to an emission time dependence which can be given as the difference between two exponential functions of time, similar to the situation found with the formation of NO(B) from N_2O [13]. The NO_γ emission curves experimentally found correspond to this shape and a previously developed fitting routine[4] offers itself to be adapted and applied in due course to obtain rough reaction rates.

Fig.9 shows NO_γ emission from the N_2/NO mixture ("BOX") in dependence upon added water vapour content and demonstrates the strong quenching of the precursor(s) by water. - Addition of CO_2 to the moist gas produces little change to the time dependence of the NO_γ emission, but increases the tendency of the streamer discharge to branching (gap spacing 20 mm).

Further work

The modified configuration (Fig.1b) should permit - by suitable choice of observed streamer segment - utilization of shorter time spans for the primary electron process[2,3] , thus approaching much more closely the δ-pulse assumption in the simple model mentioned above. In the same arrangement, the averaging along the streamer track in the measurements presented here is replaced by a certain degree of spatial resolution permitting access to processes in particular zones of the discharge.

CONCLUSIONS

Light emission measurements with good spectral and time resolution - utilizing time-correlated single-photon counting - permit observation of certain conditions (e.g. energetic electron concentrations) and processes (generation and destruction of excited states) in repetitive corona discharges. Apart from indicating *early* processes in the discharge, such emission measurements could help to verify the validity of simulations[14,15] of the early processes in treatment discharges.

ACKNOWLEDGEMENTS

The authors wish to thank H. Wymann for construction of essential apparatus, B. Waldesbühl for building auxiliary electronics and K. Lehmann for provision of data processing programs and assistance with data presentation.

REFERENCES

1. B.M. Penetrante and S.E. Schultheis (Eds.). "Non-Thermal Plasma Techniques for Pollution Control", NATO ASI Series G **34,** Part A and Part B, Springer, Berlin (1993)
2. N. Ikuta and K. Kondo, A spectroscopic study of positive and negative coronas in N_2 - O_2 mixture, *Proc. 4th Intern. Conf. on Gas Discharges, Swansea; IEE Conf. Public. No.* **143**, 227-230, IEE London (1976)
3. K. Kondo and N. Ikuta, Highly resolved observation of the primary wave emission in atmospheric positive-streamer corona, *J.Phys.D* **13**, L33-L38 (1980)
4. T.H. Teich, Emission spectroscopy of corona discharges, *in* Ref.1, 231-248 (1993)
5. P.C. Johnson, G. Berger and M. Goldman, The temporal behaviour of emission from excited metal atoms on the cathode region of a pulsed discharge in air, *J.Phys.D* **10**, 2245-2256 (1977)
6. G. Hartmann. "Spectroscopie de la décharge couronne: Étude des mécanismes de collisions dans le dard (streamer)", Thèse U. Paris Sud, No.1783 (1977)
7. G.H. Dieke and H.M. Crosswhite, The ultraviolet bands of OH, *J.Quant.Spectrosc.Radiat.Transfer* **2**, 97-199 (1962)
8. R. Peyrous. "Simulation de l'évolution temporelle de diverses espèces gazeuses créées par l'impact d'une impulsion électronique dans de l'oxygène ou de l'air sec ou humide", Thèse U. de Pau (1986)
9. B. Eliasson and U. Kogelschatz, N_2O formation in ozonizers, *J.Chimie Phys.* **83**, 279-282 (1986)
10. K. Schofield, Critically evaluated rate constants for gaseous reactions of several electronically excited species, *J.Phys.Chem.Ref.Data* **8**(3), 723-798 (1979)
11. T.H. Teich and R. Bräunlich, UV and other radiation from discharges in artificial air and its constituents, *Proc. 8th Intern. Conf. on Gas Discharges and Their Applications, Oxford*, 441-444, Leeds University Press, Leeds (1985)
12. I.M. Campbell and R.S. Mason, The kinetics of β and γ band quenching in the nitric oxide afterglow, *J.Photochem.* **8**, 321-336 (1978)
13. T.H. Teich, Determination of reaction data from corona discharge observation, *Proc. 3rd Intern. Symp. on High Pressure Low Temp. Plasma Chem.*, Strasbourg (1991), 77-83
14. P.A. Vitello, B.M. Penetrante and J.N. Bardsley, Multi-dimensional modelling of the dynamic morphology of streamer coronas, *in* Ref.1, 249-271 (1993)
15. M. Jacob, W. Lowell Morgan and E.R. Fisher, Simulation of the plasma chemistry of a pulsed corona discharge in dry and humid air, *46th Annual Gaseous Electronics Conference,* Montreal (1993), and private communication

DISCUSSION:

E. MARODE: (1) You showed there is a difference on some of the spectra when you add water. Is the change proportional to the amount of water added, in other words can we use this technique to calibrate the amount of water you inject? (2) Regarding the electron energy, you said we have to excite high levels. Do you think all excitation is from the ground state or from already excited levels?

T. TEICH: In some cases the proportionality works well but you also have to take into account the additional self-quenching which may occur with the additional concentration of water. (2) I tried some variation of electron number in the pulses so the voltage was varied over quite a range and observed linearity between the current injection and emission but this is not yet a safe proof because it will increase repetition rate. There is only a moderate change in electron density with significant change in current.

R. BARTNIKAS: The comparison between the electrical and optical pulse is quite interesting. Did you find that there was any time delay between the two pulses?

T. TEICH: There is only one diagram which shows the relation to the current. I also make a sort of time delay calibration by looking at a very short lifetime emission from the oxygen Hopfield bands. On the basis of that the current and the emission were lined up. In Fig. 5 for nitrogen it is fairly clear that one has this relation between the very sharp streamer front rise and the emission from states of high electron energy. These are very closely related although there can be a difference between the primary and secondary wave of the streamer. For NO the emissions only start after the rise of the current and in fact the emissions increase while the current has long been forming and in some cases persist long after the current.

ON THE IONS IN AN ARGON-HEXAMETHYLDISILOXANE RADIO-FREQUENCY-DISCHARGE

Rüdiger Foest, Martin Schmidt, Mario Hannemann and Ralf Basner

Institut für Niedertemperatur-Plasmaphysik e.V.
an der Ernst-Moritz-Arndt-Universität
D 17489 Greifswald Germany

INTRODUCTION

The RF generated anisothermal low pressure plasma is widely used in plasma technology for thin film deposition, etching and surface treatment. For plasma polymerization, the silicon organic compound hexamethyldisiloxane $(CH_3)_3$-Si-O-Si-$(CH_3)_3$ (HMDSO) plays an important role in the field of surface protection, optics, electronics and biomedicine[1].

The plasma wall interaction is of special interest in plasma technology. It is essentially determined by the properties of the boundary sheath. They control the energy of the ions striking the wall. This report presents results of mass spectrometric investigations on the ion currents, the kinetic ion energy distribution function (IEDF) and the neutral gas composition at the wall in a capacitively coupled RF discharge in an Ar plasma with 1% HMDSO admixture. Information on electron energies and electron densities is given by Langmuir probe measurements. Diagnostics of the internal plasma properties contribute to the understanding of the plasma behavior and are necessary for a more precise deposition process control in technological applications as well.

EXPERIMENTAL

The RF discharge burns in a stainless steel vacuum chamber (Fig. 1) between two planar steel electrodes of 128 mm diameter and spacing of 40 mm. The top electrode is capacitively coupled through a matching network to a 13.56 MHz RF generator (ACG-5, ENI) while bottom electrode and chamber wall are grounded. This asymmetrical discharge configuration causes a DC self-bias voltage of about -100 V_{DC} to -750 V_{DC} depending on chamber pressure and RF amplitude (280 V_{PP} to 1500 V_{PP}). A commercial power meter (NS 663 PA, DAIWA) was used to determine the RF power which varied from 15 W to 120 W.

The vacuum system is evacuated by a turbomolecular pump backed by a rotary pump The residual pressure is lower than $1\cdot10^{-4}$ mbar as measured by ionization vacuum gauge (IG 3, INFICON). Chamber pressure (0.01 - 1.0 mbar) and gas flow rate are regulated

Gaseous Dielectrics VII, Edited by L.G. Christophorou
and D.R. James, Plenum Press, New York, 1994

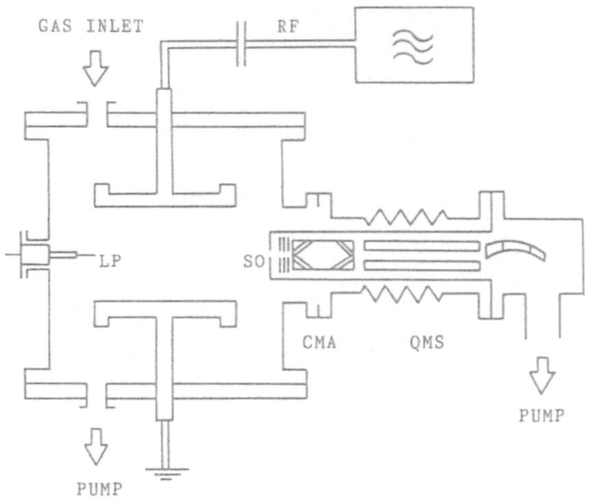

Figure 1. Schematic diagram of the experimental assembly (abbreviations: see text)

using the Multi Gas Controller 147 (MKS Baratron) with mass flow controller (MFC), capacitance vacuum meter and feedback controlled throttle valve. The argon (purity 99.999%) flow rate is 10 sccm, the HMDSO flow rate is 0.1 sccm. The HMDSO vapor is fed into the vacuum vessel via vapor source flow controller and heated tubes.

The quadrupole mass spectrometer (QMS) (SXP 300H, VG Instruments) is equiped with a cylindrical mirror analyzer (CMA) to obtain IEDFs. Its sampling orifice (SO) of 0.1 mm diameter fits to the chamber wall, equidistant to the electrode surfaces, and serves

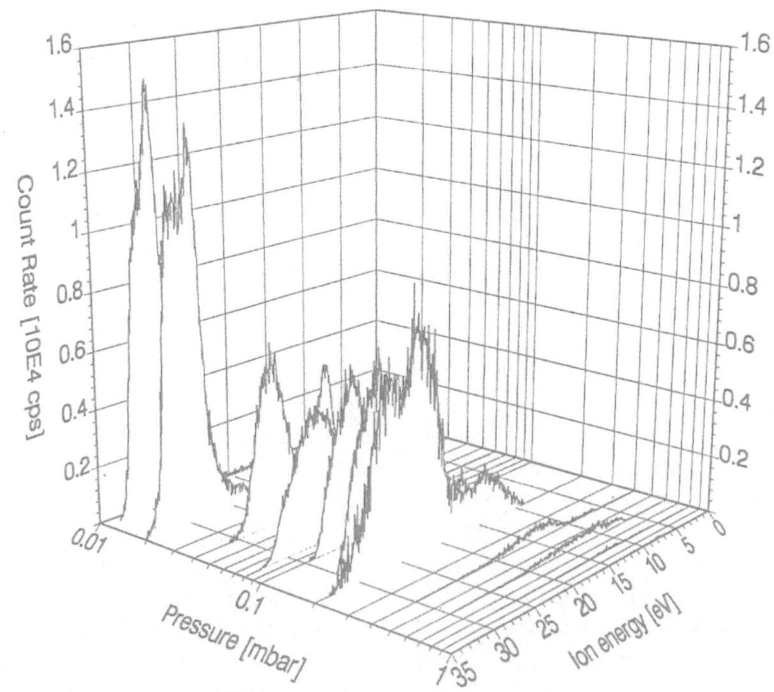

Figure 2. IEDF of the ion H_3^+ at the chamber wall in an Ar-HMDSO-RF discharge (1% HMDSO), RF power: 90 W

as a pressure stage. The residual pressure inside the mass analyzer is lower than $1 \cdot 10^{-7}$ mbar. It did not exceed $1 \cdot 10^{-5}$ mbar during the measurements. The arrangement allows the mass analysis of neutrals ionized by electron impact in the internal ion source as well as the analysis of ions generated in the plasma with the internal ion source switched off.

The HMDSO admixture was held at 1% to keep the deposition rates low. At higher admixtures (5%) the orifice region, transfer optics and ion source were rapidly contaminated by a dielectric layer which resulted in crucial signal instabilities. These problems could only be corrected by a repeatedly mechanical cleaning of the inner parts of the analyzer.

The Langmuir probe (LP) is an external heated tungsten-wire loop (0.025 mm diameter, 4 mm length). The use of a heated probe is necessary to avoid thin film deposition on the probe surface. A tuned LC network protects the probe circuit against disturbances by the sustaining RF signal. The digitally measured probe characteristic is numerically evaluated by means of nonrecursive digital filters to obtain the electron energy distribution function (EEDF), electron density (n_e), mean electron energy ($\bar{U}$) and space potential.

RESULTS

The IEDFs of H_3^+, Ar^+, ArH^+ and $(CH_3)_3SiOSi(CH_3)_2^+$ for a RF power of 90 W in dependence on pressure are presented in Figs. 2-5. The relative ion currents at 0.1 mbar received by integration over the IEDF are shown in Fig. 6. In addition to these ions an ion (m/z = 221) with small intensity was identified. The measurement of the neutral gas mass spectrum shows a strong decay of the intensity of the HMDSO and the increase of the decomposition products as CH_4 and H_2 after ignition of the discharge. The EEDFs (Fig. 7) are evaluated from voltage current characteristics measured with the Langmuir probe and

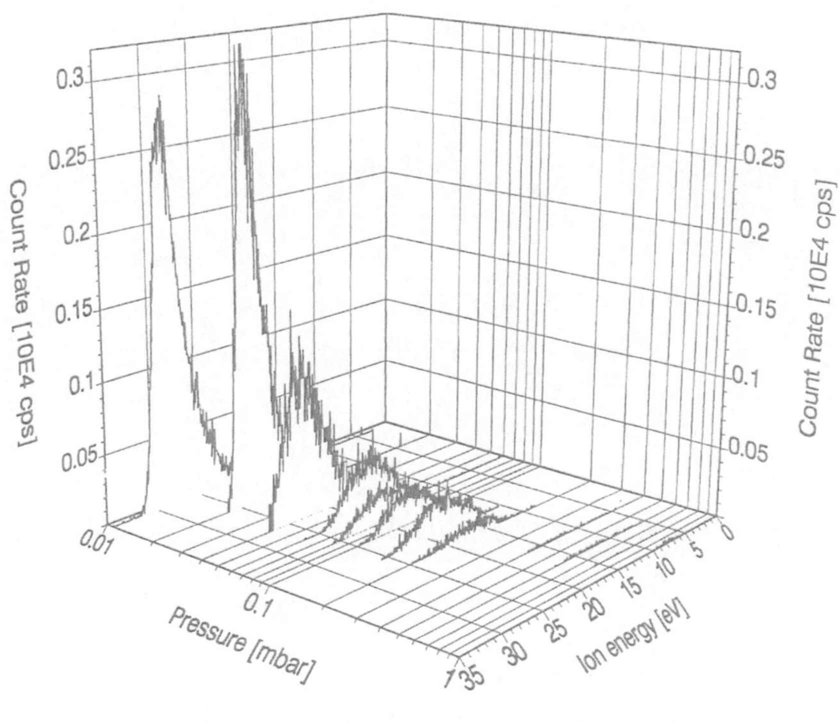

Figure 3. IEDF of the ion $(CH_3)_3SiOSi(CH_3)_2^+$ with m/z 147 at the chamber wall in an Ar-HMDSO-RF discharge (1% HMDSO), RF power: 90 W

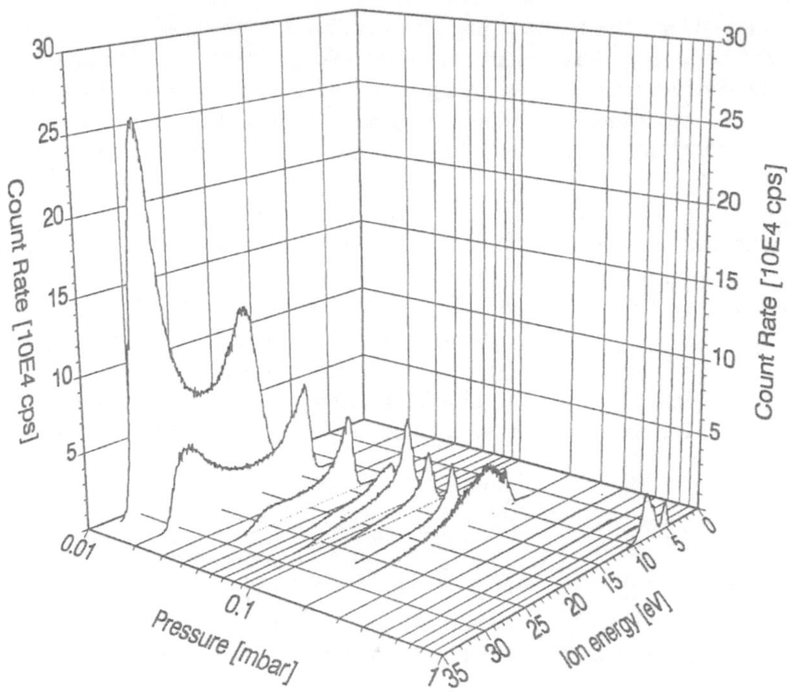

Figure 4. IEDF of the ion Ar^+ at the chamber wall in an Ar-HMDSO-RF discharge (1% HMDSO), RF power: 90 W

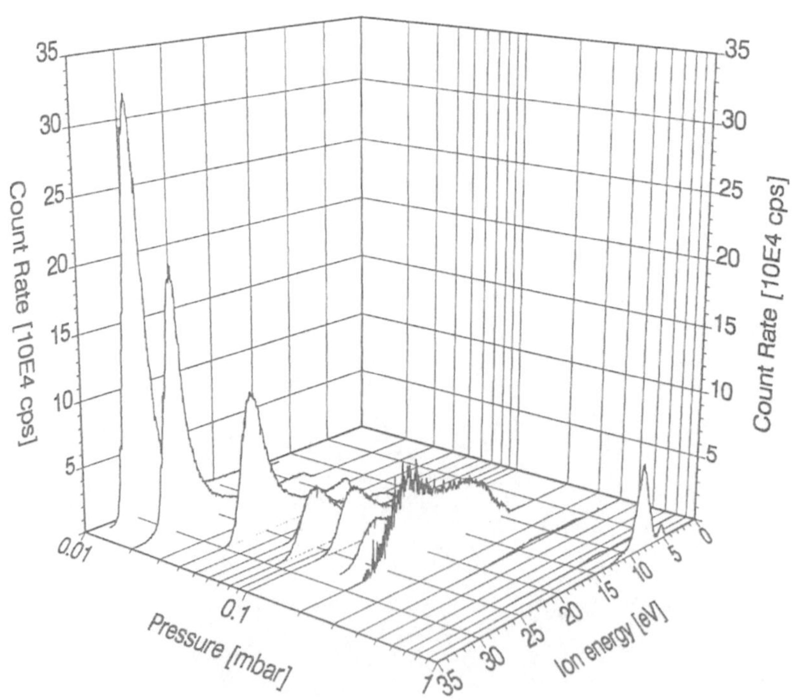

Figure 5. IEDF of the ion ArH^+ at the chamber wall in an Ar-HMDSO-RF discharge (1% HMDSO), RF power: 90 W

indicate the change of the distribution function of a pure Ar discharge (n_e=1.1·10^{10} cm^{-3}, $\bar{U}$=1.45 eV) by the admixture of HMDSO (n_e=8.2·10^9 cm^{-3}, $\bar{U}$=2.34 eV).

DISCUSSION

The mass spectrum of the ion current to the reactor wall is characterized by the dominance of Ar^+, ArH^+, smaller parts of H_3^+ and carbon hydrogen ions CH_3^+, $C_2H_2^+$, $C_2H_4^+$ (mixed with CO^+). Moreover, Ar_2^+, the base peak of the HMDSO $(CH_3)_3SiOSi(CH_3)_2^+$ and the $Si(CH_3)_4^+$ ion were identified. The ion with the mass number 221 corresponds to the base peak of octamethyltrisiloxane.

The IEDF of H_3^+ is modulated by the RF field in the boundary sheath, because the transfer time of this ion is comparable with the period of the RF field[2]. For higher pressures, the IEDFs are shifted to lower energies due to collisions of the ions with neutral particles[3]. In the IEDF of Ar^+ the resonant charge transfer collisions of the Ar^+ ions with Ar atoms induce a loss of energetic ions already at lower pressures. The maximum energy of the ions nearly corresponds to the space potential measured by the Langmuir probe.

The ions may be generated by electron impact of ground state atoms or molecules, stepwise ionization, collisions of heavy particles in excited states (pair collisions, Penning ionization and Hornbeck-Molnar process) and by ion molecule reactions. As we do not know concentrations of metastable and excited Ar atoms, evaluation of such processes was not performed. The high gas temperature (near 1500 K)[4] in this discharge reduces the concentration of metastable atoms[5]. With the EEDF (Fig. 7) the rate coefficients for direct electron impact ionization (Tab. 1) are obtained, approximating the EEDF by two maxwellian distributions with temperatures of 0.99 eV and 2.63 eV. The experimental ionization cross sections of Rapp[6] (H_2, Ar), Biaocchi[7] (CH_3), Orient[8] (CH_4), Seefeldt[9] with corrections according to Deutsch[10] (HMDSO) and calculations corresponding to Deutsch[10] ($Si(CH_3)_4$) are used for the calculations. The atomic ionization cross sections of C and Si are taken

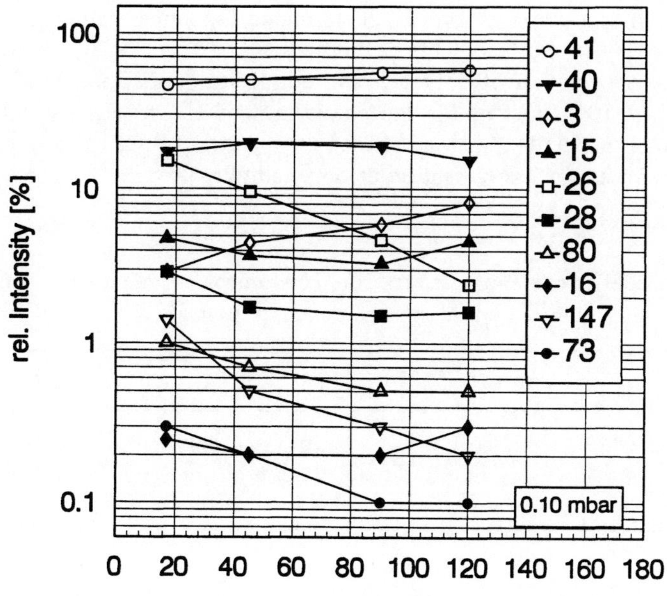

Figure 6. Relative ion currents of the Ar-HMDSO discharge, received by integration over the IEDF. Indicated are the m/z values of the different ions.

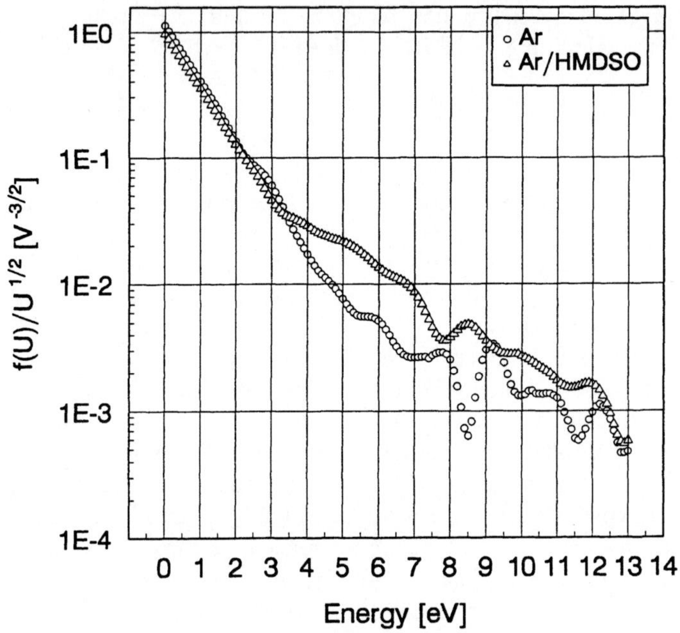

Figure 7. EEDF of the Ar-HMDSO discharge and a pure Ar discharge (0.1 mbar, 45 W) derived from Langmuir probe measurements.

from Freund[11]. An estimation of the relative concentrations of the neutral components in the plasma (45 W, 0.1 mbar, 1% HMDSO admixture) based on the measured ion currents (Fig. 6) and the rate coefficients for electron impact ionization leads to the values listed in Tab. 1. The calculation of the relative concentrations with the values taken from the neutral gas mass spectrum leads to similar results for the stable molecules. This correspondence supports the assumption of ion production by electron impact ionization of ground state molecules as a first approximation. The determined H_2 concentration is sufficient for the formation of ArH^+ by ion molecule reaction (k=$1.1 \cdot 10^{-9}$ cm^3s^{-1})[12] of Ar^+ ions with H_2 molecules. The observation of only H_3^+ and not H_2^+ is understandable by the large rate coefficient (k=$2.0 \cdot 10^{-9}$ cm^3s^{-1})[12] for the conversion of H_2^+ to H_3^+ by collisions with H_2 molecules. An ion molecule reaction between the ion $(CH_3)_3SiOSi(CH_3)_2^+$ and the HMDSO molecule is proposed for the formation of octamethyltrisiloxane[13].

Table 1. Rate coefficient k of electron impact ionization calculated with the measured EEDF (Fig. 7) and relative neutral species concentrations.

Gas	H_2	CH_3	CH_4	Ar	$Si(CH_3)_4$	HMDSO
k [m^3s^{-1}]	$1.8 \cdot 10^{-17}$	$1.7 \cdot 10^{-16}$	$9.0 \cdot 10^{-17}$	$4.0 \cdot 10^{-17}$	$5.1 \cdot 10^{-16}$	$9.0 \cdot 10^{-16}$
rel. concentr.	0.1	(0.01)	0.001	1	0 0002	0 0003

An important question in plasma assisted deposition processes is the portion of ions compared with neutrals for the thin film formation. The significance of monomer ions was shown for inert gas DC plasmas with small admixtures of organic compounds[14]. In our experiments the deposition rate (0.1 μm/h, 0.1mbar, 45 W) was measured in the vicinity of the ion sampling orifice. The distinct siloxane structure of the film proved by IR spectroscopy shows that it must be essentially formed by siloxane bond containing particles and not only by products of a strong decomposition of the monomer molecules. Assuming a sticking coefficient of 1, the film formation with the HMDSO ions (m/z 147) would demand a particle current of at least $1.1 \cdot 10^{13}$ $cm^{-2}s^{-1}$. With a Bessel distribution of charge carrier density in the space between electrode edge and grounded wall and the measured electron density of $8 \cdot 10^{9}$ cm^{-3} in the centre of the discharge an ion flux density at the wall of $8 \cdot 10^{13}$ $cm^{-2}s^{-1}$ is evaluated. A value of $5 \cdot 10^{11}$ $cm^{-2}s^{-1}$ results for the particle flux density of the HMDSO ions (see Fig. 6), which is more than one order of magnitude too small. So the conclusion may be drawn that for these conditions the transport of mass for thin film deposition occurs also by smaller ions and neutral siloxane particles.

ACKNOWLEDGMENTS

This research was supported by the BMFT of Germany, contract N° 13N 5939.

REFERENCES

1. W. Kulisch and C. Rau, Mechanisms of plasma polymerization of different silico organic monomers, in "11th Int. Symp. Plasma Chem. Symp. Proc." J. Harry, ed., Int. Org. Comm. of the 11th ISPC, Loughborough (1993), 1010.
2. M.J. Kushner, Distribution of ion energies incident on electrodes in capacitively coupled rf discharges, J. Appl. Phys. 58:4024 (1985).
3. J. Liu, G.L. Huppert, H.H. Sawin, Ion bombardement in rf plasmas, J. Appl. Phys. 68:3916 (1990).
4. J. Röpcke, A. Ohl and M. Schmidt, Comparison of optical emission spectrometric measurements of the concentration and energy of species in low-pressure microwave and radiofrequency plasma sources, J. Anal. At. Spectrom. 8:803 (1993).
5. A. Rutscher and S. Pfau, Zum Mechanismus der positiven Niederdrucksäule im Grenzfall kleiner Stromdichten III, Beitr. Plasmaphys. 7:187 (1967)
6. D. Rapp and P. Englander-Golden, Total cross-sections for ionization and attachment in gases by electron impact. I. Positive ionization, J. Chem. Phys. 43:1464 (1965)
7. F.A. Biaocchi, R.C. Wetzel and R.S. Freund, Electron impact ionization and dissoziative ionization of the CD_3 and CD_2 free radicals, Phys. Rev. Lett. 53:771 (1984)
8. O.J. Orient and S.K. Srivastava, Electron impact ionization of H_2O, CO, CO_2 and CH_4, J.Phys. B: At. Mol. Phys. 20:3923 (1987)
9. R. Seefeldt, W. Möller and M. Schmidt, Zur Elektronenstoßionisierung des Hexamethyldisiloxans, Z. phys. Chemie, Leipzig, 266:797 (1985)
10. V. Tarnovsky, K. Becker, H. Deutsch and T.D. Märk, Electron impact ionization of CF_x and NF_x (x=1-3), in "45th GEC Program and Abstracts", Boston, 1992.
11. R.S. Freund, R.C. Wetzel, R.J. Shul and T.R. Hayes, Cross section measurements for electron impact ionization of atoms, Phys. Rev. A41: 3575 (1990)
12. E.W. McDaniel, V. Cermak, A. Dalgarno, E.E. Ferguson and L. Friedman. "Ion-Molecule Reactions", Wiley Interscience, New York (1970)
13. A.M. Wrobel and M.R. Wertheimer, Plasma-polymerized organosilicones and organometallics, in "Plasma deposition, treatment and etching of polymers" R. d`Agostino, ed., Academic Press, San Diego (1990), 163.
14. H. Deutsch and M. Schmidt, On the quantitative treatment of the growth rate of thin polymer films produced in glow discharges, Beitr. Plasmaphys. 21:279 (1981)

DISCUSSION

J. S. CHANG: (1) How do you evaluate mass discrimination effects in the ion sampling processes and inside the mass spectrometer? (2) How are the Ar_2^+ formed?

M. SCHMIDT: (1) The mass discrimination is small in the investigated mass region according to the information of the factory. Our own measurements of neutral gas mass spectra in comparison with known mass spectra confirm this. (2) The Ar_2^+ ions are formed by the Hornbeck-Molnar process in collisions with Ar atoms in high excited states.

L. G. CHRISTOPHOROU: What is responsible for the structure in your electron energy distribution function?

M. SCHMIDT: The fine structure of the electron energy distribution function may be a result of the residual disturbances by the RF sustaining voltage in the probe circuit in this low current part of the probe characteristic. The second derivative is very sensitive against disturbances of the probe characteristic.

R. J. VAN BRUNT: (1) In the case of the ion energy distribution measurements, we have found that the observed ion energy distribution will depend on whether or not the sampling probe extends into the plasma, that is, on whether or not a sheath forms around the probe. Where was the probe located relative to the electrodes and chamber wall? (2) Could you also comment about the cutoff in the ion energy distribution at 5 eV? It is expected that ion energies should extend down to 0 eV. Do you believe that this is an instrumental energy shift?

M. SCHMIDT: (1) The orifice was situated near the reactor wall. (2) The gap in the ion energy distribution in the energy range between 0 and 5 eV may be connected with a low efficiency of ion collection in this region. The energy scale was tested by a thermionic ion source (Na^+) in vacuum. In the low energy range there is a shift of $\sim$ 1 eV.

ELECTRICAL BREAKDOWN AND OPTICAL EMISSION PROPERTIES OF HIGH PRESSURE PULSED RF GAS DISCHARGES

Scott R. Hunter[1,2], William A. Gibson[1], G. Samuel Hurst[1,2], James E. Turner[3], Robert N. Hamm[3], Harvel A. Wright[2]
[1]Pellissippi International Inc.
110 Ridgeway Center, Oak Ridge, TN 37830, USA
[2]Consultec Scientific Inc.
725 Pellissippi Parkway, Knoxville, TN 37932-3300
[3]Health Sciences Research Division
Oak Ridge National Laboratory, Oak Ridge, TN 37831-6123

I. INTRODUCTION

A fundamentally new technique for the measurement of the number of subexcitation electrons produced by the passage of ionizing radiation (electromagnetic and high energy charged particles) through a gas has been developed by us during the past few years [1-5]. The new detector is based on the digital characterization of the electrons in the particle track produced by the ionizing radiation where the charged particle track is registered by measuring the number of electrons found in given subvolumes of the gas in the ionization chamber [6-10]. The track is thus characterized by a set of integers in each volume element, from which parameters such as the track length and energy can be measured. We have specifically developed this technique for use in the field of microdosimetry, where the study of the fluctuations in the energy deposition in the charged particle track on the order of the molecular to cellular (i.e., nanometers to microns) dimensions is important in understanding the chemical and biological effects of ionizing radiation [2,4,6]. If the charged particle is a recoil nucleus produced by a neutron interaction, the energy deposited and the linear energy transfer can be inferred from this information, allowing the detector to act as neutron spectrometer [3,11]. In principle, the detector can be used for any application that requires knowledge of the three dimensional locations of single electrons in a gas [1,12].

In this paper the experimental apparatus used to make these measurements and the optical and pulsed RF electric field optimization studies we have performed are described. Images of particle tracks produced by α and β particles, and recoil protons from the interaction of high energy neutrons with hydrogenous materials, are also presented to illustrate the sensitivity and track imaging resolution we have obtained with this detector. The quality of the information gained from the detector (i.e. the uncertainty in the locations of each secondary electron produced by the ionizing particle) is a very sensitive function of the gas mixture constituents, purities and pressures used in the chamber, along with the magnitude, duration, and frequency of the pulsed RF electric field. These considerations are discussed in detail.

Gaseous Dielectrics VII, Edited by L.G. Christophorou
and D.R. James, Plenum Press, New York, 1994

II. DETECTOR DESIGN

The detector optically images the secondary electrons produced by the passage of ionizing particles through a gas, and is shown in Figure 1. The technique is an adaption of an experimental method originally developed by Cavalleri [13] to determine the electron density in a low electron concentration gas discharge. More recently the technique has been used to measure thermal electron diffusion coefficients and electron drift velocities in Townsend discharges in several gases [14]. We have previously discussed the use of this technique for the measurement of the locations of the secondary electrons produced by the passage of ionizing particles in a gas [1,4,5] and the method is only briefly outlined here. A pulsed, high-voltage (peak voltages up to 30-40 kV), high frequency RF field (f = 27.125 MHz stabilized by a crystal controlled 3000 watt oscillator-amplifier) is temporarily applied to the chamber electrodes, immediately following the passage of the ionizing particle through the chamber (Figure 1). In the present measurements, the chamber was either filled with a 0.5-10 % mixture of N_2 in Ar or a 0.1-0.5% mixture of TMAE (tetrakis[dimethylamine]ethylene) in Ar with total gas pressures in the range 2.5-100 kPa, and a corresponding E/N range of ≈10-100 V-cm^2.

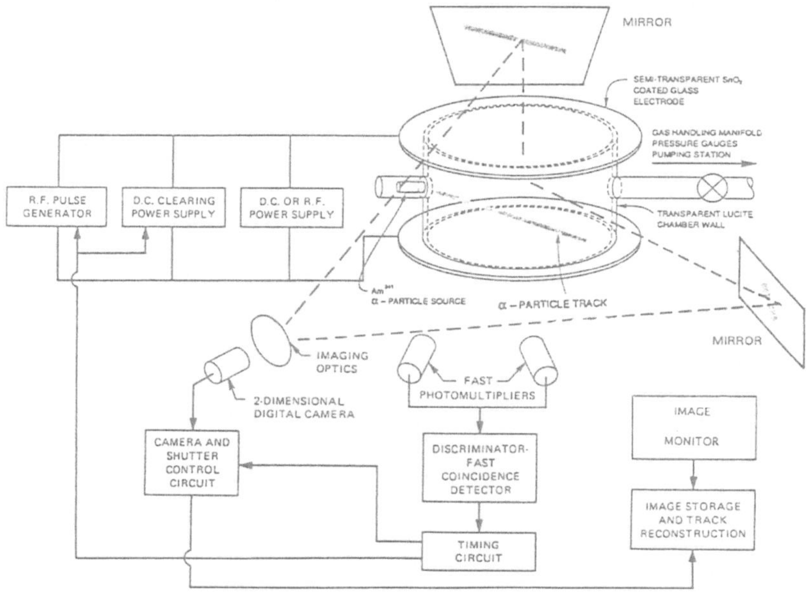

Figure 1. Schematic diagram of the Optical Ionizing Radiation Detector. The imaging optics in front of the camera were replaced by a spectrometer for the pulsed gas discharge spectroscopy measurements.

The prompt fluorescence from the decay of the gas molecules excited by the passage of the ionizing particle through the ionization chamber is used to trigger the detector. The light is detected by two fast photomultipliers operating in coincidence and the output from the coincidence detector is used to trigger the RF field circuit where a short RF voltage burst (pulse duration ≈ 1-3 μs) is applied to the chamber electrodes. Alternatively, the RF field can be operated in a pulsed duty cycle mode (the duty cycle for these measurements was usually 0.1% - 10%) with a pulse duration of typically 0.1-10 ms. The electric field is switched off when the photomultipliers detect a burst of light from the discharge produced by the excited electrons in the particle track. Although both techniques were successfully used to obtain images of the particle tracks in the gas, the particle track results presented in this paper were obtained by the latter technique due to the minimization of the electron diffusion before the application of the high voltage RF field when this technique is used. The maximum RF excitation pulse width is this case was ≈ 0.5 μs.

The pulsed RF field excites the subexcitation electrons left along the particle's path leading to ionization of the surrounding gas and to the emission of light in the immediate vicinity of the particle track. Under **low** electron amplification, the amount of light emitted from a given sub volume of the gas will be directly proportional to the initial number of electrons in that sub volume due to the particle track. The detection of the light emission by one or both of the photomultipliers is used to turn off the RF field in the pulsed duty cycle mode before space charge problems or gas breakdown occurs in the chamber. The emitted light is imaged by a fast microchannel plate intensified digital camera (shutter speed ≈ 0.01-5 μs) and recorded on a 512 x 512 CCD pixel array within the camera (Photometrics Series 200 camera). The image is subsequently digitized and transferred to a computer (PC 486 clone) for later analysis.

III. TRACK IMAGING RESULTS

Images of two α particle events are shown in Figures 2 and 3, and were produced by a ^{232}Th source located at the top electrode of the chamber. These α particle tracks were produced in a gas mixture of 2% N_2 in Ar at a total gas pressure of 55 kPa and show two of the general types of α particle tracks that we have detected with this chamber. That is, conventional tracks with a high degree of ionization near the stopping point of the track (Figure 2), and more unexpectedly, double ended tracks with a high degree of ionization at both the beginning and end of the track (Figure 3). The reason for the track structure observed in Figure 3 is presently not understood, but may be related to charge stripping (and therfore a lower rate of gas ionization) of the α particle during part of its passage through the gas. The half width of the track image shown in Figure 3 ≈ 0.7-1.0 mm, giving a corresponding unit density resolution of ≈ 0.6-0.9 μm.

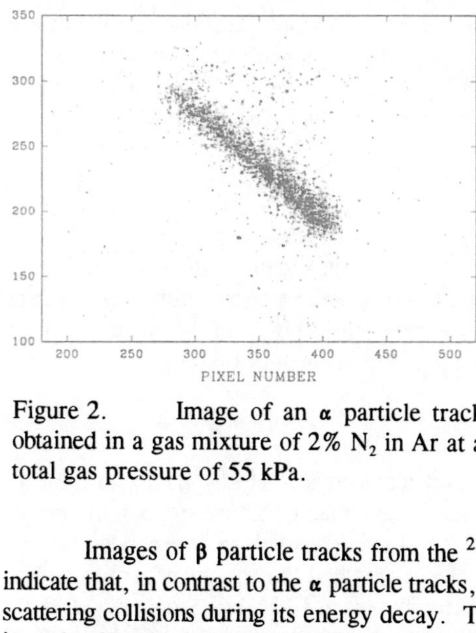

Figure 2. Image of an α particle track obtained in a gas mixture of 2% N_2 in Ar at a total gas pressure of 55 kPa.

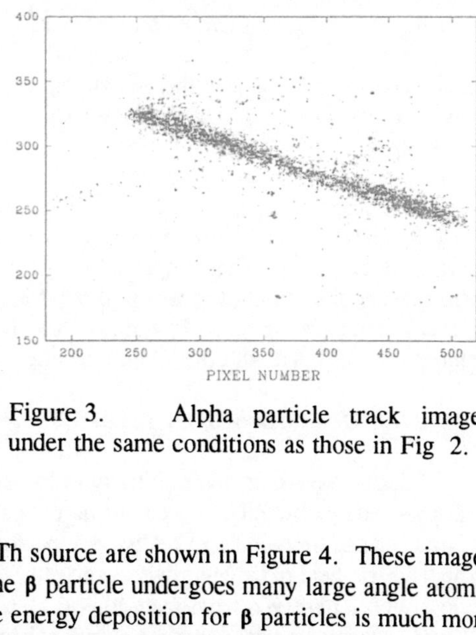

Figure 3. Alpha particle track imaged under the same conditions as those in Fig 2.

Images of β particle tracks from the ^{232}Th source are shown in Figure 4. These images indicate that, in contrast to the α particle tracks, the β particle undergoes many large angle atomic scattering collisions during its energy decay. The energy deposition for β particles is much more irregular than that for the α particles showing regions of dense electron production (and hence high energy deposition) and other regions with very little ionization of the gas (and hence low energy deposition). A Monte-Carlo simulation of an electron track produced by a 20 keV β particle in liquid water is shown in Figure 5. The three tracks shown in Figure 5 are the result of electron diffusion at (a) 10 ps, (b) 28 ns and (c) 280 ns after the passage of the β particle through the water [6]. The β particle measurements shown in Figure 4 were made several hundred nano- seconds after the passage of the β particle through the gas and correspond approximately to the track in Figure 5c. The similarity between the observed β particle track shown in Figure 4b and the calculated track shown in Figure 5c is, although coincidental, remarkable.

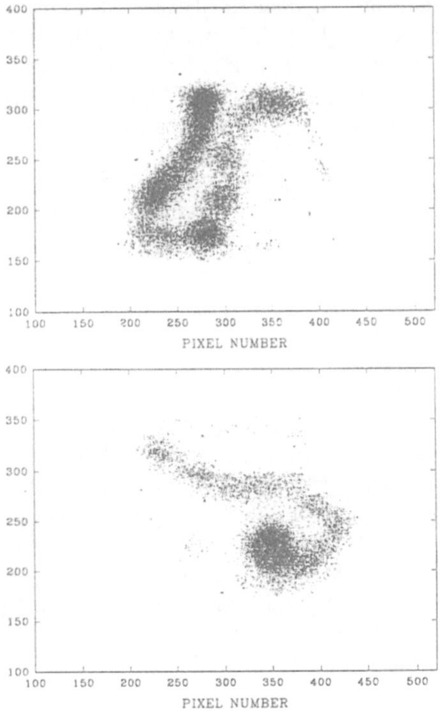

Figure 4. Images of two β particle tracks obtained from a [232]Th source located within the ionization chamber. The chamber contained a 2% N_2 / Ar gas mixture at a total gas pressure of 55 kPa.

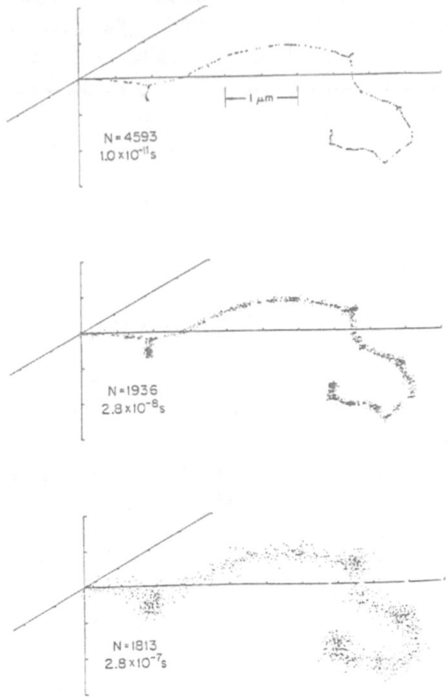

Figure 5. Monte Carlo simulation of a 20 keV β particle in liquid water including electron diffusion at (a) 10 ps, (b) 28 ns, and (c) 280 ns after the passage of the β particle through the water.

The detector has been shown to have the ultimate sensitivity of being able to image single electrons, and can thus be used to image any low density gas discharge containing one or more electrons. Figure 8 shows the amplified image of a single electron and an "old" α particle track, where most of the initial electrons in the particle track have been lost by electron attachment to impurities in the gas mixture. In addition, the three dimensional locations of the electrons in the chamber can be found with the addition of the imaging mirrors shown in Figure 1.

IV. GAS OPTIMIZATION STUDIES

Ideally, we require the sensitivity of the optical detector to be such that the exact location of each secondary electron produced during the passage of the ionizing particle through the gas in the chamber can be found. In reality, the performance of the detector is very dependent on the gas and electric field parameters with which the detector is operated. Strong optical emission has previously been observed from the $C^3\Pi_u$ - $B^3\Pi_g$ transition in pulsed 10 MHz RF discharges in N_2 [15], and our initial studies focussed on this gas in dilute mixtures with Ar due the high emission coefficients previously observed from this system [16]. The emission spectrum from a 1% N_2/Ar pulsed gas discharge is shown in Figure 6 in comparison with the spectral response of the intensified CCD camera with a quartz and a standard 50 mm objective lens. The spectrum given in Figure 6 indicates that emission in this gas mixture occurs predominantly at 337.1 nm [v = 0-0], 357.7 nm [v = 0-1], 380.5 nm [v = 0-2] and other higher vibrational levels from the second positive system. No emission from the first negative system ($B^2\Sigma^-_u$ - $X^2\Sigma^+_g$) was observed in the present studies, while Sato and Haydon [15] have observed strong emission at 391.4 nm [v = 0-0]. Plane parallel electrodes were used to create a uniform electric field in the present studies, while the measurements performed by Sato and Haydon were made with a point-plane electrode

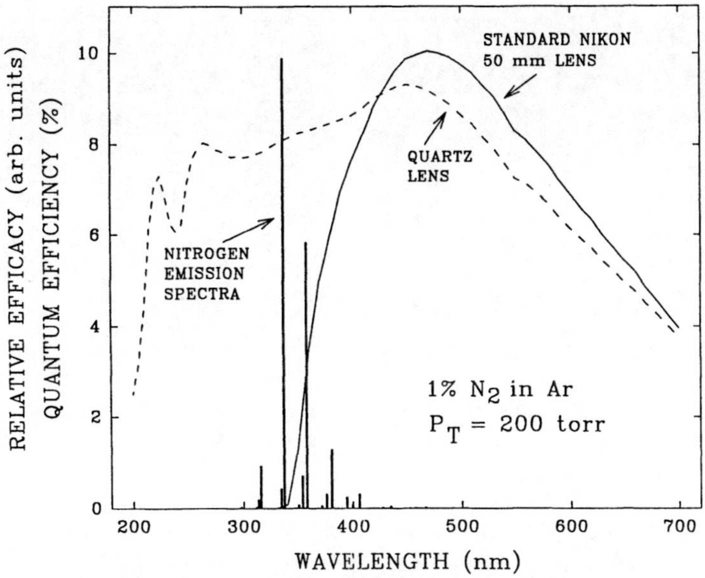

Figure 6. The quantum efficiency of the intensified CCD camera with a standard optical 50 mm objective lens and a 50 mm quartz objective lens. Also shown are the major lines from the nitrogen emission spectra in a pulsed N_2/Ar gas discharge.

arrangement under corona discharge conditions. The higher electric fields near the tip of the rod electrode are expected to lead to excitation of higher lying states than for uniform electric field excitation, possibly leading to the observed spectral lines from N_2^+ [15]. The present measurements show that the quartz lens is required if the full optical emission from the discharge is to be used to image the particle tracks in this gas mixture. In practice, the standard crown glass/flint glass lens must be used even though light below about 350 nm is not transmitted. Spherical aberration problems (leading to a loss in image resolution) were found to be too severe with the quartz objective. As a result, only approximately 10% of the total light generated by the N_2/Ar gas discharges is used to obtain the images in Figure 2-4.

Typical RF electric field voltage waveforms for a gas mixture containing 1.0 torr of N_2 in Ar are shown in Figure 7a at a total gas pressure of 200 torr, and at a total gas pressure of 300 torr in Figure 7b. Also shown is the light output from the 337.1 nm line in N_2 in these two gas mixtures. These waveforms illustrate the major problem with applying the RF voltage after the passage of the ionizing particle through the detector. With the resonant tank excitation circuit we use to produce the pulsed RF electric field [5], it takes 2-4 μs for the voltage level across the discharge chamber electrodes to rise to a point where significant ionization and excitation in the gas can occur and thus, before significant light emerges from the discharge. Significant thermal diffusion of the secondary electrons on the particle track can occur during this time leading to a loss of resolution of the initial track and to reduced information on the energy deposition in the particle track (see Figure 5).

The fluorescence decay times observed in the light output measurements in Figure 7 were 6.4×10^{-7} sec. at 200 torr and 4.4×10^{-7} sec. at 300 torr. These decay times are much longer than the expected 9 ns decay time for the $C^3\Pi_u$ state of N_2 at these gas pressures [15, 17]. A significant fraction of the emission from the N_2/Ar gas mixtures is expected to come from the collisional quenching of high lying metastable states in Ar by the N_2 leading to additional emission from the second positive group [18]. The apparent decay times decrease with increasing gas pressure primarily due to the loss of Ar metastables from the formation of Ar excimers in three body collisions with Ar and N_2 [17].

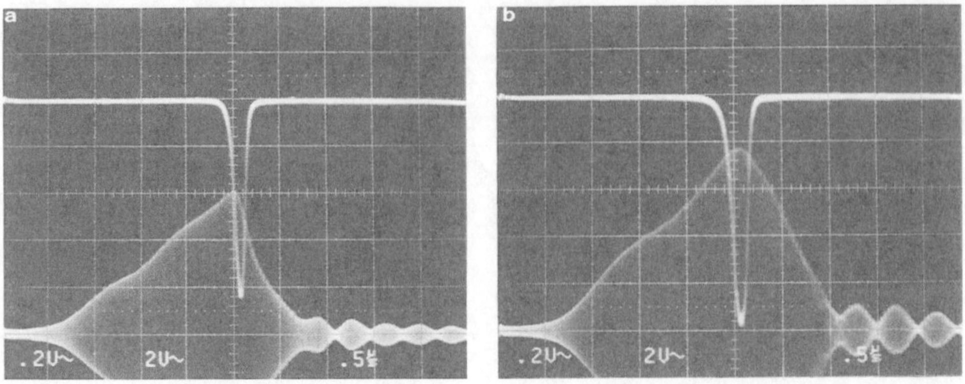

Figure 7. Typical RF voltage waveforms and 337.1 nm light output from pulsed RF 27 MHz discharges in a 1.0 torr N_2 gas mixture with Ar at a total pressure of (a) 200 torr and (b) 300 torr.

The large delay between the formation of the particle track in the gas and the generation of light from the discharge is the primary reason for the loss of resolution with the present detection scheme [1]. This is demonstrated by the results shown in Figure 9 where the apparent half width of the particle track is plotted as a function of the time required for the RF excitation voltage to reach its peak value. Excitation times considerably shorter than the present minimum value of ≈ 3 μs must be used to reduce the track half width. We are presently exploring this possibility using a shock excitation scheme where delay times < 0.5 μs are possible.

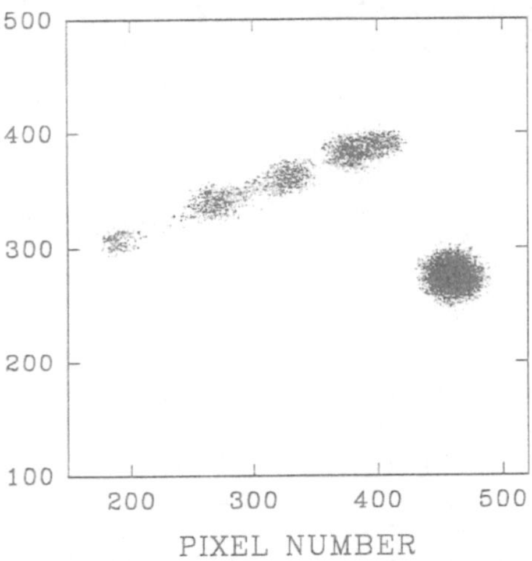

Figure 8. An amplified image of a single electron and an "old" α particle track in a 1% N_2 / Ar gas mixture.

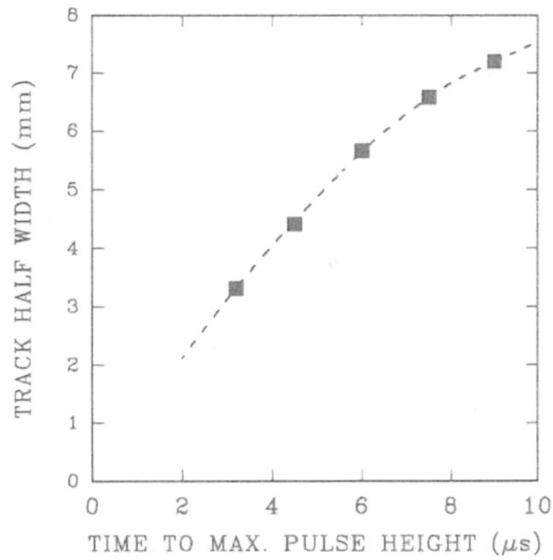

Figure 9. The apparent half width of an α particle track as a function of the delay time to reach the peak excitation voltage.

Optical emission measurements have been performed in N_2 /Ar gas mixtures as a function of the partial pressure of N_2 and as a function of the total gas pressure to find the optimum gas mixture which maximizes the light output from the discharge, thus giving the highest detector sensitivity. These results are shown on Figures 7 and 8 for several of the strongest emission lines in N_2. At very low partial N_2 pressures, the second positive band emission is a simple function of the N_2 concentration. At higher N_2 concentrations, energy loss due to vibrational excitation of N_2 alters the high energy tail electron energy distribution function in the discharge, reducing the $C^3\Pi_u$ excitation coefficient. At the low total gas pressures, the nitrogen emission increases with the N_2 (direct excitation) and Ar (collisional quenching) concentrations. At the higher gas pressures, Ar excimer formation and N_2 collisional quenching compete with N_2 emission to reduce the rate of increase in the $C^3\Pi_u$ emission. Gas mixtures composed of ≈ 0.5% N_2 at total gas pressures above 200 torr are required to optimize the light output in N_2 / Ar gas mixtures.

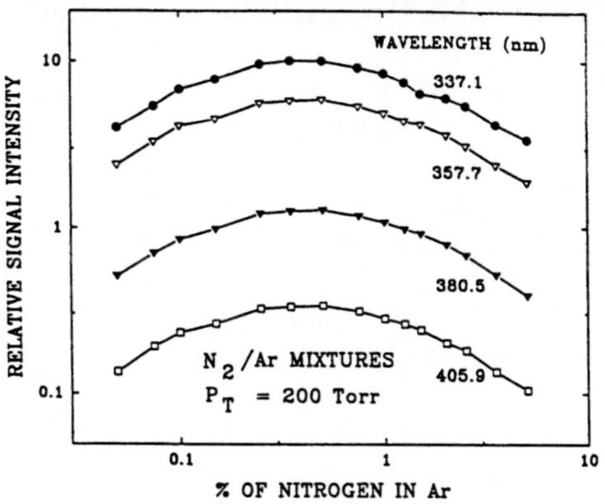

Figure 10. Prominent N_2 emission lines from the RF discharges in N_2/Ar gas mixtures as a function of the percentage of N_2.

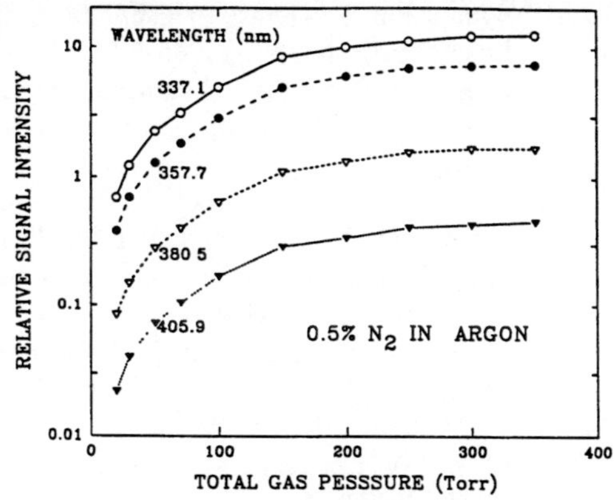

Figure 11. Prominent emission lines from N_2 in a 0.5% N_2/Ar gas mixture as a function of the total gas pressure.

V. CONCLUSIONS

We have developed a new type of ionizing radiation detector which can be used to image the particle tracks produced by an ionizing particle as it traverses an ionization chamber. The detector has been shown to be capable of imaging α and β particle tracks with a resolution of < 1 mm, and to have the sensitivity to resolve single electrons. Measurements of the detector's performance have been made in N_2/Ar and TMAE/Ar gas mixtures (only the results for the N_2/Ar gas mixtures have been presented in this paper). Future improvements will give a detector resolution down to ≈ 0.1 mm by reduction in the electronic delays, and the proper choice of gas mixtures, and RF electric field strength, frequency, and duration.

ACKNOWLEDGEMENTS

The present work was supported by the National Cancer Institute, National Institutes of Health, under Grant No. CA45869-03 to Pellissippi International, Inc.

REFERENCES

1. S. R. Hunter. "Evaluation of a Digital Optical Ionizing Radiation Particle Track Detector," *Nucl. Instru. Meth.* A260:469 (1987).
2. J. E. Turner, S. R. Hunter, R. N. Hamm, H. A. Wright, G. S. Hurst, and W. A. Gibson. "Development of an Optical Digital Ionization Chamber," *Radiat. Protect. Dosim.* 29:9 (1989).
3. J. E. Turner, S. R. Hunter, R. N. Hamm, H. A. Wright, G. S. Hurst, and W. A. Gibson. "Digital Characterization of Recoil Charged-Particle Tracks for Neutron Measurements," *Nucl. Instr. and Meth.* B40/41:1219 (1989).
4. S. R. Hunter, W. A. Gibson, G. S. Hurst, J. E. Turner, R. N. Hamm, and H. A. Wright. "Optical Imaging of Charged Particle Tracks," *Radiat. Protect. Dosim.* 52:323 (1994).
5. W. A. Gibson and S. R. Hunter. "Technique for the Optical Imaging of Charged Particle Tracks in a Gas," *Rev. Sci. Instr.* - submitted for publication.
6. H. A. Wright, J. L. Magee, J. E. Turner, A. Chatterjee, R. N. Hamm, R. H. Ritchie, Physical and Chemical Evolution of an Electron Track in Liquid Water, *in*: "Radiation

Protection," Proc. Eighth Symposium on Microdosimetry, J. Booz and H. G. Ebert, eds., CEC, Brussels, pg. 101 (1983).

7. J. E. Turner, R. N. Hamm, G. S. Hurst, and H. A. Wright, A Digital Approach to Neutron Dosimetry and Microdosimetry, *in*: "Radiation Protection," Proc. Fifth Symposium on Neutron Dosimetry, H. Schraube, G. Burger, and J. Booz, eds., CEC, Luxembourg, pg. 675 (1985).

8. J. E. Turner, R. N. Hamm, G. S. Hurst, H. A. Wright, M. M. Chiles. "Digital Characterization of Particle Tracks for Microdosimetry," *Radiat. Protect. Dosim.* 13:45 (1985).

9. W. E. Bolch, J. E. Turner, R. N. Hamm, H. A. Wright, and G. S. Hurst. "A Method of Obtaining Neutron Dose and Dose Equivalent from Digital Measurements and Analysis of Recoil-Particle Tracks," *Health Physics*, 53:241 (1987).

10. J. E. Turner, R. N. Hamm, H. A. Wright, R. H. Ritchie, J. L. Magee, A Chatterjee, and W. E. Bolch. "Studies to Link the Basic Radiation Physics and Chemistry of Liquid Water," *Radiat. Phys. Chem.* 32:503 (1988).

11. J. E. Turner, R. N. Hamm, T. E. Huston, H. A. Wright, W. A. Gibson, and G. S. Hurst. "Development of an Optical Digital Ionization Chamber," *Radiat. Protect. Dosim.* 32:157 (1990).

12. R. N. Hamm, S. R. Hunter, G. S. Hurst, J. E. Turner, and H. A. Wright. "Ionizing Radiation Detector System," United States Patent No. 4,931,653 June 5, (1990).

13. G. Cavalleri. "Measurements of Lateral Diffusion Coefficients and First Townsend Coefficients for Electrons in Helium by the Electron-Density Sampling Method," *Phys. Rev.* 179:186 (1969).

14. D. K. Gibson, R. W. Crompton, and G. Cavalleri. "Measurements of the Thermal Diffusion Coefficient for Electrons in Helium," *J. Phys. B* 6:1118 (1973).

R. Hegerberg, and I. D. Reid. "Electron Drift Velocities in Air," *Aust. J. Phys.* 33:227 (1980).

I. D. Reid and R. W. Crompton. "The Drift Velocity of Low Energy Electrons in Oxygen at 293 K," *Aust. J. Phys.* 33:215 (1980).

R. W. Crompton and G. N. Haddad. "Thermal Electron Attachment to SF_6 and $CFCl_3$," *Aust. J. Phys.* 36:15 (1983).

15. N. Sato and S. C. Haydon. "Nanosecond Time-Resolved Spectroscopic Investigations of RF Corona with High Spatial Resolution," *J. Phys. D* 17:2023 (1984).

N. Sato and S. C. Haydon. "Ionization Growth between Hyperboloidal Needle and Plane Electrode under R.F. Electric Fields in N_2," *Aust. J. Phys.* 39:271 (1986).

16. W. Legler. "Anregung Von UV-Strahlung in Stickstoff und Wasserstoff Durch Einen Elektronenschwarm," *Z. Physik.* 173:169 (1963).

17. P. Grimm, F. J. Hambsch, M. Mutterer, and J. P. Theobald. "Fast Decay Components in the Scintillation Gases Ar, Ar-N_2, Xe, and He-Xe Excited by Heavy-Ion Impact," *Nucl. Instru. and Meth.* A262:394 (1987).

18. T. D. Strickler and E. T. Arakawa. "Optical Emission from Argon Excited by Alpha Particles: Quenching Studies," *J. Chem. Phys. 41*:1783 (1964).

DISCUSSION

E. MARODE: The intensity of light emission seems to increase with pressure, while generally an increase in pressure leads to pressure quenching and light reduction. Could you comment about that?

S. R. HUNTER: As I mentioned in the paper, a significant fraction of the light from the N_2/Ar discharges is expected to come from collisional quenching of high lying metastable states of Ar by N_2, leading to additional emission from the second positive group. This is a three-body process and increases with increasing total gas pressure (Ref. 17 of the paper).

R. BARTNIKAS: What is the limitation of going to frequencies higher than 50 MHz?

S. R. HUNTER: A tank circuit is used for generating the high voltage. We are operating in a resonant circuit mode and it is very hard to go above 50-100 MHz because of component size limitations. We could go to microwaves but we would have to go to a different excitation scheme.

EXHAUST GAS TREATMENT USING ELECTRICAL DISCHARGES

Adriano Nasciuti, Timm H. Teich

High Voltage Engineering Group
Swiss Federal Institute of Technology
CH - 8092 Zürich

INTRODUCTION

Interest in the use of non-thermal electrical discharges for the conversion of toxic or otherwise undesirable gas components has been building steadily over more than a decade[1]. Both direct and indirect treatment methods have been applied. With *indirect* methods, a discharge-exposed gas stream is added to the exhaust flow to be treated[2]; with *direct* methods[3], the exhaust gas stream itself is exposed to the discharge; corona, barrier and surface discharges have been investigated[1]. Besides modest capital outlay, discharge treatment has the advantage (in common with electron-beam techniques) that it can reduce certain pollutants practically to zero concentration and that it can still be relatively effective with quite small initial concentrations. A disadvantage is a relatively high energy investment, and therefore numerous laboratories concentrate their present efforts on finding the most energy-efficient operating conditions for the particular treatment problem. One has the choice of basic type of discharge, discharge geometry, type of voltage application to promote particular discharge properties, selection of additives etc.. The results of various investigations seem to indicate the usefulness of application of steep-fronted pulses of short duration; however, one has to realize the difficulties of providing such pulse voltages for industrial-scale systems, taking into consideration the associated electromagnetic compatibility problems and the energy efficiency of *complete* systems including the power conditioning. Another point that has to be addressed is the possibility of the formation - as a consequence of the discharge - of highly undesirable new products not necessarily detected with standard gas analysis; in this context, the more exotic input gas compositions may require special attention.

Nitric oxide from various combustion processes is a central problem because of its direct physiological effects and its involvement with undesirable subsequent atmospheric chemistry. This paper reports on discharge chemistry operated with the purpose to convert hazardous gases to less harmful substances utilising radicals produced by electrical discharges. Examples of gas discharges which can be driven even at atmospheric conditions are the barrier discharge, the corona discharge and some kinds of surface discharges. The investigations presented here are based on experiments made with a barrier discharge

Gaseous Dielectrics VII, Edited by L.G. Christophorou
and D.R. James, Plenum Press, New York, 1994

reactor. A problem arises from the fact that in every combustion exhaust gas CO_2 is present in large quantities. This causes quite a high CO production. The purpose of this work is to determine the working conditions of the reactor with regard to the products and the energy efficiency in reducing or producing NO, NO_2 and CO. The parameters varied are the temperature and the content of oxygen and water in the synthetic exhaust gas. The gas analysis is mainly made using a FT-IR spectrometer (3 to 20 μm) which allows the measurement of the main nitrogen oxides.

To compare the efficiencies of the reactor at different working conditions, a quantity Θ is introduced:

$$\Theta = \frac{\Delta c}{E} = \frac{\Phi \cdot \Delta c}{P} \quad \left[\frac{l \cdot ppm}{Wh}\right]$$

which represents the change in specific concentration Δc [ppm] per unit energy input / volume, E [Wh/l]; E can also be given in terms of electrical power [W] and flow rate Φ [l/h]: E=P/Φ. The changes Δc in concentration are specific for the particular gas mixtures considered; we compare concentrations $\Sigma\ [X]_o$ without discharge ("reactor off") to concentrations with the reactor active, $\Sigma\ [X]'$:

$$\Delta c = \{\Sigma\ [X]_o - \Sigma\ [X]'\}$$

and correspondingly have specific efficiencies Θ for reduction of NO, NO_x ($NO_x = NO + NO_2$) and production (Δc negative) of CO.

EXPERIMENTAL SET-UP

The experimental set-up consists of a gas mixing and heating unit, interchangeable reactors and a FFT-IR spectrometer for the gas analysis. As is shown in Fig. 1, the gases coming from compressed gas cylinders flow into a gas mixing chamber. There, if needed, they can also be heated. Afterwards the synthetic exhaust gas flows to the reactor to be exposed to the discharge and then proceeds to the gas cell for the analysis. The total gas flow is typically of *6 l/min* (referred to STP) at *100 kPa (750 Torr)*.

In the present series of experiments four different synthetic gas mixtures have been investigated. The compositions relate to typical combustion flue gases, though the water contents are here always lower (see Table 1).

Table 1. Target compositions of the test gases and target powers.

						target power [W]	
	CO_2	NO	H_2O	O_2	N_2	P1	P2
Gas A	13%	100 ppm	0%	0%	87%	8,8	12
Gas B	13%	100 ppm	0%	6%	81%	3	4,5
Gas C	13%	100 ppm	1,72%	0%	85,28%	8,8	12
Gas D	13%	100 ppm	1,72%	6%	79,28%	3	4,5

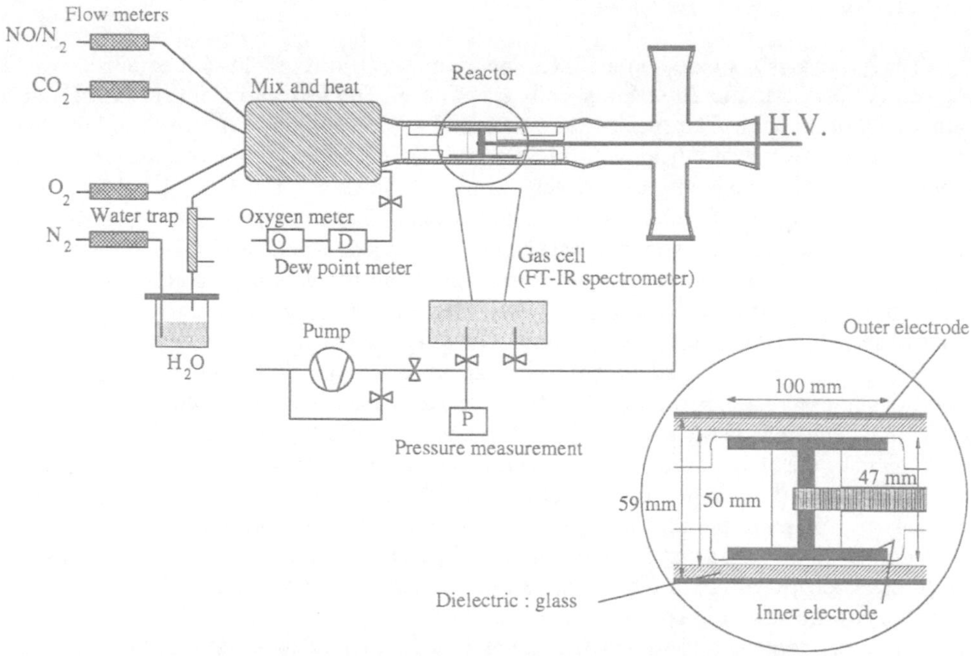

Figure 1. Experimental set up with a barrier discharge reactor.

The investigation of these four test gases shows the influence of the oxygen and water content. The water content in the gases C and D is small compared to that of real flue gases, but this is at present the highest acceptable value for reliable gas analysis by means of the instrumentation used here. The experiments were carried out at two different gas temperatures: *30°C* (T1) and *100°C* (T2).

The energy consumption *of the discharge* is measured using the "Lissajous figure" method[4]. The voltage at the reactor is measured with a high voltage probe and the charge with a condenser between outer electrode and ground. The signals of voltage and charge are digitised on an oscilloscope and transferred by GPIB to a PC. There the data can be analysed with a specific program to calculate the area of the Lissajous figure.

Every test gas was treated at two fixed discharge powers (see Table 1) to better compare the results with those of other experiments. It was to be expected that the efficiencies are to some extent power dependent.

Every measuring series consists of the collection of a spectrum with the reactor off (to determine the inlet gas composition) and two spectra taken at different reactor powers (Table 1).

The Reactor

The reactor consists basically of a stainless steel inner electrode, a dielectric (borosilicate glass) and a copper outer electrode (see Fig. 1). The gas flows through the gap between inner electrode and the glass. On the inner electrode a sharp edged thread was cut to create a small local field enhancement. The high voltage (500 Hz AC) is brought into the reactor from the right (in the Fig. 1) using a high voltage leadthrough.

Gas analysis

The gas analysis is based on a FT-IR spectrometer (Bio-Rad FTS-40) equipped with a gas cell (Wilks Scientific 20-m Long Path Gas Cell of The Foxboro Company EMO) at a path length of 11,25 m. The spectra were recorded at a resolution of 0,5 cm^{-1} and the total gas pressure was always held at 100 kPa at 24°C. The gas components, which can simultaneously be measured quantitatively, are the following: CO_2, CO, NO, NO_2, N_2O, O_3, N_2O_5, HNO_2 and HNO_3. The quantitative analysis was carried out by means of a computer program (PLSplus), which makes use of a special algorithm PLS-1. This method is very suitable for gas mixtures of many components because it takes into account the cross sensitivities (overlapping of absorbance bands) between the single gases. This quantitative analysis needs a calibration. A set of spectra of different but known concentrations of the different gas components had to be collected, as far as possible, under closely related conditions (pressure, temperature, resolution, etc.) to those encountered with the discharge-treated gas. The uncertainty of the measurement over the whole span (generally 0-500 ppm, 0-100 ppm for the higher nitrogen oxides and 0-20% for CO_2) is of about ±1 % of full scale. For that reason the smaller concentrations (0-10 ppm) had always to be determined manually by comparison with the calibration spectra. When measuring wet gases a water spectrum has to be subtracted from the measured spectrum. The spectrum obtained is what would have been measured if the gas had been without any water content. The best results are obtained when the "wet spectrum" and water spectrum have the same water content.

The water content is measured with a condensation type dew point instrument and the oxygen content with a paramagnetic oxygen meter.

RESULTS AND DISCUSSION

Influence of the gas composition

The "exhaust" gas composition has an important influence on the efficiency of NO_x removal. In our experiments - the findings of which are summarized in Fig. 2 - we noticed a positive effect of oxygen content. The radicals, produced in the presence of oxygen (gas B), cause the stepwise oxidation of nitrogen oxides. At low power input the NO is converted to NO_2. Only at higher power NO_2 can be transformed into N_2O_5 and HNO_3. The production of HNO_3 is due to the small water content (some tens of ppm) of the gases in the compressed gas cylinders. At this point some ppm of ozone can be measured, too. Larger quantities of water may reduce those effects dramatically. Water seems to promote the production of O_xH_y radicals, leading to HNO_2 instead of HNO_3. The NO reduction and the CO production are decreased by a factor of about 1,5 compared to a dry gas (see for example Fig. 3). Water reduces the efficiencies of the discharge relating to NO_x, too. This effect was already noticed in the earlier work of Heer and Loepfe[5], performed with a different gas analysis method. Scheuermann and Bloss[6] and Masuda and Wu[3] (the latter with pulsed corona discharge) observed the opposite influence of water content; in their case, water seems to promote the NO_x oxidation. In the gas D (even at higher power up to 16 W) we could achieve only the oxidation of the NO to NO_2, besides a small production of HNO_2. When neither O_2 nor H_2O are present in the gas containing NO (gas A), the composition of the treated gas seems to us to indicate that a decomposition of NO with formation of N_2 and O_2 is then possible. No higher oxides are then detectable in the treated gas A (see Fig. 4).

Depending on the power and on the gas composition, a few ppm of N_2O are produced in the discharge. Typical concentrations are between 2 and 6 ppm.

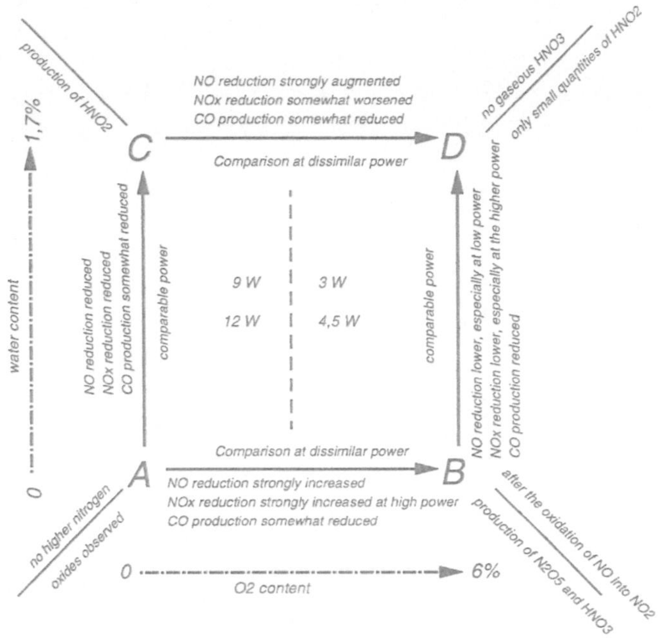

Figure 2. Influence of the gas composition on the discharge products.

Influence of the temperature

The temperature effects observed are not very great (over the step from 30°C to 100°C) except for gas B where the [NO_x] reduction efficiency collapses at the higher temperature (see Fig. 5). A similar though much less pronounced effect is also observed with gas D where, however, the NO_x reduction was very small anyway. The production of radicals from the oxygen seems to be reduced because in the gas B no higher oxides are measured. In all the gases investigated, there is a slightly reduced NO conversion at the higher temperature.

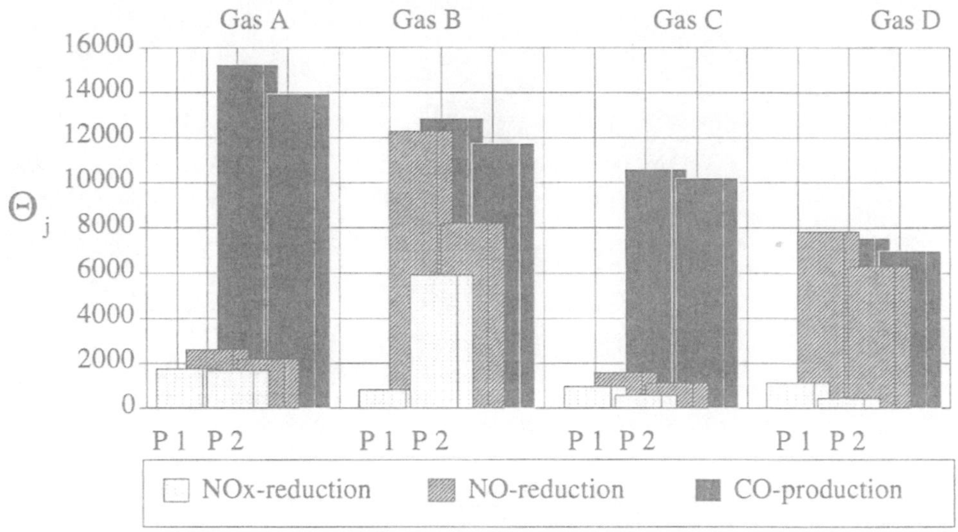

Figure 3. Effect of gas composition on the efficiencies Θ (30°C).

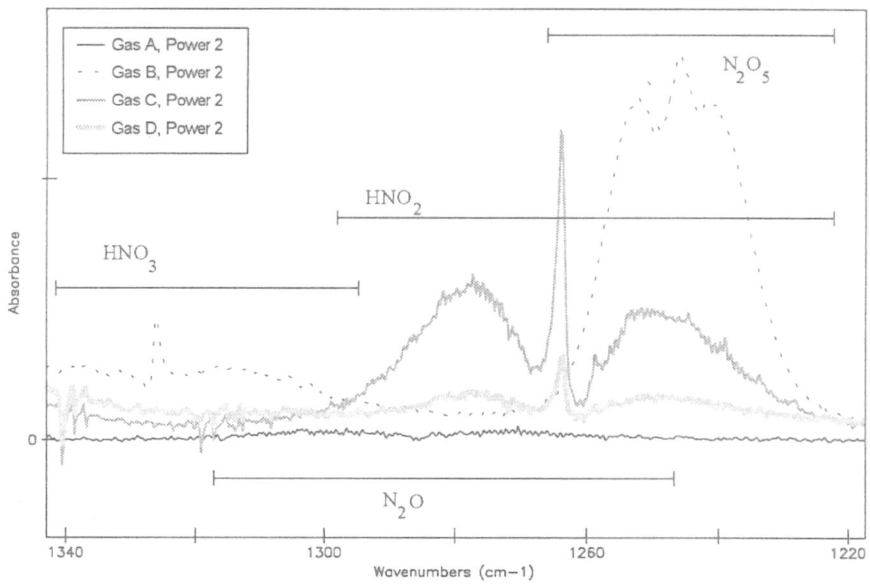

Figure 4. Absorbance bands of the higher nitrogen oxides and N_2O.

Gas A: No higher nitrogen oxides are visible. The barely visible small bands are caused by about 2 ppm N_2O.

Gas B: The main band is N_2O_5 (about 22 ppm). On the left of this band about the same quantity of HNO_3 is evident.

Gas C: In this spectrum the significant production of HNO_2 is observable (about 120 ppm).

Gas D: Only a small quantity of HNO_2 is noticeable. No gaseous HNO_3 is detectable.

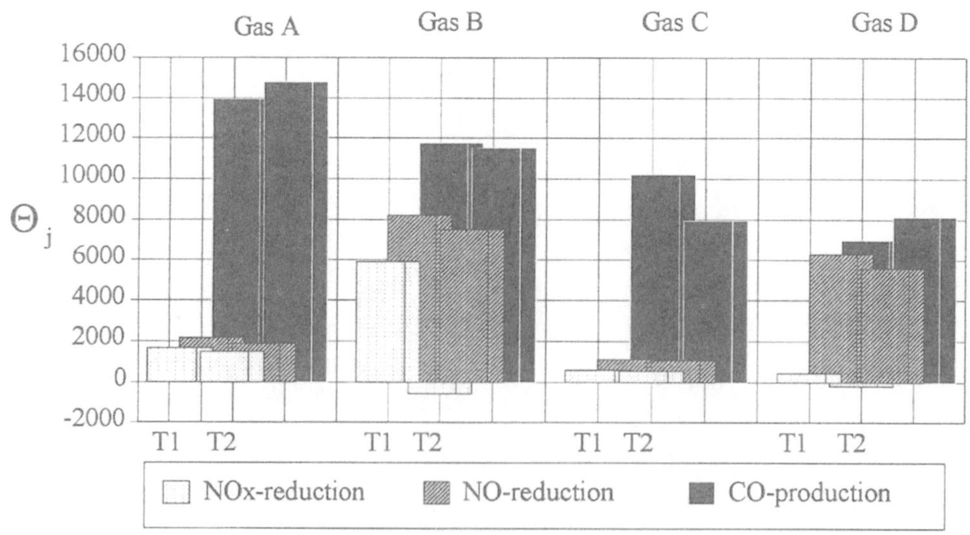

Figur 5. Effect of the gas temperature on the efficiencies Θ at power P2 (30°C (T1) and 100°C (T2)).

CONCLUSIONS

The experiments show the considerable dependence of the removal and generation efficiencies Θ and the reaction products on the gas composition. The results presented here are in general agreement with earlier work of Masuda and Wu[3] and of Scheuermann and Bloss[6], except the influence of the water content on the efficiencies Θ. In our work it is evident that water represents an obstacle to the removal of NO_2, under the conditions studied here. Furthermore, it proves necessary to take the CO production associated with the discharge into account. Both problems can be avoided by *indirect* gas treatment, adding ozone to the gas containing NO[2,7] instead of exposing the gas directly to the discharge. However, in this way no *direct* effects of the discharge on other harmful components of an exhaust gas can be exploited.

Future work will also include measurements on gases with water contents more representative of combustion processes (8-20 %) and with additives such as NH_3.

REFERENCES

1. B.M. Penetrante and S.E. Schultheis (Eds.), "Non-thermal plasma techniques for pollution control", NATO ASI Series G volumes 38A and 38B, Springer, Heidelberg (1993).
2. S.S. Novoselov, A.F. Gavrilov, V.A. Svetlichnyi, V.E. Chmovzh, V.Yu. Simachev and I.M. Zaplatinskaya: "The ozone method of removing SO_2 and NO_x from the flue gases of thermal power stations. Thermal Eng. **33**, 496-498 (1986).
3. S. Masuda and Y. Wu; "Removal of NO_x by corona discharge induced by rising nanosecond pulse voltage", Electrostatics 1987: Seventh International Conference on Electrostatic Phenomena, St Catherine's College, Oxford, 8-10 April (1987).
4. U. Kogelschatz; "Advanced Ozone Generation", Process Technologies for Water Treatment, edited by Samuel Stucki, Plenum Press, New York (1988).
5. L. Heer, B. Loepfe; "Die Beeinflussung der Gaszusammensetzung durch eine stille Entladung", Semesterarbeit WS 1990/91, Fachgruppe Hochspannungstechnik, ETH Zürich (1991).
6. A. Scheuermann and W.H. Bloss; "Verfahren zur Entstickung und Entschwefelung von Verbrennungsabgasen mittels Corona-Entladung", Projekt Europäisches Forschungszentrum für Maßnahmen zur Luftreinhaltung (PEF), Kernforschungszentrum Karlsruhe, KfK-PEF 47 (1988).
7 A. Nasciuti; "Stickstoffoxidreduktion mit Hilfe elektrischer Entladungen", Diplomarbeit SS 1990, Fachgruppe Hochspannungstechnik, ETH Zürich (1990).

SECTION 7: PULSED POWER SWITCHING/GAS RECOVERY

RECENT ADVANCES IN HIGH POWER GAS SWITCHES

K.H. Schoenbach[1] and A.H. Guenther[2]

[1]Physical Electronics Research Institute
Old Dominion University, Norfolk, VA 23529
[2]Sandia National Laboratories
Albuquerque, NM 87185

ABSTRACT

Primarily because of a shilft from defense-related applications to those more commercial, the emphasis in recent developments of high power closing gas switches has been on lifetime and reliable operation at medium average power, rather than on improvements in single shot performance at very high peak power. Most of the work has focused on pseudosparks and very high pressure spark gaps. The lifetime of pseudosparks is now approaching that of thyratrons. Recovery times of less than one millisecond in high pressure hydrogen spark gaps allow operation with repetition rates exceeding 1 kHz. Anode voltage fall times of less than 100 ps have been achieved in strongly overvolted, high pressure spark gaps. In the field of opening switches, Plasma Opening Switches (POS) have received the most attention. The increase in peak power into the terawatt range and the controllability of the opening effect in POS were recent major achievements.

KEYWORDS

Pulsed power, closing switches, and opening switches.

INTRODUCTION

The amplification of electrical power by orders of magnitude up to the terawatt level in short pulses has been a principal and central topic of Pulsed Power Physics and Technology. Most of the rapid development of pulse power technology in the past decade was stimulated by the needs of the Department of Defense. With changing priorities, the applications of pulse power systems are now directed more to commercial and environmental uses. With the new applications, the emphasis in the development of pulsed power systems is changing; there is less emphasis on single shot, very high peak power

Gaseous Dielectrics VII, Edited by L.G. Christophorou
and D.R. James, Plenum Press, New York, 1994

devices and increasingly more emphasis on reliable, repetitive, high average power, long lifetime systems. It is in this area, and in the area of extremely short pulses, that pulsed power systems have advanced most in recent years. The improvements are due to intense research and development on the most critical component in pulse power systems: the pulse power switch.

There are two basic types of pulse power circuits, based on the type of energy storage system used. Most of the pulsed power systems utilize capacitive storage systems where, in order to transfer the energy to the load, closing switches are required. The second type of pulse power circuit is based on inductive energy storage. For the energy to be transferred to the load, opening switches (circuit breakers) need to be employed. Hybrid systems, with capacitive and inductive storage systems arranged in a pulsed power chain, contain both types of switches. Regardless of the circuit type, pulse power operation requires the switches to operate at high hold-off voltages (> kV) and to be capable of transferring large amounts of electrical charge on time scales approaching the order of picoseconds, corresponding to currents in excess of kA.

CLOSING SWITCHES

Gas discharge closing switches are the classic type of switches in pulsed power systems.[1] They are the most thoroughly investigated and most widely used. The closing mechanism is the growth of ionization (electrical breakdown) at high electric fields. They cover a wide range of pressures, from vacuum to high pressure (> atmospheric pressure). A classification of the various switches can be based on their position in the Paschen curve (pressure, p, times breakdown gap, d, range versus breakdown voltage) with respect to the Paschen minimum (Fig. 1). Pseudosparks, thyratrons, and Crossatrons operate (in the breakdown phase) on the left hand side of the Paschen minimum. Diffuse discharge switches and spark gaps are located on the right hand side of the minimum. Advances have been made in all types of gas discharge closing switches in recent years, the most impressive ones being in pseudospark switches and high pressure spark gaps.

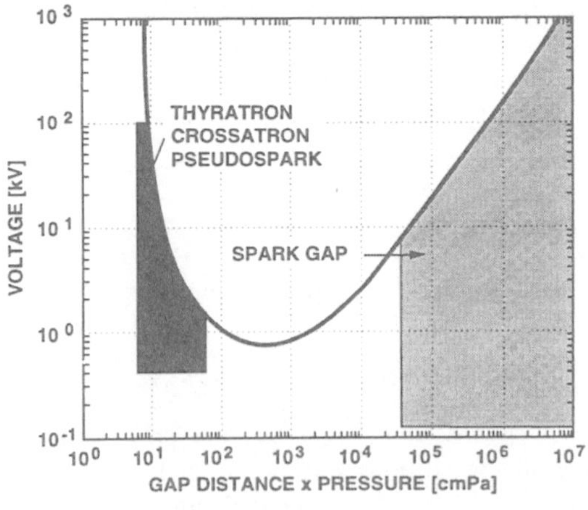

Figure 1. Range of operation of various types of switches in a Paschen diagram.

THYRATRONS

Besides the spark gap, there is no switch on which more engineering studies have been performed and which is used so commonly as the hydrogen thyratron. The hydrogen thyratron was developed in the nineteen forties by K.J. Germeshausen[2] and has undergone little change in the past decade. It is a gridded gas switch with the control grid close to the anode. There are two reasons for the use of hydrogen: one is the fact that the damage of the cathode, which is due to ion bombardment, is minimal with hydrogen; secondly, the deionization time of hydrogen is shorter than for mercury-filled or rare gas-filled thyratrons, except for helium, and therefore allows operation of the switch at the maximum repetition rate. The gas pressure is below 1 Torr which causes the grid-anode gap to operate on the left hand side of the Paschen curve. Application of a voltage to the control grid therefore causes breakdown across the longer gap between grid and cathode. The glow discharge between cathode and grid brings the grid potential almost to the cathode potential, causing the anode-grid gap to break down. Typical performance parameters are hold-off voltages of up to several tens of kV, kA's of peak current, and a reliable operation with lifetime approaching 10^9 shots.

The rise time of conventional thyratrons is on the order of 10 ns, largely due to the inductance of the switch. Reduction of the inductance has been achieved by using an annular device (POLOTRON) where a microwave discharge between cathode and grid was generated by means of a slow wave structure, shown in Fig. 2.[3] The longitudinal electric field in the tube decreases with distance from the helix, reaching a minimum on axis. The plasma which is generated by the annular discharge, formed close to the outside walls of the tube, diffuses into the high voltage gap at the anode and causes a rapid breakdown. Voltage fall times of less than 3 ns have been measured with a 30 kV cold cathode POLOTRON.

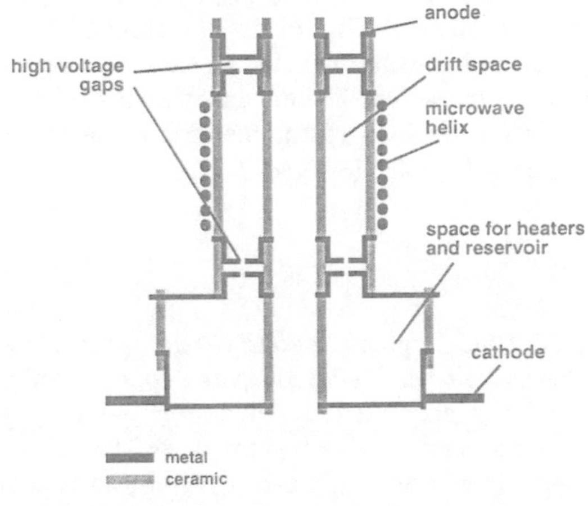

Figure 2. Schematic diagram of a POLOTRON.[3]

CROSSATRON

The Crossatron is a low-pressure discharge device which is similar to the thyratron in closing behavior but can also be used as a grid-controlled opening switch.[4] The electrons in the Crossatron are provided by a hydrogen plasma in a crossed electric and magnetic field, between a cold cathode and a source grid (Fig. 3). The gas pressure is 200 to 900 microns, which is slightly below that of thyratrons. This is due to the shift in the Paschen curve in a crossed field configuration to lower pressure values. It allows, therefore, an increase in the hold-off voltage of the anode-control grid gap.

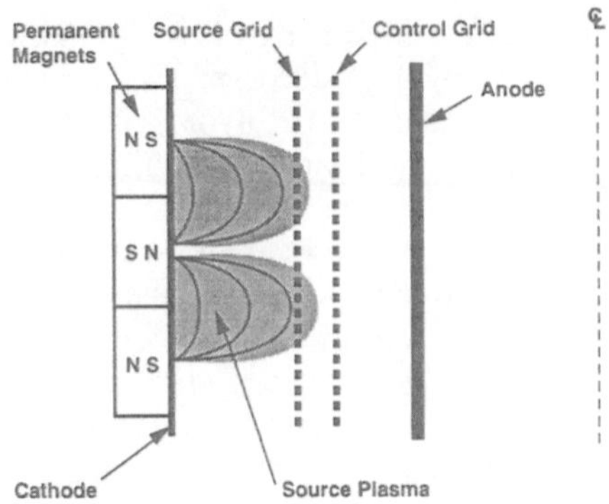

Figure 3. Crossatron switch electrode configuration.[5]

A low-current discharge (20 mA typical) between the cathode and the source grid is maintained in order to minimize the turn-on jitter. Closing of the switch is obtained by raising the second grid, the control grid, above the potential of the cathode. This allows the plasma to diffuse into the anode-control grid gap and to establish low impedance conduction. The advantages compared to thyratrons are longer life and faster recovery. Typical recovery time is 10 microseconds. Recent advances in the design of the Crossatron, such as modifications in the cathode surface and grid transparency, have lead to higher hold-off voltage and higher peak current.[5] For Crossatrons used as closing switches, the hold-off voltages and peak currents are comparable to thyratrons.

Interruption of the current in Crossatrons is achieved by pulsing the control grid negative with respect to the cathode. This blocks the flow of plasma from the cathode region into the anode space, and the plasma in the anode space erodes with a time constant given by the thermal velocity of the ions and the dimensions of the gap. Typical opening time is one microsecond. Compared to Crossatrons operated as closing switches, the peak current is an order of magnitude lower (~1 kA).

PSEUDOSPARK/BLT

Pseudosparks or BLTs (Back Lighted Thyratrons) are the most recent addition to types of high power gas discharge switches.[6] Pseudosparks operate at about the same pressure as normal thyratrons. They are filled as thyratrons with hydrogen or deuterium. Advantages of pseudosparks compared to thyratrons are the smaller size, higher peak currents, and fast start-up. In addition, it is the possibility to operate with 100% current reversal which makes the pseudospark attractive for application where bipolar operation is required. The present lifetime of pseudosparks is lower than that of thyratrons; however, there have been strong improvements in this area within the past two years.

The structure of a pseudospark is shown schematically in Fig. 4 (lower part), which also illustrates the various phases in the discharge development.[7] The discharge develops as a Townsend discharge along the longest distance between cathode and anode, because the pd-value is on the left hand side of the Paschen minimum. The field distortion generated by this filamentary discharge causes the development of a hollow cathode discharge, with radial current components. With increasing current, a superdense glow discharge develops in the central region around the cathode hole. The development up to the hollow cathode discharge seems to be well understood.[8] During the immediate past

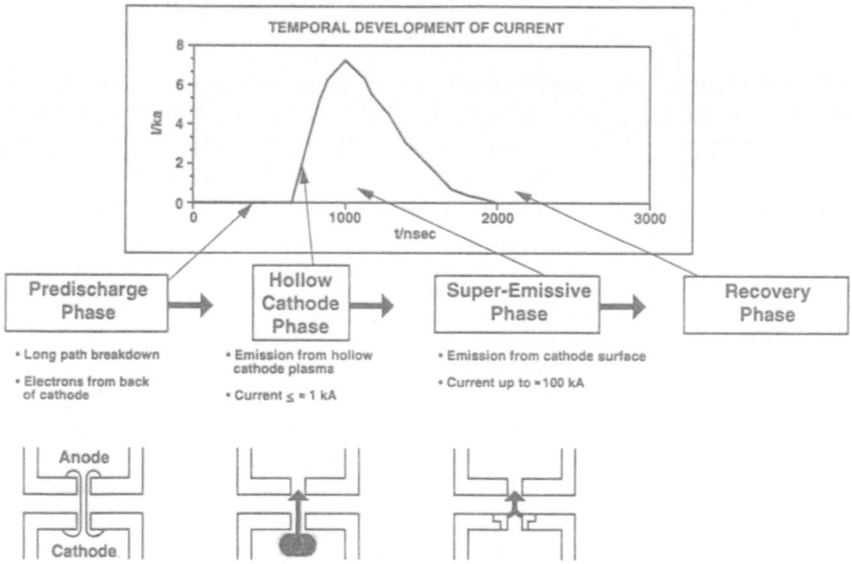

Figure 4. The temporal development of a pseudospark discharge.[7]

years, efforts have concentrated on the understanding of the physics of the superdense glow, the phase of high current flow.

A model developed by Puchkarev[9] and by A. Anders, S. Anders, and Gundersen[10] describes the "super-emission" of electrons from the cathode as a field-induced thermal emission. Heating of the cathode through ion impact is not sufficient to explain the high current density observed in pseudosparks. However, the formation of a virtual plasma anode in close proximity to the cathode surface causes a dramatic increase in electric field intensity at the cathode surface. From vacuum breakdown studies, it is known that for electric fields, exceeding a critical value, microtips at the cathode surface heat up explosively and form a metal plasma. At currents of some kA, the cathode spots form groups and eventually merge into a "unified" spot of much larger diameter. The molten surface of this spot is able to provide the measured electron current density in the superdense glow phase. Observations of merging cathode spots during the breakdown of the switch voltage[11] confirm the assumptions made in this model. Experiments at Siemens using laser induced fluorescence have indicated that spots with temperatures in excess of 7000 K exist on the cathode and are sufficient to provide the required electron emission for the high current phase.[12]

The engineering development of these switches has proceeded very quickly. Integrated Applied Physics, Inc. (IAP) in the United States and Siemens and CERN/Thompson in Europe already produce a line of pseudospark devices for applications such as pollution control[13], RADAR[14], lasers[15], and accelerators[16]. Most recent efforts have concentrated on an increase in lifetime. In excess of 10^8 pulses has been obtained by using multichannel pseudosparks, a value which is close to the lifetime of thyratrons.[17] Another approach to the increase in the lifetime of BLTs is rotation of the discharge through a circular multihole device.[14] Triggering of pseudospark switches is generally performed electrically, using trigger electrodes; however, optical triggering has also been used in the BLT. Another successful trigger method applied in the CERN switches is electron-beam triggering using ferroelectric electron emitters.[18]

SPARK GAPS

Spark gaps seem to be the least sophisticated of switches, certainly in design. In their simplest form, as self-breakdown gaps, they consist of two electrodes separated by an insulating gas medium. However, although the geometry is simple and the mechanisms seem to be less complex than in other switches, there is still room for much improvement and innovation. In recent years, research on spark gaps, particularly high pressure spark gaps, has led to impressive results. Several areas are receiving attention: triggering of spark gaps; rise of current or anode fall time, respectively; recovery (which determines maximum repetition rate); and lifetime. Triggering of spark gaps, by means of a third electrode, optically or using electron-beams is described in ref. 1. However, most of the work in recent years has focused on the other important related issues.

The pulsed power group at Texas Tech University has studied electrode erosion with various electrode materials dependent on current and transported charge. Fig. 6 shows erosion rates and lifetimes versus current and charge, respectively, for graphite and copper as electrode materials.[19] The copper pattern is typical for most of the electrode materials. In the lower range of charge transferred per discharge, Q (or current, current density, etc.), the erosion is determined by evaporation due to a single, current-carrying filament. The dramatic increase in erosion is caused by collective filament melting and ablation, when a critical number of filaments is formed. The third phase, at high Q, is typical for evaporation (or sublimation) over the entire arc cross-section. For a graphite electrode where no ablation of molten material occurs, the erosion transition from region 1 to region 3 is primarily due to an increase in the sublimation rate due to collective filament interaction.[19]

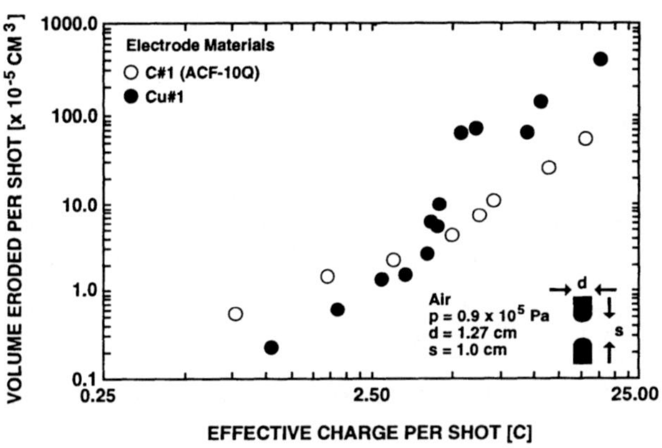

Figure 5. A comparison of Cu and graphite electrode versus transferred charge.[15]

Studies on very high pressure spark gaps (p ≫ 1 atm) have received the most attention in recent years. The research group at NSWC, Dahlgren has concentrated on high repetition rate spark gaps.[20,21] For gases such as air, nitrogen, argon, oxygen, and SF_6, the recovery time for high pressure spark gaps is typically on the order of 10 ms. With hydrogen, however, this time could be reduced by one order of magnitude. It was assumed that this reduction in recovery time is due to fast channel expansion of the hydrogen plasma. Besides the gas, the pressure plays an important role in shortening the recovery time. Increasing the pressure up to 7 MPa allowed reduction of the gap spacing down to 0.1 mm. The recovery time was also reduced, probably due to closer gas contact to the surface of the electrodes.

The recovery of the gas occurs in two steps as shown in Fig. 6.[20] The first stage, up to the plateau, is determined by the recovery of the gas density to pre-breakdown values (cooling of the gas). The second stage of the switch recovery, where the switch can be overvolted again, is strongly influenced by the presence of negative ions (or positive ions) which are created in the gap by the initial discharge.[22] It could be shown that it is possible to reduce the recovery time by sweeping out the residual charges in the interpulse period through application of electric fields.[22] Another method to obtain high repetition rates is by undervolting the switch, i.e., trigger the switch in stage one of the recovery curve. Repetition rates of 1 kHz at voltages up to 500 kV have been demonstrated with this method.[21] Ten kHz repetition frequencies seem to be feasible with properly triggered switches operating in the undervolted mode. Measurements in a burst mode indicate that the jitter for successive pulses can be kept in the nanosecond range for operation down to 60% breakdown.

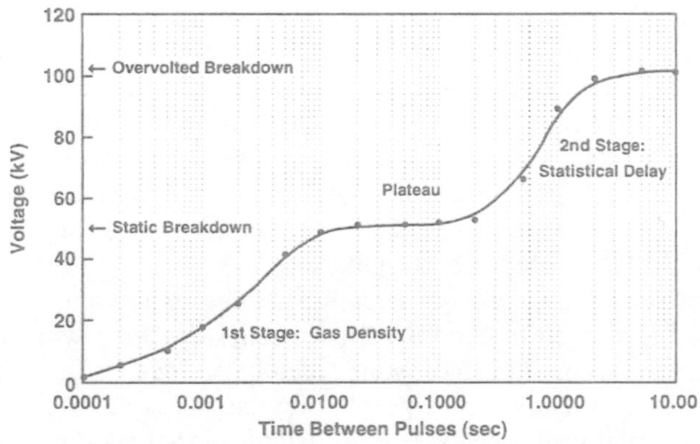

Figure 6. Two stages of over-volted spark-gap recovery. Charge voltage risetime is 1 kV/ns in an air gap of 450 kPa.[20]

Another advantage of operating spark gaps at very high pressures is the speed of gas breakdown which occurs when the switch is overvolted. Fig. 7 shows an experimental setup which is used to test ultrafast switches.[23] A fast pulser with conventional switch generates a fast rising voltage pulse across the high pressure, miniature switch. In order to minimize the impedance of the switch, it is made part of either a coaxial line, as shown in Fig. 7, or a stripline.[24] Pulses with 50 to 100 ps risetime, to voltages of 75 kV to 170 kV at pulse repetition frequencies of 1 kHz, were achieved.[23]

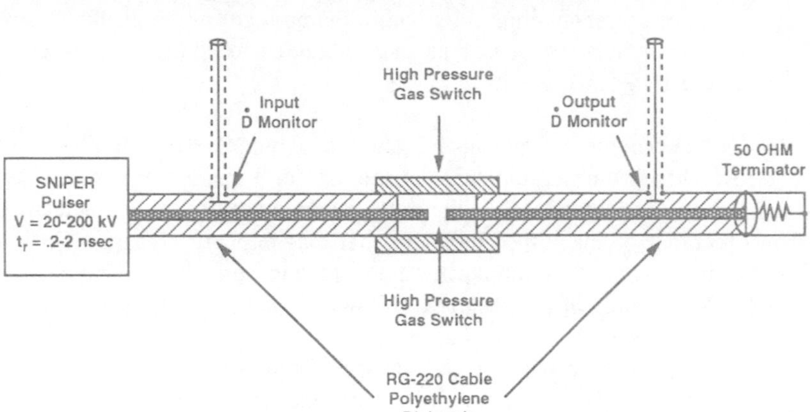

Figure 7. The experimental apparatus is based on an in-line cable switch in a 50 Ohm system.[23]

Gases used in the experiment with a coaxial line were hydrogen, nitrogen, and sulphur hexafluoride and mixtures thereof. The risetime was found to be inversely proportional to E/p in any gas. Typically, the highest pressure with the shortest gap spacing produces the fastest output risetime and minimum switch loss. The initial fields, before breakdown of the gap, are such that field emission from the electrode can be expected. Fields of 6 MV/cm were measured in a 0.05 cm hydrogen gap at pressures of 165 atm. Comparisons of experimental with modeling results indicate that the discharge in the initial phase is diffuse (or at least multifilamentary) initiated by the copious source of electrons field-emitted from the cathode.

OPENING SWITCHES

In the field of opening switches at very high power levels (> GW), Plasma Opening Switches (POS), also called Plasma Erosion Switches, of late have dominated. Therefore, in the following section, we have concentrated on this type of switch. Advances, however, have also been made in other opening switches. In the lower power range up to 100 MW, the Crossatron is commercially available. The opening mechanism of this switch was briefly discussed in the section on Closing Switches. Research on most of the other plasma opening switches (e.g., diffuse discharge switches, reflex switches, and plasma focus) has slowed not because they were unsuccessful, but rather because of the shrinking range of applications and advances in alternative approaches.

An opening effect which deserves attention relates to current quenching in glow discharge switches. Hydrogen thyratrons show current quenching when the peak current or pulse duration exceed a certain value. More recently, a similar effect has been observed in pseudosparks[25] and in diffuse discharges[26]. Attemps to explain this effect have been made for thyratrons[27] and pseudosparks[17]. An overview of the various types of opening switches can be found in ref. 28.

PLASMA OPENING SWITCHES

The geometry of a typical Plasma Opening Switch is shown in Fig. 8.[29] A plasma is injected in a vacuum between the electrodes of a coaxial line (generally by using a flash board as carbon plasma source) to conduct a temporally increasing current. The electron density is on the order of several 10^{15} cm^{-3}. The developing current sheath in the plasma is predominantly radial. Due to its azimuthal magnetic field, it is driven towards the load, generally by a distance corresponding to its width, thereby storing magnetic energy in the space behind it. The switch opens, i.e. the plasma ceases to conduct, when the current through the plasma has reached a critical value.

A model for the conduction and opening phase of microsecond POS was recently developed at NRL, based on measurements of the temporal change in electron density.[30] The measured reduction of the axial line-integrated electron density in the conduction phase by more than one order of magnitude indicates that magnetohydrodynamic (**JxB**) effects displace the plasma in radial direction during the conduction time which is typically 0.5 to 1 microsecond. The thinning of the plasma is followed by a rapid plasma erosion. Plasma erosion, in the collisionless plasma, is due to the limited number of injected ions. With increasing current, the ion density at the cathode decreases, and the current in the developing gap is carried only by electrons which are field emitted from the cathode. With increasing electron current density, the voltage across the gap increases according to Child-Langmuir's law, and the switch begins to open. Part of the current is now

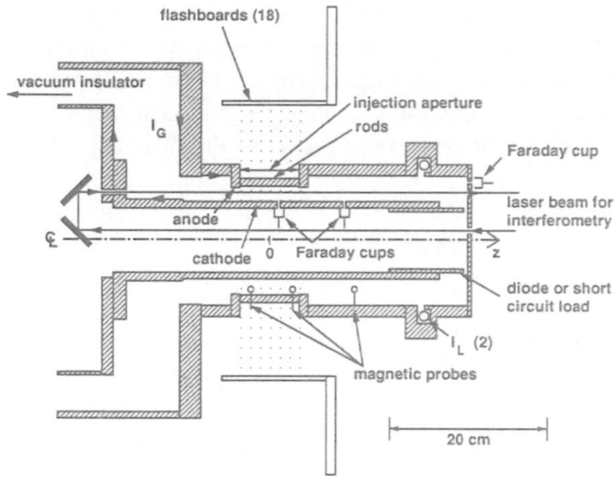

Figure 8. Drawing of the vacuum section of the pulsed power system, illustrating the POS geometry and indicating the location of diagnostics. The diagnostics include generator and load current monitors, I_a and I_n, magnetic probes, Faraday cups, and He-Ne laser interferometry. The arrows in the conductor indicate the direction of the current flow.[29]

commutated to the load, and the azimuthal magnetic field which develops in the gap due to this current flow accelerates the speed of opening, and eventually magnetically insulates the switch. For this to happen, the gap must be large compared to the electron Larmor radius.

For optimum operation as an opening switch in an inductive energy storage system, the conduction time should be as long as possible and the opening time as short as possible. The conduction time is dependent on factors such as device geometry, temporal development, and amplitude of current, time, and type of plasma injection. The injection of a large mass of plasma, required for long conduction time, reduces the opening speed. A method to command trigger the opening of a microsecond POS and to keep the opening time short has been developed at Sandia.[31,32] The modified circuit is shown in Fig. 9. It consists of two POS, one used to trigger the main opening switch.

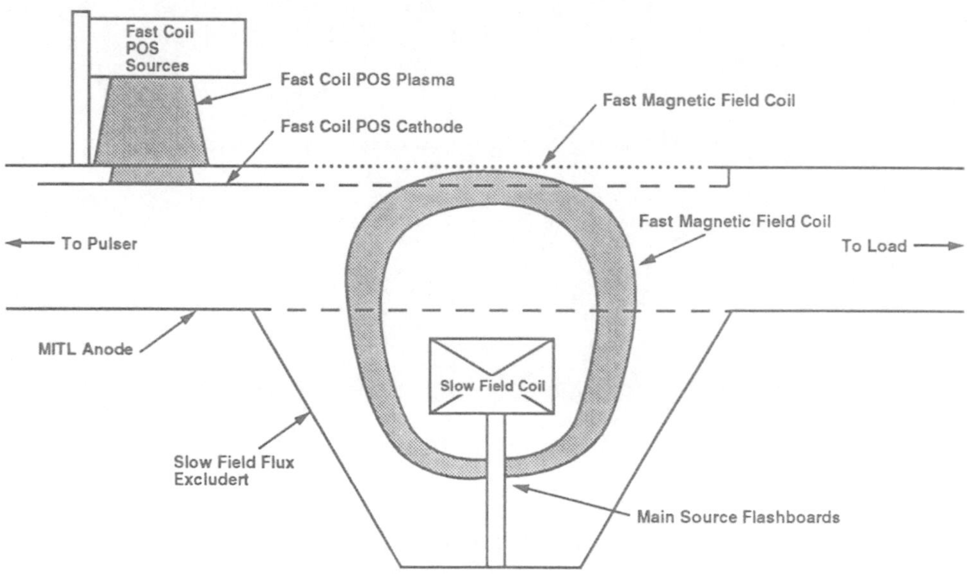

Figure 9. The Sandia long Conduction Plasma Opening Switch.[32]

The temporal sequence is as follows: The slow coil capacitor bank is fired and establishes a magnetic field in the transmission line gap which holds the injected plasma in place. During conduction, the magnetic field coil is shunted by means of trigger POS (Fast Coil POS plasma) with conduction time less than one microsecond. The current flows through axial vanes to the downstream side of the fast coil. Before the current has reached its maximum, the trigger POS opens, the fast magnetic coil is energized, and its increasing axial field pushes against the slow magnetic field, thereby opening the main switch through a fast erosion process.

The results obtained with the various POS are impressive. Recent results at the Sandia "Current Toggled" POS are: voltage > 1 MV, with peak switch current of 500 kA, generated after a 0.65 microsecond conduction time. The switch opens in 30 ns or less. The trigger ability makes this system unique; however, it requires a considerable current - up to one third of the switched current. Efforts continue to concentrate on the reduction of this current.

Even higher power was switched with a POS at Physics International.[33] Peak opening voltages of more than 2 MV, at a current of approximately 1 MA, were achieved. Typical current and voltage waveforms for the PI switch are shown in Fig. 10. The improvement in peak voltage is due to modifications at the injection region, particularly at the anode. The "bird cage" metal structure (which allows the injected plasma to enter the transmission line) was eliminated, and the injected plasma itself serves as anode. In addition, the downstream part of the anode was moved radially outward. Both modifications were done to reduce the electron damage to the anode. The reasons for the dramatic improvement in performance caused by these relatively simple geometry changes awaits further study.

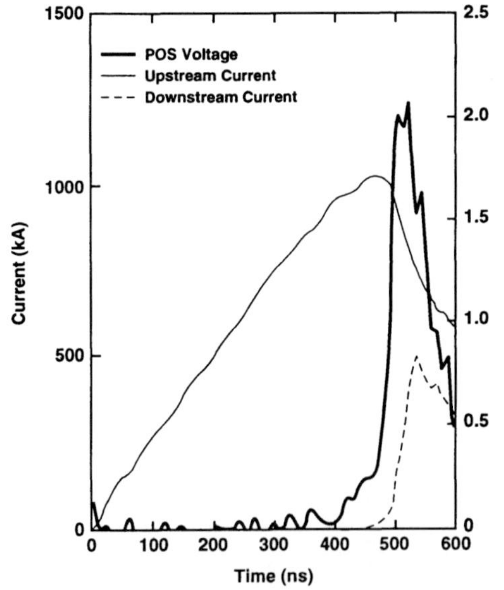

Figure 10. Representative POS-voltage and current waveforms.[33]

SUMMARY

The ongoing efforts to convert the primarily defense-related pulsed power technology for civilian use have had a strong impact on the recent development of closing switches. Affordability, lifetime, maintainability, and reliability are the main issues, rather than highest performance for single-shot operation and cutting edge, but unproven technology.[34] Most of the recent advances in closing switches are therefore related to an increase in lifetime and to more reliable operation at high average power, rather than on high peak power. Because of their reliability and lifetime, thyratrons still head the list of pulsed power switches. Efforts to increase their speed are underway. Crossatrons and pseudosparks, however, are gaining ground. Pseudosparks, which are a hybrid between thyratron type discharges and spark discharges, are particularly attracting attention. The physics of the high current phase in these switches, the "super-emissive" phase, seems to be understood - a fact which might lead to further improvement of pseudospark operation. In regard to repetitive operation, spark gaps have made tremendous progress. The use of high pressure hydrogen and the triggering at voltages below the self-breakdown voltage have moved the repetition rate into the kHz range. In addition, there is a better understanding of erosion processes, with impact on lifetime of spark gaps. Another important result which was also achieved with very high pressure, overvolted spark discharges was the reduction of the gas breakdown time to less than 100 ps. This brings spark gaps into a temporal range which so far has been dominated by photoconductive solid state switches. In the field of opening switches, most of the work done is on Plasma Opening Switches (POS). Power levels of terawatt have been achieved with pulsed power systems using POS, and attempts to control the opening phase were successful. Plasma Opening Switches still seem to be in the early phase of development; lifetime and repetition rate are not yet an issue. However, there are tremendous efforts to advance the performance of these truly high power switches. Research on most of the other opening switch types which were studied in the nineteen eighties[35] has halted. However, the challenge to develop a reliable, controllable, high average power opening switch still exists.

ACKNOWLEDGEMENT

We would like to thank our colleagues R. Commisso (Naval Research Laboratory), K. Frank (U. Erlangen), C. Frost (Sandia National Laboratories), D.M. Goebel (Hughes), J.R. Goyer (Physics International), M. Gundersen (USC), W. Hartmann (Siemens), G. Kirkman (IAP), M. Kristiansen (TTU), A. Maitland and N. Ridge (U. St. Andrews), S.L. Moran (NSWC, Dahlgren), P. Osmokrovic (U. Belgrade), W. Pfeiffer (TH Darmstadt), H. Riege (CERN), and M.E. Savage (Sandia National Laboratories) for providing us with their latest and most pertinent results on high power switching for this state-of-the-art review.

REFERENCES

1. G. Schaefer, M. Kristiansen, and A. Guenther, eds., "Gas Discharge Closing Switches," Plenum Press, New York, 1990.
2. K.J. Germeshausen, "The Hydrogen Thyratron," in "Pulse Generators," G.N. Glasoe and J.V. Lebacqz, eds., Radiation Laboratory Series, McGraw-Hill Book Company, Inc., New York and London, 1948, p. 335.
3. N.A. Ridge, B.P. Condon, P.F. Hirst, A. Maitland, D.M. Parkes, and P.D. Smith, "Microwave Triggered Annular Thyratron (POLOTRON)", Proc. 9th Intern. IEEE Pulse Power Conf., Albuquerque, NM, 1993.
4. R.W. Schumacher and R.J. Harvey, "Low Pressure Plasma Opening Switches," in "Opening Switches," A. Guenther, M. Kristiansen, and T. Martin, eds., Plenum Press, New York, 1987, p. 93.
5. D.M. Goebel, R.L. Poeschel, R.M. Watkins, and J.G. Elverum, "Recent Advances in Crossatron Switches," Proc. 9th Intern. IEEE Pulse Power Conf., Albuquerque, NM, 1993.
6. M.A. Gundersen and G. Schaefer, eds., "Physics and Applications of Pseudosparks," NATO ASI Series B: Physics, Vol. 219, Plenum Press, New York, 1990.

7. M.A. Gundersen, USC, private communication.
8. L. Pitchford, J.P. Boeuf, V. Puech, R. Liou, and M. Gundersen, "The Hollow Cathode Phase of Pseudospark Operation," Proc. Intern. IEEE Pulse Power Conf., Albuquerque, NM, 1993.
9. V. Puchkarev, "Fast Processes on Cathode Surface Resulting from Pseudospark Discharge," IEEE Trans. Plasma Science 21, 725 (1993).
10. A. Anders, S. Anders, and M.A. Gundersen, "Model for Explosive Electron Emission in a Pseudospark "Superdense Glow"," Phys. Rev. Lett. 71, 364 (1993).
11. K. Frank, J. Christiansen, A. Goertler, H. Granzer, A. Schwander, M. Stetter, and R. Tkotzk, "Forward Voltage Drop Measurements in Pseudospark Switches," Proc. 9th Intern. IEEE Pulse Power Conf., Albuquerque, NM, 1993.
12. G. Lins and W. Hartmann, "Metal Vapor Densities and Excitation Temperatures in a Pseudospark Switch," IEEE Trans. Plasma Science 21, 511 (1993).
13. G. Kirkman, N. Reinhardt, B. Jiang, J. Hur, and Y. Yampolski, "Power Conditioning for Corona Discharge Treatment of Effluent Gases," in "Non-Thermal Plasma Techniques for Pollution Control," B.M. Penetrante and S.E. Schultheis, eds., Springer Verlag, Berlin and Heidelberg, 1993, p. 379.
14. N. Reinhardt, G. Kirkman, J. Hur, B. Jiang, and T. Reinhardt, "Rapid Start, High Power BLT Switch for Anti-Missile RADAR," Proc. Intern. IEEE Pulse Power Conf., Albuquerque, NM, 1993.
15. D.W. Branston, W. Hartmann, G. Lins, H.D. Rohde, and J. Stroh, "Pseudospark Switch Development at Siemens - Compact High PowerSwitches for Gas Discharge Lasers and as Spark Gap Replacement," Proc. 9th Intern. IEEE Pulse Power Conf., Albuquerque, NM, 1993.
16. P. Faure, H. Riege, K. Schmidt, and G.H. Schroeder, "High-Power Gas Switches Triggered by Ferroelectrically Generated Electron Beams," Proc. 9th Intern. IEEE, Pulse Power Conf., Albuquerque NM, 1993.
17. K. Frank, O. Almen, P. Bickel, J. Christiansen, A. Goertler, W. Hartmann, C. Kozlik, A. Linsenmeyer, H. Loscher, F. Peter, A. Schwandner, and R. Stark, "Pseudospark Switches for High Repetition Rates and High Current Applications," Proc. IEEE 80, 958 (1992).
18. H. Riege, "New Ways of Electron Emission for Power Switching and Electron Beam Generation," in "Physics and Applications of Pseudosparks", M. Gundersen and G. Schaefer, eds., NATO ASI Series: Physics, Vol. 219, Plenum Press, New York, 1990, p. 349.
19. A.L. Donaldson and M. Kristiansen, "The Erosion Performance of Graphite Electrodes in High Current, High Coulomb, Spark Gaps," Proc. 8th IEEE Intern. Pulsed Power Conf., San Diego, 1991, p. 874.
20. S.L. Moran, L.W. Hardesty, and M.G. Grothaus, "Hydrogen Spark Gap for High Repetition Rates," Proc. 8th Intern. IEEE Pulse Power Conf., San Diego, CA, 1991, p. 336.
21. M.G. Grothaus, S.L. Moran, and L.W. Hardesty, "Recovery Characteristics of Hydrogen Spark Gap Switches," Proc. 9th Intern. IEEE Pulse Power Conf., Albuquerque, NM, 1993.
22. S.J. MacGregor, S.M. Turnbull, F.A. Tuema, and O. Farish, "The Operation of Repetitive High Pressure Gas Switches," J. Phys. D: Appl. Phys. 26, 954 (1993).
23. C.A. Frost, T.H. Martin, P.E. Patterson, L.F. Rinehart, G.J. Rohwein, L.D. Roose, J.F. Aurand, and M.T. Buttram, "Ultrafast Gas Switching Experiments," Proc. 9th Intern. IEEE Pulse Power Conf., Albuquerque, NM, 1993.
24. W.R. Cravey, E.K. Freytag, D.A. Goerz, P. Poulsen, and P.A. Pincosy, "Picosecond High Pressure Gas Switch Experiment," Proc. 9th Intern. IEEE Pulse Power Conf., Albuquerque, NM, 1993.
25. A. Goertler, J. Christiansen, S. Doellinger, K. Frank, U. Herleb, S. Insam, T. Mehr, M. Schlaug, A. Schwandner, and R. Tkotzk, "Pseudospark Switches for High Power Applications," Proc. ISDEIV'94, Moscow - St. Petersburg, 1994.
26. W. Pfeiffer and M. Junker, "Comparison of Two Different Concepts to Realize an Opening Switch with High Repetition Rate," Proc. Intern. IEEE Pulse Power Conf., Albuquerque, NM, 1993.
27. B.M. Penetrante and E.E. Kunhardt, "Fundamental Limitations of Hydrogen Thyratron Discharges," in "Closing Switches," G. Schaefer, M. Kristiansen, and A. Guenther, eds., Plenum Press, New York, 1990, p. 451.
28. A. Guenther, M. Kristiansen, and T. Martin, "Opening Switches," Plenum Press, New York, 1987.
29. R.J. Commisso, P.J. Goodrich, J.M. Grossman, D.D. Hinshelwood, P.F. Ottinger, and B.V. Weber, "Characterization of a Microsecond-Conduction-Time Plasma Opening Switch," Phys. Fluids B4, 2368 (1992).

30. D. Hinshelwood, B. Weber, J.M. Grossmann, and R.J. Commisso (1992), "Density Redistribution in a Microsecond-Conduction-Time Plasma Opening Switch," Phys. Rev. Lett. 68, 3567 (1992).
31. C.W. Mendel Jr., M.E. Savage, D.M. Zagar, W.W. Simpson, T.W.Grasser, and J.P. Quintenz, "Experiments on a Current-Toggled Plasma Opening Switch," J. Appl. Phys. 71, 3731 (1992).
32. M.E. Savage, W.W. Simpson, G.W. Cooper, and M.A. Usher, "Long Conduction Time Plasma Opening Switch Experiments at Sandia National Laboratories," Proc. 9th Intern. Conf. on High-Power Particle Beams, Washington, DC, 1992, p. 621.
33. J.R. Goyer, D.Kortbawi, and P.S. Sincerney, "Improved Performance of a Plasma Opening Switch Using a Novel Anode Configuration," to appear in IEEE Trans. Plasma Science.
34. G. Frazier, "Pulsed Power for Civilian Applications," Presentation at the Joint Workshop: Power Semiconductor Coordination Committee and Inter-Agency Power Group, EPRI, Palo Alto, CA, February 1993.
35. K.H. Schoenbach, M. Kristiansen, and G. Schaefer, "A Review of Opening Switch Technology for Inductive Energy Storage," Proc. IEEE 72, 1019 (1984).

DISCUSSION

S. W. ROWE: For hollow cathode closing switches field emission was stated to be the main cause of transfer to the high conducting phase between the main electrodes. Would not the heating of the electrodes due to the lower current phase play a role? Could you give an explanation of the operating mode of modern high pressure spark gaps?

K. H. SCHOENBACH: According to models by Puchkarev (Ref. 9 in the paper) and Anders et al. (Ref. 10 in the paper), the high current phase in pseudosparks is initiated by field emission from microtips on the cathode surface. The current after ignition, however, is sustained through thermal emission of electrons from cathode spots at the position of the exploded microtips which merge at very high current densities in a "unified" spot. Heating of the cathode due to ion impact does not seem to be significant in the transfer to the high current mode.

L. G. CHRISTOPHOROU: Is H_2 unique as a gaseous medium for high pressure fast closing switches?

K. H. SCHOENBACH: Hydrogen, nitrogen, argon, oxygen, SF_6, and mixtures thereof have been used as media in high pressure fast closing switches. The best performance, i.e., fastest current rise and shortest recovery time, was obtained with hydrogen as fill gas. The reason for the fast recovery is assumed to be the high thermal diffusivity (Ref. 20 in the paper) which allows for fast cooling of the gas. The fast current rise is, according to Frost et al. (Ref. 23 in the paper), due to the low E/N required for breakdown in hydrogen. In order to reach the same hold-off voltage as in other gases, the hydrogen pressure (density) must exceed that of these gases. Since, according to a breakdown model by T. H. Martin, the switching time is inversely proportional to the gas density, the highest rate of current rise can be expected in hydrogen.

D. W. BRANSTON: One of the main reasons for using hydrogen as filling gas is that it is the only gas which can be readily stored in a solid reservoir. This is absolutely necessary for a sealed-off device if reasonable lifetimes are to be achieved.

K. FESER: Could you comment on developments in laser-triggered spark gaps?

K. H. SCHOENBACH: I have not seen any really new results in this area. This is a technology which works nicely but is pretty expensive in that it requires a very high power solid state laser which may limit its commercial use. It allows extremely fast rise times however.

DIELECTRIC RECOVERY OF SF_6 DURING THE POST-ARC PHASE

I. Spiliopoulos[1], G. Berger[1], E. Marode[1], S.W. Rowe[2], G. Bernard[2]

[1] Laboratoire de Physique des Décharges
Ecole Supérieure d'Electricité
Plateau de Moulon - 91190 Gif Sur Yvette
France

[2] Merlin Gerin
Centre de Recherche - A2
38050 Grenoble Cedex
France

INTRODUCTION

In SF_6 circuit breakers the success of current interruption depends on the capability of the arcing region to withstand the applied recovery voltage during the dielectric post-arc phase. In fact in this region, where the arc burnt, a hot low density channel remains. The recovery of such a gaseous channel to its initial dielectric strength can not be immediate, but has to pass through successive steps related to different chemical activities and recombination processes. The study of the breaking process is difficult due to the complexity of the physical phenomena taking place inside the breaker chamber and it has to be split into various phases. We chose to focus our attention on the dielectric recovery phase following arc extinction during the application of the transient recovery voltage (TRV). Indeed, this phase has already been studied elsewhere under various different conditions. For instance, an industrial unit has been investigated[1,2,3] using gas flow, to cool the arc column. Elsewhere an experimental simulation of the interruption phenomena has been set up by means of a rotating arc between fixed contacts[4]. In this case however the metallic vapour contamination of the gas can not be avoided. In another approach[5] the measurement of the dielectric withstand of SF_6 exposed to high temperatures, below the dissociation limit of the molecule (of about of 2000 K), is studied. The comparison with the recovery performances of other insulating gases such as air[6], nitrogen[7], argon and hydrogen[8] shows a clear advantage for SF_6 which presents excellent quenching and recovery properties. This justifies the choice of SF_6 as a good industrial interrupting gas.

For the experiment presented in this paper, the power arc phase, which appears during contact separation in a breaker, is simulated by a high current electrical arc running between two rails. The dielectric strength of the recovering gas is evaluated by means of the application of a high voltage step impulse to a test gap, after a forced arc current interruption close to its crest value and not, as is the case in other approaches, after a natural zero of the current. The main objective here is to separate out the contribution on the recovery performances of the gas itself from various other parameters which are present in a real breaker, such as to gas pollution by metallic vapour formation, influences due to the stagnation of arc roots (hot points), and finally due to gaseous flow. Consequently, only the phenomena related to the gas chemistry and the effect of a reduced density hot channel at the place of the arc column are taken into account.

Gaseous Dielectrics VII, Edited by L.G. Christophorou
and D.R. James, Plenum Press, New York, 1994

This paper presents the recovery voltage characteristics of SF_6 and SF_6 mixtures with gaseous additives such as air, nitrogen and oxygen. Influences on the SF_6 recovery performances of various parameters, such as arc current value, gas contamination by water vapour and also test gap geometric parameters have been investigated. The recovery curves show distinct phases and rates. Using these data an estimation of the mean temperature of the recovering gas is obtained. The statistics of the post-arc time to breakdown t_b as a function of the gas nature is also presented.

EXPERIMENTAL SET-UP

An idealised model breaker experiment with two independent electrical circuits (figure 1a) has been designed. The first circuit generates the high current electrical arc, which runs between two divergent runners inducing a lengthening of the arc during its development. The arc simulates the breaking arc, dissociating and exciting the gas molecules. The second circuit delivers a high voltage impulse which will be applied to a test point-to-plane gap, in order to evaluate the SF_6 dielectric strength, after the arc current interruption.

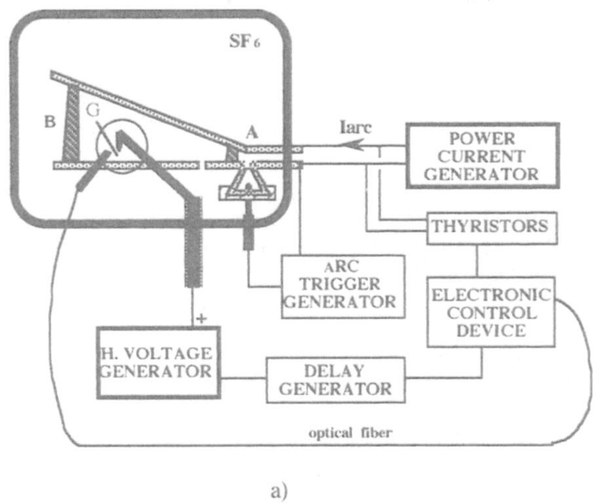

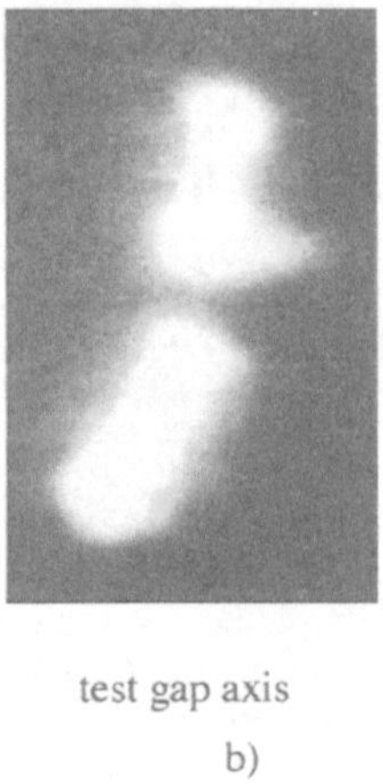

Figure 1 a) Experimental set-up. b) Pictures of an electrical arc of 880 A propagating along the runners, obtained respectively at 5.5 ms (upper picture) and 6 ms (lower picture) after the arc ignition.

The electrical arc is generated inside the experimental vessel by the discharge of a capacitor bank, previously charged to voltages up to 500 V. The arc current is a sine wave, with a frequency of 42 Hz, able to reach a peak value of 900 A. The arc is initiated at position (A on figure 1a) well removed from the test gap (G), by a plasma-jet system which generates a conducting channel between the metallic runners. This approach avoids arc stagnation and consequent gross metal vapour production. This arc, under the combined action of its self induced magnetic field and that of an external transverse field generated by strong magnets (0.1 T), propagates rapidly towards the end of the runners (B), crossing the test gap region. A study of the arc dynamic[9] shows that the arc is not filamentary but diffuse. Indeed, Figure 1b presents still photographs of a 860 A arc, using neutral density filters, obtained in the frame mode with a high speed STL electronic camera, the picture being in the plane defined by the runners. As seen, at t=5.8 ms corresponding to the current crest value, an apparent column width of 15 mm was measured while the arc column is practically vertical. The recovery tests are performed after the interruption of the arc, close to its current crest value, and it was found that under such conditions the test region is fully bathed by the hot arc plasma.

The arrival of the arc column at the test gap is detected by an optical fiber which commands an electronic control device. This device switches the arc current towards a pair of thyristor, at a time t_0 used as a zero temporal reference, and triggers a high voltage generator with a variable delay Δt from t_0. Full current commutation requires about 9 μs, which gives a value of di/dt of about 100 A/μs. This rate corresponds to the slope, at current zero, of a 220 kA 50 Hz current. The positive high voltage impulse (30 ns / 20 ms) is applied to the point to test the dielectric strength of the arced SF_6 gas and to evaluate its recovery characteristics. The electrode arrangement is composed of a copper point of small radius of curvature (100 μm) and a plane, made of a copper-tungsten, with a diameter of 10 cm. The absolute SF_6 pressure is one bar. The arc current is measured by a magnetic current transformer while the high voltage pulse is recorded through resistive voltage divider.

SF_6 RECOVERY CHARACTERISTICS

The high voltage "step impulse" is applied to the test point after a given delay Δt (or recovery time) with respect to arc extinction. Figure 2 shows two typical recovery tests, with current commutation at its crest value of 860 A, for a test gap distance of 6 mm and a delay of 500 μs. On the first example (figure 2a) the recovering gas withstands the applied voltage, while on the second and for the same arc current (figure 2b, expanded on 2c) breakdown takes place with a formative time t_b of 180 ns. This time is defined as the time between the high voltage front and voltage collapse.

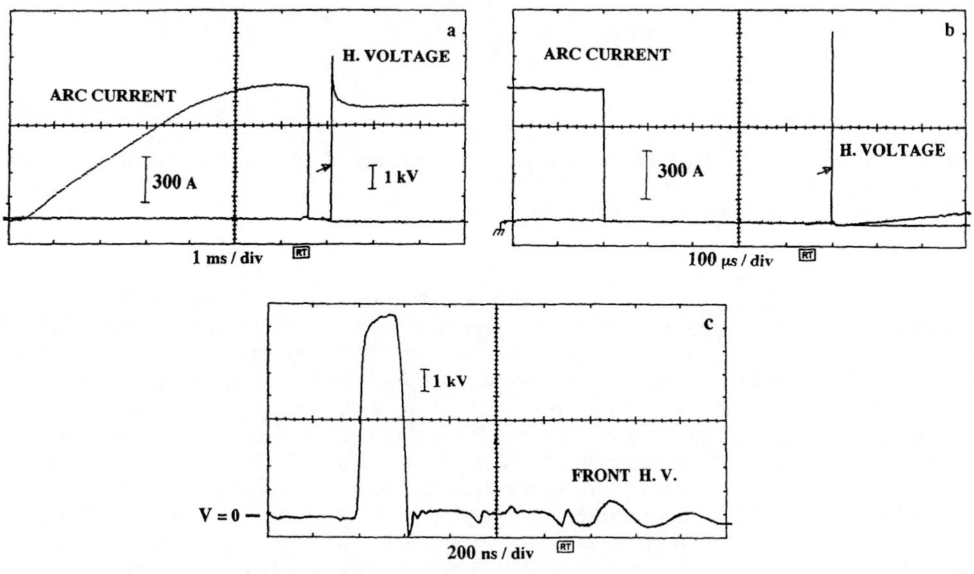

Figure 2 Typical results of dielectric recovery tests performed 500 μs after the interruption of 860 A arc current : (a) Dielectric withstand - (b), (c) Dielectric post-arc breakdown, with a formative time of 180 ns.

Figure 3 presents a typical recovery result, obtained for a 6 mm point-plane gap and an arc current crest value of 860 A ($\pm$ 20 A). The minimum breakdown voltage V_{bm} (dotted line) is plotted as a function of the delay Δt. The dielectric recovery characteristics will be referred to this lowest breakdown value V_{bm}, which determines the limit for practical industrial applications. In order to define a good reference potential, the minimum breakdown value V_{bf} has been measured in the absence of a power arc, i.e. for a non arced cooled gas. For the above described experimental conditions this value was 28.3 kV, at room temperature. The reduced (or normalised) minimum breakdown voltage V_{bm}/V_{bf} is an expression of the degree of gas recovery during the post-arc phase. Figure 3 presents also, in %, the ratio V_{bm}/V_{bf} (as continuous line) versus time. According to this data, the recovery rate is of the order of 2.75 kV/ms during the last phase, so that it predicts a complete gas recovery time of about 6.75 ms. This large value of the recovery time, is in agreement with the assumption that the recovery is governed by the arc column cooling directly related to the axial heat flow towards the electrodes and the radial heat flow towards the surrounding cold gas. It should be stressed, that the recovery curve features great similarity with experimental curves previously found for the SF_6[2,3,4] and features the same general tendency as measured in other insulating gases recovering more slowly[6,7,10].

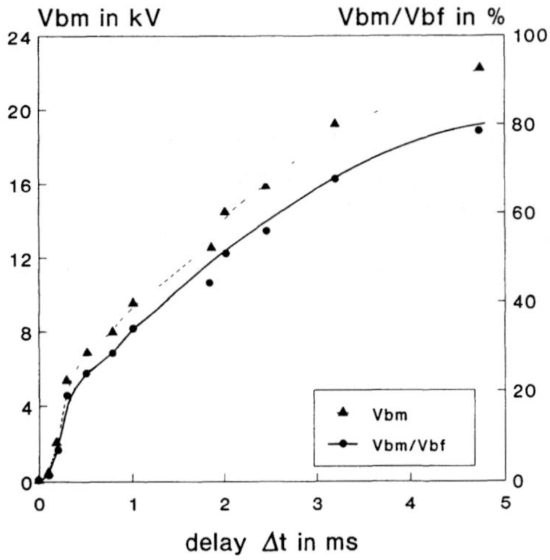

Figure 3 Recovery results for a 6 mm test gap and a 860 A arc. Recovery curve V_{bm} (dotted line) and ratio V_{bm}/V_{bf} (continuous line) versus Δt.

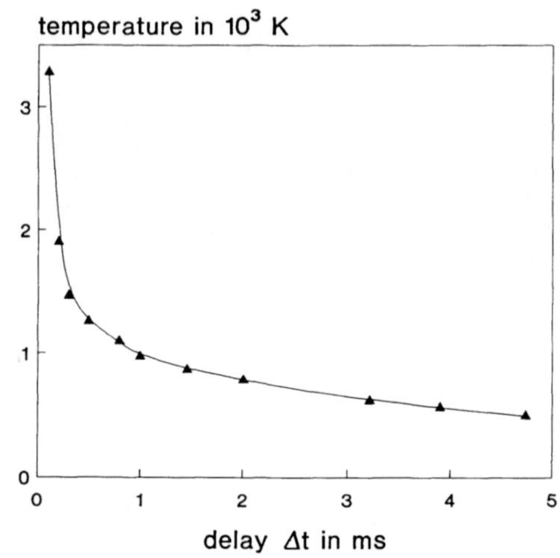

Figure 4 Gas mean temperature decay during recovery phase, versus delay Δt.

Just after arc extinction, the SF_6 breakdown strength can not be explained in terms of gas density decrease alone because the SF_6 remains strongly dissociated. Theoretical[2,11] work of the recombination processes of the decomposed SF_6 suggests that after about 100 μs the value of the density of SF_6 determines the strength of the mixture. More precisely, evidence[5] that the temperature affects the SF_6 dielectric strength only above 2000 K, limit for which strong SF_6 dissociation occurs, has been obtained, while below this limit, the hot gas presents a dielectric strength proportional to the gas density. The breakdown process is triggered as soon as the conditions for streamer formation are fulfilled. If one supposes that the dielectric strength of the recovering hot plasma is proportional to gas density, this density may be estimated from the ratio V_{bm}/V_{bf}. From the experimental data of figure 3 and using the SF_6 gas properties at high temperatures[12] relating the gas density to its temperature, an average value of the gas channel temperature may be calculated. Figure 4 shows the evolution of the mean value of the plasma temperature as a function of the time following arc extinction. One may notice that initially the decay is fast so that the temperature falls under 2000 K after 200 μs. Actually, as far as dielectric recovery is concerned, this initial period is critical and will be the period where our future investigation will be developed.

Figure 5 shows the recovery characteristics of a 6 mm point-plane gap for different arc currents. It is important to note that the propagation velocity depends on the current crest value. The instant of arc interruption being governed by its physical arrival at the test gap, it is therefore the arc current value at this moment which is reported in the figures, and not its crest value. The curves obtained for arc current values (at interruption) of 430, 560 and 860 A, show that the hot channel related to the smaller arc current recovers fastest.

For all investigated conditions, the same general recovery tendency is observed. In particular, three consecutive recovery phases may be identified, according to a slow-fast-slow sequence of recovery rates. In particular, the recovery tests performed after the interruption of a 860 A arc with a 6 mm gap, and 0.1 mm point radius, show that during the first slow phase ($\Delta t<100$ μs) the recovery rate is 4 V/μs while during the second phase ($100<\Delta t<300$ μs) it is of the order of 23 V/μs and during the third phase ($\Delta t<1000$ μs) it is some 6.5 V/μs. The same general recovery shapes are obtained for different test gap distances, so that the recovery rates are not far from those given for 6 mm.

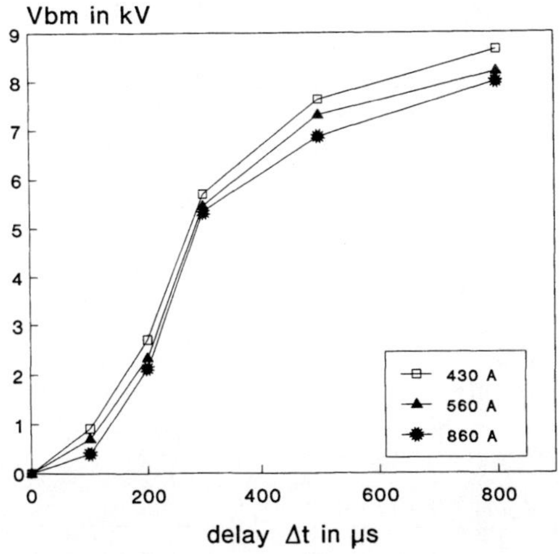

Figure 5 SF_6 recovery curves V_{bm} as a function of the interrupted current value and the delay Δt (gap distance 6 mm).

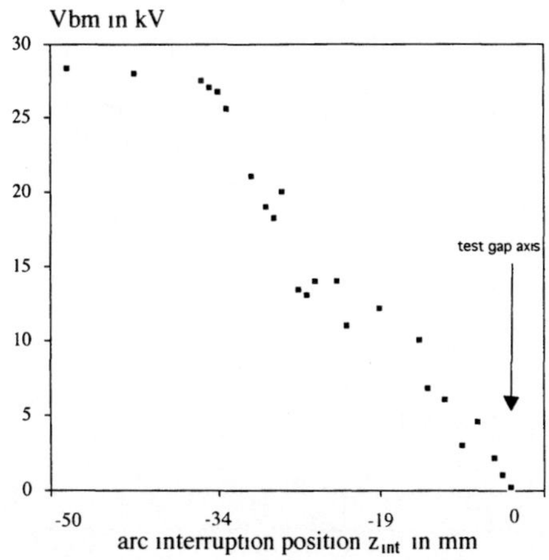

Figure 6 Minimum breakdown voltage V_{bm} for a 6 mm gap, when the arc is interrupted in various positions before the test gap axis.

Clearly, a power circuit breaker would not interrupt correctly if it depended uniquely on the natural recovery characteristics of SF_6, even in absence of gross pollution due to metal vapour or electrode hot-spots. The remarkable ability of rotating arc / auto expansion breakers and puffer breakers, to improve the recovery rate during the initial 100 μs by at least a factor of 20 is hence highlighted.

Interrupting the arc, before its physical arrival at the test gap region, allows the investigation of the possible influence, on the test gap dielectric strength, of the distant arc plasma. The minimum breakdown voltage V_{bm} of a 6 mm gap is measured for arc interruption at various positions ahead of the test gap axis. The arc current is set so that its maximum value should be 860 A. The value of the position is not precisely known, but the time t_{int} at which the arc is interrupted can be easily obtained from the oscilloscope measurements, t_{int} being less than 5.8 ms, at which time the moving arc is known to reach the test gap. From the arc kinematics study[9], it is possible to derive the arc position z_{int} in space knowing the time t_{int} of its interruption. Figure 6 presents the results obtained plotting the value V_{bm} as a function of z_{int}. A progressive fall of dielectric strength with increasing z_{int} is registered, i.e. when the arc approaches the test gap. So, at 34 mm before test gap axis the breakdown voltage is almost equal to the cold condition value of 28.3 kV, while at 19 mm before test gap axis this value is reduced to 12 kV. For t_{arc}=5.8 ms the arc touches the test gap. This behaviour can be explained by assuming that the initially running arc puts the surrounding gas in movement. When the driving force is suddenly removed, at arc extinction, the gas continues to move forward, cooling and recombining as it goes towards the test gap. The further ahead of the test gap the arc is extinguished, the greater the degree of recovery of the gas is and the less the dielectric strength is influenced on eventual arrival at the gap.

Effect of Humidity on the SF_6 Recovery Characteristics

The contamination of the SF_6 due to water vapour inside a G.I.S. may lead to a degradation of the insulating performance. Even if the initial moisture level, at the time of equipment installation, is low, it is known that moisture may increase later on due to the desorption of various components. Humidity levels as high as 2000 ppm_V have been measured inside actual G.I.S. under operation[13]. However, during the breaking high current phase, the instantaneous value of the humidity concentration in the circuit breaker is not known, because a transient increase of the humidity may appear due to local temperature increment. In order to investigate the influence of water vapour on the post-arc recovery performances, dielectric recovery tests have been performed at various humidity concentrations. The results presented so far were performed with commercial quality "dry" SF_6, for which the concentration of water vapour is of the order of 270 ppm_V. In these experiments the water vapour concentration $[H_2O]$ ranges from 800 to 9800 ppm_V, the concentration being measured by a General Eastern Hygrocomputer 1500 probe. Three different delay Δt values of 100, 200 and 300 μs after the interruption of an arc current of 860 A are investigated in a test gap of 6 mm.

The measurements of the minimal recovery breakdown voltage for wet conditions V_{bmw}, shows an important reduction of the recovery capability compared to the corresponding value for dry conditions V_{bmd} ($=V_{bm}$). Figure 7 summarises the results obtained, presenting the % loss of recovery strength for the wet SF_6, expressed as the ratio $R_H=(V_{bmd}-V_{bmw})/V_{bmd}$, against the humidity concentration in ppm_V. As seen, performances degrades for wetter conditions. Furthermore, the influence of the humidity is more pronounced for the smaller delay value. For only 1000 ppm_V we note a performance loss of about 10%.

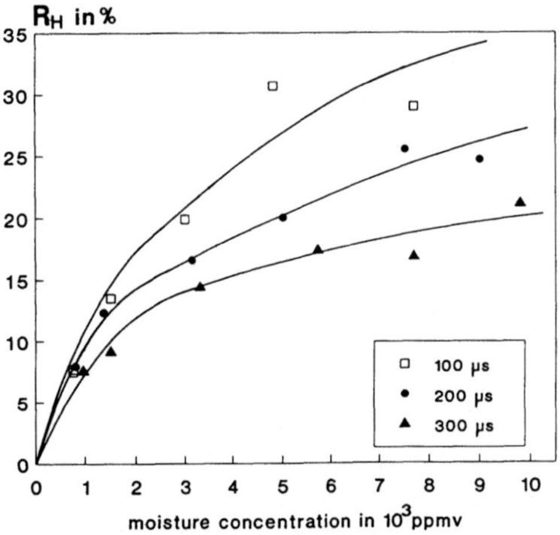

Figure 7 Evolution of the ratio R_H as a function of the water vapour contamination and the delay Δt (6 mm gap distance and 860 A arc current).

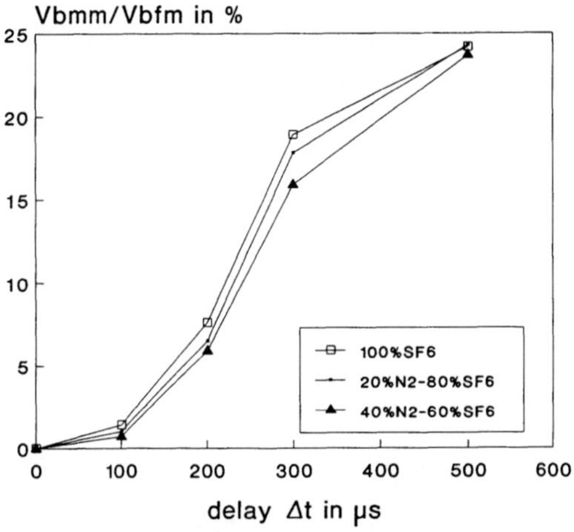

Figure 8 Normalised recovery curves V_{bmm}/V_{bfm} versus delay Δt, for SF_6-N_2 mixtures (6 mm gap, 860 A arc current).

Effect of Gaseous Additives on the SF_6 Recovery Characteristics

Recovery strength measurements for gases other than the SF_6 have already been performed by other researchers. In particular, the recovery characteristics of nitrogen[7], air[6], argon[14] or hydrogen[8] have been studied, as a function of the time following the extinction of a power arc. Our experimental results show that at atmospheric pressure the SF_6 recovers faster than these gases. One of the aims of our experimental work is also to investigate the recovery behaviour of SF_6 gaseous mixtures, in order to define the influence of gaseous additives to SF_6 recovery performances. The concentration of each additive is calculated in percent per volume. The gaseous additives, used for this experimental study, were atmospheric air, oxygen and nitrogen. As previously, the minimal dielectric breakdown voltage of each mixture V_{bmm}, during recovery, is measured as a function of the delay Δt. On the other hand the reference minimal breakdown voltage V_{bfm} is measured at room temperature for each mixture. Then, using the same analysis as in the case of humidity mixtures, it is possible to estimate the dielectric strength reduction, due to the additive, in comparison to the SF_6 value. This reduction is expressed as $R_m = (V_{bf} - V_{bfm})/V_{bf}$ for cold conditions or $R_{bm} = (V_{bm} - V_{bmm})/V_{bm}$ for post-arc conditions. For each mixture the recovery level V_{bmm}/V_{bfm} is also measured versus the recovery time.

The recovery curves of mixtures show distinct phases as for the case of pure SF_6. For SF_6-air and SF_6-O_2 mixtures the recovery levels at a given Δt seems to be only slightly affected by the gaseous additive concentration. In the case of SF_6-N_2 mixtures, the differences are more pronounced. As seen on figure 8, while the same recovery tendency is observed, the increment of the nitrogen concentration induces a more evident reduction of the recovery level at a given Δt. On the other hand, the results show that a general feature is that the presence of a gaseous additive degrades the recovery performances (figure 9). The degradation depends on the gaseous additive nature, its concentration and also on the value of the recovery time. In particular, the worst recovery performances are obtained for a mixture with strong (40%) nitrogen concentration, specially during the first 200 microseconds. Since the 50% air mixture, as well as the 25% air mixture, gives about the same behaviour as the 40% nitrogen mixture and 20% mixture respectively, it seems quite clear that degradation induced by the air-SF_6 mixture to the SF_6 properties are mainly due to the nitrogen component and not to the oxygen.

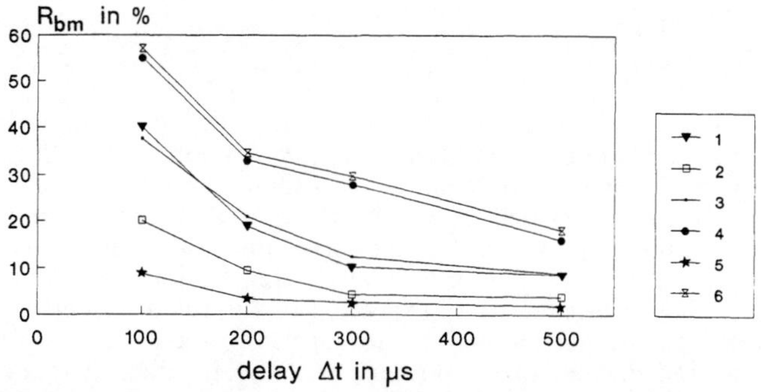

Figure 9 Evolution of the ratio R_{bm} as a function of the delay Δt and the gaseous additive concentration. : 1 - 25% air, 2 - 10% ai'r, 3 - 20% N_2, 4 - 50% air, 5 - 10% O_2, 6 - 40% N_2 (6 mm gap and 860 A arc current).

Measurements of the Breakdown Formative Time t_b in the Test Gap.

The breakdown formative time t_b is measured for all the gaseous conditions investigated for various value of V_H applied to the test gap. For cold conditions in dry SF_6, the time t_b, varies inversely to the applied V_H values (and consequently to the breakdown probability). This general behaviour is also observed during recovery measurements in dry, arc exposed, SF_6 and its mixtures. However, in the case of wet SF_6 the formative t_b time do not present a distribution with clear correlation with voltage or water vapour concentration. Now, using $V_H = V_{bm}$, and fixing a delay of 100 μs, the time t_b becomes comparable to the impulse rise time (of the order of 30 ns). For greater delays, t_b increases to values which depends on the gas composition. In fact, referring to the value of t_b corresponding to V_{bm} values, figure 10 summarises the results obtained showing the maximum, mean and minimum t_b measurements for any delay greater than 100 μs (200 μs, 300 μs....). According to these results, a rapid breakdown is observed in the presence of humidity, effect which may be due to the easy presence of initiatory electrons, triggering streamer formation which, in turn, leads to a quick breakdown, or to the creation of unstable oxyfluoride ions or metastables.

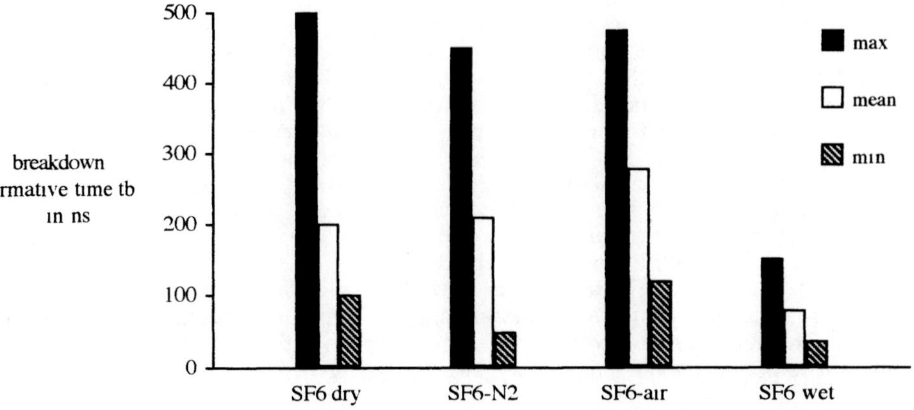

Figure 10 Maximum, mean and minimum measured values of breakdown formative time t_b, for different gaseous conditions.

CONCLUSIONS

The SF_6 recovery characteristics after submission to an electrical power arc, simulating a power arc breaker has been studied. Using a sweeping arc allows to separate the effect on recovery of the gas itself from metal vapour contamination due to hot point stagnation, or arc cooling by a strong forced gas flow. The SF_6 recovery performances are influenced by the gap spacing, the arc current value, the test point radius and by gaseous additives, especially nitrogen. The main parameter which seems, after some 100 microseconds, to explain the recovery is the value of the neutral density, reduced by the temperature increase in the gas. Remains the question of the recovery during the first temporal steps after the current interruption, where the physics of recovery is not well understood. Another important result is that contamination by water vapour leads to a large reduction of the gas recovery capability, related to short time to breakdown t_b values. Experiments are in progress to characterise the gas recovery, not only by the breakdown potential of the test gap, but also by the complete analysis of the test breakdown formation. Knowledge of the electrical and optical characteristics of this streamer breakdown formation may lead to a better understanding.

REFERENCES

1. E. Schade, K. Ragaller, "Dielectric recovery period of axially blown SF_6-switchgear arcs", 7th International Conference on Gas Discharges and their Applications, London, pp. 100-103, 1982.
2. E. Schade, "Similarity of the dielectric recovery characteristic of axially blown arcs in SF_6", 8th Intrnational Conference on Gas Discharges and their Applications, Oxford, pp. 50-53, 1985.
3. H. Schoetzau, H. Meili, E. Fischer, Ch Sturzenegger, H. Graf, "Dielectric phase in an SF_6 model breaker", IEEE Transactions on Power Apparatus and Systems, vol. 104, n° 7, pp. 1897-1902, July 1985.

4. D. Birtwhistle, A. Stokes, "Interruption of low currents by short arcs in SF_6", 10th International Conference on Gas Discharges and their Applications, Swansea, pp. 226-229, 1992.
5. B. Elliason, E. Schade, "Electrical breakdown of SF_6 at high temperatures (<2300K)", 13th International Conference on Phenomena in Ionised Gases, Berlin, pp. 409-410, 1977.
6. K. Tsuruta, A. Kato, A. Sato, G. Watanabe, "Dielectric voltage recovery characteristics of small air gaps for a variety of electrode geometries", 8th International Sumposium on High Voltage Engeneering, paper 37.01, pp. 433-436, 1993.
7. G. Farral, J. Cobine, "Recovery strength measurements in arcs from atmospheric pressure to vacuum", IEEE Transactions on Power Apparatus and Systems, vol. 86, n° 8, pp. 927-932, 1967.
8. S. Moran, L. Rinehart, "Voltage recovery time of small spark gaps", IEEE Transactions on Plasma Science, vol. 10, n° 4, pp. 277-280, 1982.
9. I. Spiliopoulos, G. Berger, E. Marode, S. W. Rowe, G. Bernard, "Dielectric recovery characteristics of SF_6 after the extinction of an electrical power arc", 8th International Sumposium on High Voltage Engeneering, paper 31.05, pp. 295-298, 1993.
10. K. Tsuruta, I. Takahashi, Y. Kanzaki, T. Ojima, Y. Komuro, "Experimental study of the voltage recovery characteristics of small air gaps", IEEE Transactions on Plasma Science, vol. 17, n° 3, pp. 560-564, 1989.
11. E. Schade, K. Ragaller, W. Egli, K. Brand, L. Niemeyer, "Dielectric recovery of axially blown SF_6 arc after current zero", IEEE Transactions on Plasma Science, vol. 10, pp. 141-172, 1982.
12. L. Frost, R. Liebermann, "Composition and transport properties of SF_6", Proceedings IEEE, vol. 59, n° 4, pp. 474-485, 1971.
13. Y. Doin, S. W. Rowe, J. P. Hours, G. Riquel, "The control of humidity in G.I.S. Quality control aspect", 7th International Sumposium on High Voltage Engeneering, paper 33.04, pp. 125-128, 1991.
14. J. Perkins, A. Parker, "Mechanism of dielectric recovery in uniform and nonuniform field gaps after high-current high-pressure spark discharges", Proceedings IEE, vol. 117, n° 7, pp. 1460-1466, 1970.

DISCUSSION

L. G. CHRISTOPHOROU: It seems that water plays an important role in your studies. What do you see its role to be?

E. MARODE: Figure 10 shows that the breakdown formative time is clearly reduced with the presence of water. Actually in dry SF_6, the sequence leading to a breakdown begins with some streamers whose space charge prevents, for a while, the spark to be formed, while in wet SF_6, there is a continuous sequence leading to breakdown. The absence of the streamer prevented section in wet air may explain why the minimum breakdown potential is lower in wet SF_6 . Now, the reasons which explain this water effect are still not known. Is H_2O easier to photoionise? Does H_2O change the surface electrical properties? (See the paper by Lalmas et al., page 617, which shows how sensitive the surface condition on breakdown ability may be.)

E. ILLENBERGER: Can you comment on the present knowledge concerning the relation between recovering time and gas composition on the molecular scale?

E. MARODE: This knowledge remains poor and more research should be devoted to the first 500 μs after the arc interruption. During that period the detachment of negative ions, formed within the decaying arc plasma, should be one of the main parameters which governs the insulation recovery. The fast initial recovery rise should be related to the plasma recombination. It is probably not the only parameter since transport coefficients as well as source terms due to ionizing collisions and photoionising radiation induced by the remaining or detached electrons are all very sensitive to the nature of the molecules. After 500 μs, the recovery, as is said in the paper, should be related mainly to the hot cloud left by the initial arc, which cools down inducing the insulation recovery.

INSTABILITIES OF DISCHARGES AND THEIR APPLICATION FOR OPENING SWITCHES IN INDUCTIVE ENERGY STORAGE SYSTEMS

M. Junker, W. Pfeiffer

Technische Hochschule Darmstadt
Institut für Hochspannungs- und Messtechnik
Landgraf-Georg-Strasse 4
64283 Darmstadt
Germany

ABSTRACT

In pulse power applications, where high energies have to be commutated from the energy storage system to the load in a very short time interval, there is a great interest in using inductive energy storage systems. The basic condition for use of inductive energy storage systems is the availability of an opening switch.

During the last few years different concepts were persued to realize such an opening switch. For that purpose diffuse discharges have been a very promising candidate because of the reasonable feasibility to influence the conductivity of the discharge volume by external sources. However diffuse discharges typically have a rather low current density.

For that reason another concept has been recently persued to overcome these limitations. This new concept is based on instabilities of the high current discharge channel which may occur in special attacher gases.

Our investigations have been focused on the variation of the most important parameters influencing the development of this special kind of discharge. The aim is to find out the interrelations between the influencing parameters and the extinction of the discharge current, so that an external control of the opening switch will be possible.

INTRODUCTION

For high voltage switching in pulse power applications spark gaps are often used as switches. The following characteristics are the reason:

1) easy to realize:
   Compared to other kind of switches it is rather easy to build a spark gap.

2) easy to match the dielectric strength to the operating voltage:

Gaseous Dielectrics VII, Edited by L.G. Christophorou
and D.R. James, Plenum Press, New York, 1994

The required dielectric strength can be achieved by suitable choice of gas pressure and electrode spacing.

3) very fast switching (closing switch):
   Very short switching times are possible because the breakdown occurs within a view nanoseconds.

There are several possibilities to trigger a spark gap with very short delay time. For example it is possible to trigger the spark gap by a short voltage impulse at a trigger electrode. Also optical triggering is possible by ionisation of the used gas with an UV-laser system.

Recently inductive energy storage systems have attracted considerable interest because of their high energy density and the very high lifetime. In figure 1 the basic function of such an energy storage system is demonstrated.

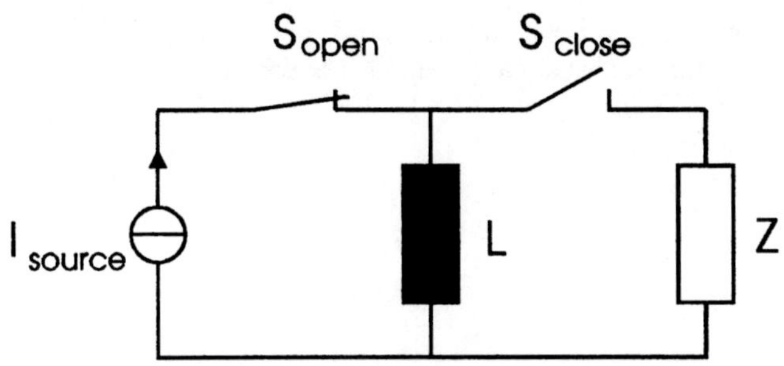

Figure 1. Equivalent circuit of an inductive energy storage system.

It is obvious, that besides of a closing switch, an opening switch is necessary in order to commutate the current flow through the energy storage inductor from the current source to the load. So different concepts are pursued to realize an opening switch.

In this paper some results with a commutating switch based on instabilities of the high current discharge channel are presented.

EXPERIMENTAL SETUP

The experimental setup being used (figure 2) consists of the following components:

1) Pulse generator
   The high voltage pulse generator produces square wave pulses with an adjustable amplitude of up to 250 kV. The generated pulses are characterized by a pulse duration of 400 ns and a rise time of 3 ns.

2) The test gap
   To investigate the commutating behaviour a coaxial construction with three electrodes was chosen.

3) The load resistors
   The circuitry consists of two loads, each realized by a high-voltage high-power resistor. The discharge current is commutated between the two loads R_1 and R_2.

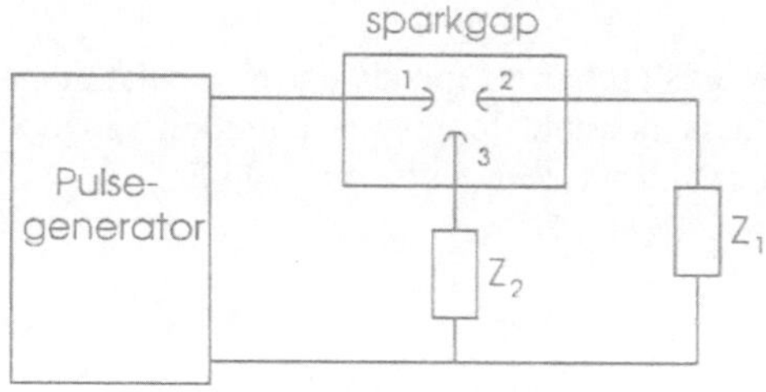

Figure 2. Experimental setup.

The gap between electrode 1 and electrode 2 breaks down induced by the voltage pulse delivered by the generator. The result is a fast rising current in the first load circuit (R_1). Because of instabilities in the high current spark channel a second discharge channel between electrode 1 and electrode 3 can be formed. The development of this current may lead to an extinction of the first spark channel. The current flow in both load circuits as well as the voltage drop across R_1 and R_2 are measured.

EXPERIMENTAL BASIS

The investigations described here are based on instabilities in gas discharges, which may lead to both, the formation of a second spark channel and the extinction of the first spark channel. To reach this characteristic no external control mechanism is required.

The special kind of discharge can be obtained in the following electronegative gases (attacher gases): N_2O, CF_4, $CClF_3$, CCl_2F_2 and SF_6. Investigations with neutral gases have shown that N_2 to some extent also shows this behaviour. Two essential requirements have to be fulfilled to realize the special kind of discharge:

1) The overvoltage

The overvoltage marks the ratio between the applied impulse voltage amplitude and the DC breakdown voltage. Only when the overvoltage exceeds a certain limit being characteristic for the used insulating gas, the special kind of discharge can be obtained. Figure 3 shows the required overvoltage in order to reach the commutation for different gases.

The maximum overvoltage is only limited by the dielectric strength of the switch chamber

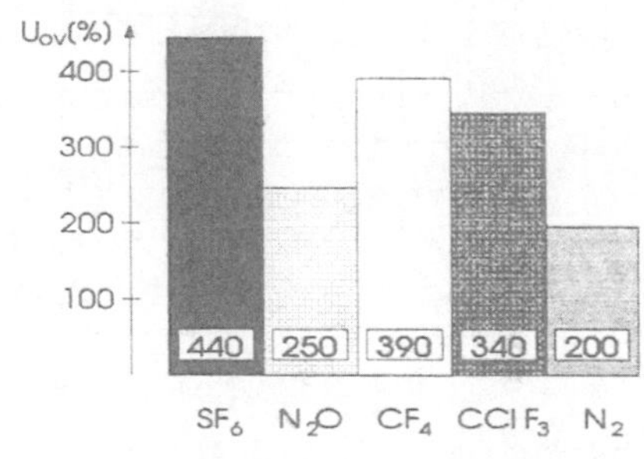

Figure 3. Required overvoltage for the investigated gases.

2) The gas pressure

Although the required gas pressure depends on the insulating gas used it is only a very weak effect. Therefore, it is possible to give a common range of gas pressure for all investigated gases, i.e. the range between 5 kPa and 20 kPa.

INVESTIGATIONS

The investigations were concentrated on the evaluation of a mechanism to obtain an external control of the discharge development. Consequently the most important parameters which affect the discharge characteristics were varied. Thereby, the following parameters were investigated:

1) The impedance of the second load resistor
2) The rate of overvoltage

EXPERIMENTAL RESULTS

1) The second load resistor R_2 is varied to investigate the influence to the discharge process. In this contribution measurements with a 50 Ω and a 5 Ω resistor are compared.

Figure 4 shows the current course in the load circuit for a second load resistor (R_2) of 50 Ω and 5 Ω respectively. It is obvious that the use of a second load resistor of 50 Ω has two essential disadvantages:

a) $R_2 = 50\ \Omega$

b) $R_2 = 5\ \Omega$

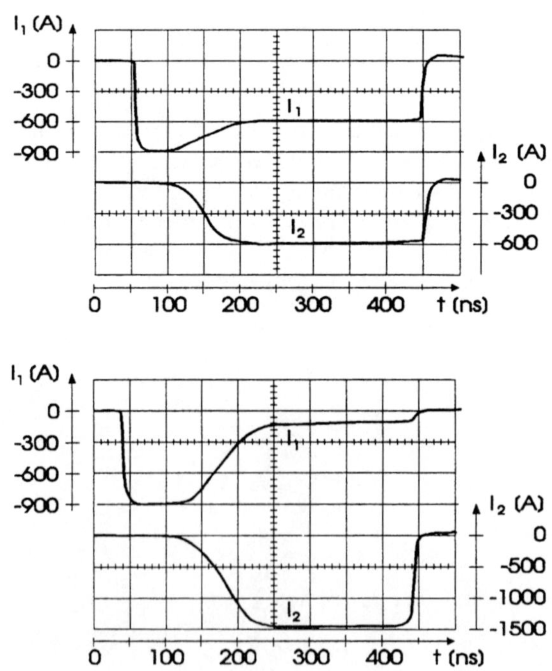

Figure 4. Current course in both load circuits a) $R_2 = 50\ \Omega$ b) $R_2 = 5\ \Omega$. The testvoltage is 70 kV ($U_{ov} = 410\%$). The insulating gas is N_2.

a) The very small decrease of the discharge current I_1 through the first load circuit.
b) The relative small amplitude of the second load current I_2 (590 A).

In contrast to this behaviour, a second load resistor of 5 Ω results in a strong decrease of the first load current. Besides, the amplitue of the second load current amounts 1470 A. It is obvious, that an effective commutation of the discharge current requires for a small second load resistance.

2) Variation of the overvoltage rate
The investigations show that the beginning of the current commutation depends strongly on the applied overvoltage. Hence, this effect can easily be used to control the moment of commutation externally. The current in both, the first (I_1) and the second (I_2) load circuit is shown in figure 5.

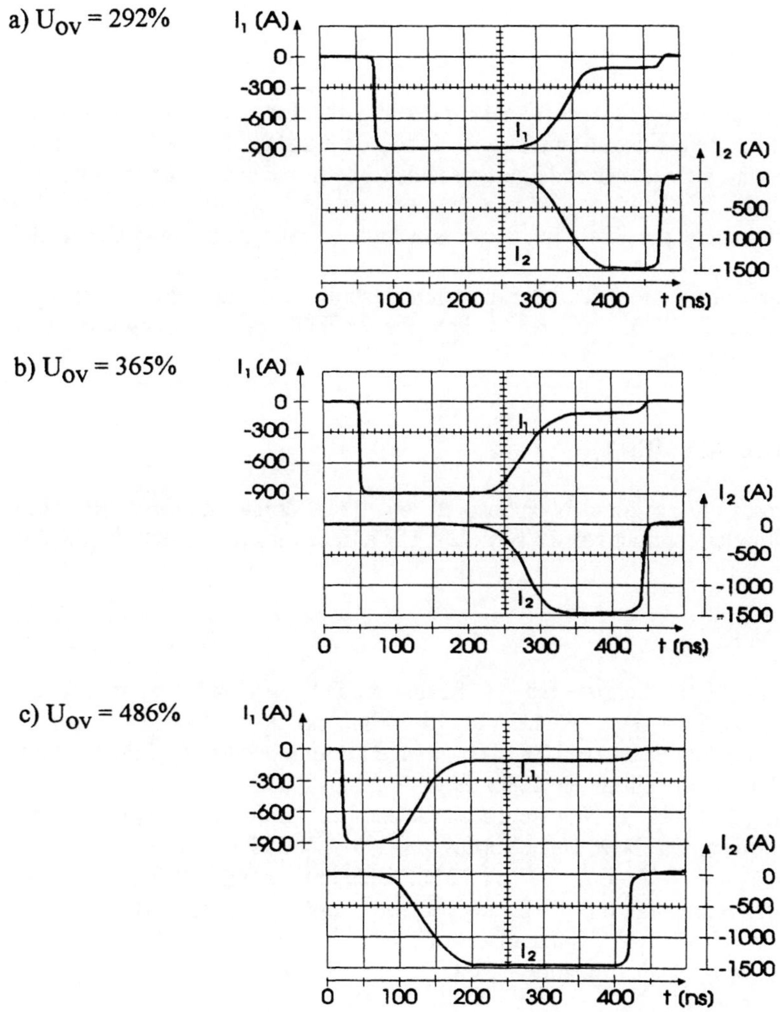

Figure 5. Development of the discharge current in the first and the second load circuit for different overvoltages. The test voltage is 70 kV. The used insulating gas is N_2.

An overvoltage of 292 % corresponds to a current flow during 275 ns in the first load circuit (Fig. 5a). An increase of the overvoltage reduces the duration of the current flow (Fig. 6).

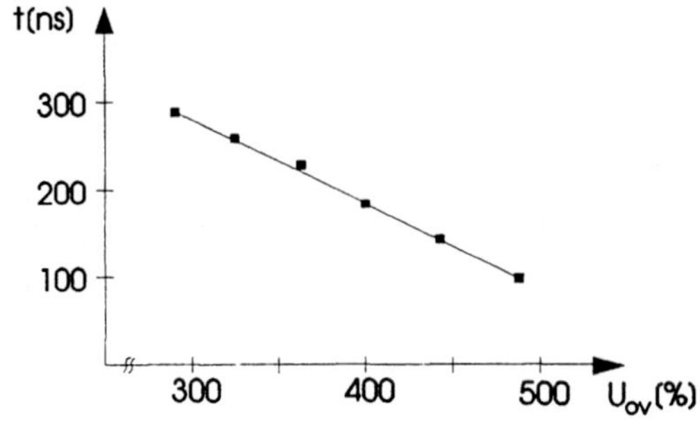

Figure 6. Duration of the current flow (I_1) versus the applied overvoltage

CONCLUSION

The special discharge type seems to be suitable for application in a commutating switch due to its self extinguishing behaviour based on instabilities in the high current spark channel. So with a three electrode gap it seems to be possible to combine a closing and an opening switch.

An additionally possible application is a high voltage pulse generator with variable pulse duration.

The next step is to use the experimental setup with an inductive energy storage element. Further investigations have to show the feasibility of this arrangement in inductive energy storage systems.

ACKNOWLEDGEMENT

This research program is sponsored by the German ministry of research and technology (under contract No. 13 N 5992). The authors are responsible for the contents.

REFERENCES

1. G. Schaefer, K. H. Schoenbach, "A Review of Diffuse Discharge Opening Switches", IEEE Trans. On Plasma Science Vol. PS-14 No., pp. 564, 1986
2. H. Shimamori, R.W. Fesseenden, "The temperature dependence of electron attachment to N_2O in the gas phase", J. Chem. Phys. 70, Feb. 1979
3. W. Pfeiffer, D. Stolz, "A New approach for an opening switch for repetitive operations", 6th Int. Symp. On Gaseous Dielectrics, pp. 341-347, 1990
4. M. Giesselmann, W. Pfeiffer, J. Wolf, "Short time optical and electrical diagnostics of N_2 and SF_6 Discharges", 6th IEEE Pulsed Power Conference, 182, Arlington/VA, 1987
5. M. Junker, W. Pfeiffer, "Comparison of two different concepts to realize an opening switch with high repetition rate", 9th IEEE Pulsed Power Conference, Albuquerque/NM, 1993

DISCUSSION

L. L. HATFIELD: How can this scheme be used to switch an inductive store onto a load?

W. PFEIFFER: Precondition for a sufficient switching efficiency of the commutating switch is that the current is switched from a higher load to a lower load. However, practical measurements with inductive stores have still to be performed, which will be done in the near future.

INVESTIGATION FOR A SUITABLE SHAPE OF LIGHTNING CONDUCTOR AIR TERMINAL ELECTRODE

Rajiv Sawhney[1] and Ravindra Arora[2]

[1]Faculty of E/M Engineering
CME, PUNE - 411 031
India

[2]Professor in Dept of Elect Engg
IIT Kanpur - 208 016
India

INTRODUCTION

Apart from being an interesting natural phenomenon that serves to maintain a balance in the global electrical charge system, lightning has serious destructive effects on property and life. Any effort to reduce this destruction by even a fraction is worthwhile. 'Lightning Protection' is a vast area of research and plenty of work has been done to study the phenomenon and provide adequate shielding against its ill effects.However, while designing the protection system against lightning, no emphasis is made for the suitable shape and size of the Air Terminals of lightning conductors. Most of the specifications recommend a simple 'rod' shape Air Terminal.

The probability of 'Natural Lightning' striking the objects on ground is a statistical phenomenon. A particular area seen as 'Attractive Area' by a lightning stroke is a matter of design of the lightning protection system, if it exists. In this work effort is made to investigate the effects of shape and size of lightning conductor Air Terminals to provide a greater 'attraction' for lightning to strike. It is taken for granted that by increasing the probability of lightning strike on a particular Air Terminal, it renders other objects in the vicinity safer.

To the best of information available, the idea of determination of 'Lightning Protection Zone' for a given set of lightning conductors by 'Rolling Sphere Method' was first floated by Armstrong et al. in 1968 [1]. Although this concept is widely accepted,its practical validity on a scaled model in the laboratory investigated in this work is not known to have been performed earlier.

Gaseous Dielectrics VII, Edited by L.G. Christophorou and D.R. James, Plenum Press, New York, 1994

BREAKDOWN STRENGTH OF AIR WITH LIGHTNING IMPULSE

When lightning strikes on the ground, it chooses the passage and the object for which electric strength of air under the prevailing conditions is weakest. Breakdown strength of air is a complicated phenomenon which not only depends upon its atmospheric conditions but also upon the field configuration determined by the shape and size of the electrode and the gap distance. Besides, the breakdown strength also depends upon the type of voltage applied and its polarity. For long air gaps where an extremely non-uniform field prevails, the breakdown strength when measured with ac and long duration impulse voltages depends upon the type and extent of stable PD which precede the breakdown.

The phenomenon of breakdown with lightning impulse is different. Because of a short rise time of the impulse as well as its small duration, no stable PD are able to take place in this case. The breakdown is directly accomplished with an instable 'leader' followed by an arc. The breakdown strength in this case depends on the Statistical Time Lag which is a function of the applied voltage. As Statistical Time Lag depends upon the availability of initiatory electrons the breakdown strength not only depends upon the shape, size and irradiated condition of the electrodes but also upon the extent of pre-ionisation available in the gas. This is the reason that Power System installations attract the lightning most.

For a lightning conductor it has been followed in the past that its attractive effect for the lightning to strike is merely a function of the height of the air terminal above ground or above the conducting component of the structure to be protected [2]. It has also been thought that the possibility of lightning strike is higher at sharp point electrodes where the electric field concentration is high to initiate a discharge process at the cathode. In order to study the effect of the air terminal, that is its 'shape', 'area' and the 'field intensity' effects, a small model of electrode system was set up in the laboratory for investigations.

ZONE OF PROTECTION BY ROLLING SPHERE

Having originated from the clouds, among numerous other factors, the probability of a lightning stroke striking the ground or an object on the ground depends upon the extent or the reach of the last stage of the stepped leader discharge. For a given condition of clouds when the gap distance between any ground object and the extent of the last stretch of discharge is minimised, the lightning is able to strike the object. This gap distance is known as the 'striking radius' or 'attractive distance' (r_s). Another important condition for the lightning to strike is that the length of this so called last stretch should be equal to or smaller than the 'length of the return stroke step' (l_{rs}) of the original leader, shown in Figure 1(a).

For the construction of protection zone from lightning, Armstrong and Whitehead [1] suggested to roll a sphere of radius r_s over the objects providing a three dimensional sense of protection, shown in Figure 1(b). A total perspective of the protected zone suggested by Naumann [3] with four lightning conductors installed at equal heights of 'H' is shown in Figure 1(c). The heights x_1, x_2 and x_3 represent the minimum heights of the protective zone at these locations.

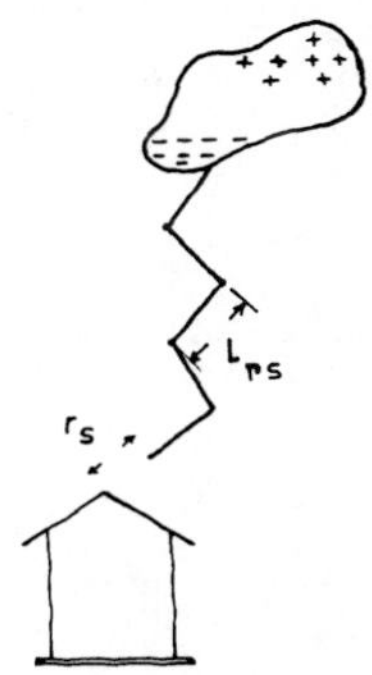

Fig. 1(a). Lightning discharge with Return Stroke.

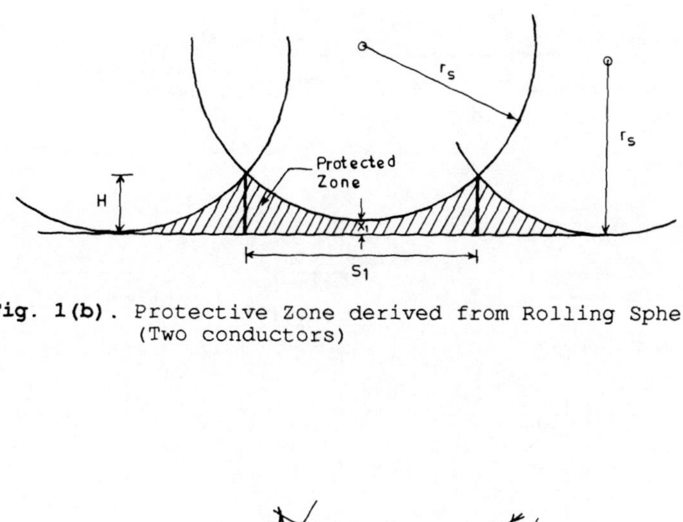

Fig. 1(b). Protective Zone derived from Rolling Sphere (Two conductors)

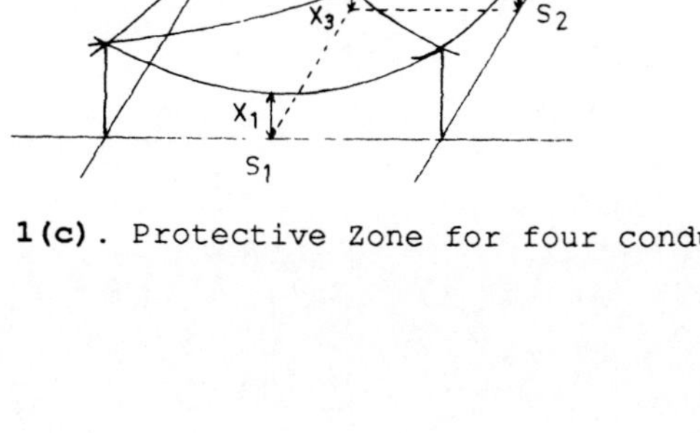

Fig. 1(c). Protective Zone for four conductors.

EXPERIMENTAL SETUP

To simulate a cloud (anode),a large inverted bowl shaped aluminium electrode of circa 75 cm in diameter was suspended from the ceiling with the help of a toughened glass insulator string as shown in Figure 2. The grounded aluminium plate of 75 cm square (cathode) had provision to screw the test electrodes vertically at any desired location and height. Clearance between the inverted bowl (cloud) and the grounded electrodes could be adjusted as per requirement. Figure 3 shows the four shapes of lightning conductor air terminals investigated. Experiments were performed with different combinations of the shape and size of these electrodes.

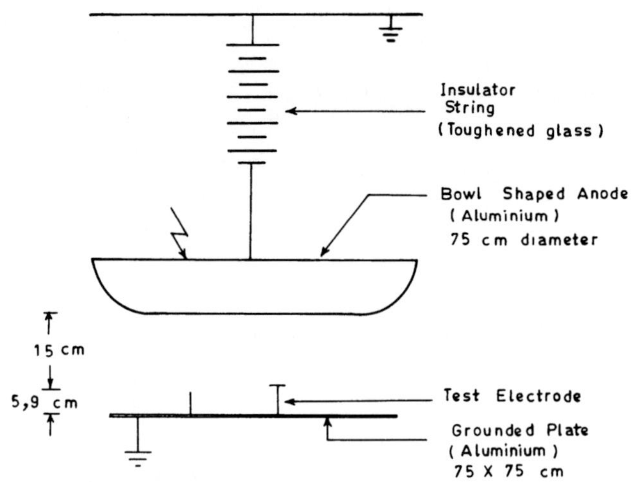

Fig. 2. Set-up for the investigation of the shape of air terminal.

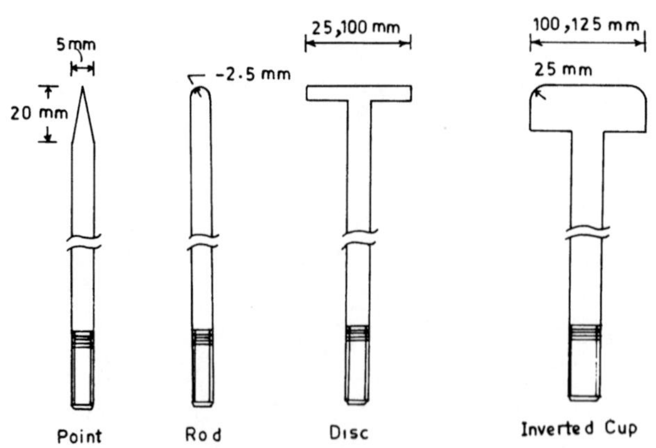

Fig. 3. Air terminal test electrodes.

Two scaled models of electrode setup were prepared for the investigation of validity of protection zone derived by rolling sphere. In the first setup, a disc shaped electrode,

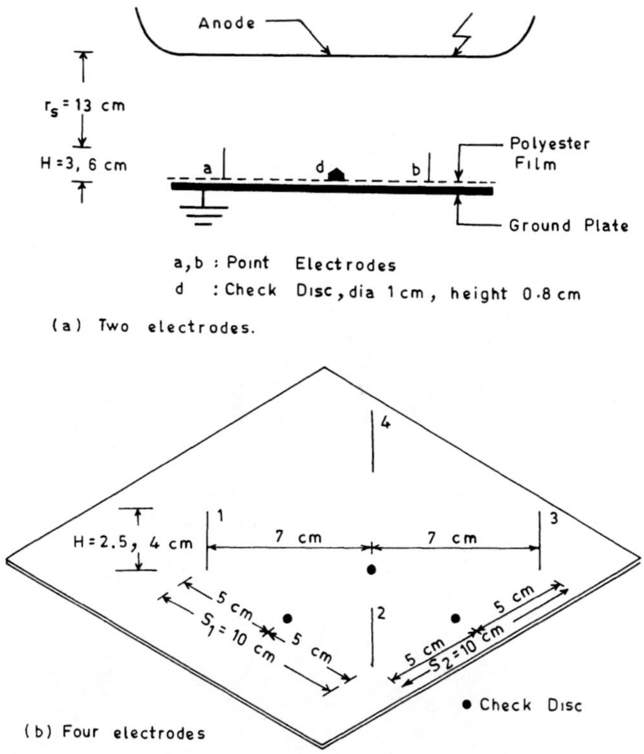

Fig. 4. Electrode lay-out Rolling Sphere

1 cm in diameter and 0.8 cm high was placed in the middle of straight line joining two tall point shaped electrodes. The disc electrode served as a check electrode between two air terminal electrodes, as shown in Figure 4(a).

The second setup prepared for investigation of three dimensional protection zone consisted of four point electrodes as air terminals which necessitated three check disc electrodes, as shown in Figure 4(b). Since the height of the electrodes on the ground had to be maintained rather low, the grounded plate was completely covered with an insulating film of polyester for the investigations on rolling sphere protection zone. It prevented the surge striking on the plate, leaving the only possibility to strike on any grounded electrode.

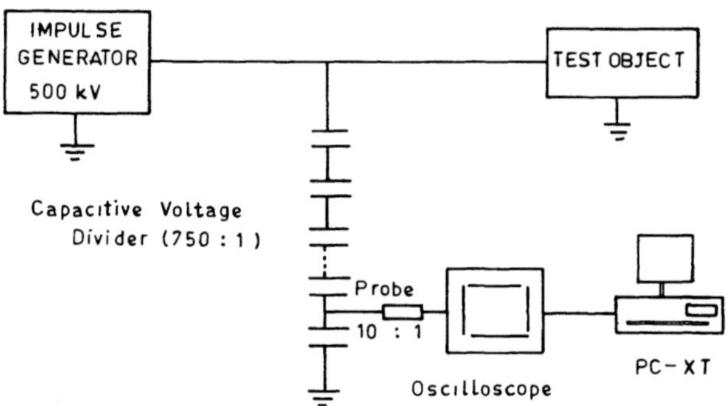

Fig. 5. Schematic Connection

Figure 5 shows the schematic of electrical connections. A 500 kV, 4.4 kJ Impulse generator was used to produce the standard shape of lightning impulse (li). The actual wave shape was recorded over a pure capacitive voltage divider having division ratio of 750:1. A passive probe (10:1 or 50:1) was connected at the end of the 25 m long measuring cable having an impedance of 75 Ω before connecting it to the digitizing oscilloscope which had a frequency range of 50 MHz and maximum sampling rate of 200 Mega samples/sec. Experiments were performed with positive polarity li having actual wave shape of 0.9/54 μs which is well within the tolerance limits for 1.2/50 μs wave. The oscilloscope, provided with a GPIB port, was interphased with a PC-XT in order to take the printouts of the wave shape.

MEASUREMENTS AND RESULTS

Investigation on different Shapes of Air Terminal

Although difficult, experiments were performed measuring 100% breakdown voltages U_{b-100}, of atmospheric air for a gap distance of 15 cm between the cloud electrode and all the air terminals investigated. First the breakdown voltages were measured for each one of nine individual electrodes of four shapes but of different sizes, shown in Figure 3. Their heights varied between 5 to 9 cm above the ground plate.These results are brought together in Table 1. The last column gives the mean potential gradient or breakdown strength measured for different electrodes.

Table 1 Breakdown(individual electrodes)

SER NO.	CONFIGURATION	$U_{(b-100)}$ (kV)	GRADIENT (kV/cm)
1.	Point 15cm 5cm	372	24.8
2.	Point 15cm 7cm	362	24.1
3.	Point 15cm 9cm	354	23.6
4.	Rod φ=5mm 15cm 5cm	381	25.4
5.	Rod φ=24mm 15cm 13cm	288	19.2
6.	Disc φ=25, 45,65mm 15cm 5cm	297, 289, 290	19.8, 19.3, 19.3
7.	Disc φ=65mm 15cm 7cm	287	19.1
8.	Disc φ=100mm 15cm 9cm	276	18.4
9.	Cup φ=100, 125mm 15cm 16cm	276	18.4

For determining each U_{b-100}, a few tens and even upto hundreds of strikes were necessary.to measure the correct magnitude and observe confident striking behaviour. Following observations on their striking behaviour could be made.

When a point electrode was placed 5 cm above the ground plate, only 40% flashover struck the point and rest the ground plate. This striking characteristic increased to 90% when the point electrode was replaced by a 5 mm diameter rod at the same height (S.No.1 & 4 of Table 1). One third of the strokes on the point electrode were at the wave front or at the crest but remaining two third were at the wave tail with a time lag beyond the crest between 2 to 17 μs. On the contrary, for the rod all strokes were on the wave tail with a time lag of only about 1.5 μs. The average breakdown strengths did not vary much in the two cases.

When the height of the point electrode was raised to 7 cm, 100% strokes were on the electrode. For all the

Table 2 Breakdown (grouped electrodes)

SER NO	CONFIGURATION	$U_{(b-100)}$ kV	GRADIENT kV/cm
1.	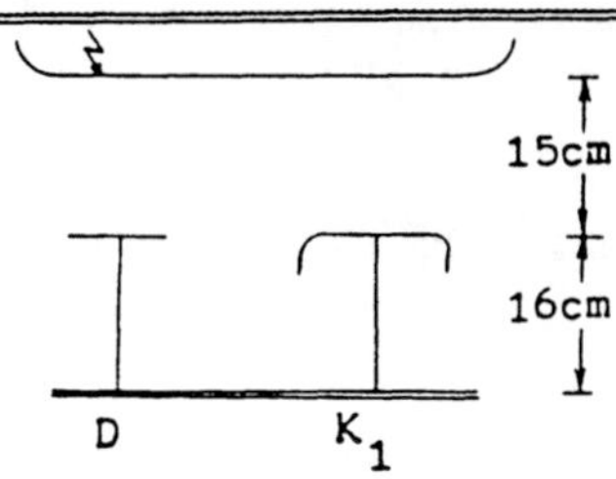 φ=10cm	276	18.4
2.	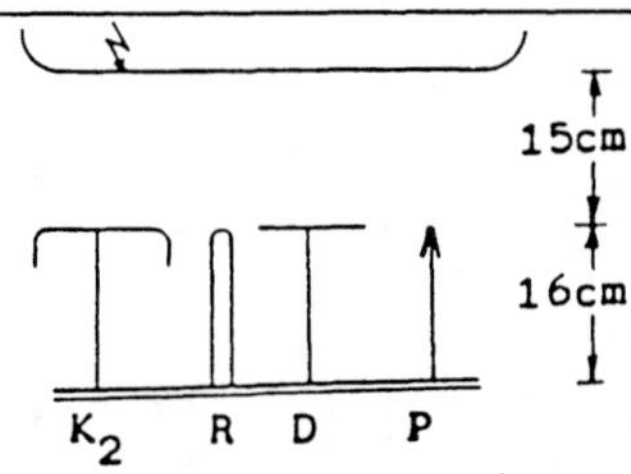 K_2:Cup φ=12.5cm P:Point D:Disc φ=6.5cm R:Rod φ=2.4 cm	276	18.4
3.	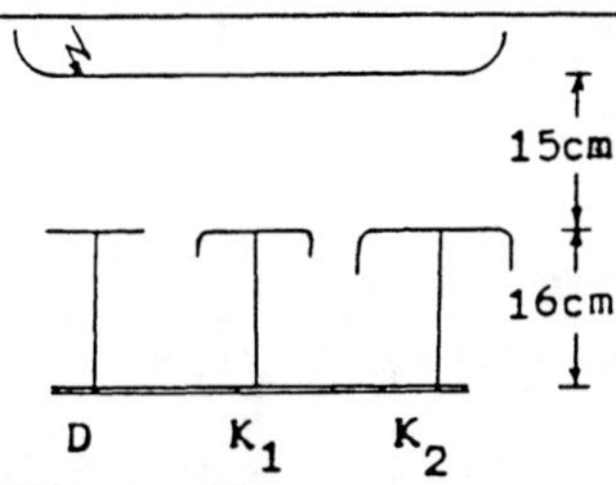 D:Disc φ=10cm K_1:Cup φ=10cm K_2:Cup φ=12.5cm	276	18.4

investigations with electrode at a greater than 5 cm height and with broader surface area electrodes placed even at 5 cm above the ground electrode, 100% strokes were on the air terminals and on the wave tail for which the time lag could be measured. Average breakdown strength for each case were calculated and are given in the last column. It can be observed from this table that as the surface area of the air terminal electrode is increased, not only the probability of lightning striking on the electrode increases but the breakdown strength of air also reduces.

A typical striking behaviour could be observed in case of flat disc and inverted cup electrodes. On examining these electrodes closely after hundreds of strikes, small spots of 'electrode pitting' could be observed all along their periphery. Rest of the electrode surface were found spotless.

In the second phase of experimentation with air terminal electrodes, these were grouped together, as shown in Table 2. When a flat disc and an inverted cup, both having diameter of 10 cm were placed together 16 cm above the ground, lightning pulse struck the cup every time. When a 12.5 cm diameter inverted cup was placed with these two at the same height (S.No.3), the lightning preferred to strike on this largest area electrode. Obviously, when this inverted cup was placed along with a point, a rod and a disc electrodes together, it could be observed that the lightning struck only on this large cup again (S.No.2).

Investigation on Protection Zone derived by Rolling Sphere

A protected zone under more than one lightning conductors is determined by the smallest height x above the ground between any two conductors derived by rolling the sphere of radius r_s. For lightning conductors located at the same height H above the ground (Figure 1), considering the simple geometry, x is given by the following equation,

$$x = H - \left[r_s - \sqrt{r_s^2 - \left(\frac{S}{2}\right)^2} \right] \qquad (1)$$

where H is the height of air terminals, r_s the striking radius and S is the distance between two air terminals under consideration.

Length of the lightning return stroke step l_{rs} is recommended by German standard TGL 200-0616/02 to be calculated by the equation,

$$l_{rs} = 2\,\hat{i} + 30 \left[1 - \exp\left(-\frac{\hat{i}}{6.8} \right) \right] \qquad (2)$$

where l_{rs} is in m and $\hat{i}$, the peak value of lightning impulse current is in kA. A safe value of 40 m is recommended for the striking radius r_s by TGL 200 after a number of considerations. An average intensity lightning stroke may accompany a current of 20 kA. The length of return stroke l_{rs} for this value works out to be 71 m.

As suggested by Naumann [3], point electrodes were chosen as air terminals. The check disc electrodes were 1 cm in diameter and having a sharp pointed shape in the middle with a total height of 0.8 cm.

Due to the constraint of the magnitude of impulse voltage in the laboratory, a scaled model of 300:1 was constructed as shown in Figure 4a. Accordingly 40 m of striking radius works out to be 13 cm and a gap distance of 60 m to be 20 cm. First two 3 cm tall point electrodes as air terminals were placed 20 cm apart and the check disc in the middle. On applying 76 strokes of U_{b-100}, it was observed that about 18% of them struck on the check disc. This conveys that for the given dimensions the disc is not falling in the safe zone. With the distance S maintained at 20 cm (60 m) and raising the heights of the air terminals to 6 cm (18 m) above ground, 100% strokes hit the air terminals. This verified that now the check disc was within the safe zone of li protection. As mentioned, these experiments were performed covering the ground plate with an insulated barrier. The protection zones derived analytically by this method for the scaled dimensions are shown in Figure 6 for the two heights of the air terminals investigated. Thus a practical validity of the zone of protection could be made.

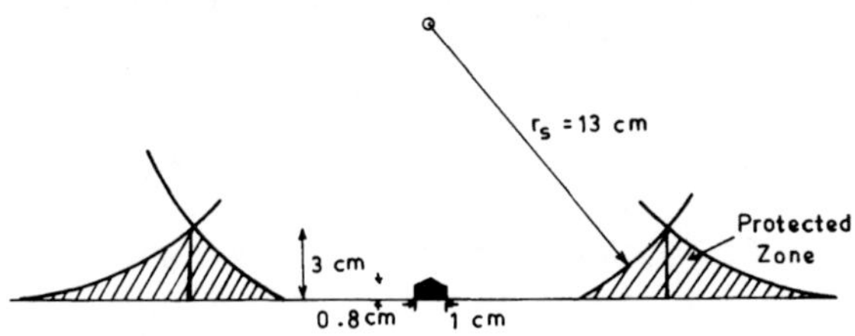

Fig. 6(a). Disc in unprotected area.

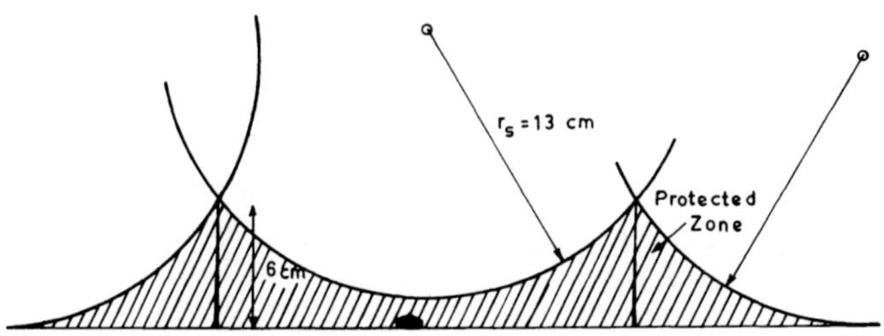

Fig. 6(b). Disc under protection.

Another model of four air terminals was constructed also with a scaled ratio of 300:1 shown in Figure 4b with actual dimensions. Investigations were made with air terminal heights of 2.5 and 4 cm which simulates the air terminals of 7.5 and 12 m above the ground on a field of 30x30 m. The striking radius of li, that is, the gap distance between the cloud electrode and the air terminals was kept at 13 cm representing actual distance of 40 m. Substituting these dimensions in Equation 1, for H=2.5 cm, the minimum heights of safety $x_1 = x_2$ and x_3 are obtained to be 1.5 and 0.45 cm respectively. For H=4 cm these work out to be 3.3 and 2 cm respectively. Since the check disc was 0.8 cm in height, it thus projected above the protected zone in the middle of the diagonal for

H=2.5 cm. On applying a total of 75 li of U_{b-100} of magnitude of 318 kV for this configuration, about 12% of them struck on the check disc placed in the middle of the diagonal and rest 88% on the four air terminals. None of the impulse struck on check discs placed on the sides of the quadrangle. When the heights of the air terminal electrodes was raised to 4 cm, it was remarkable to observe that 100% strokes hit the four air terminals. Thus the validity of protection zone formed by rolling sphere could be confirmed practically on this scaled model of four air terminal electrode system also.

CONCLUSION

Experimental investigations in the laboratory performed on comparatively smaller gap distance of 15 cm reveal that for a lightning impulse the average breakdown strength of air strongly depends upon the surface area of the air terminal of the lightning conductor. During the measurements, a lightning preferred to strike not only the air terminals with largest surface area in the vicinity but also the location where high field intensity existed on it. There exists a likelyhood for a sharp point electrode air terminal located at a low height to be missed by the lightning. The most suitable shape of electrode for air terminals that could be recommend through these investigations is in the form of an inverted cup.

Laboratory investigations on a scaled model confirmed the lightning protection zone derived by the rolling sphere method even when provided with sharp point electrodes as lightning conductors.

REFERENCES

[1] H.R. Armstrong et al., "Field and Analytical Studies of Transmission Line Shielding", IEEE Trans. Vol. PAS-87, No.1 (1968), pp. 270-281.

[2] R.H. Golde, "Protection of Structures against-Lightning", Proc. IEE, Vol.115, No.10 (1968), pp. 1523-1529.

[3] W. Naumann, "Koustruktion der Schutzräume in Blitzschutz", Elektropraktiker, Berlin, 40 (1986)8, pp.258-263 (in German).

SECTION 8: GAS DECOMPOSITION/PLASMA CHEMISTRY

INFLUENCE OF ELECTRODE MATERIALS AND COVERINGS ON THE GASEOUS BY-PRODUCTS GENERATED BY AIR CORONA DISCHARGES

N.FOULON-BELKACEMI, M-P.PANAGET
M.GOLDMAN, A.GOLDMAN

Laboratoire de Physique des Décharges, CNRS / ESE,
Plateau de Moulon, 91192 Gif/Yvette Cedex, FRANCE

INTRODUCTION

It is well known that, by electron / molecule collisions, corona discharges in atmospheric air produce reactive species which can be enumerated as followed: primary charged products which should mainly be O_2^-, O^- and OH^- ions for the negatively charged species and N_2^+, N^+, O_2^+, O^+ ions for the positively charged species; neutral active products N, N_2^*, N^*, O, O_2^*, $O^*$and OH.

Most of these active species rapidly lead to secondary species hydrated ions , neutral compounds and aerosols, the chemical behaviour of which most generally reflects oxidising and acidifying properties. This is in particular supported by a lot of observations on surface effects. For example, on a polymer surface, let say polypropylene, these species induce oxidised products through a complex reaction scheme[1], or on a copper surface, they can enhance corrosion rates[2]. Then, physical and chemical alterations on electrodes appear as a consequence of reactions induced by reactive gaseous species provided by the discharge. But the surface processes may in their turn bring modifications to the discharge as well as to the nature and the evolution of the corona gas and of its gaseous by-products through, for example, electrocatalytical processes[3,4].

In this paper, our objective is to study the influence of electrode materials and coverings on the gaseous species in the gap. For this purpose, are carried out simultaneous analyses of the evolution of the gaseous species in the discharge gap and of the alterations produced on the electrodes.

EXPERIMENTAL

The experimental device consists in a point-to-plane gap of constant length fixed to 8 mm. The point electrode (Fe-Ni) is connected to a 20 kV / 50 Hz voltage supply through a 25 MΩ resistance. Experiments are carried out on a copper plane electrode and on 15 μm thick polypropylene foils placed on this plane electrode, in synthetic air at controlled temperature and humidity (25°C and 60% RH). To understand the physico-chemical processes induced by the discharge in air, it is necessary to recall some basic data on the energy distribution between the reactive chemical products. The production of gaseous species by the discharge is associated to the different components of the discharge current under alternating voltage which are[5]:

- the current component synchronous with the applied voltage, which corresponds to the drift of ions through the gap, including more than half of the current flow in the positive half-cycles and all the current flow in the negative ones. Corresponding current densities are quite low (some tens μA/cm^2). The gaseous species produced by this way are for 99% electrically neutral and their kinetic energy does not exceed a fraction of eV.

Gaseous Dielectrics VII, Edited by L.G. Christophorou
and D.R. James, Plenum Press, New York, 1994

- the impulsive current component which is associated to the formation of streamers developing with high speeds (of the order of 10^7 cm/s) exclusively in the positive half-cycles above a critical field. Corresponding current densities are in this case much higher, of the order of 2000 to 3000 $\mu A/cm^2$. The positive ions provided by these streamers to the surface have kinetic energies which can reach tens of eVs and thus not only break chemical bonds but also produce sputtering effects. The streamers concerned with this paper are repetitive streamers, all concentrated on the gap axis and superposed with the glow components.

The chemical and morphological modifications of the plane electrode are studied by means of a Scanning Electron Microscope (SEM) equipped for Energy Dispersive X-ray analysis (EDX) and of capillary electrophoresis. The evolution of the gaseous products generated by the discharge and by its interaction with the plane electrode is followed in situ by use of a Fourier Transform Infra Red Spectrometer (Fig 1).

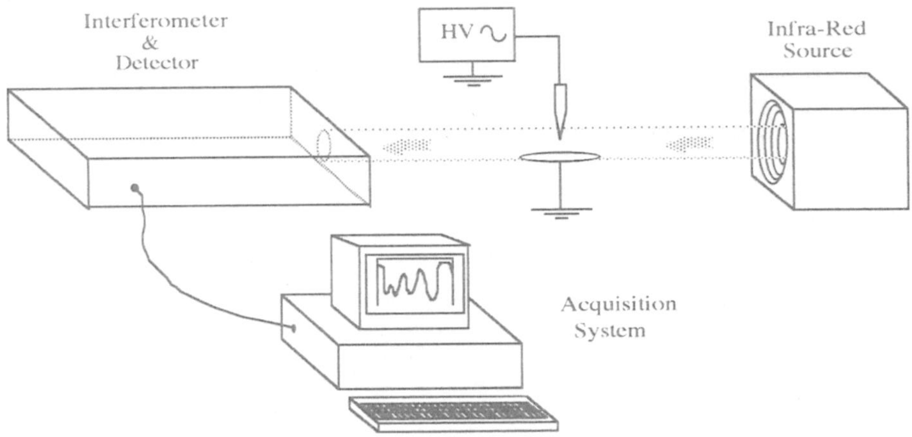

Fig 1: Experimental set-up used for infra-red in situ analyses of the gas phase.

RESULTS

Infra Red in situ Analysis

First, the experiment is carried out with the naked copper electrode. During the discharge, several infra-red spectra are acquired and it appears clearly (Fig 2) that the discharge components produce: ozone O_3, in the region of 1030 cm^{-1}; nitrous acid HNO_2 and nitric acid HNO_3 at about 800 and 1270 cm^{-1}; sulphur dioxide SO_2 as an atmospheric pollutant at 1110 and 1185 cm^{-1}. These species are produced with a constant rate along the discharge duration.

Then the discharge is faced to the electrode covered with a polypropylene foil, and several infra-red spectra are acquired before the dielectric breakdown of the polypropylene occurs (Fig 3). The analysis of the gas then permits to characterise the presence of -OH and $-CH_2-$ / $-CH_3$ functions and of nitrogen dioxide NO_2 and to study the variations of their concentrations with time during the discharge (Fig 4). The results first show the removal of oligomers from the polymer foil from the first seconds of the treatment. The -OH functions linked to the organic groups of the polymer are seen to appear only after a few minutes treatment and then to increase to a maximum value maintained till the foil breakdown; their delayed apparition gives to think that they should be produced by second step reactions acting on the oxidised by-products of the surface which are in turn, after a given time, submitted to the same aggressions than the substrate. The reported results so bring to light the competition which develops between the mechanical degradation processes induced by the streamers and the chemical reactions of the reactive species generated by the discharge in the gas. The simultaneous change observed after ~ 10 min treatment on all curves shows that the polymer surface gets a new equilibrium through this period of time but, at this stage, we have no definitive explanation for it.

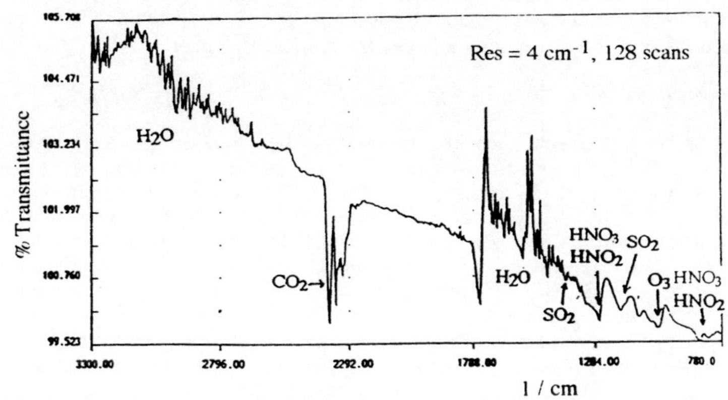

Fig 2: Infra red spectrum of the gaseous species produced by the discharge on the copper plane electrode.

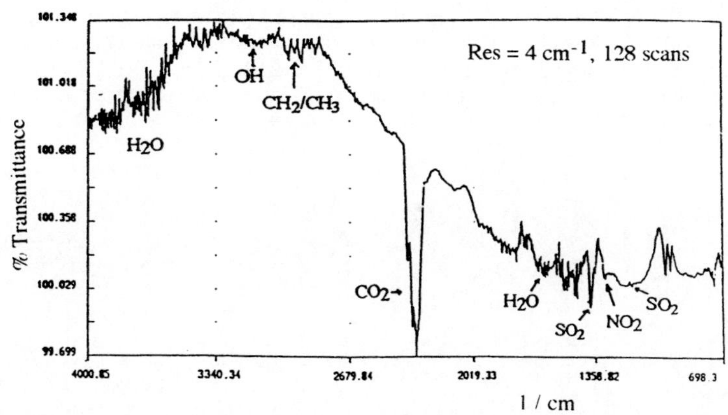

Fig 3: Infra red spectrum of the gaseous species produced by the discharge on an electrode covered with a polypropylene foil.

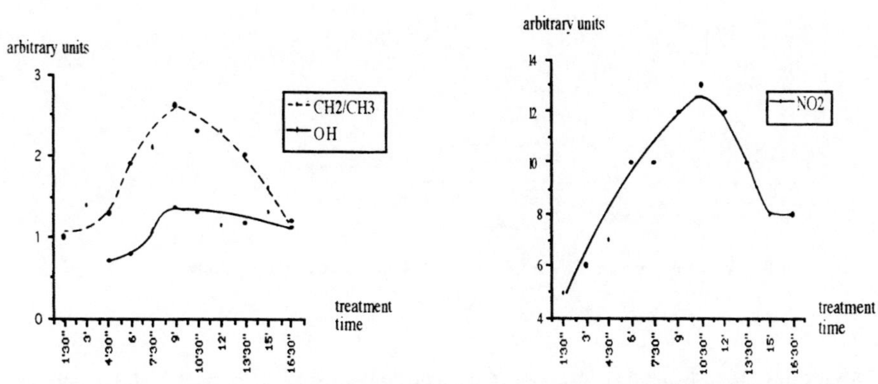

Fig 4: Evolution of CH_2 / CH_3 functions, of OH functions and of NO_2 observed in the gas by in situ infra-red analysis.

At last, from the comparison of the results obtained with and without a polypropylene foil on the plane electrode, it comes out that nitrous acid HNO_2 and nitric acid HNO_3 have reacted with the polymeric surface; effectively these products are no more detected in the gas when the plane electrode is covered with a polypropylene foil. These acids may acidify the polypropylene surface and participate to the alteration of its properties. In the same manner, ozone appears only when the copper electrode is naked. It may react with the polymer by ozonolysis reactions described in another paper by Foulon-Belkacemi et al.[1].

Capillary Electrophoresis

The by-products generated by the interaction between the gaseous products and the surfaces were collected by depositing a drop of distilled water on the site of the impinging streamers with a syringe and by reusing it to take off the drop of water with the surface products dissolved in it. Injected for analysis into the capillary of an equipment for capillary electrophoresis, the negative ionic species are identified according to their migration speeds directly proportional to their mobilities along the capillary to which a drifting potential is applied. An electro-osmotic flow increases the migration times of the species which are identified by UV detection.

First analyses are carried out on the products taken off from the naked copper electrode. The electroferrogram of Figure 5 shows the species identified: chlorides (Cl^-) and sulphates (SO_4^{2-}) originating from the gas used for the experiments, and nitrites (NO_2^-) and nitrates (NO_3^-) generated in the discharge gap, all species transferred to the plane electrode probably with the assistance of the well-known "electric wind". As for the acids which appear with higher retention times, they may be provided by organic pollutants always present on metallic surfaces.

In a second step, analyses are performed on the products taken off from a polymer surface. The electroferrogram of Figure 6 first shows that there are numerous final oxidised by-products which were already detected by Fourier Transform Infra Red Spectroscopy[6].They result from the breaking of polymer chains occurring randomly on the point where streamers are impinging.

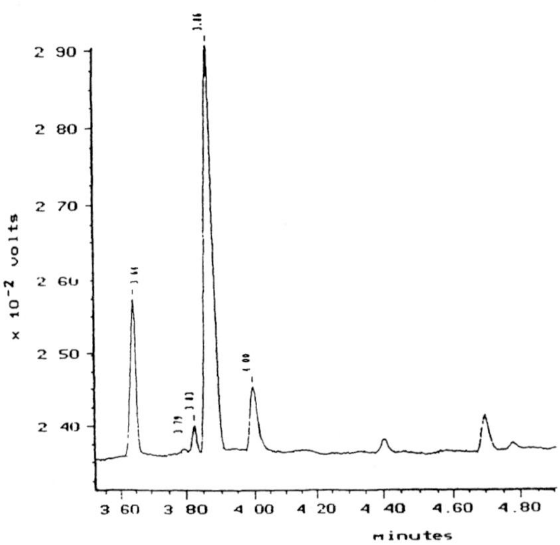

Cl^-	3 64 min
SO_4^{2-}	3.79 min
NO_2^-	3.83 min
NO_3^-	3 86 min
$^-OOC\text{-}COO^-$	4 37 min
$HCOO^-$	4 71 min

Fig 5 Electroferrogram of a sample of products collected on the copper electrode

Again are also observed chlorides (Cl^-), sulphates (SO_4^{2-}), nitrites (NO_2^-) and nitrates (NO_3^-) as in the case of the naked copper electrode and again with nitrates in a greater rate than nitrites. This is easily explained by considering the two systems HNO_2 / NO and NO_3^- / HNO_2 with their respective standard potentials E°=0.99 V and E°=0.94 V vs NHE: not stable in acid aqueous solutions, nitrite acids are transformed by dismutation reactions in the form of:

$$3\,HNO_2 \rightarrow 2\,NO + NO_3^- + H_3O^+$$

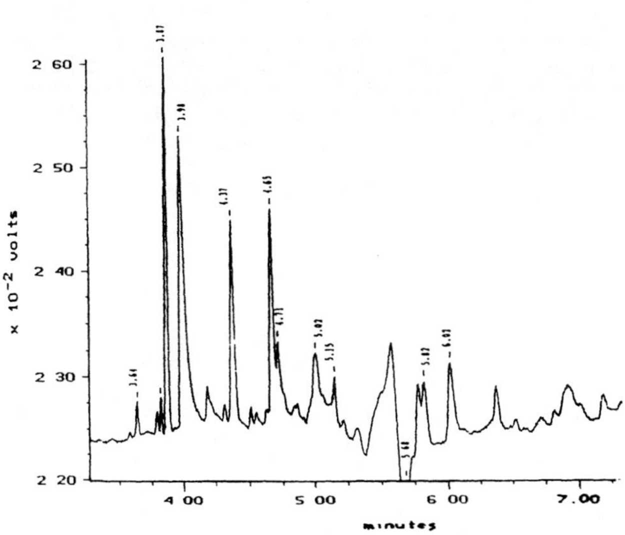

Cl^-	3.64 min
SO_4^{2-}	3.72 min
NO_2^-	3 85 min
NO_3^-	3 87 min
$^-OOC\text{-}COO^-$	3 98 min
$^-OOC\text{-}CH_2\text{-}COO^-$	4,25 min
$^-OOC\text{-}CH{=}CH\text{-}COO^-$	4.37 min
$^-OOC\text{-}(CHOH)_2\text{-}COO^-$	4.52 min
$^-OOC\text{-}(CH_2)_2\text{-}COO^-$	4.65 min
$HCOO^-$	4.71 min
$(^-OOC)_2\text{-}C_3H_5$	5.02 min
$^-OOC\text{-}C(CH_3)_2\text{-}COO^-$	5.15 min
CO_3^-	5.40 min
$CH_3\text{-}CHOH\text{-}COO^-$	5.82 min
$CH_3\text{-}CH_2\text{-}COO^-$	6.02 min
$CH_3\text{-}CH{=}CH\text{-}COO^-$	6.37 min
$CH_3\text{-}(CH_2)_2\text{-}COO^-$	6 90 min
$CH_3\text{-}(CH_2)_4\text{-}COO^-$	7 20 min

Fig 6: Electroferrogram of a sample of products collected on the polymer electrode covering

Scanning Electron Microscopy associated with Energy Dispersive X-Ray Analysis

The observation of the surface of the copper electrode by SEM shows agregates, while EDX analysis gives access to the oxygen fixation in the forms of CuO and Cu_2O as main products, as identified by Sobek et al.[2]. On the Figure 7 which shows the distribution of the oxygen found on the copper electrode after its exposure to the discharge, it is noticeable that, at the point where the streamers were impinging, there is little oxygen on the surface. This again shows the mechanical effects of the discharge; here, the mechanical degradation processes induced by the streamers acted on the oxidised surface products. No oxidised copper was detected in the gaseous phase by Infra Red Spectroscopy; it was probably in the form of aerosols.

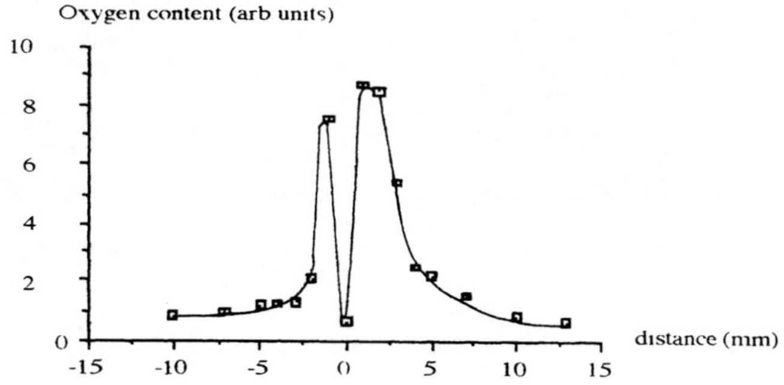

Fig 7· Oxide formation rate on the copper along the transversal axis of the electrode.

At last, the observation of the polymer surface treated by corona streamers shows as well the formation of nodules well-known in this case since 1971[7] and if the observation is carried out at low temperatures, on a sample kept at -150°C in the chamber of the microscope, one can see small particles of less than 1μm deposited on the nodules (Fig 8). These particles may correspond to aerosols, as described by Borra et al.[8] who show that such aerosols produced by corona discharges can originate from metallic oxides provided by pulverisation processes at the electrodes or by gas to particule conversion processes which should be triggered by the clustering of hydrated ions such as from nitric acid[9] or from volatile organic compounds (VOCs), produced by the discharge in air; nitric acid was not then observed by infra red spectroscopy in the gaseous phase, but the formation of a great number of carboxylic acids was observed on the polypropylene: this acid may have reacted with the polypropylene surface in terms of acid-base reactivity, according to a model proposed by Amouroux et al.[10] which refers to the desactivation of the non-charged plasma products through collisions with water molecules in the water saturated gas layer close to the surface increasing the surface acidity.

CONCLUSIONS

The paper shows that the reactive species formed in the gas by point-to-plane corona discharges (but it should be the same for all kinds of gas discharges) are species aged by chains of reactions; in air, these are typically acids such as HNO_2 and HNO_3 identified while working with naked metallic electrodes. But moreover, it was shown, by simultaneous analyses of the plane electrode, naked or covered with a polymer foil, and of the gas submitted to the discharge that the surfaces undergoing corrosion and/or sputtering processes through gas-surface interactions will participate to the pollution state of the gas; some active species primarily formed in the gas phase may, through these interactions, disappear from the gas while others, originating from surface reactions, may be injected in. So, in the case treated in the paper of a polypropylene foil submitted to the discharge in air, it occured a consumption of the nitrous acid HNO_2 and of the nitric acid HNO_3 formed in the gas phse, with a release of -CH_2- and -CH_3 functions from the material.

The products formed by the gas-surface interactions can, with the corona species produced and remaining in the gas phase, further react directly with materials and provoke ageing effects on them or they can undergo further reactions leading to the creation of long life species with different degrees of reactivity.

At our knowledge, it does not exist any material able to remain completely passive in front of a cold discharge plasma, so that any surface in contact with the discharge must be regarded as a source of pollution changes for the gas, but the consequences of their effects on the gas dielectric strength are, at this stage, quite unknown, while indeed strongly dependent on the initial gas composition.

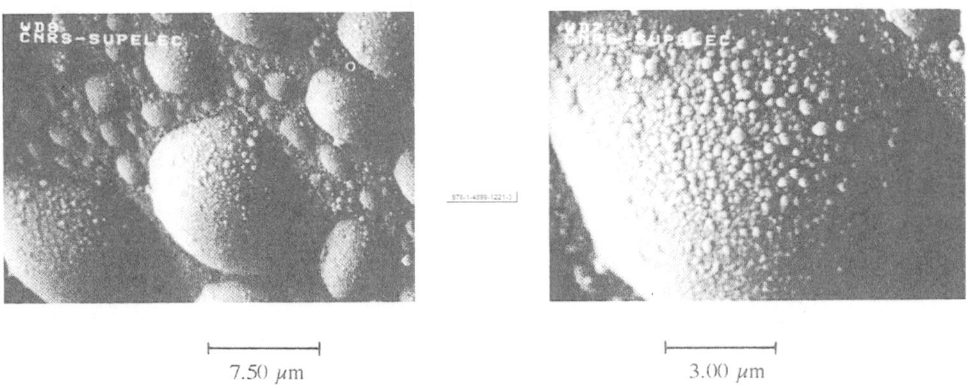

Fig 8: Observation by SEM of a polypropylene surface exposed to corona streamers.

REFERENCES

1. N.Foulon-Belkacemi, M.P.Panaget, M.Goldman, A.Goldman, Cinétique des phénomènes de dégradation d'un isolant soumis à des décharges couronnes filamentaires, to be published in *Journal de Physique* (1993).

2. V.Sobek, A.Goldman, M.Goldman, S.Bureau, S.Bouquet, C.Fiaud, Experimental simulation of atmospheric corrosion by use of a corona discharge, *4th International Symposium of High Pressure and Low Temperature Plasma Chemistry, Hakone IV*, Slovakia (1993).

3. M.Morvova, M.Kurdel, I.Morva, The plasmachemical and electrocatalytical effects of corona discharge of both polarities on CO and CO_2 mixtures with air for various materials of electrodes, *4th International Symposium of High Pressure and Low Temperature Plasma Chemistry, Hakone IV*, Slovakia (1993).
4. D.Rapakoulias, J.Amouroux, M.P.Bergougnan, A.Gicquel, Processus catalytiques dans les réacteurs à plasma hors équilibre III-Décomposition de NH_3, *Revue Physique Appliquée*, 17:95 (1982).
5. M.Goldman, A.Goldman and R.S.Sigmond, The corona discharge, its properties and specific uses, *Pure and Applied Chemistry*, 57:1353 (1985)
6. N.Foulon-Belkacemi, M.Goldman, A.Goldman, H.Dejean, J.Amouroux, The mechanism of degradation of polymers under corona streamers: relative humidity influence, *IEE Internatinal Conference of Partial Discharges*.(1993).
7. C.Y.Kim, D.A.I.Goring, Surface morphology of polyethylene after treatment in a corona discharge, *Journal of Applied Polymer Science*, 15:1357 (1971).
8. J.P.Borra, A.Goldman, M.Goldman, D.Boulaud, Generation of condensation nuclei by dc electrical discharges in point-to-plane configuration, to be published in *Journal of Aerosol Science* (Juin 1994).
9. J.S.Chang, Energetic electron induced plasma processes for reduction of acid and greenhouse gases in combustion flue gas, Non Thermal Plasma Techniques for Pollution *Control NATO ASI Series*, Vol.G34, Part A (1993).
10. J.Amouroux, J.L.Brisset, A.Doubla, A.Goldman, M.Goldman, Acid-base properties and reactivity towards surfaces of the gaseous uncharged activated species created in a DC corona discharge, *International Symposium of plasma chemistry n°8*, DVII-02 ,(1987).

DISCUSSION

R. BARTNIKAS: Prof. Goldman has presented some interesting data obtained with dielectric surfaces exposed to electrical discharges. We have also observed the formation of droplets and crystals on the surface and have identified a number of acids (C. Hudon et al., *IEEE Trans. on Electrical Insulation*, Vol. 28, 1993, pp. 1-8). However, some of the higher molecular weight substances are more difficult to identify and we would like to know whether he has encountered similar problems.

M. GOLDMAN: The results presented on dielectric materials are published. I tried to show that nodules are easy to evaporate and will become aerosols or dust in the discharge gap. Large chains were identified in the past by different authors in particular by the work of Prof. Amouroux. We have not identified higher molecular weight substances but Prof. Amouroux has. The chromatogram shows that the sensitivity decreases as you increase the weight so we are able to show here only the smaller fragments. There is no statistical reason to cut the polymer in only small chains - it can also be cut in big chains. Amouroux has shown this by SIMS. In the beginning he had large molecular weights which were then cut into smaller chains. In the beginning of a discharge you cut the polymer into big chains because statistically you cut everywhere and then these fragments get cut into pieces. The more you wait, the more you will see smaller chains. Amouroux has measured chains with a molecular weight of more than 400, but I think these big chains will also be acidic.

J. KENNEDY: Regarding Fig. 4 of your paper, does the evolution of OH, CH_2/CH_3, and NO_2 eventually reach a steady-state condition?

M. GOLDMAN: The experiment stopped when the polymer broke down.

W. KINSNER: Since the surface is so important in the study of gas kinetics, how do you characterize the surface systematically?

M. GOLDMAN: We have to use different complementary methods: electron microscopy, X-ray, ESCA, Auger, FTIR, surface-potential, and others.

DECOMPOSITION OF SULFUR HEXAFLUORIDE BY X-RAYS

J. K. Olthoff and R. J. Van Brunt

Electricity Division
National Institute of Standards and Technology*
Gaithersburg, Maryland 20899

INTRODUCTION

The decomposition of gaseous sulfur hexafluoride (SF_6) by exposure to high-energy photons, and the subsequent formation of toxic and corrosive oxyfluoride by-products, is of interest due to the use of SF_6 as a high-voltage insulator near sources of radiation, such as particle accelerators and X-ray units. Additionally, information about by-product formation due to radiation exposure can be compared with data obtained for the decomposition of SF_6 in electrical discharges.[1,2] This comparison is interesting since the volume in which the decomposition occurs is usually many orders of magnitude larger for radiation exposure than for electrical discharges. However, little previous work has been done to determine the effects of radiation upon gaseous SF_6.[3]

In this paper, we present results of by-product formation in gaseous SF_6 exposed to high-energy X-rays. The identity and concentration of the decomposition by-products are determined by gas chromotography/mass spectrometry techniques that were developed to investigate the decomposition of SF_6 exposed to corona discharges.[1,4] The production curves of SOF_2 and S_2F_{10} are determined for a range of SF_6 gas pressures, X-ray energies, and X-ray fluxes. Evidence for the presence of other by-products, such as SOF_4, SO_2F_2, and $S_2O_2F_{10}$ is also presented. The decomposition data for the SF_6 exposed to X-rays are compared with previously published data for SF_6 exposed to corona discharges.

EXPERIMENTAL DESCRIPTION

The beam of high-energy X-rays used to irradiate samples of SF_6 was produced by a commercial X-ray unit. X-rays were generated in the unit by a high-energy beam of electrons striking a tungsten surface at a 20° incident angle. The energy and intensity of the X-rays were varied by changing the accelerating voltage (V_x) and current (I_x) of the incident electron beam. For the experiments presented here, electron energies of 100 kV and 195 kV were used, with electron currents of 4.0, 2.5, and 1.0 mA. The X-rays then pass through approximately 1 mm of glass and 4 mm of aluminum before exiting the X-ray unit.

*Electronics and Electrical Engineering Laboratory, Technology Administration, U. S. Department of Commerce

Gaseous Dielectrics VII, Edited by L.G. Christophorou
and D.R. James, Plenum Press, New York, 1994

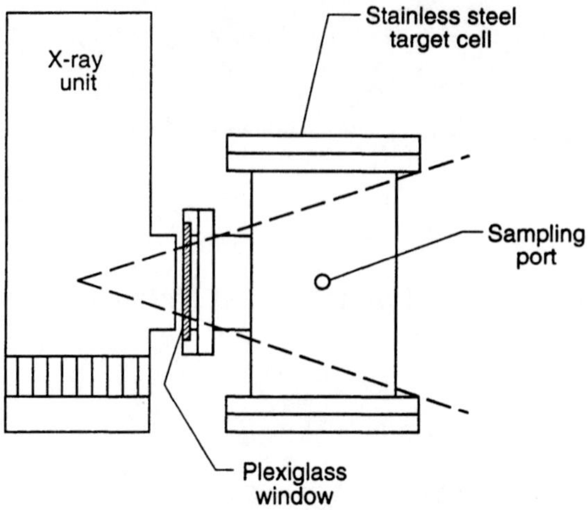

Figure 1. Schematic diagram of the apparatus used to investigate the decomposition of SF_6 by exposure to X-rays. The dashed lines show the approximate spatial spread of the X-ray beam.

The SF_6 that was exposed to the X-ray beam was contained in a stainless steel vacuum chamber with a volume of 6.2 ℓ. Prior to filling with SF_6, the vacuum chamber was evacuated by a mechanical vacuum pump down to approximately 1 Pa, and then flushed several times with SF_6. During the experiments, the gas pressures in the cell were maintained at 300 or 600 kPa. The vacuum chamber was equipped with a 10-mm-thick, 100-mm-diameter plexiglass window, through which the X-rays were directed, and a sampling port that allowed the sampling of the compressed SF_6 from inside of the cell by a gas-tight syringe through a rubber septum. The X-ray unit and target cell were placed in a lead-lined concrete bunker, with the two chambers sufficiently close together such that the entire X-ray beam intercepted the target cell. This arrangement is shown schematically in Fig. 1.

To determine the effects of irradiation by X-rays on an SF_6 sample, the X-ray unit was operated for a set period of time at a particular beam current and voltage setting. During the experiment, the X-ray unit was periodically turned off, and the target cell was removed from the radiation bunker for gas analysis. The gas was analyzed using gas chromatography/mass spectrometry (GC/MS) techniques that are described in detail elsewhere.[1,4] In brief, 0.5 ml samples of decomposed SF_6 were removed from the target cell by gas-tight syringe. The samples were then injected into the GC/MS where the various components of the gas are separated in time as the gas flows through the chromatographic column. The compounds SOF_2, SO_2F_2, and SOF_4 were detected using a Poropak[†] column in the GC, whereas detection of S_2F_{10} and $S_2O_2F_{10}$ required use of a Chromosorb[†] column. Both columns were operated at room temperature with helium carrier gas. The various compounds were then detected by monitoring ions created by electron impact in the source of a mass spectrometer. For example, ion 86 (SOF_2^+) was monitored to detect SOF_2 and SOF_4, and ion 83 (SO_2F^+) was observed to detect SO_2F_2. Ion 86 was also used to detect S_2F_{10} due to a conversion of S_2F_{10} to SOF_2 that occurs at the heated heated interface between the GC column and the mass spectrometer.[4]

[†]Certain commercial equipment, instruments, or materials are identified in this paper to specify adequately the experimental procedure. Such identification does not imply recommendation or endorsement by the National Institute of Standards and Technology, nor does it imply that the materials or equipment identified are necessarily the best available for the purpose.

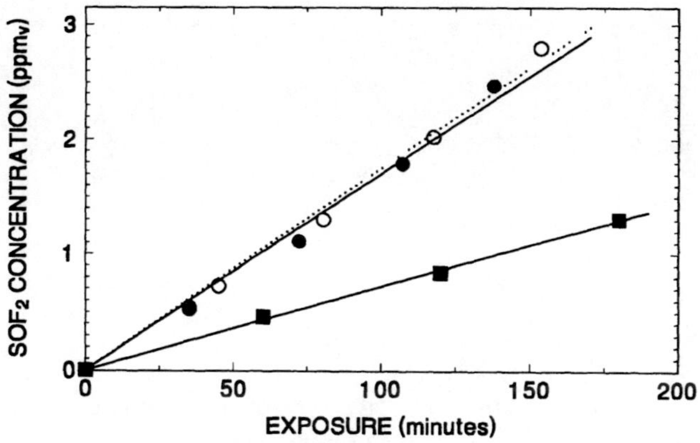

Figure 2. Production curves for SOF_2 in a SF_6 sample exposed to X-rays. The X-ray voltage and SF_6 gas pressures used to obtain these data are: (•) V_x = 195 kV, P = 600 kPa; (■) V_x = 100 kV, P = 600 kPa; (○) V_x = 195 kV, P = 300 kPa. The beam current (I_x) of the X-ray unit was 2.5 mA for all data in this figure. The solid lines are the least square fit for the 600 kPa data, while the dotted line is the least square fit for the 300 kPa data.

To calibrate the response of the GC/MS to the presence of the decomposition products, injections from reference samples containing known concentrations of SOF_2, SO_2F_2, SOF_4, or S_2F_{10} were made immediately before and after injections from the target cell. Uncertainties in the relative concentrations of by-products in the target cell due to variations in injection sizes and changes in GC/MS sensitivity were calculated to be less than ±4%. However, the absolute concentrations of oxyfluorides in the prepared reference samples used to calibrate the GC/MS were observed to vary by as much as ±25%, which significantly increases the uncertainties of the measured production rates.

RESULTS

The GC/MS analysis of the SF_6 exposed to the X-ray beam indicates that SOF_2 is probably the primary gas-phase decomposition product from the interaction of the radiation with the gas. Figure 2 shows the production curves (solid lines) for SOF_2 inside the cell for V_x = 195 kV and 100 kV, with I_x = 2.5 mA and a SF_6 pressure of 600 kPa. The SOF_2 concentrations are given in parts in 10^6 by volume (ppm_v). A clear voltage dependence is observed indicating that the production rate is dependent upon the energy of the irradiating X-rays. The production curve represented by open circles and a dotted line in Fig. 2 is for an SF_6 pressure of 300 kPa with I_x = 2.5 mA and V_x = 195 kV. The agreement between this data and the data obtained at 600 kPa, clearly shows that the production rate of SOF_2 by exposure to X-rays is not dependent upon the SF_6 gas pressure.

Figure 3 illustrates the dependence of SOF_2 production upon the intensity of the X-ray beam, by showing the production curves for different setting of I_x with V_x held constant at 195 kV. As expected, the production rate of SOF_2 increases with increasing I_x, corresponding to higher X-ray beam intensities. However, the observed dependence does not scale linearly with I_x.

Production of SO_2F_2 and SOF_4 was also observed for all of the experimental conditions investigated. Unfortunately, SOF_4 and SO_2F_2 have nearly the same GC retention times which results in significant signal interference if both are present in the gas sample.[1] Therefore, production rates for SO_2F_2 and SOF_4 inside the target cell could

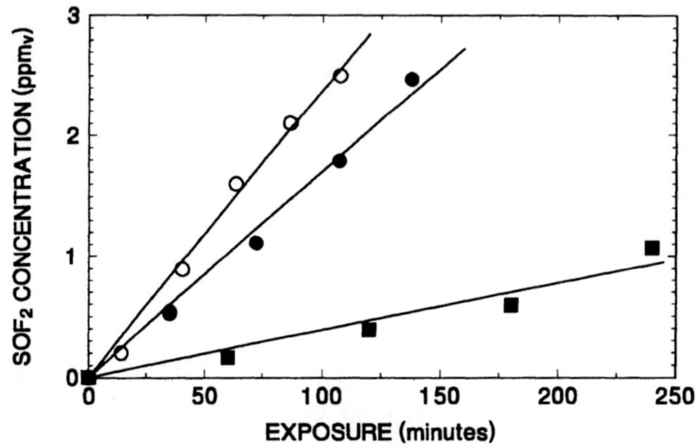

Figure 3. SOF_2 production curves for 600-kPa SF_6 samples decomposed by exposure to X-rays produced with V_x = 195 kV and (■) I_x = 1.0 mA, (•) I_x = 2.5 mA, and (○) I_x = 4.0 mA.

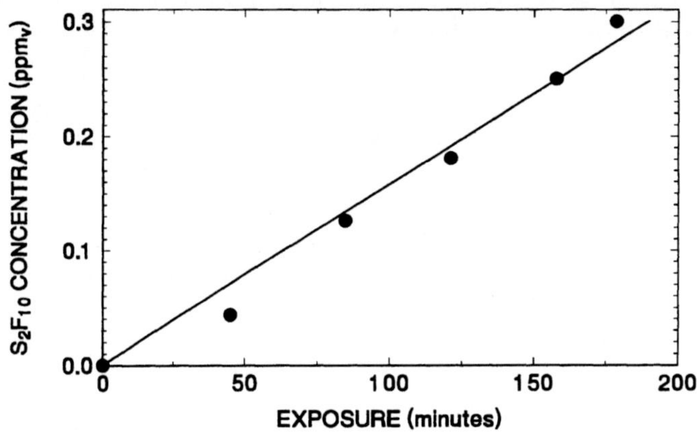

Figure 4. Production curve for S_2F_{10} formed in a 300-kPa SF_6 sample exposed to X-rays produced with V_x=195 kV and I_x=2.5 mA.

not be determined. However, the magnitude of the combined signal observed in these experiments does not preclude the possibility that concentrations of SO_2F_2 and SOF_4 in the decomposed SF_6 may be of the same order as those measured for SOF_2.

Production of S_2F_{10} and $S_2O_2F_{10}$ was also observed in the target cell, and a production curve for S_2F_{10} is shown in Fig. 4 for a 300-kPa SF_6 sample irradiated by the X-ray beam with V_x = 195 kV and I_x=2.5 mA. The production rates of S_2F_{10} are nearly a factor of ten lower than those observed for SOF_2. No quantitative analysis was performed for $S_2O_2F_{10}$ due to the lack of a reliable reference sample.

DISCUSSION AND CONCLUSIONS

The by-products produced in SF_6 during exposure to X-rays are essentially the same as the by-products produced by corona,[1,2] sparks,[5] and arcs[6] in SF_6. This implies that the chemistry of by-product formation in decomposed SF_6 is essentially independent of the dissociation mechanism. Interestingly, the relative production rates of S_2F_{10} and SOF_2 for dissociation of SF_6 by X-rays are nearly the same as for cases where SF_6 is decomposed by exposure to corona discharges.[1,2] This implies that the SF_5 radical, the precursor to S_2F_{10} formation, must possess a relatively long lifetime under the conditions used here, because of the large volume in which S_2F_{10} formation occurs during exposure to radiation (a factor of 10^9 larger than in a corona discharge).[7] The data also show that a significant concentration of toxic by-products can be produced by X-ray exposure in a relatively short period of time (hundreds of minutes). This fact should be taken into account when dealing with the safe handling of SF_6 that has been exposed to radiation.

The dissociation of SF_6 in the presence of X-rays may be caused by either of two processes: (1) direct photodissociation of the gas by K-shell absorption by the sulfur and the fluorine, or (2) dissociation or dissociative attachment induced by secondary electrons generated by X-rays striking the back wall of the target cell. Direct photodissociation of SF_6 is due to K-shell absorption by the sulfur and fluorine, with absorption edges at 2.3 keV and 0.68 keV, respectively.[8] Dissociative attachment is due to the capture of a free electron to form a temporary negative ion that dissociates into a stable negative ion and a neutral fragment.[9]

The fact that the concentration of decomposition by-products in the target cell is independent of pressure (see Fig. 2) suggests that the primary dissociating process under these conditions is photodissociation, since the number of secondary electrons produced by the X-rays will not change significantly with gas pressure. In order to verify this conclusion, a curved, perforated stainless steel plate was installed inside the target cell, 2 mm from the back wall of the chamber. A dc voltage (20 kV) was then applied to the plate in order to produce an electric field inside the target gas while the sample was exposed to the X-ray source. Since the dissociation or dissociative attachment processes are highly dependent upon the energy of the free electrons, the rate of by-product formation would vary with the applied electric field if these processes played a significant role in SF_6 dissociation. However, the production rates of the decompostion by-products were observed to be unaffected by the applied field, thus indicating that secondary electrons do not play a major role in the dissociation of SF_6 under these conditions. This confirms that photodissociation is the primary means of decompostion of SF_6 in the presence of high-energy photons.

REFERENCES

1. R. J. Van Brunt, Production rates for oxyfluorides SOF_2, SO_2F_2, and SOF_4 in SF_6 corona discharges, *J. Res. Natl. Bureau Standards*, 90:229 (1985).
2. R. J. Van Brunt, J. K. Olthoff, and M. Shah, Rate of S_2F_{10} production from negative corona in compressed SF_6, *in*: "Conference Record of the 1992 IEEE Symposium on Electrical Insulation," IEEE, Piscataway, NJ (1992).
3. R. Hergli, J. Casanovas, A. Derdouri, R. Grob, and J. Mathieu, Study of the decomposition of SF_6 in the presence of water, subjected to gamma radiation or corona discharges, *IEEE Trans. Electrical Insul.* 23:451 (1988).
4. J. K. Olthoff, R. J. Van Brunt, J. T. Herron, and I. Sauers, Detection of trace disulfur decafluoride in sulfur hexafluoride by gas chromotography/mass spectrometry, *Anal. Chem.* 63:726 (1991).
5. I. Sauers, By-product formation in spark breakdown of SF_6/O_2 Mixtures, *Plasma Chem. Plasma Process.* 8:247 (1988).
6. B. Belmadani, J. Casanovas, A. M. Casanovas, R. Grob, and J. Mathieu, SF_6 decomposition under power arcs, *IEEE Trans. Electrical Insul.* 26:1163 (1991).
7. R. J. Van Brunt and J. T. Herron, Plamsa chemical model for decomposition of SF_6 in a negative glow discharge, in preparation.
8. CRC Handbook of Chemistry and Physics, R. C. Weast, Ed., CRC Press, Boca Raton, FL (1986).
9. H.-X. Wan, J. H. Moore, J. K. Olthoff, and R. J. Van Brunt, Electron scattering and dissociative attachment by SF_6 and its electrical-discharge by-products, *Plasma Chem. Plasma Process.* 13:1 (1993).

DISCUSSION

J. CASTONGUAY: Can you comment on the reaction schemes that produce a similar relative ratio of S_2F_{10}/SOF_2 in your "diffused" reaction volume compared to the "minute" and very localized corona discharges?

J. OLTHOFF: The fragmentations of SF_6 by X-rays and by corona both produce SF_5 radicals. The surprising aspect of S_2F_{10} formation in the large volume exposed to X-rays, is that the SF_5 radicals exist long enough to find each other and form S_2F_{10}.

PRODUCTION OF S_2F_{10}, S_2OF_{10}, AND $S_2O_2F_{10}$ BY SPARK DISCHARGES IN SF_6*

I. Sauers, S. M. Mahajan[1], and R. A. Cacheiro[2]

Health Sciences Research Division
Oak Ridge National Laboratory
Oak Ridge, Tennessee 37831-6123

INTRODUCTION

Decomposition of SF_6 in electrical discharges has been the subject of extensive study and review over the past decade.[1-4] Because of toxicological concerns recent[5] research has focused on the formation of S_2F_{10}, an SF_6 byproduct of high toxicity.[6,7] As a result of a concerted effort to establish the formation of S_2F_{10} in practical systems the results of laboratory experiments[8,9] have shown that S_2F_{10} is formed over a wide range of discharge conditions including power arc, corona, and spark discharges. In previous work in our laboratory the yield of S_2F_{10} formation in SF_6 spark discharge experiments was measured by conventional gas chromatography with a sensitivity of only around 50 ppm, requiring the deposition of large amounts of energy (generally requiring hundreds to thousands of sparks) into a small volume breakdown cell to develop large enough concentrations for detection. In addition to the formation of S_2F_{10}, other disulfur compounds are known to be formed including the oxyfluorides S_2OF_{10} and $S_2O_2F_{10}$. These compounds are of interest for several reasons. Both byproducts may be potential interferents in the detection of S_2F_{10}. Although their toxicities are not well characterized they both exhibit toxicity, with $S_2O_2F_{10}$ having toxicity approaching that of S_2F_{10}. Understanding the chemistry governing formation of these compounds is important because they may involve reactions with the SF_5 radical, a precursor to S_2F_{10} formation.

---

[1]Department of Electrical Engineering, Tennessee Technological University, Cookeville, TN 38505

[2]Department of Physics, University of Tennessee, Knoxville, TN 37996

Gaseous Dielectrics VII, Edited by L.G. Christophorou
and D.R. James, Plenum Press, New York, 1994

EXPERIMENT

Analytical Technique

Gas samples (typically 2mℓ at 100kPa) before and after sparking were taken directly from the spark chamber via a stainless steel sampling port (with a silicone rubber septum) using a gas-tight syringe. Samples were then injected into a cryogenic enrichment-gas chromatograph with an electron capture detector shown in Figure 1. The apparatus which had been described in Ref. 10 and 11 had been modified to improve the sensitivity to below 10 ppb. The gas sample, introduced at A (Figure 1), is entrained in a helium carrier gas stream which sends the sample to a trap filled with Porapak Q (Supelco, Inc.) operated at T=-20°C. At that temperature the S_2F_{10} component of the sample is selectively retained, while the unretained SF_6 as well as some of the other major by-products (SOF_2, SO_2F_2, SOF_4, and SO_2) are vented, thereby enriching the original sample in S_2F_{10} content relative to SF_6. Following this enrichment stage the trap is slowly warmed sending the trapped S_2F_{10} and other trapped components to a cryofocusing injector (CFI [Tekmar Model 1000]) where the sample is collected at T=-196°C. Flash heating, with a temperature rise of 50°C/s for 10s, injects the sample into the gas chromatograph (Varian Model 3600) for analysis. The chromatographic column is 50-m long fused silica with an inside diameter of 0.25 mm and a 0.6 μm inside coating of SPB-1 (Supelco, Inc.) operated at T=-50°C isothermal (He carrier gas flow rate=4.2 mℓ/min). This column has similar separating properties as that used by Janssen.[12]

The elution order of all three disulfur compounds studied was S_2OF_{10}, S_2F_{10}, and $S_2O_2F_{10}$. The same elution order was found using a packed chromasorb WAW column. We should also note that S_2OF_{10} elutes before S_2F_{10}, opposite to the elution order reported by Janssen[12] who used a capillary column similar to that used in the present work. The elution order of these species is consistent with other gas chromatographic data.[8,11,14,15]

Reference samples of S_2F_{10}, S_2OF_{10}, and $S_2O_2F_{10}$ were synthesized at Clemson University and received in stainless steel cylinders as liquid, each under its own vapor pressure. S_2F_{10} was found to be relatively stable as a liquid but decays in the gas phase by reactions associated with the cylinder walls.[16] S_2OF_{10} was found to be stable both in the liquid and gas phase showing no indication of decay. The sample of $S_2O_2F_{10}$ however appears to be the least stable of these compounds in either gas or liquid form. Because of its relatively short shelf life, we were unable to obtain and maintain reproducible reference sample mixtures over the course of this work. Although no quantitative data on $S_2O_2F_{10}$ will be reported, preliminary data indicate that $S_2O_2F_{10}$ is produced in sparked SF_6.

Spark Cell and Energy Measurement

Spark discharges were produced in a 1.1-ℓ stainless steel chamber as shown in Figure 2. For most of the data reported here the energy was supplied from a 0.4 μF capacitor which was charged until breakdown occurred across the gas gap. A reproducible breakdown voltage was obtained by illuminating the high voltage electrode (producing initiating electrons) with continuous UV light, using a deuterium lamp, through a sapphire viewing port. Both electrodes were made of stainless steel. The spark energy was measured by simultaneously sampling the voltage waveforms V_A and V_B using a dual channel digital oscilloscope (Nicolet 320) and high voltage probes at points A and B in Figure 2. The spark current is given by V_B/R_s where R_s is the shunt resistor.

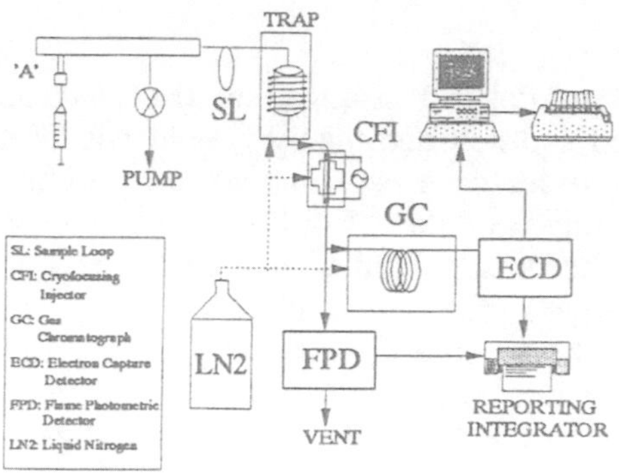

Figure 1. Cryo-GC analytical system.

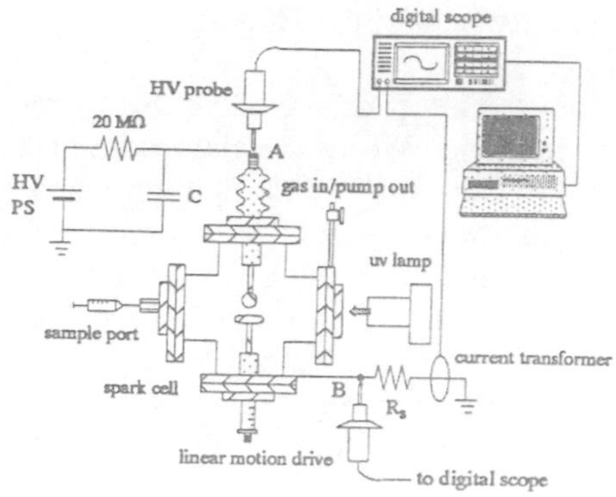

Figure 2. Spark cell and energy measurement system.

The energy, ε, deposited into the spark is determined by integrating the product of the voltage across the spark gap, V_A-V_B, and the current V_B/R_s, given by $\varepsilon = \int((V_AV_B-V_B^2)/R_s)dt$. Typical spark energies for the experiments reported here are ε=80 J/spark.

RESULTS AND DISCUSSION

Byproduct production in 'pure' SF_6

In previous work using small spark cell volumes and large number of discharges in SF_6, yields of S_2F_{10} were found to fall in the range 0.04-0.37 nmole per joule, depending on factors including moisture content of the gas.[17] Because of the large number of sparks required to produce measurable concentrations of S_2F_{10} using conventional gas chromatography it was not possible to carefully control and measure the moisture content of the gas during the experiment. Qualitatively, however, it was found that the addition of water to SF_6 resulted in decreased production of S_2F_{10}, so that the maximum yield corresponded to experiments conducted under extremely dry conditions. In the experiments reported here the S_2F_{10} yield was found to be lower than the yields found in the earlier work indicating that the large spark chamber used in the present studies is not as dry, which is to be expected. Repeated sparking into a small volume in the previous work effectively dries the surfaces through production of SF_4 which reacts with water. Spark yields of S_2F_{10} for 7, 20, 30, and 80 J per spark are given in Table 1. No significant dependence on spark energy was observed for the yield. For these measurements an average yield of 1.6±0.5 x 10^{-11}mol/J was obtained.

Table 1. Spark yield of S_2F_{10} in SF_6 at different spark energies.

Byproduct		Yield (10^{-11}mol/J)				
	ε (J/spk)	7 J	20 J	30 J	80 J	Avg
S_2F_{10}		1.2±0.1(2)[a]	1.0±0.2(2)	2.1±0.1(2)	1.7±0.3(5)	1.6±0.5(11)

[a] Number of data sets in parentheses.

Electrode Surface Effects

The production of S_2F_{10} was found to depend strongly on the electrode condition. When the electrodes were cleaned and polished the S_2F_{10} yield was found to be substantially lower than when the electrodes have been subjected to multiple breakdowns, generally in excess of one hundred sparks. Insufficient control and characterization of the electrode surface for the present studies, make interpretation of electrode effects on S_2F_{10} speculative. Since gas phase water and oxygen were found to influence the yield of S_2F_{10} production, these same contaminants arising from electrode surfaces could also influence the yield. Whether the smoothness of the surface affects the yield through surface catalyzed reactions has not been investigated.

Effect of Oxygen Addition

When oxygen is added to SF_6 the production of S_2F_{10} decreases while the production of S_2OF_{10} increases. Figure 3(A) and (B) shows the concentration of S_2F_{10} and S_2OF_{10} as a function of spark number for oxygen concentration over the range 0-10% in SF_6. Yields (in moles per joule) were determined from the slopes obtained from linear fits to the data in Figure 3 for each O_2/SF_6 mixture studied and are plotted in Figure 4.

A simple model is offered here to explain the dependence of the S_2F_{10} yield on the oxygen content. First it is assumed that SF_5 is the precursor to S_2F_{10} (reaction 1) and S_2OF_{10} (reactions 2 and 3) and that in the absence of species added to SF_6, the same amount of SF_5 is produced in every spark, that is, each spark is independent of each other.

$$SF_5 + SF_5 \xrightarrow{k_1} S_2F_{10} \quad (1)$$
$$SF_5 + O \xrightarrow{k_2} SF_5O \quad (2)$$
$$SF_5O + SF_5 \xrightarrow{k_3} S_2OF_{10} \quad (3)$$

In addition to the above reactions, SF_5 can react with other species such as F or OH (from background water) to produce products other than S_2F_{10} and S_2OF_{10}. The fraction, f, of the SF_5 concentration, $[SF_5]$, that results in S_2F_{10} formation is expressed as the ratio

$$f = k_1[SF_5]/\sum k_i[X_i] \quad (4)$$

where X_i and k_i are reactants and corresponding rate constants respectively for all reactions involving SF_5. The concentration of S_2F_{10} is given by

$$[S_2F_{10}] = f[SF_5] \quad (5)$$

When oxygen is added, an additional reaction path is included so that f becomes

$$f = k_1[SF_5]/(\sum k_i[X_i] + k_2[O]) \quad (6)$$

Converting concentration to yield by dividing the concentration by the spark energy per volume (ε/V), the yield, Y, assumes the form

$$Y_{S_2F_{10}} = c_1/(c_2 + [O_2]) \quad (7)$$

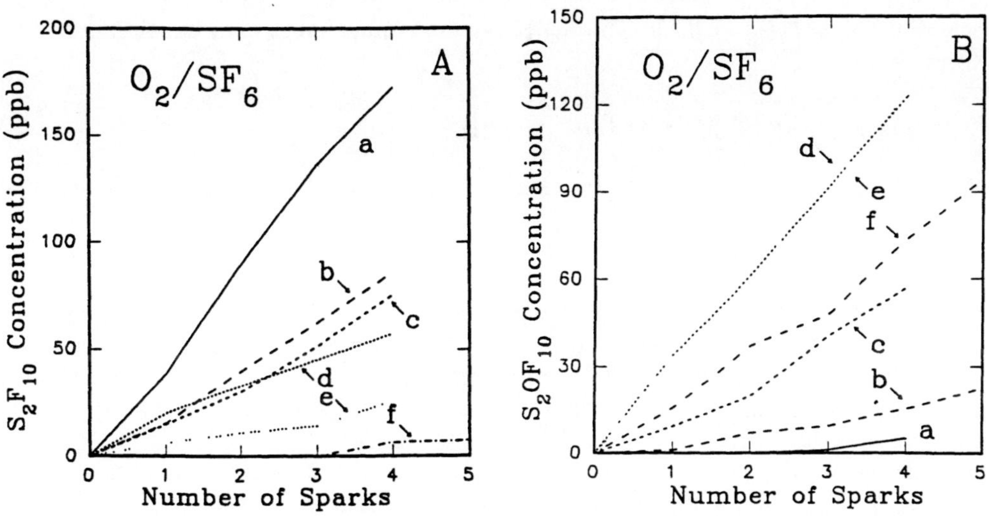

Figure 3. Production of (A) S_2F_{10} and (B) S_2OF_{10} in sparked O_2/SF_6 mixtures for oxygen concentrations (in percent): (a) 0, (b) 0.1, (c) 0.5, (d) 1.0, (e) 5, (f) 10. Pressure=100kPa and energy per spark=80J.

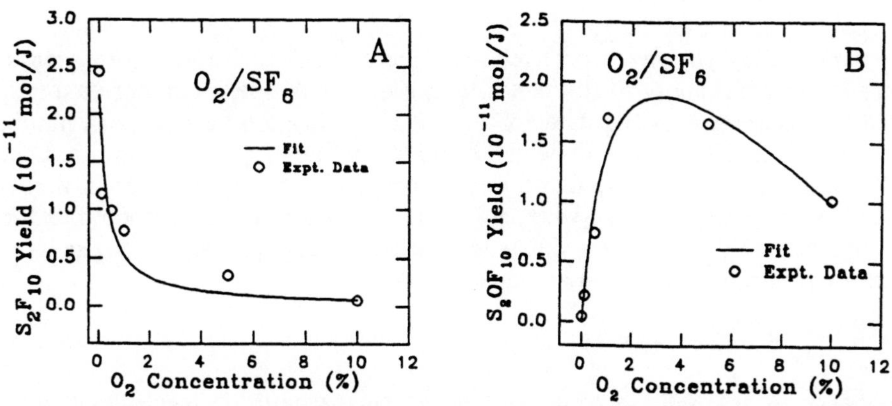

Figure 4. Spark yield of (A) S_2F_{10} and (B) S_2OF_{10} as a function of added oxygen concentration.

where [O] is proportional to $[O_2]$. A fit of eqn. (7) to the experimental data is shown (solid line) in Figure 4(a).

The production of S_2OF_{10} is viewed as proceeding through the intermediate step, reaction (2), leading to formation of SF_5O. As before it is assumed that each spark is independent of each other and that the fractional amount of SF_5 leading to formation of SF_5O is the ratio of the reaction (2) rate to the sum of all the rates times the SF_5 concentration

$$[SF_5O] = \{k[O]/(\sum k_i[X_i] + k[O])\}[SF_5] \tag{8}$$

For large [O] the concentration of SF_5O approaches that of SF_5. Reaction of the remaining SF_5 radicals with SF_5O results in formation of S_2OF_{10} given by

$$[S_2OF_{10}] = \{k[SF_5O]/(\sum k_i[X_i] + k[SF_5O])\}[SF_5]_r \tag{9}$$

where $[SF_5]_r = [SF_5] - \gamma[O_2]$. The S_2OF_{10} yield assumes an oxygen dependence of the form

$$Y_{S_2OF_{10}} = c_3[O_2](c_4 - [O_2])/(c_5 + [O_2]) \tag{10}$$

A curve fit of this expression to the data is indicated by the solid line in Figure 4(B).

Effect of Water Addition

The influence of moisture on S_2F_{10} production was investigated by injecting 1-5 $\mu\ell$ of water directly into the spark chamber. After allowing the water to equilibrate with the walls and the water vapor pressure to approach a relatively constant value, SF_6 was added to a pressure of one atmosphere. One to eight sparks were then produced and the gas was immediately analyzed after each spark. In Figures 5(A) and (B) production of S_2F_{10} and S_2OF_{10} is given as function of the number of sparks for 1, 2, 3, 4, and 5 μl of water injected into the spark chamber. The corresponding water vapor concentrations (for 1-4 μl) were 630, 1190, 1840, and 2750 ppm. A small air leak in the 5 μl experiment did not permit a valid measurement of the water vapor partial pressure only for that experiment. For these experiments SF_6 was sparked at 20 kV, 100 kPa, and 80 J per spark.

The dependence of the yield of S_2F_{10} and S_2OF_{10} on the amount of water added to the spark cell is shown respectively in Figure 5(A) and (B). The yields were determined from a linear fit to the concentration versus spark number data in Figure 5, with the exception of the data for S_2OF_{10} concentration versus spark number for pure SF_6 (Figure 5(B) curve a), which exhibits a nonlinear increase in production.

Analyses similar to those made for the oxygen data can be made in the case of H_2O addition to SF_6. Adding water to the gas adds additional reaction paths for the S_2F_{10} precursor, SF_5. The S_2F_{10} yield should follow a $[H_2O]$ dependence similar to that found for $[O_2]$ in eqn. (7), and assumes the form,

$$Y_{S_2F_{10}} = c_1'/(c_2' + N_{H_2O}) \tag{11}$$

where N_{H_2O} is the amount of water injected (in $\mu\ell$) and is proportional to the gas phase water concentration, $[H_2O]$. A fit of this expression to the experimental data is shown in Figure 6(A). In comparing the water data to the oxygen data, water addition of 5 $\mu\ell$ corresponds in concentration to approximately 0.3%. Water is therefore more effective in decreasing the S_2F_{10} yield than O_2 for comparable concentrations in SF_6. This suggests that possibly OH rather than O is responsible for SF_5 depletion when water is added.

In an analysis similar to that used for the O_2/SF_6 data, a fit of the S_2OF_{10} yield data is shown in Figure 6(B) along with the experimental data. The yield is significantly higher for oxygen addition than for an equivalent concentration of water. Yield data for S_2F_{10} and S_2OF_{10} for water and oxygen are listed in Table 2.

Table 2. Spark yield (in 10^{-11}mol/J) of S_2F_{10} and S_2OF_{10} in O_2/SF_6 and H_2O/SF_6 mixtures

	O_2 Concentration (%)					H_2O Content ($\mu\ell$)				
	0.1	0.5	1.0	5.0	10.0	1	2	3	4	5
S_2F_{10}	1.16	0.989	0.778	0.323	0.063	1.03	0.753	0.672	0.639	0.316
S_2OF_{10}	0.219	0.738	1.70	1.66	1.01	0.020	0.030	0.043	0.047	0.054

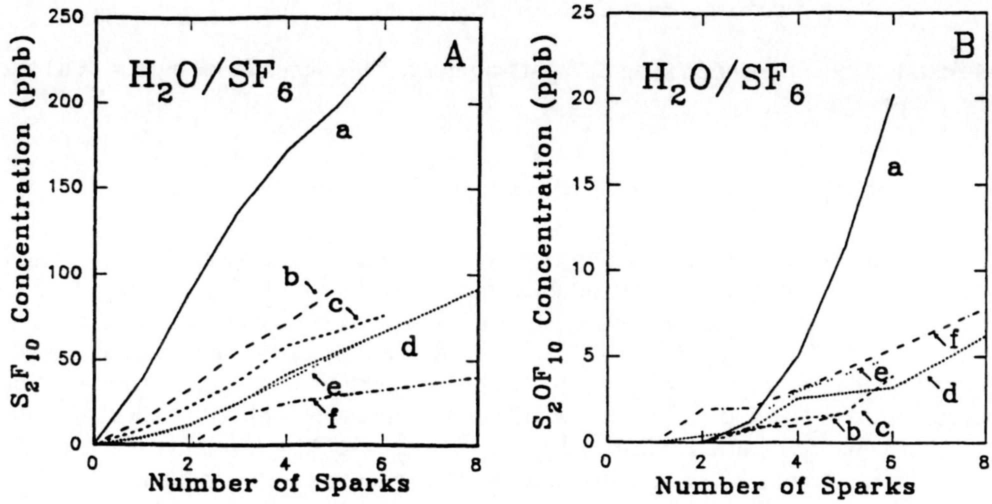

Figure 5. Production of (A) S_2F_{10} and (B) S_2OF_{10} in sparked H_2O/SF_6 mixtures for water additions (in $\mu\ell$) into a 1.1-ℓ spark cell: (a) 0, (b) 1, (c) 2, (d) 3, (e) 4, (f) 5. Pressure=100kPa and energy per spark=80J.

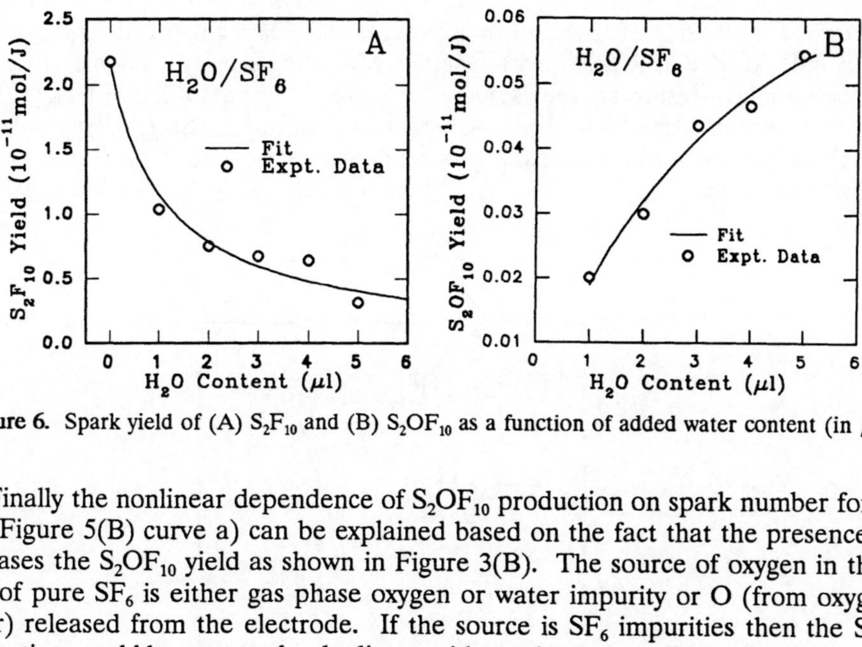

Figure 6. Spark yield of (A) S_2F_{10} and (B) S_2OF_{10} as a function of added water content (in μl).

Finally the nonlinear dependence of S_2OF_{10} production on spark number for pure SF_6 (Figure 5(B) curve a) can be explained based on the fact that the presence of O increases the S_2OF_{10} yield as shown in Figure 3(B). The source of oxygen in the case of pure SF_6 is either gas phase oxygen or water impurity or O (from oxygen or water) released from the electrode. If the source is SF_6 impurities then the S_2OF_{10} production would be expected to be linear with spark number. Rate equations in units of spark number, n, in place of time, t, can be written for S_2OF_{10} (reaction 3), SF_5O (reaction 2), and O (released from the electrode) as

$$d[S_2OF_{10}]/dn = k_3'[SF_5O][SF_5] \quad (12)$$
$$d[SF_5O]/dn = k_2'[SF_5][O] \quad (13)$$
$$d[O]/dn = \alpha \quad (14)$$

where each spark produces a constant addition of [O] to the gas and "rate" constants are in units of volume per spark. Assuming that the amount of SF_5 produced is independent of n, the concentration of S_2OF_{10} is given by

$$[S_2OF_{10}] = cn^3, \text{ where } c=\alpha k_2' k_3' [SF_5]^2/6 \quad (15)$$

indicating a cube dependence on the spark number. This dependence agrees well with the data (Figure 7).

CONCLUSIONS

Measurements have been made of the spark yield of S_2F_{10} and S_2OF_{10} as a function of O_2 and H_2O concentration. In addition the formation of $S_2O_2F_{10}$ is reported. The spark yield of S_2F_{10} is reduced when either oxygen or water is added to SF_6. On the other hand the yield of S_2OF_{10} is increased when either oxygen or water is added, oxygen having the greater effect. A simple model is presented based on a common SF_5 precursor to S_2F_{10} and S_2OF_{10} showing good agreement with the yields of both byproducts in O_2/SF_6 and H_2O/SF_6 mixtures. The yield of S_2F_{10} is found to be greatest when the spark chamber is very dry and the electrodes are well conditioned.

ACKNOWLEDGMENTS

This work was conducted under a Cooperative Research Development Agreement supported by the Office of Energy Management, U.S. Department of Energy (under contract DE-AC05-84OR21400 with Martin Marietta Energy Systems, Inc.), the Tennessee Valley Authority, the Bonneville Power Administration, Electric Power Research Institute, Canadian Electrical Association, Empire State Electric Energy Research Corporation, National Institute of Standards and Technology, Ontario Hydro, and Hydro Québec.

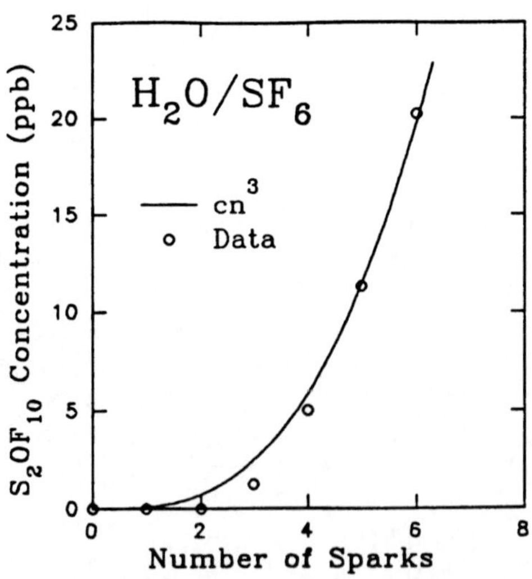

Figure 7. Nonlinear increase in S_2OF_{10} production with number of sparks: data (open circles), curve fit (solid line).

REFERENCES

1. F. Y. Chu, SF_6 decomposition in gas-insulated equipment, IEEE Trans. Elec. Insul. **21**, 693 (1986).
2. R. J. Van Brunt, Production rates for oxyfluorides SOF_2, SO_2F_2, and SOF_4 in SF_6 corona discharges, J. Res. Nat. Bur. Stand. **90**, 229 (1985).
3. I. Sauers, H. W. Ellis, and L. G. Christophorou, Neutral decomposition products in spark breakdown of SF_6, IEEE Trans. Elec. Insul. **21**, 111 (1986).
4. R. J. Van Brunt and J. T. Herron, Fundamental processes of SF_6 decomposition and oxidation in glow and corona discharges, IEEE Trans. Elec. Insul. **25**, 75 (1990).
5. D. R. James, I. Sauers, G. D. Griffin, R. J. Van Brunt, J. K. Olthoff, K. L. Stricklett, F. Y. Chu, J. R. Robins, and H. D. Morrison, Investigation of S_2F_{10} production and mitigation in Compressed SF_6-insulated power systems, IEEE Elect. Insul. Magazine 9(3), 29-40, 51 (1993)
6. L.A. Greenberg and D. Lester, The toxicity of sulfur pentafluoride, Arch. Indust. Hygiene and Occupat. Med. 2, 350-353 (1950).
7. G.D. Griffin, K. Kurka, M.G. Nolan, M. D. Morris, I. Sauers and P.C. Votaw, Cytotoxic activity of disulfur decafluoride (S_2F_{10}), a decomposition product of electrically-stressed SF_6, In Vitro 25(8), 673-675 (1989).
8. I. Sauers, G. Harman, J. K. Olthoff, and R. J. Van Brunt, Gaseous Dielectrics VI, Proc. 6th Int. Symp. Gaseous Dielectrics (Plenum, New York, 1991) pp. 553-562.
9. R. J. Van Brunt, J. K. Olthoff, I. Sauers, H. D. Morrison, J. R. Robins, and F. Y. Chu, Detection of S_2F_{10} produced by electrical discharges in SF_6, Proceedings of the Tenth International Conference on Gas Discharges and Their Applications, W. T. Williams, ed., 418-21 (1992)
10. I. Sauers and R. A. Cacheiro, A Cryogenic enrichment technique for gas chromatographic detection of S_2F_{10} in SF_6 discharges, Conference Record of the 1992 IEEE International Symposium on Electrical Insulation, 340-44 (1992).
11. I. Sauers and S. M. Mahajan, Detection of S_2F_{10} produced by a single spark discharge in SF_6, J. Appl. Phys. Comm. **74**(3), 2103-2105 (1993).
12. F. J. J. G. Janssen, Measurements of the sub-ppm level of sulfur-fluoride compounds resulting from the decomposition of SF_6 by arc discharge, *KEMA Scientific and Technical Report* **2**(2), 9-18 (1984).
13. R. J. Van Brunt, J. K. Olthoff and M. Shah, Rate of S_2F_{10} production from negative corona in compressed SF_6, *Conference Record of the 1992 IEEE International Symposium on Electrical Insulation*, IEEE Cat. No. 92-CH-3150-0, pp. 328-331 (1992).
14. J. Gutbier and J. Luy, *Anal. Chemie* **231**, 329-339 (1967).
15. J. M. Hanrahan and A. R. Paterson, *J. Chromatog.* **193**, 265-270 (1980).
16. J. K. Olthoff, R. J. Van Brunt, J. T. Herron, I. Sauers, and G. Harman, Catalytic decomposition of S_2F_{10} and its implications on sampling and detection from SF_6-insulated equipment, Conference Record of the 1990 IEEE Int. Symp. on Elect. Insul., IEEE Cat. No. 90-CH2727-6, New York, 248-252, (1990).
17. I. Sauers, G. Harman, J. K. Olthoff, and R. J. Van Brunt, S_2F_{10} formation by electrical discharges in SF_6: comparison of spark and corona, *Gaseous Dielectrics VI* (L. G. Christophorou and I. Sauers, eds) Plenum Press, New York, pp. 553-562 (1991).

DISCUSSION

J. CASTONGUAY: Your data in Table 2 show that the S_2F_{10} yield is reduced by ~ 25 to 35% with the addition of water in quantities of 1 to 4 μL. This seems to prove that water has not that large effect that is usually claimed. Can you comment on that and the much lower value obtained with the addition of 5 μL of H_2O?

I. SAUERS: Five microliters of water injected corresponds to about 0.3% (3000 ppm) water vapor concentration added to SF_6. This water addition results in reduction of S_2F_{10} from about 2.2 x 10^{-11} mol/J for no added water to 0.32 x 10^{-11} mol/J for 0.3% water. This is an extremely large effect for such a small water concentration. Water is much more effective in reducing S_2F_{10} than oxygen. Figure 6 of our paper shows a relatively smooth (with some scatter) decay in S_2F_{10} agreeing in oxygen dependence with the analytical expression from the simple model. It should also be pointed out that even in the 0 μL case there already exists some water in the gas of unknown concentration, possibly on the order of a few hundred ppm, making the effect of added water even more significant.

J. CASTONGUAY: When you added water you waited for its equilibrium to be reached but did not wait long for sample analysis. Do you think that by-product equilibrium was reached after such a short waiting time?

I. SAUERS: Each spark results in rapid dispersal of the decomposition products throughout the chamber. This is easily checked by taking several samples after a spark. These tests indicate that mixing of the by-products in the chamber occur faster than the sampling time. On the other hand when making H_2O/SF_6 mixtures, water is injected into an evacuated chamber where water must first reach equilibrium with the walls before a relatively constant concentration of water is established.

R. J. VAN BRUNT: It should be noted that the reaction SF_5 + O has been investigated by others (Ryan and Plumb), and at low pressures the most likely products are SOF_4 + F. In general, there will be competing reactions that can complicate the model.

I. SAUERS: The reactions used in this simple model probably occur at some unknown temperature. Also at high pressure the SOF_4 reaction may not be favorable. It is possible that SF_5O is produced in collisionally stabilized reactions. The data indicated that S_2OF_{10} accounts for the S_2F_{10} loss suggesting that the SOF_4 reaction may not be important with regard to SF_5 depleting processes.

PRODUCTION AND DECAY OF S_2F_{10} IN A DISCONNECT SWITCH

H.D. Morrison[1], V.P. Cronin[1], F.Y. Chu[1], M. Eygenraam[1]
I. Sauers[2], and M.J. Dallavalli[3]

[1]Ontario Hydro Technologies
800 Kipling Avenue, Toronto, Ontario
Canada M8Z 5S4

[2]Oak Ridge National Laboratory
Oak Ridge TN 37831-6123

[3]COPEL-LAC, Curitiba 80230
Brazil

INTRODUCTION

The production of S_2F_{10} by electrical discharges in SF_6 has been demonstrated in a variety of laboratory conditions, including arcs, sparks and corona [1]. The question arises whether S_2F_{10} is produced by electrical discharges that occur under normal operating conditions of gas-insulated substation (GIS) components. To address this question we have extended the laboratory studies to investigate the production and decay of S_2F_{10} in a typical GIS component, a disconnect switch. During a disconnect operation, multiple sparks are created and there may be some corona. The energies of the discharge were not measured, but they are considerably less than in a breaker, which more closely resembles an arc fault. Thus, we expect to see decomposition of SF_6 in a disconnect switch to be more similar to spark discharges than to power arcs as studied in the laboratory.

At Ontario Hydro, our research facility includes an entire GIS, rated for 138 kV and operated at 115 kV. For this study we simulated typical operating conditions with one disconnect switch of the GIS, but at a rate of switching about 100 times more frequent than normal. The decomposition products of the SF_6, including S_2F_{10}, were analyzed in our laboratory from gas samples taken from the switch compartment. As S_2F_{10} is known to decay in sample cylinders on the scale of several hours [2], the location on site of a sensitve technique for rapid analysis of samples from a discharge experiment is essential for collecting meaningful results. Through active participation in the Co-operative Research And Development Agreement (CRADA, see Ref. 1) to study S_2F_{10} in SF_6-insulated power systems, we at Ontario Hydro have developed a technique using infrared absorption spectroscopy over a long pathlength in a Fourier transform infrared (FTIR) spectrometer to

Gaseous Dielectrics VII, Edited by L.G. Christophorou and D.R. James, Plenum Press, New York, 1994

analyze decomposed SF_6 for S_2F_{10} and many other possible byproducts. Our detection limit for S_2F_{10} is less than 0.10 parts per million (ppm) by volume, even in the presence of several tens of ppm of SOF_2 and SO_2F_2, usually the most abundant decomposition byproducts found in SF_6.

In this paper we present the analysis of samples taken during a week of operation of the disconnect switch and for several weeks thereafter. From these measurements we calculate average rates of production and decay of S_2F_{10} and four other byproducts. Separate analysis at Oak Ridge National Labs (ORNL) of two of the samples confirms the presence of S_2F_{10} and the accuracy of the FTIR technique. The implications of these results for operating GIS will be discussed in relation to the permissible limits of exposure to the more well-known decomposition byproducts of SF_6, such as SOF_2 and SO_2F_2.

DISCHARGES IN A DISCONNECT SWITCH

A section of a single phase of the GIS including a disconnect switch was prepared according to the manufacturer's manual: it was evacuated, filled with nitrogen to take up moisture, evacuated again, and then filled with commercial grade, virgin SF_6 to 465 kPa at 7 °C. No adsorbents were placed in the 216-liter volume of the switch compartment. Just before the first discharge sequence, the dewpoint of the gas in the compartment was about -30 °C, as measured with a chilled-mirror hygrometer. That corresponded to a water concentration of about 80 ppm by volume.

A transformer supplied power at 70 kV line-to-ground into a gas-to-air bushing and along about 6 m of bus duct to the disconnect switch. Because of the configuration of the GIS, another 10-m section of bus duct contributed to the capacitance on the source side of the switch, to a total of 925 pF. About 12 m of GIS components were connected to the load side of the disconnect switch to provide a load capacitance of 690 pF.

Typical operation of such a switch could be up to 1 or 2 open-close cycles per day. To facilitate the measurement of byproducts, however, we operated the switch through 200 open-close cycles per day, a total of 1000 cycles over one week. The switch was controlled by a computer through a relay to the switch linkage motor to keep track of the number of cycles and to provide continuous automatic operation. The average switching rate was 3 open-close cycles per minute, sustained for 200 cycles each day of the test. We monitored the light produced by the discharge in the disconnect switch with a photomultiplier detector to determine the number and duration of the sparks in each cycle. We observed about 1000 sparks per cycle, each about 100 ns in duration, varying in intensity with the gap separation in the switch.

To monitor the decomposition of the SF_6 in the compartment of the disconnect switch, we sampled the gas about 2 hours before and then immediately after each day's operation of 200 open-close cycles. After a total of 1000 cycles, four more samples were taken over the next two months without otherwise disturbing the gas in the switch compartment. The samples were taken in 1-liter stainless steel cylinders and then analyzed within 1 hour by infrared absorption spectroscopy in a Fourier transform infrared (FTIR) spectrometer. Figure 1 shows a spectrum of the gas sampled after 1000 cycles, with a spectrum of clean SF_6 for comparison. All the samples were taken under essentially similar conditions. After 14 samples taken over 58 days, the pressure in the switch compartment was only reduced to about 92% of the initial pressure. During the test, the ambient temperature at the GIS varied from 4 to 12°C.

ANALYSIS OF DECOMPOSITION BYPRODUCTS

With the FTIR we recorded the infrared absorption spectra of each gas sample at as high a pressure as possible, about 60 kPa, in a cell of 19.8 m pathlength. Reference spectra of the four major byproducts, SOF_2, SO_2F_2, SF_4, and SOF_4, as well as S_2F_{10}, were obtained with pure samples in the same cell on the same instrument. The spectrum of the first sample, taken before any discharges in the switch, was used as the reference for SF_6. Figure 1 shows that the presence of the four major byproducts are easily identified in the infrared region near 550 cm^{-1}, even in the presence of SF_6 at high pressure. The presence of S_2F_{10} is not immediately apparent (see Ref. 3 for a spectrum), but can be determined by subtraction of the spectra of the other components in the gas mixture.

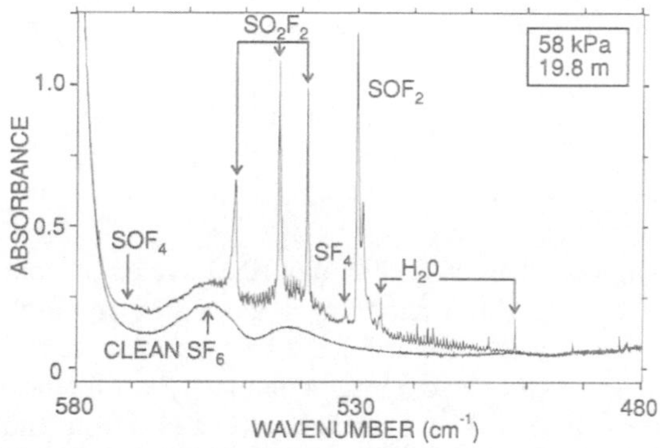

Figure 1. The infrared absorption spectrum of a sample of SF_6 from a disconnect switch compartment after 1000 open-close cycles, in comparison with a spectrum of clean SF_6. Both spectra were recorded in a cell of 19.8 m pathlength at a pressure of 58 kPa. S2F10 is not apparent in the upper spectrum, but it can be seen by subtracting reference spectra of the other byproducts.

The concentration of all five byproducts in the samples taken from the disconnect switch were determined by spectral subtraction, i.e. simultaneously subtracting a proportion of each byproduct reference spectrum from a sample spectrum to produce a flat baseline. In every case, a flat baseline was achieved, neglecting absorption lines of water, with only seven components: SF_6, SOF_2, SO_2F_2, SF_4, SOF_4, S_2F_{10}, and S_2OF_{10}. The latter compound, present at concentrations less than 1 ppm, is more stable and less toxic than S_2F_{10} [4]. From the proportions obtained by subtraction and the conditions of the reference spectra, we calculated the concentration of the byproducts in each sample. The results of this procedure for 13 of the 14 disconnect samples are illustrated in Fig. 2, beginning with the first sample of clean SF_6 containing no detectable concentration of byproducts. The SOF_2 and SO_2F_2 are most abundant and decay at the slowest rate. The S_2F_{10} is least abundant, but decays more slowly than either SF_4 or SOF_4. From the measurements, we estimated a detection limit of less than 0.10 ppm for S_2F_{10}.

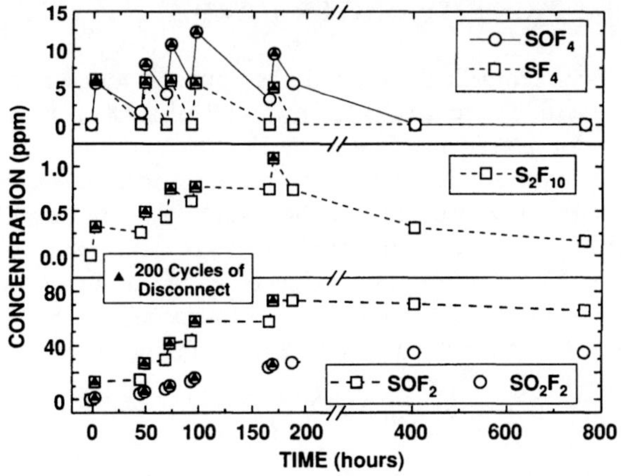

Figure 2. The concentration of five decomposition byproducts of SF_6 as a function of time in the disconnect switch. The triangles indicate the samples taken just after every 200 cycles of the switch. Note the change of scales for the three panels. The connecting lines are provided only as a guide.

Portions of the samples taken after 800 and 1000 cycles of operation were also sent to ORNL for analysis by cryogenic enrichment gas chromatography (GC) [5]. Those samples confirmed that S_2F_{10} was produced in the disconnect switch, and also demonstrated a problem in sampling. The S_2F_{10} decomposed completely within the 150-ml stainless steel sampling cylinders in one week. These results are compared in Table 1 with the results obtained by FTIR. One final sample in a 1-liter cylinder, taken 58 days after the last operation of the disconnect switch, was only analyzed at ORNL. This sample, analyzed

Table 1. Comparison of byproduct analysis by two methods: cryogenic-GC at ORNL, and FTIR at Ontario Hydro.

	Cryogenic-GC at ORNL			FTIR at Ontario Hydro		
Sample (cycles)	SOF_2 (ppm)	SO_2F_2 (ppm)	S_2F_{10} (ppm)	SOF_2 (ppm)	SO_2F_2 (ppm)	S_2F_{10} (ppm)
800	57	33	0.15[a]	57.7	16	0.77
1000	75	65	0.69[a]	73.2	26	1.09
1000 (after 58 days)	72	57	<0.010	—	—	—

[a]Initial values measured. Samples decayed with time.

within 24 hours of removal from the switch compartment, contained less than 0.010 ppm of S_2F_{10}.

RESULTS AND DISCUSSION

From the analysis of all the samples as shown in Fig. 2, we calculated the production and decay rates as shown in Fig. 3 and Table 2. The accumulated production of the five byproducts for each 200 cycles of operation is shown in Fig. 3. These values represent the production only during the operation of the disconnect switch, and not the effect of any production or decay processes between the sequences of 200 cycles. Note that the data form roughly straight lines, indicating a linear dependence of production for all five byproducts proportional to the number of cycles, and hence to the energy injected into the discharge with each cycle. The rates of production calculated from these data are listed in Table 2. If we had an accurate estimate of the energy discharged per cycle then we could convert the production rates to more familiar units of mol/J. That is a worthwhile problem for later study.

As shown in Fig. 2, the most abundant compound at all times is SOF_2. The production rates for SF_4 and SOF_4 are about half that of SOF_2, but they react rapidly with water to form SOF_2 and SO_2F_2, respectively, and HF. Thus, the SF_4 decays to zero within 24 hours; SOF_4 reacts more slowly than SF_4, but it too disappears within a few days. Note that the decay of SOF_4 is the main source of SO_2F_2, as indicated by the steadily rising curve in Fig. 2 and by a production rate during switching only about 30% of that of SOF_4. SO_2F_2 accumulates and approaches the concentration of SOF_2 because it decays at a slower rate.

The relative proportions of the five byproducts is similar to that observed in spark discharge experiments, and not at all similar to that seen in arc discharge experiments. For

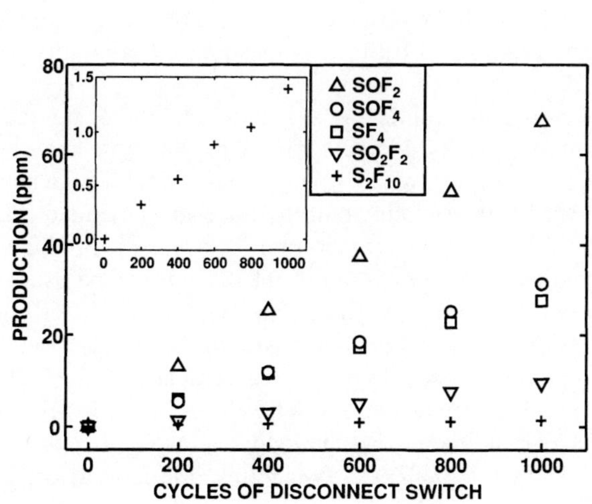

Figure 3. The accumulated production, measured in ppm concentration, of five byproducts for every 200 cycles of operation of the disconnect switch. The inset shows the production of S_2F_{10} with an expanded scale.

Table 2. Rates of production and decay of SF_6 decomposition byproducts.

Byproduct	Production Rate (nmol/cycle)	Decay Rate (sec^{-1})
S_2F_{10}	59 ± 16	7.0×10^{-7}
SO_2F_2	402 ± 86	6.2×10^{-9}
SOF_2	2849 ± 268	5.2×10^{-8}
SF_4	1171 ± 104	$> 4.5 \times 10^{-5}$
SOF_4	1323 ± 98	8.1×10^{-6}

example, in sparks, the production rates of SOF_2 and S_2F_{10} are within two orders of magnitude. In arcs, we found that the production rate for SOF_2 is at least 10,000 times greater than for S_2F_{10}.

As noted above, the production rate of S_2F_{10} during the discharge cycles is also roughly linear with the number of cycles. The chief difficulty in assessing the true production rate of S_2F_{10} is its propensity for decay in sample cylinders at low concentrations, on the order of 1 ppm or less. Analysis is not the problem. Table 1 indicates that the analysis techniques of ORNL and Ontario Hydro are comparable; the correspondence of the SOF_2 results is excellent, and the factor of 2 between the SO_2F_2 results is consistent. Detailed measurements by ORNL of the 800-cycle and 1000-cycle samples over a period of one week demonstrated an exponential decay of the samples in the cylinders at a rate 10 times greater than observed in the switch compartment. Extrapolation back to the time of sampling gives values consistent with the results obtained at Ontario Hydro. Note that the decay rate listed in Table 2 is for the S_2F_{10} in the switch compartment, not in the cylinders sent to ORNL. Using this decay rate to extrapolate from the last three samples measured by FTIR to the time of the last sample analyzed at ORNL, we predict a concentration of 0.020 ppm of S_2F_{10} in the compartment at that time. Although this is consistent with the observation of less than 0.010 ppm measured at ORNL within 24 hours of sampling, a doubt still remains about the reliability of sampling such low concentrations of S_2F_{10}.

Finally, let us consider the implications of these results for operating GIS equipment. Under normal operation, typically 1 or 2 cycles a day at most, the production and decay rates of S_2F_{10} in the disconnect switch suggest that S_2F_{10} would never be a significant safety hazard. One cycle of the switch would produce an immediate concentration of about 0.0014 ppm in SF_6 in the compartment (see Fig. 3); decay will reduce this value. That is less than the Threshold Limit Value (TLV) of 0.010 ppm for exposure to S_2F_{10} set by the American Conference of Governmental Industrial Hygienists (ACGIH) [6] (see Ref. 1 for discussion). Any release of the decomposed SF_6 to the atmosphere would immediately dilute the S_2F_{10} to even lower concentrations. For comparison, the TLVs for SOF_2, SO_2F_2 and SF_6 are 1.6, 5, and 1000 ppm, respectively. The TLV for SF_6 implies that any gas released from a GIS compartment should not be allowed to exceed 1000 ppm in the ambient air, a dilution factor of 1000. With this dilution factor, even the maximum concentration of S_2F_{10} achieved in the disconnect switch in this study would be below its TLV.

It is interesting to note that the decay rates of the five byproducts listed in Table 2 cover five orders of magnitude, with S_2F_{10} right in the middle. Thus, within a few days following any disconnect operations, we would find only SOF_2, SO_2F_2, and S_2F_{10} in the SF_6. Until now, the limits of exposure to decomposed SF_6 were determined mainly by the presence of SOF_2. In the disconnect switch, the production rate for SOF_2 was 48 times greater than for S_2F_{10}, and its decay rate was at least 10 times slower. Therefore, when handling SF_6 from a disconnect switch, SOF_2 is probably still the dominant contributor to the hazard of exposure, but S_2F_{10} should also be considered.

CONCLUSIONS

Production and decay rates for S_2F_{10} and four other decomposition byproducts of SF_6 have been measured in a disconnect switch, a typical GIS component, without -adsorbents present. From anlaysis by FTIR, the production rate of S_2F_{10} is about 2% of that of SOF_2, and its decay rate is more than 10 times faster. The switch was operated under normal conditions except for the accelerated frequency of cycling to facilitate the measurement of the byproducts. An independent analysis at ORNL confirmed the presence of S_2F_{10} in the samples. When considered all together, the results establish the validity of using an FTIR for quantitative analysis of S_2F_{10} to less than 0.10 ppm even in the presence of several 10's of ppm of SOF_2 and SO_2F_2. This detection limit is more than sufficient for assessing the limits of exposure to decomposed SF_6.

Although this study examined a typical set of conditions for GIS, the problem of S_2F_{10} decaying more rapidly in sampling cylinders illustrates that the factors affecting the decay rate of S_2F_{10} provide many opportunities for further study. Another area worthy of more detailed work is the dissipation of energy during a disconnect operation. Then the production rates found here can be related to other studies of SF_6 byproducts in electrical discharges.

ACKNOWLEDGEMENTS

This report was prepared as an account of work sponsored by agencies of the United States Government, the Electric Power Research Institute, the Empire State Electric Energy Research Corporation, the Bonneville Power Administration, the Tennessee Valley Authority, Ontario Hydro, Hydro Quebec, and the Canadian Electrical Association. None of the Parties nor any agency thereof, nor any of their employees, makes any warranty, express or implied, or assumes any legal liability or responsibility for the accuracy, completeness, or usefulness of any information, apparatus, product, or process disclosed, or represents that its use would not infringe privately owned rights. Reference herein to any specific commercial product, process, or service by trade name, trademark, manufacturer, or otherwise, does not necessarily constitute or imply its endorsement, recommendation, or favoring by the Parties or any agency thereof. The views and opinions of authors expressed herein do not necessarily state or reflect those of the sponsoring Parties including the United States Government or any agency thereof.

REFERENCES

1. D.R. James, I. Sauers, G.D. Griffin, R.J. Van Brunt, J.K. Olthoff, K.L. Stricklett, F.Y. Chu, J.R. Robins, and H.D. Morrison, Investigation of S_2F_{10} production and mitigation in compressed SF_6-insulated power systems, *IEEE Elec. Insul. Mag.* 9:29 (1993).
2. J.K. Olthoff, R.J. Van Brunt, J.T. Herron, and I. Sauers, Detection of trace disulfur decafluorides in sulfur hexafluoride by gas chromatography/mass spectrometry, *Anal. Chem.* 63:726 (1991).
3. F.Y. Chu, I. Sauers, and G.D. Griffin, A review of the formation of S_2F_{10} in gas insulated equipment, *in*: "Conference Record of the 1988 IEEE International Symposium on Electrical Insulation," IEEE (1988).
4. Private communication with G.D. Griffin. See also Ref. 1.
5. I. Sauers and S.M. Mahajan, Detection of S_2F_{10} produced by a single-spark discharge in SF_6, *J. Appl. Phys.* 74:2103 (1993).
6. "1992-1993 Threshold Limit Values for Chemical Substances and Physical Agents and Biological Exposure Indices," American Conference of Governmental Industrial Hygienists, Cincinnati (1992).

DISCUSSION

L. NIEMEYER: Your measurements under accelerated operation conditions indicate that the S_2F_{10} concentration to be expected under normal operation conditions (~2 operations per day) will be roughly 2 orders of magnitude lower, i.e. in the 10 ppbv range. Hence it can be concluded that there is no health risk associated with GIS disconnectors even under the worst case assumption that no absorber is provided in the disconnector compartment.

S. W. ROWE: The main question with which we must not loose contact is: "Is there a risk to operator health linked to S_2F_{10} for disconnector operation?" The important information is thus the maximum concentration which can be built up in a work place atmosphere. The conditions used in these experiments are at least one hundred times more severe than those of normal operating conditions due to the extremely high operation rate and the lack of adsorbents. However, even in these conditions one can calculate the maximum work place concentration by scaling the value measured inside the GIS compartment of a few hundred liters to the switchgear room volume of several hundred thousand liters. Doing this, and using the results presented here, one finds a concentration of about 0.3 ppb. This is about 30 times less than the value in which full-time work is permitted (TLV). The same treatment for SOF_2 shows this also to create a concentration far below the TLV. The conclusion seems to be thus that even under these extremely unfavorable conditions there should be no risk to operator health due to these gases. In real life situations this conclusion is even more true. Do you agree with this?

H. MORRISON: I agree with your analysis that there is practically no risk to operator health under normal operating conditions of a disconnect switch. Proper gas handling procedures will still need to be followed to minimize exposure to any by-products. I would also like to point out that the concept of TLV is a guideline to assist in the control of health hazards. In the case of S_2F_{10}, the TLV is a ceiling limit--it should never be exceeded. The best practice is to maintain concentrations of all airborne contaminants as low as is practical.

J. CASTONGUAY: You took the FTIR spectra of many by-products of SF_6. Can the IR spectrum data base you collected be used on other FTIR without the need to record again the standard spectra?

H. MORRISON: There exists an international standard for the exchange of IR spectra, known as JCAMP-DX. All FTIR control software, as far as I know, recognize this standard for importing and exporting data files. I propose to make my FTIR spectra available in DX format eventually.

E. KUFFEL: The results you presented today are reassuring that perhaps we need not panic. Specifically your Fig. 3 in the paper which summarizes the concentrations of the various by-products produced shows that while other by-products increase with the number of cycles of operation of disconnect switch, the concentration of S_2F_{10} remains nearly constant and at a low value.

H. MORRISON: I agree that the results are reassuring. In the case of all by-products other than S_2F_{10}, however, there is a good scientific understanding of the chemical reactions leading to decomposition. In the case of S_2F_{10}, we observe decomposition under many different conditions but we do not yet understand the processes involved.

PROCEDURE FOR MEASURING TRACE QUANTITIES OF S_2F_{10}, S_2OF_{10}, AND $S_2O_2F_{10}$ IN SF_6 USING A GAS CHROMATOGRAPH-MASS SPECTROMETER

R. J. Van Brunt, J. K. Olthoff, K. L. Stricklett, and D. J. Wheeler*

Electricity Division
National Institute of Standards and Technology†
Gaithersburg, Maryland 20899

INTRODUCTION

The compounds S_2F_{10}, S_2OF_{10}, and $S_2O_2F_{10}$ are formed by decomposition of gaseous sulfur hexafluoride (SF_6) in electrical discharges.[1,2] The species S_2F_{10} is known to be highly toxic to humans,[3] and there is recent evidence that $S_2O_2F_{10}$ may also be very toxic.[4] There is, therefore, an interest in having analytical methods to detect these compounds in compressed SF_6 at trace levels down to 10 parts in 10^9 by volume (10 ppb_v). Two chromatographic methods have been used to detect these compounds at the 10 ppb_v level or lower. The first method developed by Sauers and coworkers[5] is based on a cryogenic enrichment procedure first proposed by Janssen,[6] and uses a gas chromatograph with an electron-capture detector. The second method, which is the focus of the present work, utilizes a gas chromatograph-mass spectrometer (GC/MS) with a thermal-chemical converter.[7]

The purposes of the present work are: 1) to define a procedure for using the GC/MS method to make fast, reliable measurement of trace S_2F_{10}, S_2OF_{10}, and $S_2O_2F_{10}$ in pressurized SF_6; 2) to discuss the major sources of error that can be encountered in using the method; and 3) to discuss the factors that limit the sensitivity such as interference from other compounds.

MEASUREMENT SYSTEM

The GC/MS measurement system, shown diagrammatically in Fig. 1, has previously been described in detail by Olthoff and coworkers.[7] The gas chromatograph connects to a quadrupole mass spectrometer through a jet separator that is heated to a temperature of about 180 °C. A significant fraction of the S_2F_{10} that enters the jet separator tube from the GC column is converted to SOF_2 by a surface-catalyzed reaction with H_2O adsorbed on the walls of the separator. Peaks therefore appear in

*Summer student, University of Illinois.

†Electronics and Electrical Engineering Laboratory, Technology Administration, U. S. Department of Commerce.

Gaseous Dielectrics VII, Edited by L.G. Christophorou and D.R. James, Plenum Press, New York, 1994

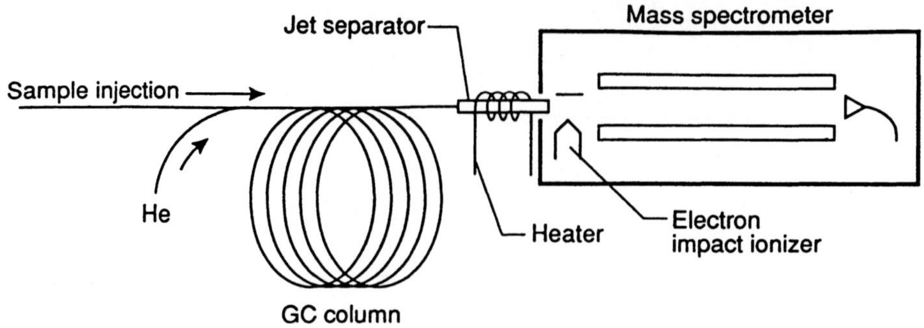

Figure 1. Diagram of the gas chromatograph-mass spectrometer system.

Table 1. Mass-to-charge ratios of ions in atomic units (u) selected for observing S_2F_{10}, S_2OF_{10}, and $S_2O_2F_{10}$ and typical GC-column retention times.

	S_2F_{10}	S_2OF_{10}	$S_2O_2F_{10}$
Ions observed (m/z)	(86,67,48)	(86,48)	(121,102,86,67)
Column retention time (min)	4.48	3.99	5.72

the "single-ion" chromatograms at mass-to-charge ratios (m/z) of 48, 67, and 86 u (characteristic of the SOF_2 mass spectrum) at the time that S_2F_{10} elutes from the column. The conversion to SOF_2 avoids the interferring effects of SF_6 on the S_2F_{10} signal and enormously enhances the sensitivity of detection. The interference of SF_6 is a consequence of the fact that it has a mass spectrum identical to that of S_2F_{10} at the standard 70 eV electron-impact ionization energy.[8] This method allows S_2F_{10} to be detected in SF_6 down to the 1 ppb_v level.

It is not necessary for the species S_2OF_{10} and $S_2O_2F_{10}$ to undergo a chemical conversion in the jet seperator in order to detect them with high sensitivity, because the predominant ions in their mass spectra are different from those in the SF_6 mass spectrum. The observed GC-column retention times for S_2F_{10}, S_2OF_{10}, and $S_2O_2F_{10}$ and the m/z values of the ions that were used to detect these compounds are given in Table 1.

The retention times given in Table 1 apply to a 7.3 m Chromosorb WAW column[‡] operated at room temperature (20–23 °C). No attempt has yet been made to determine the lowest concentrations of S_2OF_{10} and $S_2O_2F_{10}$ that can be detected with this system.

ANALYTICAL PROCEDURE

A quantitative analysis of SF_6 gas using the GC/MS method to determine the concentrations of S_2F_{10}, S_2OF_{10}, and $S_2O_2F_{10}$ at trace levels is most accurately performed by making immediate comparisons with reference gas samples that contain known concentrations of these compounds. Using the programmable features of the GC/MS, it is possible to make a sequence of sample injections that give responses which appear

[‡]Certain commercial equipment, instruments, or materials are identified in this paper to specify adequately the experimental procedure. Such identification does not imply recommendation or endorsement by the National Institute of Standards and Technology, nor does it imply that the materials or equipment identified are necessarily the best available for the purpose.

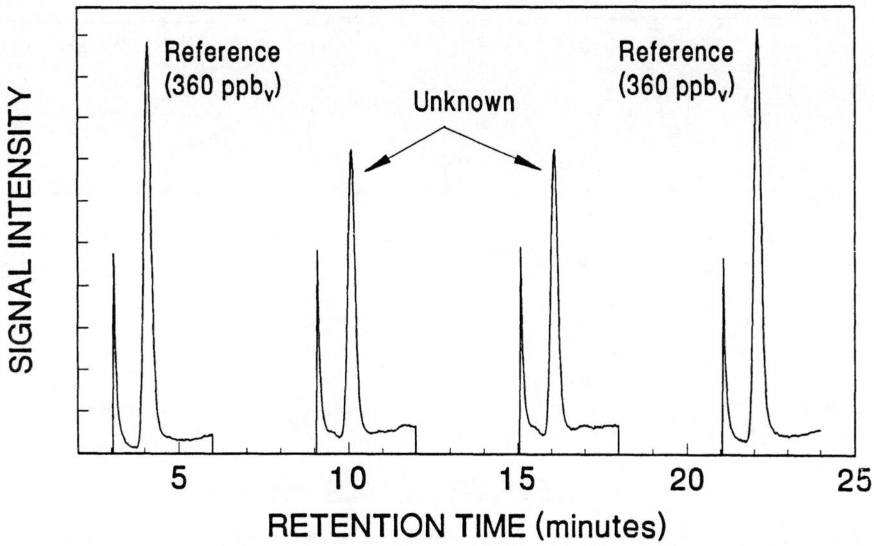

Figure 2. Single-ion chromatogram for m/z = 86 u showing responses from a sequence of four 1.0 ml injections: two unknowns and two references containing the indicated amount of S_2F_{10} in 300 kPa SF_6.

on one single-ion chromatogram as shown in Fig. 2. In obtaining the chromatogram displayed in Fig. 2, the mass spectrometer was not operated for the first three minutes following the injection of a sample into the GC column. This is the time during which most of the SF_6 elutes from the column. Deactivation of the mass spectrometer during this time reduces the background and inhibits exposure of the detector to the large ion flux associated with SF_6.

In following this procedure, it is desirable to bracket one or more injections from the unknown sample with at least two injections from a reference containing a known concentration of the species of interest. In this way, it is possible to eliminate errors due to drift in the GC/MS response such as associated with GC-column conditioning effects, changes in the jet separator surface composition, and variations in the electron-impact source conditions. It is also desirable to use the same injection size (quantity of gas) for both the reference and the unknown and to adjust the concentration of the reference to give a response (peak height) that closely matches that of the unknown. These precautions reduce the errors associated with nonlinearities in the response which tend to be more of a problem with S_2F_{10} than with the other species.[7] The results of analysis for S_2F_{10} content shown in Fig. 2 were obtained using a 1.0 ml×300 kPa injection size. The unknown concentration is determined from the ratio of the average responses from the unknown and reference injections based on either peak heights or areas under the peaks. If there are no interference effects associated with other compounds or with previous injections, the area and peak-height determinations should yield the same result to within the normal statistical variability which will tend to increase as the response (concentration) decreases. Interference from preceding injections can usually be eliminated by appropriate adjustment of the sample monitoring times (MS on) in the injection sequence, which is 3.0 min for the case shown in Fig. 2. Before setting up the injection sequence, it is desirable to make is single "long-time" chromatogram from the unknown sample to determine the possible presence of other peaks at times longer than that for the species of interest.

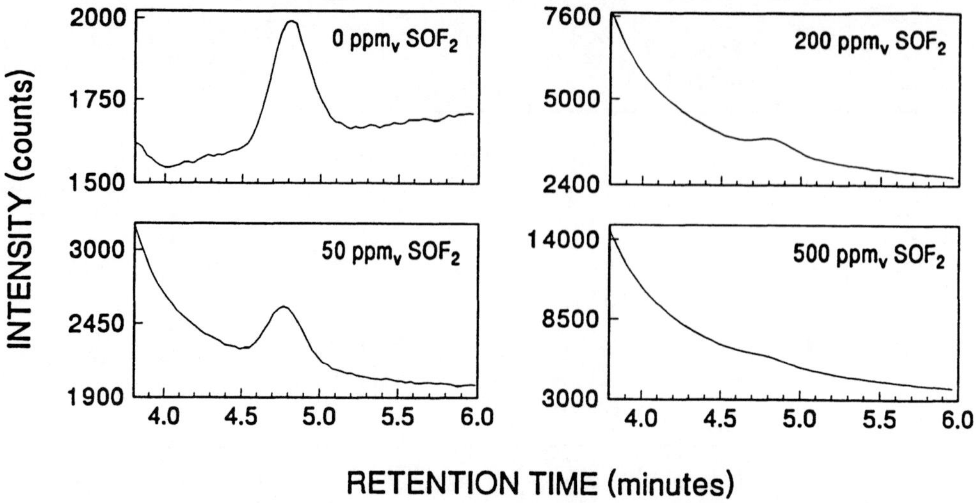

Figure 3. Single-ion chromatograms at $m/z = 86$ u that show the effect of SOF_2 on the detection of 20 ppb_v S_2F_{10} in 200 kPa SF_6. Shown are the concentrations of SOF_2 that were present and the peak for S_2F_{10} that appears at a retention time of about 4.8 min.

SOURCES OF ERROR

Interference Effects

When the GC column is operated under "normal conditions" at a temperature of 20 °C and a helium carrier gas flow rate of 20 to 30 ml/min at atmospheric pressure (100 kPa), the compounds S_2F_{10}, S_2OF_{10}, and $S_2O_2F_{10}$ are sufficiently well separated in time that they do not interfere with each other. It is thus possible to observe all three species on a single chromatogram for the common ion at $m/z = 86$ u (see Table 1). The compound SOF_2 has a retention time of less than 2.4 min under the conditions that give the results listed in Table 1; therefore it is sufficiently well separated from the compounds of interest that it does not ordinarily cause significant interference. On the other hand, if its concentration greatly exceeds that of the compounds of interest, particularly S_2F_{10}, it can have the effect of increasing the background level enough to reduce the detection sensitivity. This is illustrated by the results displayed in Fig. 3. Shown in this figure are the responses for samples that all contain 20 ppb_v S_2F_{10} and differing amounts of SOF_2 from 0 to 500 parts in 10^6 by volume (ppm_v). It is seen that the S_2F_{10} detection sensitivity is reduced to about 20 ppb_v when the SOF_2 concentration reaches the 500 ppm_v level. Over this range of SOF_2 concentrations, the lowest concentration of S_2F_{10} in ppb_v that can be detected is estimated from the data to be $(2.0 + 3.6 \times 10^{-2}[SOF_2])$ where $[SOF_2]$ is in ppm_v.

Of the contaminants that are often present in commercial SF_6, it was found that the compound CCl_2F_2 can be a source of interference in the detection of S_2F_{10} under some conditions. This is because it has a retention time that is only 0.10 min shorter than that for S_2F_{10}. In most cases it is possible to distinguish CCl_2F_2 from the SOF_2 formed from S_2F_{10} by examination of the mass spectrum for the feature attributed to S_2F_{10}. The relative abundances of the major ions that appear in the mass spectra of both SOF_2 and CCl_2F_2 for $m/z \geq 50$ u are given in Table 2. Interference from CCl_2F_2

Table 2. Relative abundances of ions in the mass spectra of SOF_2 and CCl_2F_2 for $m/z \geq 50$ u.

	Abundances			Abundances	
m/z (u)	SOF_2	CCl_2F_2	m/z (u)	SOF_2	CCl_2F_2
50	0.012	0.118	70	0.017	0.002
51	0.036	0.001	85	0.000	1.000
66	0.000	0.036	86	0.556	0.010
67	1.000	0.000	87	0.005	0.326
68	0.010	0.012	101	0.000	0.087
69	0.046	0.000	103	0.000	0.057

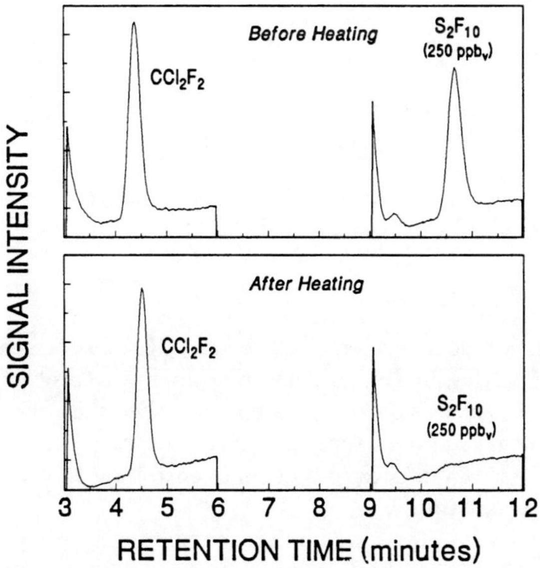

Figure 4. Single-ion chromatograms at $m/z = 86$ u from two different SF_6 gas samples containing either CCl_2F_2 or S_2F_{10} before and after heating at 250 °C for 30 min.

is only a problem if the mass spectrometer is slightly detuned to enhance the detection sensitivity at $m/z = 86$ u. Under conditions where the mass resolution is insufficient to allow complete separation of the adjacent ions at $m/z = 85, 86, 87$ u, the CCl_2F_2 interference can become significant. However, even with poor mass resolution, CCl_2F_2 can usually be distinguished from SOF_2 at $m/z = 67$ u.

Independent of how the mass spectrometer is operated, it is possible to perform a pyrolysis test to confirm that the observed peak in the chromatogram is due to S_2F_{10}. This is illustrated by the chromatograms shown in Fig. 4 that were obtained from two different SF_6 gas samples, one containing CCl_2F_2 at an unknown trace level and one containing 250 ppb_v S_2F_{10}. Indicated in this figure are the chromatograms that were obtained before and after these samples were heated for 30 min at a temperature of 250 °C in a 150 cm^3 stainless-steel cylinder. It is seen that the CCl_2F_2 feature remains unchanged after heating, whereas the S_2F_{10} disappears, presumably by the same process that converts it into SOF_2 in the heated jet separator of the GC/MS. The compound $S_2O_2F_{10}$ is also thermally unstable at 250 °C and will disappear from the sample after heating as illustrated by the results shown in Fig. 5. It is also seen from this figure that S_2OF_{10} is unaffected by the heating. This compound is relatively more stable than either $S_2O_2F_{10}$ or S_2F_{10}, and is frequently observed to be a contaminant in commercial SF_6.

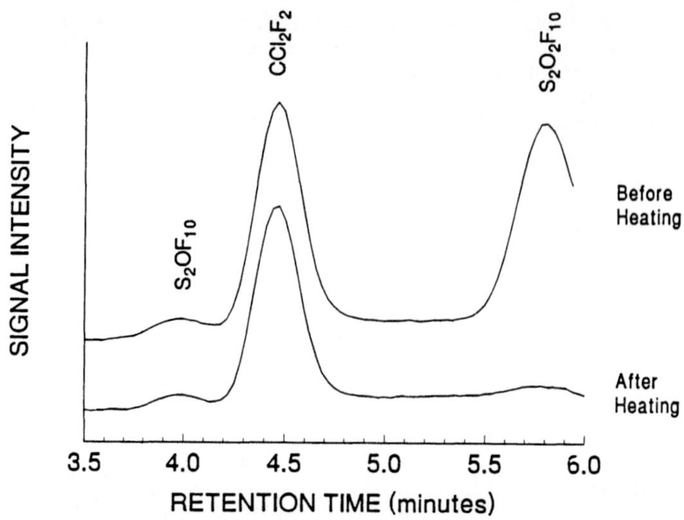

Figure 5. Single-ion chromatograms at m/z = 86 u before and after heating an SF_6 gas sample containing trace amounts of CCl_2F_2, S_2OF_{10}, and $S_2O_2F_{10}$.

Reference Sample Preparation

The most significant source of error in using a GC/MS to measure the content of S_2F_{10}, S_2OF_{10}, and $S_2O_2F_{10}$ in SF_6 is likely to occur in the process of preparing reference samples. The decomposition of S_2F_{10} and $S_2O_2F_{10}$ that occurs at temperatures above 180 °C also occurs at room temperature at a slower rate. This decomposition process can introduce errors in determining and controlling the concentration of S_2F_{10} or $S_2O_2F_{10}$ in a reference sample.

The usual procedure in preparing a reference sample is to inject a known quantity of the trace gas of interest into a sample cylinder under vacuum and then introduce the SF_6 to the desired pressure. In the process of doing this, the trace gas is usually retained initially in tube of relatively small diameter. In the case of S_2F_{10}, some of the gas will decompose during this retention time as illustrated by the chromatogram shown in Fig. 6. The data in this figure correspond to a sequence of injections made after pure S_2F_{10} was introduced into a 0.6 cm diameter stainless-steel tube under vacuum. It can be seen that a significant decay of the sample occurs in a matter of minutes. There are two precautions that can be taken to minimize the error due to S_2F_{10} decay during sample preparation. The first, is to remove as much H_2O as possible from the surfaces of the sample cylinder and associated tubing, either by direct heating or by flushing with a gas that readily hydrolyzes such as SF_4. The second is to minimize the time that the sample container is under vacuum by introducing the diluent gas (SF_6) immediately after the S_2F_{10} is injected into the system. Generally it is better to start by preparing a sample that contains a relatively high concentration of S_2F_{10} ($\geq$ 1000 ppm$_v$) and then use successive dilutions with SF_6 to reduce the concentration down to the desired level.

It has been found[7] that the rate of S_2F_{10} decomposition in a reference sample increases with decreasing S_2F_{10} concentration, total gas pressure, and sample cylinder size. It also increases with increasing temperature and water vapor content. These factors must be considered in maintaining reference samples. To avoid problems of sample decay, it is advisable to prepare a new reference each time that an analysis is to be performed.

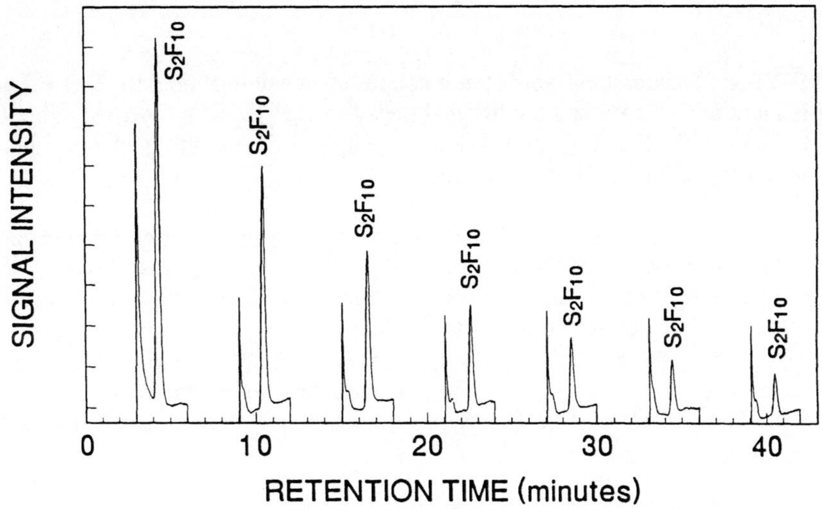

Figure 6. Chromatogram showing the decay of S_2F_{10} after being injected into a 0.6 cm diameter stainless-steel tube under vacuum at room temperature.

ACKNOWLEDGMENTS

This work was conducted under a Cooperative Research and Development Agreement supported by the Office of Energy Management, U.S. Department of Energy, Martin Marietta Energy Systems/Oak Ridge National Laboratory, the Bonneville Power Administration, Canadian Electrical Association, Electric Power Research Institute, Empire State Electric Energy Research Corporation, Hydro-Québec, National Institute of Standards and Technology, Ontario Hydro, and Tennessee Valley Authority.

REFERENCES

1. R. J. Van Brunt, J. K. Olthoff, and M. Shah, Rate of S_2F_{10} production from negative corona in compressed SF_6, *in:* "Conference Record of the 1992 IEEE Symposium on Electrical Insulation," IEEE, Piscataway, NJ (1992).
2. A. M. Casanovas, J. Casanovas, F. Lagarde, and A. Belarbi, Study of the decomposition of SF_6 under dc negative polarity corona discharge (point-to-plane geometry): Influence of the metal consituting the plane electrode, *J. Appl. Phys.* 72:3344 (1992).
3. L. A. Greenberg and D. Lester, The toxicity of sulfur pentafluoride, *Arch. Indust. Hygiene and Occupat. Medicine* 2:350 (1950).
4. D. R. James, I. Sauers, G. D. Griffin, R. J. Van Brunt, J. K. Olthoff, K. L. Stricklett, F. Y. Chu, J. R. Robins, and H. D. Morrison, Investigation of S_2F_{10} production and mitigation in compressed SF_6-insulated power systems, *IEEE Elec. Insul. Mag.* 9:29 (1993).
5. I. Sauers and S. M. Mahajan, Detection of S_2F_{10} produced by a single-spark in SF_6, *J. Appl. Phys.* 74:2103 (1993).
6. F. J. G. Janssen, Measurements of sub-ppm level of sulfur-fluoride compounds resulting from the decomposition of SF_6 by arc discharge, *Kema Sci. Tech. Rep.* 2:9 (1984).
7. J. K. Olthoff, R. J. Van Brunt, J. T. Herron, and I. Sauers, Detection of trace disulfur decafluoride in sulfur hexafluoride by gas chromotography/mass spectrometry, *Anal. Chem.* 63:726 (1991).
8. J. K. Olthoff, R. J. Van Brunt, and I, Sauers, Electron-energy dependence of the S_2F_{10} mass spectrum, *J. Phys. D: Appl. Phys.* 22:1399 (1989).

DISCUSSION

L. NIEMEYER: I understand you have a couple of decay channels for $S_2O_2F_{10}$ and one of them is thermal. We know how thermal decay of S_2F_{10} works. Are you able to state how it would work for $S_2O_2F_{10}$ - do you know the appropriate binding energies that would need to be overcome thermally to decompose it?

R. J. VAN BRUNT: Unfortunately I do not know the answer to that question because that compound has not been studied very thoroughly. There is much less known about this compound than S_2F_{10} and we are not aware of any published data on thermal decay of $S_2O_2F_{10}$. We now have evidence that $S_2O_2F_{10}$ spontaneously decomposes at a low rate at room temperature (300 K). From our preliminary tests, it appears that one of the decomposition products is SO_2F_2. At the present time, for example, we know nothing about the mechanism responsible for the decomposition whether it is a gas-phase or a surface reaction.

I. SAUERS: I would like to point out that we also observed the thermal instability and decay of $S_2O_2F_{10}$ to be similar to the situation with S_2F_{10}. Also, from preliminary work done by Dr. Guy Griffin at Oak Ridge National Laboratory on the toxicity of $S_2O_2F_{10}$ using cell culture bioassays, it appears that the toxicity of $S_2O_2F_{10}$ approaches that of S_2F_{10}, i.e. it is less toxic than S_2F_{10} but more toxic than the other SF_6 by-products studied (SOF_2, SO_2F_2, SF_4, SO_2 and HF). No other toxicity data are currently available in the literature.

S. W. ROWE: In switchgear insulating spacers will constitute a quasi-permanent source of humidity even if a filter is fitted. This is because the humidity must diffuse through the gas to reach the filter and a local high humidity zone will probably build up at the spacer-gas interface. What sort of reactions with S_2F_{10} and S_2OF_{10} would be expected under these conditions? Would decay be promoted with respect to an "all-metal" situation? Would any catalytic type reactions be expected with the diverse components of the spacer?

R. J. VAN BRUNT: We have evidence that the rate of S_2F_{10} decomposition in SF_6 at room temperature (300 K) is increased when cast epoxy materials are introduced. Presently there is no evidence that the mechanism of S_2F_{10} decomposition on epoxy surfaces is any different from that which occurs on metal surfaces. The enhancement in S_2F_{10} decomposition rate when epoxy is present may simply be due to the fact epoxy is more efficient than metals in adsorbing H_2O.

H. MORRISON: The probable source of CCl_2F_2 (Freon 12) in commercial SF_6 is a leak from the cooling compressors used by the manufacturers in the production of SF_6. As well, CCl_2F_2 has been used by some GIS manufacturers and some utilities as a tracer to test for pressure leaks.

R. J. VAN BRUNT: It should be noted that CCl_2F_2 generally appears at trace levels (less than 200 ppb) in nearly all commercial-grade SF_6 that we have analyzed.

J. CASTONGUAY: You did try to condition the internal surface of your small introduction tube to limit the surface conversion of S_2F_{10}. Was this effective?

R. J. VAN BRUNT: We have found that predrying of the sample cylinder, either by heating under vacuum or by introducing SF_4 that readily reacts with H_2O to form HF and SOF_2, helps to reduce the rate of S_2F_{10} decomposition in gas mixtures that are subsequently introduced into that cylinder. We would recommend the heating method for drying. The formation of HF from SF_4 could have undesirable side effects.

S_2F_{10} AND OTHER HEAVY GASEOUS DECOMPOSITION BYPRODUCTS FORMED IN SF_6 AND SF_6-GAS MIXTURES EXPOSED TO ELECTRICAL DISCHARGES

Jacques Castonguay and Isabelle Dionne

Sevice Chimie des Matériaux
Institut de recherche d'Hydro-Québec
Varennes, Québec
Canada, J3X 1S1

ABSTRACT

Mixtures of SF_6 with nitrogen or perfluoromethane are needed for equipments to operate correctly at temperature as low as -50°C. Improvement of experimental setups and analytical techniques have permit the detection and identification of new gaseous byproducts formed when these mixtures are exposed to electrical discharges. Formation of NF_3, N_2F_2 and some perfluoromethylsulfanes has been revealed in addition to other *heavy* compounds containing the SF_5• radical. The paper also presents results obtained with SF_6-air (dry and wet) mixtures.

INTRODUCTION

Study of the elusive but most potent SF_6 decomposition byproducts, S_2F_{10}, has received much attention in the last decade[1-8]. However, many difficulties were encountered, due to its peculiar chemical properties and stability[2,3]. In addition to S_2F_{10}, Janssen[1] and Sauers[2] have observed the presence of S_2OF_{10} but there was some controversy regarding its identification and its relative elution time in their respective articles.

Van Brunt[6] then brought some light on that matter and tentatively identified as $S_2O_2F_{10}$ a new gaseous byproduct formed in partial discharges and which eluted under his analytical conditions, after S_2F_{10}. The overall picture on the identification of the SF_6 decomposition heavy byproducts was still in need of clarification.

This work is thus a logical complement to a previous study[10] that resulted in the development of a simplified analytical method[4,9] for S_2F_{10}. Some of the results found previously suggests that the SF_6 byproduct domain is more complex and wider than first visualized. Consequently, this investigation was essentially focused on the exhaustive analysis and positive identification of the heavy SF_6 decomposition byproducts formed in low energy electrical discharges (coronas and some sparks). Therefore, some special attention was devoted to S_2F_{10}. Four gas mixtures which are of practical interest to the electrical equipment operation were studied in these conditions: pure SF_6, SF_6 + 5% air, SF_6-N_2 (50-50) and SF_6-CF_4 (50-50).

Gaseous Dielectrics VII, Edited by L.G. Christophorou
and D.R. James, Plenum Press, New York, 1994

EXPERIMENTAL

The SF_6 degradation tests were mostly carried under partial discharges using two experimental cells differing by their internal volume: 30 and 120 mL. The smaller one consists of a main cylindrical Teflon body and two stainless steel (s.s.) end plates, supporting each s.s. electrode (point and plane). The second cell has a Plexiglass body covered on the inside by a thin Teflon sheet. It is sealed with aluminum plates, one of which including the s.s. point electrode.

A negative HV-DC source was employed to generate coronas in a point-plane electrode geometry (gap ≈3 mm). In general, all the corona tests were done at 200 kPa gas pressure with currents of 10 and 50 µA. Typically, their durations were limited to less than 1h and most often to 10-15 min.. Consequently, the total charges injected into the gas phase were less than 150 mC and most frequently in the range of 5 to 20 mC. This is much lower than those generally reported by other authors for similar studies[2,5-7]. In addition, few AC spark experiments (15kV and 60 mA) were carried out at atmospheric pressure with exactly the same cell-electrode setups. However, these tests were lasting only 30 s.

It should also be noted that most discharges experiments were performed without the need to open periodically the test cell for cleaning. This was probably a consequence of the very mild discharge conditions chosen. Negative coronas are known not to affect much the electrode conditions, resulting into stable electrical discharge parameters over long period of time. Thus, the opening of a cell was only required, after a long series of AC sparks, to polish the tip of the point electrode that has been eroded. This approach helps minimize some unpredictable conditioning effects.

Two different gas chromatograph/mass spectrometer (GC/MS) instruments were used in combination to identify the heavy byproducts resulting from the SF_6 decomposition in low energy electrical discharges. Standard Electron-Impact Ionization (EI) spectra were recorded with a Saturn II (VARIAN) operated under analytical conditions described elsewhere[10].

Negative Chemical Ionization (NCI) spectra using CH_4 gas were obtained on a HP-5988 (Hewlett-Packard) GC/MS. The chromatograph of this system was equipped with similar capillary columns but not identical to the first one: HP-5 (25m, 0.3mm, 0.5µm film tickness) in tandem with a Ultra-1 (50m, 0.2mm, 0.5µm film tickness). The HP-5 and Ultra-1 are Hewlett-Packard designation of DB5 and DB1 types of stationary coating inside the columns. In both setups, the outlet of the capillary column was directly coupled to the MS ion source.

The temperature of the critical instrument zones were adjusted sufficiently low (<60°C) to limit the thermal decomposition of the byproducts[10]. Similar cryogenic chromatographic conditions and oven temperature programming were chosen for both instruments. In addition, all the byproduct gaseous samples were injected manually using gas-tight syringes. Finally, both MS were operated in scan mode and typical analytical results were recorded in the form of total ion chromatograms (TIC).

RESULTS

The present section is devoted to the qualitative analysis and identification of the new degradation byproducts of SF_6. We will also designate as "heavy byproducts" those we consider different from the usual ones: SO_2F_2, SOF_4 and SOF_2.

The analytical results presented in the followings were obtained from GC/MS data obtained from the two instruments operated in EI and NCI modes, respectively. The first mode is recognized to induce a large number of ion fragments from molecules smashed with electrons of 70 eV. On the other hand, the NCI can be consider a rather gentle mean of ionization, being some sort of electron capture. Consequently, it produces a much simpler mass spectrum.

Often, new ion fragments are found in such spectrum. They can provide helpful clues for the identification of unknown compounds. On the other hand, NCI analysis cannot as easily serve for quantification purposes, as EI[4,6,10]. In fact, NCI-MS is rarely used, related data are seldom reported in the literature and no data base exists. In addition, very little is known concerning NCI molecular response factors which seem to depend largely on the chemical composition. Indeed, they may vary by orders of magnitude from compound to compound, in analogy with the electron capture detector for GC.

Formation of Heavy Byproducts in Pure SF_6

Typical analytical results obtained with samples of pure SF_6 decomposed in contact with partial discharges are illustrated in figure 1. The EI-TIC chromatogram shows a large peak at R.T. ≈4 min. which correspond to the SOF_2. This figure also presents four other smaller peaks eluting between 5 and 10 min. and some very small ones at R.T. >10 which are only visible on the 10x trace. These are the object of the present study.

The informations gathered from the selective ion chromatograms (SIC) are quite instructive regarding the probable molecular composition of the corresponding heavy byproducts. The reader can consult the annex where are tabulated the ion intensities of the EI and NCI mass spectra of the principal new heavy byproducts identified during this work. It also summarized the chemical formula of most of the ion fragments encountered in either EI and NCI mass spectra.

However, to facilitate the understanding, one can find below the identify of a few ion species which characterize important chemical groups found in most of the heavy byproducts:

m/z 67:	SOF	m/z 83:	SO_2F	m/z 105:	SOF_3
m/z 121:	SO2F3	m/z 127:	SF_5	m/z 143:	SOF_5

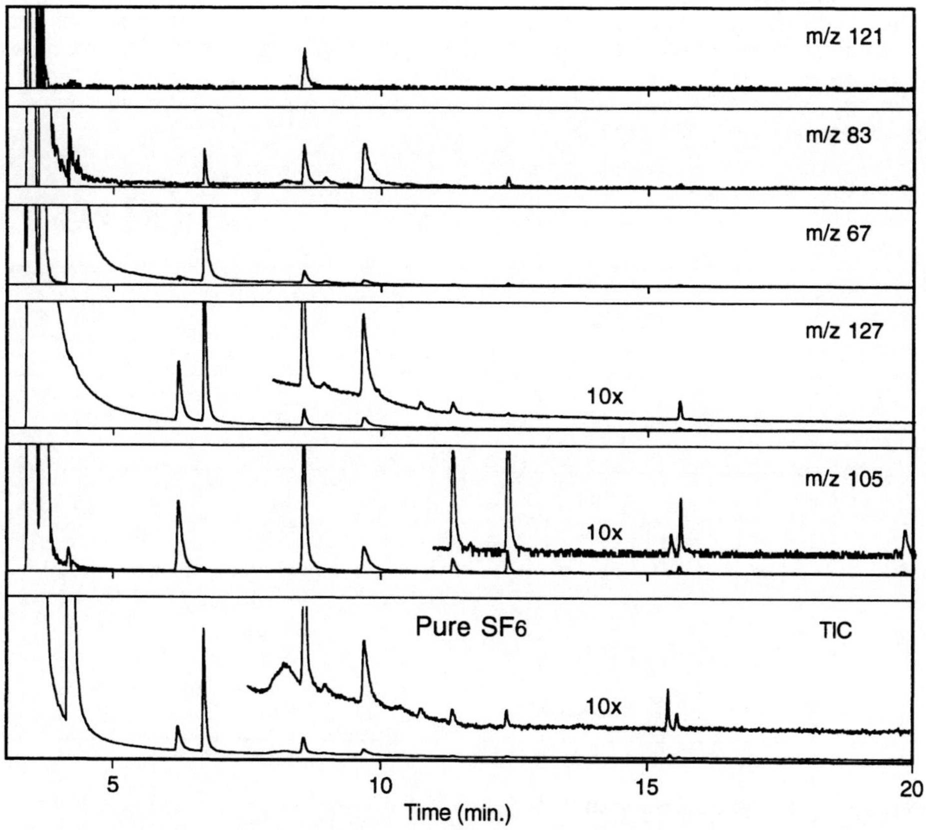

Figure 1. Typical TIC chromatogram and selective ion chromatograms of heavy byproducts of pure SF_6 after degradation by partial discharges.

The other pertinent informations contained in the appropriate mass spectra assured us of the appropriate and positive identification of the first three heavy byproducts separated by the analytical system:

S_2OF_{10} (R.T. ≈6.2 min.), S_2F_{10} (R.T. ≈6.7 min.) and $S_2O_2F_{10}$ (R.T. ≈8.4 min.)

It is worth mentioning that the S_2F_{10} concentration corresponding to the peak in figure-1 was estimated to 40-50 ppmv.

The EI mass spectra of the two oxy-disulfur decafluorides are very similar. Only the presence of an ion of small intensity at m/z 121 permits to distinguish between them. This is also illustrated in the upper trace of figure-1. As already shown by Olthoff[4], the mass spectrum of S_2F_{10} is identical to that of SF_6. Thus, its identity was difficult to prove solely from this information.

The mass spectrometric data recorded in the NCI mode, shown in figure-2, help resolving this ambiguity. The TIC chromatogram may not look ideal, but it shows a sharp S_2F_{10} peak while those attributed to other O-containing byproducts have a rather broad and tailing appearance.

The NCI mass spectral informations collected here compare well with those published by Janssen[1]. Table 1 summarizes those data for S_2OF_{10} and S_2F_{10} and also tabulates the major ion intensities measured for SF_6 and $S_2O_2F_{10}$. One can easily notice the simplicity of the SF_6 and S_2F_{10} NCI mass spectra. Indeed, the former one comprises essentially the molecular ion of SF_6 while that of S_2F_{10} is almost limited to those of SF_4 and SF_6 which are its two main unimolecular dissociation products (1).

$$S_2F_{10} \longrightarrow SF_4 + SF_6 \qquad (1)$$

The identification of the fourth heavy byproduct (R.T. ≈9.7 min.) was not as simple as the three previous ones. In particular, the presence of a SO_2F^+ ion fragment was intriguing and we envisaged the possibility of its secondary formation within the MS ion source, as it was observed[10] for the SOF^+ ion in the S_2F_{10} spectrum. At first and without the availability of the NCI instrument, we thought this fourth heavy byproduct have the following composition: $FO_2S\text{-}SF_5$.

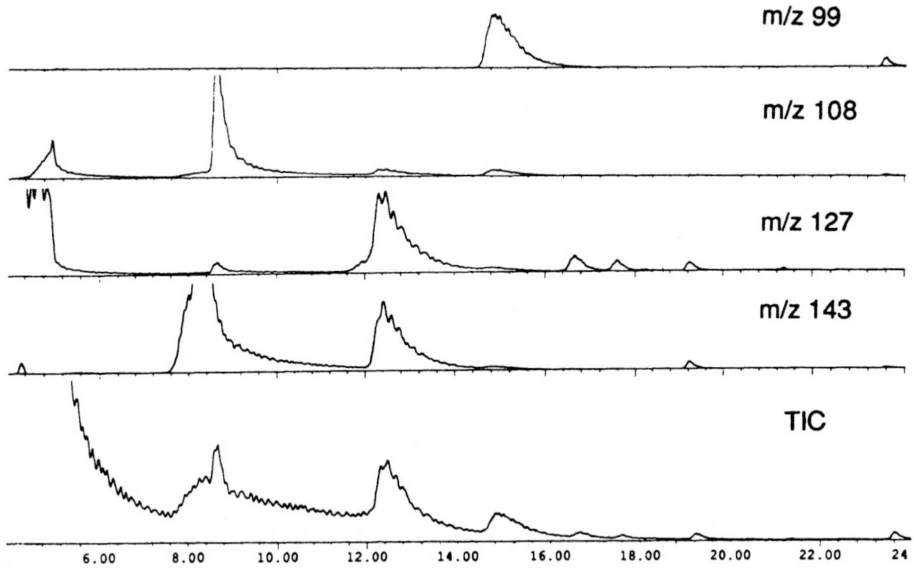

Figure 2. Typical NCI chromatogram of heavy byproducts of pure SF_6 exposed to partial discharges.

What a surprise to find that its NCI mass spectrum (R.T. ≈15 min.) contains solely one single ion at m/z 99! The only logical ionic species which can have such a mass is SO_3F^-. Consequently, the molecular formula of this new byproduct was revisited to $FO_2S\text{-}O\text{-}SF_5$. This compound will be referred from now on as $S_2O_3F_6$.

A second compound characterized by the important presence of a SO_3F^- ion was discovered in the same chromatogram, at R.T ≈23.7 min.. Fortunately, its identification was simplified because its NCI mass spectrum also include an ion fragment at m/z 223.

We found this mass to correspond to the lost of a F• atom, a usual fragmentation process, from the molecular weight (MW) of 242. The following chemical composition accounts for such a molecule: FO_2S-O_2-SF_5. The peak due to this new byproduct is seen in the EI chromatogram (figure 1) at R.T. ≈12.5 and also in the m/z 83 SIC.

Many other low-intense peaks can be observed in chromatograms of figures 1 and 2. Unfortunately, their corresponding mass spectral data lack of distinctive fragment ions to permit further identification with enough confidence. Nevertheless, from the previous discussion, one can certainly infers that the picture of the SF_6 decomposition byproducts is becoming more complex and intriguing.

Table 1. Principal ions from the NCI mass spectra of S_2OF_{10} and S_2F_{10}. Comparison with negative ions intensities reported by Janssen (reference 1)

		S_2OF_{10}		S_2F_{10}		SF_6
m/z	Ion Species	Janssen	This work	Janssen	This work	This work
146	SF_6	2.5		100	64	100
143	SF_5O	100	100	1.4	0.9	
127	SF_5	1.4	1.2	5.6	9.0	
124	SF_4O	1.6	2.5			
108	SF_4	3.4	3.1	12.6	100	
105	SOF_3	35	9.2	0.5		

Heavy Byproducts Formed in SF_6 Containing 5% Humid Air

The addition of 5% humid air at 40% RH increases by ≈700 ppm the moisture level in the SF_6 gas. This level is estimated to be 8-10 times the effective humidity content of a pure SF_6 gas sample after its introduction into the test cell.

The heavy byproducts generated by the corona degradation of such contaminated SF_6 samples are quite similar to those formed in pure SF_6. This is illustrated in figure 3. One significant difference, though, is the much taller $S_2O_2F_{10}$ peak (R.T. ≈8 min.). Indeed, an approximative 10-fold increase is observed. It clearly indicates that the formation of this by product is greatly promoted by the presence of oxygen (1% in 5% air) gas in the discharge environment. Reaction (2) can account for this behavior.

$$2\,SF_5\bullet \quad + \quad O_2 \longrightarrow SF_5\text{-}O\text{-}O\text{-}SF_5 \qquad (2)$$

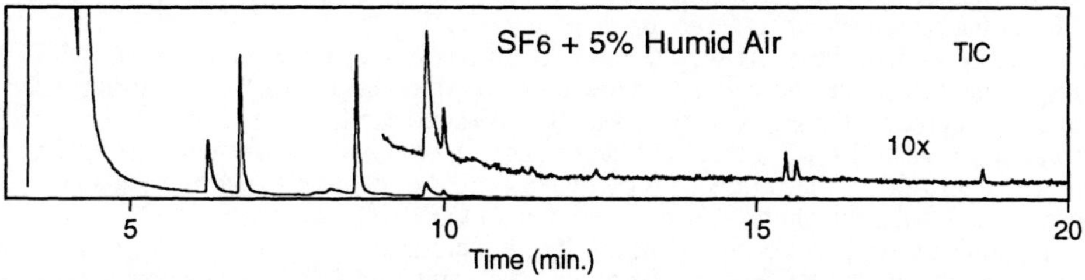

Figure 3. Typical TIC chromatogram of heavy byproducts of SF_6 containing 5% humid air after exposition to partial discharges.

The same EI-TIC chromatogram reveals another peak increase: the one eluted at R.T.-≈10 min. The amount of this byproduct is relatively small, as clearly seen on the 10x trace. In addition, its mass spectrum does not comprise an ion at m/z 105, and thus, does not contain any oxygen atom. If so, its molecular composition should otherwise contain some nitrogen atom(s). Indeed, its chemical identity will be discussed in the next section.

Heavy Byproducts Formed in SF_6 - N_2 mixtures

Mixtures of SF_6 and N_2, typically 50-50 in composition, are generally used in HV equipment specified to operate at temperatures as low as -50°C. Exposed to partial discharges, these gas mixtures generate some light nitrogen fluorides. As illustrated in figure 4, the most important one is NF_3 but some very small quantities of N_2F_2 are also detected. The significant peak of the former one indicates that the discharges are able to totally split the N_2 molecule.

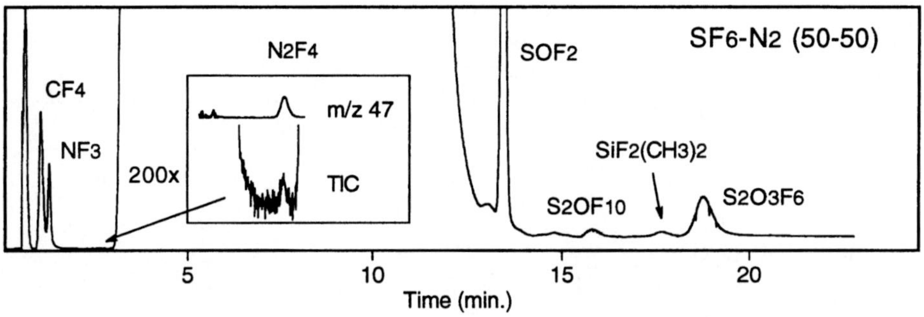

Figure 4. TIC chromatogram and selective ion chromatograms of the byproduct of SF_6-N_2 (50-50) mixtures decomposed by partial discharges and analyzed on Porapak N packed column (reference 10).

The analytical performance of the Porapak N columns may be good for light byproducts, but was not efficient for heavy ones and especially S_2F_{10} which has never been observed with such columns[10]. A better analysis is certainly done with capillary columns.

Compare to figure 1, the EI TIC chromatogram of figure 5 reveals that two new heavy byproducts are produced in significant quantities in corona-decomposed SF_6-N_2 mixtures. Indeed, new peaks (*) are clearly seen at R.T. ≈8 and 10 min.. The first one elutes simultaneously with the broad SO_2 and $SiF_2(CH_3)_2$. Its intensity is comparable to that of the S_2F_{10}.

The EI mass spectrum of this first new product contains the same SF_x fragments observed for SF_6 with the only exception at m/z= 46. This mass might corresponds to the SN^+ or N_2O^+ entities. Based on the chromatograms of figure 5, one may suppose that this byproduct contains no O• atom. The best estimate of its chemical composition would then be: SF_5-NN-F or, more simply SF_5N_2F.

The second new heavy byproduct (R.T. ≈10 min.) is eluted just after the $S_2O_3F_6$ peak. It also exhibits a simple "SF_6" mass spectrum with an additional ion of low intensity at m/z 141. This corresponds uniquely to SF_5N^+ and suggests the following chemical structure for this compound: $(SF_5)_2NF$. This hypothesis was confirmed by a close examination of the NCI chromatogram shown in figure 6.

A close look at the mass spectra recorded in the area of the $S_2O_3F_6$ peak (R.T. ≈15 min.) is needed since there is no evidence of another important peak eluting after it. Indeed, obscured by the signal of the broad $S_2O_3F_6$, one can find ions at m/z 141 and 268 in the mass spectra collected in that time range. This last ion is very informative regarding the molecular formula of the product. It corresponds to the lost of a F• atom from a chemical having a molecular weight of 287, precisely that of $(SF_5)_2NF$.

In addition to the two N-containing byproducts already discussed, the EI and NCI chromatograms show the presence of many new compounds. Closer examination of the m/z 127 selective ions data in both figures 5 and 6 reveals a new byproducts eluting near the end of the tail of the SF_6 peak. The mass spectra (plain or subtracted) obtained then did not give any more clues to help identify the corresponding byproduct. However, conditions in which it is detected suggest that it could be SF_5NF_2.

The NCI chromatogram allowed us to identify two other N-containing byproducts eluting at R.T. ≈6.8 and 19 min., respectively. The first is clearly visible at the end of the large SF_6 tail. Its NCI mass spectrum contains one single ion: m/z 84 = 100%. This mass probably correspond to the following formula: NSF_3. This is certainly a valid but

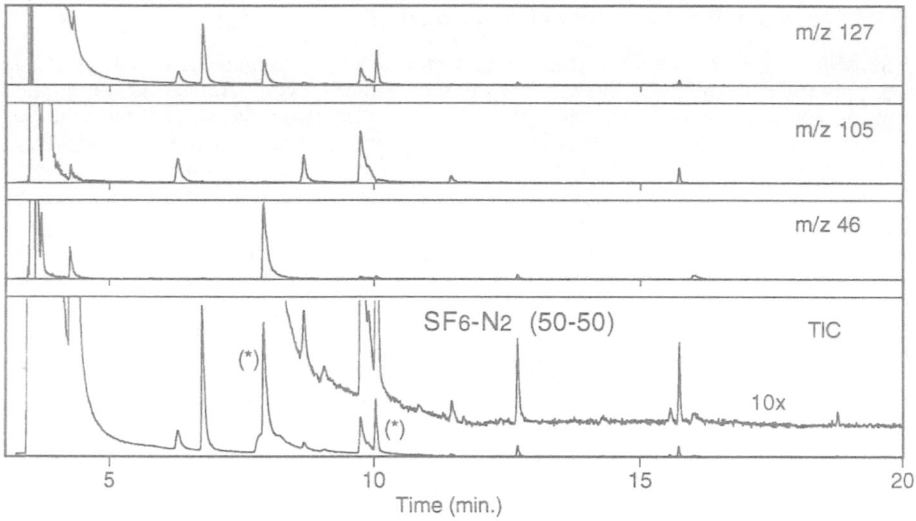

Figure 5. Typical TIC chromatogram and selective ion chromatograms of heavy byproducts of SF_6-N_2 (50-50) mixtures decomposed by partial discharges.

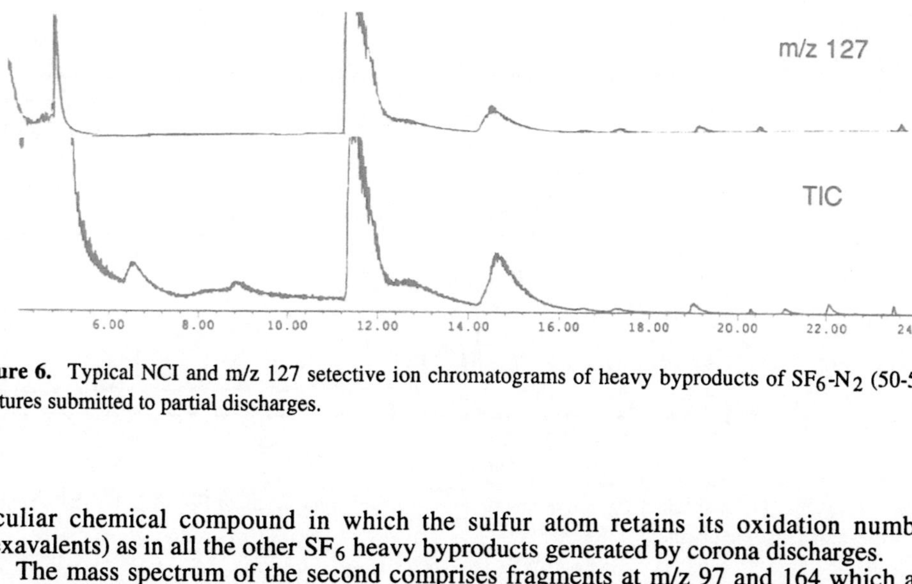

Figure 6. Typical NCI and m/z 127 setective ion chromatograms of heavy byproducts of SF_6-N_2 (50-50) mixtures submitted to partial discharges.

peculiar chemical compound in which the sulfur atom retains its oxidation number (hexavalents) as in all the other SF_6 heavy byproducts generated by corona discharges.

The mass spectrum of the second comprises fragments at m/z 97 and 164 which are most compatible with the following chemical structure: SO_2F-NF-SOF. Again, the ion at m/z 164 comes from the lost of F• from the molecule. Its retention time seems compatible with the suggested molecular composition.

Another peak eluting at R.T. ≈12.6 min. in the EI-TIC and SIC of m/z 46 in figure 5 must be attributed to the addition of N_2 in the SF_6. Its counterpart in the NCI-TIC trace is seen at R.T. ≈19 min.. Informations gathered from both EI and NCI mass spectra are not too explicit but enable us to propose the following molecular structure: $SF_5N_2SF_5$.

Finally, it is interesting to note that N_2 seemed to promote the synthesis of two peculiar byproducts containing the SO_2F- group: the last one discussed and the $S_2O_3F_6$. Moreover, it is also intriguing to notice that atomic N• is found in many byproducts of partial discharge decomposition of SF_6-N_2 mixtures, while none that could contain -SF_4- or -SF_2- entity is detected. It is possible then that such sulfur fluoride gaseous polymers have no characteristic fragments in their mass spectrum to reveal their nature.

Heavy Byproducts Formed in SF_6 - CF_4 mixtures

SF_6-CF_4 gas mixtures were introduced in the electrical apparatus to satisfy the same low-temperature purposes as the SF_6-N_2 mixtures. They even offer better arc quenching properties which are very appreciate for their use in circuit-breakers. Submitted to partial discharges, these gas mixtures are able to produce, again, some other new byproducts.

This is particularly well illustrated in the resulting EI-TIC chromatogram of figure 7. This recording clearly reveals the presence of a new product eluting in the tail of the SOF_2 peak. Moreover, its intensity is about twice that of the S_2F_{10}. The SIM chromatograms show that its mass spectrum comprise ions at m/z 69 and 85, in addition to the usual 105 and 127. The first two new ions correspond to CF_3^+ and CF_3O^+, respectively. Usually, an ion having a m/z of 85 is attributed to SiF_3^+. Two reasons seem to invalidate that possibility:

- no SiF_3^+ was ever observed before at that retention time,
- the parallel appearance of other characteristic ion fragments, especially at m/z 69

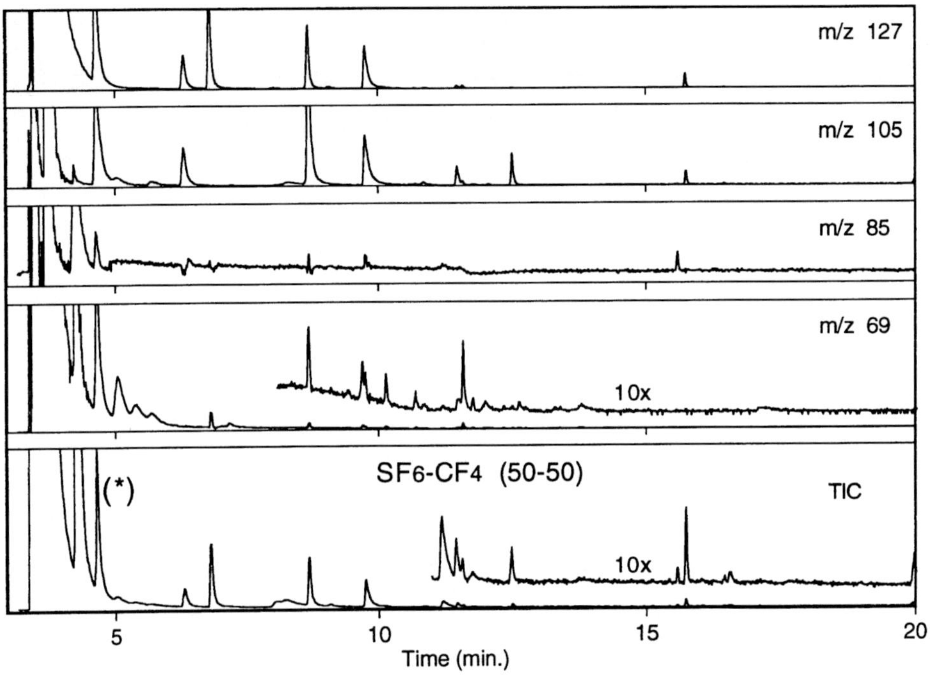

Figure 7. Typical TIC and selective ion chromatograms of heavy byproducts of SF_6-CF_4 (50-50) mixtures after exposition to partial discharges.

At first, the above spectrum data convinced us to propose the following molecular formula for this new compound: CF_3OSF_5. However, the presence of an ion of small intensity at m/z 121 make us believe that the chemical structure could also be $CF_3O_2SF_5$. In addition, an ion at m/z 209 was observed occasionally. This fragment would then corresponds to the lost of F• from the following molecule: SF_5-C_2F_3. One can see that it is not an easy task to distinguish between these multiple plausible choices. Maybe two of these possible compounds elute together, but the informations given by the EI mass spectrum is too scarce to confirm any of those choices.

The NCI chromatogram shown in figure 8 also presents a relatively intense peak in the tailing of the SF_6 matrix. Its mass spectrum includes only four ions: m/z 85, 105, 127 and 143. Unfortunately, this gives us not much supplementary informations to help with the composition of that new byproduct.

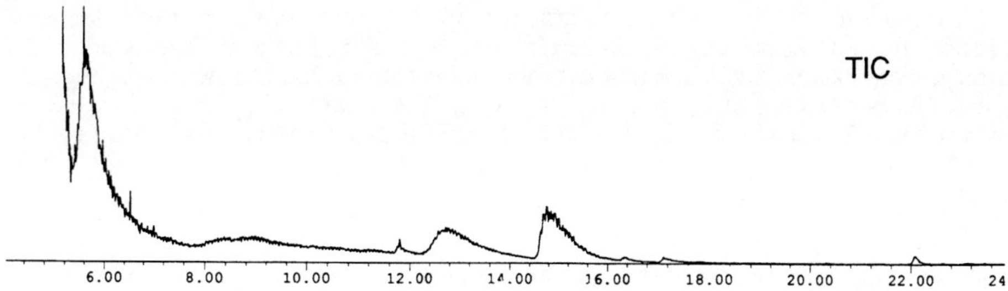

Figure 8. Typical NCI chromatogram of heavy byproducts of SF_6-CF_4 (50-50) mixtures decomposed by partial discharges.

No other pertinent informations are available from the NCI-TIC chromatogram. On the contrary, the EI-TIC chromatogram (10x trace) shows some other new decomposition products at R.T. ≈11.5 min.. However. their quantities must by quite small and only the second peak of the doublet seems to comprise a CF_3 group (m/z 69).

DISCUSSION

This section mainly presents and discusses semi-quantitative results calculated for a relatively wide spectrum of experimental conditions: types and current intensities of discharges, presence of added gas and humidity in SF_6, change of test cell. The calculations used a simple quantification method based on the assumption that all the byproducts considered have identical EI-TIC response factors[10]. That was the only logical step to take, because of the impossibility of having reference gas standards for the majority of those usual and new byproducts.

Furthermore, the study of such a large variety of operating parameters was not intended to be exhaustive. Therefore, the numbers presented below must be considered as such: an rough picture of some general trends of the behavior of SF_6 systems exposed to low energy electrical stresses. Under these conditions, we estimated a confidence limit of ±25% for the result reproducibility of similar discharge experiments. In addition, because S_2OF_{10} is already present (≈4-6 ppm) as impurity in our original SF_6, many of its results are too uncertain to be reported.

Estimated Formation Rates of S_2F_{10} and Other Heavy Byproducts

Data published on the formation of S_2F_{10} in various environments indicate that its yield may vary within a 1:10 range in similar test conditions. This is understandable, considering the many influencing factors reported: electrode and vessel wall materials, gas pressure, gas mixtures, humidity level, nature and volume of sampling bottles, analytical conditions, time delay before analysis, current intensities and power of electrical discharges, types of discharges, experiment duration, electrode conditioning...

Surely, one can think of some other items to add to this list, such as the test cell conditioning that we found to play an important role, especially in the amount of certain heavy byproducts detected. For example, the first series of preliminary tests performed with the Plexiglass cell that has been left on the shelf for many years (>5) gave yields twice as low as those measured later (0.06-0.08 compared to 0.1-0.15 nmole/J). Thus, it appears that internal surface adsorption and conversion is an important issue to consider and it will certainly have a great influence in the case of real electric apparatus.

Discharge Type and Cell Effects. Analytical data obtained from corona and spark experiments are reported in table 2. They reveal a wider dispersion of the AC sparking data which can be expected from such unstable discharges. Spark tests done in the teflon cell were those showing the largest variation. The spreads of the calculated formation yields range from 1:3 to 1:20, with the widest observed for the last two new heavy byproducts. Indeed, the difficulty of assuring the air tightness of the teflon cell is one possible factor that could explain this behavior.

Nevertheless, this did not seem to influence the formation rate of S_2F_{10}. The tabulated data show that sparks are slightly more efficient in forming this toxic byproducts. On the contrary, there is some evidence that the opposite is true for the last two byproducts which are just barely detected after sparking SF_6 in the Plexiglass cell. Otherwise, the results shown in table 2 indicate that the cell body material did not play any significant role in the decomposition product processes. The relative byproducts distribution is quite similar in both cells.

Table 2. Comparative byproduct formation rates (nmole/J) in corona and spark discharges (gas pressure= 200-kPa) (*)

Byproducts	Coronas in Teflon cell	Coronas in Plexiglass cell	Sparks in Teflon cell	Sparks in Plexiglass cell
$SOF_4 + SO_2F_2$	1.0 - 1.8	1.2 - 1.8	1.5 - 14	2.0
SOF_2	0.45 - 0.7	1.0 - 2.1	2.5 - 10	1.4
S_2F_{10}	0.16 - 0.2	0.06 - 0.34	0.3 - 0.9	0.5
$S_2O_2F_{10}$	0.01 - 0.02	0.01 - 0.03	0.005 - 0.1	<0.001
$S_2O_3F_6$	0.01 - 0.018	0.01 - 0.02	0.01 - 0.1	0.003
Discharge Energy (J)	100 - 400	100 - 200	≈7000	7000 - 13000
Injected charges (mC)	10 - 40	10 - 20	-	-

(*) The spark tests were done close to atmospheric pressure(≈110 kPa), due to limited capacity of the AC HV-source.

Influence of gas mixtures. The formation yields of all the usual and new heavy byproducts of SF_6 formed in partial discharges and discussed in this work are summarized in table 3. These interesting results clearly allow us to formulate of the following observations (one must not forget the limit of uncertainty (±25%) stated earlier:

- only the addition of CF_4 seems to significantly lower the formation of S_2F_{10} but it does not have any effect on the other byproducts,
- the presence of 1% O_2 in the 5% air mixture promote more than 10 times the formation rate of $S_2O_2F_{10}$,
- the addition of ≈700 ppm, with respect to less than 100 ppm in the *dry* SF_6 sample, of humidity really increases the amount of S_2OF_{10} measured and double that of $S_2O_3F_6$,

Table 3. Relative byproduct formation (nmole/J) in various gas mixtures exposed to corona discharges (Plexiglass cell, total pressure= 200 kPa, I= 50µA, duration= 1h)

Byproducts	Pure SF_6	SF_6 + 5% Humid Air	SF_6 - N_2 (50 - 50)	SF_6 CF_4 (50 - 50)
$SOF_4 + SO_2F_2$	1.50	2.25	2.45	1.90
SOF_2	1.35	1.75	1.75	1.25
S_2OF_{10}	-	≈0.025	-	-
S_2F_{10}	0.145	0.150	0.130	0.065
$S_2O_2F_{10}$	0.007	0.081	0.004	0.004
$S_2O_3F_6$	0.013	0.022	0.048	0.022
$S_2O_4F_6$	0.001	<0.001	<0.001	<0.001
SF_5N_2F	-	0.003	0.095	-
$SF_5N_2SF_5$	-	0.003	0.030	-
$SF_5O_2CF_3$	-	-	-	0.045

- a higher content of humidity does not change, in either way, the production rate of S_2F_{10},
- the sum of the two new heavy byproducts generated in the SF_6-N_2 mixture nearly equals that of the S_2F_{10},
- a similar behavior is found with the unique new heavy decomposition of the SF_6-CF_4 mixture,
- part of the $S_2O_3F_6$ formation rate increase is probably due to the contribution (unknown) of the coeluting $(SF_5)_2NF$.

Finally, if the amount of some new heavy byproducts appears to be low, we can not tell if it is not due to the low vapor pressure of the corresponding compounds.

Influence of the presence of dry and humid air and of humidity. The corona discharges results just discussed did not reveal any clear influence due to the presence of significant amount of water vapors in the SF_6. However, on other occasions evident effects have been found, as it is shown by the calculated data reported in table 4. Here, the addition of 5% air, either dry or humid (it does not matter), noticeably diminished the formation of S_2F_{10}. A decrease by a factor or 3-4 is observed.

Similar contradictory results were found by different authors, however. Indeed, Casanovas[7] reported a relatively minor influence of the addition of 2000 ppm of humidity whereas Sauers[5] measured an even higher decrease of the S_2F_{10} formation than we did, with comparable addition (600 & 1200 ppm) of water to his test samples. Beside this peculiar finding that will need to be explore more closely in the future, the influence or either dry of humid air added cannot be differentiate.

Table 4. Effects on the byproduct formation rates (nmole/J) of the addition of 5% of dry and humid air in SF6 exposed to partial discharges (teflon cell filled with gas at P= 200-kPa) (*)

Byproducts	Pure SF_6	SF_6 + 5% Dry Air	SF_6 + 5% Humid Air
SOF_2	0.54	0.39	0.38
S_2F_{10}	0.108	0.026	0.027
$S_2O_2F_{10}$	0.016	0.059	0.062
$S_2O_3F_6$	0.01	0.007	0.003

(*) Data presented are the averages of two distinct series of tests; standard deviation was < 25%

CONCLUSIONS

This study has permitted the positive identification of the three anticipated disulfur decafluorides $S_2O_xF_{10}$, where x= 0, 1 and 2. In addition it gave the opportunity to discover and identify a certain number of new heavy SF_6 decomposition byproducts which are generated in low energy electrical discharges (coronas and sparks). Especially in absence of added N_2, the majority of most important ones were the oxy-sulfur fluorides, as revealed by the presence of characteristics ion fragments in both EI (SF_3O^+, m/z 105) and NCI (SF_5O^-, m/z 143) mass spectra. The most significant exception to this picture is the S_2F_{10}.

As expected, the molecular formula of these new heavy byproducts was directly influenced by the gas composition. Those of which were identified with enough evidence are summarized below:

- pure SF_6: $S_2O_3F_6$ and $S_2O_4F_6$
- SF_6-N_2 mixtures: SF_5N_2F, $(SF_5)_2NF$, $SF_5N_2SF_5$, with possible SF_5NF_3
- SF_6-CF_4 mixtures: CF_3OSF_5 or $CF_3O_2SF_5$ or $C_2F_3SF_5$

Many other heavy byproducts were also detected in small quantities. They could not been identified because their mass spectrum contains no distinctive ion fragments, except in one case where it was not possible to find any valid molecular formula.

Semi-quantitative analysis of the results allows us to formulate the following observations:

- the presence of air (1% oxygen) increased the yield of $S_2O_2F_{10}$ by about 10-fold,
- in corona discharges, the relative formation rates of S_2F_{10} vary in the range of 0.2-0.4 that of SOF_2; in sparks, this relative yield is roughly reduced by 2 (0.07-0.2),
- increase of the level of humidity (+700 ppm) in the SF_6 exposed to coronas has produced contradictory effects on the formation of S_2F_{10}, thus, our results reflect what two different authors have published

In conclusion, this study has revealed a broader picture of the heavy decomposition byproducts of SF_6 than expected at the beginning. It was also surprising to find that nitrogen gas can act as a real promoter for the production (in significant quantities) of 4-6 new byproducts. In all cases, the S• atoms forming the new byproducts retains its hexavalents characteristic and were only present as either SF_5- and SF_5O- groups. Finally, the problem of the toxicity of SF_6 heavy byproducts seems left with many more questions that will have to be addressed.

REFERENCES

1. F.J. Janssen, "Decomposition of SF_6 by Arc Discharge and the Determination of the Reaction Product S_2F_{10} ", Gaseous Dielectrics V (Proceedings of the Fifth International Symposium on Gaseous Dielectrics, Knoxville, 1986), pp. 157-162, Pergamon Press, NY, 1987.
2. I. Sauers, M.C. Siddagangappa, G. Harman, R.J. Van Brunt and J.T. Herron, "Production and Stability of S_2F_{10} in SF_6 Corona Discharges", Proceedings of the 6th International Symposium on High Voltage engineering,, New Orleans, Aug. 28-Sept. 1, paper 23-8, 1989.
3. J.K. Olthoff, R.J. Van Brunt, J.T. Herron and I. Sauers, G. Harman, "Catalytic Decomposition of S_2F_{10} and its Implications on Sampling and Detection From SF_6-Insulated Equipment", Conference Record of the 1990 IEEE International Symposium on Electrical Insulation, pp 248-252, Toronto, 3-6 Juin 1990.
4. J.K. Olthoff, R.J. Van Brunt, J.T. Herron and I. Sauers, "Detection of Trace Disulphur Decafluoride in Sulfur Hexafluoride by Gas Chromatography/Mass Spectrometry", Analytical Chemistry, Vol 63 No. 7, pp. 726-732, 1991.
5. I. Sauers, G. Harman, J.K. Olthoff and R.J. Van Brunt, "S_2F_{10} Formation by Electrical Discharges in SF_6 Comparison of Spark and Corona", Gaseous Dielectrics VI, New York: Plenum Press, pp. 553-562, 1991.
6. R.J. Van Brunt, J.K. Olthoff and M. Shah, "Rate of S_2F_{10} Production from Negative Corona in Compressed SF_6 ", Conference Record of the 1992 IEEE International Symposium on Electrical Insulation, Baltimore, June 7-10, pp. 328-331, 1992.
7. A.M. Casanovas, J. Casanovas, F. Lagarde and A. Belarbi. "Study of the Decomposition of SF_6 under DC Negative Polarity Corona Discharges (Point-to-Plane Geometry) : Influence of the Metal Constituting the Plane Electrode", *J. Applied Physics*, vol. 72, no. 8, pp. 3344-3354, October 1992.
8. D. R. James (Editor), I. Sauers, G. D. Griffin, R.J. Van Brunt, J.K. Olthoff, K., K. L. Strickett, F. Y. Chu, J. R. Robins and H. D. Morrisson, "Investigation of S_2F_{10} Production and Mitigation in Compressed SF_6-Insulated Power Systems", *IEEE Electrical Insulation Magazine*, vol. 9, no. 3, pp. 29-40, May/June 1993.
9. I. Sauers and R.A Cacherio, "A Cryogenic Enrichment Technique for Gas Chromatographic Detection of S_2F_{10} in SF_6 Discharges", Conference Record of the 1992 IEEE International Symposium on Electrical Insulation, Baltimore, pp. 340-344, June 7-10 1992.
10. J. Castonguay and I. Dionne, "S_2F_{10} and Other Heavy Gaseous Decomposition Byproducts Formed in SF_6 and SF_6-Gas Mixtures Exposed to Electrical Discharges", presented at the Seventh International Symposium on Gaseous Dielectrics, Knoxville, Tennessee, April 24-28, 1994.

ANNEX 1

Table A.1 EI mass spectra of the major new SF_6 decomposition heavy byproducts (important positive ion fragments observed) (relative ion intensities are expressed in %).

		SF_6 and SF_6 + 5% Humid Air				SF_6 - N_2 (50-50)		SF_6 - CF_4 (50-50)
m/z	Positive Ion Fragments	$S_2O_2F_{10}$	$S_2O_3F_6$	$S_2O_4F_6$	Unknown (RT ≈ 10.5 min.)	SF_5-N_2-F ?	SF_5-N_2-SF_5 ?	$SF_5O_2CF_3$ + ? SF_5OCF_3, ?
209	CSO_2F_7							5.6
157					49			
141	SF_5N						1.2	
139					29			
127	SF_5	60	100	46	100	62	100	100
122	SF_4N						0.5	
121	SO_2F_3	1.8						4.8
108	SF_4	1.2				3.9	1.6	
105	SOF_3	100	37	100	92			32
89	SF_3	15	26	15	43	65	40	12
86	SOF_2	9		18	2			
85	CF_3O							1.6
83	SO_2F	3.3	7.0	11				
70	SF_2	6.0	10	8	14	20	8.5	
69	CF_3							31
67	SOF	7.2	10	23	7.7	5.5		
51	SF	5.0	4.7	6	8	12	3.9	
48	SO	5.6	9.0	11	4.2			
46	SN					100	3.2	

Table A.2 NCI mass spectra of the major new SF_6 decomposition heavy byproducts (important negative ion fragments observed) (relative ion intensities are expressed in %).

		SF_6 and SF_6 + 5% Humid Air				SF_6 - N_2 (50-50)		SF_6 - CF_4 (50-50)
m/z	Positive Ion Fragments	$S_2O_2F_{10}$	$S_2O_3F_6$	$S_2O_4F_6$	Unknown (RT ≈ 10.5 min.)	SF_5-N_2-F ?	SF_5-N_2-SF_5 ?	$SF_5O_2CF_3$ + ? SF_5OCF_3, ?
223	$S_2O_4F_5$			33.2				
146	SF_6	7.1	1.1	4.8	6.0			
143	SF_5O	81	56	13.5	57			100
141	SF_5N							
127	SF_5	100	100		100	100	100	49
124	SF_4O	13	8.0	6.9	5.5			3.6
108	SF_4	5.7	1.9	16	3.4	2.9		0.4
105	SOF_3	47	28	86	23	0.2		13.5
99	SF_3			100				
86	SOF_2							
85	CF_3O							7.0
84	SO_2F						3.6	

Table A.3 Chemical formula of the principal ionic fragments observed in the EI and NCI mass spectra of the new SF_6 decomposition heavy byproducts.

m/z	Species	m/z	Species
46	NS, NO_2	47	N_2F
48	SO_2	50	CF_2
51	SF	52	NF_2
54	SF_4	60	COS
64	S_2, SO_2	65	NSF
66	N_2F_2	67	SOF
69	CF_3	70	SF_2
77	$SiF(CH_3)_2$	81	$SiF_2(CH_3)$, C_2F_3
83	SO_2F, S_2F	84	NSF_2
85	SiF_3, CF_3O	86	SOF_2
89	SF_3	96	$SiF_2(CH_3)_2$
99	SO_3F	100	C_2F_4, $SiF_3(CH_3)$
102	SO_2F_2, S_2F_2	103	NSF_3
105	SF_3O	108	SF_4
121	SF_3O_2	122	SF_4N
124	SF_4O	127	SF_5
139	?	141	SF_5N
143	SF_5O	146	SF_6
155	$Si_2F_2O(CH_3)_3$	207	$S_2O_3F_5$, CSO_2F_7
209	CSO_2F_7	223	$S_2O_4F_5$
226	$S_2O_3F_6$	268	NS_2F_{10}

DISCUSSION

L. NIEMEYER: What are the known physicochemical properties of the new decomposition products $S_2O_3F_6$, $S_2O_4F_6$, SF_5N_2F, $(SF_5)_2NF$, $SF_5N_2SF_5$ and $SF_5O_2CF_3$? Do you have calibration samples of these gases?

J. CASTONGUAY: I do not know what the physicochemical properties of those new SF_6 decomposition by-products are. Neither do I know the properties of ($S_2O_2F_6$) found since then and mentioned during my oral presentation. I do not know if some of these have ever been synthesized. As was indicated in the paper, the chemical composition of some of the new by-products could not clearly be established from their EI and NCI mass spectra, so it is stated as a probable or logical one. This is the case of all the N-containing compounds, except the $(SF_5)_2NF$ which was clearly identified, and the new by-products formed with the SF_6-CF_4 mixtures. Otherwise we are quite confident of the identification of the three members of the new SF_6 heavy by-product family: FO_2S-SF_5, FO_2S-O-SF_5 and FO_2S-O_2SF_5. Finally, considering the above arguments and the recent discovery of these new compounds, one cannot expect to find available sources of these to allow the preparation of reference gas standards.

R. J. VAN BRUNT: In the case of SF_6/N_2 mixtures, did you observe formation of nitrogen oxides?

J. CASTONGUAY: No, we did not observe the formation of NO_x in our SF_6-N_2 mixture experiments. One would expect to find NO_x in the discharge tests involving SF_6 with 5% humid air added. In that case also, no nitrogen oxides were found, neither other by-products containing -N=O or -O-NO groups. Finally, if NO was formed, it might have eluted under the large SF_6 peak without being detected.

I. SAUERS: (1) With regard to the disulfur compounds, were you able to verify the number of sulfur atoms from the $^{34}S/^{32}S$ isotope ratios which should be about 9% when two sulfur atoms are present? (2) What was the energy per spark discharge in your spark experiments and how was the spark produced? (3) What is the effect of SOF_2 on S_2F_{10} sensitivity?

J. CASTONGUAY: (1)Yes, as a mass spectroscopist, looking at isotopic ratio regarding peak intensities is the first thing I do after estimating a possible fragment composition. The $^{34}S/^{32}S$ ratio matched well for the two by-products for which their negative chemical ionization respective mass spectrum contains such ions (for example: $(SF_5)_2N^-$ at m/z 268 and $S_2O_4F_5^-$ at m/z 223). (2) The spark conditions are briefly explained in the paper. We used the same cells, the same electrodes and applied the high voltage from the AC transformer rated (15 kV and 60 mA) 900 watts. Since the gas breakdown occurred at ≈ 10 kV, not all the available source power is consumed in the sparks. Some oscillographic observations have been made and confirm this hypothesis. Thus, we choose to use half of the source power as the basis of our spark energy calculations. If your concern is with our reported higher formation yields (SOF_2) than yours, please consider that, at most, they can be divided by 2, but can be higher if the relative efficiency of the energy transfer to the spark discharges is less than 50% which we used. (3) I do not expect a large influence on the S_2F_{10} sensitivity due to the presence of high concentrations of SOF_2 and I explain briefly why. In the EI chromatograms presented in the paper, the SF_6 peak is 1-1.2 min. wide and its tail extends to ≈ 11 min. This is the condition in which we estimated our S_2F_{10} limit of detection, using the three ions (those of SF_6, for instance: m/z 127, 89 and 70). Thus, since at very low concentration of S_2F_{10}, I expect its mass spectrum will contain probably only ions of SOF_2, and since the SOF_2 is only eluting slightly less than one minute after SF_6, I do not anticipate detrimental reduction of the S_2F_{10} sensitivity of our system for SOF_2 concentrations up to 5% (100-125 times the SOF_2 concentrations usually found in our most contaminated samples for which analytical results are presented).

ANALYSIS OF S_2F_{10} IN DECOMPOSED SF_6 BY GAS CHROMATOGRAPHY : VARIOUS ASPECTS

Jacques Castonguay and Isabelle Dionne

Sevice Chimie des Matériaux
Institut de recherche d'Hydro-Québec
Varennes, Québec
Canada, J3X 1S1

ABSTRACT

This paper presents the results of experiments performed to find some practical and simplified chromatographic method for analyzing S_2F_{10} in SF_6. A mass detector was used to achieve the necessary sensitivity and to help with the identification of the SF_6 decomposition byproducts. The discussion includes the influence of various instrument parameters and temperature conditions. The use of packed columns (Porapak types) gave proper resolution but they induced S_2F_{10} decomposition. Bonded-silicon capillary columns were finally chosen for their overall performances. Indeed, the usual gaseous byproducts (SOF_4, SO_2F_2, SOF_2) can be analyzed and quantify. Also, other heavy SF_6 byproducts (S_2OF_{10}, $S_2O_2F_{10}$ and other unknowns) have been detected. However, some chemical interactions still take place in the column with the SF_6 byproducts formed by partial discharges, forming various fluoromethylsilanes. Finally, sensitivity in the tens of ppbv can be obtained with 50 µL injected, while quantification relying on total ion current, (due to the lack of standard S_2F_{10} solution) seem to give satisfactory results.

INTRODUCTION

The use of sulfur hexafluoride as an insulating gas in electric-power systems is widely spread. Consequently the decomposition byproducts formed in such equipments are of interest. In particular, the formation of disulfur decafluoride (S_2F_{10}) by electrical discharges has led to concern because of its known high level of toxicity. Thus, a good separation of S_2F_{10} from SF_6 and a sensitive detection technique is needed. This paper presents the development of a practical, direct and simplified chromatographic method using a mass detector to achieve the necessary sensitivity to identify S_2F_{10} in SF_6 as well as the usual gaseous byproducts (SO_2F_2, SOF_4, SOF_2). Thus, the method differs from that published by Olthoff[1]. It also allows the identification of other heavy gaseous byproducts (S_2OF_{10}, $S_2O_2F_{10}$) without the use of a more complex cryogenic enrichment procedure[2-3].

EXPERIMENTAL

GC separations were performed on a Varian 3400 gas chromatograph equipped with a Septum Programable injector (SPI) maintained at 50°C and coupled to a Saturn II (Varian) mass spectrometer. Gas samples were introduced by a gas-tight syringe. The columns used were operated with helium as the carrier gas with a flow rate of 2-3 mL/min to the MS

Gaseous Dielectrics VII, Edited by L.G. Christophorou
and D.R. James, Plenum Press, New York, 1994

source. Throughout this work, a number of stationary phases and column materials were tested and the conditions employed in the GC analyzer were optimized to achieve good separation and resolution. Thus, different aspects of experimental conditions are given in more details in the appropriate section.

Data acquisition and data analysis were performed by the provided software. Retention time (R.T.) was measured from the time of sample injection. All acquisitions were performed in full scan (m/z: 36-300; 1-1.5 scan/s.) giving total flexibility for further data treatment. The latter was done by means of single (SID) or multiple (MID) ions detection and by total ion current (TIC) detection.

Samples of decomposed SF_6 were produced in a 30 mL cylindrical teflon cell at room temperature by submitting the SF_6 (Accudri, Allied Chemicals Canada Inc.) gas to partial discharges. A point-to-plane electrode geometry with a 3 mm gap spacing between the stainless steel electrodes was used. The point was connected to (dc, (-) polarity) high voltage (Bertan Associates Inc. model 205A-50R). The corona current and voltage varied from 1-15 uA and 6-15 kV, respectively. A SF_6 gas pressure of 200 kPa was generally used; with some experiments done at 100 kPa SF_6.

RESULTS

The principal goal of developing a different analytical method for S_2F_{10} was to simultaneously simplify the overall procedure to a minimum and positively identify the S_2F_{10} peak in every case. Many chromatographic conditions were evaluated.

Analysis with Packed-Columns (Porapak)

The first approach used for S_2F_{10} detection employed Porapak type packed columns. A 3.2 mm (1/8") diameter stainless steel column filled with Porapak N over a length of 2 m was coupled to the MS via a 1:10 capillary splitter. Under our GC conditions, with a carrier gas (He) flow rate of 20 mL/min, an oven temperature programing from 30 to 150°C at 20°C/min and a 1 mL injection, the SF_6 peak was long to elute (≈7 min. wide at R.T.= 3 min.).

Although large peaks were observed, well-separated peaks for CF_4, SOF_2, S_2OF_{10}, and $SiF_2(CH_3)_2$ were obtained with R.T. of 1.2, 10.2, 14.5 and 17 min., respectively. In addition, the tail of the extremely large SF_6 signal obscures the SO_2F_2, and SOF_4, signals but not that of SOF_2. Peaks due to fluoromethylsilane impurities arising from reactions with HF were always recorded and are discussed later. Some others are also detected, such as NF_3 [4].

Of more importance, using Porapak columns, there was no sign of the elution of S_2F_{10}. and only the S_2OF_{10} peak was recorded. A typical chromatogram is shown in Figure 1. Lowering the initial oven temperature down to -20°C gave, as expected, a better separation but widen the peaks of interest. The MS response is proportional to the instant concentration of the eluted compounds but since broad peaks were observed, a low detection limit cannot easily be obtained under these conditions.

Improvement of the peak widths was possible with shorter columns (25 and 60 cm) filled with either Porapak N or Q. However, proper resolution, with respect to the longer column, could only be achieved by lowering the initial oven temperature to -30°C, reducing the analysis time from 20 to 15 min.. For similar column length, the initial oven temperature needs to be set 10°C lower for the column filled with Porapak Q to obtain similar performance. Finally, shorter columns permitted to lower the maximum oven temperature to 100°C but again no S_2F_{10} peak has ever shown up in any of the chromatograms.

Analysis using other Packed-Columns (Filled with Silicone Gum Stationary Phase)

It was shown that it was not possible to detect S_2F_{10} using any of the Porapak columns, even the shorter ones. Either, its decomposition occurs from direct reaction with the stainless steel surface of the column or because of the high temperature (up to 150°C) needed for the analysis. Therefore, the next column tried for S_2F_{10} detection was a 6 m x 3.2 mm teflon coated stainless steel tubing containing 20% SE-30 on 80/100 Chromsorb WAW.

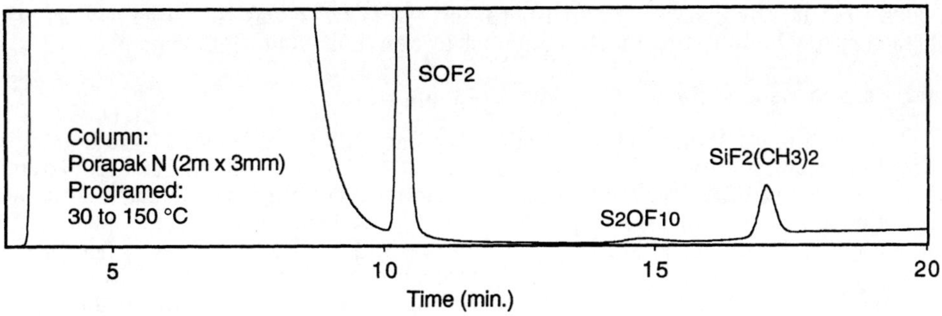

Figure 1. Typical TIC chromatogram recorded using a stainless steel packed column (2m x 3.2mm) filled with Porapak N.

This type of column is similar to that previously used by Van Brunt[1]. The conditions employed were similar to those mentioned above: He flow rate = 20 mL/min, 1:10 split, oven
temperature programed from -40 to 0°C at 20°C/min. For a 500 µL to 1 mL injection of the decomposed SF_6 sample, it was possible to separate S_2F_{10} (R.T.: 6.2 min) completely from the SF_6 matrix, as illustrated in Figure 2a. However, a signal due to $SiF_2(CH_3)_2$ was observed under the S_2F_{10} peak. It did not interfere with the qualification of the relatively more intense S_2F_{10} peak but could easily impair its measurement at concentrations lower than 1 ppm. Indeed, the presence of the silicon impurity depends on the elapsed time between the end of the discharge and the analysis. This will be discussed further. Finally, under these experimental conditions, the analysis time was relatively short (8 min.) but a rather poor resolution was obtained and the large SF_6 signal obscured the usual byproducts (SO_2F_2, SOF_4, and SOF_2).

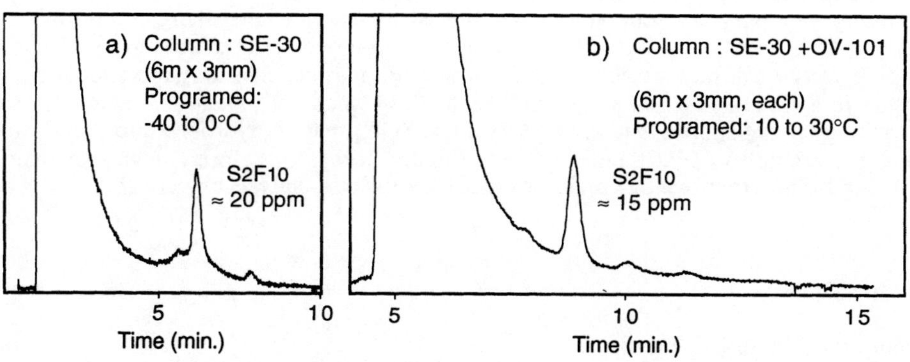

Figure 2. Typical TIC chromatograms recorded with teflon-coated stainless steel packed columns (6m x 3.2mm) using OV-101 and SE-30 as stationary phase

In hope of increasing the separation, the column was lengthen by coupling in tandem the above column with a similar column using instead 20% OV-101 as the stationary phase (similar column used by Janssen[5]). The system was optimized giving a 10-12 minutes analysis for a 100 µL injection (He flow rate = 15 mL/min, 1:10 split and oven temperature programed from -30 to 10°C) and resulting in an almost room temperature operation. Figure 2b illustrates a chromatogram obtained with these conditions, where the resolution has indeed improved. However, the peaks are still quite large (peak width of ≈18 s) and the SF_6 tail still obscures the usual byproducts. The peaks at 7.4, 8.4, 9.4, 10.4, and 11.5 minutes can be ascribed to S_2OF_{10}, S_2F_{10}, $SiF_2(CH_3)_2$, $S_2O_2F_{10}$ and $S_2O_3F_6$, respectively, as identified in reference 4. Furthermore, the S_2F_{10} peak was too large to achieve a satisfactory low limit of detection (minimum detection limit attained: 0.25 ppm)

Analysis with Bonded-Silicon Capillary Columns

A 30 m long, 0.25 mm ID. fused silica capillary with an inside 0.25 μm coating of DB5 phase was then tried. Under our GC conditions (-60 to -40°C oven temperature programing, 2 mL/in He flow rate), S_2F_{10} was found to elute from the column ≈2.1 minutes after injection (5 μL) and was separated from the well eluted (15 s) SF_6 peak. The chromatogram thus obtained (4 min.) showed sharp and well-separated peaks but they are too thin for an accurate quantification, being formed of only 3-4 scans. In addition, a quick analysis (2 min.) can easily be performed with this column maintained (isothermal) at -30°C.

Improved performance. In order to increase the separation, a second fused silica capillary column (25 m x 0.20 mm coated with 0.33 μm Ultra 1 (DB1 type from Hewlett Packard) stationary phase, similar in characteristics to the column used by I. Sauers[2], was coupled to the one mentioned above. The usual discharge byproducts, S_2F_{10} and other sulfur oxyfluorides were well separated using an oven temperature programing of -60 to 20°C at 20°C/min. and an He flow rate of 2 mL/min. With this relatively short analysis (8.5 min.), the retention of S_2F_{10} is 6.1 min. and elutes after S_2OF_{10} (R.T.: 2.8 min.). No interferences from the various silane impurities are observed. The optimum estimated (due to the lack of standard) limit of quantification is 20 ppb (S/N= 3) for a sample volume of 250 μL injected.

Choice of oven temperatures. Figure 3 shows the effects of initial oven temperatures on chromatograms for a 5 μL injection of the decomposed SF_6 sample. The analysis performed at 30°C gives an ideal total elution time (less than 5 min.) but bunches the peaks corresponding to S_2OF_{10}, S_2F_{10} and $S_2O_2F_{10}$ and the SF_6 tail completely obscures the SO_2F_2, SOF_4, and SOF_2 signals. As expected, the separation efficiency of the column system improves as the initial oven temperature is lowered. The peaks of interest, except those for the usual monosulfur-oxyfluorides, are well eluted from the SF_6 signal when the oven temperature is programed from -20 to 30°C at 20°C/min.

SOF_2 can only be separated from the SF_6 matrix when using an initial oven temperature of -40°C or lower. By lowering the initial oven temperature to -60°C, SO_2F_2 and SOF_4 peak can now be separated from the SF_6 matrix. These two compounds have similar retention times and may only be distinguished by selective ion monitoring. However, their separation is not necessary since SOF_4 readily hydrolyzed to SO_2F_2 and it seems preferable to count them together. Finally, it was therefore shown that one can easily adjust the experimental conditions to satisfy various analysis objectives.

DISCUSSION

Evolution of Sample Contents

As stated above, few fluoromethylsilanes were always observed in any of the chromatograms and warranted further studies on the "aged" decomposed SF_6 samples. Therefore, a given SF_6 sample submitted to partial discharges was analyzed at regular time intervals while being left in the test cell at room temperature.

A close look to Figure 4 indeed reveals that the amount of $SiF_2(CH_3)_2$ (R.T.: 8.1 min.) present in the decomposed sample increases drastically over a short period of time. That of $SiF(CH_3)_3$ (R.T.: 11.1 min.) is also significant after 2 hours. In addition, SiF_3CH_3 (R.T.: 3.4 min., covered by SF_6) and $(CH_3)_2FSiOSi(CH_3)$ (R.T.: 19.4 min.), not shown here, are sometimes observed; the latter also slowly increases with respect to the time spent in the cell. This is a partly fluorinated dimer fragment of a typical dimethylsilicone polymer. It suggests some reactions involving HF and other reactive F-containing compounds with silicones (septum, oil traces) present in our inert teflon cell.

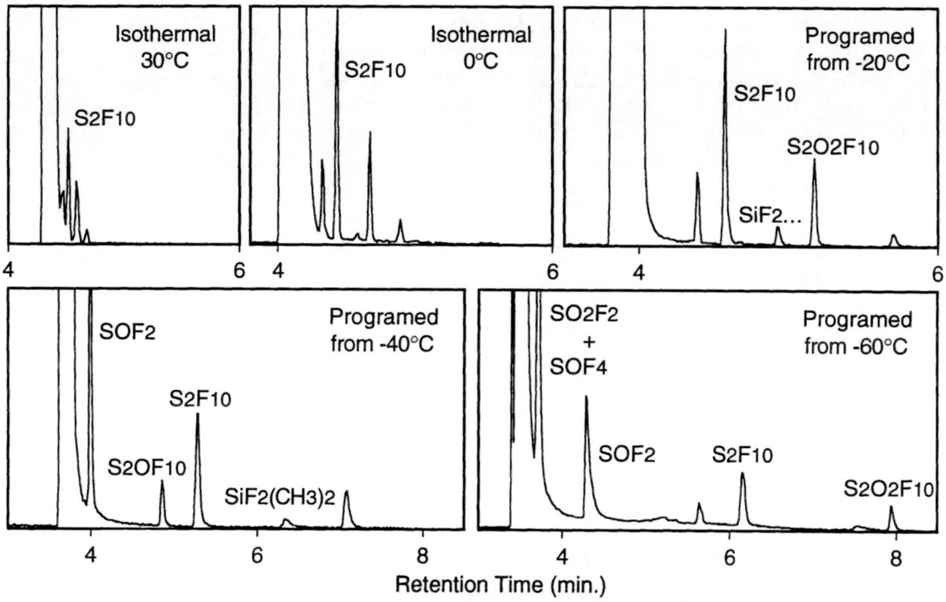

Figure 3. Typical TIC chromatograms recorded with DB5 and DB1 (55m x 0.2mm) bonded-silicon capillary columns, showing the influence of the initial oven temperature on the byproduct separation.

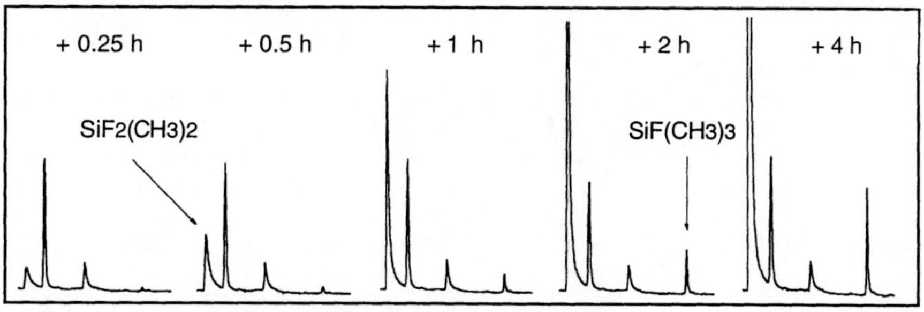

Figure 4. Increase of the amount of fluoromethylsilane impurities with residence in the test cell.

Fortunately, a short residence time of the sample prior to analysis does not affect the detection of S_2F_{10} since the developed chromatographic method allows complete separation of all components of interest. It was also noted that immediately after column conditioning and for a freshly prepared SF_6 decomposed sample, the fluoromethylsilane peaks are less intense and that SiF_3CH_3 is not always present. This seems to indicate that some slow degradation of the chemically bonded silicon gums of the column occurs in the long-term when exposed repetitively to reactive fluorine compounds.

S_2F_{10} Stability during analysis

Further investigation on the stability of the decomposed sample reveals that some SOF_2 ion fragments are present in the S_2F_{10} peak, as well as in other byproducts[4] . The gas-phase unimolecular decay of S_2F_{10} into SOF_2 has been reported in the past[1,6] . Data reported in Table 1 compare the relative importance of unreacted S_2F_{10} (m/z 127) with that of SOF_2 (m/z 67) over a concentration range of three decades. It is observed that the intensities of the SOF^+ ion vary in inverse proportion to the amount of S_2F_{10} present in the byproducts.

In fact, these data are only compatible with a disulfur decafluoride decomposition that occurs in the MS source, after its elution from the analytical column. Indeed, the total contribution of ions at m/z 67 and 127 appears to be relatively constant, as shown in the last row of Table 1. This demonstrates the roughly similar MS signal response of both byproducts. In consequence, SOF_2 ions are always present in the mass spectrum of S_2F_{10}, even with the MS source maintained close to room temperature (≈50°C). However, their contributions become quite small at the highest S_2F_{10} concentrations.

Table 1. Importance of the SOF^+ (m/z 67) ion in S_2F_{10} mass spectrum as a function of the quantity (TIC) of S_2F_{10} present in the decomposed SF_6 samples.

S_2F_{10} TIC (000s)	5.8	46	217	2300	7350
m/z 67 / m/z 127 (%)	49	29	20	5.2	1.7
m/z 127 / TIC (%)	35	38	39	47	47
Σ m/z (67 & 127) / TIC (%)	52	49	47	50	48

Effects of Instrument Zone Temperatures

Analysis of S_2F_{10} for SOF_2 content was also performed as a function of temperature of the different zones of the GC/MS system: injector, transfer line and MS ion source. This allowed the evaluation of the influence of the S_2F_{10} thermal decomposition prior to detection on its quantity finally measured. The relative amount of S_2F_{10} was evaluated using CF_4 (normal impurity in SF_6 gas) as internal standard. All the representative ions for SOF_2 (SOF_2^+, SOF^+ and SO^+; m/z 86, 67 and 48) were used in assessing its presence in the S_2F_{10} (SF_5^+; m/z 127) mass spectrum, in contrast with Olthoff[1] who did not considered the principal SOF_2 ionic fragment at m/z 67.

All data are listed in Table 2. In general, increasing the temperature of any zone results in a decrease of measured S_2F_{10}. This effect is accentuated in the case of the injector where a relative decrease of ≈45% is noted in going from 50 to 250°C. For a similar temperature variation of the transfer line and the MS-ion source, a comparable effect is observed (Table 3). Thus, the nature of the glass insert itself and a longer residence time might play a role on the decomposition of S_2F_{10} in the injector.

Table 2. Signal intensities (%) of the representative ions for SOF_2 (m/z 86, 67 and 48) in the S_2F_{10} (SF_5^+: m/z 127) mass spectrum as a function of temperature of certain zones (injector, transfer line and MS-ion source) of the GC/MS analytical system.

Zone Temperature (°C)			Fraction of SOF_2 fragments in S_2F_{10} (%)			Relative quantity of S_2F_{10}
SP Injector	Transfer Line	MS Source	48	67	86	S_2F_{10}/CF_4 (*)
50	50	50	10	19	5.3	1.07
150	50	50	10	19	5.1	0.95
250	50	50	10	20	2.4	0.47
50	150	50	9	18	4.4	1.12
50	250	50	5	14	2.8	0.73
50	100	50	1.5	2.1	0.7	1.44
50	100	150	1.4	1.8	0.8	1.34
50	100	250	0.7	1.2	0.4	1.00

(*) CF_4 is an impurity in the SF_6 and can therefore serve as internal standard to evaluate the relative amount of S_2F_{10}.

Interestingly, the percentage of SOF_2 (ion fragments) found in S_2F_{10} is relatively constant in all cases tabulated. The percentage of SOF_2 fragments in S_2F_{10}, when the temperature of the source is varied, differs by one order of magnitude from that of the injector or the transfer line and is not accounted for at the present time. We have no explanation for this yet, since all the gas samples were obtained from partial discharge tests done exactly under identical conditions.

Nonetheless, these results indicate that the partial surface decomposition of S_2F_{10} does not seem to form SOF_2 quantitatively. Indeed, the presence of its three main ions in the S_2F_{10} mass spectrum is rather constant, even in the case of the heated transfer line in which the column passed through before reaching the MS ion source. What is obvious, however, is the net lost of S_2F_{10} which is not compensated by the relatively constant amount of SOF_2 measured by its 3 major fragments. Thus, the only suitable explanation is the irreversible reactions of S_2F_{10} with the column material (thin bonded silicone coating). Finally, the net result of having higher zone temperatures is lowering S_2F_{10} quantities eventually measured, as depicted in Table 3.

Table 3. Relative amount of measured S_2F_{10} with respect to the increase of the zone temperatures (relative to 50°C).

Zone Temperature (°C)	Injector	Transfer Line	Source
150	0.89	1.00	0.93
250	0.44	0.68	0.70

One final but interesting observation was made concerning the S_2F_{10} irreversible lost during the elution in the analytical column. After a severe and exhaustive conditioning of the column, the absolute concentration of the disulfur decafluoride was found to increase by almost three fold. As seen in Table 4, S_2F_{10} was always measured in higher quantities than the SOF_2 . The experimental conditions for producing the S_2F_{10} were exactly similar to previous ones where the S_2F_{10} / SOF_2 ratios were found to be 0.25-0.5, instead of >>1 . After a while, this ratio returned to the *normal* range (0.2-0.5).

In addition, the data in Table 4 reveal also that some part of the SOF_4 also disappears into the column during a *usual* analysis. On the contrary, SOF_2 did not seem to be affected by such irreversible interactions within the column.

Table 4. Relative formation of S_2F_{10} and SO_2F_2 + SOF_4 compare to that of SOF_2 measured before and after exhaustive conditioning of the analytical columns (*)

Byproducts	Test results before cond.	Test results after cond.	Test results after cond.	Test results after cond.	Average of 3 tests	Standard Deviation
(SO_2F_2 + SOF_4) / SOF_2	2.7	4.2	3.4	5.2	4.3	± 21 %
S_2F_{10} / SOF_2	0.39	2.30	3.45	2.90	2.9	± 20 %
Formation rate of SOF_2 (nmoles/J)	0.75	0.89	0.96	0.70	0.85	± 16 %

(*) Each test is a distinct discharge experiment done in identical conditions

Calibration, Quantification and Sensitivity

Calibration analysis with the present GC/MS system shows a linearity of three decades. Calibration of the MS response was done in the concentration range of 1 to 100 ppm for SF_6 in nitrogen and with 500 ppm of the same gas mixture. A slight deviation from linearity is probably due to uncertainties in the preparation of standards of low concentrations. The detector response versus the amount of pure SF_6 injected (1-10 μL) was also measured. The response is linear and reproducible with an estimated accuracy of 10 %. It is noteworthy that our calibration study indicates that 5-6 ppm of S_2OF_{10} is initially present in the commercial SF_6 gas we used. This should be taken into account during quantification.

With the present analytical method, employing 55 meters of capillary column, it was found that a large injected volume (>300-350 μL) overloads the column. A detection limit of ≈50 ppb was attained for a 50 μL sample injected. Caution should be exercised since the presence of a great concentration of one component influences the detector exhibiting its resolution limits. Indeed, for any given intense fragment ion, ions at m/z ± 2-3 are observed. The problem is probably associated with the small volume of the trap and its ion extraction and intensity measurement characteristics.

The MS response was found to be stable over a nine months period and was periodically checked with standards. However, on occasion, a higher response (10x) for SF_6 in nitrogen or in dry air was noted. The cause of this peculiar behavior, probably due to the high quantity of nitrogen or dry air matrix introduced at once in the MS trap, is still not quite understood at this point, but is easily picked up and was taken into account when it occurred (50% reproducibility in the measurements).

A 15% reproducibility, within each others, for CF_4, SOF_2 and SF_5Cl standards, was observed. Whenever TIC measurement was not possible, the gas components were quantified based on the sum of their three most abundant fragment ions, which was found to correspond to 75-80% of the integrated surface based on the TIC. Without the availability of a specific gas reference standard, this still allows a relative quantitative analysis of S_2F_{10} with a confidence range better than 50% (i.e. concentration error: factor of 2). In addition, 25% in reproducibility, for the combination of the discharge test plus the analysis, of repeated injections of freshly decomposed SF_6 samples was usually found and is shown in Table 5. This gives a margin error of ± 25%. It is also illustrated by data of Table 4.

Table 5. Reproducibility of the S_2F_{10} formation and analysis: TIC signal measured for various samples taken from distinct discharge experiment done in identical conditions

Injection Volume (μL)	5	10	50	200
S_2F_{10}	298	566	1 981	7 328
TIC signal	237		2 246	
(000s)			2 008	

CONCLUSIONS

This paper has described the development of a simple chromatographic method for the analysis of S_2F_{10}. The mass detector was found to be a reliable and sensitive technique for the direct identification and determination of S_2F_{10} analysis in SF_6. Further, the simplified one-step analysis does not require removal of SF_6 from the gas sample with tedious cryogenic enrichment procedure.

In the optimum conditions it also allows the separation of all the known principal SF_6 decomposition byproducts with the initial cooling of the oven to -60°C. Analysis of S_2F_{10} at the ppm level can even be done in less than 5 min. with the oven temperature almost maintained at room temperature (0°C), but at the expense of a loosing resolution. In the best analytical conditions, the quantification limit was estimated at ≈20 ppb for an injected sample volume of 300 μL which is slightly less than the column saturation capacity.

In addition, the use of a MS detector has the great advantage of permitting semiquantitative measurements of the various byproducts, even those with no calibration gas standard available, due to their relative similar response factor (TIC). This property is of particular interest in the case of S_2F_{10} since reference gas samples of this compound are difficult, if not impossible, to obtain. However, it will eventually be necessary to verify at least once the relative MS response with a known quantity of S_2F_{10}.

REFERENCES

1 J.K. Olthoff, R.J. Van Brunt, J.T. Herron and I. Sauers, "Detection of Trace Disulphur Decafluoride in Sulfur Hexafluoride by Gas Chromatography/Mass Spectrometry", Analytical Chemistry, Vol 63 No. 7, pp. 726-732, 1991.

2. I. Sauers and R.A Cacherio, "A Cryogenic Enrichment Technique for Gas Chromatographic Detection of S_2F_{10} in SF_6 Discharges", Conference Record of the 1992 IEEE International Symposium on Electrical Insulation, Baltimore, pp. 340-344, June 7-10 1992.

3. R.J. Van Brunt, J.K. Olthoff, J.K.,I. Sauers, H. D. Morrisson, J. R. Robins and F. Y. Chu, "Detection of S_2F_{10} Produced by Electrical Discharge in SF_6", Proceedings of the X International Conference on Gas Discharges and Their Applications,, Sept. 13–18, 1992, Swansea, England, pp. 418-421.

4. J. Castonguay and I. Dionne, "S_2F_{10} and Other Heavy Gaseous Decomposition Byproducts Formed in SF_6 and SF_6-Gas Mixtures Exposed to Electrical Discharges", presented at the Seventh International Symposium on Gaseous Dielectrics, Knoxville, Tennessee, April 24-28, 1994.

5. F.J. Janssen, "Decomposition of SF_6 by Arc Discharge and the Determination of the Reaction Product S_2F_{10} ", Gaseous Dielectricsc V (Proceedings of the Fifth International Symposium on Gaseous Dielectrics, Knoxville, 1986), pp. 157-162, Pergamon Press, NY, 1987.

6. I. Sauers, M.C. Siddagangappa, G. Harman, R.J. Van Brunt and J.T. Herron, "Production and Stability of S_2F_{10} in SF_6 Corona Discharges", Proceedings of the 6th International Symposium on High Voltage engineering,, New Orleans, Aug. 28-Sept. 1, paper 23-8, 1989.

DISCUSSION

S. W. ROWE: Can you identify SF_4?

I. DIONNE: Not with these analytical techniques. Actually, SF_4 might be injected to condition the column. It takes a very large quantity injected before seeing any SF_4 eluted peak.

DECOMPOSITION OF SF_6 AND PRODUCTION OF S_2F_{10} IN POWER ARCS

H.D. Morrison[1], F.Y. Chu[1], M. Eygenraam[1], I. Sauers[2], R.J. Van Brunt[3]

[1]Ontario Hydro Technologies
800 Kipling Avenue, Toronto, Ontario
Canada M8Z 5S4

[2]Oak Ridge National Laboratory
Oak Ridge TN 37831-6123

[3]National Institute of Standards and Technology
Gaithersburg, MD 20899

INTRODUCTION

Decomposition of SF_6 in electrical discharges produces many toxic solids and gases. S_2F_{10} is the most toxic of the gaseous byproducts and has been found in arcs, sparks and corona [1]. Of these, S_2F_{10} production in arcs is the least understood, in part because S_2F_{10} is known to decay rapidly at temperatures above 250°C [1]. As temperatures in an arc are considerably higher, it is believed that S_2F_{10} cannot be formed directly by an arc. The first experiments where S_2F_{10} was detected in SF_6 decomposed by a power arc employed a burn-through configuration into another chamber containing SF_6 at a lower pressure [2,3]. In those experiments the S_2F_{10} may have been formed during the volume expansion and cooling of the SF_6 decomposition products into the second chamber. We have conducted a series of tests of a power arc discharge contained completely within a bus duct configuration. Among the many other gaseous byproducts, we have detected S_2F_{10} at or below the part per million (ppm) by volume level, proving that S_2F_{10} can be formed directly by a power arc within SF_6-insulated equipment. The relative production rate of S_2F_{10} with respect to that of SOF_2 and SF_4, however, implies that S_2F_{10} is not a significant contributor to the hazard of exposure to decomposed SF_6.

EXPERIMENT

The power arc tests were performed in an arc cell designed to simulate a section of bus duct. The arc cell consisted of two coaxial aluminum pipes with plastic end plates to support the inner pipe as the inner conductor, as indicated in Figure 1. The inner diameter

Gaseous Dielectrics VII, Edited by L.G. Christophorou
and D.R. James, Plenum Press, New York, 1994

of the outer conductor was 154 mm, and the outer diameter of the inner conductor was 72.5 mm. The end plates were separated by 1.83 m, enclosing a volume of 26.5 L. Connections on the end plates provided for gas filling and sampling, and for pressure relief. The arc cell was designed to be re-usable. The original cell was used for the first two tests, both pipes were replaced for the the third test, and finally a new inner pipe was used for the last test. Any re-used components were thoroughly cleaned before being used in another test.

At least 24 hours before each test, the arc cell was evacuated and then filled with commercial grade, virgin SF_6 to a pressure of about 200 kPa. Just before each test, the equilibrium humidity of the gas in the cell was measured by flowing some of the gas through a chilled-mirror hygrometer. Then, the final pressure in the cell was read from the portable rack and the valve at the cell was closed.

The tests were performed in an outdoor, explosion-proof room attached to the high current laboratory at Ontario Hydro. The current source was connected to the inner

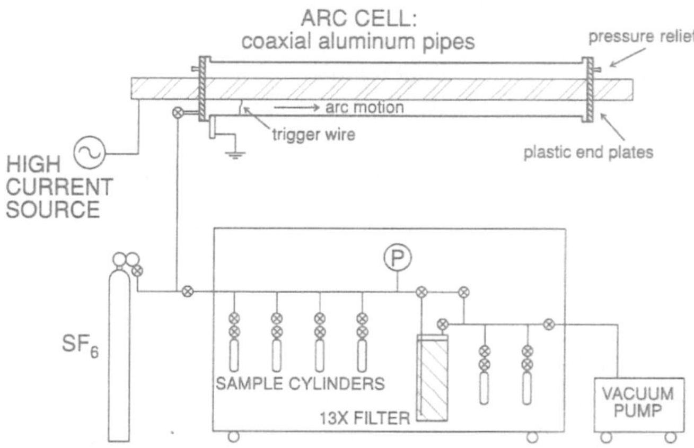

Figure 1. Schematic diagram of the arc cell and portable sampling rack used in the power arc experiments. The internal volume of the arc cell is 26.5 litres.

conductor at one end only, and the outer conductor was grounded at the same end. A fuse wire between the two conductors, inside the cell and near one end, served to initiate the arc discharge when the current was applied.

The top part of Table 1 lists the experimental conditions for each of four tests: SF_6 pressure and quantity, water vapour concentration, arc current (rms) sustained, and energy dissipated in the cell. The energy dissipated is expressed as energy per unit volume of SF_6 at atmospheric pressure (101.3 kPa). For each test, the duration of the arc current was fixed between 50 and 150 ms set to provide a range of energy discharged into the cell without burning through the conductors or the end plates. For a single-ended current, the arc is not stationary; we measured current-normalized arc velocities of about 1 m/(s.kA) [4].

After each test, the solid byproducts, mostly AlF_3, were allowed to settle for at least 15 minutes before withdrawing any gas samples. Several gas samples were taken from each test in stainless-steel cylinders, either 150 or 1000 ml in volume. At least one sample from

Table 1. The test conditions of four power arcs, labelled A, B, C, and D, and the production rates of the SF_6 decomposition byproducts of interest.

		A	B	C	D
SF_6: Pressure	(kPa)	158	210	208	195
Quantity	(moles)	1.74	2.47	2.26	2.25
Water	(ppm)	>800	100	860	340
Current	(kA)	9.5	7.8	7.5	7.2
Energy Dissipated	(kJ/L)	1.59	0.97	2.53	2.47
SOF_2 & SF_4	(mol/J)	1.74×10^{-6}	5.31×10^{-7}	7.31×10^{-7}	8.43×10^{-7}
SO_2F_2	(mol/J)	5.3×10^{-10}	9×10^{-11}	5.3×10^{-10}	8.7×10^{-10}
S_2F_{10}	(mol/J)	4×10^{-11}(a)	$<2\times10^{-10}$(b)	7.5×10^{-12}(a)	1.38×10^{-11}(a)

(a) Measured by cryo-GC at ORNL.
(b) Measured by GC-MS at NIST.

each test was analyzed by infrared absorption spectroscopy in a Fourier Transform infrared (FTIR) spectrometer at Ontario Hydro [1]. A typical infrared absorption spectrum of the gas in the arc cell, before and after the arc, is shown in Fig. 2. Another sample from each test was analyzed either by gas chromatography-mass spectrometry (GC-MS) at the National Institute for Standards and Technology (NIST) [5] or by cryogenic enrichment gas chromatography (cryo-GC) at Oak Ridge National Labs (ORNL) [6].

RESULTS AND DISCUSSION

The example spectra in Fig. 2 show absorption bands due to SF_6 and other absorption bands, appearing in various windows of the SF_6 spectrum, due to a variety of decomposition byproducts. The main byproducts of interest here are SF_4 and SOF_2. As Fig. 2 indicates, analysis of these two compounds individually by FTIR can be straightforward. However, as SF_4 reacts rapidly with water to produce SOF_2 and HF, most GC analysis techniques cannot readily distinguish SF_4 from SOF_2. Thus, many reports of SF_6 decomposition in arcs relying on GC analysis have lumped SF_4 in with SOF_2 production rates. To facilitate comparison with previously published results, we report SF_4 production and SOF_2 production, measured separately by FTIR, together in the bottom part of Table 1. Table 1 also lists the production rates of SO_2F_2 and S_2F_{10}. The analysis of SO_2F_2 was performed by GC-MS at all three sites. The S_2F_{10} measurements were split between the GC-MS at NIST and the cryo-GC at ORNL, as indicated in Table 1.

The combined production rates of SF_4 and SOF_2 observed in the last three arc tests are each comparable to the average production rate for SOF_2 in arcs of 6.0×10^{-7} mol/J reported in Ref. 7. The much higher than average production rate of SF_4/SOF_2 in arc test A has not been adequately explained. Pettinga [3], however, noted a correlation between lower vacuum attained in preparation of an arc cell and lower production of SF_4/SOF_2. In our case, the cell for test A was pumped a shorter time than the cells for the subsequent tests.

In contrast to the SOF_2 production, the very low production rates of SO_2F_2 are typical of power arcs. SO_2F_2 production is enhanced in cooler discharges, such as sparks and corona. Thus, the ratio of SOF_2 to SO_2F_2 in decomposed SF_6 is a simple measure of the type of discharge. For equal concentrations, the absorption peaks of SOF_2 and SO_2F_2 are

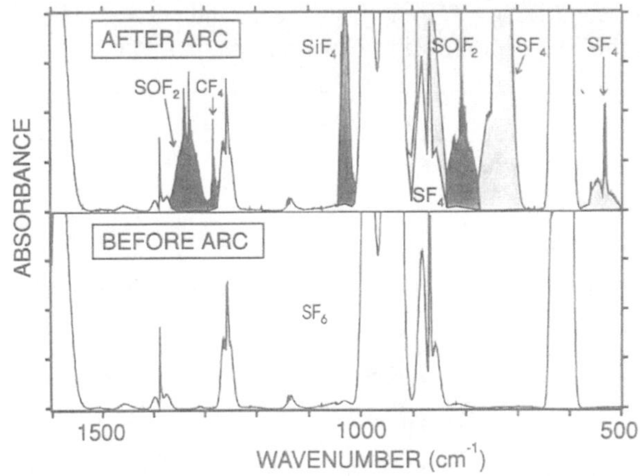

Figure 2. The two panels show the infrared absorption spectrum of gas samples from the arc cell before and after a power arc. In the lower panel, the spectrum is entirely due to SF_6. In the upper panel, absorptions due to major byproducts of the arc are indicated: SF_4, SOF_2, SiF_4, and CF_4. The region near 550 cm^{-1} is of greatest interest for detecting S_2F_{10}.

about the same height in FTIR spectra such as Fig. 2. Absorption peaks due to SO_2F_2 are not apparent in Fig. 2 because the concentration is about 0.1% of that of SOF_2.

In the absence of SOF_2 and SF_4, our detection limit for S_2F_{10} is about 0.10 ppm by FTIR and is less than 0.010 ppm by GC-MS. The concentration of S_2F_{10} in the virgin SF_6 used to fill the arc cell was less than 0.010 ppm as measured by GC-MS. The high concentrations of SOF_2 and SF_4 found in the arc samples created problems for detecting S_2F_{10} either by FTIR or GC-MS. The S_2F_{10} production rates listed in Table 1 for arc B represents a lower limit of 5 ppm for detection by GC-MS. Fortunately, the cryo-GC method at ORNL for detecting S_2F_{10} is not affected by high concentrations of either SOF_2 or SF_4. The detection limit for S_2F_{10} with the cryo-GC was about 0.1 ppm for arc A and less than 0.010 ppm for the analysis of arcs C and D. Thus, analysis of the samples C and D revealed S_2F_{10} concentrations of 0.43 and 0.76 ppm, respectively, which correspond to the production rates shown in Table 1.

A problem frequently encountered in the sampling of S_2F_{10}, at concentrations on the order of 1 ppm or less, is its decay with time in sample cylinders [5]. In general, higher pressures, larger volumes and dryer surfaces result in slower decay. Thus, 1000-ml cylinders are usually superior to 150-ml cylinders for retaining S_2F_{10}. The samples from arc A were taken in 150-ml cylinders, before we realized fully the problem of sample decay. The samples for arcs B, C and D were taken in 1000-ml stainless-steel cylinders. The measurement of sample C, however, was delayed about two months beyond the time of the arc test. In that time, the S_2F_{10} within the sample may have decayed.

Sample D was measured within 24 hours of sampling, and then re-measured several times over a period of several months. These measurements showed that S_2F_{10} was stable over a 50 day period, and decreased to about 50% after 200 days. We believe that the stability of the S_2F_{10} in the cylinder may be attributed to the dryness caused by the large amount of SF_4 present. As arcs C and D were very similar, the measurement of arc C, delayed as it was, probably represents more than 50% of the initial, unknown value.

The guidelines for exposure to many compounds are defined by the Threshold Limit Values (TLV) set out by the American Conference of Governmental Industrial Hygienists (ACGIH) [8]. The TLV for SF_6 is 1000 ppm, for SOF_2 is 1.6 ppm, for SF_4 is 0.10 ppm, and for S_2F_{10} is 0.010 ppm (see Ref. 1 for discussion). In all four arc tests, the combined production rate of SOF_2 and SF_4 was about 10,000 times greater than the production rate of S_2F_{10}. Thus, the limit for exposure to SF_6 decomposed by a power arc is governed by the presence of SF_4 and SOF_2, and not by the presence of S_2F_{10}.

CONCLUSIONS

We have performed four arc tests enclosed in a bus duct configuration with aluminum electrodes and have confirmed the presence of S_2F_{10} as a decomposition byproduct of SF_6. In two of the tests, we measured S_2F_{10} concentrations of 0.43 and 0.76 ppm, with a detection limit of 0.01 ppm, by a cryogenic-enrichment-GC technique. Of the three techniques investigated for measuring S_2F_{10}, the cryo-GC was the only one unaffected by the large concentrations of SOF_2 and SF_4 in the samples. The detection limits for S_2F_{10} by an FTIR technique and by a GC-MS technique were degraded to the range of 5-10 ppm. The concentrations of SOF_2 and SF_4 were measured separately by the FTIR technique. In every test, the combined production rate for SOF_2 and SF_4 exceeded the production rate for S_2F_{10} by a factor of about 10,000. The limits for exposure to SOF_2 and SF_4 are only about 160 times and 10 times more, respectively, than for S_2F_{10}, as specified by the Threshold Limit Values set out by the ACGIH. Therefore, the hazard of exposure to SF_6 decomposed by a power arc is determined by the SOF_2 and SF_4 and not by the presence of S_2F_{10}.

ACKNOWLEDGEMENTS

This report was prepared as an account of work sponsored by agencies of the United States Government, the Electric Power Research Institute, the Empire State Electric Energy Research Corporation, the Bonneville Power Administration, the Tennessee Valley Authority, Ontario Hydro, Hydro Quebec, and the Canadian Electrical Association. None of the Parties nor any agency thereof, nor any of their employees, makes any warranty, express or implied, or assumes any legal liability or responsibility for the accuracy, completeness, or usefulness of any information, apparatus, product, or process disclosed, or represents that its use would not infringe privately owned rights. Reference herein to any specific commercial product, process, or service by trade name, trademark, manufacturer, or otherwise, does not necessarily constitute or imply its endorsement, recommendation, or favoring by the Parties or any agency thereof. The views and opinions of authors expressed herein do not necessarily state or reflect those of the sponsoring Parties including the United States Government or any agency thereof.

The authors thank Dr. Jeffrey R. Robins for his assistance in establishing and continuing the arc testing program, and for his work in setting up the FTIR.

REFERENCES

1. D.R. James, I. Sauers, G.D. Griffin, R.J. Van Brunt, J.K. Olthoff, K.L. Stricklett, F.Y. Chu, J.R. Robins, and H.D. Morrison, Investigation of S_2F_{10} production and mitigation in compressed SF_6-insulated power systems, *IEEE Elec. Insul. Mag.* 9:29 (1993).
2. F.J.J.G. Janssen, Measurements at sub-ppm level of sulfur-fluoride compounds resulting from the decomposition of SF_6 by arc discharges, *Kema Scientific & Technical Reports* 2:9 (1984).
3. J.A.J. Pettinga, Full-scale high-current model tests on busbar constructions, Proc. of CIGRE Symp. High Current in Power System, 506-585 (1985).
4. F.Y. Chu, F. Lutz, J.M. Braun, and H.A. Stuckless, Effects of power arc faults in gas-insulated substations, *in*: "CIGRE Symposium on High Current in Power Systems", Brussels (1985).
5. J.K. Olthoff, R.J. Van Brunt, J.T. Herron, and I. Sauers, Detection of trace disulfur decafluorides in sulfur hexafluoride by gas chromatography/mass spectrometry, *Anal. Chem.* 63:726 (1991).
6. I. Sauers and S.M. Mahajan, Detection of S_2F_{10} produced by a single-spark discharge in SF_6, *J. Appl. Phys.* 74:2103 (1993).
7. F.Y. Chu, SF_6 decomposition in gas-insulated equipment, *IEEE Trans. Elec. Insul.* EI-21:693 (1986).
8. "1992-1993 Threshold Limit Values for Chemical Substances and Physical Agents and Biological Exposure Indices," American Conference of Governmental Industrial Hygienists, Cincinnati (1992).

DISCUSSION

J. CASTONGUAY: Your test A shows a much larger production rate of SOF_2, that is, twice the usual "maximum" yield ever measured by any other experimenter. Any comment?

H. MORRISON: Yes. I mention this point in the paper. The higher yield may be related to the fact that the cell for test A was not pumped down as long as was done for the other tests.

J. CASTONGUAY: Do you think that the level of moisture has some effect in your power arc tests?

H. MORRISON: I believe that the humidity level in the gas must have some effect on the by-products, but the effect is not very strong. Water adsorbed on the surfaces of the chamber where the arc passes may be just as important as the water vapor in the gas.

SECTION 9: DC GIS/CHARGING OF SOLID INSULATORS

MEASURING THE DC STRESS DISTRIBUTION ACROSS DISC INSULATORS

J.S. Pearson, B.F. Hampton, O. Farish and C. Alderton

Centre for Electrical Power Engineering
University of Strathclyde
Glasgow G1 1XW
UK

INTRODUCTION

Switching operations in a gas insulated substation (GIS) on an AC transmission system may leave a trapped charge on an isolated section of busbar, thereby stressing the solid support barriers by a DC voltage. This condition may persist for several hours or days, because it is not normal practice to earth the EHV conductor and drain the charge unless the chamber is to be opened for some purpose.

An opening disconnector restrikes many times at progressively higher voltages as the contacts separate, until the gap can withstand the voltage across it. The final discharge often occurs close to the peak of the AC wave, and a positive or negative charge of up to 1pu could be left on the busbar.

Barriers for use on an AC system are also designed to withstand a high DC voltage, because, for example, a cable feeding the GIS will need to be tested with DC and any barriers between it and the first disconnector must also withstand this voltage. Nevertheless it is suspected that under some conditions trapped charge may cause problems, and perhaps be connected with the low energy flashover marks on barriers reported previously[1]. Some small scale laboratory tests were therefore undertaken to investigate this possibility.

Gaseous Dielectrics VII, Edited by L.G. Christophorou
and D.R. James, Plenum Press, New York, 1994

THE VOLTAGE DISTRIBUTION ACROSS GIS BARRIERS

Barriers in GIS are designed to have a low surface stress, so that in the absence of severe contamination the risk of them flashing over under normal service voltage is very small indeed. The maximum design stress for a barrier in a 400kV GIS (240kV phase-ground) is typically 4kV(pk)/mm, and its AC flashover voltage is at least 700kVrms - that is, in excess of 10kV(pk)/mm.

Ionisation cannot occur in SF_6 unless the peak local field (peak AC, or DC) exceeds 8.9kV/mmbar. This corresponds to almost 50kV/mm at 4.5bar(g) working pressure, and so under trapped charge conditions a field enhancement of at least 12 would be needed at the most highly stressed part of the barrier before any discharge could start. There are various mechanisms which might lead to stress enhancement under DC, and the most relevant of these are discussed below.

Bulk Conductivity

The AC stress distribution across a barrier may readily be calculated. On applying a DC stress, the displacement current dominates and since some dipoles become oriented only very slowly, the small conduction current may be swamped and not seen for a considerable time.

The permittivity (ϵ) of a resin system is known precisely and typically has a value of 5. Against this the resistivity (ρ) of the material probably varies quite widely, since it depends on the low levels of ionic impurities which may be completely ignored for all other properties of the material. It is likely to be controlled by factors such as the level of trace impurities and preparation of the filler, the curing process, temperature, and so on. One manufacturer quotes the resistivity as approximately 10^{13} Ω-m, and then the material has a relaxation time constant (τ), where:

$$\tau = \epsilon\epsilon_0\rho = 4.5\times10^2 s$$

In practice the time constant could well vary from minutes to several hours. The important point, however, is that after a sufficiently long time the stress distribution through the bulk of the barrier material will be controlled by its resistivity - and if the resistivity is non-uniform due to differences in processing or temperature across the barrier, the stress will be non-uniform too.

Surface Conductivity

It is very difficult to make a surface retain the physical and electrical properties of the bulk material, because its behaviour is almost always changed by adsorbed gases or contamination of one form or another. When the most precise laboratory measurements have to be made on samples of material which will withstand a high temperature, the favoured means of adequately cleaning the surface is to bake it under an ultra-high vacuum. In comparison with this standard of preparation, wiping a barrier with a cloth and solvent does no more than remove the gross contamination. Of course the traces of contamination remaining have no effect at all on the AC performance of the barrier, but under DC could well play a significant rôle.

The values of surface conductivity likely to be met in practice could be in the region of 10^{12}-10^{15}Ω, but again it is the uniformity of the surface resistivity which controls the stress distribution and apparently insignificant changes in the cleaning technique or materials could well be important.

Surface Charge Deposition

Free ions in the gas can be attracted to the barrier surface, and once there will not easily be removed. There are several possible sources of ions, but it is likely that they originate from high surface stresses on the busbar. Although the average stress there may be only, say, 3kV/mm due to the trapped charge, it can be increased locally either by surface roughness or the oxide layer which is inevitably present on aluminium, and might lead to field emission.

If the emission were in the re-entrant part of the barrier, ions would move in the overall field to the barrier and build up a patch of charge there. In theory the maximum surface charge density which can be sustained on an infinite plane surface in SF_6 at 4.5bar(g) is $720\mu C/m^2$, but in this non-uniform case sparking would occur at values lower than this. In addition, the accumulated surface charge would reduce the stress at the busbar, and the emission could well be self-limiting. The time needed to generate a significant level of charge depends on the emission current, which is very unpredictable. At low stresses of 1kV/mm or so on an oxide covered surface the emission can be 10^5 electrons/sec^2, and it would take some hundreds of hours to build up a patch of charge. This effect is only likely to be significant if some emission mechanism leading to higher currents is present[2].

MEASURING THE DC STRESS DISTRIBUTION EXPERIMENTALLY

Insufficient data is available for the DC stress distribution to be calculated, and it is necessary to measure the surface stresses experimentally. Measurements have been made of the charge left on an insulating surface when the applied voltage has been removed[3,4], but to our knowledge such a measurement has not been made of the resistively controlled stress with the voltage still applied. However this is at the heart of understanding any DC flashover mechanism, and is the objective of the work described here. It has been undertaken on a model insulator, because at the outset it appeared too difficult to attempt the measurement on a full sized barrier.

The average surface stress on a conical barrier in a 420kV GIS due to a trapped charge of 1pu on the busbar is likely to be in the region of 1kV/mm. For the present work use is made of a plain coaxial disc insulator, 130mm in overall diameter and with an insulating surface 50mm wide. To stress the disc realistically 50kV needs to be applied to the centre conductor. It was therefore not practical to measure the stress distribution by scanning the surface with a normal charge-measuring probe at earth potential, because at any useful separation breakdown to it would certainly occur.

Measurement Technique

The approach is to scan the disc with a probe, but to continually adjust the potential of the probe to equal that of the surface it faces. The probe is connected to a variable HVDC supply, which has to be adjusted progressively so that the probe and surface potentials become equal.

Such a scheme depends on the probe being able to sense any difference V between its own potential and that of the closest part of the insulating surface. This is achieved

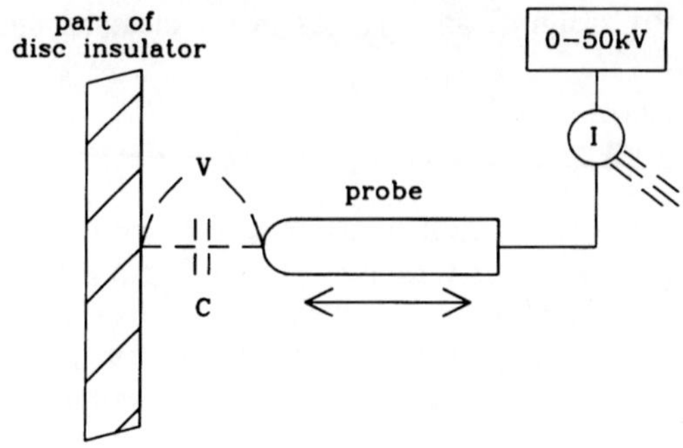

Figure 1. Principle of the oscillating probe.

by continuously oscillating the probe perpendicularly to the surface (Fig 1), thereby changing the capacitance C between the two. The displacement current in the probe lead is then given by:

$$i = d(CV)/dt$$

$$= V*dC/dt$$

since the voltage remains constant during the measurement.

The magnitude of the current indicates the voltage difference between the probe and the surface, and its phase whether the probe potential should be raised or lowered in order to reduce the difference. Having adjusted the probe potential to minimise the current, and so the voltage difference, the probe can be brought closer to the surface. The capacitance between the two increases, giving a higher probe current and a more sensitive measurement - so the process can be repeated until the probe is near enough to give to give a good resolution. In the limit, it is possible for the probe to touch the surface without disturbing its potential.

Apparatus

The sample disc is contained in a large glass vessel fitted with several ports, as shown in Fig 2. It is capable of withstanding both vacuum and a slight overpressure, but not the normal working pressure in a GIS. However the stress distribution on the disc is unlikely to be affected significantly by the gas pressure, and this limitation is more than offset by the convenience of working with a completely visible system.

The vessel contains a means of scanning the probe radially and axially with respect to the disc, using stepping motors so that the probe position relative to set limit switches is known. A third stepping motor is used to rotate the disc.

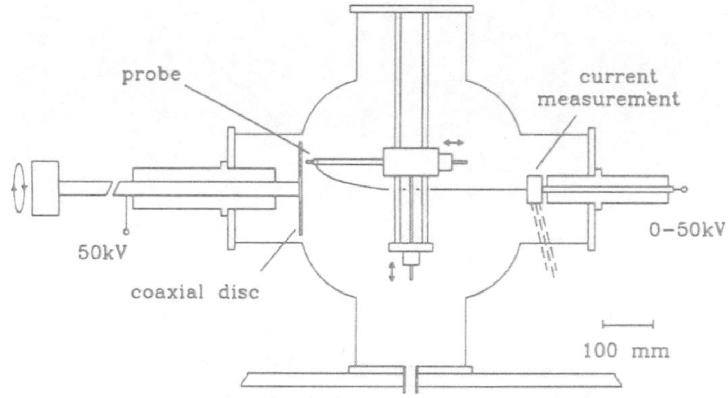

Figure 2. General arrangement of the test vessel.

Probe Current Measurement

Currents of some nanoamperes are induced in the probe lead and measured by a battery operated unit mounted on the HV bushing. The block diagram of the circuit (Fig 3) shows a high impedance buffer amplifier at its input, followed by two stages of amplification and an RMS to DC converter. From this the signal is passed to a voltage to frequency converter, and the frequency signal to an infra-red transmitter. A receiver on the outside of the vessel detects the transmitted frequency, and passes it to the controlling PC. The measured current/frequency calibration is given in Fig 4.

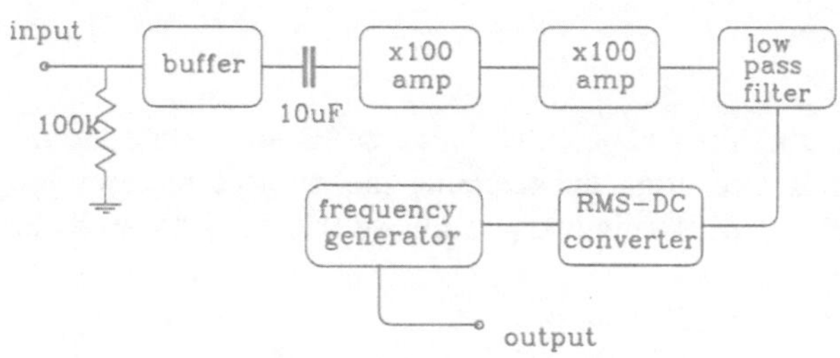

Figure 3. Block diagram of the current measuring circuit.

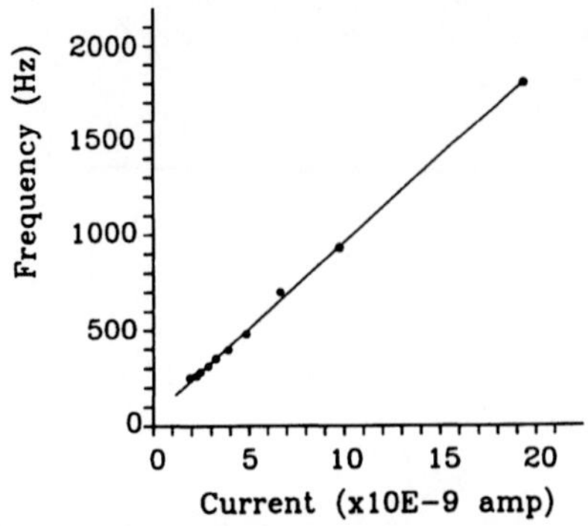

Figure 4. Frequency calibration of the current measurement at a probe oscillation of 25Hz.

Procedure for Scanning the Disc

The potentials of both the central electrode of the disc and the probe are set to 50kV, and the probe moved close to the electrode. Having moved the probe radially on to the insulating surface there is an imbalance of potentials and a current is induced in the probe lead. Its magnitude (expressed as a frequency) is fed to the PC, which varies the probe HV supply until the current is a minimum. The position of the probe and its potential are then stored by the PC.

The PC may be programmed to move the probe stepwise over the insulating part of the disc, repeating the measurement until the surface has been covered. Having obtained a complete set of potential data it can be displayed in various ways - plan, elevation, 3-dimensional - by the PC, and so allow any highly stressed areas on the disc to be identified.

CONCLUSIONS

A method of measuring the DC potential on an insulating surface has been demonstrated, and a programme of measuring the scanning the surfaces of discs prepared in various ways will soon be started. Preliminary results of this work are expected to be presented at the Symposium.

ACKNOWLEDGMENTS

This work was funded jointly by the National Grid Co and Scottish Power, and the authors are grateful to the directors of these companies for permission to publish the paper.

REFERENCES

1. B M Pryor. Tracking of in-service GIS barrier insulators, *IERE Workshop on GIS*, Toronto, 1990.
2. F LLewelyn Jones. "Reports in Progress in Physics," The Physical Society, London, 1953.
3. O Farish and I. Al Bawy. Effect of surface charge on impulse flashover of insulators in SF_6, *IEEE Trans on Elec Ins*, vol 26, no 3, June 1991.
4. I. Al-Bawy and O Farish. Charge deposition on an insulating spacer under impulse voltage conditions, *Proc IEE-A*, vol 138, no 3, May 1991.

DISCUSSION

S. M. MAHAJAN: The tip of the probe may induce charges on that part of the dielectric surface (in the vicinity of the tip) where the voltage of the probe does not match the voltage at the dielectric surface. Will this not be even more important as you try to improve the spatial resolution of the measurement system which incorporates an "unshielded" probe?

O. FARISH: It will be necessary to increase the measurement sensitivity (by using a better input buffer amplifier and/or by increasing the oscillation frequency if a small probe is used). A reduced voltage of less than 10 V in (say) 1000 V should be readily achieved.

D. L. SCHWEICKART: (1) Are the probe oscillations internal (like a vibrating membrane) or external (mechanically driven fixture)? (2) What is the oscillation frequency? (3) Is there a lower limit to the probe diameter, i.e. spatial resolution, for which this technique will give satisfactory results?

O. FARISH: The probe tip is mechanically oscillated in a direction normal to the surface at a frequency of 25 Hz and with a displacement of $\sim$2mm. The lower limit to size will depend on the sensitivity of the measurement system and the maximum voltage on the sample. Measurements at, say, 1 kV would allow a considerable reduction in the size of the probe without corona occurring and hence in increased spatial resolution.

E. MARODE: You report about possible local charge variation along the surface. Would the probe be able to follow these variations? What is actually the space resolution of your probe?

O. FARISH: The spatial resolution depends on the physical size of the probe. For the present case (where the system is designed to realistic levels of stress) it will be a few mm, but greater resolution can be obtained if the probe is reduced in size. This would in turn involve working at lower applied voltage in order to avoid corona from the probe housing. Such measurements will be useful in mapping the evolution of the potential distribution with sub-mm resolution.

R. BARTNIKAS: To what degree does your described measurement system differ from the usual vibrating capacitance methods used for surface charge measurement? Do you have any data on the nature of the charge traps in the insulators which you propose to examine?

O. FARISH: Vibrating reed probes, field mills, etc., have been used to measure surface charge density with absence of any applied voltage. The present system is designed to measure the potential distribution with high-voltage applied to the electrodes. In the present context the term "trapped charge" refers to the condition which leaves an isolated bus section charged at up to 1 pu following a switching operation. Our study aims to follow the evolution of the stress distribution over time, with different spacer conditions. These might include local variations in bulk or surface resistivity as well as surface charge deposition. Depending on the results, it may be necessary to investigate charge mobility and only then would it be necessary to consider the nature of charge traps in the insulator.

K. NAKANISHI: As you well know, the charges can accumulate on both sides of the barrier. In such a situation, can your probe system separate the individual charge in the form of the charge density on the barrier?

O. FARISH: Our probe cannot separate individual charges on both sides of the barrier and this would introduce an error for the thin samples being used in the study. However, our intention is to treat one side of the barrier so as to create a nonuniform potential distribution and this will be the dominant effect in our study, whose aim is to achieve an insight into the source of low-energy discharge actually on barriers.

CHARGE ACCUMULATION AND DIELECTRIC CHARACTERISTICS OF DC-GIS

Yoshio Yoshida[1], Koji Yamaji[2], Masatoshi Sampei[3] and Koji Ibuki[4]

[1]The Kansai Electric Power Co. Inc.
[2]Shikoku Electric Power Co. Inc.
[3]Electric Power Development Co. Ltd.
[4]Mitsubishi Electric Co.

INTRODUCTION

It is widely recognized that the charge accumulation on the surface of epoxy resin barrier during the operation of GIS under a DC voltage gradually modifies the electrical field distribution. The mechanism of the charge accumulation is classified into some categories.[1] Anomalies in charge accumulation by conduction through bulk of the insulator is usually with a long time constant and the distortion from the non-disturbed condition is relatively limited. It is because the inhomogeneity of the conductivity of the material is the main cause of the deviation of the distribution and the inhomogeneity of the conductivity is usually small when the material is processed under a reasonable control. The non-uniformity of the surface conductivity is another cause of the distortion of the electrical field distribution. However, the barrier manufactured under a strict quality control is usually with clean surface and it is rather unlikely to have this kind of gradient of the conductivity. Therefore, the most common cause of the unexpected distortion of the field is the charge accumulation induced by the current through the gas space such as the partial discharge caused by some sort of irregularity on the conductor or the enclosure.

The behavior of the charge on the epoxy resin either by the regular conduction through the insulator body or the accumulated charge through the gas space is more critical and the extent of the effect of the static charge on the insulator is more important for DC-GIS than for AC-GIS because the accumulation of the charge continues while the charging mechanism works and it becomes to the steady state only when the charging mechanism is balanced by the applied voltage. Therefore, the understanding of the steady state of the charge accumulation which is dependent on the nature of the emitting point is necessary to derive a reasonable configuration of the insulator to fulfill the given requirement.

Present work is mainly directed to obtain the understanding of the charge build up process and the likely limit of the charge build up depending on the location of the charge emitting site by numerical simulation. Some experimental effort to demonstrate a behavior similar to the simulation results is, also, reported.

Gaseous Dielectrics VII, Edited by L.G. Christophorou
and D.R. James, Plenum Press, New York, 1994

NUMERICAL SIMULATION

A cone type epoxy barrier model intended for a use in a ±125 kV GIS was used for the present work. It is not a simple step to calculate an electrical field affected both by the conduction current and the space current because the field cannot be described by a single Poisson's equation. Electrical field distribution in the current field determined by the conduction current can be described by eq.(1).

$$\mathrm{div}\ (E_r/r + \partial/\partial t\ (\epsilon\ E_r)) = 0 \qquad (1)$$

E_r : electrical field in the current field
r : volume resistivity given as a function of the point in space

The second term on the left corresponds to the displacement current and this is affected both by the charge which appears due to the current conduction and by the static charge caused by other reasons such as the space current by partial discharge. The steady state electrical field is obtained by the eq.(2) as $\partial/\partial t=0$.

$$\mathrm{div}\ (E_r/r) = 0 \qquad (2)$$

The field by this equation can be calculated by a program of "surface charge method". When the electrical field in the conduction current field is obtained, the charge to reproduce the same field distribution under a given distribution of dielectric coefficient can be obtained by the eq.(3)

$$\sigma_r \equiv \mathrm{div}\ (\epsilon\ E_r) \qquad (3)$$

σ_r : electrical charge necessary to reproduce the electrical field determined by conduction current

The solution of this equation is used as the initial condition for the numerical simulation of the distortion of the field. Emission of the electrons from corona discharge on a protrusion located either on the enclosure or on the conductor of the GIS is assumed to add the charge expressed by σ_s. It was, also, assumed that the surface or bulk conduction of the insulator is small enough in comparison with the current through the gas space. It is expected that the electrons emitted from a emitting site will run through the gas space along the electrical field line passing the emitting point. When the electrons hit the insulator it was assumed that the electrons are captured on the spot where they hit and the total charge σ appears as the result of the trapping of the additional charge is given by eq.(4).

$$\sigma = \sigma_s + \sigma_r \qquad (4)$$

The electrical charge trapped on the insulator distorts the electrical field as given by eq.(5) and this distortion changes the charge accumulating point.

$$\mathrm{div}\ (\epsilon\ E) = \sigma \qquad (5)$$

This process of numerical calculation is repeated until the electrical field line crosses the electrode of the opposite polarity.

Result of the simulation

A typical example of the numerical simulation, that shows the distortion of the field such as a shifting of the highest field point, is shown in Fig. 1. When there is no charge caused by the corona current is added—the steady state electrical field—which is decided by the current conducted through the insulator body, obtained by the calculation is rather flat in the space across the gap. However, the charge accumulation caused a significant distortion of the field and the highest field point is pushed toward the center conductor

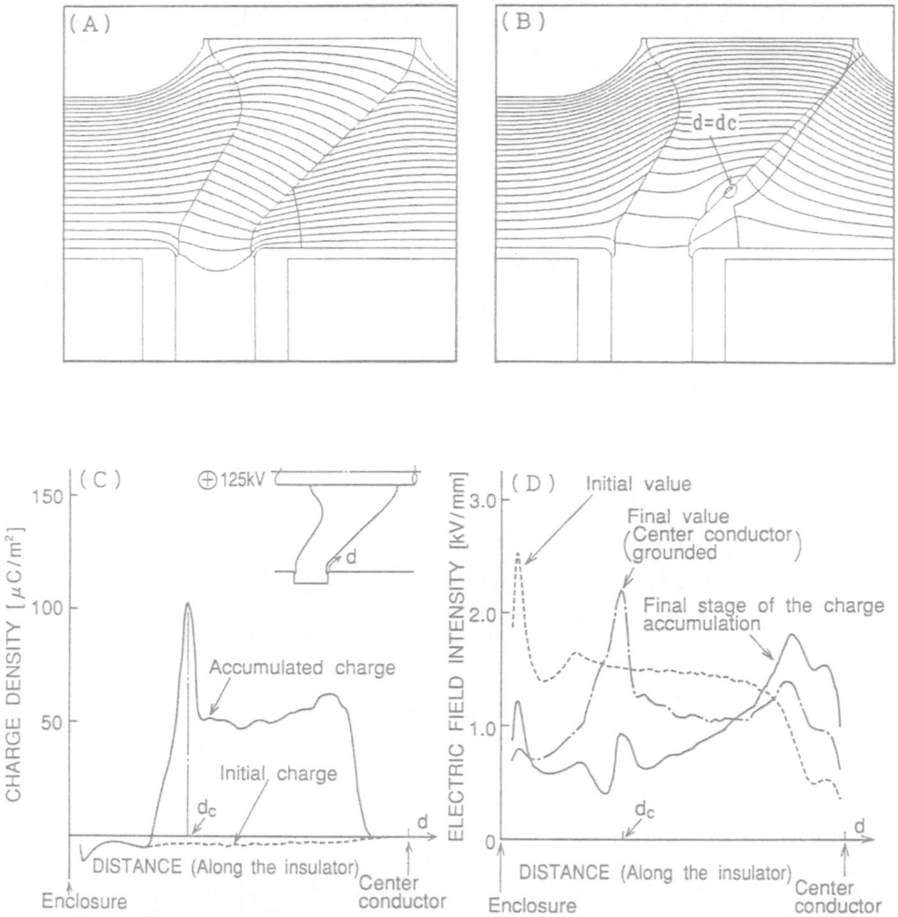

Fig. 1. Simulation results of the charging process on the surface of the epoxy barrier. (A) shows the geometry of the model insulator and the electrodes and the position of the emission point fixed on the enclosure. The equi-potential lines before the start of the simulation are also shown. The electrical field line passing the emitting point is crossing the insulator immediately across the gas gap. Electrons hitting this point is assumed to stay there and the field distribution with the charge is calculated. (B) shows the field line in the final stage of the calculation. All the electron emitted from the emitting site is expected to be captured by the center conductor. (C) shows the charge accumulated on the surface of the insulator compared with the charge at the steady state under the assumption that the field is fully dominated by the resistivity of the insulator. (D) shows the electrical field intensity at the initial condition (resistivity dominated field distribution) and the intensity at the final stage of the charge accumulation both with the voltage at the center conductor. The field intensity with the center conductor grounded is also plotted. The field intensity in (D). shows the absolute value of the intensity at the surface of the insulator as a function of the distance along the surface.

side. When the charge is accumulated to the extent shown in Fig. 1(b), the field line passing through the emission point is deformed and it hits the center conductor without touching the insulator any more. Therefore, it is expected that the further charge accumulation by this anomalies does not continue any more. It can be seen that this distortion itself does not give any serious effect as long as the same DC voltage is applied because the highest field is not significantly higher than the initial values. However, the field configuration is significantly affected when the applied voltage is changed—one typical example is the polarity reversing. Fig. 2. shows the electrical field when the polarity of the center conductor is reversed after the preceding process of charge accumulation. The comparison of the highest electrical field point between the field values with and without the charge accumulation show a significant difference. The highest field intensity becomes 90% higher than before reversing.

One of the possibilities of the anomalies assumed above is the existence of the free metal particles in the enclosures. The numerical simulation can clarify the influence of these particles and it becomes possible to estimate the favorable or unfavorable geometry of the barrier under the regular operating voltage or under special cases such as the polarity reversing.

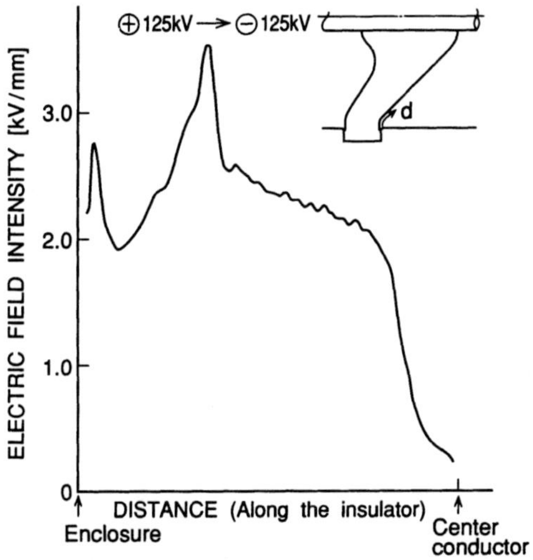

Fig. 2. A simulation result with a voltage of reversed polarity with the accumulated charge by the process shown in Fig. 1.

EXPERIMENTAL WORK

The behavior of the electrical field can be estimated in a qualitative manner as shown above. However, it is not easy to measure the field intensity or the accumulated charge especially in the SF_6 gas space. Two provisional experiments were carried out to provide evidences to support the basic idea introduced in the simulation.

The experimental set up is shown in Fig. 3. Two means (partial discharge detector and dust figure charge visualizing setup) are provided to show the charge accumulation and the occurrence of the partial discharge as well as a protrusion on the enclosure, which is expected to work as emitting site, as shown in the figure. The basic object of the experiment is to show (1) that the zone in which the electrical charge is accumulated agrees to the result obtained by the calculation and (2) that the order of the accumulated charge agrees with the calculated value when it is estimated by the initiation of the discharge by the accumulated charge itself. The process of the experiment is described as follows.

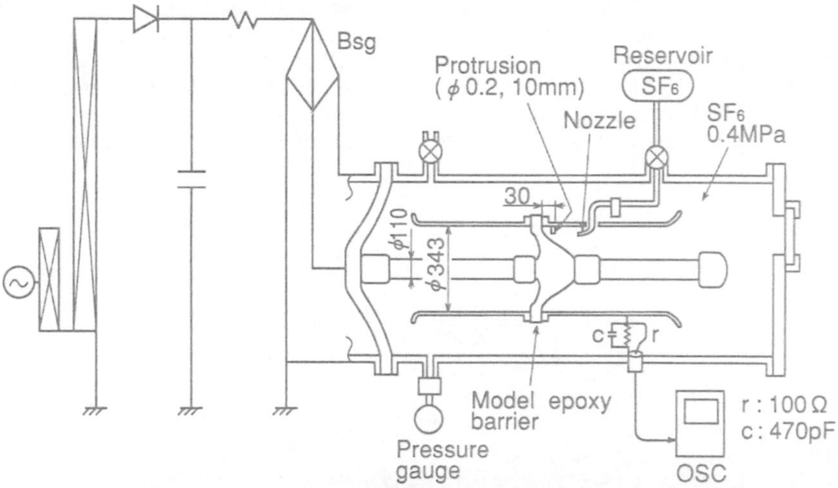

Fig. 3. Experimental setup. The outer electrode described as "enclosure" in the text is set in an outer enclosure and the resistor is provided between inner and outer enclosure to detect the electrical discharge signal with a high sensitivity. The "dust" for visualizing the charge distribution is provided in the container and the nozzle is directed to the insulator.

Experimental process and the results

A DC voltage is applied on the center conductor for more than 1 hour to allow the partial discharge at the protrusion to charge up the epoxy barrier. The applied voltage is reduced to zero after this process and the center conductor is grounded. The SF_6 gas is evacuated until the accumulated charge on the insulator is discharged through the gas gap. The discharge is monitored by observing the voltage across the resistor shown in Fig. 3. The electrical field intensity on the surface of the insulator when both the center conductor and the enclosure are grounded is calculated as shown in Fig. 1(d). The maximum electrical field appears at the midway between two electrodes and it is approximately 2 kv/mm. (A reduced pressure of the order of 200 to 500

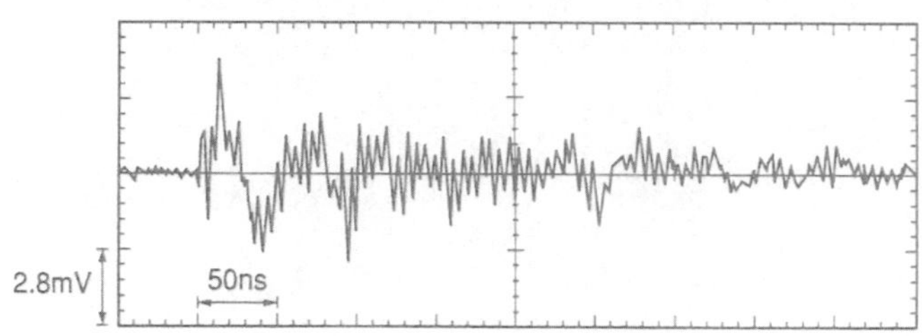

Fig. 4. A typical electrical signal across the resistor. It was detected at a pressure 250 Torr. The pressure at which the discharge took place was from 230 to 300 Torr and the average was 255 Torr in 7 trials of the same experimental process.

Torr is likely to lead the discharge when this order of electrical field is on the electrode[2].) A typical electrical signal is shown in Fig. 4. This gas pressure can be interpreted as the likely electrical maximum field and it is possible to estimate the order of the electrical charge accumulated on the insulator.
Another process of the experiment is to visualize the zone in which the charge is accumulated by "dust figure method". A container of the toner to visualize the charged up surface is provided. A diverging nozzle directed toward the insulator surface is fitted to the container and the toner can be driven by SF_6 gas pressure controlled by a valve. By opening the valve, the toner in the container is blasted toward the insulator and it is trapped by the negative charge on the insulator. It can show the area in which the charge is accumulated. Two examples of the "dust figure" are shown in Fig. 5 and Fig. 6. One is the figure immediately after reducing the applied voltage to zero.

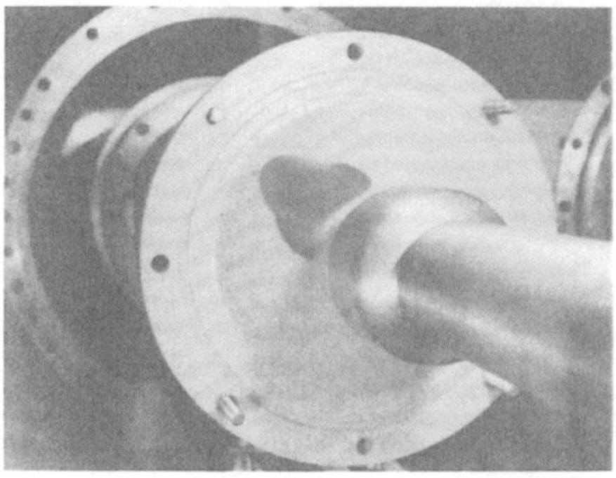

Fig. 5. The "dust figure" observed after applying the rated DC voltage for 1 hour. The needle was fixed at an azimuthal position of the fitting bolt. The figure is slightly non-symmetric and this slight deviation was reproducible in all the trials.

Fig. 6. The "dust figure" obtained after the observation of the partial discharge shown in Fig. 4. The point corresponding to the highest charge density in Fig. 1(c) has lost the charge.

(Fig. 5) This is showing that 70% of the gap between the enclosure and the center conductor is charged up negatively. This is qualitatively and quantitatively in agreement with the calculation in spite of the axi-symmetric modeling of the simulation and the non-axi-symmetric arrangement of the experimental set up. Fig. 6. shows a dust figure immediately after the detection of the partial discharge at 250 Torr. It, obviously, shows that the charge in an area of an elliptic shape has disappeared and the location of the charge-missing-part is in agreement with the place of the highest charge density and at the same time the place of the highest field strength.

DISCUSSION

The experimental results, although it is not fully interpreted quantitatively, can support the process assumed in the numerical simulation and it can provide a background for the usage of the simulation for practical purposes such as deciding the geometry of the barrier or evaluating the extent of the influence of the foreign particles which can act as emitting spots. Some examples of the study about the polarity reversing process is included in the simulation results and it is showing that the expected electrical field exaggeration by reversing is of the order of 90%. Optimization of the geometry of the barrier will be a practical target for this simulation and the reason of the insensitivity of the flat type insulators to the charging up will be obvious from the point of view of charge accumulation.

A question left without being discussed is the classification of the emitting points. An aluminum needle of .2 mmΦ with a 10 mm length fixed on the enclosure (the electrical field strength at the top of the needle by a field calculation was 50 kv/mm) clearly showed that it is an efficient emitting site. "What kind of foreign particle or surface anomalies on the electrode work as emitting site?" is also a question for practical design process. These questions are going to be studied and the study will be tied with the simulation mentioned here to complete a practical design of a DC-GIS.

REFERENCES

1. T. Nitta and K. Nakanishi, Charge accumulation on insulating spacers for HVDC GIS, IEEE Trans. EI1-26:3 (1991)
2. T. Nitta, Y. Shibuya, Electrical breakdown of long gas gaps in sulfur hexafluoride, IEEE Trans. PAS-90:3(1971)

DISCUSSION

S. W. ROWE: The situation simulated uses an axi-symmetrical geometry and the charge "accumulated" takes on the form of a ring charge. In real cases the situation is 3D and a localized "packet" of charge will be built up. What would be the expected effect on calculated breakdown voltages and charge distribution?

K. IBUKI: The 3D effect in simulation is not properly taken into account at present. The expected effect by 3D simulation is the extension of the charge accumulated zone but the maximum electric field value will not be affected so much. The experimental result is not greatly different from the axi-symmetric simulation and this fact encourages us to apply this technique for practical purposes.

G. C. CRICHTON: In GIS under service conditions temperature gradients will occur. Thus, with DC-stressing, charge accumulation inherently related to the gradient (∇) of the volume and surface conductivity (γ) could become important. γ is strongly dependent on T. Have you considered such temperature dependent charging in your on-going studies?

K. IBUKI: So far, we have not tried to give an evaluation of the effect of the temperature. However, we have the impression that the effect of conductivity is very small in comparison with the current through gas space. Therefore, the effect of the temperature on the resultant electrical field will be small enough although the time constant of redistributing of the charge will be affected to a significant extent.

G. N. TRINH: (1) Under condition of voltage reversal, what is the time necessary for the charges to redistribute themselves? (2) Normally, the distribution time is long, and therefore the most severe stress would occur immediately following the potential reversal. Have you evaluated this stress?

K. IBUKI: (1) The time constant on an actual spacer has not been measured in our work so far but we expect that it is longer than a few hours. (2) The stress obtained in the simulation of the reversed polarity corresponds to the polarity reversed instantaneously and, therefore, the calculated value corresponds to the highest stress that is even higher than the value reproduced in the practical experimental reversing.

CHARGING AND BREAKDOWN CHARACTERISTICS OF VARIOUS SURFACE-TREATMENT SPACERS UNDER DC VOLTAGE IN SF_6

T. Hasegawa[1], T. Fujiwara[2], H. Ooi[3], F. Endo[4], T. Rokunohe[4] and T. Yamagiwa[4]

[1]The Kansai Electric Power Co., Inc.
[2]Shikoku Electric Power Co., Inc.
[3]Electric Power Development Co., Ltd.
[4]Hitachi, Ltd.

INTRODUCTION

Recent progress in DC transmission projects has promoted of DC gas insulated switchgear (DC GIS) to ensure better reliability, compactness and economy.[1, 2] One difficulty with DC GIS is charge accumulation on solid insulating spacers under DC stress. A number of studies have looked at the effects of the charge accumulation,[3,4] but little has been done on countermeasures for accumulated charge. In this study, we investigated the effects of surface-treatment conditions of spacers treatments included use of a low permittivity material coating, an anti-charging material coating, and a metal-sputtering coating. A final condition was use of coated electrodes with a non-coated spacer. The test results are described along with our evaluation of the spacer for DC GIS.

EXPERIMENTAL SET UP AND TEST PROCEDURE

Test Circuits

Figure 1 (a) shows the test circuit used in this investigation. The DC voltages for the DC withstand test were supplied from a 1000kV, 10mA DC generator and positive polarity switching impulse breakdown test were supplied from a 2400kV impulse voltage generator.

Model Spacers

Cylindrical post spacers, shown in Figure 1 (b), (diameter : 50mm, height :

Gaseous Dielectrics VII, Edited by L.G. Christophorou
and D.R. James, Plenum Press, New York, 1994

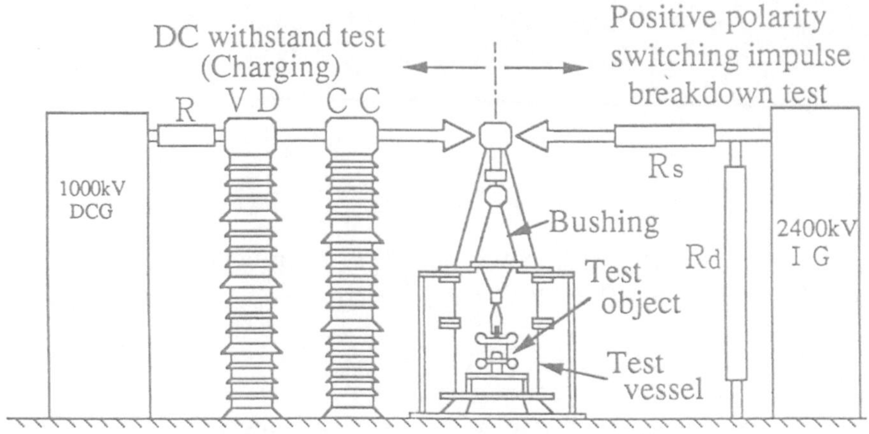

(a) Test circuit

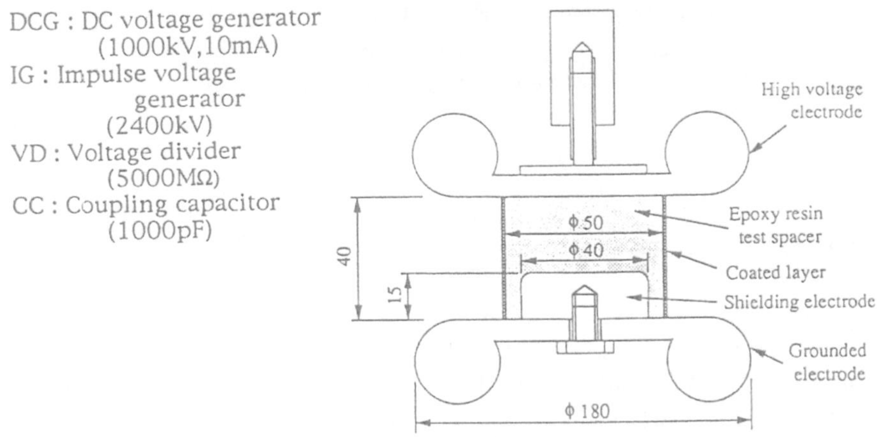

(b) Test object

Figure 1. Test circuit and object

Table 1. Test conditions and results

Set No.	Spacer coating material	Electrode coating material	Surface resistivity (Ω)	Thickness of coating (μm)	DC Breakdown voltage (BDV)
1	Without	Without	10^{16} - 10^{17}	–	-200kV, withstood
2	Epoxy resin (low permittivity)	Without	10^{14} - 10^{15}	300 - 800	-200kV, withstood
3	Anti - charging material	Without	10^{11} - 10^{12}	20 - 50	-150 - 180kV
4	Metal sputtering coating	Without	10^{11} - 10^{13}	0.03 - 0.08	-200kV, withstood
5	Without	Epoxy resin		80 - 150	-200kV, withstood

40mm) were used to investigate the effects of the surface-treatments as shown in Figure 1 (b). These spacers were made of epoxy resin using AC gas insulated switchgear and the shielding electrode (diameter : 40mm, height : 15mm) was molded on the grounded side of the spacer.

Table 1 lists test conditions for the surface-treatments on tested spacers. Five sets of conditions were investigated: low permittivity material coating, anti-charging material coating, or metal-sputtering coating on the spacer, and no coating on the spacer. The final condition used was of an epoxy resin coated electrode with an uncoated spacer.

Measurement Equipment for Accumulated Charge

The accumulated charge on the model spacer surface was detected using an electrostatic probe set inside the test vessel as shown in Figure 2. We could measure the whole surface of a test spacer in SF_6 immediately after a DC voltage was applied. To observe the residual charge after removal of the DC voltage, an electrostatic probe was extended to a position 2mm away from the spacer surface, by a robot, controlled from a remote control system. The charge quantity was measured using a vibrating reed electrometer connected to the probe.

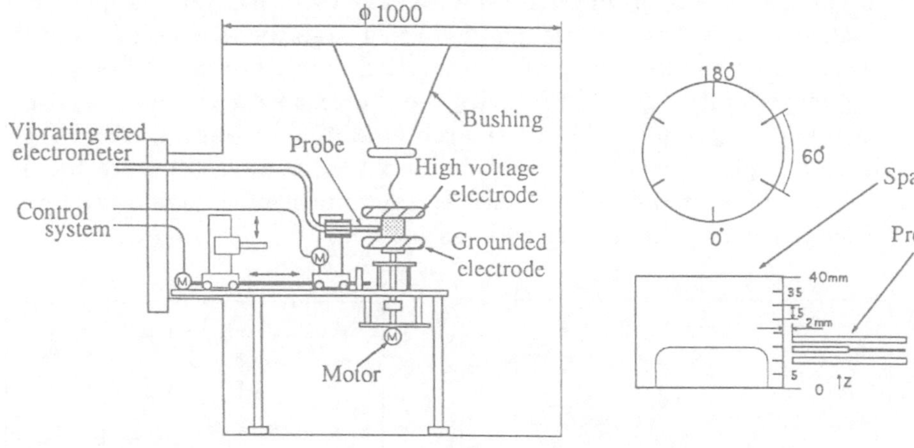

Figure 2. Measurement equipment for accumulated charge on model spacer

Test Procedure

Figure 3 outlines the test procedure. Two kinds of tests were performed to study the effect of charge accumulation. The first test was a DC 2-hour withstand test which was used for evaluation of the coating materials and accumulation of space charge. We chose a negative polarity DC voltage for easy charging. The second test was a switching impulse (250/2500μs) breakdown test. In it, we applied a reversal polarity voltage (positive polarity) against the DC test to investigate the effect of the charge accumulation. For some tested spacers, the distributions of the accumulated charge along the surface after the DC withstand test were measured.

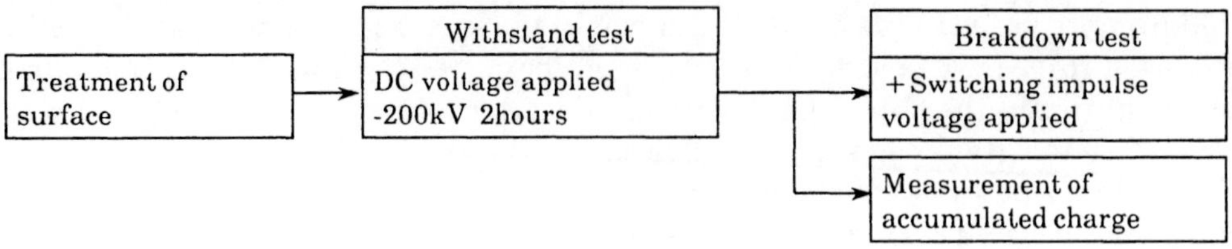

Figure 3. Test procedure

RESULTS

DC Breakdown Characteristics

Table 1 also lists results of the 2-hour DC-200kV withstand test. Only for the anti-charging material (set No. 3) did breakdown occur (150-180kV). The surface resistivity of set No.3 was the lowest (10^{11}-$10^{12}\Omega$). It seems to be that the current through the coated layer causes the thermal breakdown occurred before dielectric breakdown did. For all other sets, spacers withstand DC-200kV for a two hour period.

Effects of Accumulated Charge (Polarity Reversal Characteristics)

The four cases which withstood the DC-200kV were next subjected to positive polarity switching impulse voltage. The relative switching impulse breakdown voltages and the effects of accumulated charge are shown in Figure 4 (a) and (b), respectively. For Figure 4 (a), the coated electrode (No. 5) had the highest breakdown voltage. We considered that there was no effect from accumulated charge because (1) there was no triple junction; (2) there was no border surface between the epoxy resin spacer and coated layer; (3) fielde mission from the high voltage electrode was avoided.

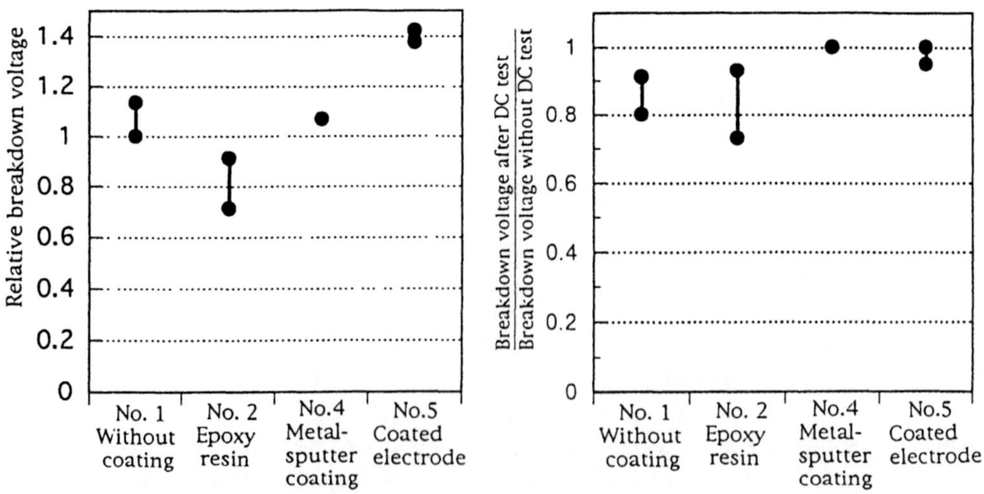

(a) Relative switching impulse breakdown voltage

(b) Effect of accumulated charge

Figure 4. Results of switching impulse breakdown voltage

Figure 4 (b) shows the effect of accumulated charge. Nos. 4 and 5 had better performance regarding accumulated charges than No. 1 (without coating). In the case of No. 4 (metal-sputtered) it would be necessary to improve the dielectric strength of the sputter coating layer.

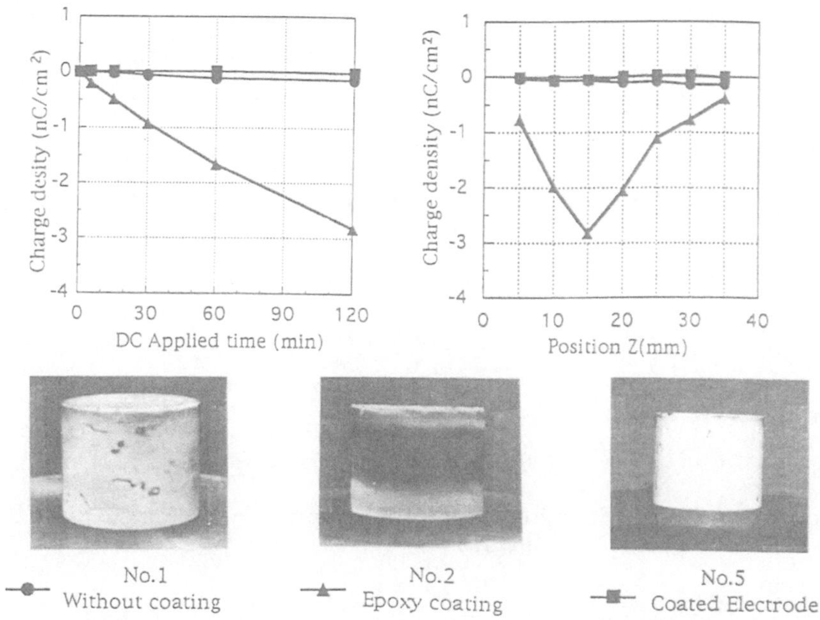

Figure 5. Examples of accumulated charge distribution

DISCUSSION

Accumulated Charge Distribution

Figure 5 shows accumulated charge distributions for Nos. 1, 2 and 5. For Nos.1 and 5, few charges were observed, however, some spot charges (several nC/cm^2) remained for No. 1. This is one of the reason that No.1 breakdown voltage is lower than that of No. 5. On the other hand for No. 2, many negative polarity charges accumulated at the surface near the top of the inside shielding electrode. It is assumed that these charges are supplied through the coated layer. Therefore, it is important that the resistance of coated layer isn't reduced (thickness is as thin as possible) and the surface resistivity of the coated layer is reduced.

Figure 6 shows the potential distributions and electric field strength along the surface for AC and DC applied conditions. For the DC analysis, non-linear resistivities (bulk of epoxy resin : ρ_1 (E), surface layer of spacer : ρ_2 (E)) were used., while only permittivity was used for AC analysis. The accumulated charge distribution of No.2 (Figure 5) agree with that of the normal component of the field strength for the AC analysis (Figure 6 (a)). The field distribution for AC application is an important point for DC-spacer design.

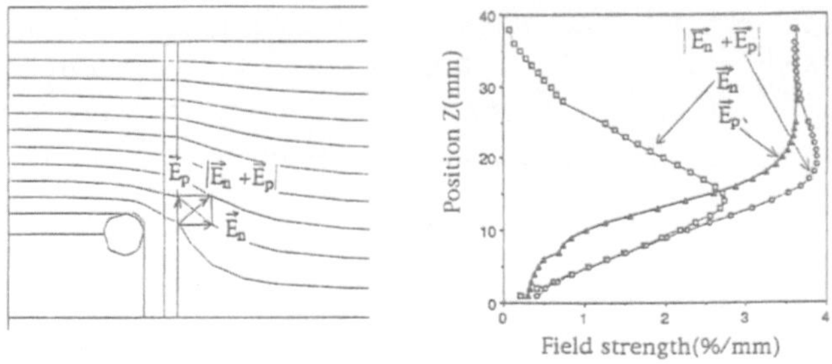

(a) AC applied condition (used only permittivity)

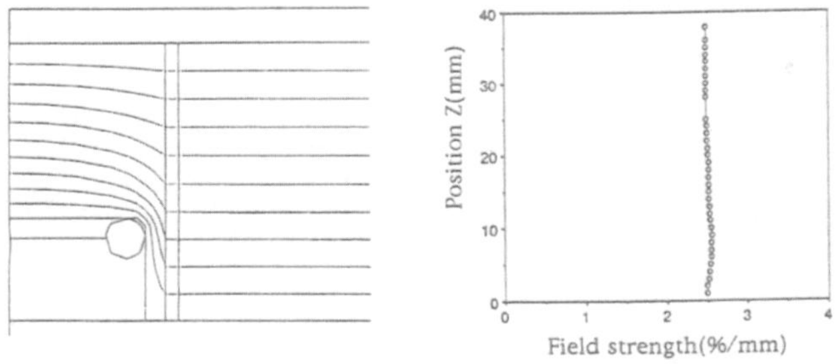

(b) DC applied condition (used only resistivity)

Figure 6. Variation of conditions of field mapping and potential distribution along the spacer surface

Characteristics of Sputter Coated Surface

In the case of No. 4 (metal-sputter coating), the coated layer was rather thin and surface resistivity was relatively low, however, the relative breakdown voltage was about the same value as No. 1 (no coating). Figure 7 shows an SEM photograph of the sputter-coating surface same spots with no coating were observed. These defects seemed to be one of cause for the drop in breakdown voltage.

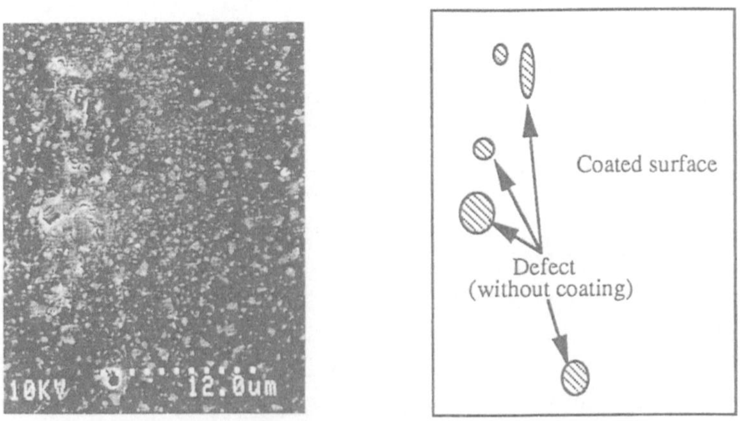

Figure 7. SEM photograph of sputtering coated surface

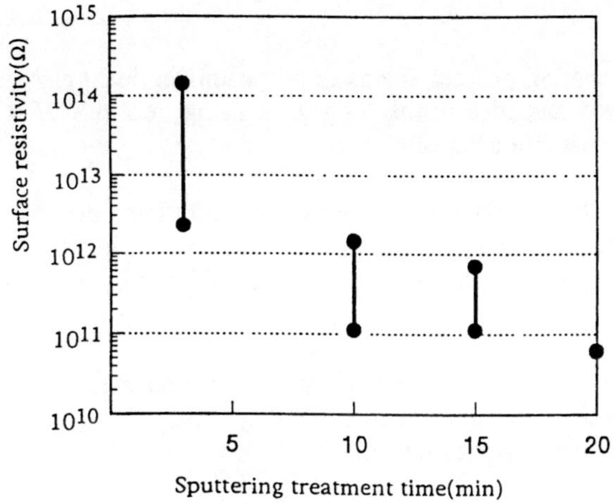

Figure 8. Relation between surface resistivity and sputtering time

Figure 8 shows the relation between the surface resistivity and the sputtering time. Longer sputtering time gave lower surface resistivity. We are investigating this point further.

CONCLUSION

We investigated the effect of surface-treatment on spacers for DC GIS. The results obtained are summarized as follows.

(1) A thick coating layer of a low resistivity material was not very good for DC spacer coating.

(2) An uncoated spacer with coated electrodes had the best performance regarding reversal polarity breakdown characteristics.

(3) The metal-sputter coating might possibly be a good treatment method for DC spacers. We are studying it more.

REFERENCES

1. F. Endo, T. Kichikawa, R. Ishikawa and J. Ozawa, Dielectric characteristics of SF_6 gas for application to HVDC systems, IEEE PES Summer Meeting, F 79 620-6 (1979)
2. T. Nitta and Y. Arahata, Development of ±500kV HVDC gas -insulated switchgears, in: "Gaseous Dielectrics V", L.G. Christophorou and D. W. Bouldin, eds., Pergamon Press, New York (1990), pp.471-482
3. K. Nakanishi, A. Yoshioka, Y. Arahata and Y. Shibuya, Surface charging on epoxy spacer at DC stress in compressed SF_6 gas, IEEE Trans. on PAS, Vol. PAS-102, No. 12 (1983)
4. M. Yashima, H. Fujinami and T. Takuma, Measurement of accumulated charge on dielectric surfaces with an electric probe, in : "Gaseous Dielectrics V", L. G. Christophorou and D. W. Bouldin, eds., Pergamon Press, New York (1990), pp.242-248

DISCUSSION

S. W. ROWE: The coating process seems to be promising in improving performance of DC spacers. Have you any idea of the long term aging behavior of this type of surface treatment for the 25 year life span of GIS?

T. YAMAGIWA: This is one of the important problems for this type of surface treatment. Our study is now in a fundamental stage and we are selecting the coating material for the DC spacer. Therefore, we have to investigate the aging behavior of coating materials.

C. X. WANG: One of your conclusions is that an uncoated spacer with coated electrodes had the best performance regarding reversal polarity breakdown characteristics. Can you comment on this increase in performance?

T. YAMAGIWA: The reasons may be that the field emission is reduced from the high field electrode.

THE INFLUENCE OF CHARGING OF GIS EPOXY INSULATOR ON DC/IMPULSE SUPERPOSED BREAKDOWN CHARACTERISTICS

T. Hasegawa[1], T. Fujiwara[2], H. Ooi[3]
S. Yanabu[4], H. Aoyagi[4], Y. Kanno[4]

[1]The Kansai Electric Power Co., Inc.
Osaka, Japan
[2]The Shikoku Electric Power Co., Inc.
Takamatsu, Japan
[3]Electric Power Development Co., Inc.
Tokyo, Japan
[4]Toshiba Corporation
Kawasaki, Japan

INTRODUCTION

Because of their high reliability, various models of gas insulated switchgear (GIS) are finding wide application, their voltage classes ranging up to 800kV on AC lines[1]. However, if any of such models is used (as DC GIS) on a DC line, the electric field impressed on it is directed in one way, which involves a peculiar tendency (kind of electrification) for charges to be accumulated in insulating spacers as supports for the high voltage conductors. For example, the impression of a reversed-polarity surge voltage may, because of the electrification just mentioned, cause the insulating capacity of the insulating spacers to decrease. This has encouraged extensive studies on the insulating characteristics of insulators at the time of voltage polarity reversal[2,3]. However, such studied have more often than not investigated the effect of electrification at very high level of DC voltage.

Using basic models designed with the internal (embedded electrode interface) and surface insulation of the insulating spacer of DC GIS respectively taken into account, the present study examined the relation between prestressed DC and lightning impulse breakdown voltages in detail to investigate the effects of bulk and surface electrification due to a prestressed DC voltage on puncture breakdown and surface flashover characteristics. For the prestressed DC voltage, a value larger than the equivalent GIS operating voltage was chosen. In accordance with the test result, investigation was made of the dielectric performance of the DC GIS.

Gaseous Dielectrics VII, Edited by L.G. Christophorou
and D.R. James, Plenum Press, New York, 1994

EXPERIMENTAL MODELS AND METHODS

Figure 1 shows the experimental models, the puncture breakdown model in (A) and the surface flashover model in (B). Two puncture breakdown models were used: their embedded electrodes made of aluminum and conductive plastics respectively. On these models, dielectric breakdowns are caused between embedded electrodes. With the surface flashover model, both maximums of normal component electric field strength (En) and tangential component electric field strength (Et) of the epoxy molding were comparable and both En and Et are lower than the electric field strength of the high-voltage electrode. This causes a dielectric breakdown to take place with discharges starting at the high-voltage electrode.

The experimental procedure began to obtain the breakdown voltage with L.Imp only and proceeded, while varying the level of prestressed DC voltage and its impressing time, to obtain the DC/L.Imp breakdown voltage. The dielectric characteristics of the surface flashover model was investigated with two surface conditions: bare and planted with aluminum wires. Table 1 shows the experimental conditions.

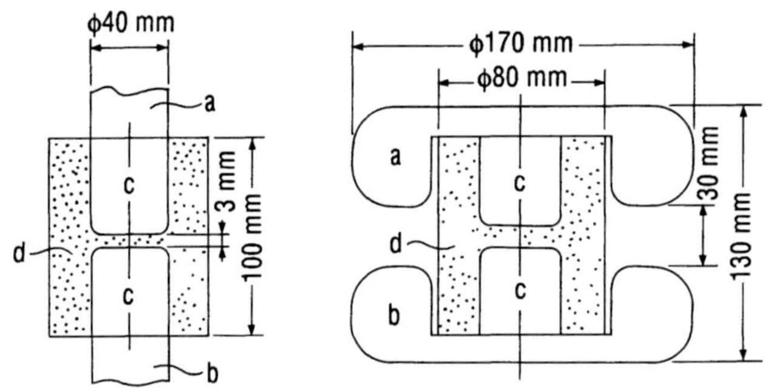

(A) Puncture breakdown model (B) Surface flashover model

Figure 1. Sectional view of experimental models.
(a: high-voltage electrode, b: grounding electrode, c: embedded electrode, d: epoxy molding)

Table 1. Test conditions.

	Puncture breakdown model	Surface flashover model
Model condition	Embedded electrode material • Aluminum • Conductive plastics	Surface condition • Clean • With aluminum wires (0.25mm in dia. and 5mm long)
Prestressed DC voltage impressing time	1 – 1000 h	1 – 10 h
SF_6 gas pressure	0.4MPa	0.5MPa

EXPERIMENTAL RESULTS AND DISCUSSIONS

Breakdown Characteristics of Puncture Breakdown

Figure 2 shows the dielectric breakdown characteristics of the puncture breakdown model. The impressing time of prestressed DC voltage ranged up to 1000 h. Without prestressed DC voltage impression, the L.Imp breakdown voltages (initial values) were obtained with 10 model units and the breakdown voltages with a prestressed DC voltage impression were obtained with 3 model units. Although showing some dispersion, the L.Imp breakdown electric field strengths of the two models each having an aluminum or conductive-plastic embedded electrode depend on the levels of prestressed DC voltage and its impressing time. In a prestressed DC voltage impressing time of 1000 h, the L.Imp breakdown voltage shows little decrease with a prestressed DC voltage of 20kV but suffers a 10% decrease with 50kV and a 15% decrease with 100kV. Decreases in dielectric breakdown electric field strength begin to take place at about 3 h of prestressed DC voltage impression.

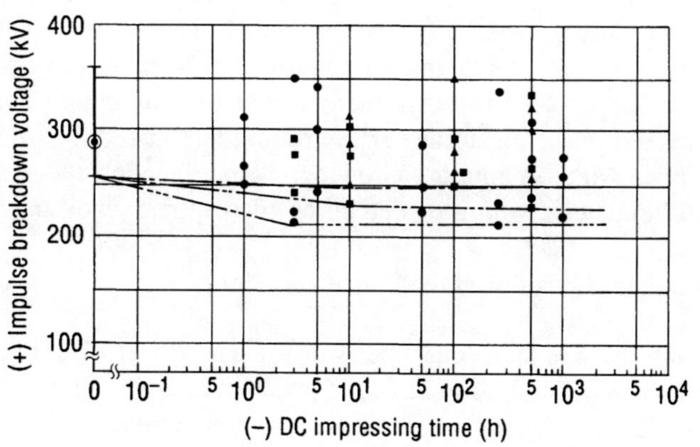

(a) Embedded electrode material: aluminum

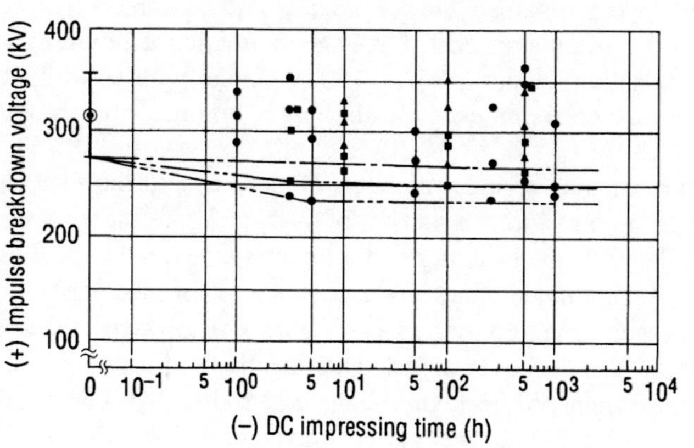

(b) Embedded electrode material: conductive plastics

Figure 2. Dielectric breakdown characteristics of puncture breakdown model.
(Prestressed DC voltage: ◉ zero, ▲ 20kV, ■ 50kV, ● 100kV)

Time constant τ for the transition of the bulk potential of a solid insulator from capacity to resistance distribution is generally given by

$$\tau = \varepsilon s \cdot \varepsilon o \cdot \rho v \quad (1)$$

where εs: specific dielectric constant, εo: vacuum dielectric constant and ρv: volume resistivity. εs of the puncture breakdown model here is 5.8 and ρv at room temperature is 1.5×10^{16} (Ω·cm). From these, τ of the puncture breakdown model is calculated as about 2.1 h. This value nearly agrees with the prestressed DC voltage impressing time at which the L.Imp breakdown voltage start to decrease.

The L.Imp breakdown voltage of the puncture breakdown model, on the other hand, is about 10% higher with the conductive-plastic embedded electrode than with the aluminum one. Ordinarily, puncture breakdown strength depends greatly on the electrode surface condition. Main indices to electrode surface condition may be surface roughness and adhesion to epoxy resin. The embedded electrodes for the present model were sand blasted to increase adhesion to epoxy resin. This made the maximum roughness of the electrode surface about 20–30μm. Since both electrodes are comparable in surface roughness, the puncture breakdown strength is assumed to be dependent on adhesion to epoxy resin. From this point of view, the thermal expansion coefficient of the conductive plastics is closer to that of the epoxy moldings than to that of aluminum. This suggests that residual stresses caused in the electrode surface during thermosetting are lower with conductive plastics than with aluminum, allowing the surface of the conductive-plastic electrode to be less roughened with such defects as microscopic voids and cracks[4].

Breakdown Characteristics of Surface Flashover Model

Figure 3 shows the flashover characteristics of the surface flashover model with a bare electrode surface. The impressing times of prestressed DC voltage ranged up to 10 h. The initial values of L.Imp breakdown voltage in the figure were obtained with 10 model units and are shown by error bars and 50% values of flashover voltage. The values with 3 h and 5 h of voltage impression were obtained with 5 model units and those with 10 h of voltage impression were obtained with 2 model units. Decreases in flashover voltage started to appear at 1 h of prestressed DC voltage impression, amounting to about 3% with a DC voltage of 100kV and about 6% with 150kV. Either with or without prestressed DC voltage impression, dielectric breakdown discharges originated in the high-voltage electrode.

About the mechanisms of electrification of the insulator surface, there are several theories, one of which points out the transfer of charges from gaseous space[5]. The quantity of charges accumulated on the insulator surface corresponds to the level of surface electric field strength En. Accumulated charges are contrary to En in polarity. As electrification proceeds up to a saturation, En decreases. Figure 4 shows the electric field strength distribution of the surface flashover model. The eclectic field strengths are standardized on the basis of the maximum on the high-voltage electrode as 100% and Et as of positive polarity. The 100% electric field strength is equivalent to 0.42 kV/cm/kV. In the present experiment, a positive DC voltage was impressed on the high-voltage side. This causes positive charges to be accumulated on that part of the epoxy molding surface near the high-voltage electrode. If in this condition, a negative L.Imp is impressed, it causes the electric field strength of the high-voltage electrode to rise equivalently. However, decreases in L.Imp breakdown voltage due to the impression of a DC voltage of 150kV are about 6%.

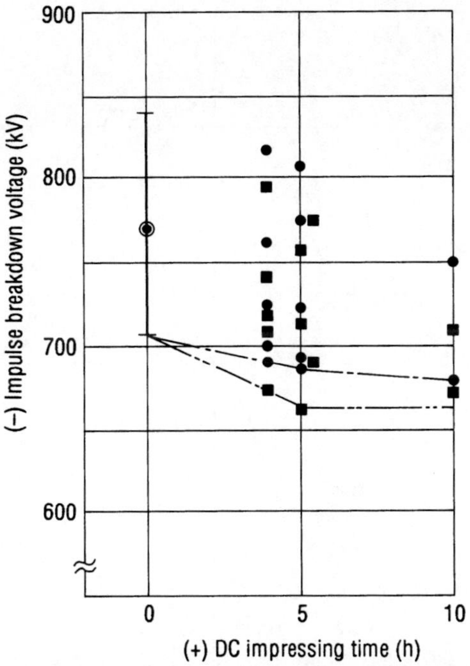

Figure 3. Dielectric breakdown characteristics of surface flashover model.
(Prestressed DC voltage: ◉ zero, ● 100kV, ■ 150kV)

Figure 5 shows the flashover characteristics of the surface flashover model with a wire-planted surface. The flashover voltage with some deposits on the insulator surface depends on electric field strength Et. In the present test, wires were planted in the center of the surface of the epoxy molding having a maximum value of Et. The same number of model units as with bare electrodes were used. Both mean and minimum values of flashover voltage were about 5% lower with a prestressed DC voltage of 100kV and about 10% lower with 150kV. Figure 6 shows fine discharges at the tip of an aluminum wire with a prestressed DC voltage of 150kV impressed. These discharges probably cause the model surface near the tip of the aluminum wire to be electrified, which causes the reverse polarity L.Imp flashover voltage to decrease. However, such flashover voltage decreases resulting from electrification due to fine discharges are about 10%.

Reliability of Insulating Spacer for DC GIS

From the results of the experiments with the puncture breakdown and surface flashover models, the insulating reliability of insulating spacers for DC GIS can be reviewed as follows: Generally, with the DC operation electric field strength of an insulating spacer assumed, the surface of the embedded electrode corresponds to 20kV of prestressed DC voltage on the puncture breakdown model and the surface of the insulating spacer to 100kV on the surface flashover model. For Et, however, a prestressed DC voltage of 100kV is equivalent to the test voltage for the insulating spacers.

Even with the assumed DC GIS exposed to a surge voltage under an operating voltage, decreases in the dielectric performance of the insulating spacer are as small as

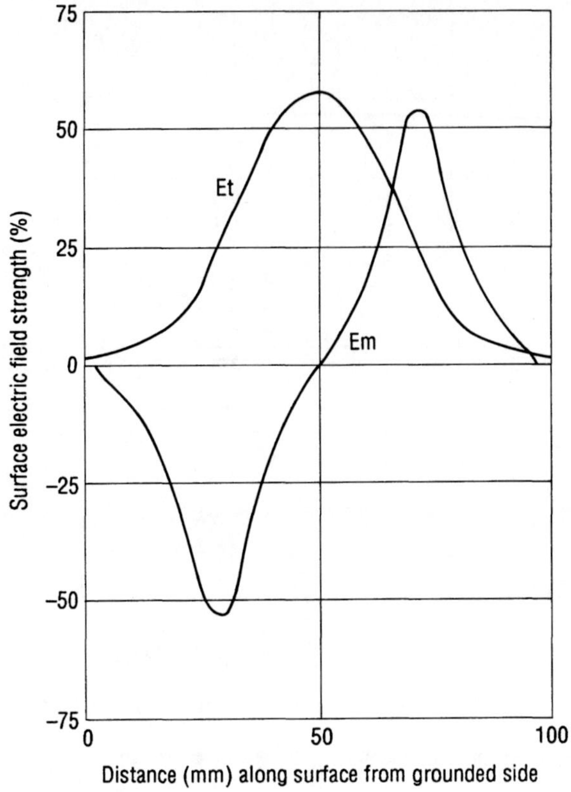

Figure 4. Surface electric field strength distribution of surface flashover model.

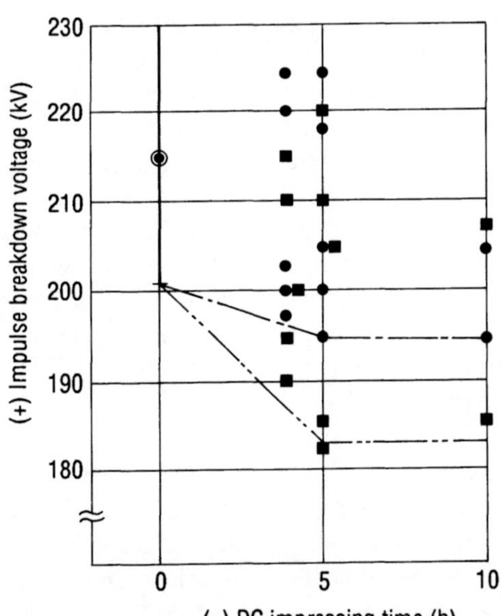

Figure 5. Dielectric breakdown characteristics of surface flashover model. (Surface planted with aluminum wires)
(Prestressed DC voltage: ◉ zero, ● 100kV, ■ 150kV)

Figure 6. Fine discharges from aluminum wire tip.

about 5%. However, if there are some deposits on the insulator surface, the breakdown voltage is much lower than without such deposits, as shown in the present experiment. This suggests that to increase the insulating reliability of DC GIS, controlling inclusions of depositing particles is important.

CONCLUSIONS

To investigate the surge voltage dielectric performance of insulator spacers, in operation, for DC GIS, DC/L.Imp breakdown electric field strengths were examined using two kinds of models for verifying surface insulation and interface insulation between insulating spacer and embedded electrode. The results proved the following: The breakdown voltage starts the decrease in a few hours of prestressed DC voltage impression and the higher the DC voltage impressed, the larger the breakdown voltage decrease. At a voltage level equivalent to the operating voltage, however, decreases in surge voltage dielectric performance of the insulating spacer were as small as about 5%, which is tolerable in practical operation.

However, if the insulating spacer has some deposits on its surface, its dielectric performance, with or without DC voltage impression, might be greatly reduced. To increase the insulating reliability of DC GIS, therefore, sufficient control of deposits is necessary.

REFERENCES

1. H. Toda, Y. Ozaki, I. Miwa, S. Nishiwaki, Y. Murayama, and S. Yanabu, "Development of 800kV Gas-Insulated Switchgear," *IEEE Transactions on Power Delivery*, Vol. 7, January 1992.
2. A. Bargigia, R. Brambilla, A. Pigini, and G. Rizzi, "Influence of Solid Insulation on the Dielectric Performance of SF6 Configurations," *4th Gaseous Dielectrics*, p. 437, 1984.
3. S. Takahashi, H. Okubo, H. Aoyagi, H. Murase, Y. Kawaguchi, and S. Yanabu, "Dielectric Characteristics of GIS for Combined Voltage," *IEEE/PES 88 SM625-2*, 1988.
4. M. Honda, H. Aoyagi, M. Koya, N. Kobayashi, and M. Tamura, "V-T Characteristics of Epoxy Mold Insulation for Sustained AC Voltage," *IEEE Transactions on PAS*, Vol. PAS-103, No. 5, 1984.

5. H. Fujinami, T. Takuma, M. Yashima, and T. Kawamoto, "Mechanism and Effect of DC Charge Accumulation on SF_6 Gas Insulated Spacers," *IEEE Transactions on Power Delivery*, Vol. 4, No. 3, 1989.

PHYSICAL MODEL FOR SURFACE CHARGE SUPPORTED FLASHOVER

Chunxiao Wang

Technology and Science Division
National Grid Company
Kelvin Avenue
Leatherhead
Surrey KT22 7ST
U.K.

INTRODUCTION

It has been reported by some utilities[1] that low energy track marks were found on in-service GIS insulators during normal inspection. In the past these have been attributed to fast transients generated by switch operations. However, recent work has shown that these low energy flashovers always occur sometime after switch open operations when transients should long have disappeared. The time delays range from a few minutes to several hours.

The switch open operations leave trapped charges on de-energised side of GIS, which, because of the ultra high insulation leakage resistance of GIS, can remain there for up to several days as shown by both site measurement and indicated by operational experience. An AC GIS under this situation has to endure electrical stress that are only present in a DC GIS. Track marks are left on an insulator when trapped charge voltage (equal or less than system peak voltage) is observed to undergo a step change with concurrent and clearly audible sound. Figure 1 shows an example of such an event. The blue phase had +1.0 per unit voltage trapped charge. The coupled voltage through the circuit breaker's grading capacitors lowered the effective DC voltage to about +0.8pu. The yellow and red phases all had negative trapped charges with voltages less than 1pu. A low energy flashover occurred on the blue phase 8 minutes and 31 seconds after de-energisation.

Although low energy flashover marks on an insulator do not seem to affect the insulator's system AC withstand capability after successful re-closure, the processes that are active under DC and responsible for initiating low energy flashovers are under intensive investigation to see whether under certain conditions subsequent re-energisations of GIS can be influenced. As the structure, choice of material and manufacturing process for GIS used by different manufactures are basically very similar, the issue calls for immediate and full attention as there have been occasions when failures occurred shortly after closing although stringent site cleaning procedure had been adopted during and after GIS erection and

Gaseous Dielectrics VII, Edited by L.G. Christophorou
and D.R. James, Plenum Press, New York, 1994

installed partial discharge monitoring equipment revealed that no detectable levels of discharge were present before the failures. The cause of these failures can no longer be casually associated with the possibility of consequential particles being present. Other utilities have also reported[2] unexplained switching related failures.

Trapped charge related problems in an AC GIS can be alleviated by ensuring that no trapped charges are left on GIS during de-energisation, e.g. by arranging a transformer, or a set of surge arresters connected to the section of GIS that is to be de-energised. However, these measures cannot be applied to majority GIS operations, if no undue operation restrictions are to be imposed. Frequently, configurations of GIS is such that trapped charges are unavoidable for open operations. Therefore, the understanding of trapped charge related phenomena is important for the reliable design and safe operation of AC GIS.

It is known that significant surface charge accumulation can occur under DC stressing[3,4]. There are many mechanisms that can lead to the transfer of electrostatic charges from busbars on to insulators. The charge carriers could be ions or dust-like microparticles. The ions could be naturally present, or generated by field emission as a result of roughness on busbar or chamber wall. Net charges can also arise on the surface or in the bulk of insulators if there is conductivity inhomogeneity in the insulator material. This effect, however, can be mitigated by adopting a suitable curing process in the manufacturing process[5]. For correctly manufactured insulators with homogeneous resistivity, the dominant mechanism of insulator charging is through charge transportation in the gas media and not through inhomogeneous conduction in the solid media. Particulate contamination generated PD can quicken, but does not constitute a prerequisite for, the process of charge transfer.

Surface charges can cause local field enhancement at the triple junctions and thus providing a triggering mechanism for surface flashover. However, this does not seem to represent the whole picture. Flashover marks have been found on both convex and concave side of cone shaped support barriers. The flashover paths usually adhere closely to the insulator surface rather than developing in the radial direction, which is the shortest route, resulting in breakdown in the gas media. In addition, it has been observed experimentally that sometimes surface flashover can propagate a long distance along a direction that is perpendicular to that of the applied electric field lines, apparently unaided by the external field. These observations all suggest that surface charge not only play an important part in providing the initiating mechanism for flashovers but also in supporting their materialisation and propagation.

Clearly, a physical model is needed to address the following important engineering question: what is the tolerable level of surface charge density on an insulator that will not sustain a flashover? Since the maximum surface charge density is determined totally by the geometry of insulator and conductor configuration, such knowledge will lead to the design of optimal insulator shapes with the best trade-off between insusceptibility to charge accumulation, mechanical strength, ease of casting, savings in material, long creepage length, etc.

A surface discharge criterion has been proposed[6] that is based upon experimental results of chopped DC tests. The criterion centres on electrostatic surface energy density rather than electric field stress, which constantly changes during the surface discharge process and is hard to formulate. By reasoning on the level of energetics, economy of thought is achieved and no assumptions, such as the diameters of the discharge filaments, needs to be made to follow the process of surface discharge mathematically. This paper will explore this criterion further by simulating experimental results. The aim is to establish a numerical procedure to assess the safety margin of an insulator design against trapped charge induced problems. The results can provide input to product approval and operational procedure considerations.

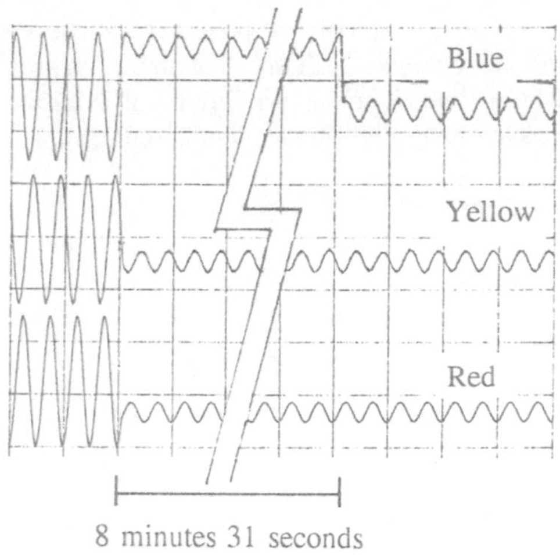

Figure 1. Low energy flashover under trapped charge stressing on the blue phase of a de-energised section of GIS. Time delay after circuit breaker open: 8 minutes and 31 seconds.

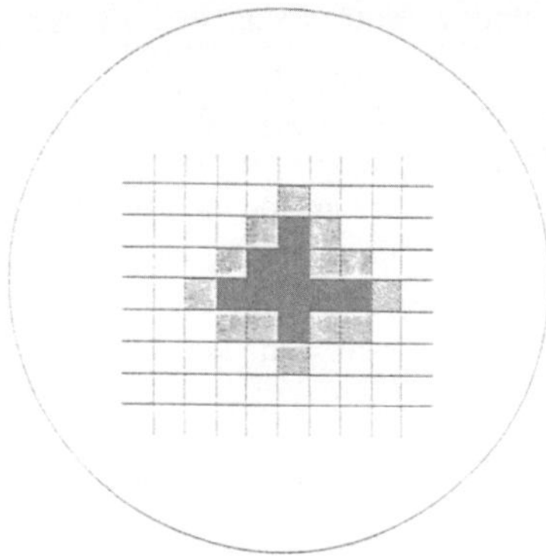

Figure 2. Schematic simulation grid. Discharge develops from black nodes to adjacent cross-hatched ones. The probability of a candidate node is determined by surface energy density.

FORMULATION OF A COMPUTER SIMULATION ALGORITHM

When surface charges accumulate on the surface of an insulator, electrostatic energy is stored. The surface density of the stored energy E_s depends on both the surface charge density σ and on the potential distribution ϕ as expressed in the following:

$$E_s = \frac{1}{2}\sigma\phi \qquad (1)$$

The potential ϕ is determined by the geometry, the surface charge on the insulator and the voltage of electrodes. Surface charge can support self-sustained surface discharge when the electrostatic surface energy has attained a critical level E_{lim} to support renewed ionisation activities.

$$E_s \geq E_{lim} \qquad (2)$$

During the growth of discharge, the surface potential distribution constantly changes and with it the electrostatic surface energy changes during each step of the surface discharge process. The complicated interaction between geometry, surface potential, surface charge density, surface energy density and SF_6 pressure can be examined by developing a computer model.

For the benefit of comparison with experimental observations, simulation is carried out with an insulator-conductor configuration that is the same as that used experimentally[7]. A circular disk is divided into many small squares. The surface charge density and the potential on each square is represented respectively by σ and ϕ at the centre of the square. The potential around the disk boundary is held at zero. The simulation starts by imposing zero potential on a square at the centre of the lattice. This "initiation site" corresponds to the contact point between the hemispherically tipped electrode and the insulator disk for DC chopping tests[7], in which the potential at the contact point suddenly drops to zero when the applied DC voltage is chopped. Surface discharge then spreads radially outward. The area already covered by surface discharge is illustrated by black squares in Figure 1. All the neighbouring cross-hatched squares are possible candidates for subsequent discharges.

The following procedure is used to determine the probability of a particular step of surface discharge propagation. The probability that a candidate square is chosen is assigned to be proportional to the number of free electrons produced in the previous discharge step, which in turn should be proportional to the energy released in the neighbouring black square. The probability is zero when the energy released is below a threshold and no renewed ionisation takes place.

$$\begin{aligned} p &= 0 && \left(\tfrac{1}{2}\sigma\phi < WhP\right) \\ p &\sim \tfrac{1}{2}\sigma\phi && \left(\tfrac{1}{2}\sigma\phi \geq WhP\right) \end{aligned} \qquad (3)$$

where p is the probability, σ the surface charge density distribution, ϕ the potential distribution, P the SF_6 pressure, h the grid size and W the distance scaled critical surface energy density. The critical surface energy density for supporting self-sustained surface flashover was found macroscopically to be about $9.6 mJm^{-2}bar^{-1}$ over a distance of 4.5mm from the triggering site[7]. It is assumed that this value is valid with different grid sizes and the value of W adopted here is therefore $2.13 Jm^{-3}bar^{-1}$.

In cases when a candidate square borders with more than one black square, the probability is added up. It is assumed that there is no potential drop along the ionised channels of discharge and zero potential is imposed on all the black squares.

The potential distribution is calculated with finite difference method using an optimum successive over-relaxation scheme.

$$\phi^{(n+1)}_{(i,j,k)} = \phi^{(n)}_{(i,j,k)} + \alpha\left(\overline{\phi_{(i,j,k)}} - \phi^{(n)}_{(i,j,k)}\right) \qquad (4)$$

If we assume that the process of charge transfer occurs in the gas media as mentioned in the introduction, an equilibrium surface charge density is reached when the normal electric field is cancelled out to zero by the surface charge itself. This means that an upper-bound value of surface charge density is determined by the insulator's geometry and the applied voltage. The density can be calculated by imposing a boundary condition of zero normal electric field strength along insulator surfaces. The simulation input data for surface charge density distribution was obtained from a finite element method package. A "natural boundary condition" (zero normal field strength) is imposed along the insulator surface. The calculated result is the highest surface charge density distribution that an insulator surface can possibly collect. Only one such calculation at unit DC voltage is needed. At higher voltages, the input data is scaled up by multiplications with the voltage value. SF_6 pressure has no effect on the surface charge density distribution.

Simulations were carried out with different voltage and pressure combinations. At several intermediate stages of a simulation, matrix outputs are stored of the surface potential distribution, the tree structure of all the nodes that have been discharged, and the energy involved at each node.

SIMULATION RESULTS AND DISCUSSIONS

In order to visualise the simulated discharge patterns and to compare discharge length measured from static photographs taken from tests, the data of discharge energy at each growth step are converted into a digital raster image file. The greyness level at a certain node is taken to be directly proportional to the total energy involved in all descending branches connected to that node. The greyness levels at all the nodal points are found to be approximately of Gaussian distribution. Lower and upper cutoff limits that correspond to zero and 255 greyness levels are chosen to be at three times the standard deviation below and above the average.

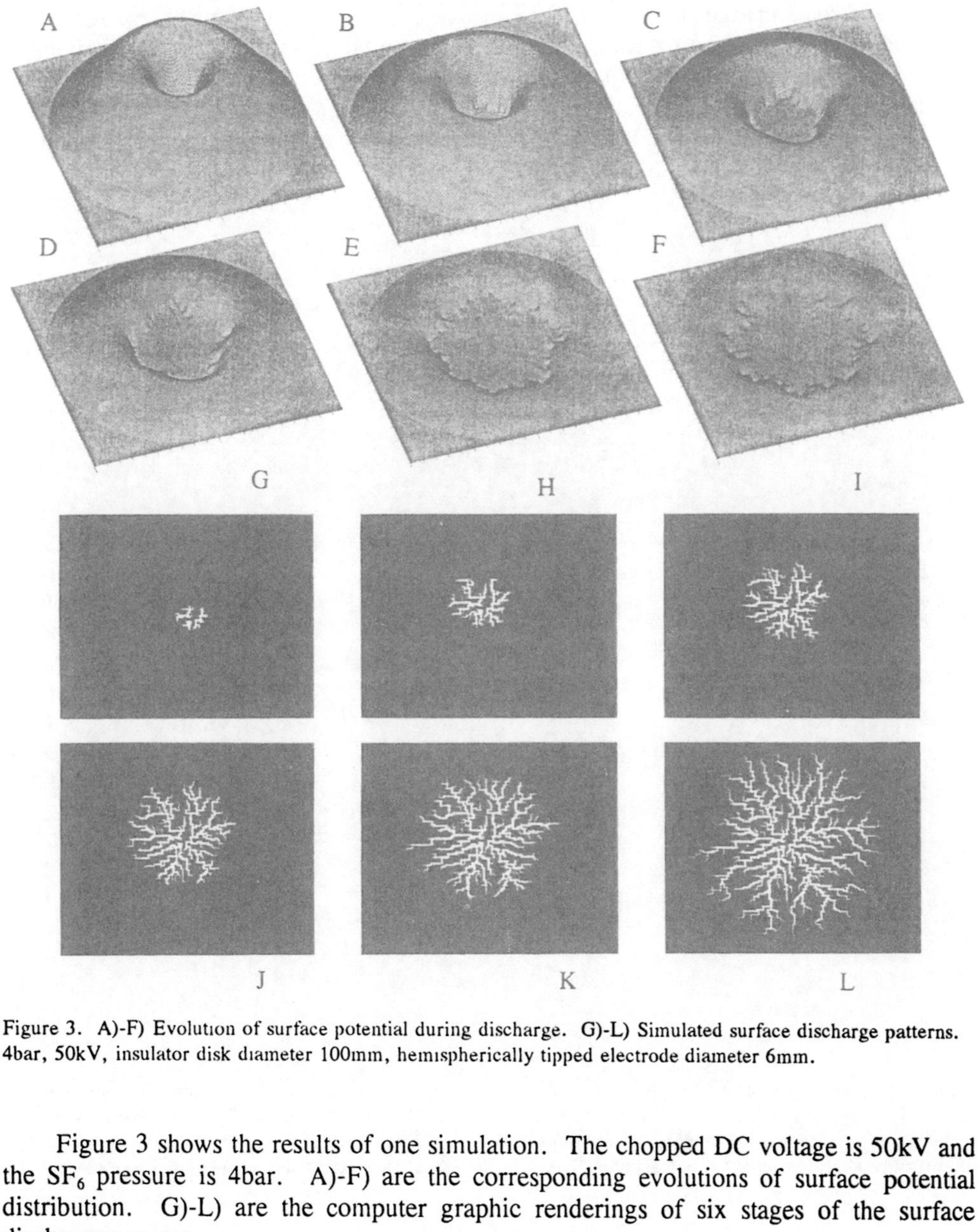

Figure 3. A)-F) Evolution of surface potential during discharge. G)-L) Simulated surface discharge patterns. 4bar, 50kV, insulator disk diameter 100mm, hemispherically tipped electrode diameter 6mm.

Figure 3 shows the results of one simulation. The chopped DC voltage is 50kV and the SF_6 pressure is 4bar. A)-F) are the corresponding evolutions of surface potential distribution. G)-L) are the computer graphic renderings of six stages of the surface discharge process.

The dendritical structure of surface discharge photographs as in Figure 3 show varied brightness among different filaments and along each filament. The varied brightness of discharge filaments is a result of different amounts of energy involved in ionisation at different positions. The simulated patterns bear strong resemblance to static photographs, which show that filaments close to the root of the discharge are brighter and tributary branches have lower and lower luminosity, indicating a diminishing and more transitory intensity of radiative ionisation activities further out from the discharge initiation site.

The simulation results show that not all surface charge can be discharged. The discharge process stops when the surface energy density falls below a threshold level and the discharge activity can no longer be sustained. This has been confirmed by experiment. Figure 4 shows an experimental scan of residual surface charge distribution using an inductive surface charge probe. The SF_6 pressure was 4bar and the DC voltage before chopping was 70kV. The measured ratio of total amount of surface charge remaining on the disk after discharge to the initial value tends to drop with increase of voltage. In other words, the higher the applied voltage, and therefore the higher the surface charge density, the more complete the surface discharge can neutralise the surface charge. This point has been borne out by simulations very well.

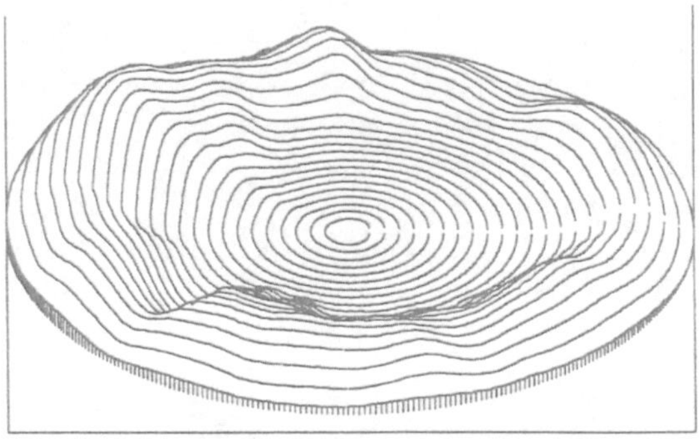

Figure 4. Measured residual surface charge distribution on the surface of an epoxy resin sheet disk after DC chopping. Disk diameter 100mm. 4bar SF_6. Positive DC voltage 70kV applied for 1 hour before chopping.

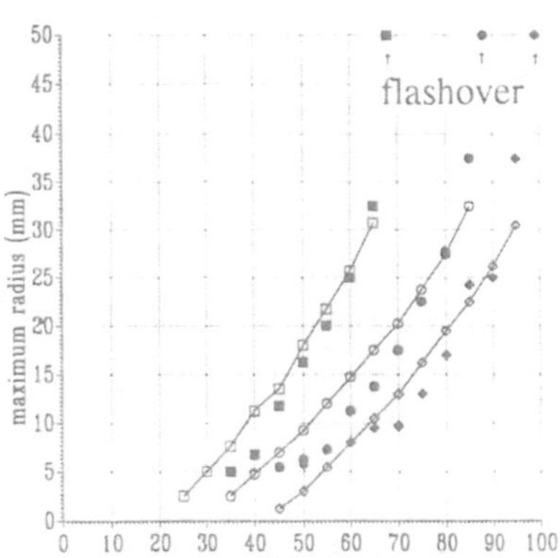

Figure 5. Positions of sequentially added discharge steps. This shows that there is a clearly defined "active front" behind which discharge activity stops. SF_6 pressure 4bar. Positive DC voltage 70kV.

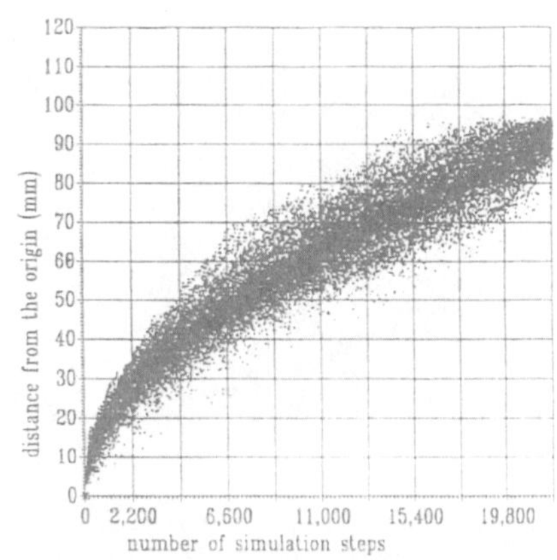

Figure 6. Simulated and measured discharge lengths.
□ 1bar, calculated ■ 1bar, measured
○ 3bar, calculated ● 3bar, measured
◇ 5bar, calculated ◆ 5bar, measured

Figure 5 shows that there is an active front in the discharge. The discharge activity tends to be concentrated at the front of the advancing filaments while the centre region only maintain passive conduction in the wake of the expanding active ring.

To see whether the model can address the engineering question of how far surface discharge can propagate, the maximum radii of the simulated discharge patterns are compared with those measured from static photographs. Three simulations with different random number seedings for choosing discharge stepping according to the discharge development probability as defined in Equation (3) were carried out with each voltage and pressure combination. The maximum radius of the discharge structure is found to vary very little between different runs. Figure 6 shows the average of the simulated maximum radius of surface discharge.

The results shown in Figure 6 show that the agreement with measured values is reasonable. Above 68KV, 88kV and 99kV while SF_6 pressure is 1bar, 3bar and 5bar respectively, bright arcs bridging the insulator disk were seen on static photographs. This is believed to be a result of the 10ns fall time of the chopped DC voltage[7]. It is possible that discharge can reach the edge of the disk before the chopped DC voltage has dropped to zero. Therefore, at higher voltages, the simulated discharge lengths become considerably shorter that those taken from photographs. Now that the algorithm has been established, it is straight forward to impose non-zero potential on the black squares as shown in Figure 2 to simulate real situations when flashover occur while the busbar voltage is not zero. Work as such is being carried out on real geometries of GIS.

SUMMARY

The condition responsible for the mysterious low energy flashover marks found on the surface of GIS insulators have been revealed to be trapped charge stressing on the de-energised sections of GIS.

A physical model for surface flashover has been successfully developed that incorporates only physically meaningful parameters. It is based on a minimum electrostatic surface energy criterion. Computer simulations have yielded results that are very similar to those observed experimentally, such as the dendritical appearance and physical scale of surface discharges, and the presence of residual surface charge after flashover.

It is now possible to calculate the extent that the surface charge supported flashover can spread for any given conductor-insulator configurations when surface charge accumulation reaches saturation under any busbar voltage. This paves the way for assessing the safety margins of GIS insulators against trapped charge stressing.

REFERENCES

1. Pryor, B. M. "Tracking of In-service GIS Barrier Insulators". IERE Workshop on GIS, Toronto. (1990)
2. Fujimoto, N. "Conduction Currents in Gas-Insulated Switchgear for Low Level DC Stress". Fifth International Symposium on Gaseous Dielectrics. (1987)
3. Cook, C. M. "Charging of Insulator Surfaces by Ionisation and Transport in Gases". IEEE Trans. Electric Insulation. Vol. EI-17, No. 2 pp172-178 (1982)
8. Nakanishi, K. "High Voltage DC Gas Insulation". Gaseous Dielectrics VI. Edited by L. G. Christophorou and I. Sauers. Plenum Press. (1990)
5. Cook, C. M., Nakata, R., Ouyang, M., Dale, S. J. and Garrity, T. F. "Compressed Gas Insulation for Advanced HVDC Transmission Equipment". CIGRE 1982 Session, Paper 15-14. (1982)
6. Wang, C. X., Wilson, A. and Watts, M. W. "Surface flashover sustained by electrostatic surface charge on epoxy resin insulator in SF_6". IEE Proceedings-A, Vol. 140, No.5, September (1993)
7. Wang, C. X. "Surface Discharge on Epoxy Resin Insulators in SF_6 under Chopped DC". 10th International Conference on Gas Discharges and Their Applications. Swansea. (1992)

DISCUSSION

I. D. CHALMERS: (1) Trapped charge implies charge in the busbar. Why does this necessarily imply charge trapped on the barrier surface? (2) Can you give some indication of the physical mechanism whereby energy surface density results in ionization to cause breakdown?

C. X. WANG: (1) There are many mechanisms, as mentioned in the presentation, that can cause charge transfer from busbars onto insulators. (2) To put it briefly, the release of surface energy can support and sustain surface discharge development.

A. PEDERSEN: You say that correctly manufactured insulators have a homogeneous resistivity. However, the resistivity of even the very best dielectric depends dramatically on very small changes in the temperature. One should not discard the resistivity as a possible source of trouble.

C. X. WANG: Thanks for a very valid comment. The only reservation that I have is that many trapped charge related problems reveal themselves under no load situations. In other words, when the conductor is still cold and is in thermal equilibrium with the environment.

K. NAKANISHI: (1) My first question is concerned with the starting locations of the low energy track marks. Do they vary from case to case for the actual spacers, or are they nearly the same? (2) Have you also experienced the low energy track marks on the post-type spacer?

C. X. WANG: (1) The appearance of track marks seems to be random. There is no consistency in their locations either in relation to a particular part of an insulator, or to the insulator's position in GIS. (2) Tracks have also been found on post-type insulators although in this case we have no firm evidence that they occurred under trapped charge stressing.

SECTION 10: DEVELOPMENTS IN GAS-INSULATED EQUIPMENT

GIS UTILITY RESEARCH - A HISTORICAL PERSPECTIVE

N. Fujimoto, J.M. Braun, R.M. Morra, H.D. Morrison, G.L. Ford , F.Y. Chu

Ontario Hydro Technologies
Toronto, Ontario, Canada

INTRODUCTION

Gas-insulated substations (GIS) have been in use for many years worldwide, especially at the lower voltage classes. In more recent years, the promise of high reliability, low maintenance and compact size has led to the extension of the technology to higher voltage classes. GIS at 550 kV, for instance, was first introduced about 15 years ago. However, the application to EHV levels (>300 kV) was accompanied by a number of problems related to the use of an emerging technology and resulted in dielectric failures and costly outages [1]. This spurred much research activity by manufacturers, universities and utilities. While manufacturer research was primarily devoted to the fundamentals of equipment design and circuit breaker development, utilities were forced to address operational and maintenance problems and, especially, to develop an understanding of dielectric failure mechanisms. Although the manufacturers contributed to the resolution of many of these problems, the effort was spearheaded by utilities. The combined research efforts have, in general, allowed for better specifications, testing and design evaluation resulting in modern GIS which is more reliable and more trouble-free. While many of the problems related to the older, first-generation GIS have been solved and overcome in modern equipment, many utilities continue to operate older installations and must therefore continue to investigate the related technical issues.

This paper reviews some of the major challenges encountered by this developing technology over recent years, the major utility research accomplishments and how the technology has been implemented in practice. A large component of the utility research work was performed at Ontario Hydro (Figure 1) and often with sponsorship by the Electric Power Research Institute (EPRI) or the Canadian Electrical Association (CEA). The paper also describes, from the utility perspective, the state of GIS technology today and the remaining issues requiring further study.

THE EARLY YEARS

The application of GIS to EHV levels, and especially to 550 kV, was based, in most cases, on extrapolated equipment designs used with success in lower voltage class equip-

Gaseous Dielectrics VII, Edited by L.G. Christophorou
and D.R. James, Plenum Press, New York, 1994

Figure 1 - Ontario Hydro's GIS Research Facility includes an entire 138 kV class GIS dedicated to research purposes. The above photo shows some of this equipment with fiberglass, air/SF_6 bushings A great deal of the research described in this paper was performed in this facility.

ment. In addition, as a result of rationalization of product lines by many manufacturers, much of the 550 kV equipment was essentially the same as equipment used at 420 kV but operated at approximately 30% higher stress. While some of the early 550 kV installations operated without many problems, others using prototype or near-prototype equipment suffered from numerous inservice failures and other forced outages, prompting active research programmes.

The early research efforts addressed the basic phenomena associated with dielectric failure and its consequences, and formed the basis for subsequent research activities. The investigations indicated that GIS was indeed a technology which required different and advanced approaches to solving problems. For instance, many of the concepts, including fault arc motion, nanosecond-risetime travelling waves/transmission line theory, grounded components which appear to be momentarily not "grounded", etc., were not new concepts to many sectors of the scientific community but, were essentially new concepts to power engineers. Power engineers were forced to think innovatively using concepts borrowed from other disciplines to develop appropriate solutions.

Major areas of research at Ontario Hydro included the following topics.

GIS Fault Analysis and Fault Location [2] - Research work sponsored by EPRI studied techniques for fault location using thermal and electromagnetic techniques as a means of quick location and reducing costly outage time. Prototype instrumentation was also developed. The origins of fast transient phenomena resulting from the initial breakdown of a fault or between disconnector contacts during operation, including the establishment of the fast, nanosecond-risetime nature of these transients, were also investigated. The dynamic behaviour of power arcs in the coaxial GIS geometry was also investigated as was fault-induced pressure rise and enclosure burnthrough. The consequences of faults, and damage assessment techniques were also addressed.

Transient Groundrise [3] - This phenomenon of short duration high voltages on the grounded components of the GIS (later called TEV - Transient Enclosure Voltage [4]) as a result of SF_6 breakdown and/or disconnector operation was investigated and the mechanisms identified and understood. The results of this study became instrumental in addressing the problems related to the electromagnetic compatibility of auxiliary circuits in GIS. Some safety concerns were also identified although reports of electrical shocks from this phenomena are rare. Methods for mitigating the effects of TEV through equipment layout and design were also established. This investigation was sponsored by CEA.

GIS Testing and Particle Behaviour - Investigations aimed at the dynamic behaviour of metallic particles in the GIS were undertaken to assist in the development of proper onsite high voltage test techniques. Since particles were identified as a major contaminant contributing to the reduction in the GIS insulation integrity, understanding their behaviour was critical to successful testing and ultimately reliable performance. These investigations were performed by several laboratories worldwide, including manufacturers and universities. One EPRI-sponsored project addressed the issue of locating the partial discharges (PD) generated by particles by electromagnetic means [5]. This project introduced several concepts, including the sub-nanosecond risetime nature of PD pulses in GIS and the use of capacitive couplers as PD sensors, which became the basis of subsequent research developments.

LATER DEVELOPMENTS

As indicated above, the early research addressed immediate concerns and established the groundwork to understand various phenomena. Failures and problems continued to occur but the focus now shifted to investigating issues affecting the operation and maintenance of the equipment although failures remained an issue. Several major accomplishments are described below.

SF_6 Decomposition Byproducts - Early work had shown that SF_6 gas, while nontoxic in itself, would decompose in the presence of arcs and sparks to form a number of solid and gaseous byproducts, some of which are toxic (the byproducts of particular concern were SOF_2, SO_2F_2 and HF). Work was performed to identify what byproducts were generated, under what conditions and in what quantities to assess the potential hazard and to develop appropriate safe work procedures [6]. These work procedures (or its variations) are essentially in use worldwide today. In addition, the effect of these byproducts on gas-handling equipment and on the GIS itself (especially on the surfaces of epoxy insulators) were studied to develop recommendations for cleanup and repair [7].

Moisture Assessment - Early operational experience with GIS indicated great difficulty in interpreting measurements of the moisture in SF_6 gas. A CEA-sponsored study has shown that moisture in GIS is a dynamic system and seasonal (i.e. temperature) fluctuations are normal as moisture migrates from the gas phase to being adsorbed in solid materials and back to the gas phase [8]. Condensation related to sudden changes in temperature in view of this type of mechanism has been suspected in several failures.

Very Fast Transient Overvoltages and Breakdown - As indicated above, disconnector operation was identified as a source of very fast transient (VFT) overvoltages [4] resulting from nanosecond-risetime travelling waves. As a travelling wave phenomenon, these transients have waveshapes determined by the physical configuration of the GIS and could also vary as a function of position within the GIS. However, since they are generated by disconnector operation, which by nature is statistical, the VFT overvoltages were also statistical in nature. These VFT stresses were part of the normal operating environment of the GIS but the fast risetime, and therefore, high frequency content, of these transients raised concerns over the insulating behaviour of the SF_6 gas and GIS systems stressed by VFT. Techniques were developed to perform computer simulations of VFT waveforms in GIS [9]. In addition, considerable work worldwide focussed on insulation behaviour under this type of stress. The results of these investigations suggested a decrease in insulation strength with a decrease in risetime in the presence of some types of defects. This lead to the greater prominence of lightning or oscillating lightning impulse waveforms as an onsite test.

Control Wiring Transients - As indicated above, the transients generated internally as VFT can couple onto the external portions of grounded equipment as Transient Groundrise or Transient Enclosure Voltage (TEV). TEV shares many of the same characteristics as VFT and becomes a source of concern with regard to the Electromagnetic Compatibility (EMC) of secondary systems. Wiring practice and coupling mechanisms were studied under CEA sponsorship and recommendations were made to reduce the coupling of these high frequency transients into secondary systems [10]. While EMC is an issue in nearly all power station environments, interference found in GIS was found to be one to two orders of magnitude higher in frequency, requiring that much greater care be taken in shielding, routing and termination of secondary wiring. Provided these recommendations are considered, EMC is generally not a problem in modern GIS installations.

Disconnector Statistical Operating Behaviour - In the early years of EHV level GIS, disconnectors were prone to failure during operation. This was found to be the result of an inadequate design (for the level of operating stress) combined with extra stresses generated by the self-induced transients. Studies were undertaken to study the statistical behaviour of disconnectors during operation, in order to identify characteristics which provide information on the likelihood of failure [11]. Manufacturers have since used many of these concepts to design disconnectors which are much more reliable. This work was also used recently in the development of a new IEC standard for type testing of GIS disconnectors.

PRESENT INTERESTS

As many of the early problems have been solved or at least explained, the focus of utility GIS research has shifted to long-term behaviour. Many utilities continue to operate earlier generation equipment which may be prone to failure. In addition, modern GIS, despite being (in general) much more reliable than older generation equipment, continues to experience failure. Operating and maintenance issues, such as the safe work procedures related to decomposition byproducts, continue to require attention, especially in view of increasing environmental sensitivity. Some of the major areas of effort in recent years are described below.

Onsite PD Measurement and Diagnostics - As faults continue to occur, researchers have looked to advanced PD measurement techniques as a way of detecting harmful defects before they can cause a premature failure [12]. PD measurements using acoustic emission (AE) sensors and by measuring electrical quantities at very high frequencies (400-1500 MHz) using built-in capacitive couplers have been shown to be a sensitive means of detecting harmful defects. These techniques are now used by many utilities to improve the efficacy of onsite dielectric tests. In the most important cases, diagnostic measurements are used in conjunction with AC and impulse waveforms to achieve the best possible dielectric onsite test.

S_2F_{10} - The early studies on decomposition byproducts of SF_6 provided the basic information on which safe work procedures were based. Since that time, another highly toxic substance, S_2F_{10}, was found in specific circumstances for which the toxicity limit was below detection limits. Sensitive techniques to detect its presence have now been developed and measurements have been performed to assess the production and decomposition rates of S_2F_{10} under actual operating (as opposed to laboratory) conditions [13]. This information will be used to modify, if necessary, the safe work practices now in use by utilities.

Ageing of Epoxy Insulators - While SF_6 gas insulation does not "age" *per se*, the mineral-filled epoxy insulators are subject to long-term degradation, ageing and possibly premature failure. Studies under EPRI sponsorship were undertaken to study the long term ageing of epoxy insulators [14]. These studies have shown that insulator failures result from internal electrical treeing starting at defects within the insulator, such as small voids, metallic inclusions, metal/epoxy interface defects, etc. Studies have also shown that conventional PD tests, performed by manufacturers as part of the quality assurance process, do not necessarily detect all of the defects which can cause premature failure. A more sensitive test using xray irradiation as an ionizing source to enhance electrical PD measurements was developed and shown to be much better at detecting the small defects which could be missed by conventional tests (Figure 2). This Xray-Induced Partial Discharge (XIPD) test is now being extended to insulators up to 550 kV class.

Effect of VFT on Other Power System Components - While VFT phenomena has been extensively studied from the point of view of its effect on the insulating properties of GIS systems, its effect on other power system components is less certain. As indicated above, VFT is a part of the normal operating environment of GIS but, for other power equipment, represents a new category of stress. Insulation systems external to the GIS need to be designed to accommodate this type of stress. Problems in the area of direct GIS/cable terminations were identified [15] and some work sponsored by CEA is presently underway to examine VFT effects on transformer insulation.

Figure 2 - The X-ray-Induced Partial Discharge (XIPD) test facility developed for high sensitivity PD measurements on GIS epoxy insulators. The photo shows an overall view of the XIPD system with a portion of the test cell enclosed in a lead-lined room and the aerial connection to the high voltage transformer. At present, a second version of the XIPD system is being developed for use on 550kV class insulators.

FUTURE PROSPECTS AND NEEDS

The research efforts by utilities combined with parallel efforts by manufacturers and by universities have led to the many improvements evident in modern GIS designs and installations. As the technology matures, many of the phenomena unique to GIS have become increasingly well understood. Some of the research results have now found their way into standards and normal engineering practice. However, failures continue to occur and research efforts continue to be necessary, especially in the areas of diagnostics and condition monitoring as GIS installations age and become subject to longer term deterioration. Other research needs are based on remaining operational problems and potential problems resulting from the widespread use of SF_6 gas and environmental concerns. Particular areas in which more research is required include the following.

GIS Condition Monitoring - In view of recent economic pressures, utilities are increasingly forced to make better use of limited resources in an effort to reduce costs. Consequently, diagnostics have received much attention in recent years. Although techniques for diagnostic measurements are well-advanced, the interpretation of data needs more attention. The object of monitoring systems is not only to detect the presence of possible defects (which is possible today), but to obtain sufficient information about the defect to achieve a positive economic benefit (e.g. by avoiding failure and/or unnecessary outages). The knowledge required to judge the severity of the defect and to estimate the time to failure is lacking for ageing-related defects. More studies of the time-evolution of diagnostic signals leading to failure are necessary.

Recycling and Reuse of SF_6 Gas - SF_6 is generally known to be a greenhouse gas although it is not an ozone depleter [16]. Although the relative contribution of SF_6 to the greenhouse effect is extremely small [17], any unnecessary discharge of SF_6 to the atmosphere must be avoided. Consequently, as some GIS are repaired, replaced or decommissioned, issues related to large scale recycling and reusing of "old" SF_6 gas need to be addressed. Issues related to the gas-handling equipment, filter systems used, testing and the development of adequate technical standards for recycled gas also need to be addressed.

SUMMARY & CONCLUSIONS

Gas-insulated substation technology has been used worldwide for many years. Promises of high reliability, low maintenance and compact size have made GIS an attractive option in many cases. However its application to EHV levels has resulted in problems, especially in the early years, and has prompted extensive research programmes. Many of the research results reviewed in this paper have been implemented in practice and, as a result, modern GIS designs available today are much improved over their predecessors.

GIS failures continue to occur in both new and old equipment, requiring further investigations. The focus of investigations at Ontario Hydro, present and future, are on issues related to the long-term performance of the equipment such as diagnostics and condition monitoring, and on developing the underlying knowledge of ageing mechanisms needed to provide useful information from diagnostic data. Other issues, such as SF_6 recycling, are related to the widespread use of GIS technology and its potential environmental impact.

Overall, many of the basic problems associated with GIS are now either solved or understood sufficiently. GIS, in general, has proven to be a technology which is sensitive to minor imperfections and contaminants, especially at higher stress (voltage) levels. However, there are technical limits to what defects can be detected through advanced testing and condition monitoring techniques. In view of the utilities' desire to maintain high levels of reliability at a reasonable cost, a continuing goal of research work should be to find the maximum tolerable stress for reliable operation (which determines the degree of sensitivity to minor defects), within the context to our technical ability to find the critical defects and contaminants.

GIS has brought to utilities many concepts and ideas unfamiliar to many power engineers. Consequently, for continuing successful research, strong links to the general scientific community are essential.

REFERENCES

1. CIGRE WG 23-10 "A Twenty-five Year Review of Experience with SF_6 Gas Insulated Substations (GIS)" 1992 General Session of CIGRE, paper 23-101.
2. "Fault Analysis in Gas-Insulated Equipment." EPRI Final Report EL-2248, Project 1360-1, February 1982.
3. N. Fujimoto, E.P. Dick, S.A. Boggs and G.L. Ford. "Transient ground potential rise in gas-insulated substations - Experimental studies." IEEE Trans. on PAS, *PAS-101,* No.10, October 1982.
4. CIGRE Working Group 33/13-09 "Very Fast Transient Phenomena Associated with Gas-Insulated Substations." 1988 Session of CIGRE, paper 33-13.
5. S.A. Boggs. "Electromagnetic Techniques for Fault and Partial Discharge Location in Gas-Insulated Cables and Substations." IEEE Trans on PAS, *Vol. PAS-101*, No. 7, July 1982. Page 1935.
6. F.Y. Chu. "Decomposition in Gas-Insulated Equipment", IEEE Trans on Electrical Insulation, *Vol EI-21*, No 5, October 1986.
7. F.Y. Chu. "Degradation Mechanisms for Epoxy Insulators Exposed to SF_6 Arcing Byproducts." IEEE International Symposium on Electrical Insulation, 1986.
8. J.M. Braun, L.A. Gonzalez and F.Y. Chu. "A Model of Moisture Ingress and Distribution in Gas-Insulated Switchgear." CIGRE Symposium on New and Improved Materials for Electrotechnology. Vienna, Austria, May 1987. Paper 400-04.
9. N. Fujimoto, H.A. Stuckless and S.A. Boggs. "Calculation of Disconnector Induced Overvoltages in Gas-Insulated Substations." In L.G. Christophorou and M.O. Pace (Eds.) *Gaseous Dielectrics IV*, Pergamon Press, New York, 1984.
10. S.M. Harvey. "Control Wiring Transients and Electromagnetic Compatibility in GIS." In S.A. Boggs, F.Y. Chu and N. Fujimoto (Eds.) *Gas-Insulated Substations - Technology & Practice*, Pergamon Press, New York 1986.
11. S.A. Boggs, N. Fujimoto, M. Collod and E. Thuries. "The Modeling of Statistical Operating Parameters and the Computation of Operation-Induced Surge Waveforms for GIS Disconnectors." CIGRE 1984, paper 13-15.
12. CIGRE WG 15.03 "Diagnostic Methods for Gas Insulating Systems." 1992 General Session of CIGRE, Paper 15/23-01.
13. D.R. James, I. Sauers, G. Griffin, R. Van Brunt, J. Olthoff, K. Stricklett, F. Chu, J.R. Robins, H.D. Morrison. "Investigation of S_2F_{10}." IEEE Electrical Insulation Magazine, *Vol.9*, No.3, May/June 1993. Page 29.
14. J.M. Braun, N. Fujimoto, G.C. Stone, S.A. Boggs, G. Addis, K.J. Diederich, A. Diessner, G. Luxa and A. Girodet. "Long Term Reliability of Cast Epoxy Insulators in Gas-Insulated Equipment". 1990 Session of CIGRE, paper 15/33-07.
15. N. Fujimoto, S.J. Croall and S.M. Foty. "Techniques for the Protection of Gas-Insulated Substation to Cable Interfaces." IEEE Transactions on Power Delivery, *Volume 3*, Number 4, October 1988. Page 1650.
16. L. Niemeyer and F.Y. Chu. "SF_6 and the Atmosphere." IEEE Trans. on Elect. Insulation, *Vol. 27*, No.1, February 1992. Page 184-187.
17. M.K.W. Ko, N.D. Sze, W.C. Wang, G. Shia, A. Goldman, F.J. Murcray, D.G. Murcray and C.P. Rinsland. "Atmospheric Sulfur Hexaflouride: Sources, Sinks and Greenhouse Warming." Journal of Geophysical Research, *Vol. 98,* No. D6, June 20, 1993. Pages 10,499-10,507.

DISCUSSION

T. NITTA: To my understanding, Ontario Hydro is the utility company that has been working for so many years intensively on the fundamental aspect of gaseous insulation. And you said you have a good reason for that. In spite of that, the statistics you presented for Ontario Hydro systems show many times higher failure rate in comparison to the systems working in other utility companies. I believe that this is not the kind of figure you want to see for your system. What is your direction to go from here to improve these statistics? The list of your current interests are all familiar subjects for years and looks like not fulfilling the requirement for this improvement.

H. D. MORRISON: You are correct in your assumption that our high failure rate is unacceptable to our company. However, we believe that the main reason for the high failure rate is our unfortunate early application of an immature technology. As indicated in the paper, many of these problems have been addressed and solved. However, we continue to experience problems. Many of the recent problems have been, unfortunately, recurring "old" problems for which the failure cause is understood but, for various reasons, remedial actions have not yet been or could not be implemented. We fully expect that the most recently installed equipment will operate much more reliably than the older equipment. Statistics might influence our decision to use GIS technology in the future but our present line of interest is driven more by our need to address issues in operating and maintaining our existing installations.

E. KUFFEL: Reference was made by Dr. Nitta to Japanese statistics relating to GIS failures. The reported information indicates that their failures are far less frequent than Ontario Hydro's in North America. Could the difference be the result of quality control during manufacture, assembly, maintenance, etc.? In such case perhaps part of the blame should be borne by the suppliers.

H. D. MORRISON: Today we know the importance of high quality as GIS is a technology which is very sensitive to minute defects and contaminants. However, our understanding is that the Japanese took a very conservative approach while on the "learning curve" with GIS designs which were considerably overdesigned and therefore, relatively trouble-free. Unfortunately, as a result of the "competitive bidding" process used in many parts of the world, economic pressures did not permit the same degree of overdesign. This obviously resulted in a greater number of failures but also assisted in the learning process both in terms of equipment design and in the requirements for quality. Both manufacturers and utilities have gained (hopefully) from these experiences.

E. MARODE: Compared to the purity of the SF_6 when initially introduced in the GIS, have you an idea of the loss of SF_6 purity when, after a long period of service, the old SF_6 has to be removed from the equipment and recycled?

H. D. MORRISON: Recently, in the course of extending some of our GIS, we have sampled gas from compartments, both bus duct and bushings, that has been in service for 15 years and essentially untouched over that time. Our analysis shows that the gas purity is practically as good as new. Concentrations of by-products have been at most a few ppm.

A. H. COOKSON: Are there plans to use the X-ray technique for screening solid insulators for GIS as a regular part of testing?

H. D. MORRISON: Ontario Hydro already has provisions in its purchase specification for using the X-ray induced partial discharge test for insulators. However, at this point, no manufacturer can yet offer such a test in their production. Assuming the manufacturers develop such a capability, its use as a routine production test will very much depend on the cost/benefit determined for that particular purchase. Alternatively, the test may be adopted as a type test, as a check on the manufacturing process and the ability to consistently produce good insulators.

DEVELOPMENT OF 275kV GAS COOLED TYPE GAS-INSULATED POWER TRANSFORMER

C.T.Wan*, A.Kudo, M.Asakura, Y.Murakami, K.Itoh, T.Nishitani, T.Yoshikawa

*The Hongkong Electric Co., Ltd.
Mitsubishi Electric Corp. Ako Works, Ako, Japan

ABSTRACT

A 275kV power transformer with a low noise, using SF6 gas alone for both insulation and cooling, was developed with new technology and put into service in 1990 in Hongkong. This 275/11kV, 30MVA transformer is the first gas-insulated power transformer in the world to reach the voltage level of 275kV. And the new technology is transferred to the manufacturing of a 275 kV, 60MVA transformer. The paper presents the design philosophy, the new technology, the rating and the feature of the 275kV gas cooled type gas-insulated power transformer with the low noise ideal for urban areas.

KEYWORDS

Gas-insulated transformer; SF6; PPS film;On-load tapchanger;Gas blower;Insulation spacer

INTRODUCTION

SF6 gas cooled type gas-insulated power transformer, using SF6 gas alone for both insulation and cooling, has been highlighted in these days with many technical advantages prevailing over conventinal oil filled transformer. The advantages of this type transformer are fire-safety, higher reliability, flexible layout and they require little maintenance. Already a large number of this type transformers up to class 77kV, 30MVA have been manufactured so far.
Furthermore, it would be advantageous to develop this type 275kV large capacity transformer to meet the increasing electricity consumption in urban areas. For such reasons this type 275kV (BIL 1050kV),30MVA transformer was developed with new technology and put into service in 1990. The new technology has been also transferred to the manufacturing of a 275kV,60MVA transformer, which has been placed in service since 1993.

Gaseous Dielectrics VII, Edited by L.G. Christophorou
and D.R. James, Plenum Press, New York, 1994

The paper presents the design philosophy, the new technology, the rating and the features of this type 275kV transformer with the low noise ideal for urban areas.

DESIGN PHILOSOPHY FOR 275kV GAS COOLED TYPE GAS-INSULATED POWER TRANSFORMER

In Figure 1 the world's first 275kV gas cooled type gas-insulated power transformer is shown. The ratings and the features are also shown in Table 1. The overload rating (112.5%) is additionally required to cope with any emergency operation in substation. This 275kV transformer is developed with the following design philosophies.

1) Adoption of SF6 gas alone for both the insulation and the cooling to realize little maintenance
2) Self cooled capacity of 50% continuous maximum rating
3) Low transformer noise to realize environmental harmony, especially for the substation in a densely populated area
4) Adoption of 275kV SF6 gas insulated vacuum switch on-load tapchanger
5) Adoption of compact size cable connection box to obtain the high space efficiency in the substation

Table 1 Rating and features of 275kV,30MVA transformer

Phase	Three	
Capacity	15/30/(33.75)*MVA	
Cooling type	GNAN/GFAN/(GFAN)*	* 8h rating
Primary voltage	$288.75 \sim {}^{R}275 \sim 206.25$kV(19 taps)	
& insulation level(BIL)	1050kV	
Secondary voltage	11kV	
& insulation level(BIL)	95kV	
Connection	Yd11	
Frequency	50Hz	
Thermal insulation class	E(coil temperature rise limit 75℃ by resistance method)	
Impedance voltage	31% at 30MVA base	
Gas pressure(main tank)	240kPa at 20℃	
Noise level	55dB(A) at 2m	
Dimensions	Inside tank W2.3×L6.5×3.4m External W4.1×L12.0×7.5m (excluding soundproof wall)	
Weight of core & coils	45,000kg	
Total weight	92,000kg(excluding soundproof wall)	

Figure 1 External view of 275kV, 30MVA transformer

MAJOR INSULATION STRUCTURE

Since the dielectric strength of SF6 gas has a close dependency on the electric field strength and significantly falls under non-uniform electric field distribution, static shields are arranged on the coil ends to improve the electric field distribution on those parts. Also, the structures with round edges or shields are employed to the parts adjacent to the coil assembly to avoid the non-uniform electric field.

To cope with the dual ratings for cooling, GNAN (natural cooling) and GFAN (forced cooling), the cooling structure of HV and LV coils is designed with zigzag flow system, which is achieved by the gas flow stopper alternately inserted between some disk coil sections and outer or inner barrier tubes. Furthermore the vertical ducts between the coil and the outer or the inner barrier tubes is designed to be larger than horizontal duct of each coil section so that the cooling gas can flow uniformly into each coil section.

The magnetic shields are arranged on the upper and lower sides of coil assembly to reduce the stray loss and avoid local over-heating due to the high impedance voltage. (Figure 2, Figure 3)

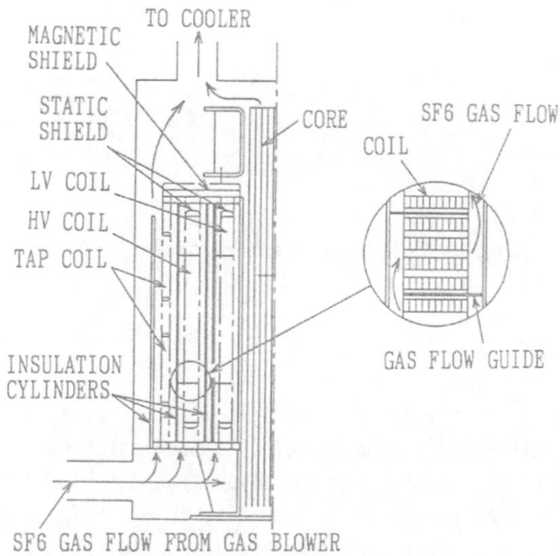

Figure 2 Section of coil assembly

Figure 3 Internal assembly of 275kV, 30MVA transformer

HIGHER THERMAL-RESISTANT PLASTIC FILM

Polyphenylene sulphide (PPS) film is used as conductor insulation material, instead of polyethylene terephthalate (PET) film which was previously in use, because of the higher thermal insulation grade and the exceptional resistance to hydrolysis even in vapour at high temperature. Figure 4 is an Arrhenius plot showing the relationship between the temperature and the time required for the tensile strength of PPS film to be half of its initial values. This result shows the thermal durability of PPS film is much superior to that of PET film[1]. PPS film can be used at higher temperatures and considered suitable for large capacity transformers and transformers with heavy overloading duty.

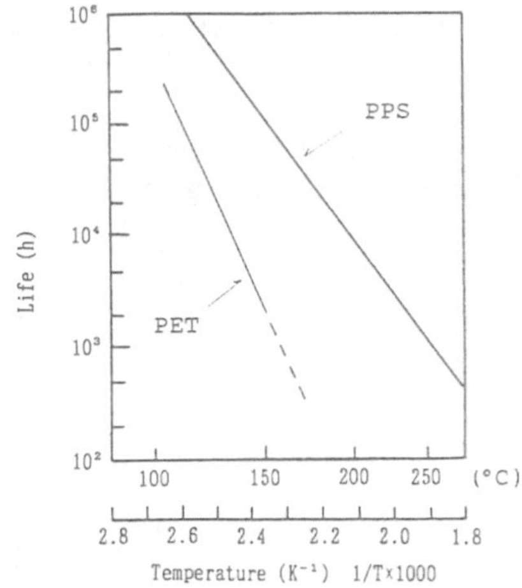

Figure 4 Lifetime characteristics of PPS film

LOW NOISE TYPE GAS BLOWER

A low noise type gas blower is essential for a low noise gas cooled type gas-insulated transformer, although the transformer operating noise can be lowered by external soundproof wall over the main tank. The low noise type gas blower employs a multi-blade impeller with a small diameter. Thereby the noise level is improved to be 10 dB(A) and more,compared with the conventional one. The two low noise type gas blowers are provided for the 275kV, 30MVA transformer (Figure 5,Table 2).

Table 2 Ratings of gas blower

Phase	Three	
Frequency	50	Hz
Voltage	380	V
Output	3.7	kW
No.of pole	6	
Rated gas volume	35	m^3/min.
Static pressure	3	kPa

Figure 5 External view of gas blower

GAS-INSULATED VACUUM SWITCH ON-LOAD TAPCHANGER

Vacuum switches are used as the switching element in the gas-insulated vacuum switch on-load tapchanger, and the diverter insert is installed in the cylinder with SF6 gas, which is separated from the transformer main tank. In addition, roller contact and lubricant-free bearings are adopted for the selector contact portion. Electrical(200 000 operations) and mechanical (800 000 operations) performance tests are performed according to Japanese standard JEC-2220 "On-load tapchanger" and good results are obtained (Figure 6, Table 3).

Table 3 Ratings of on-load tapchanger

Switching system	One-resister type
Number of vacuum switches	Three/phase
Connection	⅄
Max. Step voltage	3000 V
Max. Through current	550 A
Rated Step capacity	1100 kVA
Frequency	50/60 Hz
Number of tapping positions	Max. 23
Insulation level to earth	BIL 550 kV

Figure 6 External view of on-load tapchanger

RATED GAS PRESSURE

The transformer is provided with three different gas pressure ratings, each corresponding to the main tank (240kPa at 20℃), the HV cable box (500kPa at 20℃) and the diverter compartment of on-load tapchanger (130kPa at 20℃). The differential-pressure tight insulating spacer and the insulation cylinder are provided for the separation of the HV cable box and the diverter compartment of on-load tapchanger respectively from the transformer main tank.

The insulation spacer and the cylinder are designed sufficiently to withstand not only the maximum pressure difference, but also high top gas temperature.

The application of the insulation spacer enables the cable box to be compact and thereby improves the space efficiency in the substation (Figure 7).

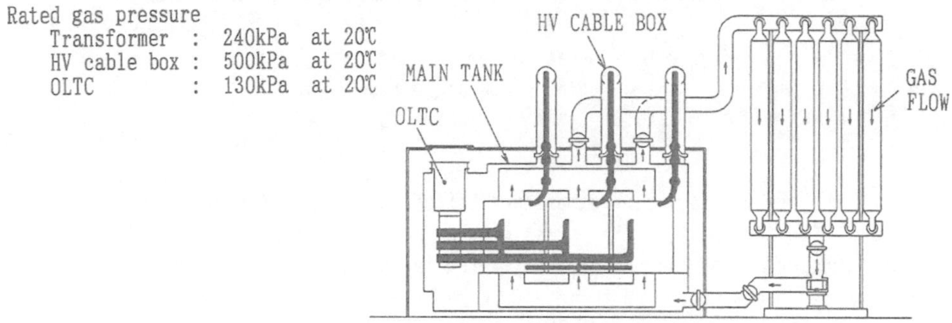

Figure 7 Section of 275 kV 30MVA transformer

PERFORMANCE TESTS

All performance tests specified in IEC-76 and BS-171 "Power transformers" were performed, including lightning impulse withstand test (BIL 1050kV), switching impulse withstand test (850 kV), induced over-voltage withstand test (ac 460kV), partial discharge measurement, temperature rise test and noise level measurement.

From the temperature rise test, the maximum gas temperature 70.5°C and the maximum gas pressure 270 kPa were obtained under the overload rating 33.75MVA. All performance tests made in the factory showed satisfactory results (Figure 8, Table 4).

Table 4 Performance test results of 275kV, 30MVA transformer

Lightning impulse	1050kV
Switching impulse	850kV
Induced over-voltage	460kV
Partial discharge measurement at 1.3E($1.3\times300/\sqrt{3}$=225.2kV)	Background
No load loss	14.0kW
Load loss at 75℃ 30MVA(275kV tap)	162.0kW
Top gas temperature rise at 30MVA(GFAN)	51.3℃
Average winding temperature rise at 30MVA(GFAN)	HV:63.0℃;LV:65.7℃
Noise level by IEC-551(at 2m)	52.6dB(A)

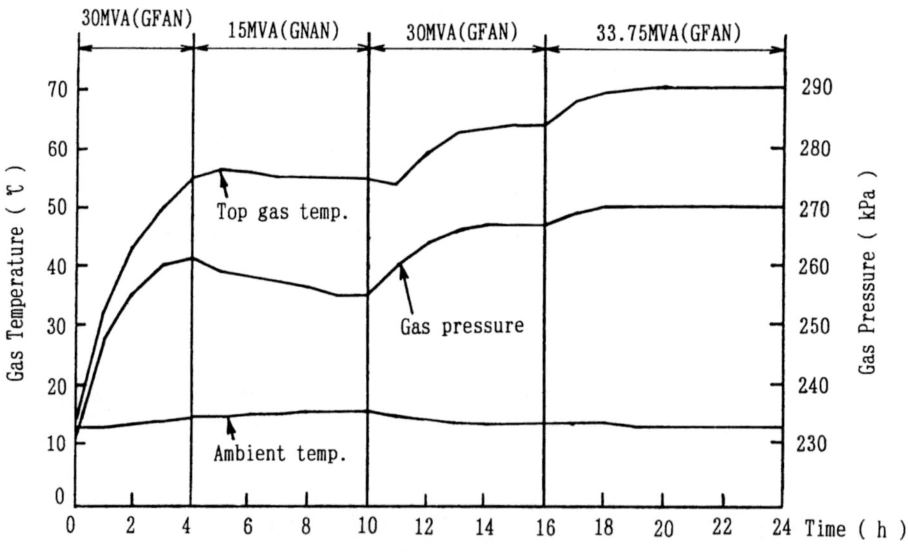

Figure 8 Top gas temperature and gas pressure under temperature rise test

Based on the manufacturing experience of the 275kV, 30MVA transformer, the 275kV, 60MVA and also the 132kV, 65MVA transformers have been manufactured in succession. The new technologies established are reflected in the design and manufacturing of high voltage and large capacity transformers (Figure 9, Figure 10).

The radiator banks are placed above the transformer main body to suit the limitation of the transformer bay in the substation.

Figure 9 External view of 275kV 60MVA transformer

Figure 10 External view of 132kV 65MVA transformer

CONCLUSIONS

The advantages of the gas cooled type gas-insulated power transformer are fire-safety, high reliability, flexible layout and little maintenance. Therefore the application of this type transformer is the most suitable for the substation in large city. Furthermore, since the application voltage range is extended to 275kV, the application of this type 275kV transformer will be increased more and more in the future to meet the increasing electricity consumption in urban areas.

REFERENCE

1. H.Muramoto and others " Developing separate-cooling / sheet-winding gas insulated transformer ", IEEE,Trans. Power Deliv.,Vol.3, No.3, PP.1067-1073,July 1988

DISCUSSION

G. N. TRINH: Can you tell us about the service experience at Hong Kong Electric with this new type of gas-insulated power transformers?

T. YOSHIKAWA: Hong Kong Electric Company has a four-year operating experience on this new type of gas-insulated power transformer. Since 1990, we've had no failures.

Y. QIU: Is it possible to use the same gas pressure for both the transformer and the on-load tap changer to make the structure of GIT simpler?

T. YOSHIKAWA: No it is not. The rated gas pressure is set lower than that of the transformer due to the limited pressure that can be applied to the vacuum switch system.

DEVELOPMENT OF 275kV 300MVA LIQUID COOLED GAS INSULATED TRANSFORMER

Yoshiaki Mino Kouji Shimada (Kansai Electric Power Co.)
Toyokazu Kouan Hiroshi Nakazawa (Mitsubishi Electric Co.)

ABSTRACT

Following the development of a 77kV SF6 gas insulated transformer in 1979, a 275kV 300MVA liquid cooled gas insulated transformer was developed and put into commercial operation in March, 1991, for the first time in the world as 300MVA gas insulated transformer. The main features of the transformer are that perfluorocarbon(C8F16O) liquid is utilized for cooling, and that mixture of SF6 gas and perfluorocarbon vapor forms the insulation.

The transformer has been operating continuously without trouble for nearly three years including three hours of overload operation up to 126% of rated capacity. As a result of actual operation it was confirmed that the performance of the gas insulated transformer was equal to that of a conventional oil immersed transformer even under severe thermal condition.

KEYWORDS

Gas insulated transformer; SF6; Perfluorocarbon (C8F16O).

INTRODUCTION

Since the power demand is increasing steadily in large cities, extension of 154kV and 275kV power systems to urban districts has been promoted for stable power supply. Furthermore, an introduction of 500kV power systems into large cities is under consideration. It is hard to secure enough space for constructing substations in densely populated urban districts, and also to meet the growing demand for fire prevention and environmental preservation.

Gas insulated transformers using SF6 gas are increasingly adopted for 77kV distribution substations located in densely populated urban areas, in large buildings and at underground sites, because the incombustible property is highly evaluated for safety, fire prevention, and environmental preservation.

Following the development of a 77kV gas insulated transformer in 1979, a 275kV 300MVA liquid cooled gas insulated transformer was developed(1),(2) and put into commercial operation in March, 1991, for the first time in the world.

This paper describes the basic technology for development, the condition of commercial operation for three years, and the results of overload operation.

Gaseous Dielectrics VII, Edited by L.G. Christophorou and D.R. James, Plenum Press, New York, 1994

CONCEPT OF LIQUID COOLED GAS INSULATED TRANSFORMER

Gases have much smaller thermal conductivities and heat capacities than liquids so the use of liquid coolant becomes necessary for a gas insulated transformer of several tens MVA or more. The selection of suitable coolant is based on cost and electrical, physical and chemical properties. The liquid coolant used at present is perfluorocarbon(C8F16O). This coolant has important advantages as far as dielectric strength, thermal and chemical stability, incombustibility and safe properties are concerned.

Fig.1 shows the internal construction of a liquid cooled gas insulated transformer (shell type). For cooling, coolant is supplied from the distributor to the coil and core as the heating unit through the route shown by dotted line, and SF6 gas and perfluorocarbon vapor are used for insulation.

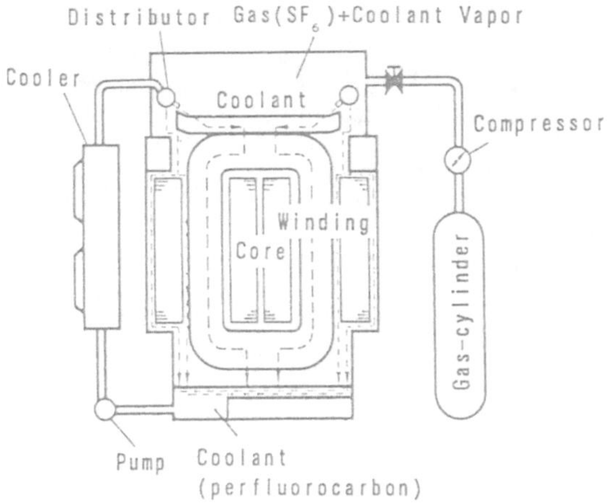

Fig.1 Internal construction of liquid cooled gas insulated transformer

INSULATION TECHNOLOGY

In comparison with oil, insulation in the mixture of SF6 gas and perfluorocarbon is featured by the following points;

- Dielectric constant of oil is about 2.3 while that of gas is 1.0. With a gas insulated transformer, therefore the electric field strength in the gas which corresponds to oil in a oil immersed transformer is relatively higher and insulation strength is primarily dependent on the electric field strength in gas.
- The ratio of insulation strength to lightning impulse and commercial frequency, i.e. impulse ratio is smaller than of oil immersed insulation. For gas insulated transformers, therefore, insulation to lightning impulse is considered more important.
- For that reason it is very important to reduce the electric field in the gaseous region. Fig.2 shows an example of the calculated electric field in a 275kV class gas insulated structure.
- Under a fixed temperature, insulation strength of gas is functional to pressure, and is higher where pressure is higher.
- Mixture of SF6 gas and perfluorocarbon vapor has better insulation strength than pure SF6 gas.

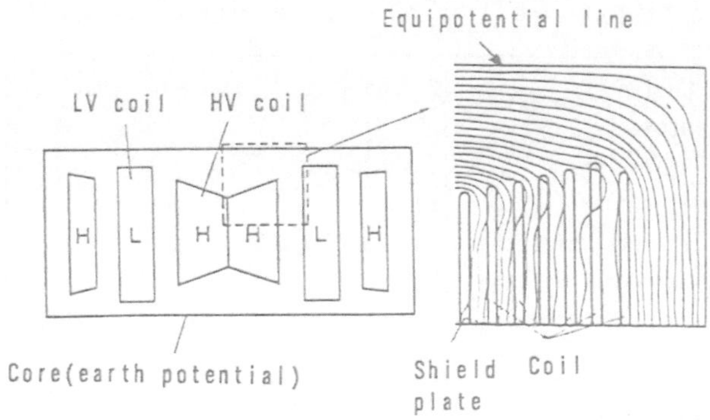

Fig .2 Distribution of Electric field in transformer coils

COOLING TECHNOLOGY

The perfluorocarbon liquid coolant has a lower viscosity than a transformer oil and so it is desirable to design a new cooling structure to make the best use of this characteristic. As perfluorocarbon is more expensive than oil at present an effective cooling system which minimizes the quantity of perfluorocarbon is required.
In order to obtain an optimum duct structure for cooling, coil models were designed using the actual transformer dimensions to allow observation of coolant flow distribution on a coil surface through acrylic covers.
To prove the coil cooling structure, temperature distribution to each part of the coil was measured by a test flowed 100% of rated current in the coil. As shown in Fig.3, the temperature difference between the highest and the average is only about 5% of the average temperature, which indicates that the cooling performance is good enough.

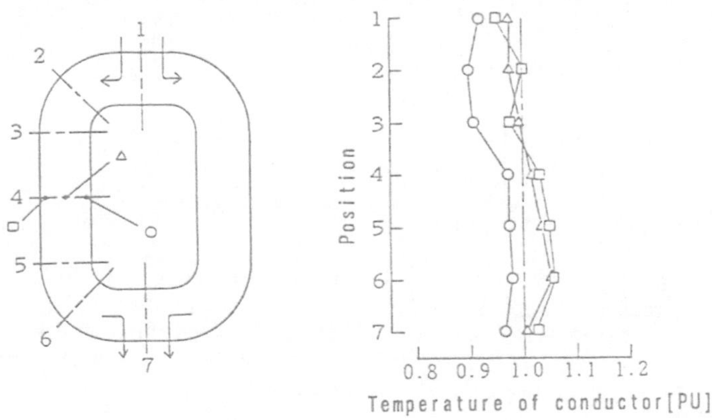

Fig .3 Temperature distribution in a coil

OPERATING CONDITION OF TRANSFORMERS IN THE FIELD

After proving the overall performance by the prototype, the liquid cooled gas insulated transformer was put into commercial operation in March, 1991, for the first time in the world as 300MVA gas insulated transformer in Hirakata substation, The Kansai Electric Power Company. Table 1 shows the ratings of the transformer and Fig.4 shows the gas insulated transformer in operation.
It has been operated continuously without trouble for nearly three years now in favorable condition of load (average:35%～45%, maximum:64%), ambient temperature(-4～37℃), liquid coolant temperature(19～50℃), gas pressure(0.1～0.2MPa), tap changer operation (about 17000 times), and operation of the auxiliaries.

TABLE1 Rating of actual transformer

Cooling	Forced-liquid, forced-air cooling
Capacity	300 MVA
Phase	3 phase
Frequency	60 Hz
Primary voltage	262.5±25 kV
Secondary voltage	77 kV
on-load tap changer	Vacuum switch type with implanted
% IZ	18 %
Pressure	0.1~0.2 MPa
	Pressure regulated type

Fig.4 External view of 275kV 300MVA transformer

RESULT OF OVERLOAD OPERATION IN THE FIELD

In a development process, overload performance of the gas insulated transformer was confirmed in factory by equivalent test and computer simulation. But checking overload performance by applying actual voltage and current of the power system must be a great help to improve the design of the transformer and achieve cost reduction. For this reason, overload operation test was performed.

(1)Operating method

In Hirakata substation where the gas insulated transformer was installed, 300MVA oil immersed transformer shares the load with the gas insulated transformer. The load of the oil immersed transformer was transferred to the gas insulated transformer to make overload condition.

The test was practiced on April, 1993. As the transformer was connected to the commercial transmission system, load of the gas insulated transformer fluctuated. But overloading was kept for 3 hours and 8 minutes with the maximum load of 126% and the average load of 113% to the rating of the transformer. Fig.5 shows the measured values(load, ambient temperature and liquid temperature) of the operation. During the test, gaschromatograph analysis of decomposition gas products was also practiced in the shortest interval for earlier detection of abnormalities. It was confirmed that there was no local overheat in the transformer.

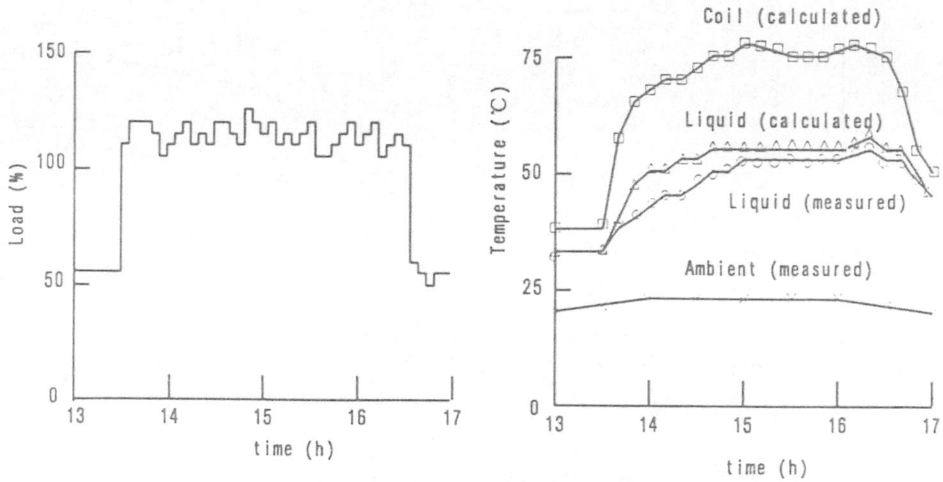

Fig.5 Results of overload operation

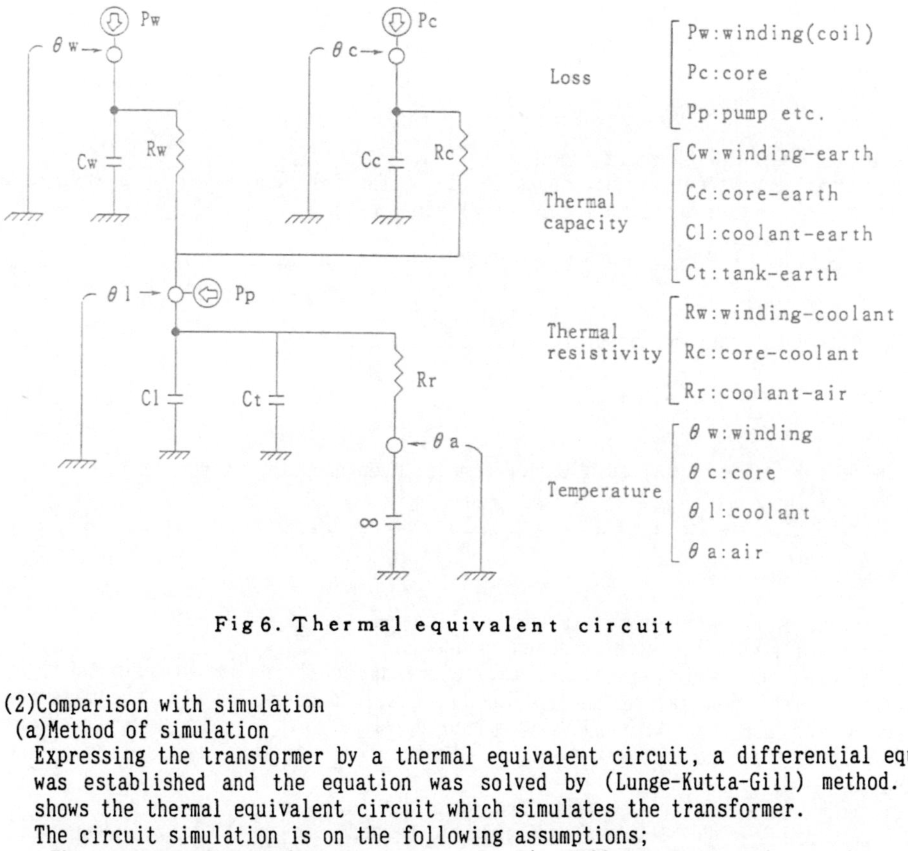

Fig6. Thermal equivalent circuit

(2)Comparison with simulation

(a)Method of simulation

Expressing the transformer by a thermal equivalent circuit, a differential equation was established and the equation was solved by (Lunge-Kutta-Gill) method. Fig.6 shows the thermal equivalent circuit which simulates the transformer.

The circuit simulation is on the following assumptions;

- The core is taken uniform as the loss doesn't differ much by each position.
- It is regarded as insulated between the coil and the core because the spaces are filled with PB (pressboard) and SF6 gas and heat can hardly transferred.
- Coil loss is considered to change according to temperature.
- Thermal capacity is to remain unchanged regardless of temperature characteristics.

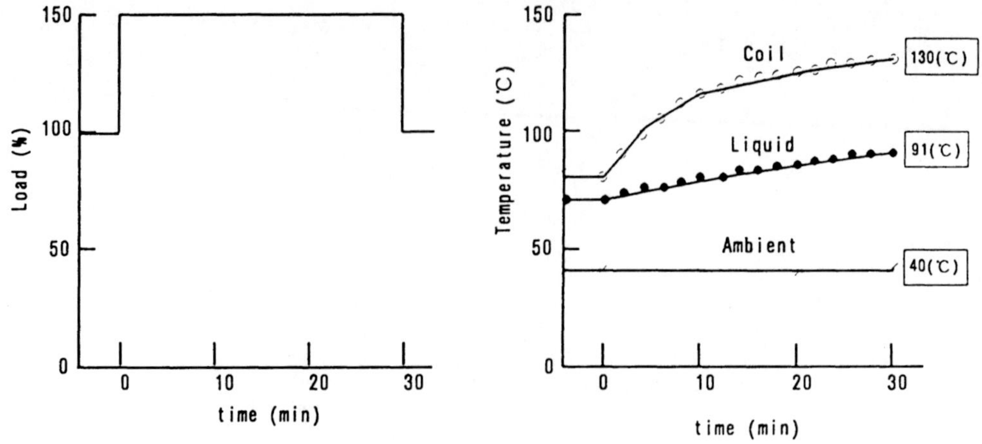

Fig.7 Results of simulation on the most severe overload condition

(b)Equation of thermal equivalent circuit

To the thermal equivalent circuit shown in Fig.6, the following equation expresses;

·With regard to loss and temperature of winding(coil):

$$C_w \frac{d(\theta_w)}{dt} + \frac{\theta_w - \theta_L}{R_w} = P_w \qquad \text{-------------------------------- (1)}$$

·With regard to loss and temperature of core:

$$C_c \frac{d(\theta_c)}{dt} + \frac{\theta_c - \theta_L}{R_c} = P_c \qquad \text{-------------------------------- (2)}$$

·With regard to loss and liquid temperature of pump and other structures:

$$C_{-} + C_{t} \frac{d(\theta_L)}{dt} + \frac{\theta_L - \theta_a}{R_r} = P_p + \frac{\theta_w - \theta_L}{R_w} + \frac{\theta_c - \theta_{-}}{R_c} \qquad \text{---------------------- (3)}$$

The equations of above(1)to(3)are solved by "Lunge-Kutta-Gill" method.

(c)Comparison of simulation with measured values

Fig.5 compares the result of liquid temperature and coil temperature simulated by using the ambient temperature and the load at overload operation with the measured values by the monitor. Although the measured values of liquid temperature are somewhat lower than the calculated values, the results are almost as expected.

(3)Evaluation of field overload operation

Liquid temperature and coil temperature were simulated on the assumption of the most severe overload condition of 40℃ ambient temperature and 150% overload for 30 minutes. The results are shown in Fig.7.

The reached maximum temperatures are;

Liquid temp.:91℃(Assured limit:95℃) Coil temp.:130℃(Assured limit:165℃)

It turned out that there were still margins to the assured limits.

By these evaluation it was confirmed that the performance of the gas insulated transformer was equal to that of a conventional oil immersed transformer even under severe thermal condition such as overload operation.

MERITS OF A GAS INSULATED TRANSFORMER

It is proved by the records of actual operation that the high voltage large capacity liquid cooled gas insulated transformer has equal performance as that of the oil immersed transformer.

Though the gas insulated transformer is more expensive than the oil immersed transformer, it has following advantages.

- Total weight is low.(about 90%)
- Noise is low.(about 2~3dB lower)
- Fire preventive devices such as fire extinguisher are not needed at all or can be simplified.

Thanks to these advantages, we can reduce the volume of underground substation considerably (see Fig.8). Thus, the construction cost can be reduced.

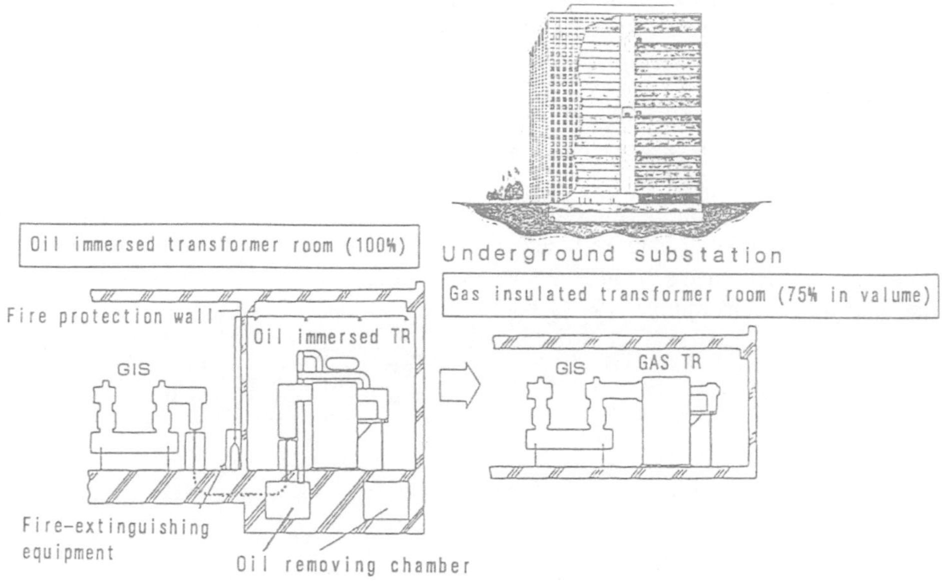

Fig.8 Underground substation design

CONCLUSIONS

The world first 300MVA SF6 gas insulated transformer, developed successfully, has been operating continuously without trouble for nearly three years including three hours of overload operation up to 126% of rated capacity. As a result of actual operation it was confirmed that the performance of the gas insulated transformer was equal to that of a conventional oil immersed transformer even under severe thermal condition.

Demand for gas insulated transformers is now increasing greatly because of the incombustible property and down-sizing effect to underground substation. We continue to make research to improve the design of a gas insulated transformer and achieve future cost reduction. Development of a 500kV gas insulated transformer is also under way.

REFERENCES

1. Y.Harumoto, Y.Kabayama, Y.Yoshida, Y.Kuroda, H.Kan, Y.Miura, E.Tamaki, T.Hakata, "Development of 275kV EHV Class Gas-Insulated Power Transformer", IEEE Trans. PAS, Vol. PAS-104, No.9, 2501-2508, September 1985
2. T.Hasegawa, K.Betsui, S.Ohnishi, K.Itoh, T.Kouan, Y.Makino, H.Nakazawa, "Development of 275kV Liquid Cooling Type Gas-Insulated Transformer", The Transactions of the Institute of Electrical Engineers of Japan Vol.110-B, No.12, 999-1005, December 1990

DISCUSSION

Y. QIU: Just two comments. First, the technical term "liquid cooled" is somewhat ambiguous because it can also mean "liquid immersed". Second, the cooling method used in your transformer makes the gas pressure, and hence the dielectric strength of the gas insulation, greatly dependent on the load. Therefore, the worst situation is to energize the transformer when it is "cold".

H. NAKAZAWA: The technical term "liquid cooled" means the cooling system in which the perfluorocarbon liquid flows down the duct, cooling the coil and core. The term might be changed to describe the right situation. The insulation design of the transformer was conducted taking the dielectric properties of SF_6 alone into consideration. The tests for the transformer were performed at the worst condition in which any contribution was not expected to the dielectric performance and showed the satisfactory results.

A. WASSERSTROM: If a mixture of SF_6 and perfluorocarbon enhances the insulating properties of pure SF_6, then why is this mixture not used in GIS?

H. NAKAZAWA: Mixing perfluorocarbon can increase the dielectric strength as shown in the paper. However, it is more practical to add SF_6 to improve the dielectric strength of GIS because it eliminates the necessity to worry about the peculiar behavior in pressure dependent on the temperature.

SECTION 11: RELIABILITY TESTING OF GIS

CALCULATION OF RESONANT FREQUENCIES IN GIS FOR UHF PARTIAL DISCHARGE DETECTION

R Kurrer,[1] K Feser,[1] and I Herbst[2]

[1] Institute of Power Transmission and High Voltage Technology
University of Stuttgart, Germany
[2] ABB High Voltage Technology, Switzerland

ABSTRACT

One of the different measurement techniques for partial discharges (p.d.) in gas insulated substations (GIS) works in the frequency range above 300 MHz called Ultra-High-Frequency (UHF) range. This paper gives a review of the GIS-relevant part of the theory of electromagnetic wave propagation and modes in this UHF-range. For the application of the UHF-narrowband method in practice typical UHF-p d.-spectra are explained. In addition to this review new computations of the reflection of waves at spacers are presented. The corresponding attenuation for different wave modes is discussed Practical conclusions are drawn from these new results

1. INTRODUCTION

The measurement of partial discharges (p d.) in GIS is an important element for quality assurance and diagnostics Practical experience with one measurement technique for p.d. in GIS at UHF has been reported[1] Fig 1 shows typical results of p.d. signals and background noise. Fig. 1a shows an example of background noise created by radio and television transmitters. As antenna a flat round coupler electrode positioned in the flange opening of a GIS is used. A spectrum analyser in combination with a preamplifier measures the frequency contents of the picked up signal The p.d source is a 25 mm needle on the inner conductor of a GIS for 362 kV rated voltage. The experimental observations (fig. 1b) show that electromagnetic waves with frequencies up to 2 GHz are excited. Peaks appear at definite frequencies in the spectrum when a p.d is detected with the UHF-narrowband method. Groups of frequency bands are recognisable and for large p d values the spectrum becomes very dense. For simplicity the UHF-narrowband method will be called "UHF-method", as it is introduced in the literature

This paper describes the theoretical background of the high frequency electromagnetic waves in GIS to evaluate and understand the measuring results with the UHF-method. Further different attenuation mechanisms of electromagnetic waves in GIS are investigated and discussed. Especially new information considering the reflection of electromagnetic waves at spacers is presented.

Gaseous Dielectrics VII, Edited by L.G. Christophorou
and D.R. James, Plenum Press, New York, 1994

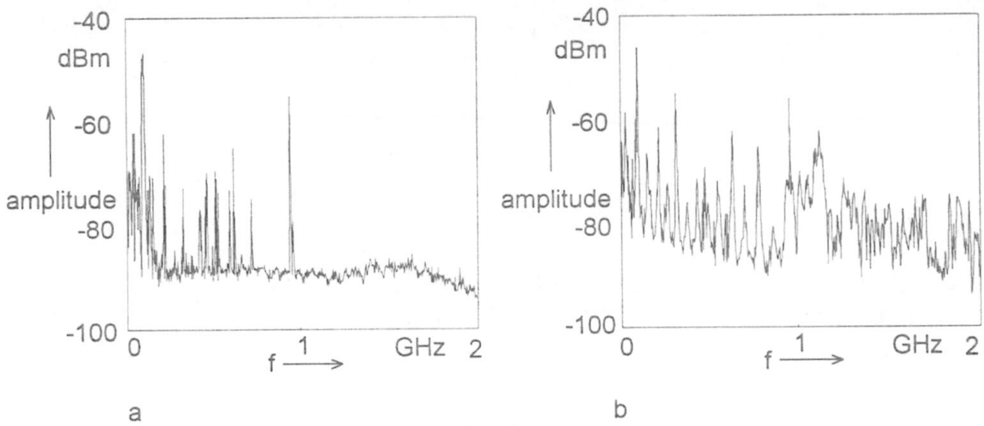

Figure 1. Spectrum of the background noise (a) and spectrum of a p.d. corresponding to 10 pC (b) (10 pC measured according to the IEC 270 method).

2. HIGH FREQUENCY MODEL FOR GIS

The GIS can be modelled as a cylindrical waveguide. The electrical field $\vec{E}$ and the magnetic field $\vec{H}$ are assumed to be periodic functions of the time with the angular frequency ω. For charge free and current free regions the Maxwell's equations result in the wave equations for $\vec{E}$ and $\vec{H}$, where μ and ε are the permeability and dielectric constant of the medium. It is possible to derive two equations (eqs.1, 2) for the longitudinal z-components E_z and H_z of the electromagnetic fields.

$$\nabla^2 E_z = -\omega^2 \mu\varepsilon E_z \tag{1}$$

$$\nabla^2 H_z = -\omega^2 \mu\varepsilon H_z \tag{2}$$

$$E_z = \left[A_E \cdot J_m(k_{mn} \cdot r) + B_E \cdot N_m(k_{mn} \cdot r)\right] \cdot e^{j(\omega t - m\varphi - \beta z)} \tag{3}$$

$$H_z = \left[A_H \cdot J_m(k_{mn} \cdot r) + B_H \cdot N_m(k_{mn} \cdot r)\right] \cdot e^{j(\omega t - m\varphi - \beta z)} \tag{4}$$

$$k_{mn}{}^2 = \omega^2 \mu\varepsilon - \beta^2 \qquad m = 0,1,2\ldots \quad n = 1,2\ldots \tag{5}$$

The solutions (eqs.3, 4) of these equations (eqs.1, 2) in cylindrical coordinates are well known and consist of the mth-order Bessel function $J_m(k_{mn} \cdot r)$ and the mth-order Neumann function $N_m(k_{mn} \cdot r)$ where n defines the order of the zeros of this functions. A_E, B_E and A_H, B_H are indefinite coefficients. The modal eigenvalue constant k_{mn} (eqs.5) has to be determined by the evaluation of the boundary conditions. Three different types of wave modes are distinguished. The TM_{mn}-mode ($H_z = 0$), the TE_{mn}-mode ($E_z = 0$) and the TEM-mode ($H_z = 0$, $E_z = 0$).

2.1 WAVE MODES IN GIS

If β of eqs.(5) is real the electromagnetic field describes a wave which can propagate in a GIS bus bar. Complex values of β result in electromagnetic waves which are attenuated, even if an ideal conducting waveguide without losses is assumed. $\beta = 0$ defines the mode-characteristic cut-off frequency f_c(m,n) below which the corresponding mode cannot exist.

$$f_c(m,n) = \frac{k_{mn}}{2\pi\sqrt{\mu\varepsilon}} \qquad (6)$$

A model for a GIS with an inner conductor is the coaxial waveguide shown in fig. 2a.

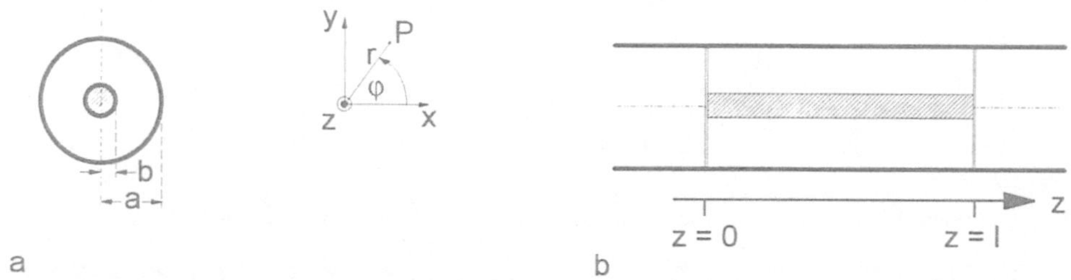

Figure 2. (a) Cross-section of a GIS (a: radius of the outer conductor b. radius of the inner conductor). (b) GIS bus bar terminated with metallic walls

For the TM-mode the field component E_z must vanish at the inner conductor (r = b) and the outer conductor (r = a). To get non trivial solutions the determinant in eqs.(7) must be zero. This is the characteristic function of the TM_{mn}-mode propagating in a coaxial GIS.

$$\begin{matrix} A_E \cdot J_m(k_{mn} \cdot a) + B_E \cdot N_m(k_{mn} \cdot a) = 0 \\ A_E \cdot J_m(k_{mn} \cdot b) + B_E \cdot N_m(k_{mn} \cdot b) = 0 \end{matrix} \quad \begin{vmatrix} J_m(k_{mn} \cdot a) & N_m(k_{mn} \cdot a) \\ J_m(k_{mn} \cdot b) & N_m(k_{mn} \cdot b) \end{vmatrix} = 0 \qquad (7)$$

The modal eigenvalue constant k_{mn} can be determined by calculating the roots of eqs.(7). The cut-off frequency is then derived from eqs.(6). For the TE-mode the characteristic function eqs.(8) must be satisfied $J'_m(k_{mn} \cdot r)$ and $N'_m(k_{mn} \cdot r)$ are the derivates of the mth-order Bessel function and the mth-order Neumann function

$$\begin{vmatrix} J'_m(k_{mn} \cdot a) & N'_m(k_{mn} \cdot a) \\ J'_m(k_{mn} \cdot b) & N'_m(k_{mn} \cdot b) \end{vmatrix} = 0 \qquad (8)$$

Both TE and TM-modes can not only exist in coaxial waveguides discussed above but also in systems without an inner conductor. Contrary to this the TEM-mode, having longitudinal components E_z and H_z equal to zero, can only exist in a waveguide with an inner conductor. As the wave number β is real for all frequencies no cut-off frequency of the TEM-mode exists. Therefore all frequencies from dc upwards will propagate in the GIS.

2.2 CALCULATION OF CUT-OFF FREQUENCIES IN GIS

The characteristic equations (eqs.7, 8) for the TE and TM-modes can be solved numerically. Fig. 3 shows the cut-off frequencies of a coaxial GIS bus bar for 362 kV rated voltage. The cut-off frequency of the TEM-mode is 0 Hz and therefore not shown. The calculations show that higher TE and TM wave modes can propagate in GIS bus bars. Therefore the travelling wave theory of the TEM-mode alone is not able to describe the theoretical background at frequencies above 300 MHz.

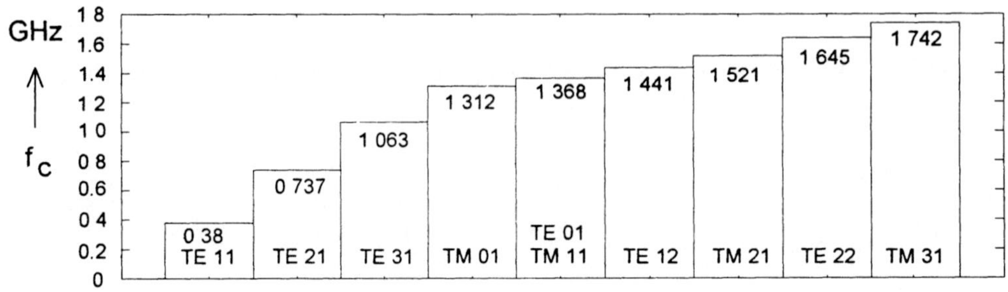

Figure 3. Cut-off frequencies for a 362 kV GIS bus bar (a = 0.185 m, b = 0.072 m).

2.3 RESONANCES IN GIS

The limitation of the consideration discussed up to now is that the cut-off frequency $f_c(m,n)$ of a mode delivers only a limit for the frequency above which the mode can propagate, but not the frequency at which the mode is predominant in a GIS. The measured UHF-spectrum is only described accurately when reflections and resonance phenomena are additionally taken into account. Electromagnetic waves which propagate in a GIS are reflected at metallic end covers, insulators and other discontinuities. Fig. 2b shows a GIS bus bar terminated with metallic walls on both sides The superposition of the incident wave and the reflected wave results in a standing wave The solution of the standing wave is given in eqs. (9, 10).

$$E_z = \left[A_E \, J_m(k_{mn} \cdot r) + B_E \cdot N_m(k_{mn} \cdot r)\right] \cdot e^{j(\omega t - m\varphi)} \cdot \cos\beta z \tag{9}$$

$$H_z = \left[A_H \, J_m(k_{mn} \cdot r) + B_H \cdot N_m(k_{mn} \cdot r)\right] \cdot e^{j(\omega t - m\varphi)} \cdot \sin\beta z \tag{10}$$

The solution fulfils the boundary conditions at the point z = 0 and z = l if the wave number β fulfils eqs.(11). It is possible to describe a GIS chamber with the length l as a cylindrical resonant cavity. The resonant frequency f_{mnp} is determined in eqs (12).

$$\beta = p \cdot \frac{\pi}{l} \qquad p = 1, 2, \ldots \tag{11}$$

$$f_{mnp} = \frac{1}{2\pi\sqrt{\mu\varepsilon}} \cdot \sqrt{k_{mn}^2 + \left(p \cdot \frac{\pi}{l}\right)^2} \qquad p = 1, 2, \ldots. \tag{12}$$

Another form of eqs (12) is given in eqs (13) where v is the velocity of light in the medium. It is possible to express the resonant frequencies of a GIS chamber with the cut-off frequencies $f_c(m,n)$ (eqs.6) of the GIS bus bar with the same cross-section.

$$f_{\mathrm{mnp}} = \sqrt{f_c^2(m,n) + \left(\frac{p \cdot v}{2 \cdot l}\right)^2} \qquad \begin{array}{l} m = 0,1,2,\ldots \\ n = 1,2,\ldots \\ p = 1,2,\ldots \end{array} \tag{13}$$

With eqs.(13) it is possible to describe an idealised spectrum. A mode with the cut-off frequency f_c(m,n) causes resonances in the GIS. The resonant frequencies of the TEM-mode appear at multiples of the first resonance because the cut-off frequency f_c is equal to 0 Hz. The TEM-mode is therefore contained in the complete frequency range of the spectrum. For n and m assumed to be constant the variation of the index p results in resonant frequencies of the TE and TM-mode which differ only 5-10 MHz from each other Therefore groups of frequency peaks near the cut-off frequency f_c(m,n) of a mode can be measured. The numerical calculation of eqs (13) for a resonant GIS cavity is shown in fig. 4. The calculation is carried out for an index p up to 3 This results in the formation of well distinguished groups of frequency bands of the TE and TM-modes The TEM-mode is shown up to the third resonance of 128 MHz For higher values of p the resonance spectrum becomes dense. This might explain the measurement results during the practical application of the UHF-method. At low p d values only a few resonances are excited and can be recognised in the measured spectrum as resonance peaks For higher p d discharge values further modes with a another index p are excited The measured spectrum is more dense and continuous.

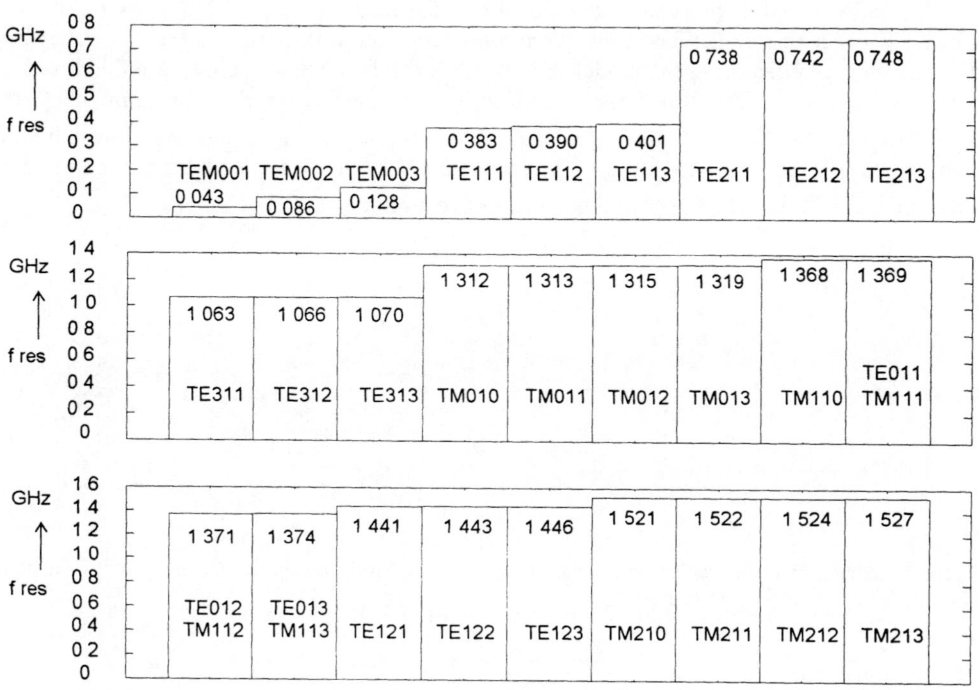

Figure 4. Resonant frequencies for a GIS cut (a = 0 185 m, b = 0 072 m, length l = 3.5 m, index p ≤ 3).

To understand the principles of the VHF and UHF-wave propagation simple models are useful. The theory of the TEM-mode alone is able to describe resonance phenomena up to 300 MHz in a GIS if a network of transmission lines and lumped elements is used as a first approach. In[2] it is shown that the simulation of a GIS as a network of transmission lines reproduces a transfer function from the p.d.-source to the p.d.-sensor of the GIS which shows the characteristic resonances. The addition of higher TE and TM-modes leads to a complete theory in the UHF-range. For both TE and TM-modes, and for TEM-modes, the transfer function will show a resonant structure.

3. ATTENUATION OF ELECTROMAGNETIC WAVES IN GIS

The waves, which propagate in GIS are attenuated because of the power dissipation in the metallic enclosure Further the reflection of waves at dielectric spacers and discontinuities result in an attenuation.

3.1 Attenuation in the Metallic Enclosure

The attenuation coefficient in the metallic enclosure is approximately calculated with the power loss method The results are known and presented in[3]. The theoretical investigations result in an attenuation coefficient of 2 dB/Km. On the other hand values around 1-2 dB/m are reported[1] The attenuation in the enclosure alone is certainly not sufficient to explain this measurement result

3.2 Attenuation by Reflections at a GIS Spacer

To study the effect of a spacer the model of a flat disc is used as a first assumption. For a conical spacer this model has to be taken with care. The reflection at a lossless dielectric disc can be computed using the model of a transmission line where sections are filled with a dielectric (ε_{r2}, μ_{r2}). The transmission and reflection coefficients of a dielectric disc are discussed in[4]. Fig. 5 shows the calculated reflection coefficients r_{TEM} and transmission coefficients t_{TEM} for the TEM-mode. The calculation for this spacer model is carried out for a dielectric disc with a thickness of 5 cm and a dielectric constant ε_{r2} equal to 6.5.

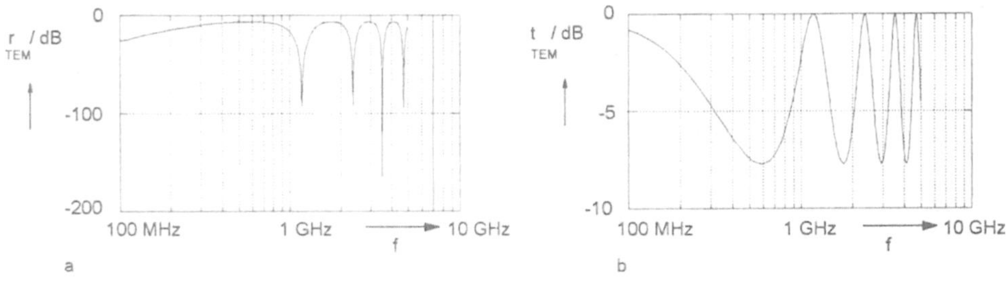

Figure 5. Magnitude of the reflection coefficients (a) r_{TEM} and transmission coefficients (b) t_{TEM} of the TEM-mode for a dielectric spacer disc.

To compare the coefficients of the TM and TE-modes the same cut-off frequency is assumed for both modes. Fig 6 shows the calculated reflection and transmission coefficients for the TM and TE-mode with a cut-off frequency f_c of 500 MHz. Because the cut-off frequency is different for each mode there will be in general different results for the reflection and transmission coefficients of the different waves. It is important that for TM-modes a frequency exist near the cut-off frequency where no attenuation occurs thus enabling to excite resonant frequencies near the cut-off frequency shown in eqs.(13) and fig. 4. On the other hand the transmission coefficient of the TE-mode shows a 5-10 dB attenuation just above the cut-off frequency The transmission coefficient of both the TE and TM-mode approaches zero in several frequency bands which means, that for every mode there exists a series of frequencies at which it is not attenuated. At all other frequencies a noticeable attenuation of about 2-3 dB occurs.

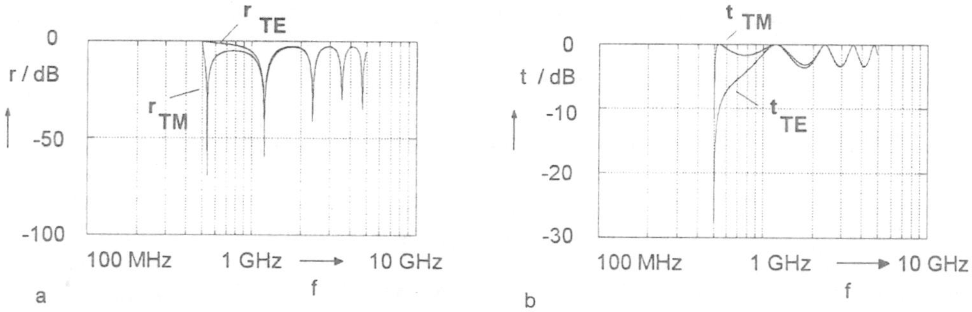

Figure 6. Magnitude of the reflection coefficients (a) r_{TE}, r_{TM} and transmission coefficients (b) t_{TE}, t_{TM} for a dielectric spacer disc.

The TEM-mode is not attenuated like the higher modes for low frequencies below the cut-off frequency of the TE and TM-modes The TEM-mode can be measured over a long distance with several spacers, and will show resonances in all frequency regions as discussed above. TE and TM-modes are able to form resonances between dielectric spacers because the reflection coefficients are high enough near the cut-off frequency. For conical spacers these results will be similar but the resonance spectrum will be more diffused because of the spacer shape.

4. CONCLUSIONS

In the presented paper the theory of high frequency electromagnetic waves is derived and discussed. It is necessary to consider the TEM-mode, the higher TE and TM-modes and resonance phenomena in GIS to describe a spectrum measured with the UHF-method. The attenuation of wave modes at a spacer can now be explained by reflections. The attenuation depends of the wave mode type. For the maximum attenuation coefficient values about 3-10 dB are possible and depend on the frequency. It will be interesting to investigate the location dependence of measurement results when using different UHF-measurement techniques with different frequency resolutions.

5. REFERENCES

1. B.F. Hampton et al., Experience and Progress with UHF Diagnostics in GIS, Paper 15/23-03, Cigre-Paris (1992).
2. I. Herbst, PD-Measurement in GIS - a Theoretical and Experimental Comparison of different Electrical Measurement Techniques, Paper 130-02, Cigre-symp. Berlin (1993).
3. N. Marcuvitz, Waveguide Handbook, McGraw-Hill Book Company Inc , New York (1951).
4. J.A. Stratton, Electromagnetic Theory, McGraw-Hill Book Company Inc.,New York (1941).

DISCUSSION

R. BARTINIKAS: You have demonstrated that multiple reflections from spacers in GIS systems may lead to substantial interpretational difficulties of PD measurements carried out at EHF. Although these problems were recognized in the past, they have not been systematically analyzed so that the work you reported constitutes a very timely contribution on the subject. The difficulties at EHF underline the importance of supplementing the PD measurements at these frequencies with acoustical PD measurements at low frequencies.

R. KURRER: We agree with Dr. Bartnikas' comment.

L. G. CHRISTOPHOROU: How does the frequency dependence of the attenuation coefficient depend on the material?

R. KURRER: For the calculation of the transmission coefficients we assumed that the permittivity of the spacer material does not depend on the frequency. The frequency dependence of the transmission factor has two reasons. First, the field impedance of the TE and TM modes is a function of the frequency. Second, the transmission of the transmission factor by the line depends on the frequency.

M. JUDD: (1) Referring to Fig. 1 of your paper, could you state the partial discharge source used and the size of the UHF coupler? (2) Have you any figure for the loss tangent of the epoxy spacers at UHF frequencies or an estimate of the extra attenuation this may cause?

R. KURRER: (1) The partial discharge source is a needle on the inner conductor of a GIS for 245 kV rated voltage. The length is 25 mm and the diameter is 0.1 mm. The spectrum is measured at a voltage of 200 kV and corresponds to an apparent charge of 10 pC. The size of the coupler is 10 cm in the diameter. (2) We did not take into account the attenuation in the spacer material. In general, the losses will increase at higher frequencies but the reflections will be further dominant for the attenuation at a spacer.

PREDICTION OF SURFACE FLASHOVER VOLTAGE FOR CONDUCTING PARTICLE ATTACHED ON EPOXY INSULATOR IN GAS-INSULATED SWITCHGEAR

A. Ieyasu, [1] M. Yumoto, [2] and M. Hara [3]

[1] Nissin Electric Co.,Ltd.
Maebashi, Japan
[2] Musashi Institute of Technology
Setagaya, Japan
[3] Kyushu University
Fukuoka, Japan

ABSTRACT

Conducting particles mingled in Gas-Insulated Switchgear(GIS) make flashover voltage decrease. In particular, the flashover voltage is severely decreased in case of GIS including the particles with a sharp tip in high pressurized SF_6 gas. It is all the worse for the particles reached epoxy insulators. Some methods for flashover voltage prediction were already proposed. However, in case of the surface flashover, few methods have been proposed.

In the present paper, a new prediction method for the surface flashover voltage with particles attached on an epoxy insulator is proposed. It is revealed through comparision with experimental values that the proposed method is useful to predict the flashover voltage of GIS including conducting particles with a sharp tip.

KEYWORDS

Gas-insulated switchgear(GIS), SF_6 gas, Epoxy insulator, Conducting particle, Flashover voltage, Electric-field

INTRODUCTION

Gas-Insulated Switchgear(GIS) has been extensively used in power systems. The GIS must fulfill the requirements of high insulation reliability. SF_6 gas which is used in the GIS, is dielectrically superior to other gases. However, it is very sensitive

Gaseous Dielectrics VII, Edited by L.G. Christophorou
and D.R. James, Plenum Press, New York, 1994

to local electric-field enhancements caused by conducting particles, etc. Though GIS is protected from contamination of particles in closed vessels, it is worried that some particles may occur by sliding components, etc. In that case, the particles are mingled in the GIS. In some cases, they are reached epoxy insulators. It can not be perfectly avoided in practical systems. The particles mingled in GIS make flashover voltage decrease. In particular, the flashover voltage is severely decreased in case of GIS including the particles with a sharp tip in high pressurized SF_6 gas. The reason is that little corona stabilizing effect is expected and that the electric-field becomes very large around the sharp tip of the particle. It is all the worse for the particles attached on epoxy insulators. The reason is that the surface flashover voltage is usually lower than the gaseous flashover voltage in comparatively high pressurized SF_6 gas. Some methods for flashover voltage prediction were already proposed [1-5]. However, the phenomena of surface flashover are so complicated that few prediction methods have been proposed.

To predict the surface flashover voltage on an epoxy insulator with particles attached, a new prediction method is proposed. Two new characteristic quantities are proposed by authors. One is a factor at a sharp tip of the particle and the other is a criterion judging whether the surface flashover happens or not. They are named "Surface Electric-Field Intensity Factor : κ_t" and "Surface Flashover Criterion : κ_{tBmin}" ,respectively. The particle attached on the insulator is predicted to have no trouble, in case that κ_t at the tip of the particle is less than κ_{tBmin} previously obtained with a simple experiment. The predicted surface flashover voltage is calculated as the critical applied voltage under which $\kappa_t = \kappa_{tBmin}$. In some conditions of particle contamination on epoxy insulators of GIS, the surface flashover voltages are determined by means of the proposed method. The predicted values are compared with the experimental results.

PREDICTION OF FLASHOVER VOLTAGE IN SF_6 GAS

Electric-Field Intensity Factor and Flashover Criterion

An electric-field profile $E(\chi)$ around a sharp tip of a conducting particle is shown as Figure 1, where χ is a distance from the tip. At the sharp tip, the electric-field is infinitely large. The tip has singularity of the electric-field. Therefore, "Electric-Field Intensity Factor : κ" characterizing the severity of the sharp tip is already defined as,

$$\kappa = \lim_{\chi \to 0} E(\chi) \cdot \chi^{\lambda} \tag{1}$$

Where, λ : singularity index of electric-field

by authors [4, 5].

If the index λ of Eqn. (1) is chosen as a suitable value, κ is expected to be a constant value independent of the distance χ from the tip shown as Figure 2. A critical value of the electric-field intensity factor κ is symbolized as κ_B, where the flashover may be occurred if $\kappa \geqq \kappa_B$. This idea is hinted by a method for prediction of crack propagation in fracture mechanics [6]. Here, the flashover voltage V_B depends on gaseous pressure shown as Figure 3. Therefore, "Flashover Criterion : κ_{Bmin}" is defined as κ_B at the gaseous pressure, where the flashover voltage is minimal in the region of high pressurized SF_6 gas.

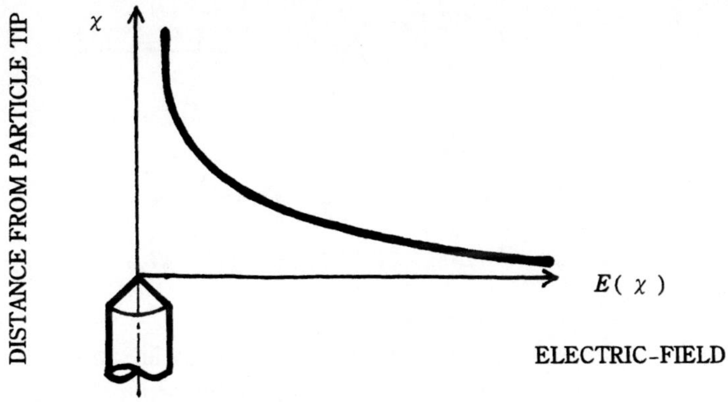

Figure 1. Electric-field around tip of conducting particle.

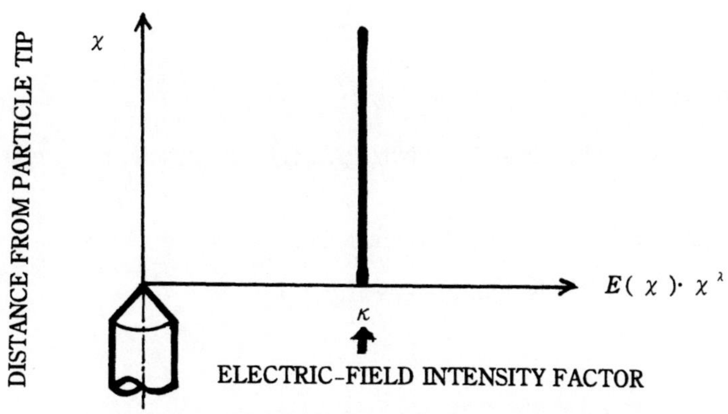

Figure 2. Concept of electric-field intensity factor.

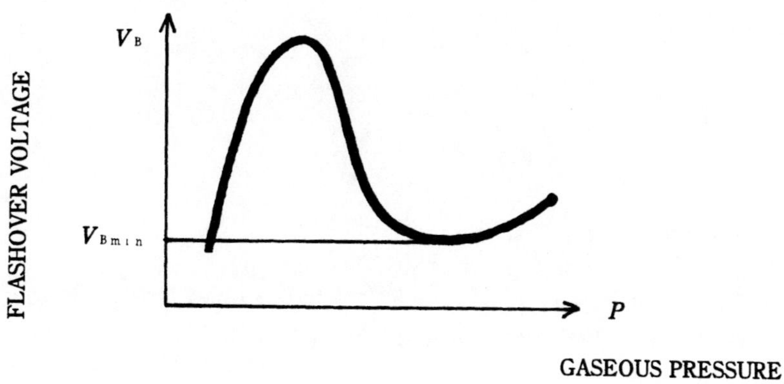

Figure 3. Relationship of flashover voltage and gaseous pressure.

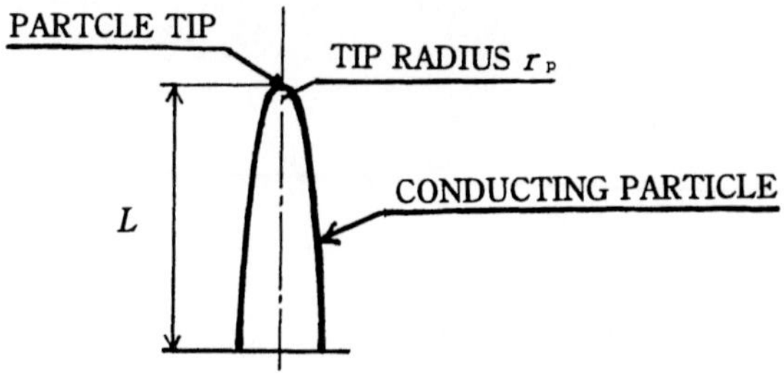

Figure 4. Conducting particle approximately represented prolate spheroid.

Electric-Field Profile and Electric-Field Intensity Factor at Sharp Tip of Conducting Particle

The electric-field profile around the tip of the conducting particle approximately represented prolate spheroid shown as Figure 4 has the formula shown as Eqn.(2) [7].

$$\left.\begin{aligned} E(\chi) &= E_0 \cdot \left[1 - \frac{\ell n \dfrac{L+\chi+C}{L+\chi-C} - \dfrac{2C(L+\chi)}{(L+\chi)^2 - C^2}}{\ell n \dfrac{L+C}{L-C} - \dfrac{2C}{L}} \right] \\ C &= \sqrt{L(L-r_p)} \end{aligned}\right\} \quad (2)$$

where, E_0 : electric-field without particle, L :length of particle, r_p : radius of particle tip

Eqn. (2) is replaced as $E(\chi)=A{\cdot}f(\chi)$ where, $A = E_0 \Big/ \left\{ \ell n \dfrac{L+C}{L-C} - \dfrac{2C}{L} \right\}$.

If r_p is very small, $f(\chi)$ is deformed as Eqn.(3).

$$f(\chi) \approx \frac{L}{\chi} \quad (3)$$

Consequently, if the singularity of the electric-field concerning the particle tip is deleted with $\lambda = 1$, the electric-field intensity factor κ has the formula shown as Eqn.(4).

$$\kappa = \lim_{\chi \to 0} E_0 \cdot \left[1 - \frac{\ell n \dfrac{L+\chi+C}{L+\chi-C} - \dfrac{2C(L+\chi)}{(L+\chi)^2 - C^2}}{\ell n \dfrac{L+C}{L-C} - \dfrac{2C}{L}} \right] \cdot \chi \quad (4)$$

Calculation of Flashover Criterion

"Flashover Criterion : κ_{Bmin}" is calculated using experimental data by Cookson, et al. [8], on condition that the conducting particle shown as Figure 5 was attached on the parallel plane electrode shown as Figure 6. The electric-field intensity factors κ_B in the instant of flashover are calculated as a function of gaseous pressure. The factors κ_B are calculated about four kinds of particle length. They are shown as Figure 7. The minimum values of κ_B are almost independent of the length of conducting particles. This value is taken as the flashover criterion κ_{Bmin}.

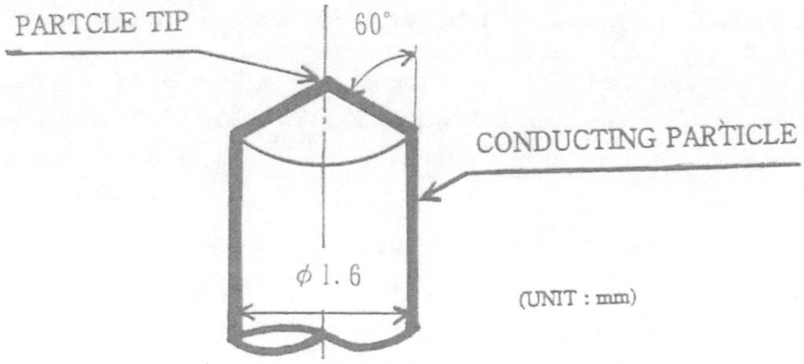

Figure 5. Conducting particle with sharp tip.

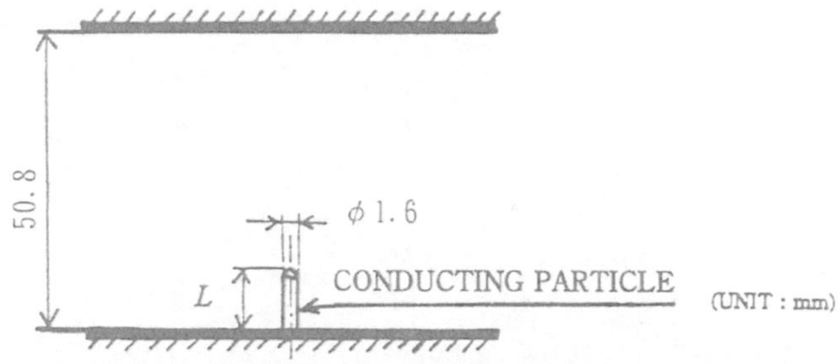

Figure 6. Parallel plane electrode with conducting particle.

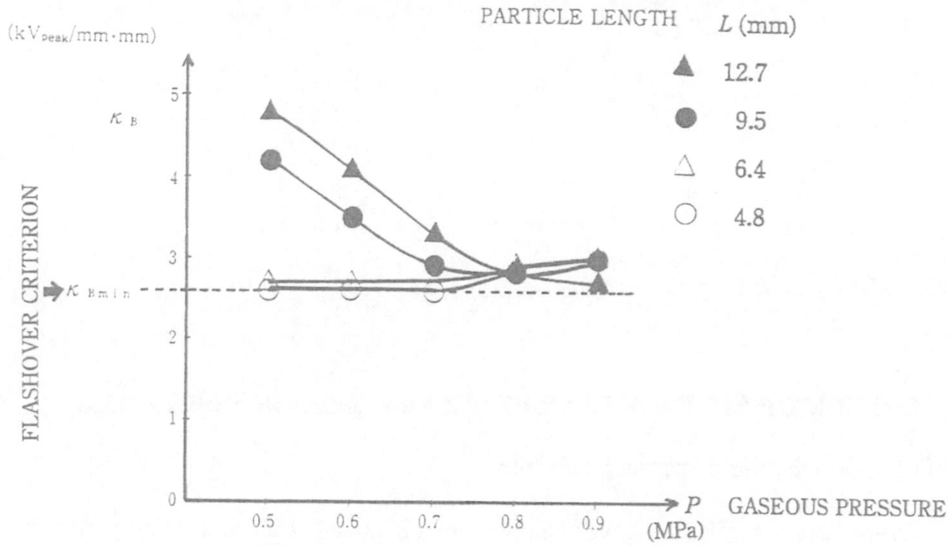

Figure 7. Flashover criterion κ_{Bmin}.

Prediction of Flashover Voltage in SF_6 Gas

The flashover voltage in SF_6 gas including a conducting particle is predicted under application of A.C. voltage using the method proposed by authors. It is calculated as the critical applied voltage under which $\kappa = \kappa_{Bmin}$. The flashover voltages are calculated about some kinds of conducting particles. They are compared with experimental values quoted from references [8-10]. They are listed as Table 1.

Table 1. Comparison of predicted and experimental flashover voltages for SF_6 gas.

Conducting particle				Predicted	Experimental	Ref.number of
Material	Diameter (mm)	Condition	Length (mm)	value of F.O.V. (kVr.m.s.)	value of F.O.V. (kVr.m.s.)	experimental value
Stainless	1.6	fixed	3.2	211	217	
−steel			6.4	115	123	
						8
			3.2	211	167	
Copper	0.43	fixed	6.4	115	125	
			12.7	62	83	
			3.2	176	141	
Copper	0.4	free	6.4	96	94	
			12.7	52	52	
			3.2	176	142	
Copper	0.1	free	6.4	96	78	9
			12.7	52	69	
			3.2	206	320	
Copper	0.4	free	6.4	113	223	
			12.7	61	139	
Aluminium	0.45	fixed	6.4	87	114	
						10
Aluminium	0.45	0.1mm from conductor	6.4	87	110	

PREDICTION OF SURFACE FLASHOVER VOLTAGE ON INSULATOR

Surface Electric-Field Intensity Factor

Prediction of surface flashover voltage on an insulator with conducting particles attached needs to include the effects of (I) and (II) in the prediction method of flashover volage in SF_6 gas.

(I) The distortion of the surface electric-field profile around the tip of the particle attached on an insulator depends on the surface position of the particle. The relationship between the enhancement of electric-field concentrations at the particle tip and the enhancement of surface electric-fields should be included in the prediction process.

(II)The dependence of the surface flashover voltage with particles attached on the size of the insulator and on the flashover length along the insulator surface should be included in the prediction process.

The effects of (I) and (II) are represented as the dielectric rate factor ε and the surface flashover length factor β, respectively. These factors are included in Eqn. (4). The new characteristic quantity is named "Surface Electric-Field Intensity Factor: κ_t", which has the formula shown as Eqn.(5).

$$\kappa_t = \left\{ \lim_{\chi \to 0} \varepsilon \cdot E_{to} \left[1 - \frac{\ln \dfrac{L+\chi+C}{L+\chi-C} - \dfrac{2C(L+\chi)}{(L+\chi)^2 - C^2}}{\ln \dfrac{L+C}{L-C} - \dfrac{2C}{L}} \right] \cdot \chi \right\} \cdot \beta \qquad (5)$$

where, E_{to}: surface electric-field without particle

Calculation of Dielectric Rate Factor. The dielectric rate factor ε is calculated about two kinds of insulators shown as Figure 8 with conducting particles attached. The surface electric-fields without particles and the flashover voltages with a particle attached at the position of Ⓐ or Ⓑ are shown as Figure 9 and Figure 10, respectively[11, 12]. Here, distortions caused by difference of the dielectric rates between SF_6 gas and the insulator are considered. The distortion must be very little in case of the particle attached at the position Ⓑ. The flashover voltage $V_{BⒶ}$ with the particle attached at the position Ⓐ is calculated by multiplying the flashover voltage $V_{BⒷ}$ with the particle attached at the position Ⓑ by a reciprocal of the ratio of the surface electric-fields shown as Figure 9. The calculated flashover voltages $\widetilde{V}_{BⒶ}$ at the position Ⓐ are also included in Figure 10 which shows the experimental flashover voltages $V_{BⒶ}$ about some kinds of voltage waveforms. The dielectric rate factors ε are obtained through the comparison of $\widetilde{V}_{BⒶ}$ with $V_{BⒶ}$. They are listed as Table 2. All of the dielectric rate factors ε are less than 1. Therefore, the enhancement of electric-field concentration at the particle tip is less than the enhancement of surface electric-fields. $\varepsilon < 1$ is at least applicable to these two kinds of insulators examined.

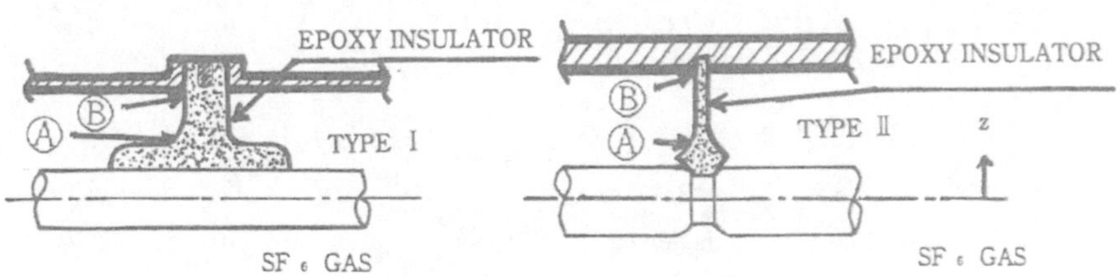

Figure 8. Two kinds of insulators.

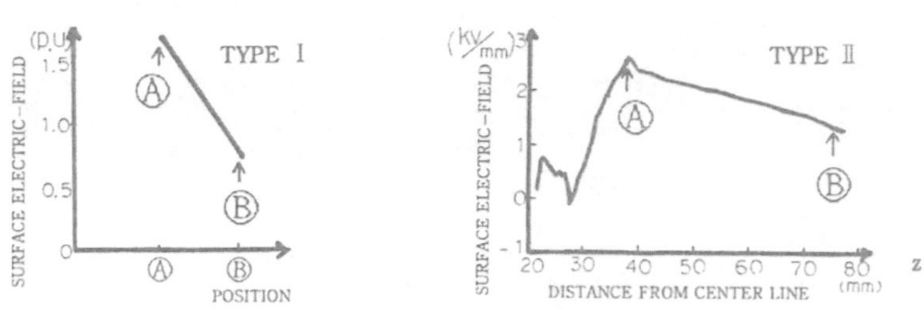

Figure 9. Surface electric-field strength.

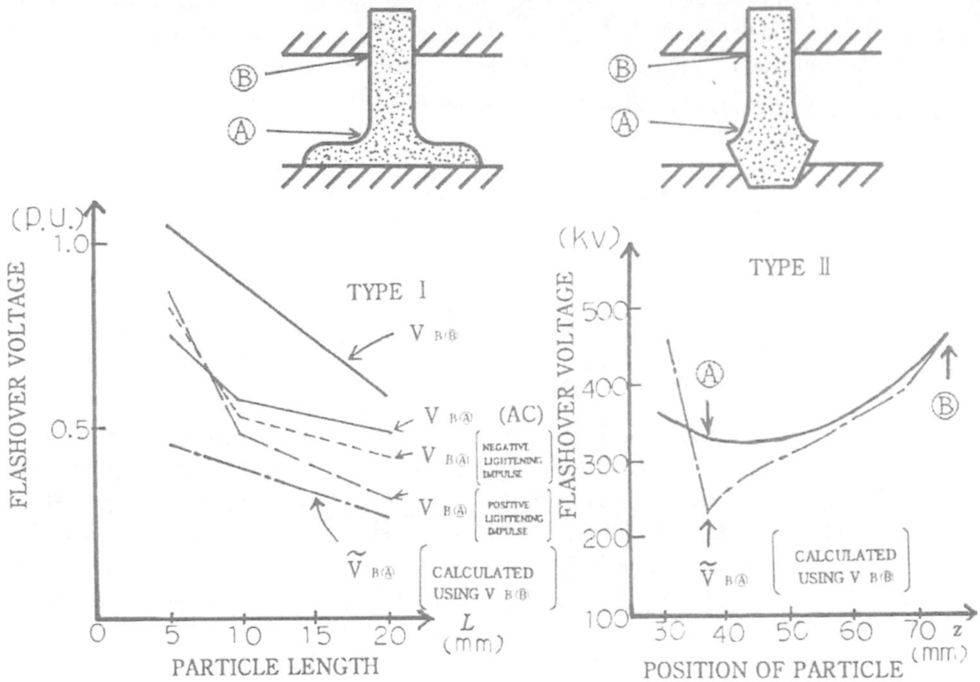

Figure 10. Experimental and predicted flashover voltage.

Table 2. Dielectric rate factors.

Insulator type	Gaseous pressure (MPa)	Applied voltage waveform	Conducting particle			Dielectric rate factor
			Material	Diameter (mm)	Length (mm)	ε
I	0.45	AC	Copper	0.2	5	0.61
					10	0.68
					20	0.53
		Negative lightening impulse	Copper	0.2	5	0.56
					10	0.74
					20	0.60
		Positive lightening impulse	Copper	0.2	5	0.53
					10	0.81
					20	0.84
II	0.3	Lightening impulse	Aluminium	0.5	5	0.61

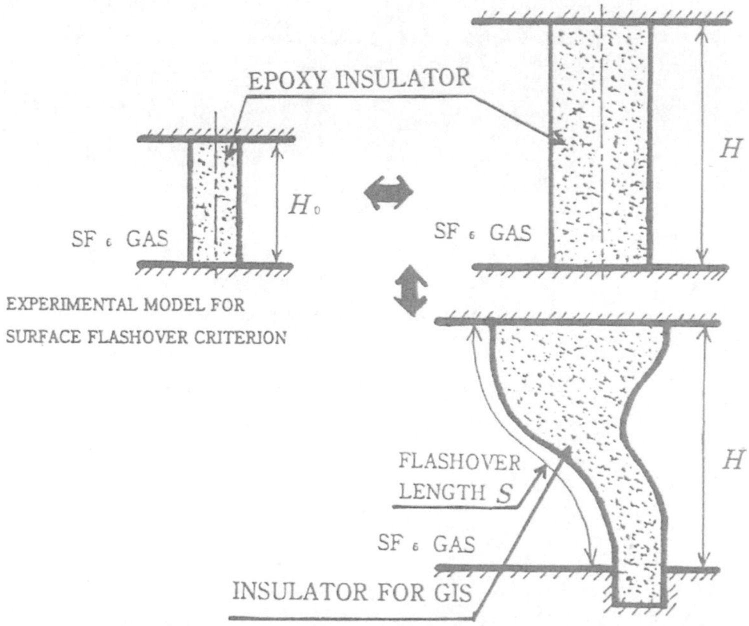

Figure 11. Distance between electrodes and flashover length along surface.

Calculation of Surface Flashover Length Factor. The flashover voltage is considerably proportional to the flashover length S along surface [13] in case of the same distance H between electrodes. In general, the distance H of the insulator for GIS differs from the distance H_0 of the simple experimental model for the surface flashover criterion, which is shown as Figure 11. The model that has the distance H equal to the insulator for GIS and that is similar to the simple experimental model for the criterion is imaged. The increase rate of the surface flashover voltage is proportional to 0.8 power of the size of insulators [11]. Consequently, the factor γ which includes the effect that the surface flashover is easy to occur according to the size of insulators, even if the electric-field strength is equal, is shown as Eqn.(6).

$$\gamma = \frac{(H/H_0)^{0.8}}{H/H_0} = \left(\frac{H}{H_0}\right)^{-0.2} \quad \textbf{(6)}$$

Consequently, the surface flashover length factor β is given by Eqn.(7).

$$\beta = \frac{1}{\gamma} \cdot \frac{H}{S} = \left(\frac{H}{H_0}\right)^{0.2} \cdot \frac{H}{S} \quad (7)$$

Calculation of Surface Flashover Criterion

"Surface Flashover Criterion : κ_{tBmin}" is calculated using experimental data [14], on condition that the conducting particle was attached on the insulator between parallel plane electrodes shown as Figure 12. The surface electric-field intensity factors κ_{tB} in the instant of surface flashover are calculated as a function of gaseous pressure. The factors κ_{tB} are shown as Figure 13. The minimum value of κ_{tB} is taken as the surface flashover criterion κ_{tBmin}.

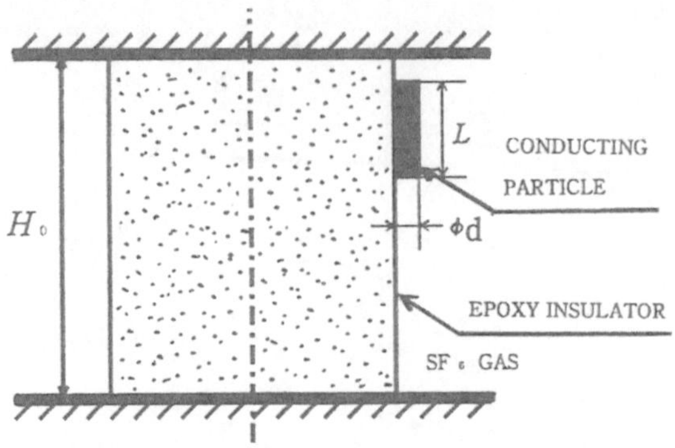

Figure 12. Insulator with conducting particle.

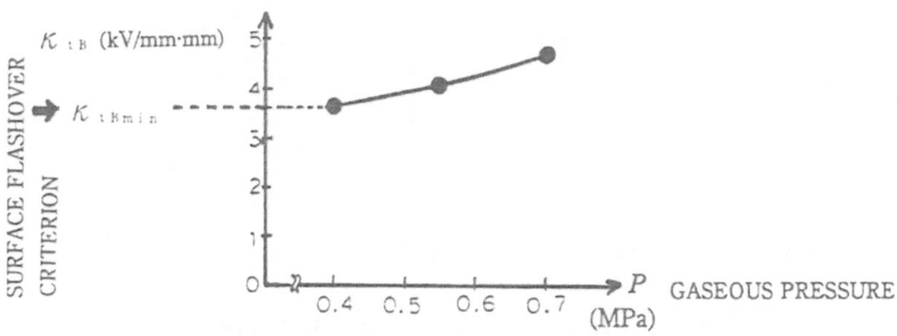

Figure 13. Surface flashover criterion κ_{tBmin}.

Prediction of Surface Flashover Voltage on Insulator

The surface flashover voltage on an insulator for GIS with a conducting particle attached is predicted under application of A.C. voltage using the method proposed by authors. It is calculated as the critical applied voltage under which $\kappa_t = \kappa_{tBmin}$. The surface flashover voltages are calculated about the insulators of disc type and post type, respectively. They are compared with experimental values quoted from references [15, 16]. They are listed as Table 3.

Table 3. Comparision of predicted and experimental surface flashover voltages on insulator.

Conducting particle attached on epoxy insulator	Predicted value of surface F.O.V (kV r. m. s.)	Experimental value of surface F.O.V (kV r. m. s.)	Ref. number of experimental value
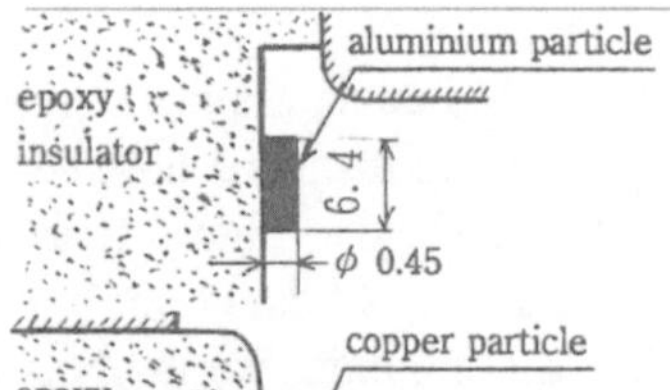	124	100 ~ 191	15
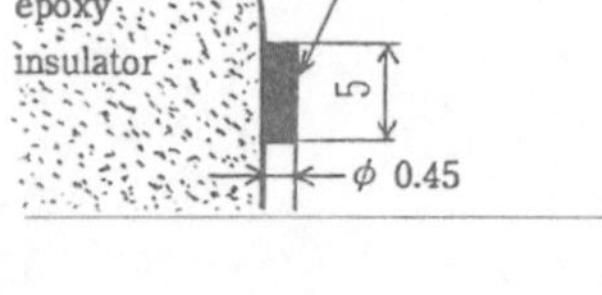	240	191 ~ 297	16

CONCLUSIONS

In order to predict the flashover voltage in SF_6 gas including the conducting particle with the sharp tip, "Electric-Field Intensity Factor" and "Flashover Criterion" were proposed by authors.After that, in order to predict the surface flashover voltage on an insulator with the conducting particle attached, "Surface Electric-Field Intensity Factor" and "Surface Flashover Criterion" were also proposed. The dielectric rate factor and the surface flashover length factor were included in the prediction process. The flashover voltages in SF_6 gas and the surface flashover voltages on the insulator for GIS including conducting particles were calculated using the proposed method. The predicted values were compared with the experimental results. It was verified that the proposed method is useful to predict the flashover voltage of GIS including conducting particles with a sharp tip.

REFERENCES

1. T.Takuma, T.Watanabe and T.Kawamoto, Proc. International Symposium on Gaseous Dielectrics, PP.190–205(1978).
2. N.Wiegart, L.Niemeyer, F.Pinnekamp, et al., IEEE Trans. on Power Delivery , Vol.3, No.3, PP.923–946(1988).
3. S.Matsumoto, H.Aoyagi, H.Okubo, N.Kobayashi and S.Yanabu, T. IEE Japan, Vol.108–B, No.4, PP.173–179(1988). (in Japanese)
4. A.Ieyasu and M.Yumoto, T.IEE Japan, Vol.113–B, No.4, PP.369–375 (1993). (in Japanese)
5. A.Ieyasu and M.Yumoto, Proc. 8th International Symposium on High Voltage Engineering, 33.07, PP.343–346(1993).
6. H.Kobayashi, Y.Arai, H.Nakamura, T.Koide, Y.Kajiyama and A.Ieyasu, T. of JSME, Vol.54, No.499, A, PP.461–465(1988). (in Japanese)
7. F.Endo, T.Yamagiwa, T.Ishikawa and M.Hosokawa, IEEE Trans. on Power Delivery, PWRD–1, No.1, PP.58–65(1986).
8. A.H.Cookson, O.Farish and G.M.L.Sommerman, IEEE Trans. on Power Apparatus Syst., PAS–91, PP.1329–1338(1972).
9. A.H.Cookson and O.Farish, IEEE Trans. on Power Apparatus Syst., PAS–92, PP.871–876(1973).
10. M.M.Morcos, K.D.Srivastaba and H.Anis, Conf. Record of International Symposium on Electrical Insulation, PP.159–162(1984).
11. T.Nitta, Y.Fujiwara, F.Endo and J.Ozawa, CIGRE Vol.26, No.15–04(1976).
12. Li Ming and Antonios E.Vlastos, IEEE Trans. on Power Delivery, Vol.4,No.1, PP.326–334(1989).
13. M.Hara, H.H.Lee, T.Egashira and K.Oyama, T.IEE Japan, Vol.110–B, No.7, PP.568–575 (1990). (in Japanese)
14. M.Hara, H.H.Lee and T.Egashira, IEEE Trans. on Electrical Insulation, Vol.26, No.1, PP.156–170 (1991).
15. A.K.Chakrabati, R.G.van Heeswijk and K.D.Srivastaba, IEEE Trans. on Electrical Insulation, EI–24, PP.549–560(1989).
16. S.Okabe, T.Yorozuya, F.Endo, T.Yamagiwa and S.Iwaasa, IEEE PES Summer Meeting, No.91SM372–3PWRD,(1991).

DISCUSSION

Y. QIU: Table 1 shows that in some cases the predicted value of FOV is far below the experimental value, e.g. 61 kV and 139 kV, respectively, for the copper free particle of 0.4 mm diameter and 12.7 mm in length (line 3 from the bottom of Table 1). Can you explain this?

A. IEYASU: The prediction is based on the assumption that the particle is attached on the conductor. Consequently, the predicted FOV is lower than the experimental FOV, because the free particle does not cross between the conductor and the vessel in the condition of this experiment.

PARTIAL DISCHARGES AND BREAKDOWN IN GIS INITIATED BY FREE MOVING PARTICLES

Michael Wohlmuth

Institute of High Voltage Engineering
Technical University of Munich
Germany

INTRODUCTION

Gas insulated substations (GIS) have proved to be very reliable in their longterm operation. However, the insulation properties of GIS are influenced mainly by the presence of free conducting particles. Several diagnostic methods allow the detection, identification and the localization of such particles during conditioning or in operation[1]. The efficiency of these methods can be considerably improved by an extensive understanding of the physical processes leading to a particle initiated breakdown, in order to qualify the risk of an insulation failure.

In the experiments particle initiated partial discharges (PD) and breakdowns are related to the phase of the applied AC voltage and to the actual position of the particle moving in the gap. Discharges at impacts with the enclosure or the internal conductor and microdischarges in mid-gap position can be distinguished.

Based on these data and with consideration of existing insulation failure models, a computer simulation program was developed. This allows the calculation of the particle motion, of particle initiated partial discharges and of breakdown voltages for different experimental conditions.

EXPERIMENTAL SETUP

The test vessel consists of single-phase 170 kV GIS components (Fig. 1) with a total length of 1200 mm[2]. It was additionally equipped with a particle input valve. The inner conductor with a diameter of 114 mm is fixed without a spacer in the horizontal part. The gap distance is 97.5 mm. The test vessel is energized by a 50 Hz 400 kV_{rms} transformer via a 16 kΩ resistor. The maximum SF_6 gas pressure is 0.4 MPa.

The whole measuring equipment is screened in a shielding case. In order to quantify the PD signals in terms of apparent charges or the energy, a conventional electrical PD measuring system MWB TEM with a wide band filter EB-1 (40 – 220 kHz) is used. This is the only method, which can be calibrated according to IEC 270. Nevertheless

Gaseous Dielectrics VII, Edited by L.G. Christophorou
and D.R. James, Plenum Press, New York, 1994

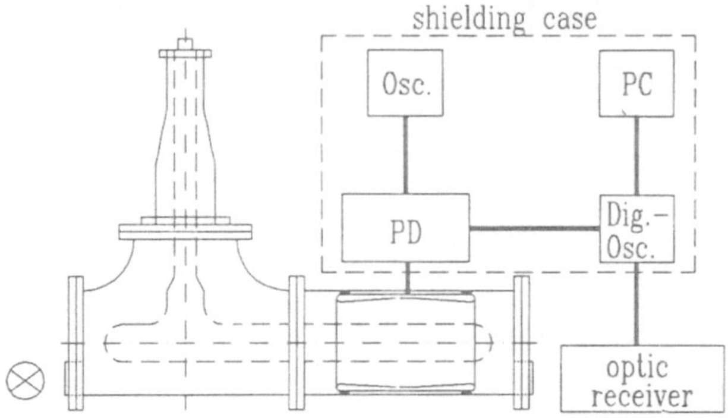

Figure 1. Experimental setup and measuring equipment

the common problems of PD measuring systems are known[3]. Additionally in case of particle initiated PD's there is no repetitive signal and the signal amplitudes show a big scatter and vary by some orders of magnitude. Due to the small length, the test vessel acts as a lumped capacitance with a constant transfer function. Therefore the apparent charges of the calibration signal and of real PD's are independent of the axial position, but depend still on the radial PD position in the gap. As coupling capacity the ground capacitance of 170 pF is used. For the signal amplitudes considered external noise under laboratory conditions is rarely a problem and no conditioning of the measuring signal, like additional filtering or locking is necessary. The particle movement is monitored through the observation window by the optic receiver device[2]. Both signals are fed into a digitizing oscilloscope, where they can be related to the phase of the AC voltage and evaluated and stored with a PC. The real PD energy can be gained by multiplying the apparent charge with the applied voltage. The time bases of the particle motion and of the discharges are totally different. Fortunately the apparent charge is given only by the amplitude of the signal. Therefore the PD signal can be measured in peak detection mode. This allows to monitor the particle motion for a longer time period with a slow time resolution.

Cylindrical aluminum wires with a length of 10.0 mm, a diameter of 1.0 mm and semispherical shaped tips are used as particles. Although a big variety of different shapes and sizes of particles can be found in GIS, this relatively simple shape is used in the experiments in order to get reproducible results, which can be numerically simulated by the computer program.

SIMULATION MODEL

Some particle parameters, which are responsible for the motion and the discharges (e.g. net charge, floating potential, tips' field) can not be measured. Therefore a computer simulation program was developed, which is based on the 3-dimensional calculation of the electric field[2, 4]. Input parameters are the particle and the gap geometry, the voltage, the gas pressure and some characteristic particle values. Physical models, which were established for discharges at fixed protrusions, are applied. These models will be described only briefly and references will be made for more detailed information.

The particle motion is calculated in a loop by solving the differential motion equation in small time steps (usually 1 ms). Each time step in the regarded time period requires a 3D-field calculation, taking into account the actual particle position, the particle net charge and the applied voltage. A correct calculation of the surface charge and the local fields enables the calculation of the driving electrostatic force. If no breakdown occurs the new particle position is calculated and the loop is continued. In comparison to the motion and the AC voltage the relevant discharge processes are much faster. Therefore they are simulated within one time step and any formation time lags are neglected.

If the streamer onset criterion according to the Pedersen law[5] is fulfilled at one or both particle tips, the corona extension can be estimated by the intersection of the internal and the external potential curve[2]. Thereby the influence of the first electron statistic and space charge effects as well as polarity dependencies are neglected. The field strength within a corona structure is balanced to the critical value. A simplified model replaces the corona, which is caused by ionization processes, by a conducting sphere, which carries the same amount of charge on its surface[6]. This charge is calculated by assigning the floating potential of the particle to the corona structure and finally the influence on the particle net charge is considered[2].

The transformation of a streamer into a leader requires an energy input, which heats the gas in the initial leader channel and dissociates it completely[6, 7]. This energy can be supplied only from the internal conductor. Therefore a (quasi-) conducting connection of the moving particle and the internal conductor is necessary, as it is the case by an impact or by a contact discharge, where the small remaining gap is bridged by a corona. Then the particle suddenly acquires almost the full potential of the internal conductor and acts as a fixed protrusion. A corona at the lower particle tip will be established and can be transformed into a leader, if the energy input is sufficient. This first leader step is equivalent to an effective elongation of the particle and reduces the inception level for the following steps. Breakdown can follow before corona stabilization has had time to become established, if the background field remains constant during the leader propagation. As this is the case for AC voltages, the leader inception criterion is equivalent to the breakdown criterion[6]. As breakdown criterion a critical energy can be used. Since the field within the corona is balanced to the critical value, this critical energy can be transformed into a critical corona charge, representing a sufficient number of electrons and ions for the energy transport. A time dependent leader channel radius resulting in a reduction of the internal field strength has not to be considered.

The initial leader channel radius is dependent on the voltage and decreases with the gas pressure. The equivalent channel radius was measured to be smaller than 40 μm and a simulation value of 30 μm was used in the precursor mechanism in case of a gas pressure of 0.1 MPa[8]. In the computer simulation a constant value of 25 μm was used, but a pressure dependent radius gives better results for higher gas pressures.

For comparison with experimental results the calculated PD charges are transformed into measurable apparent charge values, by taking into account the ratio of the capacitances C_1, C_2 and C'_P. Fig. 2 shows the equivalent circuit for a PD in mid-gap position pointing to the internal conductor. The total capacitance C_P is divided into the capacitance C_1 of the gas volume, which is bridged by the streamer corona and depends on the corona extension, and C_2, which represents the capacitance of the gas volume between the corona and the conductor. C_1 and C_2 represent a capacitive voltage divider:

$$C_i = (U_P/U_i) \cdot C_P \qquad i = 1,2$$

U_P is the potential difference between the particle and the internal conductor, U_1 the voltage drop within the streamer corona, C_P the total capacitance of the particle to

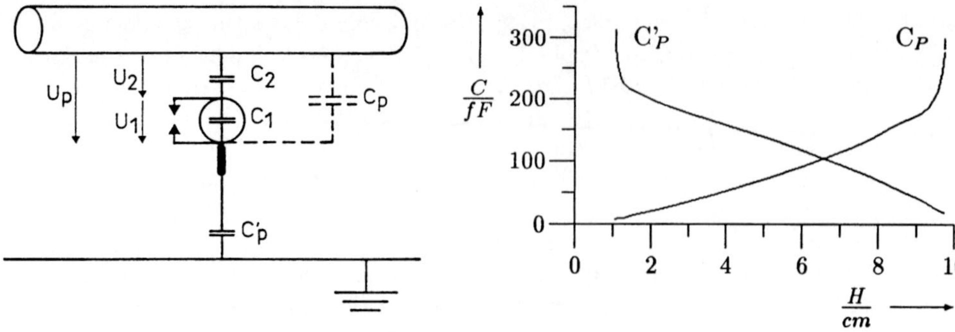

Figure 2. Equivalent circuit for a PD and total capacitances C_P and C'_P

the internal and C'_P to the external conductor. C_P and C'_P are calculated by 3D-field calculation and are shown in Fig. 2 as function of the particle height (upper tip) in the gap. In case of a contact discharge the capacitance C_2 disappears and C_1 equals C_P.

RESULTS

Fig. 3 shows on the left side the particle motion in the gap, monitored with a resolution of 16 steps by the optic receiver, and the apparent charge signal Q_a, measured by the conventional PD system. The results of the computer simulation are shown on the right side, and for comparison the real charge values Q_r are additionally transformed into measurable apparent charge values Q_a. Due to the randomness included, single particle courses in the gap cannot be simulated, but principle characteristics like the bounce height H or the time between impacts can be compared statistically. For all voltage levels and for a gas pressure above 0.1 MPa the simulation shows good results.

Clear discharge signals are measured at particle impacts with a conductor. Due to the higher field, these signals are much larger at impacts with the internal conductor, but even with the same amplification they are still clearly detectable at impacts with the enclosure. The simulation shows similar values for the measurable apparent charge, both for impacts with the enclosure and with the internal conductor. Additionally the calculated real charges are shown.

The charge values depend on the actual phase of the applied AC voltage with a maximum for an impact at the voltage crest. On the left side of Fig. 4 the apparent charge values are shown as function of the applied voltage for a gas pressure of 0.3 MPa. For lower voltages these charge signals are caused by impacts with the enclosure and above the crossing voltage of approximately 140 kV_{rms} by impacts with the internal conductor. In both cases the apparent charges increase with an increasing voltage. The charge signals for impacts with the internal conductor in the negative voltage crest show significantly higher amplitudes. As shown on the right side of Fig. 4 the apparent charges decrease for higher gas pressures for impacts with the internal conductor and also for impacts with the enclosure. In the pressure range considered the charge exchange at particle impacts in the voltage crest even with the enclosure must be combined with a contact discharge. For higher voltages or lower gas pressures the corona extension increases and this results in a bigger corona charge and according to Fig. 2 in a smaller capacitance C_1, and finally in a higher apparent charge signal.

In mid-gap position the PD's can be detected less easily, since their signal amplitudes are much lower. For lower gas pressures more PD activity is possible, and this

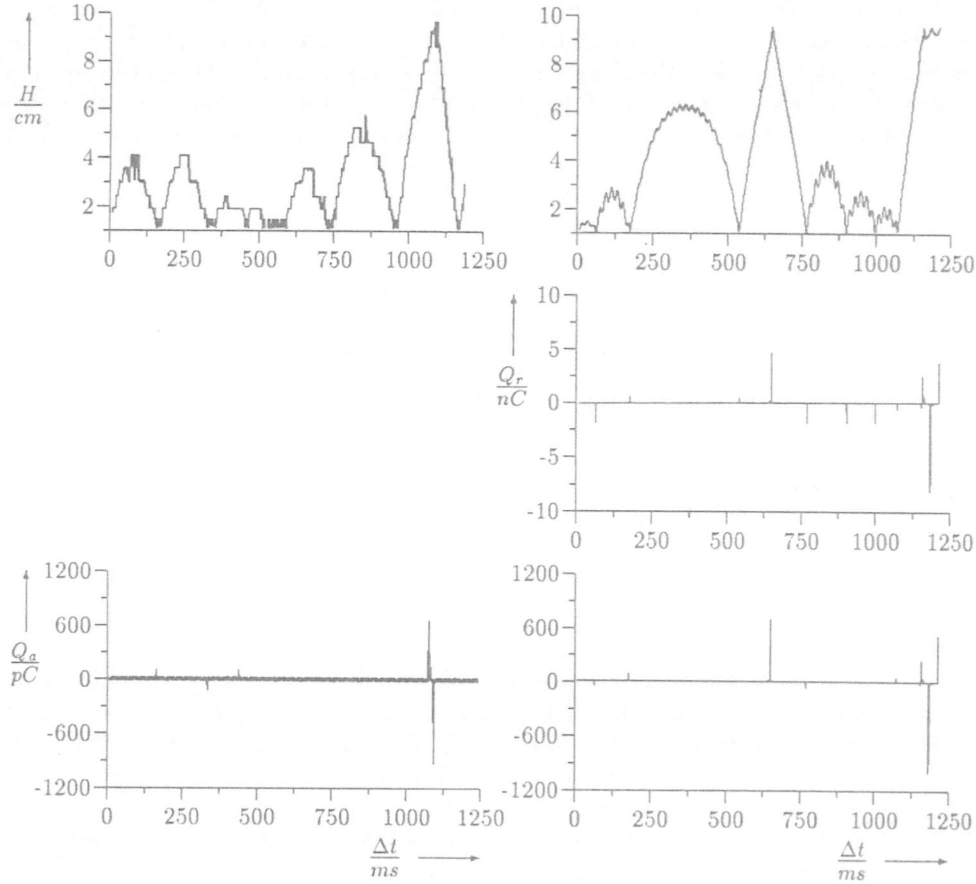

Figure 3. Particle motion and discharge signal (0.3 MPa, 160 kV$_{rms}$)

reduces the particle net charge. The superimposed motion with the power frequency disappears, whereas for higher gas pressures less PD activity is possible. In this case the net charge remains on the particle and causes a power frequent modulation of the motion. Due to the limited resolution of the optic receiver this can not be seen in the measurements, but in the simulated particle motion in Fig. 3.

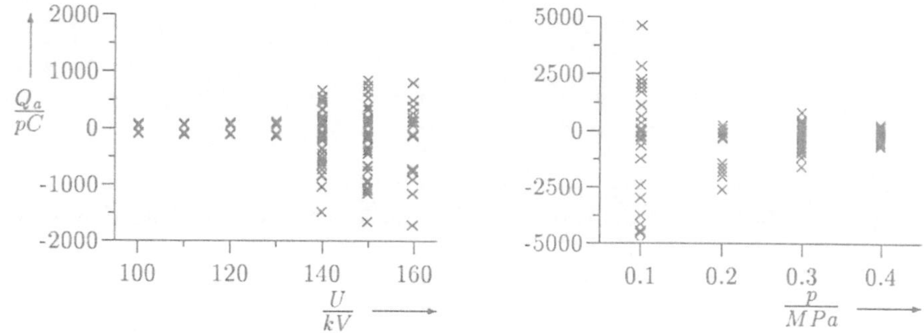

Figure 4. Maximal apparent PD signal as function of the applied voltage (0.3 MPa) and of the gas pressure (140 kV$_{rms}$)

Breakdown occurs in or right after the positive crest of the AC voltage, after the particle has passed the upper row of the optic receiver and is therefore close to the internal conductor, where the electrostatic forces cause a particle motion in upward direction[2]. In order to exclude corona stabilization the small remaining gap between the particle and the internal conductor must be bridged by a contact discharge, which gives a first PD signal. Then the particle acquires almost the full potential of the internal conductor, and the net charge can be exchanged totally. The amplitude of this charge exchange can be twice the charge amount, which is maximal induced at impacts with the internal conductor in the voltage crest. A streamer corona at the lower particle tip is established and causes a third PD component. This component represents the energy input into the gas and can cause the streamer – leader transition.

A subsequent appearance of a number of field dependent single events can lead to breakdown. The probability for this increases with time. Thus the particle initiated breakdown is strongly time-dependent.

The apparent charge signal at particle impact consists of at least two components. In case of an impact with the internal conductor there can be an additional third component, which is responsible for the streamer – leader transition. Due to the small formation time lags only one signal can be measured by the conventional method. In the PD signal a clear difference can be made between a particle impact with the enclosure or the internal conductor, but a conclusion on the charge of the corona at the lower particle tip, which can cause a leader transition is impossible.

These different charge components in the breakdown process are calculated by the computer simulation program separately. For the breakdown shown in Fig. 3 for $160\,kV_{rms}$ and 0.3 MPa similar values of 4 – 5 nC are calculated for each component and are summed up within one time step. The capacitance ratios, which are decisive for the apparent charge, are totally different. Therefore the real charge values Q_r are not constantly reduced into apparent charge values Q_a. Breakdown does not necessarily occur in the moment of the largest PD signal, because only the charge in the streamer, which is established at the lower particle tip is decisive.

Fig. 5 compares the minimal rms breakdown voltages during the measurements and the values calculated by the computer simulation for different gas pressures. For a gas pressure of 0.1 MPa the program simulates high PD activity. This prevents impacts with the internal conductor and no breakdown can take place.

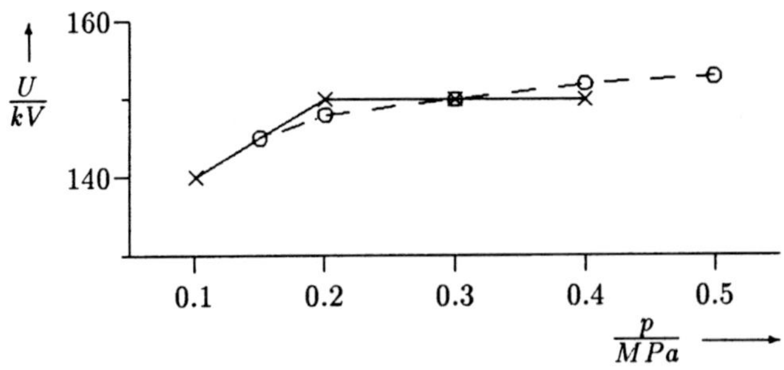

Figure 5. Measured (—) and simulated (– –) rms breakdown voltages as function of the gas pressure

For higher gas pressures there is a stabilization of the measured breakdown voltages, since the leader inception level and therefore the critical charge for the streamer – leader transition decreases with an increasing gas pressure[6, 7]. Therefore the radius of the initial leader channel must decrease with the gas pressure. Instead of the constant initial leader radius $r_L = r_{Lo} = 25\,\mu m$ a pressure dependent radius is used, and gives satisfying results.

$$r_L = r_{Lo} \cdot \left(\frac{0.1\,\mathrm{MPa}}{p}\right)^{0.7}$$

CONCLUSION

Particle initiated partial discharges and breakdowns in GIS were investigated experimentally and by a computer simulation program. The computer program simulates the characteristic motion parameters well. Existing insulation failure models have been applied and the simulation shows reasonable results for PD's and breakdown voltages.

Large PD signals are caused by particle impacts with the internal conductor and much smaller, but still clearly detectable signals at impacts with the enclosure. Even with the enclosure a particle impact in the voltage crest must be combined with a contact discharge. The apparent signal amplitudes increase with higher voltages and lower gas pressures. PD's in mid-gap position can be detected less easily, since their signal amplitude is much smaller.

The discharge component, which is responsible for a particle initiated breakdown can not be measured separately by the conventional method. Therefore particle impacts with the internal conductor, which can easily be detected due to their large charge signals can serve as an indicator for a high risk of an insulation failure, even if not every impact leads to a breakdown.

Breakdown occurs in the positive crest of the AC voltage for particle positions close to the internal conductor. The breakdown voltages are strongly time-dependent. A stabilization of the breakdown voltages for higher gas pressures does exist. This is taken into account by a pressure dependent initial leader radius in the computer model.

REFERENCES

1. CIGRE WG 15.03. "Diagnostic methods for GIS insulating systems," CIGRE Paris, 15/23-01 (1992)
2. M. Wohlmuth. "Motion and discharges of free moving particles," 8.ISH, Yokohama, 33.06 (1993)
3. I. Herbst. "PD-measurements in GIS, a theoretical and experimental comparison of different electrical measurement techniques," CIGRE Symposium, Berlin, 130-02 (1993)
4. M. Wohlmuth: "Measurement and calculation of lift-off fields and charges for free moving particles,"10. Int. Conf. on Gas Discharges, Swansea, p.414 (1992)
5. A. Pedersen. "Formulation of the streamer breakdown criterion and its application to strongly electronegative gases and gas mixtures," Archiv für Elektrotechnik 67 (1984)
6. L. Niemeyer. "Leader breakdown in compressed SF_6: recent concepts and understanding," Gaseous Dielectrics VI, Plenum Press, p.49 (1990)
7. H. Hiesinger. "The calculation of leader propagation in point/plane gaps under very fast transients," Gaseous Dielectrics VI, Plenum Press, p.129 (1990)
8. L. Niemeyer. "A model of SF_6 leader channel development," 8.Int. Conf. on Gas Discharges and their Applications, Oxford, p.223 (1985)

DISCUSSION

I. D. CHALMERS: We also have observed that breakdown with a free particle is initiated when the particle approaches the central bus bar. The fact that the bus bar potential is suddenly impressed upon the particle means that the situation may be accurately modeled by an impulse voltage applied to a fixed protrusion (which is essentially a point/plane arrangement).

O. FARISH: (1) In work reported some years ago by Wooton, experiments were carried out with particles fixed at a position close to the conductor and the critical gap for minimum breakdown voltage was determined. Is your model able to predict this condition for a given geometry? (2) In other experiments carried out at Westinghouse an argon ion laser was used to illuminate the gap and a high-speed (10,000 fps) camera was used to record particle motion at 4 bar. This shadowgraph technique clearly showed PD activity both in mid-gap and at contact with the electrode. Have you any idea why you do not see PD at high pressure in mid-gap?

M. WOHLMUTH: (1) Since the model is based on 3D-field calculation, the simulation of the particle motion, PD-onset and breakdown is possible. Therefore the critical gap can be predicted, but is strongly dependent on the geometry and the gas pressure. (2) In these investigations relatively thick particles with round tips were used. For higher gas pressures the PD-onset criterion is not fulfilled in mid-gap.

K. FESER: Can you calculate with this method the critical defect size for a certain GIS for free moving particles?

M. WOHLMUTH: The computer simulation allows the modeling of any cylindrical particle shape which is considered to be the most critical one. By comparing the calculated breakdown voltages with the required withstand voltage of a certain GIS the critical particle size can be calculated.

CHARGING MECHANISMS OF A CONDUCTING PARTICLE ON DIELECTRIC COATED ELECTRODE AT AC AND DC ELECTRIC FIELDS

T. Ooishi, M. Yoshimura, H. Hama, H. Fujii
and K. Nakanishi

Central Research Laboratory
Mitsubishi Electric Corporation
Amagasaki, Hyogo, Japan 661

INTRODUCTION

The breakdown properties of GIS can be seriously deteriorated by conducting particles which are accidentally mixed at the assembly process. The compact GIS and the gas insulated dc apparatus are expected to be one of the future directions. The particle contamination will be more crucial for both gas insulated apparatus than for the conventional GIS. Special cares against the particle contamination have to be taken for the compact GIS and dc apparatus.

Several methods which deactivate the conducting particles have been studied[1-4]. The dielectric coating to the electrode surface has shown to hamper the levitation of the particle from the electrode. The phenomenon is attributed to the effect that the dielectric coating prevents the particle acquiring net charges up to high electric stress. Two mechanisms of charging the particle have been propopsed. They are (I) partial discharges (PD) between the particle and the dielectric, and (II) conduction through the dielectric. The reference[3] has suggested that the former mechanism is effective at the situations of thin film coating and low gas pressure, and that the latter is dominant at the film thickness more than 100 um and high gas pressure.

The purpose of the paper is to reveal the mechanisms of charging the particle on the dielectric coated electrode at ac and dc stresses. We try to show that the PD is the dominant mechanism of charging the particle on the dielectric at ac and dc stresses, discussing the test results obtained by the optical measurements such as an image intensifier.

EXPERIMENTAL SETUP

In the studies, two different test equipments were used. One was a sensitive optical measurement system for detecting the PD. The other was a conventional gas tank for observing the particle lifting field.

Gaseous Dielectrics VII, Edited by L.G. Christophorou
and D.R. James, Plenum Press, New York, 1994

As the optical measurement, we used an image intensifier system with a microscope (Hamamatsu Photonics : C3230). The optical system was set to observe the faint lights of the PD initiated at the small gap between a spherical particle and a dielectric coated electrode as shown in Fig.1(a). The light signals detected by the image intensifier were taken by a computer controlled CCD camera. The measurement system was equipped with data acquisition and photon counting functions. Therefore, we could store the light emissions which took place for a certain duration and could also count the number of the light emissions.

We used a steel sphere (diameter: 2mm) as a particle to facilitate the calculations. To observe the light emission in compressed SF_6 gas, we used a transparent glass plate as a grounded electrode. The thin ITO (Indium Tin Oxide) film was evaporated on the glass plate to give conductivity and transparency. As the coating material, the Teflon film (Fluorinated Ethylene Propylene, resistivity: 5.1×10^{15} ohm cm and dielectric constant : 2.2) was used since it has transparency. The film thickness was varied 40 um to 180 um. As shown in Fig.1(b), the lower electrode was slightly dished to prevent the particle moving to the lateral direction. The gap spacing between the high voltage and the grounded electrodes was 11mm. The ac 60Hz voltage was raised with the rate of 800V/s. The gas pressure was varied 0.05 to 0.5MPa.

The particle lifting fields were measured with the same particle as used in the above test and a parallel plane electrode system (gap spacing: 1.1 cm, diameter: 10cm) set in a gas tank. Only the lower electrode was coated by the Teflon film. Lifting off from the coated electrode was observed through the glass window.

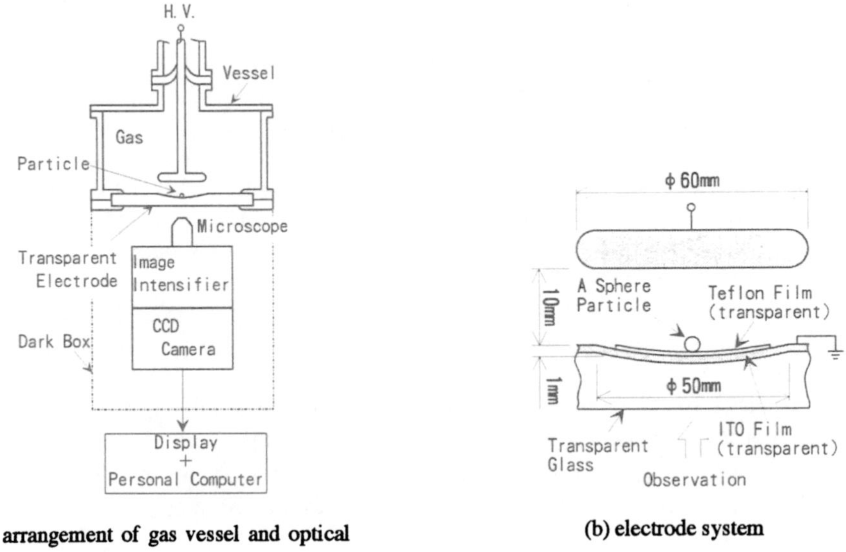

(a) arrangement of gas vessel and optical measurement system

(b) electrode system

Figure 1. Experimental Setup for detecting partial discharges at the small gap between a spherical particle and a dielectric.

EXPERIMENTAL RESULTS

Charging and Lifting of the Particle At AC Stresses

The image intensifier detected electron avalanches which happened at the small gap between the spherical particle and the dielectric. As the light emission starting stress, we defined the average stress E_e at which the light emissions began to be observed over the strayed emissions. The number and the frequency of electron avalanches at the small gap increased with the applied voltage. The particle levitated after the generation of lots of electron avalanches at higher stress than light emission starting stress, as shown in Fig.2. Both stresses almost linearly increased with the gas pressure. Moreover, the emission sites moved to the point of the particle contact with the thin film with the gas pressure as shown in Fig.3 (a) and (b).

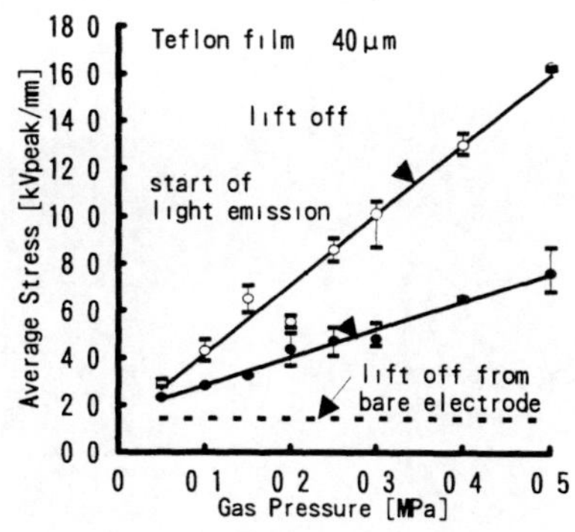

Figure 2. Dependences of light emission starting stress and particle lifting stress on gas pressure.

The number of electron avalanches was counted for the period from the beginning of the light emission up to 40% of the lifting stress. The voltage was not raised to the lifting stress in order to protect the intensifier from intense light of the flashover at the main gap. The numbers were 63986 for 94 seconds at 0.25MPa and 172884 for 157 seconds at 0.4MPa in the case of 40um thick Teflon coating. The light emissions happened at the positive and negative cycles of the ac voltage as shown in Fig.4, though the particle was charged at negative and was discharged at positive. The charge quantity Q_n of the PD at the negative cycle was much larger than the Q_p at the positive , in particular, at the low gas pressure as shown in Fig.4. The difference (Q_n-Q_p) of the charges generated at every cycle was accumulated in the particle until the total charges reached the threshold value for lifting.

Figure 5 shows the dependence of the light emission stress and the particle lifting stress on the thickness of Teflon film at the gas pressure of 0.3MPa. It is interesting that the lifting stress slightly elevated even when the film thickness was doubled.

Charging and Lifting of the Particle at DC Fields

At dc electric fields, the image intensifier system could not catch the faint lights which were always observed at ac fields before the particle lifting. However, the particle lifting stress increased with the gas pressure as shown in Fig.6, though the dependence on the gas pressure varied with the polarity of the dc. The pressure dependence may suggest that the charges were also supplied by the PD. We might not be able to detect the light emission since the frequency of the PD at dc was much smaller than that at ac.

We applied the relatively low dc voltage to the parallel plane electrode for 3 hours. In the test, we tried to observe that the charges are supplied to the particle by the bulk conduction through the dielectric and the particle lifts off from the electrode. The applied dc voltage was limited to positive 50 kV which was below the light emission starting voltage at ac. However, the particle did not levitate even after 3 hours, although the calculation showed an hour dc application was long enough to supply net charges for lifting.

Figure 3. Circular light emissions observed by image intensifier system.

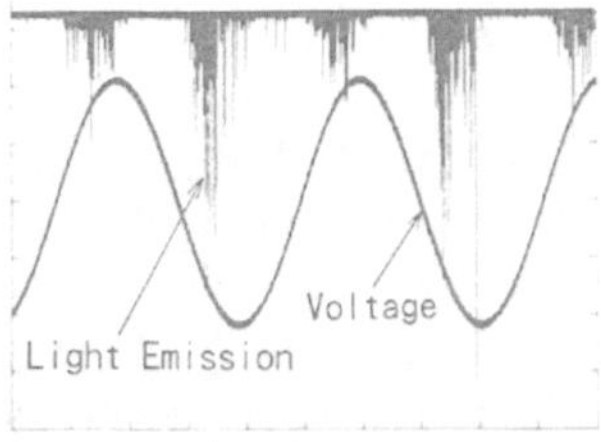

Figure 4. Partial discharges observed by photomultiplier tube at ac voltage application.
gas pressure: 0.25 MPa, applied voltage: 48kV, Teflon film: 40 um

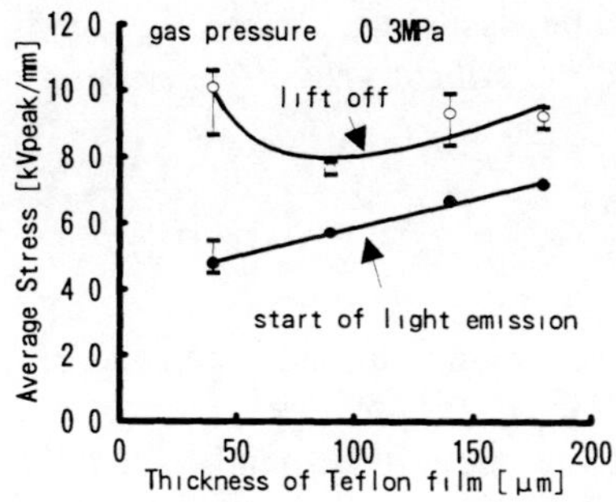

Figure 5. Influences of Teflon film thickness to light emission starting stress and particle lifting stress at 0.3 MPa.

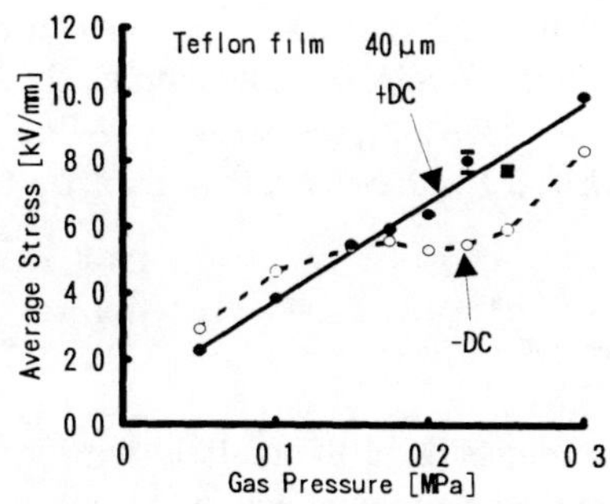

Figure 6. Dependences of particle lifting stress at positive and negative dc fields on gas pressure.

DISCUSSIONS

Charging Mechanism at AC Fields

The particles on the dielectric coated electrode cannot easily obtain charges since the dielectric film hampers charge supply from the electrode. This is the reason coating an electrode with a dielectric elevates the lifting field of the particle and eventually results in increasing the breakdown field. In this section, we limit the discussions to the mechanisms of charging the particle placed on the dielectric coated electrode.

The results shown in Figs.2-5 suggest that the PD is the dominant mechanism for charging the particle. The mechanism of charge conduction through the dielectric cannot account for the pressure dependences and light emissions. We calculated the critical number of electrons N in an avalanche, assuming that the light emission was detected when the number of electrons reached the critical value. The electron number N in an avalanche is given by equation (1)[5].

$$\ln N = K\left\{\int_{o}^{Xc} E(x)\, dX - (E/P)_{crit} \cdot P \cdot X_c\right\} \qquad (1)$$

where K is 27/kV and x_c is the length at the small gap and E(x) is the electric stress and $(E/P)_{crit}$ is 88kV/mm MPa and P is the gas pressure.

When we used lnN=18 as the critical number of the avalanche electrons for light detection, the equation (1) could account for the movement of the light emissions to the smaller gap between the particle and the dielectric with the increase of the gas pressure. Moreover, the calculation results agreed well with the observed locations of the light emissions shown in Fig.3(a) and (b). The critical value of electrons corresponded to that of the streamer initiation. Using the critical value, Fig.5 can be expressed to Fig.7 in which the ratio of E_e to E_s (streamer initiation stress) is almost constant and independent of the film thickness.

The reference[3] suggested the possibility of charge conduction through the dielectric for the 25 um anodized film (resistivity: $2x10^{10}$ ohm cm, dielectric constant: 8.9). It predicted that the lifting stress is 4.8 kV/mm. While, our calculations showed the light emission stress for the above case is 3.4 kV/mm at the gas pressure of 0.5 MPa. From these calculations, we consider the net charges are fed by the PD rather than by the conduction through the dielectric.

Charging Mechanism at DC Field

We could not detect light emission at the dc field with the optical measurement. This does not necessarily mean that the charge conduction through the dielectric was dominant at dc field. As shown in Fig.6, the lifting stress increased with the gas pressure though the properties varied, to some extent, with the polarity of the dc voltage.

In the case of dc application, the potential of the particle which is charged by the PD is close to that of the lower electrode. Therefore, the PD do not happen until the stress reaches the threshold value for electron avalanche. The number and frequency of the PD at dc stress

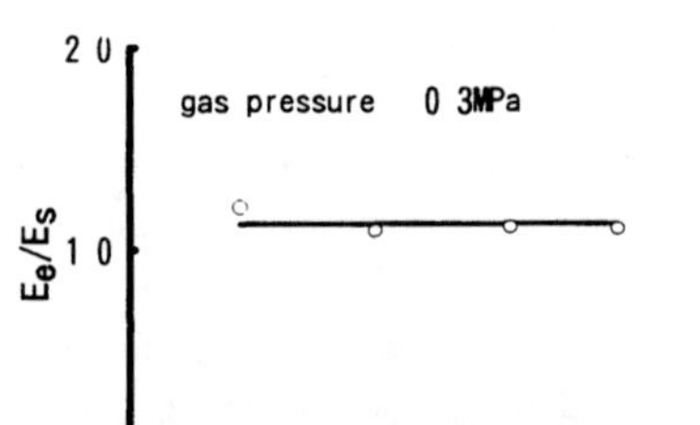

Figure 7. Relation between E_e/E_s and Teflon film thickness .E_e:light emission starting stress obtained in the experiments. E_s:streamer initiation stress calculated by equation (1) in the text.

are considered to be quite small, compared with the situations at ac. There is a possibility that the image intensifier missed the detection of infrequent light emissions.

As mentioned before, we tested the mechanism of charge conduction through the dielectric, applying the low dc stress without any avalanche for a long duration. When the contact area of the particle with the dielectric is assumed to be 1% of the cross sectional area of the particle, the calculations based on the equivalent circuit showed that the particle can acquire the necessary charges for lifting with an hour dc application. In fact, the particle did not lift off although we applied dc stress for three times duration.

CONCLUSIONS

The mechanisms of charging the particle on the dielectric coated electrode have been studied. From the experimental results and the calculations, we have drawn the following conclusions.

1. Sensitive optical measurements indicated that the particle on the dielectric coated electrode at ac stress was charged by the PD at the small gap between the particle and the dielectric.
2. The light emission starting stress which corresponded to the streamer initiation increased with the gas pressure, but was independent of the thickness of the coating film.
3. The particle levitation took place when the charges accumulated in the particle by the PD reached the threshold value.
4. The short and long term tests at dc stress suggested that the PD was the dominant mechanism of charging the particle, though the optical measurement could not detect the light emission generated by the PD.

REFERENCES

1. S.J.Dale, R.E.Wooton , and A.H.Cookson, "Effects of Particle Contamination in SF_6 Systems and Methods of Control", Gaseous Dielectrics II (edit.L.G.Christophorou), Pergamon Press, 256(1980).
2. C.M.Cooke, "Electrostatics of Particle Control in Gas Insulated Apparatus", Gaseous Dielectrics II (edit. L.G.Christophorough), Pergamon Press, 275(1980).
3. H.Parekh, K.D.Srivastava, and R.G.van Heeswijk, "Lifting Field of Free Conducting Particles in Compressed SF_6 with Dielectric Coated Electrodes", IEEE Trans. PAS, Vol.PAS-98, 748(1979).
4. K.D.Srivastava and R.G.van Heeswijk, "Dielectric Coatings-Effect on Breakdown and Particle Movement in GITL Systems", IEEE Trans. PAS, Vol-104, 22(1985).
5. T.Nitta and Y.Shibuya, "Electrical Breakdown of Long Gaps in SF_6", IEEE Trans. PAS, Vol.PAS-90, 1065(1971)

DISCUSSION

I. HERBST: What is the dependence of U_{lift} on the ϵ of the coating?

K. NAKANISHI: The main subject of the study is to make it clear whether the dominant mechanism of charging the particle is PDs or charge conduction through the dielectric. One of our conclusions showed that the PDs are the mechanism of charging the particle. Therefore, the use of the coating material with the low ϵ will increase the lift-off stress U_{lift}.

VOLTAGE-TIME CHARACTERISTICS OF BREAKDOWN IN SF_6

S.K. Venkatesh and M.S. Naidu

Department of High Voltage Engineering
Indian Institute of Science
Bangalore-560 012, India

ABSTRACT

The impulse breakdown voltage-time to flashover (V-t) characteristics of point-plane gap in SF_6 has been investigated using impulse wave fronts in the range of 1.2 μs to 250 μs. Also, field non-uniformity was systematically varied to study the effect of corona preceding breakdown on time delays. The V-t characteristics are categorized into three types depending on the field non-uniformity and gas pressure. The properties of the characteristics in these types are generalised and will be useful in the insulation co- ordination studies for abnormal voltage protection of SF_6 gas insulated power equipments and systems (GIS). It has been deduced that the lightning impulse (LI) voltage appears to be the most effective on-site test voltage to detect any defects like fixed particles or protrusions in the GIS.

INTRODUCTION

V-t characteristics of GIS are quite different from those of conventional systems. Air and porcelain insulated equipment has a rising V-t characteristic with decreasing rise time of impulse voltages, i.e. they can withstand very high voltages for very short times. However, as the duration of the voltage increases, the withstand voltage falls off considerably. On the other hand, SF_6 gas insulated equipments (without any defects) have nearly flat V-t characteristics. This is because of nearly uniform and higher field strengths used in GIS. The consequences of insulation breakdown in GIS are far more severe than in conventional substations, as GIS insulation has to be considered as non-restoring. A failure in GIS would lead to a greater disturbance

Gaseous Dielectrics VII, Edited by L.G. Christophorou
and D.R. James, Plenum Press, New York, 1994

in the operation of the system. This leads to the necessity of a detailed analysis of the stresses imposed on the GIS insulation under various conditions, as well as an economic choice of insulation levels and of means of protection to maintain the reliability of the system. An essential feature of this insulation coordination is the complete knowledge of the V-t characteristics.

V-t characteristics of SF_6, gas particularly under uniform and slightly non-uniform fields (co-axial cylinders) have been reported under various conditions[1,2]. However, studies under highly non-uniform field conditions, are important as they simulate defects that may exist within GIS.

In view of the above, V-t characteristics of SF_6 gas have been investigated in the present work using voltages having a wide range of rise times varying from standard lightning impulse to switching surge, over a pressure range of 0.1 to 0.5 MPa using hemispherically capped rod-plane electrode configuration. Field non-uniformity was systematically varied by using rods of different diameter. The effects of varying the field non-uniformity, gas pressure, and rise time of the applied impulse voltage on the breakdown have been investigated.

EXPERIMENTAL SET-UP

The electrode assembly consisted of a hemispherically capped stainless steel rod maintained at a constant distance of 15 mm from a 105 mm diameter stainless steel plane electrode. The field non-uniformity was varied by using rods of 1, 3, 6 and 15 mm diameter, having field factors (f=max. field/average field) of 20.6, 8, 4.75 and 2.55 respectively. The electrodes were housed in a pressure vessel of 12 liters internal volume. The chamber was filled with pure SF_6 gas (99.99% pure confirming to IEC 376) to the desired pressure (0.1 to 0.5 MPa) after ensuring that the chamber was clean and dry.

The impulse voltages were derived from a 10 stage, 6.25 kJ, 500 kV Marx generator. The circuit elements could be readily changed to get 1.2, 10, 60, 150 and 250 μs froı time impulse voltages. The voltage measurements were made using a damped capacitance divider and a 100 MHz digital storage oscilloscope.

The time to breakdown was measured for voltage impulses of prospective peak values corresponding to four to five different probability levels. At each level, 30 to 50 shots were applied at intervals of one minute. Only the critical values were taken to plot the V-t characteristics.

RESULTS AND DISCUSSION

Under non-uniform field conditions and large gap distances like that of conventional air insulation, statistical and formative time lags are large resulting in impulse breakdown voltages higher than the DC, AC breakdown values (Impulse ratio >1). In GIS which has a fairly uniform field, statistical and formative time lags are less and hence impulse ratios are around 1. However, when the GIS contains defects like particles or protrusions resulting in highly non-uniform fields, time lags will be significant. Also the effect of space charges of corona preceding breakdown has to be considered. Space charge can form and accumulate to affect breakdown under DC, AC and slowly rising impulse voltages. However, under fast rising voltages, there is no accumulation of space charges. It is mainly the influence of rise time on space charge accumulation and time lags which causes differences in breakdown values for impulse voltages of different rise times. The effect of space charge on breakdown for different impulse rise times is studied by varying the field non-uniformity.

Three Types of V-t Characteristics

From the results obtained for all the gap configurations and gas pressures studied, the V-t characteristics can be categorised into three types as shown in Fig.1. The distinctive features of each of these types and the conditions under which they occur are given below.

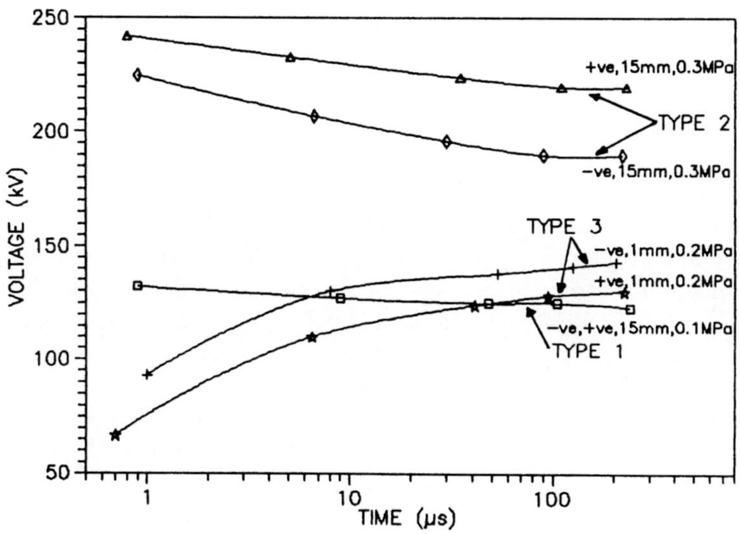

Fig.1. Three types of V-t characteristics

Type-1 The minimum switching impulse (SI) breakdown voltage coincides with the theoretical prediction based on single avalanche theory (Streamer criterion)[3]. The theoretical prediction holds good for corona free breakdown and at lower pressures where the stresses are lower, so that the effect of any surface roughness of the electrodes is insignificant. Also, the effect of polarity of voltage is negligible. The rise of V-t curve is moderate indicating lower magnitude of statistical and formative time lags. This characteristic is seen under uniform and slightly non-uniform fields, particularly at lower pressures. This type can be seen in Fig.1 for 15 mm dia rod (f=2.55) at 0.1 MPa pressure. Therefore, it can be seen that under slightly non-uniform field conditions as used in GIS, the V-t characteristic is flat down to the rise time of 1 μs.

Type-2 Here the minimum SI breakdown voltage is lower than the theoretical estimation. This type occurs particularly at higher pressures and is due to the influence of surface irregularities (weak points) distributed on the electrodes. At higher pressures, stresses will be higher, and this, further increases the number of potential weak points. This will cause an enhancement of electric field close to the surface. Normally, the extension of this enhanced field and the resulting critical volume are small. Also, as the gas pressure is increased, the critical volume for a given voltage reduces [4]. The probability of availability of the initiatory electrons in this small volume is very low and hence statistical time lag is large. Therefore, the V-t characteristic rises with decreasing rise times. Similar observations have also been made earlier[5].

For negative polarity voltages, the initiatory electrons in the small critical volume are produced easily from the electrode by field emission[2] and hence statistical time lag is reduced. This results in a breakdown value lower than that under positive polarity. This also explains the less scatter obtained under negative polarity voltages. This type is also shown in Fig. 1 for 15 mm rod dia at 0.3 MPa pressure.

Type-3 The minimum SI breakdown voltage is abnormally higher than the theoretical estimation. This type occurs under highly non-uniform field conditions and at pressures where stable corona precedes breakdown. This increase in breakdown voltage is due to corona stabilisation effect. Under AC/DC or slowly rising impulse voltages, the corona space charge accumulates around the point electrode and limits the field around it to a level near its value at onset[6]. This corona effectively shields the point electrode, thus requiring higher voltages to cause breakdown. However under fast rising impulse voltages, there is not enough time for space charge accumulation and hence breakdown occurs without corona stabilisation. This can be observed in Fig. 1 for 1 mm dia rod at 0.2 MPa pressure, where lightning impulse (LI) breakdown voltage is lower than the SI breakdown voltage by 45%. The transition to lower breakdown voltage occurs for impulses with rise times of around 10 μs and less. In this type, negative breakdown voltage is higher than the positive value and is contrary to what is observed in type 2.

Influence of Pressure On The V-t Characteristics

Figs. 2[a-c] show V-t characteristics at 0.1, 0.2 and 0.5 MPa pressures respectively. The figures also show the transition of the V-t characteristics from one type to another for different rod diameters as the pressure is increased.

At 0.1 MPa pressure (Fig.2a), corona stabilisation takes place for 1, 3 and 6 mm dia rods under impulse rise times greater than 10 μs and V-t characteristics fall under type 3. Also, due to more corona stabilisation for 1 mm rod, its breakdown voltage is higher than for 3 mm and 6 mm rods. For the less non-uniform field of 15 mm rod no corona precedes breakdown and the V-t characteristic is almost flat.

As the pressure is increased from 0.1 to 0.2 MPa, the V-t characteristics for 6 mm rod changes from type 3 to type 2 (Fig. 2b) because at 0.2 MPa there is no corona preceding breakdown for 6mm rod.

At higher pressure of 0.5 MPa (Fig. 2c) all the electrodes have rising V-t characteristics of type 2. At this pressure, the surface roughness of electrodes play a significant role and the negative breakdown values are lower than the positive values.

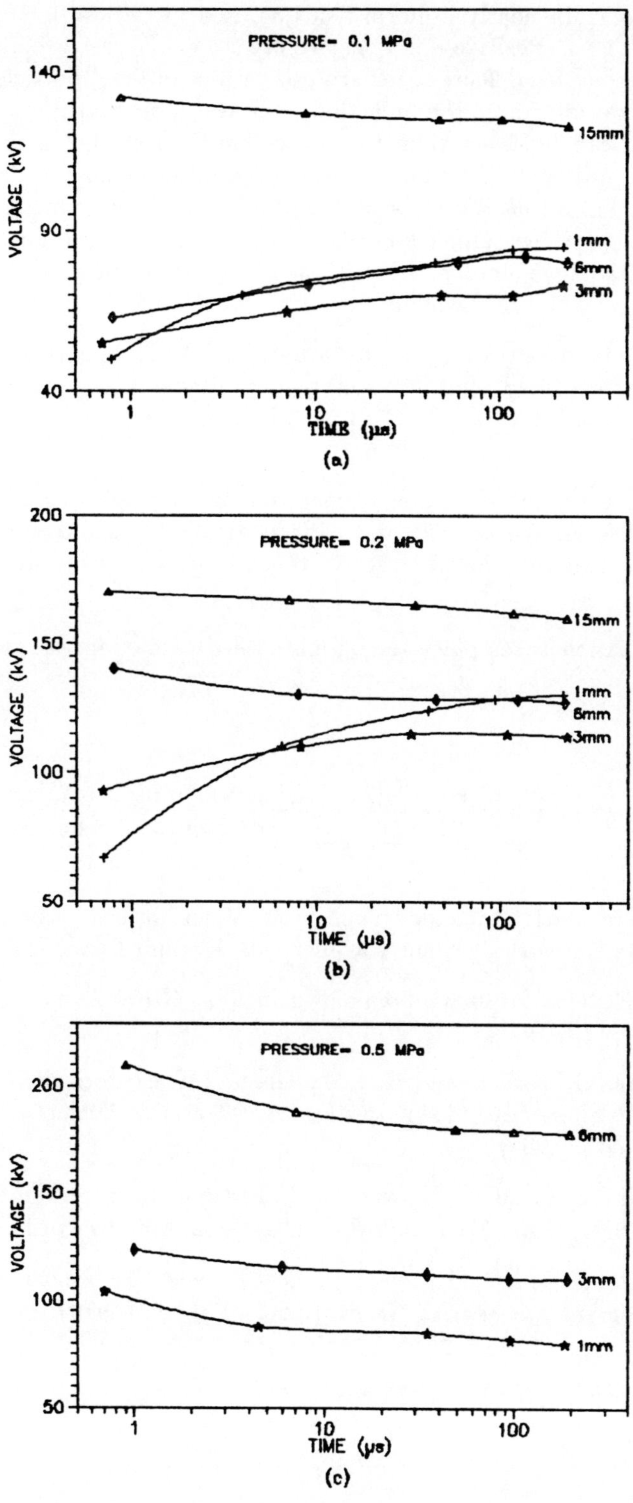

Fig.2. V-t characteristics at various pressures
(a) At 0.1 MPa (b) At 0.2 MPa (c) At 0.5 MPa

CONCLUSIONS

- V-t characteristic in nearly uniform fields as used in GIS is flat up to a time range of about 1 μs. Positive and negative breakdown values are almost equal. But if the GIS contains defects like sharp protrusions or fixed particles, resulting in high field non-uniformity, the breakdown voltage under fast impulse ($< 10\ \mu$s) is reduced and can be lower than the protective level of the arrester whereas the breakdown voltage under SI or AC is higher because of corona stabilisation effect (type 3). From this, it may be stated that in a practical system where both protrusions and particles cannot be avoided, sufficient safety margin between the arrester protective level and the LI withstand level has to be provided to ensure better reliability of the SF_6 system.
- When the GIS is installed on site, tests have to be conducted to ensure that there are no defects (field concentration) before putting it into operation. These defects can be best detected by testing with LI voltage which gives lower breakdown values.
- At higher pressures, the electrode surface roughness play a significant role in reducing the breakdown voltage and causing rising V-t characteristic due to increased statistical time lags. Negative breakdown values are lower than the positive values.
- Corona stabilisation takes place for impulse voltages having rise times greater than 10 μs.

REFERENCES

1. Farouk A.M. Rizk, Magdy B. Eteiba, Impulse breakdown voltage-time curves of SF_6 and SF_6-N_2 in coaxial-cylinder gaps, *IEEE Trans.. PAS*, Vol.101, No.12, Dec (1982)
2. W. Boeck, SF_6 insulation breakdown behaviour under impulse stresses, *in*: "Surges in High Voltage Networks," Klaus Ragaller, ed., Plenum Press, NY (1980)
3. A. Pederson, Criteria for spark breakdown in SF_6, *IEEE Trans. PAS*, Vol.89, No.8, Nov/Dec, (1970)
4. I. D. Chalmers, O. Farish, and S.J. MacGregor, The effect of impulse wave shape on point/plane breakdown in SF_6, *in*: "Gaseous Dielectrics IV", L.G. Christphorou, ed., (1984)
5. T. Nitta, Y. Shibuya and Y. Fujiwara, Voltage time characteristic of electrical breakdown in SF_6, *IEEE Trans., PAS*, Vol.94, No.2, March/April (1975)
6. O. Farish, Corona controlled breakdown in SF_6 and SF_6 mixtures, *26th Int. Conf. on Phenomena in Ionized Gases* (Invited Paper), (1983)

SECTION 12: CORONA/SURFACE DISCHARGES

INSULATION PERFORMANCE OF SF_6 MIXTURES

I. D. Chalmers, O. Farish and R. V. Shelton

Centre for Electrical Power Engineering
University of Strathclyde
Glasgow G1 1XW
UK

INTRODUCTION

Work at Strathclyde [1,2,3] and elsewhere[4] has shown that by adding small amounts of R 113 ($CClF_2$-CCl_2F) or certain other additives to SF_6, the dielectric strength of the gas can be greatly improved under certain conditions. This is at first sight surprising because in the concentrations of additive required to effect the improvement (a few percent at most), there is barely any change in the bulk dielectric properties of the gas (as measured by electron attachment or ionization coefficients). Indeed it is only under point-plane impulse conditions (or their equivalent) that the improvement is observed. Any small increases in the uniform-field dielectric strength are wholly in line with predictions from the marginally-changed net ionization coefficient[5], whereas increases of up to 50% can be obtained in the minimum impulse breakdown voltage in non-uniform-field gaps. It is worth recalling at this stage that the "minimum impulse breakdown level" is essentially the zero-probability impulse-breakdown voltage. It is derived from the well known point/plane V/p characteristics and is found to be substantially constant over a fairly wide pressure range. From an engineering viewpoint, the conditions of point-plane impulse breakdown provide an accurate model of particle-induced, power-frequency breakdown in G.I.S. and work by Farish et. al[6,13] has shown that the tolerance to conducting particles in a practical GIS is also significantly enhanced by admixing a few % of freon-R113. The present paper reports some results from a project which was stimulated by the need to investigate the physical mechanisms underlying this phenomenon in the hope that, through better understanding, an even more efficient additive and indeed the optimum conditions for improvement might be arrived at.

At the commencement of the work a considerable knowledge base relating to the discharge mechanism in SF_6 already existed from work carried out by two of the present authors[7,8] and at ABB, Baden[9,10], and therefore the aim was to devise a series of experiments to highlight the differences between the mechanism is pure SF_6 and in the mixture, in the hope of revealing which aspects of the discharge were affected by the freon, thus giving some insight into the essential physics of the processes involved.

Gaseous Dielectrics VII, Edited by L.G. Christophorou
and D.R. James, Plenum Press, New York, 1994

EXPERIMENTAL ARRANGEMENT

Details of the experimental vessel for non uniform field studies are well documented [7,8] and will therefore only be briefly commented upon here. The vessel consists essentially of a cruciform aluminium pressure vessel with a volume of about 3 liters. Two laboratory impulse generators were used, a 350kVmarx generator producing a 4/500μs waveform and an 80kV pulse-forming cable generator producing a 20ns-rise-time, flat-topped pulse 1μs long. The gas-mixing technique has also been fully described in previous publications; briefly, the vessel is pumped out by a rotary vacuum pump, a small amount of the additive is boiled into the vessel up to the required partial pressure and SF_6 is then admitted until the desired total pressure is attained.

Unless otherwise stated, experiments were carried out in commercially available SF_6 (notably containing 15ppm of water vapour and also air in concentrations up to 0.1% as impurities) in a 20mm point-plane gap. Different freons were used in parts of this study, but, to avoid unnecessary repetition, "freon" in this report refers to R113, which was the most commonly used. In all other cases the particular freon used is stated. The HV electrode used in all experiments was a paraboloidal point with a 0.5mm tip radius. In order to reduce statistical shot-to-shot variation, the gap was always irradiated with a 1 mCi ^{137}Cs γ source.

CORONA INCEPTION PROBABILITY

It now well established[7,8] that the breakdown mechanism in SF_6 under non-uniform-field conditions consists of a series of streamer and leader steps which establish a conducting path between the electrodes. The creation of the first streamer in a point/plane gap is therefore of critical importance for it is from this event that the entire sequence leading to breakdown depends. A technique previously exploited by the authors [7] has been used to identify leaders and streamers in the discharge by distinguishing their optical signatures. Leaders are "hot", the SF_6 dissociating to produce atomic fluorine which radiates spectral lines in the infra-red. No such lines are seen in the spectrum of the corona and therefore photomultiplier measurements can clearly distinguish between the two.

The inception of the first corona pulse is a statistical event which requires that there be a suitable spatial distribution of electrons in the region of the point when the voltage is applied. It was suspected at the start of the present investigation that the addition of certain additives, notably freon 113 to SF_6 might affect the availability of electrons capable of initiating a discharge and therefore it was of particular interest to determine whether the delay to corona inception changes with the presence of the additive. Accordingly, measurements were made

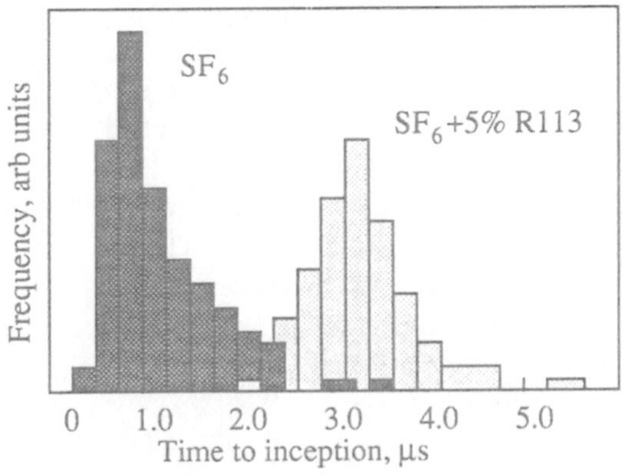

Figure 1. Distribution of corona inception times

of corona inception times using the lightning impulse. An optical photomultiplier was used to observe the gap and the time from application of the voltage wave to the occurrence of the first corona was recorded. The experimental conditions were a 0.5/20mm point-plane gap in SF_6 and in a 5% freon mixture at 1 bar with a 43kV 4/500μs wave. Figure 1 shows the distribution of corona inception times for both gases. There is clearly a large spread as would be expected, although the distribution itself was repeatable from one experiment to the next. The actual inception times recorded will obviously depend upon the particular waveform of the applied voltage and also on the irradiation level of the gap but it is possible to make comparisons for any given set of conditions. The data in figure 1 give a clear indication that the addition of R 113 significantly increases the inception time-lag (and therefore reduces the probability of inception at any given voltage level).

PROFILE OF CORONA LIGHT SIGNAL

An interesting observation was made in the course of the present work that the light from a corona discharge in the freon mixture has a distinctly different pulse-shape from that in SF_6. Figures 2a and 2b show records obtained using the "optical" photomultiplier of the light output from the whole gap with 5% R113 and with 100% SF_6. It is clear that there is a long "tail" in the light signal from the freon mixture (this is also observed in discharge-current records) persisting for some hundreds of nanoseconds. To investigate further, a quartz window was fitted to the chamber and a UV pass-band ($346<\lambda<352$ nm) filter placed between it and the photomultiplier which itself had a quartz window. The photomultiplier signal is shown in figure 2c (36 kV; 1 bar; 5% R113). The "tail" is not seen in the ultra-violet range. When the experiment was repeated using R114 ($CF_2Cl.CF_2Cl$), which also contains chlorine, again a tail in the light record was observed(figure 2d) and again it was evident only in the visible range. Although no detailed explanation for this observation can be offered at the present time, it is possible that the observed tail in these records is associated with delayed detachment of electrons from negative ions which are somewhat less stable than SF_6 ions. The fact that the phenomenon appears to be restricted to mixtures containing chlorinated freons might suggest that negative ions of chlorine could be responsible.

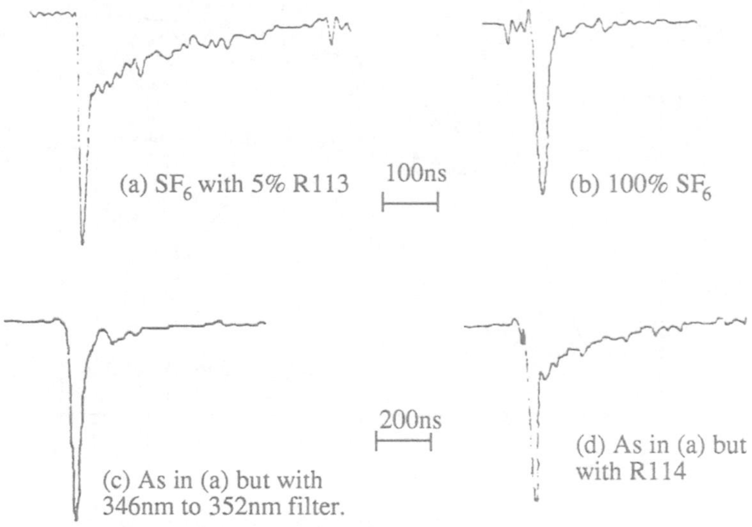

Figure 2. Photomultiplier records of the light emission from single corona pulses

MINIMUM IMPULSE BREAKDOWN LEVEL AND FREON TYPE

As argued above, the observation of the "tail" in the light signal with certain freons provides evidence which suggests that it may be the Cl^- ions which are central to whatever is happening in the freon-SF_6 mixture. Therefore a logical step in the present series of experiments was to experiment with different freons to determine if only those containing Cl had an effect on the minimum impulse level. Freon R13 ($CClF_3$), R114 ($CF_2Cl.CF_2Cl$) and R116 (C_2F_6) were used and the minimum impulse level in a 0.5/10mm point-plane gap determined using the fast-rising impulse for a 5% mixture of the various freons in SF_6. The use of the fast-rising impulse waveform reasonably ensures that corona stabilisation will not occur and that the minimum breakdown-voltage value recorded is truly that required for leader propagation. The results are shown in figure 3 and demonstrate that the mixture with R116, which is the only freon without chlorine in its molecule, behaves exactly as pure SF_6 and exhibits no enhanced minimum impulse level.

It seems fairly certain therefore that the chlorine atom in the freon molecule is associated with the enhanced electric strength of the mixture. As a further test, the minimum impulse-breakdown level can be determined as a function of the percentage of freon added. It is known that the effect of these additives saturates at very small concentrations, beyond which no further improvement in breakdown strength is observed. The question is whether the potency of the additive is a function of the number of chlorine atoms in its molecule. If the Cl^- ions are really responsible for the improvement in the dielectric strength, then one might expect that the greater the number of chlorine atoms available per molecule of additive, the more quickly will the minimum impulse level saturate and the smaller will be the amount of additive needed for the optimum improvement in dielectric strength.

Tests were carried out to determine the minimum impulse level as a function of the concentration of the additive used and figure 4 shows data for R13 and R 114 at a total pressure of 3 bar. It is clear that the R114 , which contains two Cl atoms per molecule, is more effective as an additive than R13 which contains only one Cl atom and this would tend to support the above suggestion.

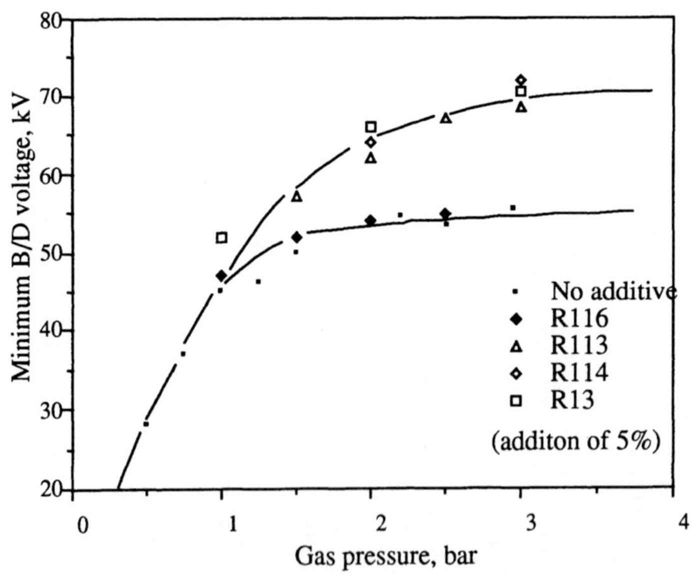

Figure 3. Minimum impulse breakdown level as a function of pressure for various additives

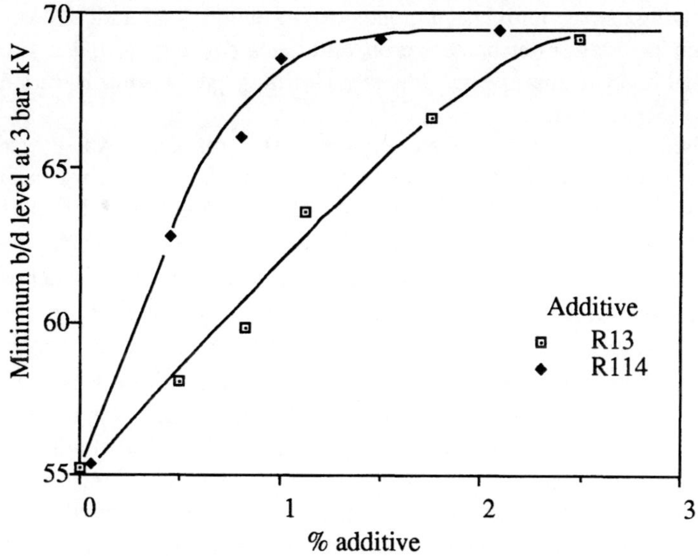

Figure 4. Minimum impulse level as a function of additive content

DISCUSSION

The results of the present work strongly suggest that the enhancement in the insulating performance of the SF_6 mixture with Freon 113, as observed in previus work, is associated with the creation of Cl^- ions from the freon molecule. One might argue that one of the effects of this will be to increase the inception voltage simply by removing free electrons but it is believed that the situation is more complex than this. In one model for the leader formation[7], the channel of the first corona discharge undergoes a series of restrikes due to the launching of ionisation waves from the region of high stress. These restrikes cause an increase in conductivity of the channel until the potential at the channel tip is sufficient for a second corona discharge to take place there. The current of the second corona discharge, which is concentrated in the primary streamer channel, provides the energy to rapidly thermalise the channel to form the first leader step. It is clearly essential in this process that there be free electrons in the region of the streamer channel tip to initiate the second and subsequent coronas. These free electrons must be formed from detachment of negative ions and the present work would suggest that the detachment processes necessary for the creation of the leader corona are changed by the freon, and that the leader propagation voltage is increased as a result. To discuss this further, the detachment processes in SF_6 and in freon must be considered.

In this context, it should be noted that the SF_6 used in this study contained significant levels of impurity water vapour and oxygen (air). Van Brunt [11] and Berger et. al. [12] have both demonstrated that the source of starter electrons for corona in SF_6 is from either O_2^- or H_2O^- ions. However the low electron affinity of these ions compared with that of SF_6 means that their mean collisional lifetime is short and therefore in the time frame of the discharge they can play no part. These authors argue, however, that hydration of the ions can stabilize them long enough for them to become involved in the discharge. Van Brunt[11] has shown that, in humid SF_6, the electron avalanches not only form with higher frequency as the above would suggest, but are also considerably larger than they are in pure SF_6. He also showed that this enhanced corona size is not explicable as a result of any increase in the ionisation coefficient of the SF_6/ water vapour mixture, but is more likely to be due to the enhanced electron detachment mechanisms giving rise to multiple, simultaneously-produced starter electrons. Applying the converse argument, a lower detachment probability would lead to a smaller streamer corona and a lower likelihood of subsequent leader steps forming; perhaps this is the situation which exists in the SF_6/freon mixture.

The key might therefore lie in the water impurity. The presence of water, as argued above, leads to the formation of hydrated negative ions with enhanced detachment but if, for any reason, there were insufficient neutral water molecules available to stabilize the oxygen ions, then their short collisional lifetimes would lead to a rapid transfer of the charge to form long lived SF_6^- ions, and this would, in turn, lead to a reduced electron detachment probability. The oxygen and water molecules are both present as impurities and there are obviously fixed numbers of each in any gas sample. However chlorine has a water-binding energy comparable with that of oxygen and hence the large numbers of highly stable Cl^- ions, even with very small concentrations of additive, will compete on at least equal terms for the available water molecules and so will reduce the number of oxygen ions which can form hydrated clusters with an associated decrease in the electron detachment probability. When the leader thermalises, this reduces the size of the leader corona and thus reduces the likelihood of the next leader propagating. A higher voltage is then required for the leader to propagate in the mixture.

The fixed amount of water vapour in the SF_6 explains the saturation in the enhancement of the minimum impulse level as more and more additive is used. Once all the water has been scavenged by the chlorine ions, hydrated oxygen ions cannot form and the detachment mechanism is then identical with that of high-purity SF_6. No further enhancement is possible.

GENERAL CONCLUSIONS

The work described and the tentative explanations given answer a number of the questions relating to the effect of additives in SF_6. Originally it was believed[3] that the effect was due only to enhanced corona stabilisation although later work indicated that under impulse conditions, (which are of direct relevance to practical situations), the stabilisation model was not really appropriate. The present work strongly suggests that the explanation for the effect lies in the ion chemistry of the mixtures. Without the additive electrons become attached to form hydrated ions of H_2O^- and O_2^- and readily detach to take part in the discharge process. Essentially the freon provides chlorine which captures electrons as stable Cl^- ions and which, in turn, hydrate leaving less water available for the formation of hydrated H_2O^- and O_2^-. The logical extension of this argument of course is that in high-purity gas where there is no significant water impurity present, hydrated ions cannot form and the additive should have no effect. This is precisely what has been found; the original experiments using amines and freon 113 as additives were carried out by the authors in Strathclyde University[3], in the Clothier Laboratory at NEI Reyrolle[13] and in Xian Jiaotong University in China[1]. In all cases, the additive was found to have a profound effect on the performance of non-uniform-field gaps and on the particle tolerance in a GIS system. In some later work[14] carried out in France, difficulty was initially experienced in producing any effect whatsoever by adding freon, the breakdown voltage being consistently high even without the additive. The results of the present study have now provided an explanation for this seemingly anomalous situation. The former three projects used commercial grade SF_6 which contains a significant level of water vapour. In the French experiments, high purity gas was used with an impurity level an order of magnitude less than that of commercial-grade gas. In fact, in these experiments, the gas had to be deliberately contaminated with water vapour before any effect of the additive could be observed. From this experience, it was then recognised that the freon somehow counteracted the effect of the water vapour although the mechanism was not at that time understood.

Thus it becomes clear that in gas-insulated equipment, freon 113, or indeed any other additive which is capable of producing negative ions of chlorine, will be beneficial in that it will mitigate the reduction in performance associated with the almost inevitable presence of water vapour. Even in systems utilising molecular filters, it is unlikely that the water vapour content could be maintained at better than a few hundred ppm at best. As a final note, it is possible that, with all the adverse publicity surrounding freons and their environmental effects at the present time, there might be an understandable reluctance to use it in GIS equipment. It is therefore of importance to investigate other more benign additives or combinations of additives which might perform the same function as the chlorine-containing freons. Here, three-gas combinations exhibiting synergistic reactions of the type described by Christophorou and colleagues[15] might well be worthy of consideration.

REFERENCES

1. I.D.Chalmers, Y. Qiu, A. Sun and B. Xu, Extended corona stabilisation in SF_6/freon mixtures, J. Phys. D. Appl. Phys. 18:L107-L112 (1985).
2. I.D.Chalmers, SF_6 mixtures, IEE Colloquium on "Latest Developments in GIS" 3/1 (1989).
3. O.Farish, I.D.Chalmers and P. Vidaud, SF_6 mixtures with enhanced corona stabilisation, Proc.7th Int.Conf.on Gas Discharges and their Applications (GD82), London: 223 (1982).
4. N.H.Malik and A.H. Qureshi, A review of electrical breakdown in mixtures of SF_6 and other gases, IEE Trans. Electr. Insul. EI-14 #1: 1 (1979).
5. N.Ijumba, PhD Thesis University of Strathclyde (1988)
6. O.Farish, I.D. Chalmers, Y.P. Feng and A. Aked, Particle initiated breakdown in coaxial systems in SF_6 and SF_6/air mixtures, Proc. 8th Int. Conf. on Gas Discharges and their Applications (GD85), Oxford: 235 (1985).
7. I.D.Chalmers, I Gallimberti, A. Gibert, and O. Farish, The development of electrical leader dischargesin point/plane gaps in SF_6, Proc. R. Soc. Lond. A412, 285 (1987).
8. I.D.Chalmers, O. Farish, A. Gibert and J. Dupuy, Leader development in short point/plane gaps, Proc. IEE, 131A #3:159 (1984).
9. L. Niemeyer and F. Pinnekamp, Leader discharges in SF6, J. Phys. D. Appl. Phys. 16: 1031 (1983).
10. F. Pinnekamp and L. Niemeyer, J. Phys. D. Qualitative model of breakdown in SF_6 in inhomogeneous gaps, J.Phys.D. Appl. Phys. 16: 1293 (1983).
11. R.J.Van Brunt, Water vapour-enhanced electron-avalanche growth in SF_6 for nonuniform fields, J. Appl. Phys. 59 #7: 2314 (1986)
12. G.Berger, B. Senouci, B. Hutzler and G. Riquel, Conditioning of sulfur hexafluoride under repetitive positive high voltage impulse conditions, Conf. Record of IEEE Int. Symp. on Electrical Insulation, Washington D.C.: 245 (1986).
13. O.Farish, I.D.Chalmers and Y.P.Feng, Particle initiated breakdown in a coaxial system in SF_6 mixtures, Conf. Record of IEEE Int.Symp. on Electricul Insulation, Washington D.C.: 206 (1986).
14. O.Farish, I.D.Chalmers, A.Gibert and J. Dupuy, The streamer/leader transition inSF_6 and SF_6/freon mixtures, Proc 6th Int Symp on High Voltage Engineering, New Orleans LA USA, Vol 2: 32.03 (1989).
15 D.L.Mc Corkle, I.Szamrej and L.C. Christophorou, Electron attachment to halocarbons of environmental interest, Chloroethanes. J. Chem.Phys. 77: 5542 (1982).

DISCUSSION

A. H. MUFTI: Why is it that the greater the number of chlorine atoms in the additive molecule the minimum is the impulse level?

I. D. CHALMERS: We believe this is because the number of hydrated Cl^- ions formed (which as a consequence leaves less water vapor available for the formation of hydrated OH^- or O_2^- ions) will be strongly related to the number of chlorine atoms available in the mixture. This should, for a given percentage of additive, be related to the number of Cl^- atoms in each molecule of additive. Thus for a given percentage of additive, there will be twice as many Cl^- atoms in an R114($CClF_2CClF_2$) mixture as there are in an R13($CClF_3$) mixture.

L. NIEMEYER: Can you explain why it takes several percent of chlorinated species to take care of several 100 ppmv of water molecules?

I. D. CHALMERS: Of course we have no information at present on the collisional cross sections for the reactions we propose. However, we do not find it particularly surprising that the number of Cl atoms required to produce apparent saturation in the increase of impulse breakdown voltage should be so much greater than the number of water molecules in the mixture.

G. C. CRICHTON: Do you anticipate any difficulties in translating results on leader suppression phenomena obtained in a macroscopic point/plane field arrangement to a perturbed field situation [i.e., particle (microscopic) superimposed or a cylindrical field (macroscopic)]?

I. D. CHALMERS: Our earlier work has shown that there is a strong correspondence between the characteristics of a point/plane gap and those of a particle-contaminated GIS system. This results from the nature of breakdown in GIS when particles are present. When the particle approaches the bus bar at or near a positive voltage peak, a discharge occurs from the bus bar to the particle. This means that the particle potential is suddenly increased to the positive peak bus bar voltage and a stepping leader discharge occurs just as in a point/plane gap under impulse application. Results obtained in a full-scale GIS system have clearly shown that a 1% concentration of additive can increase the AC breakdown voltage by approximately 50%.

R. J. VAN BRUNT: The effect of the H_2O on the breakdown of SF_6 is due to the fact the H_2O produces negative ions such as OH^- that detach electrons much more readily than any of the negative ions that result directly from SF_6, such as F^-. The effect of Cl-containing additives in inhibiting breakdown probability in SF_6 containing water vapor could be due to ion-molecule reactions that convert ions such as OH^- to other ions that do not detach as readily by collision.

I. SAUERS: We have made mass spectrometric measurements of ions sampled directly from a corona discharge and have observed the conversion of SF_6^- ions to OH^- and $OH^-(H_2O)_n$ clusters. Other additives can further change the kinds of ions that are present which in turn would influence the detachment properties. It would be interesting to look at the ions produced in these mixtures.

COMBINED OPTICAL AND ELECTRICAL MEASUREMENTS ON PULSED CORONA DISCHARGES

P.P.M. Blom, C. Smit, R.H.P. Lemmens, E.J.M. van Heesch

High-Voltage and EMC group
Eindhoven University of Technology
P.O. Box 513, 5600MB Eindhoven, The Netherlands

INTRODUCTION

To produce very intense coronas, a pulsed high voltage is applied. The peak voltage can be far above the breakdown value, if the pulse width remains below a few microseconds. Aiming at high charge production during the pulse we found that the pulse risetime is a sensitive parameter[1]. A DC bias voltage below the inception voltage is added to transport the charges to the wall[2]. The application of short pulses on top of a DC voltage instead of pure DC allows a much wider voltage range for corona operation. This wider voltage range is very advantageous in situations of serious contamination or at high operating temperatures, in the range of (400-800)°C[3]. Pulsing also permits a separate control of production (pulse) and transport (bias). Thus very high charge production rates can be achieved. Our work further aims at improving this charge production and at an understanding of these very high corona intensities. The absence of a full breakdown and the very interesting high corona current (we observed up to 600 A/m) are not fully understood however, a possible explanation might be connected to the 'bright spot' regime described later in this paper.

Previous work concentrated on the observation of corona activity using electrical diagnostics. We were especially interested in the intense corona activity near the aluminum cylinder which surrounds the corona wire. The activity occurs with both positive and negative wire voltages[4]. Two corona modes were distinguished: a moderate and an intense mode. We attributed the major contribution to the intense mode to this wall activity. A special grid sensor is used to quantify this additional high wall activity in terms of charge per meter of corona wire.

Recently we have added an image intensified CCD-camera and a photo-multiplier to the experiment. Using advanced timing techniques and the short shutter time (25 ns) of the camera, we are able to obtain semi time-dependent (one frame per corona pulse) spatial knowledge of the corona discharge. By taking proper EMC measures we now can make combined optical and interference-free electrical measurements.

Gaseous Dielectrics VII, Edited by L.G. Christophorou
and D.R. James, Plenum Press, New York, 1994

Using two, four channel high-speed oscilloscopes (HP 54542) and fast electrical diagnostics we increased the bandwidth of our electrical measurements.

EXPERIMENTAL SETUP

Electrical Diagnostics

The setup shown in Fig. 1 consists basically of a pulse source, a corona wire and a bias circuit. The 120 kV pulse is generated by the discharge of a capacitor via the main spark gap onto the corona wire. A series resistor determines the rise time (down to 20 ns) of the voltage pulse. At the other end of the corona wire a chopping gap can reduce the pulse length down to about 400 ns. At this side also a bias voltage can be applied. A pulse-sharpening spark gap is available inside the corona vessel for enhancement of the corona intensity. With two measuring cylinders inside the vessel we can detect the vacuum displacement current I_0 (upper cylinder) and the total external current I_e (lower cylinder).

A differentiating/integrating (D/I) system[5] is used to detect the voltage pulse V_p. We adjusted the pressure inside the vessel at 360 Torr to simulate the density of air at 400°C.

The pulse source induces large common mode currents in the leads of the experiment. We minimize interference problems by adequate EMC measures; we used for instance a special EMC cabinet and a D/I system.

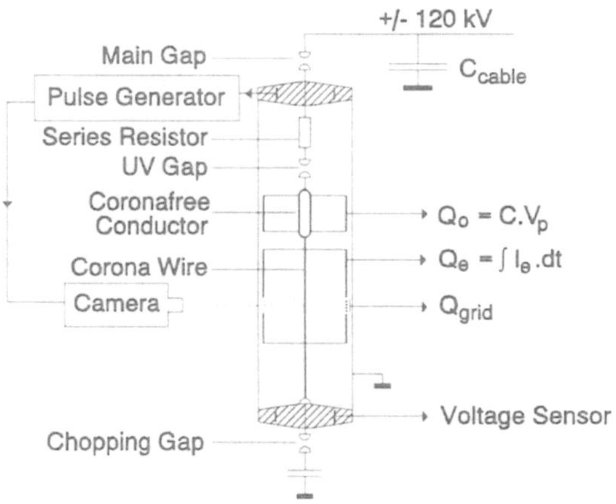

Figure 1. Experimental setup.

Imaging Diagnostics

The electrical diagnostics provide time-resolved information. The addition of an image intensified CCD-camera system allows us to investigate the spatial behavior of the corona discharge. The high sensitivity and the fast gating capability result in the desired time resolution (shutter times down to 25 ns). The readout of the CCD array takes approximately two seconds, therefore we can only make one image per corona pulse. Via a D/I system and a delay unit the camera is triggered by the high-voltage pulse on the wire. Due to the electronic delays, a minimum delay of 70 ns between the high-voltage rise on the wire and the gating of the camera is unavoidable.

The camera system is controlled by a PC, that is located in the same EMC cabinet as the oscilloscopes. The control of the experiment and data acquisition can be done from within the EMC cabinet.

MEASUREMENTS

ICCD-images; general observations

The images in this paper are presented using the following rules.
No advanced mathematical post-processing (e.g. smoothing, contrast enhancement) has been applied to images. We have chosen for a direct representation of the measured light intensity, using a linear gray-scale for the images. The images in this paper do not represent the absolute light intensity, since the offset of the linear gray-scale is adjusted to optimize the appearance of the pulsed corona structure.

Radial and axial intensity profiles are used to present the corona development process. The radial and axial profiles are calculated by summation of the measured light intensity along the vertical and horizontal axis respectively of the original CCD-image (see Fig. 2). These profiles represent the absolute measured light intensity, emitted by the corona discharge.

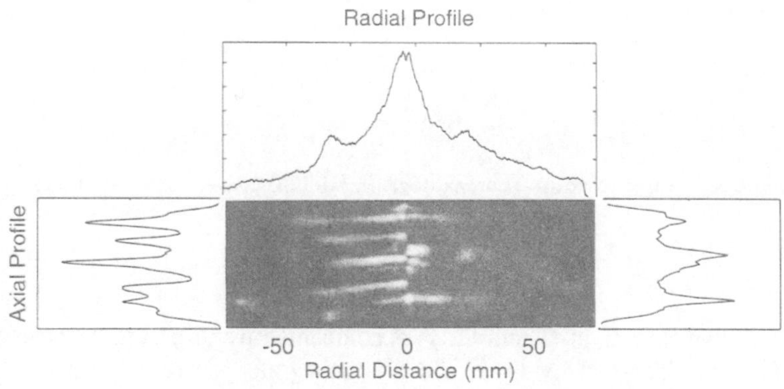

Figure 2. An image and the radial and axial profiles calculated from it.

This first optical study of our pulsed corona concentrates on the region around the wire. Optical activity is found to occur during the first microsecond of the discharge. Simultaneous short time-scale electrical measurements (1 μs) are also presented.

Measured Images and the Electrical Measurements

Figure 3(a,b) shows eight images for both positive and negative corona. These images represent two important optical regimes in the pulsed corona discharge. The images at time t1 are made at the shortest time-delay possible in our experiment (70 ns). They clearly show that the early phase of positive corona and negative corona is dominated by very interesting 'bright spots'. All images made at time t2 show regions of optical activity in contact with the corona wire: the secondary streamer regime. After the secondary streamer regime, the optical activity extinguishes and the drift phase of the corona discharge starts.

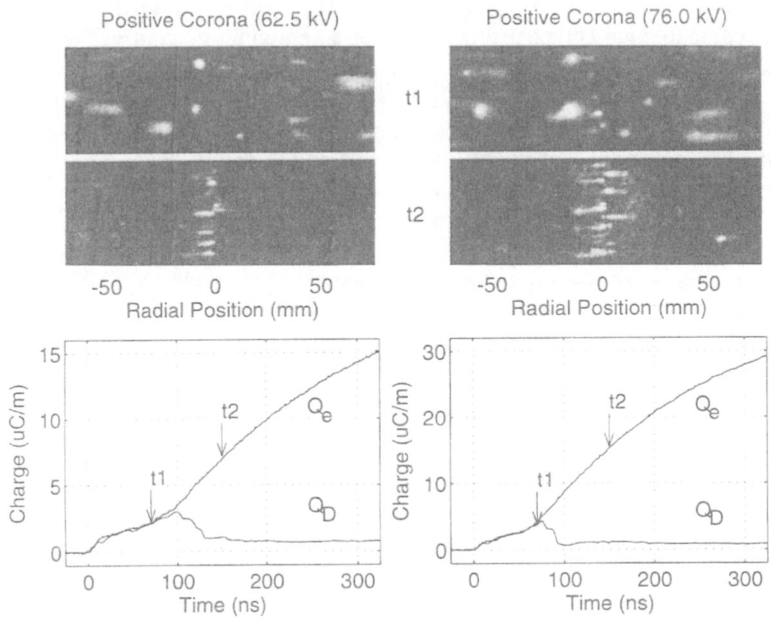

Figure 3a. Images of positive corona, together with the corresponding electrical signals.

Each set of two images at t1 and t2, is accompanied by its electrical measurements in Fig. 3. The curve denoted by Q_e is the time integrated current I_e (in the lead of the corona cylinder). Our grid sensor[6,7] measures the time-dependent change of the E-flux in front of the sensor, in the form of a displacement current I_D. Integration of the displacement current yields the curve Q_D, which represents the E-field near the cylinder electrode.

Clearly the first gradual increase of the external charge Q_e fully fits the E-field. The subsequent E-field collapse is accompanied by an enhanced growth of Q_e. The separation of Q_e and Q_D is due to material charge (ions/electrons) leaving or arriving at the cylinder electrode[6].

Streamer Development

The end of the 'bright spot' regime is reached earlier when the voltage is increased. Also the secondary streamer regime starts earlier with increasing voltage. The images and electrical measurements in Fig. 3 are an illustration of this behavior. The E-field collapse for example also comes earlier when the voltage pulse is increased.

Since the first 70 ns can not be seen, more of the early discharge stage is detected if we slow down the discharge process by decreasing the pulse voltage. The pulse voltage however must remain above the onset for wall activity (E-field collapse at the cylinder). For the observation of the early corona development we therefore had to use a relatively low voltage. Results are summarized in Fig. 4. The images are made using an increasing delay hereby time stepping through the corona process.

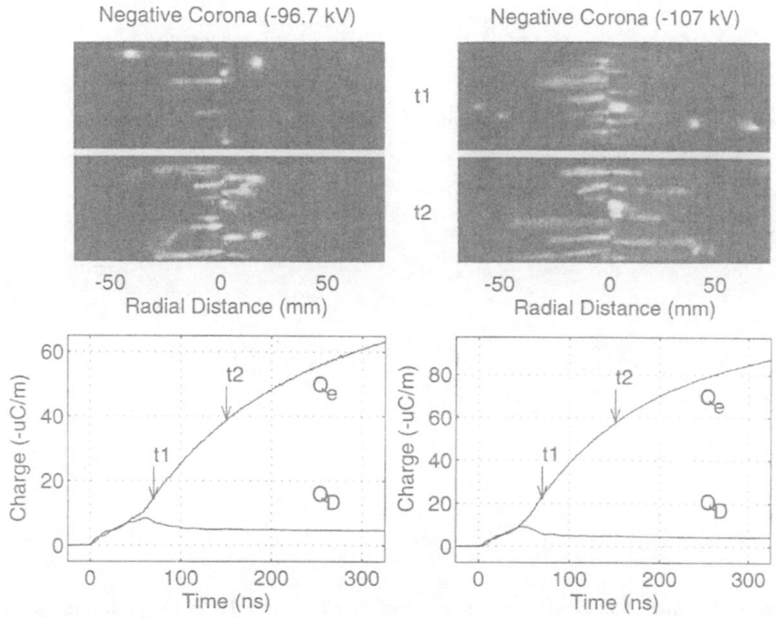

Figure 3b. Images of negative corona, together with the corresponding electrical signals.

Shown are radial intensity profiles calculated from the images, the Q_D signal and the r-t diagram of the 'bright spots'. The short horizontal baseline connected to the right side of the radial intensity profiles marks the zero-level for each of the radial profiles, and the opening time of the camera shutter.

We assume that during the 'bright spot' regime the 'bright spots' cross the cylinder volume and that the E-field collapses when the 'bright spots' reach the cylinder electrode. The high peaks seen in the radial profiles are the 'bright spots' moving outward in time, and the optical activity around the wire for $t > 120$ ns corresponds to the secondary streamer regime.

The radial positions of the separate 'bright spots' in the r-t diagram are directly measured from the image. The gray shaded area represents the possible trajectories of 'bright spots' moving towards the cylinder at r=150 mm.

The length of the bright spots increases as they move outward. From these lengths and the 25 ns gating time, a velocity increase 1.10^5 m/s near the wire to 6.10^5 m/s at r=7 cm is deduced. This velocity increase is in accordance with the marked gray shaded area.

From Fig. 4 it can be concluded that the 'bright spot' regime predominately exists during the gradual increase of the E-field (Q_D). After the 'bright spots' have reached the cylinder the secondary streamer regime, starts to develop. The secondary streamer regime is not detected when the pulse voltage is to low for the E-field to collapse. We may conclude, that the secondary streamer regime is initiated as soon as the 'bright spots' reach the cylinder electrode.

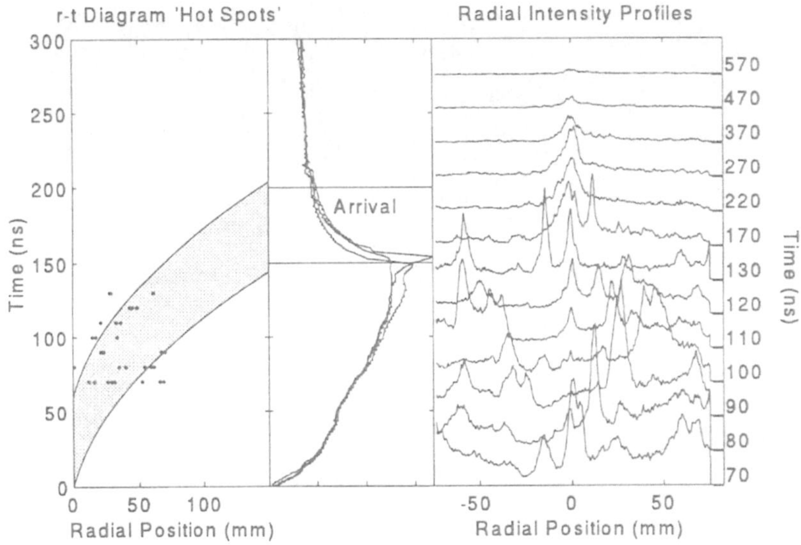

Figure 4. An r-t diagram of the 'bright spots' and the corresponding E-field Q_D and the radial profiles as a function of time.

CONCLUSIONS

Simultaneous interference free detection of E-field, external current and 2-D optical discharge activity is made at a very high time resolution. In this way a collection of perfect matching data on the early development of pulsed corona is obtained.

We observed two optical regimes during the initial phase of both positive and negative pulsed corona, which we call the 'bright spot' and the secondary streamer regime.

Our measurements of the optical activity of pulsed corona correspond well to the measured E-field and the external current.

ACKNOWLEDGEMENT

These investigations have been supported by the Netherlands Technology Foundation (STW).

REFERENCES

1. E.J.M. van Heesch, A.J.M. Pemen, J.W. van der Snoek, P.C.T. van der Laan, Pulsed Corona Experiments, Gaseous Dielectrics VI, p.391, Knoxville, 1990.
2. H.J.F. Kloth, E.J.M. van Heesch and P.C.T. van der Laan, Processes in Pulsed Electrostatic Precipitators, Proc. 5th Int. Symp. on High Voltage Engineering, Vol. 3, 91.05, Braunschweig, 1987.
3. E.J.M. van Heesch, A.J.M. Pemen, P.C.T. van der Laan, Pulsed Corona Existence up to 850°C, Proc. 6th Int. Symp. on High Voltage Engineering, Vol. 1, 14.07, New Orleans, 1989.
4. E.J.M. van Heesch, M.J.A.M. van Helvoort, P.C.T. van der Laan, High Current Pulsed Coronas of Both Polarities, Proc. Xth Int. Conf. on Gas Discharges and their Applications, Vol. 1, p.294, Swansea, 1992.
5. E.J.M. van Heesch, A.P.J. van Deursen, M.A. van Houten, G.A.P. Jacobs, W.F.J. Kersten, P.C.T. van der Laan, Field Tests and Response of the D/I High Voltage Measuring System, Proc. 6th. Int. Symp. on High Voltage Engineering, Vol. 3, 42.23, New Orleans, 1989.
6. P.P.M. Blom, E.J.M. van Heesch, P.C.T. van der Laan, Intense Pulsed Corona Due to a Highly Active Cylinder Electrode, Proc. 8th Int. Symp. on High Voltage Engineering, Yokohama, Japan, August 23-27, 1993, vol. 1, 36.08, pp. 429-434.
7. E.J.M. van Heesch, P.P.M. Blom and P.C.T. van der Laan, An E-field and an Ion Sensor for High Intensity Pulsed Coronas, Proc. ICPIG XXI, Bochum, Sept. 19-24, 1993, pp. 329-330

DISCUSSION

E. MARODE: (1) Does the origin of the streamer channels on the wire change place when making different voltage pulses at the same voltage amplitude? (2) What is the relation between the number of discharges per unit length of the central wire and the overvoltage?

P.P.M. BLOM: (1) The distribution of streamer channels along the corona wire is not determined by protrusions, and is therefore not fixed. (2) As can be seen in Fig. 3 of our paper, an increasing pulse voltage yields an increasing number of streamers. A more precise expression relating voltage and number of streamers has not yet been determined.

M. GOLDMAN: Do both positive and negative streamers move towards the cylinder electrode?

P.P.M. BLOM: For positive corona it is clearly shown in Fig. 4 that streamers move outward to the cylinder electrode. Due to the higher pulse voltage, the streamer development in Fig. 3b (negative corona) is faster than in Fig. 3a (positive corona). The minimum delay time of 70 ns and the fast development of streamers with negative corona causes the "bright spots" to move outside the boundaries of the image at $r=75$ mm (the radial position of the cylinder electrode is $r=150$ mm). Therefore it is not possible to determine the direction of the streamer displacement for negative corona.

LONG-TERM EVOLUTION OF POINT-TO-PLANE SF_6 DISCHARGES UNDER ALTERNATING VOLTAGE

M. Lalmas[1], H. Champain[1], A. Goldman[1], E. Fernandez[2]

[1] Laboratoire de Physique des Décharges (CNRS)
Ecole Supérieure d'Electricité
Plateau de Moulon, 91192 Gif sur Yvette Cedex, France
[2] Société Merlin-Gerin, 20 rue Henri Tarze, 38050 Grenoble Cedex, France

INTRODUCTION

The present paper deals with the evolution, over long durations, of SF_6 insulated systems submitted to ac high voltage conditions, able to provide ionisation and discharge phenomena in the gaseous gaps involved. Earlier studies were already carried out in the authors laboratory on such systems fed by dc voltages[1,2] and in this case it was found that the corona currents evolve with time towards new equilibrium states, very dependent on the polarity of the high field electrode, through physical and chemical changes affecting the gas-electrodes system with dominating effects on the high field electrode and with the occurrence of transient delayed spark breakdowns through mechanisms studied according to the polarity.

In the present paper, we shall see that, when a new gap (newly prepared electrodes with new gas) is put into operation under ac conditions, a conditioning period also takes place, with an increasing risk of getting transient delayed spark breakdowns but that, in this situation, currents in the two polarities half cycles are no more varying according to the polarity in the same way as under dc conditions.

EXPERIMENTAL

The experiments are carried out in a 9 litres stainless steel vessel with a point-to-plane electrode system. The point electrode made of aluminium presents a typical tip curvature radius of about 60 µm and the plane electrode is a disk of 100 mm diameter, in copper to allow easy diagnostics on the transport of matter between the two electrodes during the discharge runs. The gap length is fixed at 10 mm.

All experiments are performed with SF_6 of 99.9% purity introduced at a pressure of 2 bars in the vessel prealably evacuated to a residual pressure of $p<10^{-4}$ mbar.

Discharge runs of hours duration are carried out with a 50 Hz high voltage supply of maximum output power 7 kVA, most generally at a 60 kV peak value, independent of the

Gaseous Dielectrics VII, Edited by L.G. Christophorou
and D.R. James, Plenum Press, New York, 1994

corona current variations. Current-voltage curves are recorded over short characterization runs performed at the beginning of the long duration runs which are carried out with new sets of electrodes and new gas, (*step 1*), at their end, (*step 2*), and at last after a renewal of the gas (*step 3*) without modifying the surface state of the electrodes in order to appreciate the respective contributions of the gas and electrodes modifications on the changes appeared in the discharge behaviour, between the two preceding steps.

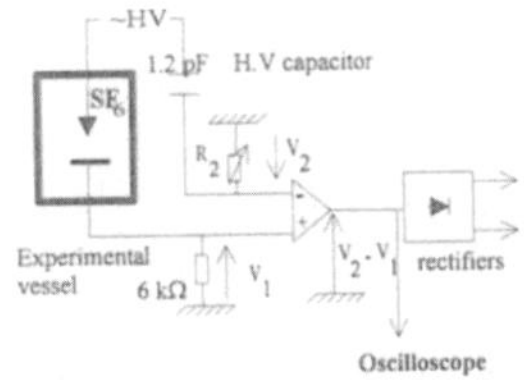

Figure 1 . Schematic view of the experimental set-up and of the device used for the separation of the active current components corresponding to the two polarities half-cycles.

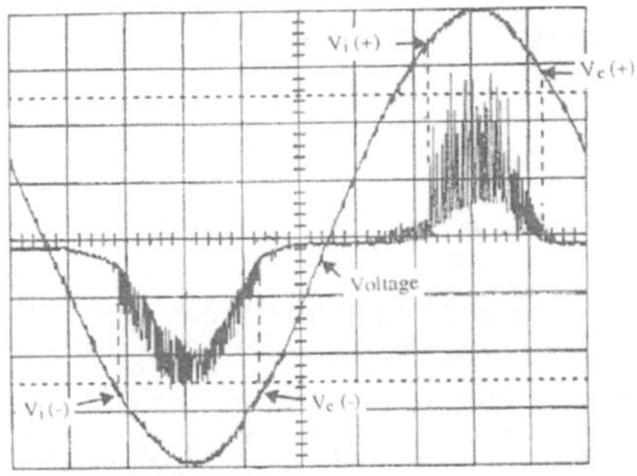

Figure 2 . Typical oscillograms of the ac voltage (13.75kV/div) and of the total current variations (33μA/div) during a cycle at the end of a discharge run (*step 2*). V_i and V_e represent the inception and extinction voltages.

The circuit shown in Fig.1 was used in combination with a 9400 Lecroy oscilloscope in order to fulfil the following requirements : (i) A correct adjustment of the resistance R_2 allows the suppression of the 50 Hz passive displacement current from the waveforms of the total discharge current and thus, at this stage, the acquisition of global data on the sum of the different active current components into play, i.e. of the voltage-synchronous components due to ionization phenomena localized in the close vicinity of the point, and of pulse components, in particular typically representative of the development of streamers[3]. The sensitivity provided by the circuit is better than 5 pC, but, because of its time constant ($\approx 1\,\mu s$), all measurements hereafter reported will be given in the form of mean values. The S^+ and S^- outputs of the circuit already provide separate mean values for the total active currents in each of the two polarities half-cycles [$\bar{I}_{tot}(+), \bar{I}_{tot}(-)$]. As an example, the oscillogram of Fig.2 shows the variations of the currents during a given cycle at the end of a long discharge run. Marks indicate, on this oscillogram, the inception V_i and extinction V_e voltages, respectively determined by the big pulse marking the onset of self-sustained discharges[4] and their end in the different half-cycles. (ii) The use of the Lecroy oscilloscope allowed to achieve a separate analysis of the different components involved in the active current waveforms get for each polarity half-cycles at the output of the bridge circuit. Mean values of the synchronous currents are obtained by using the "extrema" function of the oscilloscope which allows to get the envelope formed by the lowest absolute values over a large series of half-cycles of a same polarity ; the area delimited by this envelope represents the total ion charge which, divided by the duration of one cycle (20 ms), gives minimized mean values for the synchronous currents [$\bar{I}_{synch(+)}, \bar{I}_{synch(-)}$]. Then, by subtracting these values from the mean values of the total currents [$\bar{I}_{tot(+)}, \bar{I}_{tot(-)}$], one gets mean maximized values for the pulse currents [$\bar{I}_{imp}(+), \bar{I}_{imp}(-)$] corresponding to the half-cycles of each polarity.

RESULTS

Electrical Analysis of the Discharge Evolution with Time.

Let us consider as a typical case that one of a point-to-plane gap of d = 10 mm gap length, constituted by an aluminium point, and by a copper plane.

The electrical behaviour of the discharge at its beginning (*step 1*) and after a long run (*step 2*) are illustrated by the Fig.3 to 5 which will be discussed hereafter with particular attention to the values obtained for the same voltage (60 kV peak value) as that one used for the long run.

Results on the Total Active Current. At every half cycle, the discharge undergoes an interruption delimited by the inception and extinction voltages V_i and V_e referred to in the Fig.2. This determines dead times not taken into account for the determination of the total currents $\overline{I}_{tot(+)}$ and $\overline{I}_{tot(-)}$ reported in Fig.3 as a function of the peak voltage V_{cr}, but over long durations, the V_i and V_e values undergo an increase; so the dead times are also regularly increasing. Can their increase be sufficient to explain the decrease of $\overline{I}_{tot\,(+)}$ and $\overline{I}_{tot(-)}$ along the long discharge run (Fig.3) ? It comes out from the results that the currents are also decreasing in instantaneous values in the same time, and this in a still more important way. For instance in the case considered in Fig.3, the decreases of $\overline{I}_{tot\,(+)}$ and $\overline{I}_{tot\,(-)}$ between the steps 1 and 2 are respectively of 45 % and 27 % , due to active time decreases of respectively 10 and 3 % and to amplitude decreases of respectively 35 % and 24 %.

Finally, the results also show that, in most cases, the current in the negative half-cycles predominates over that one in the positive half-cycles at the end of the discharge run as at its beginning in the conditions of our investigations.

Results on the Voltage-Synchronous Components of the Current. A separate analysis of

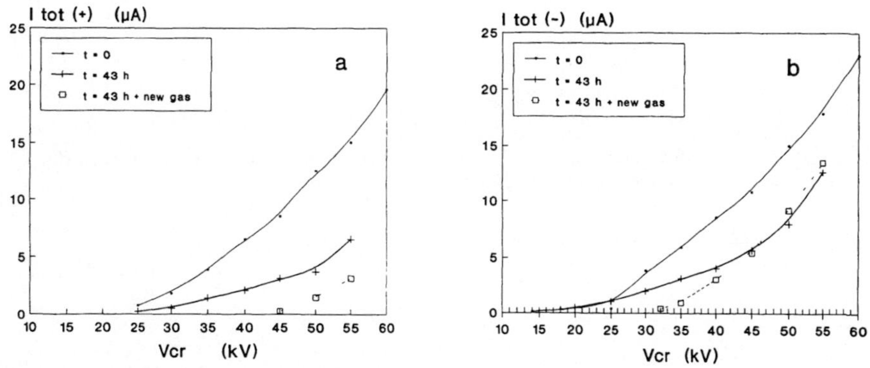

Figure 3 . Behaviour of the mean total current of a discharge, at its beginning (*step 1*), after 43 hours discharge under 60 kV peak voltage but before any gas renewal (*step 2*) and after a gas renewal (*step 3*) : (a) in the positive half-cycles, (b) in the negative half-cycles.

the $\overline{I}_{synch}$ and $\overline{I}_{imp}$ components of the active current of the discharge bring clearer ideas on the evolution of the dielectric strength of the system constituted by the gap under consideration. Let us first consider the curves of Fig.4 a and b which give informations on the mean synchronous currents $\overline{I}_{synch(+)}$ and $\overline{I}_{synch(-)}$ through the variations of the functions $\overline{I}_{synchr}$ / V_{cr} = f(V_{cr}) where V_{cr} represents the peak value of the 50 Hz high voltage waveform supplied to the discharge gap. The linearity of the curves obtained in all cases shows that a Townsend-like relation gives a good fit to describe the synchronous current variations as a function of the ac peak voltage for both polarities half-cycles. This fitting finds a physical justification by using the well known relation between the current of corona discharges and the voltage V under dc conditions $I=AV(V-V_0)$, where A is a coefficient depending on both the gap geometry and the gas and where V_0 is the corona threshold voltage. Applying this relation to the case of a sinusoidal voltage waveform :

$$V(t) = V_{cr}\sin(2.\pi.t / T) \qquad \text{with} \qquad \bar{I}_{synchr} = \frac{1}{T}\int_0^T i(t)dt,$$

where T is the cycle duration, leads to a current-voltage relation in the form of $\bar{I}_{synchr}=A'V_{cr}(V_{cr} - V_0')$, where A' depends on the inception and extinction voltages V_i and V_e of the self-sustained current in the half-cycles for the polarity under consideration, while V_0' is a function of V_0 and of both limiting voltages V_i and V_e too. This confirms that the synchronous current components are space charge limited components.

Results on the Pulse Components of the Current. As illustrated in Fig.5 a and b, for both point polarities, the pulse components show similar trends and temporal behaviour

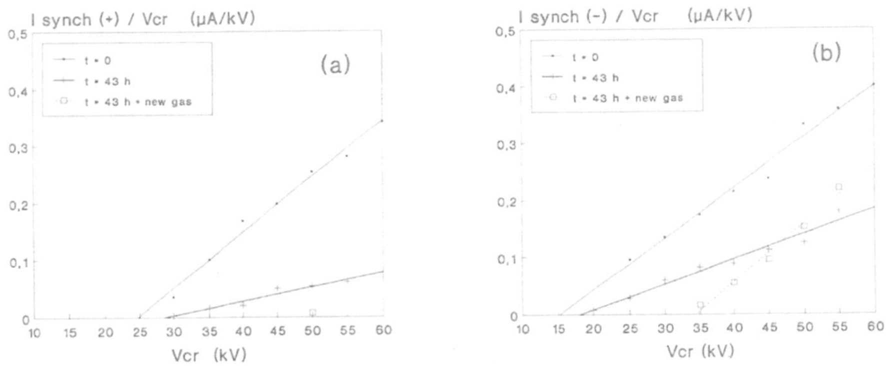

Figure 4 . Evolution of the characteristics I_{synch} /Vcr = f(Vcr) for the same experiment as in Fig.3 : (a) in the positive half-cycles. (b) in the negative half-cycles.

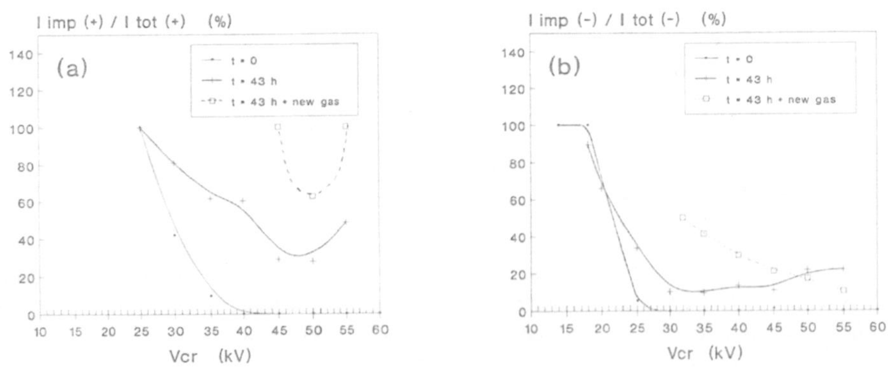

Figure 5 . Evolution of the contribution of the pulse current with regard to the mean total current versus the ac peak voltage for the same experiment as in Fig.3 : (a) in the positive half-cycles. (b) in the negative half-cycles.

with respect to the ac peak voltage variations. At the beginning of any discharge run, apart from the big pulse which initiates the self-sustained discharge in every positive or negative half-cycle, there is no pulse activity. But a pulse activity appears while time is running, as a result of the modifications produced in the gas and at the electrodes by the discharge. After a few hours, a new stabilisation state is reached with, then, a relatively important pulse activity, implying in the case considered in Fig.5 ~ 40 % of the total current for the positive half-cycles and ~ 20 % for the negative half-cycles. It is important to emphasize here that $\bar{I}_{imp(+)}$ and $\bar{I}_{imp(-)}$ variations are about equivalent in absolute values, so that the differences between $\bar{I}_{imp(+)} / \bar{I}_{tot(+)}$ and $\bar{I}_{imp(-)} / \bar{I}_{tot(-)}$ come essentially from the differences between $\bar{I}_{tot(+)}$ and $\bar{I}_{tot(-)}$.

Gas and Electrode Phenomena Involved in the Evolution of the Discharge with Time.

How phenomena in the gas volume and at the electrodes can contribute to the variations observed in the discharge behaviour with time, i.e. a loss of synchronous current simultaneously with an increase of the pulse activity, both observations going in the direction of a loss of corona stabilisation ? It is to get an answer to this question that the measurements which refer to the *step 3* (after a gas renewal) in the Fig.3 to 5 have been made. They have been achieved with the assumption that the gas renewal made after a discharge run does not affect the electrode modifications brought by this discharge run. The electrodes contribution was appreciated by the difference between the current measurements at *step 3* (after a gas renewal) and *step 1* (t=0) and the gas contribution by the difference between *step 2* (after the discharge run, but before a gas renewal) and *step 3* (after a gas renewal). The following results for the case corresponding to Fig.3 and 5 were so obtained : (i) In the positive half-cycles, a decrease of ~ 80 % on $\bar{I}_{synch(+)}$, resulting from a reduction by the electrodes contribution of ~ 100 %, partially compensated by an amplification of ~ 20 % by the gas contribution. (ii) In the negative half-cycles, a decrease of ~ 50 % on $\bar{I}_{synch(-)}$, with ~ 40 % attributable to the electrodes effects and ~ 10 % to the gas effects. (iii) In both polarities half-cycles, an increase of $\bar{I}_{imp}$ from zero at time t=0 to a few microamperes as discussed above, with only a contribution of electrodes phenomena in the positive half-cycles and with a contribution of both gas and electrode phenomena in the negative ones.

Part of these results have already received experimental confirmations. This is in particular the case for the electrode phenomena which appear to have predominant effects. We have just seen that they are responsible for reduction effects on the synchronous current in both polarities half-cycles. This can be explained : (1) By the erosion phenomena which affect the point electrodes. Fig.6 illustrates this erosion which was confirmed spectroscopically by the presence of the ray 396.15 nm of aluminium in the gap and with EDX analysis by the detection of aluminium on the copper plane electrode. The net result is an increase of the radius of curvature of the point electrode at its apex with subsequent decrease of the electric field in its vicinity. (2) Additionally, by the layers built on the electrodes as a result of the gas-surface interactions which have taken place during the discharge run. On the aluminium point electrode is found a fluorinated layer, a sectional view of which is seen in Fig.7; on the copper plane electrode is also found much fluorine distributed evenly for a part, but also often in the form of granules, generally of globular shape when combined with the metal of the point electrode (Al) and of more irregular shape when combined with the metal of the plane electrode (Cu), that can give indications on the origin of the fluorine.

In a general manner, the surface layers formed on both electrodes exhibit insulating properties demonstrated on the one hand by their charging effects which bother the observations in the apparatuses for surface analyses and also on the other hand by the time (tens of minutes or hours scale) over which an influence of preceding discharges can be felt on current measurements; furthermore, these insulating properties were controlled by electrical means (contact measurements on the point electrode, surface potential measurements on the plane electrode). A consequence of the presence of insulating layers on the electrodes is that charge accumulation can take place on them, but the permanent imbalance between $\bar{I}_{tot(+)}$ and $\bar{I}_{tot(-)}$ proves that current can permanently or quasi-permanently flow across. The microchannels which are visible across the surface layer at the point electrode (Fig.7) show how local microdischarges and/or even microsparks can develop through the layers; they may favorize the initiation of streamers and help delayed spark breakdown to occur. On the other hand, the development of retarding fields beyond the surface layer on the point electrode by charge accumulation on it may help the decrease of the synchronous current components, in the same way as the field decrease induced by the erosion of the electrode at its apex.

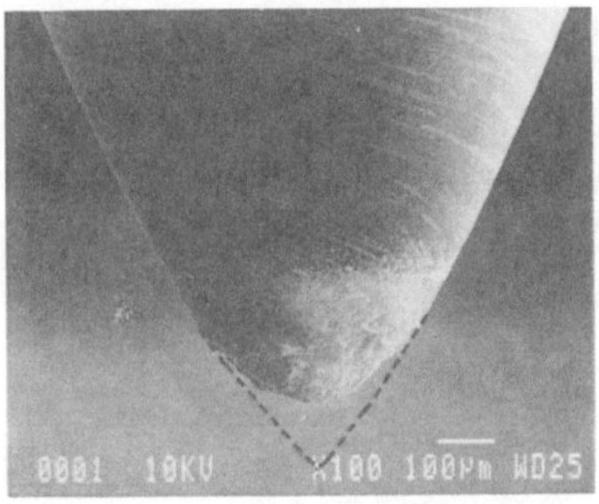

Figure 6 . Aspect of an aluminium point electrode after 7 hours ac discharge under 60 kV peak voltage. The initial profile is indicated in dashed line.

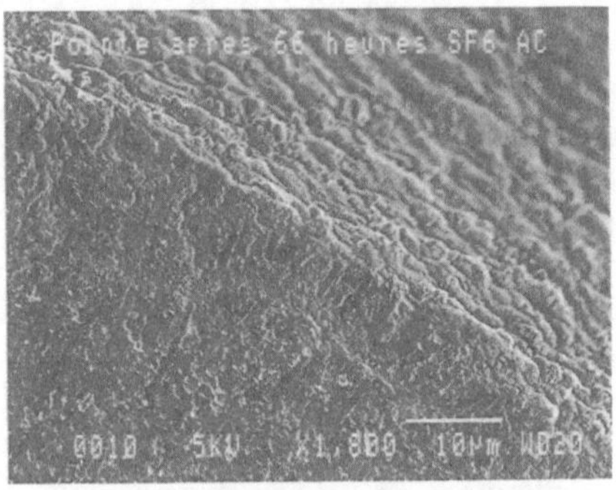

Figure 7 . Aspect of a fluorinated layer formed on a aluminium point after 66 hours under 60 kV peak voltage. Top view at the right side and oblique section showing hollow channels at the left side.

Now concerning the smaller role which should be attributable to the gas, one should have to deepen different hypotheses not yet checked. Let us just enumerate some possible causes : (1) Ionic mobilities changes (new by-products[5], clustering effects[6], fragmentation effects[7]); (2) Modifications affecting the field across the gap : the retarding fields due to charge accumulation on the surface layers should produce decreasing effects on the applied field as said above, but perhaps also gas heating effects which could have opposite effects by increasing the reduced applied field[8].

This discussion was led with the objective to get informations on the respective roles played on the variations of the discharge behaviour with time by the gas modifications and by the changes at the electrodes, but both types of contributions can be indeed tightly linked. This is for instance illustrated by the effects of electrode modifications on the applied field, acting on the discharge through both electrode and gas processes as evoked above or by surface reactions modifying both surface and gas chemistry (see for instance Casanovas et al. work[9]).

CONCLUDING DISCUSSION

The aim of the work was to characterize the electrical behaviour of SF_6 discharges in heterogeneous gaps under ac conditions, below disruptive voltages, and their evolution with time through electrodes and gas modifications in the spirit of accelerated tests for the study of the mechanisms involved in the ageing of SF_6 insulated systems. This has been achieved for a point-to-plane gap of 10 mm length in 2 bars SF_6.

In a general manner, newly prepared electrodes in new SF_6 practically only deliver positive and negative glow currents which have been referred to in the paper as voltage-synchronous currents, but after a certain evolution time (a few hours) a pulse activity arises while the stabilizing glow currents in both polarities half-cycles severely decrease in duration and amplitude with the net result of a decrease on the mean total currents of each polarity half-cycles.

Then a new equilibrium state develops with higher risks of getting delayed spark breakdowns which have been observed during this study but not systematically investigated. However, some limited observations on them can be already reported, summarized as follows : (i) The delayed spark breakdowns always occur in the positive half-cycles and only when the discharge already exhibits a pulse activity, probably correlated to surface layers, in formation or already formed on the electrodes. It is interesting to note that, in the dielectric tests carried out by S. Sangkasaad[10] in ac conditions, the breakdown was also observed to always occur in the positive half-cycles. (ii) The delayed spark breakdowns observed often appear as transient phenomena which do not alter the general behaviour of the discharge; even if a smoothing of the current signals and if cleaning effects seem to be conjugated with their occurrence. This gives to think that the delayed spark breakdowns should be initiated by burst pulses originating from microdischarges or even microsparks across the surface layers formed on the electrodes by the discharge and that these surface layers should successively go through destructive and rebuilding processes.

REFERENCES

1. K.Hadidi and A.Goldman, "Current stability of negative corona discharges in SF_6 and delayed spark breakdown" in Gaseous Dielectrics VI, L.G. Christophorou and I.Sauers, ed., Pergamon Press, New York, pp.399-406, 1991.
2. M.Lalmas, K.Hadidi, H.Champain and A.Goldman, "Corona discharges long-term behaviour and delayed spark breakdown in SF_6 under dc voltages", 4th Int. Symp. on High Pressure Low Temperature Plasma Chemistry, Bratislava, pp.189-193, 1993.
3. I.Gallimberti, G.Marchesi and R.Turri, "Corona formation and propagation in weakly or strongly attaching gases", Proceedings Gas Discharge and their Applications, pp.167-170, Oxford, 1985.
4. R. J. Van Brunt and M. Misakian, "Mechanisms for inception of dc and 60-Hz ac corona in SF_6", IEEE Trans. on Elec. Ins. Vol. EI-17, No.2, pp. 106-120, 1982
5. I. Sauers and G. Harman, "A mass spectrometric study of positive and negative ion formation in an SF_6 corona. Part II : Influence of water and SF_6 neutral by-products" J. Phys. D: Appl. Phys. 25, pp. 774-782, 1992.
6. E. W. McDaniel and E. A. Mason, "The Mobility and Diffusion of Ions in Gases", John Wiley and Sons, pp. 4-6, 1973.
7. I. Sauers and G. Harman, "A mass spectrometric study of positive and negative ion formation in an SF_6 corona. Part I : sources of sulphur-fluoride ions" J. Phys. D: Appl. Phys. 25, pp. 761-773, 1992.
8. R. S. Sigmond, "A narrow-jet model of dc corona breakdown", Proceedings Gas Discharge and their Applications, pp.167-170, London, pp. 140-142, 1982.
9. A.M. Casanovas, J.Casanovas, F.Lagarde and A.Belarbi, "Study of the decomposition of SF_6 under dc negative polarity corona discharges (point-to-plane geometry): Influence of the metal constituting the plane electrode", J. Appl. Phys. 72 (8), pp.3344-3354, 1992

10 S. Sangkasaad, "Dielectric strength of compressed SF_6 in nonuniform fields", Diss. ETH-Zürich, No. 5738, 1976.

PARTIAL DISCHARGES ON WET AND LIGHTLY CONTAMINATED INSULATING SURFACES AND RELEVANT DIAGNOSTICS

O Claus, D. König, W.-K. Park

Technical University of Darmstadt
High Voltage Laboratory
Landgraf-Georg Str.4
D-64283 Darmstadt

ABSTRACT

Partial discharge patterns of aromatic epoxy resin surfaces lightly contaminated with droplet layers were investigated under high ac-voltage stress. Furthermore relevant basic aging mechanismens are presented, with physical and chemical factors of influence. Thereby different PD measuring and evaluation systems suitable for short and long time application have been used. The influence of the surface conditions on the condensation image and the PD patterns during short and long time tests were studied.

INTRODUCTION

Partial discharges (PD) on organic insulating surfaces are an important source of aging. Depending on the environment and with respect to relevant technical application one has to distinguish between the effects on outdoor insulation, indoor insulation and insulation applied in GIS. This distinction takes into account the different types of surrounding gases and the different levels of electrical stress. Though there is a common root of the surface partial discharge phenomenon, i.e. local electrical overstress and gaseous breakdown in the interface region, the effects in detail are very different as well as the aging of the solid organic insulation surface initiated by PD[1,2,3] This contribution deals with PD occuring on wet lightly contaminated and electrochemically highly stressed insulators made of epoxy resin, thus representing worst case conditions for electrical indoor equipment Problems in this field have been reported in numerous practical orientated publications[4,5,6] and dealt with in some fundamental studies[7,2]

ROLE OF PARTIAL DISCHARGES ON WET, LIGHTLY CONTAMINATED SURFACES

Fig.1 presents a scheme including the different steps of aging on either high wet or wet and lightly contaminated organic insulating surfaces simultaneously stressed with electrical field strenght[8,9]. Due to the inhomogeneous nature of the surface layer, i e. droplets alternating with micro-dryzones as shown later on, the field strenght distribution is inhomogeneous. Thus, local electrical overstress can occur and initiate PD inside the droplet layer. This PD acts as a generator for chemical byproducts such as Ozon (O_3) and nitric oxides (NOx). Secondary reactions with the moisture layer form nitric acid. In case of sodium chloride as a contaminant hydrochloride acid, respectively sodium hydroxide can appear. The generation of all of these byproducts contributes to a "self contamination" of the moisture layer with the consequence of an increase of the surface layer conductivity and the leakage current as well as an enhancement of

Gaseous Dielectrics VII, Edited by L.G. Christophorou
and D.R. James, Plenum Press, New York, 1994

PD-activity. Though in that stage essential thermal effects do not yet happen, an ongoing of chemical and physical attacts of the surface takes place, which is termed "electrolytical PD-erosion"[10]. The resulting surface degradation leads to an ongoing loss of the surface hydrophobia towards the formation of a more or less continous film condensation. Thermal effects due to the increasing leakage currents are no unavoidable, eg the formation of thermal related dry-band zones. According to the terminology developed in Müller[8],Quint[9] the appearance of these dry-band zones accompanied by high local temperature and high energy effects indicates that the socalled "early stage of aging" is over and the "late stage of aging" starts. This later stage has been already thoroughly studied by numerous researchers[11,12] because of its importance for the application field "organic outdoor insulation".

In case of the application "indoor insulation" the late stage of aging in general is short due to the high electrical stress and the material applied. Therefore the early stage of aging is of main interest and sensitive diagnostic tools are required in order to realize and -if possible- to quantify the relevant changes of the insulating surface properties. Some important diagnostic tools are shown in Fig.2. However in this contribution mainly the relevance of PD-measurements as a diagnostic tool is investigated.

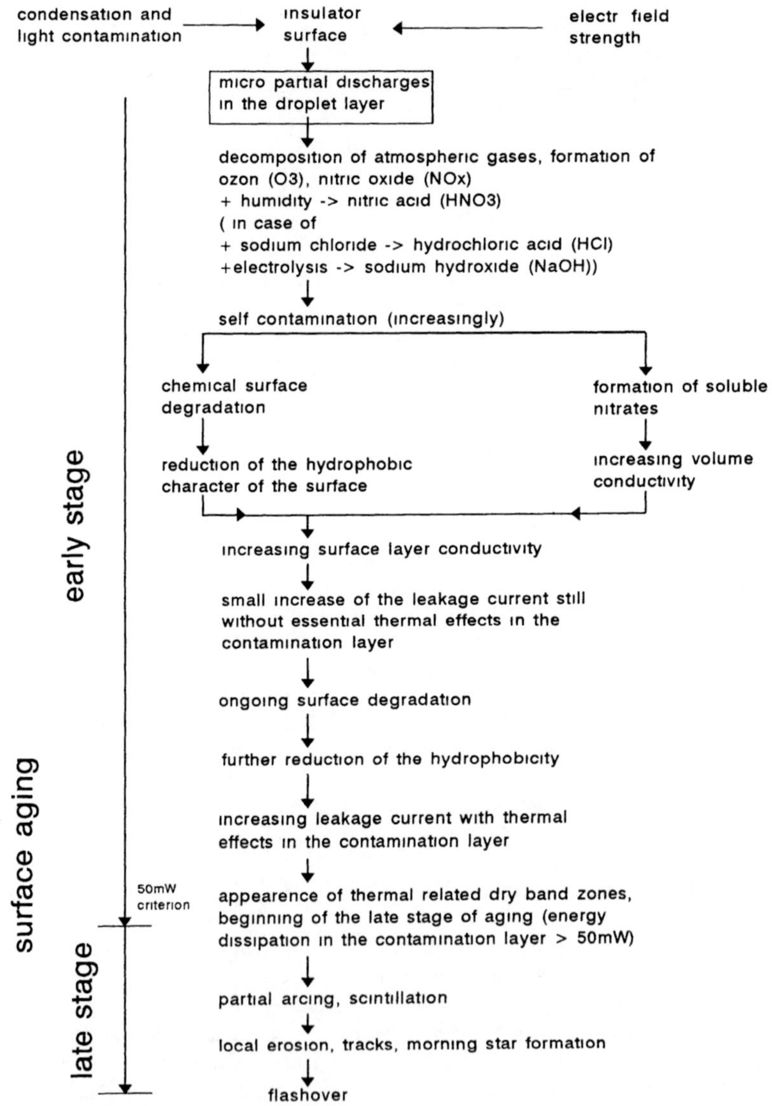

Fig.1 Basic aging mechanisms on wet and lightly contaminated organic insulation surfaces

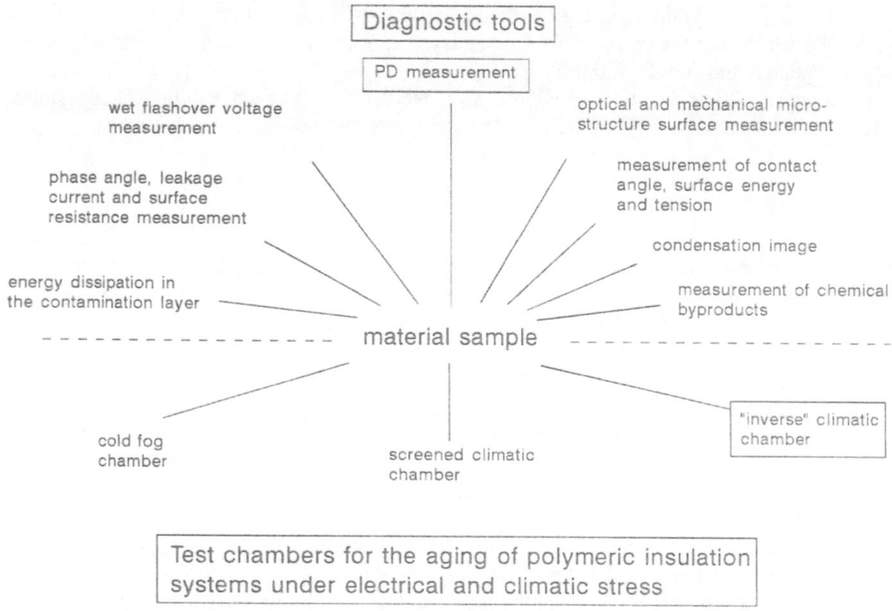

Fig. 2. Diagnostic tools and test chambers applied for aging processes

TEST ARRANGEMENT AND TEST SPECIMEN

To generate a defined droplet layer for long time investigations the test arrangenment according to Fig.3 was developed ("inverse climatic chamber")[7]. In a closed test vessel with a constant climate the test

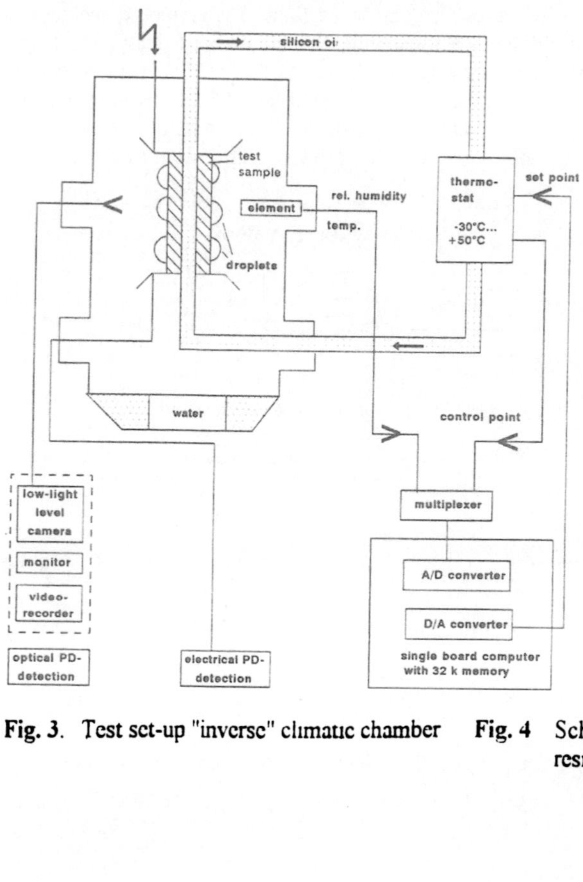

Fig. 3. Test set-up "inverse" climatic chamber

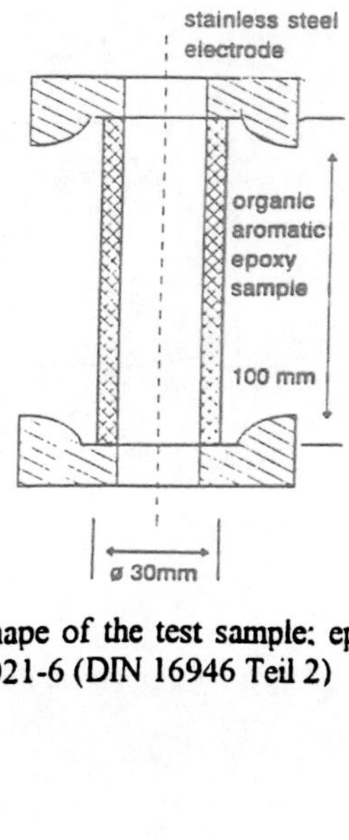

Fig. 4 Schematic shape of the test sample: epoxy resin Type 1021-6 (DIN 16946 Teil 2)

sample consisting of a hollow cylindrical model insulator (Fig.4) is arranged vertically and cooled internally in such a way that a constant droplet layer appears on the insulating surface. Secondary equipment regulates and stabilizes this state

The inverse climatic chamber operates in an electrically screened cabin, which enables PD-measurements above a noise level >0,05pC with ac-voltages up to 50kV.

PD-MEASURING SYSTEMS

A general problem with long time PD-measurements is the huge amount of data and their handling. On the one side a certain data reduction is unavoidable, while on the other side a loss of essential information has to be avoided. Often a procedure is applied, with uses PD-pattern and their comparison at defined moments, as for instance at the start of an aging cycle and during progressing times of aging. Typical changes of these patterns are expected to be indicators for changing conditions of the test objects. Up to now systematic studies on PD-pattern of wet and lightly contaminated electrically stressed organic surfaces and their changes during aging have been performed. Since the aging of these specimen is a fairly slow process, long time measurements presenting a high number of data have to be performed. The data processing has to be handled in such a way that no essential information is lost, but a reduction of data can be achieved.

Three different measuring PD-systems (Fig.5) have been developed and applied. Relevant PD-patterns have been studied during short time and long time tests All systems record the apparent charge (q_s) based on wideband or quasi-wideband measurements with integration in the time or frequency domain[13].

Short time PD-measuring system

This measuring system (Fig.5a) consists of commercial hardware components including a standardized PD-measuring instrument A (Manufactorer: Siemens/TuR, Type MTE3) The measured PD-pulse at the output of the PD-measuring instrument is transmitted to a fast digitizer, where the digitized information is stored intermediately Furthermore the divided high voltage signal is digitized. Both stored data record are transmitted to a PC, stored, processed and made available for graphic presentation. Though only the information of the peak value of a single integrated PD-pulse is needed, all parts of the measured PD-impulse are digitized and stored, which creates a high number of useless data This is also true for divided voltage signal, where only the voltage value at the time of the PD-occurrence is needed. Thus, by overloading the system with mainly unrelevant data the storage limit is rapidly reached. For example measuring of three halfwaves at a sample rate of 400ns produces an amount of data, which require a memory space of 512kByte

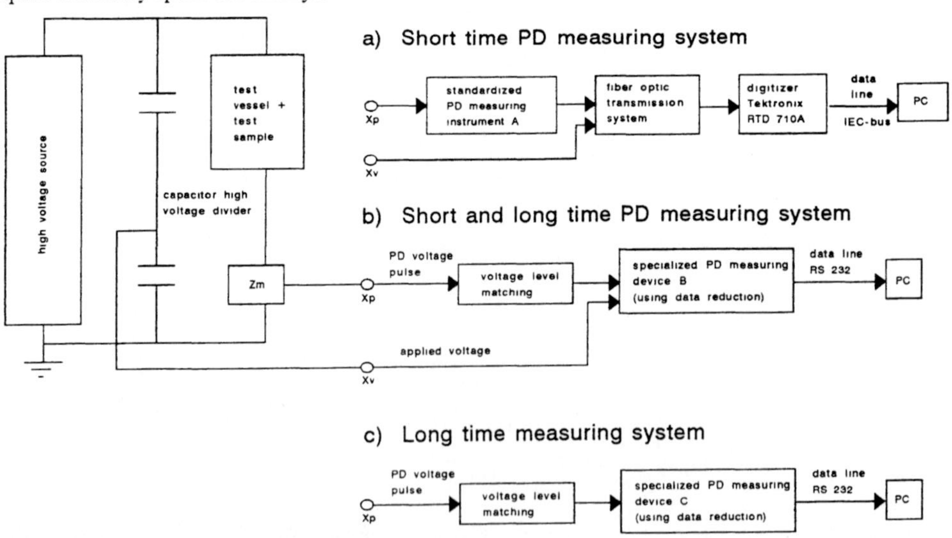

Fig. 5 Automatical PD-measuring systems applied for short and long time measurements

Short and long time PD-measuring system

PD-measuring system (Fig.5b) consists of two no commercial hardware components for PD long time tests. These are a special matching unit and a specialized PD-measuring device B. The amplified and matched single PD-pulse is transmitted to the input of the measuring device B, where a quasi-integration

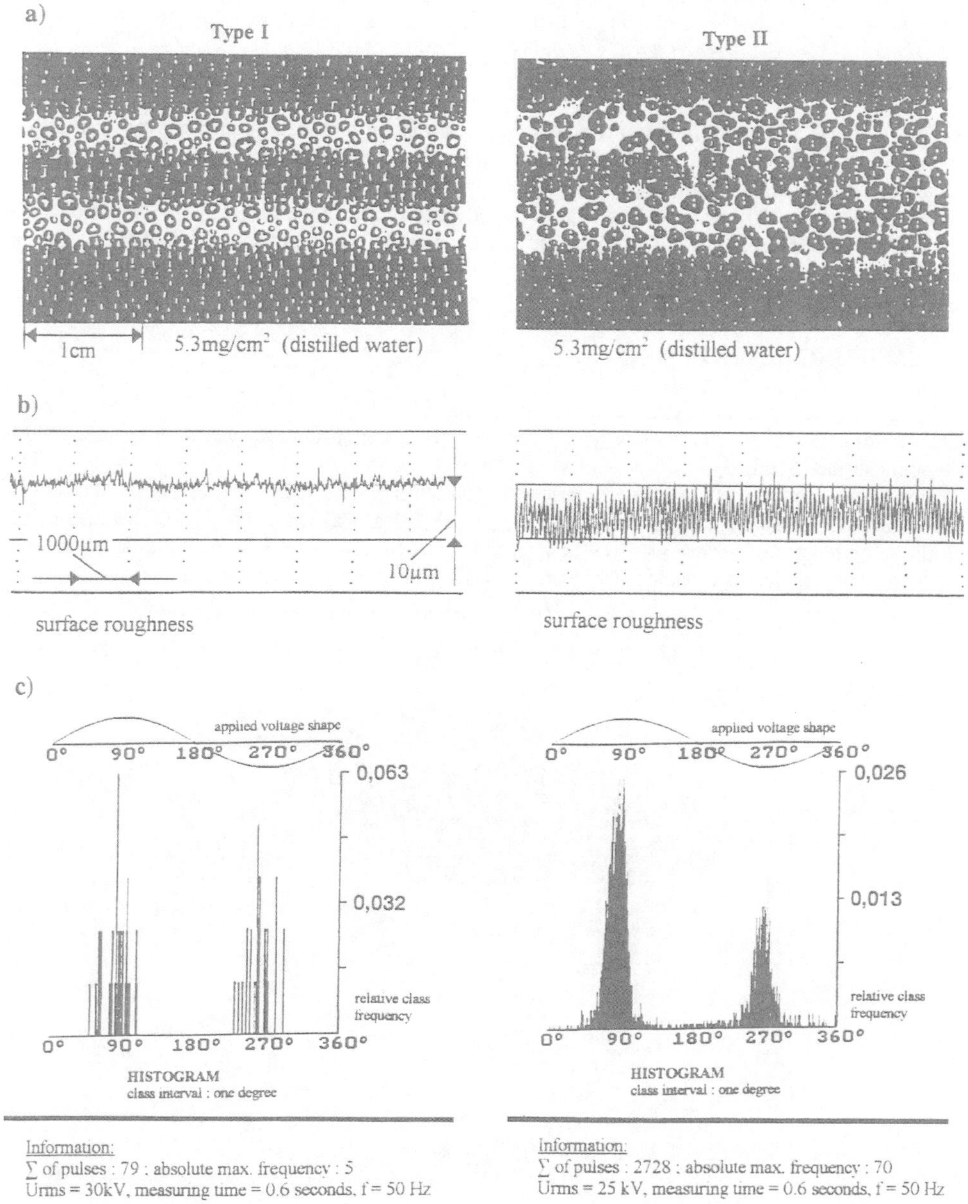

Fig. 6. PD measurement on aromatic epoxy resin specimens of cylindrical shape and different surface roughness Type I and Type II
a) condensation image
b) surface roughness
c) PD-histograms measured with PD device A

of the voltage pulse is performed. The peak value of the integrated PD-signal, which is proportionally to q_s, is digitized and stored by a random access memory (RAM). For the purpose of synchronization and determination of the phase angle of the PD the stepped down test voltage is connected to the input of measuring sytem B. Thus a counter with a high frequency can be activated to assign the correct phase angle with a solution of 0.1°. The measurement is interrupted if the storage limit is reached (32 kByte). Then the data are transmitted to a PC, where they are stored and processed for graphic output. As a result of that measuring conception only the values of the apparent charges and assigned phase angles are stored. Depending on PD activity long time measurements in the time range of hours or even days can be carried out

Long time measuring system

Fig.5c shows a PD-system designed under the conditions of high data reduction even at a loss of some relevant information. After matching the PD-pulses are quasi-integrated and digitized inside the specialized no commercial PD-measuring device C. In difference to device B only the PD-peak value during a time window of one second is evaluated and directly transmitted to a PC. A limitation with respect to the amount of stored data is given only by the memory space of the used PC. For example, a fixed disk with a relatively small memory space of 10MByte enables a continuous measuring time of about 58 days.

PD MEASURING RESULTS

The three PD-measuring systems enabled the record of PD during short and long time tests and the evaluation of patterns, which might be typical for the investigated insulation system and give hints for developing surface aging.

Fig.6 shows condensation image, epoxy insulator surface roughness and PD-patterns of two different types of surface such as a typical new one (Type I) and an artifically aged one (Type II). The quantity of moisture of both types of surface was well defined and equal. Surface Type I shows a droplet layer with a similarity of the droplet shape while the droplets on surface Type II are more different in size and shape

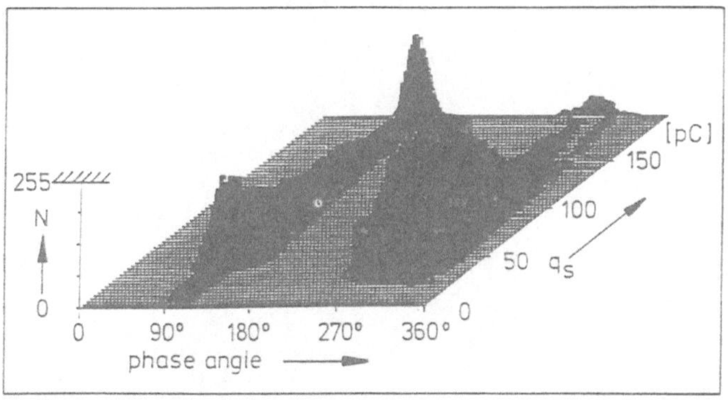

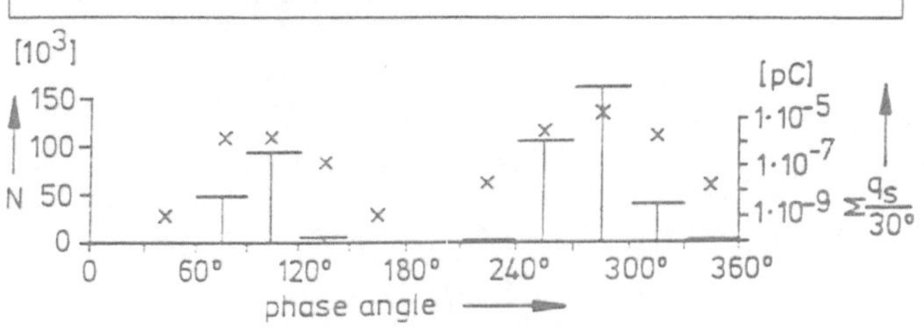

Fig. 7. PD measurement of a wet and lightly contaminated epoxy resin specimen after an aging time of 72h during a measuring interval of ten minutes;
parameters: electrical field strenght 1.4 kV/cm, moisture layer 7.9mg/cm², volume conductivity 500μS/cm (achieved with natrium chloride)

a) 3d plot as a function of the number of pulses (N), the apparent charge q_s and the phase angle

b) 2d plot of N (≙ ⊤),and $\sum q_s$ /30° (≙ x) versus phase angle

This situation is influenced by the different surface roughness (Fig 6b), which creates in case of surface Type I a hydrophob and in case of surface Type II a change to a beginning hydrophilic behaviour. These different surface conditions lead to different PD-patterns, short time measured with PD measuring system A and shown in Fig.6c. The big difference in the number of evaluated pulses is remarkable , which demonstrate a tendency towards increased PD-activity on rough surfaces, which represent an aged state even at a reduced electrical stress of about 20%. It has to be mentioned that in both cases the PD-level was mainly <1pC.

Fig 7 presents a PD-pattern taken at a time interval of 10 minutes with measuring system B at a specimen at the end of the early stage of aging (see Fig.1). The specimen was contaminated with a defined moisture layer with increased volume conductivity compared to distilled water. The three dimensional plot shows the apparent charge (q_s) and the number of all evaluated pulses (N) grouped in 1° sections as well as the relevant phase angle. An increasing frequency of pulses at about 90° and 270° is obvious. During the negative half cycle of the voltage much more PD-impulses occur. Fig.7b shows a two dimensional plot of N and q_s grouped for a section of 30°, i.e. $\sum q_s/30°$ as a function of the phase angle. This diagram confirm that the number of pulses at the negativ half cycle of the voltage is twice as large as the positive one, with the ratio $\sum(q_{spos}/30°)/(q_{sneg}/30°)$ of about 64%.

In addition to patterns recorded in the early stage of aging with the help of PD-measuring system A for about 0.6s (Fig.6) and at the end of this stage with the help of PD-measuring system B for about 10 minutes (Fig.7) measurements at a time interval of about 12h achieved with the help of measuring system C are performed (Fig.8). The apparent charge (q_s) is shown as a function of the time. While this data reductive measuring system presents peak values of the quasi-integrated PD-impulses evaluated during a time interval of 1s, these data are averaged based on different time intervals with the help of software inside the PC and plotted. For the time range Δt_1 from t=43 200s to 86.400s (Δt_1=43 200s) PD-average

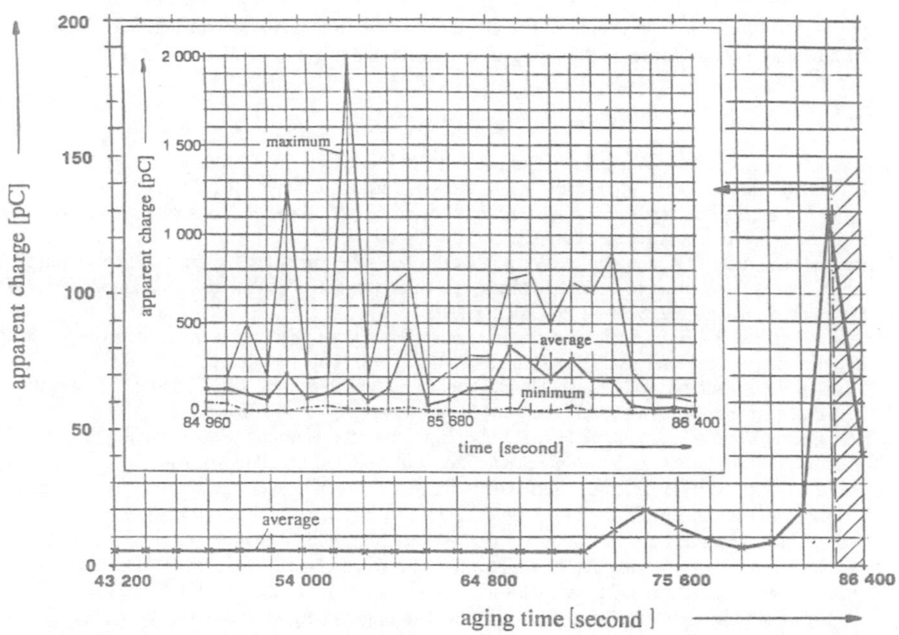

Fig. 8. Apparent PD charge measurement during a time interval of surface aging of a wet aromatic epoxy resin specimen with PD measuring device C depicted with two different time resolutions parameter: electrical field strength 1.4kV/cm, moisture layer 7.9mg/cm², volume conductivity 100µS/cm (achieved with natrium chloride)

a) Δt_1 = 43.200s - 86.400s = 43 200s ≙ 12 h

b) Δt_2 = 84.960s - 86.400s = 1.440s ≙ 0.4h

values were calculated based on a time interval of t=1.800s (Fig.8a). For the time range from Δt_2=84.960s to 86.400s (Δt_2=1.440s) PD-averaged values were calculated based on a time interval of Δt=60s (Fig.8b). In addition to the presented PD-average values in Fig.8b maximum and minimum values are shown based on measured peak values per second found in a time interval of 60s. Thus, Fig.8b is an enlarged section of Fig 8a, however with a different averaging time interval and visualizing the scatter between minimum and maximum PD-values in that time interval.

The interpretation of Fig.8 allows the following statements. Up to a time t=70.200s no essential surface changes occur. This is in accordance with findings by other diagnostic tools according Fig.2. Afterwards an increase of PD activity develops and culminates at about t=84.600s (see Fig.8a). This can be clearly diagnosed despite of a loss of detail information. An enlarged and more detailed information by using a different evaluation algorithm of the stored data including maximum and minimum values demonstrate a temendous high scatter of the intensity of the single PD-events. Some single PD have very high apparent charges of about 2000pC compared to average values of 20-140pC (Fig.8a), respectively 150-400-20pC. This behaviour seems to indicate the beginning of the formation of macro-dry zones on the surface of the specimen and their bridging by discharges with higher energies, thus enhancing the aging process and starting the late stage stage of aging.

Thus, the application of long time PD-measuring systems even with data reduction seems to supply sufficient information when used as a diagnostic tool for the judgment of surface aging.

SUMMARY

The investigations presented demonstrate that PD-measurements in the short and long time range with more or less stringent data reduction are useful tools with respect to aging diagnostics of contaminated organic insulating surfaces.

ACKNOWLEDGEMENT

The authors express their gratitude to the Deutsche Forschungsgemeinschaft (DFG), Bonn, for the support of a research programme on the aging of contaminated organic insulating surface.

REFERENCES

1 C Hudon, H Bartnikas, M R Wertheimer, Surface and Gas Phase Reactions Arising With Epoxy Exposed to Partial Discharges, Conf on Electr Insul and Dielectric Phenomena (CEIDP), pp. 725-735, 1992

2 D Konig, I Quint, Multifactor Aging Mechanisms on the surface of Model Post Insulators Conf o Electrical Insulation and Dielectric Phenomena (CEIDP),pp 664-671,1992

3 R Barsch, H Gremmel, D König, J Pilling, Verminderung des Isoliervermogens an Feststoff/Gas Grenzflächen durch fremdschichtinitiierte Teilentladungen, ETG-Fachbericht 49 Isolationskoordination in Hoch- und Mittelspannungsanlagen, S.133-149, Oktober 1993

4 N.Balsev, E Larsen, P K Birch, Pollution, condensation and discharge phenomena in medium voltage compact stations, Intern Conf on Electr Distribution (CIRED) 1983, Paper No e 11

5 P K Birch, J T Benjaminsen, K O Tangen, Climatic behaviour of substations, IEE Conference Publication No. 250, Part 1, Intern Conference on Electricity Distribution (CIRED) 1985, pp 1-5

6 D König, K -H Krefter, H Gremmel, H -J Straube, Environmental Humidity Conditions for Enclosed High-Voltage Switchgear, Intern Conf on Electricity Distribution (CIRED) 1989, pp 47-51

7 D Konig, I Quint, P Rösch, B Bayer, Surface Discharges on Contaminated Epoxy Insulators, IEEE, Trans Electrical Insulation, Vol EI-24, No 2, pp 229-237, April 1989

8 B Muller, Untersuchungen zum Oberflachenverhalten von stabformigen Isolatoren aus Epoxidharz-Formstoff bei simultaner Beansoruchung durch Feuchte und hohe Wechselspannung, Diss TH Darmstadt,1985

9 I Quint, Untersuchungen zum Einfluß von schwach leitenden Fremdschichten auf das Oberflachen-Alterungs-verhalten wechselspannungsbelasteter zylindrischer Pruflinge aus Epoxidharz-Formstoff, Diss TH Darm-stadt, 1994

10 D Konig, I Quint, Investigations on the Very Early Stage of Surface Aging of Lightly Contaminated Epoxy Resin Post Insulators, 7th Inter Symp on High Voltage Eng (ISH) Dresden, No. 24.02, August 1991

11 T C Cheng, D C Jolly, D J King, Surface Flashover of Water Repellant Insulators Under Moist Conditions, IEEE Transactions on Electrical Insulation, Vol. EI-12, No.3, pp.208-213,June 1977

12 J Pilling,R Barsh, Accelerated Ageing Tests of Housing Materials for Composite Insulators in a Fog Chamber, European Trans on Electrcal Power Engineering (ETEP) Vol 1, 2/1991, pp 89-95

13 K Schon, Fundamentals of Partial Discharge Measuring Techniques, in "Partial Discharges in Electrical Power Apparatures", D Konig, Y N Rao, VDE-Verlag, Berlin 1993

DISCUSSION

R. J. VAN BRUNT: The fact that the PD is peaked in intensity at 90° and 270° is significant. I believe that this may be a consequence of the conductivity of the surface from absorbed moisture. Ordinarily when PD occurs on an insulating surface or in a void the pattern is shifted in phase toward the zero crossing because of surface charging effects. Could you comment on this?

O. CLAUS: The reason for the position is that we talk here about PD outside, not inside. We have made such other measurements and we cannot say that it is always exactly at 90° or 270°. It can be before or behind. As aging occurs, we would expect to get more events at the zero point because we have erosion and other mechanisms.

M. GOLDMAN: I think these results are completely right because he works directly from the point to the insulating material. In this case if you use our compensating technique you find exactly the same thing - the corona and discharge are always in phase with the voltage. But in voids it is completely different and you will have a change of phase.

D. W. BRANSTON: Concerning the question by Dick Van Brunt, the water provides a conducting layer which results in the signals seen.

MODIFICATION OF CAST EPOXY RESIN SURFACES DURING EXPOSURE TO PARTIAL DISCHARGES

Helena Slowikowska, T. Łas, and J. Slowikowski
Instytut Elektrotechniki
Warsaw, Poland

R. J. Van Brunt
Electricity Division
National Institute of Standards and Technology*
Gaithersburg, Maryland 20899

INTRODUCTION

Cast epoxy resin materials are commonly used in insulating spacers of gas-insulated high-voltage power systems. It has recently been reported by Hudon and coworkers[1] that when these materials are exposed to electrical discharges in air, significant modifications occur in their insulating characteristics. In particular, it was discovered that exposure to discharges causes a decrease in surface resistivity. A decrease in surface resistivity was also reported by Van Brunt and coworkers[2,3] for ac-generated partial discharges on epoxy surfaces. They showed, moreover, that the discharge-induced decrease in surface resisitivity is responsible for dramatic changes in the stochastic behavior of the discharge.

In the present work, the nature of partial-discharge (PD) induced damage to cast epoxy resin surfaces has been investigated in more detail using various techniques to examine surface roughness and morphology, as well as resistivity. The effect of generating PD in gases other than air, such as SF_6, N_2, and N_2/O_2 mixtures, was also examined.

EXPERIMENTAL

Partial Discharge Conditions

Partial discharge was generated by applying a 50-Hz alternating voltage to point-dielectric gaps in which the dielectric was an 8×8 cm^2 planar sheet of epoxy of 1.0 mm thickness. The investigations reported here were carried out on samples consisting of two types of epoxy resins, Epidian 5 (Ep5) and Epidian 2 (Ep2) of different molecular content and on composites of these epoxies with Al_2O_3 filler, referred to here as

*Electronics and Electrical Engineering Laboratory, Technology Administration, U. S. Department of Commerce.

Gaseous Dielectrics VII, Edited by L.G. Christophorou and D.R. James, Plenum Press, New York, 1994

Ep5K, Ep2K, and Ep2K$_{mod}$.† The epoxy resins are the products of reactions between epichlorohyrite and bisphenol-A.

Most of the results reported here were obtained for discharges generated in room air of uncontrolled humidity with point-to-dielectric gaps of 0.5 mm. The relative humidity in the room was typically in the range of 60 to 70 precent at a temperature of about 20 °C. Some measurements were made in controlled atmospheres (SF_6, N_2, or N_2/O_2 mixtures) using Ep2K. In this case, the discharge was generated in a sealed vessel with a volume of 300 ml and for an absolute gas pressure of 1200 kPa. For the tests performed in the enclosed vessel, the epoxy samples were either allowed to "acclimatize" to normal conditions of the room or were dried by heating at a temperature of 105 °C for 24 hours before inserting into the vessel. The N_2/O_2 mixtures used for the tests conducted here consisted of 79% N_2 and 21% O_2 by volume.

The epoxy samples were fastened in all cases to a flat, electrically grounded metal plate. The point electrode was a stainless-steel needle with a conical tip. The cone had an angle of 60° with a tip radius of about 50 μm. A test voltage of 4 kV rms (50 Hz) was applied to the point electrode for about 22 hours. For the tests performed in SF_6, a higher voltage of 5 kV rms was applied, and the exposure times were extended to 120 hours. Under theses conditions, pulsating PD occurred on both half-cycles of the applied voltage. During the time of discharge exposure, the PD activity was monitored with a conventional PD detector (Tettex 9126).† The following PD parameters were measured: average discharge current, maximum values of positive and negative PD charge, and PD pulse repetition rate. As shown in our earlier work,[2,3] the electrical characteristics of the PD change significantly while the epoxy materials are exposed to the discharge, and this presents difficulties in any attempt to quantify the discharge exposure. The problems of quantifying discharge exposure go beyond the scope of this paper.

Characterization of Surface Damage

Observations of the dielectric surface in the vicinity of the discharge site were made using either an optical miroscope or a BS 301 (CSR) electron-scanning microscope,† and measurements were made of surface roughness parameters using a computer-controlled scanning profilometer. The profilograms provide "normal" and "upturned" views from which decrement volume (v^-), (volume of material removed), the area decrement (A), the maximum depth of decrements (d_{max}), and the density of surface pits (D) could be calculated by computer analysis.[4]

The measurements of dielectric surface resistance (R_s) were carried out using a Kiethley 617 electrometer† connected to a 100 V DC voltage supply. The resistances were measured between a pair of movable parallel brass electrodes that were 15 mm in length and always separated by a distance of 0.4 mm. Resistance values were determined at different distances (up to 16 mm) from the center of discharge exposure by moving the brass electrodes on the surface of the epoxy. Because of uncontrollable factors such as associated with relative humidity at the time of measurement, there is considerable uncertainity in the absolute resistance values reported here which may be in error by as much as an order of magnitude. However, the relative resistance values at different locations on the surface of the same material are quite reproducible and vary by less than ±20%. Most measurements of R_s were carried out under conditions where

†Certain commercial equipment, instruments, or materials are identified in this paper to specify adequately the experimental procedure. Such identification does not imply recommendation or endorsement by the National Institute of Standards and Technology, nor does it imply that the materials or equipment identified are necessarily the best available for the purpose.

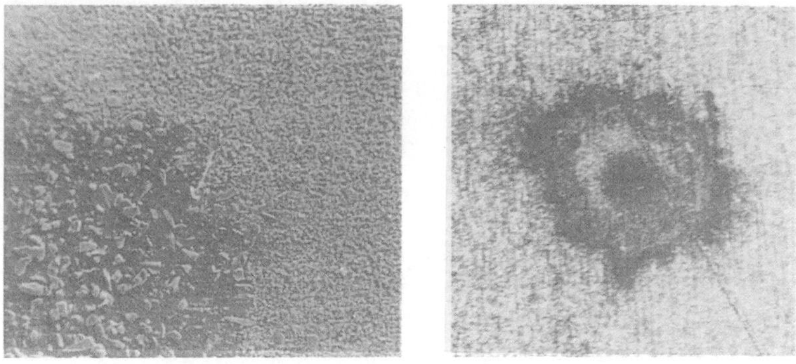

(a) (b)

Figure 1. Miroscopic image of an Ep2K sample after exposure to PD: a) scanning miroscope view of the center (×75), b) optical microscope view, full area of exposure (×20).

the relative humidity of the room air was at a constant value of about 65 percent.

For cases where the tests were carried out in an sealed vessel, measurements were also made of the gas content before and after operation of the discharge using a gas chromatograph equipped with either an electron-capture detector or a thermal-conductivity detector. Particular attention was paid to changes in the oxygen content.

RESULTS

The results shown below apply to discharges that were generated in air. Results for other gases are discussed, but because space limitations, they are not presented here. It was found that the occurrence of PD causes concentric circular features to appear on the epoxy surface centered about the location of the point electrode. Optical microscopy and scanning microscopy showed the following: 1) loss (decrement) of resin in the central area in all cases; 2) formation of bubbles (see Fig. 1a) and deposits not bound to the substrate (see Fig 1b); and 3) in the case of Ep2K, grains of Al_2O_3 devoid of enclosing resin. Consistent differences in the appearances of PD damage were noted for the composites Ep2K and $Ep2K_{mod}$. Unlike Ep2K shown in Fig. 1, $Ep2K_{mod}$ exhibited amorphous forms of resin attached to the individual grains.

Results of typical surface profile measurements in the vicinity of the discharge site are shown by the profilograms in Figs. 2 and 3. For all samples, the roughness increases significantly as the center of discharge activity is approached. The pure resin samples tend to show a greater volume decrement than the composites with Al_2O_3. Analyses of the profilograms yield the parameters such as given in Table 1 that quantify the discharge-induced surface damage.

Examples of results from surface resistivity measurements are shown in Fig. 4. For all types of epoxy, the PD activity causes a local decrease in the surface resistivity. Also, the region of reduced resistivity is seen to extend considerably beyond the area of visible damage at the discharge site. The reduction in surface resistivity is generally not permanent. Removal of the deposits that form visible rings by successive washing and drying causes R_s near the discharge site to be at least partially restored to its predischarge value (see Fig. 4b). As shown in Fig. 5, R_s will tend to increase with time since the end of discharge exposure even without washing.

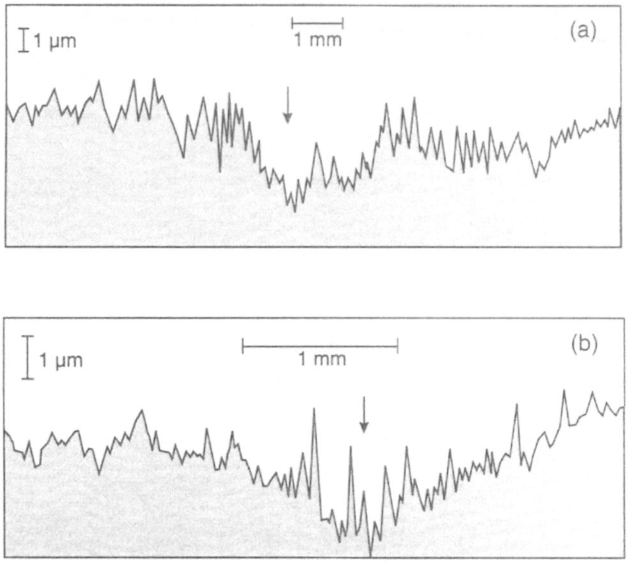

Figure 2. Profilograms of epoxy samples after exposure to PD: a) Ep5, b) Ep2. The vertical arrows indicate the location of the point electrode.

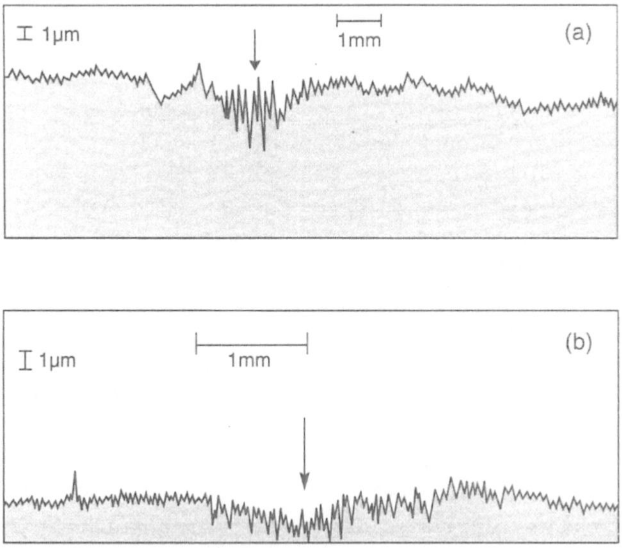

Figure 3. Profilograms of epoxy samples after exposure to PD: a) Ep5K, b) Ep2K. The vertical arrows indicate the location of the point electrode.

Table 1. Results from from determinations of surface roughness characteristics for three types of Ep2 samples after exposure to PD in room air. The relevant parameters are defined as follows: v^- - volume decrement, A - area decrement, d_{max} - maximum decrement depth, D - density of surface pits.

Type of sample	v^- $(mm)^3 \times 10^{-5}$	A $(mm)^2$	d_{max} μm	D $(mm)^{-2}$
Ep2	218.3	0.25	4.4	68.3
Ep2K	134.5	1.65	5.8	147.0
$Ep2K_{mod}$	164.7	1.37	5.6	145.0

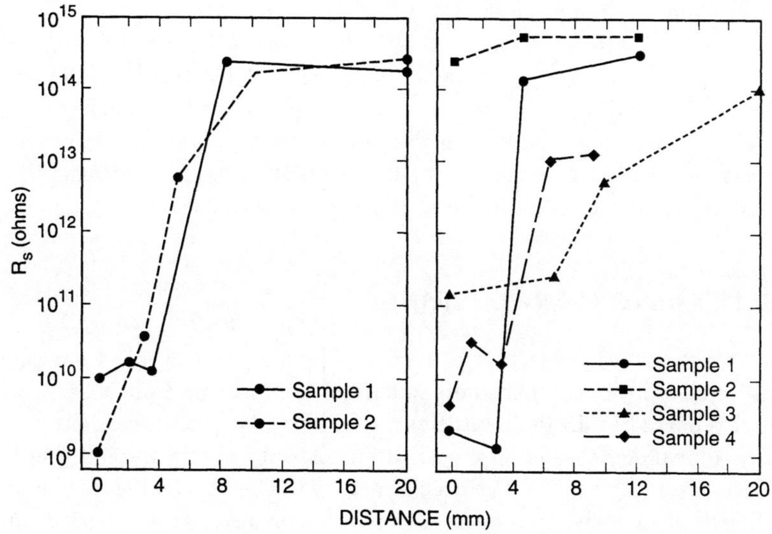

Figure 4. The measured changes in R_s as a function of distance from the discharge site: a) sample 1 - Ep5K, sample 2 - Ep5; b) sample 1 - Ep2K, sample 2 - Ep2K after washing, sample 3 - Ep2, and sample 4 - $Ep2K_{mod}$.

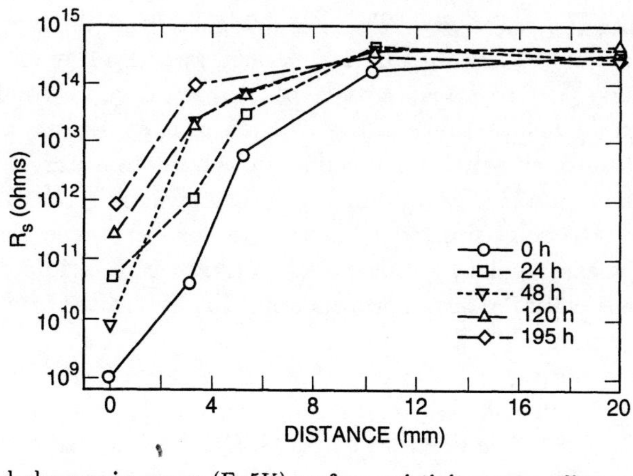

Figure 5. Measured changes in epoxy (Ep5K) surface resistivity versus distance from the center of the discharge site for different times since the termination of discharge exposure.

The results from tests performed in an enclosed vessel with different gas mixtures showed that, although the effects of PD on epoxy surface characteristics are qualitatively similar to those found in room air, there was considerable variability in the extent of discharge damage which depended not only on gas mixture but also on preconditioning of the sample, i.e., on whether or not it had been dried. Comparison of results obtained under different environmental conditions is complicated by the unresolved problem of quantifying the overall level of PD exposure. Despite this difficulty, some consistent trends were observed. For example, at comparable levels of electrical stress, it was found that the volume decrement from PD activity was consistently lower in dry N_2 or N_2/O_2 mixtures than in room air. Gas analysis showed that, in the latter mixture, O_2 was consumed during discharge activity, thus suggesting that chemical action involving this gas may play an important role in causing surface damage. It was also found that predrying of the sample tended, in most cases, to reduce the level of damage to the dielectric surface.

In the case of SF_6, it is particularly difficult to make meaningful comparisons with the results obtained for room air because the level of PD activity at any given voltage was always much lower in SF_6 than in air. To obtain comparable surface damage using SF_6, it was necessary to increase both the applied voltage and the PD exposure time. As in the case of air, local decreases in surface resistivity occurred near the discharge site which were not permanent in the sense that the resistivity could be restored after washing.

DISCUSSION AND CONCLUSIONS

It has been shown here that it possible to quantify PD-induced damage to epoxy surfaces in terms of isometric parameters derived from microscopic observations similar to those first reported by Holboll and Henrikson.[5] In all cases, it was found that PD activity causes a local decrease in epoxy surface resistivity at the discharge site consistent with previous observations.[1–3] The precise nature of the physical and chemical changes of the surface that give rise to a decrease in surface resistivity are not yet understood. The increase in surface roughness resulting from exposure to PD increases the effective surface conduction path length. This would therefore tend to produce an increase in surface resistivity which is opposite to the trend observed. It can thus be argued that the local changes in resistivity are not directly a consequence of changes in the epoxy surface profile. The results suggest instead the possible importance of deposits on the surface in modifying the resistivity. There is evidence from the work of Hudon and coworkers[6] that droplets of glycolic and formic acid appear on epoxy surfaces during exposure to discharges. These acids are known to be far more conductive than the epoxy resin matrix.[1,6] The presence of acidic deposits is consistent with the observed restoration of surface resisitivity after washing and points to a possible important role played by ionic conduction in determining the surface resistivity.[7,8] Discharge-induced changes in surface resistivity similar to those observed here have also been reported previously for polymeric insulating materials.[9–11] Evidence of acidic deposits were observed in the present work for tests performed in SF_6, even after drying of the sample prior to discharge exposure.

The observed depletion of oxygen during PD tests performed in a sealed vessel and the effect of drying in reducing the rate of dielectric deterioration suggests an important role played by reactive species such as O, OH, and O_3. In the case of SF_6, the usual oxyfluoride corona by-products such as SOF_2 reported in earlier experimental work[12] were also observed. The presence of these compounds indicates that HF and free fluorine can be involved in the relevant surface chemistry at the discharge site. It is clear that more careful and systematic investigations of the discharge-induced modifications of the surface are required to achieve a better understanding of the mechanisms responsible for changes in surface morphology and resistivity that result from exposure to partial-discharge.

ACKNOWLEDGMENTS

Investigations were conducted in the framework of cooperation between the Institute of Electrotechnics in Warsaw (Poland) and the National Institute of Standards and Technology (USA), financed by the Marie-Sklodowska Fund II. The authors are grateful to Z. Klimczak and J. Dukaczewski of the Polish Academy of Sciences for performing the roughness measurements and corresponding computer analysis.

REFERENCES

1. C. Hudon, R. Bartnikas, and M. R. Wertheimer, Spark-to-glow discharge transition due to increased surface conductivity on epoxy resin specimens, *IEEE Trans. Elec. Insul.* 28:1 (1993).
2. R. J. Van Brunt, P. von Glahn, and T. Las, Partial discharge-induced aging of cast epoxies and related nonstationary behavior of the discharge statistics, *in* "1993 Annual Report - Conference on Electrical Insulation and Dielectric Phenomena," IEEE, Piscataway, NJ (1993).
3. R. J. Van Brunt, P. von Glahn, and T. Las, Nonstationary behavior of partial discharge during discharge-induced aging of dielectrics, *IEE Proc. - A* (in press, 1994).
4. P. A. Engel and D. B. Millis, Study of surface topology in impact wear, *Wear* 75:423 (1982).
5. J. T. Holboll and M. Henrikson, Partial discharge patterns and surface deterioration in voids in filled and unfilled epoxy, *in* "Conference Record - 1992 IEEE International Symposium on Electrical Insulation," IEEE, Piscataway, NJ (1992).
6. C. Hudon, R. Bartnikas, and M. R. Wertheimer, Analysis of degradation products on epoxy surfaces subjected to pulse and glow type discharges, *in* "1991 Annual Report - Conference on Electrical Insulation and Dielectric Phenomena," IEEE, Piscataway, NJ (1991).
7. H. Slowikowska, J. Slowikowski, L. Dybowski, and J. Lampe, Some test results on the effect of SF_6 decomposition products on superficial properties of epoxy resin compositions, *in* "Conference Record - 1978 IEEE International Symposium on Electrical Insulation," IEEE, Piscataway, NJ (1978).
8. R. Rygal, H. Slowikowska, J. Slowikowski, and D. Paziewska, The influence of the filler characteristics on the surface resistivity of epoxy composition applied in SF_6 insulated systems, *in* "Conference Record - 1992 IEEE International Symposium on Electrical Insulation," IEEE, Piscataway, NJ (1992).
9. E. C. Rogers, The self-extinction of gaseous discharges in cavities in dielectrics, *Proc. IEE*, 105A:621 (1958).
10. M. Gamez-Garcia, R. Bartnikas, and M. R. Wertheimer, Synthesis reactions involving XLPE subjected to partial discharges, *IEEE Trans. Elec. Insul.*, EI22:199 (1987).
11. P. H. F. Morshuis, Partial Discharge Mechanisms, *Thesis*, Delft University Press, Delft, The Netherlands (1993).
12. R. J. Van Brunt and J. T. Herron, Fundamental processes of SF_6 decomposition and oxidation in glow and corona discharges, *IEEE Trans. Elec. Insul.* 25:75 (1990).

DISCUSSION

W. PFEIFFER: What is the aim of this research?

H. SLOWIKOWSKA: The overall objective of this research is to perform investigations relevant to the development of accelerated partial-discharge aging tests for epoxy and other solid insulating materials.

A. IEYASU: Why does the difference of surface damage occur? How does the difference depend on the applied voltage?

H. SLOWIKOWSKA: The rate of damage to epoxy is dependent on the intensity of the partial discharge. In general, the rate of material removal will increase with applied voltage because of the corresponding increase in the partial-discharge intensity. As the voltage is increased, the time required to reach a particular level of "damage" will tend to decrease. There is evidence that the rate of damage does not increase linearly with voltage; however, the exact dependence on voltage was not established in this work.

SURFACE CURRENT ON INSULATING SPACERS

M.C. Siddagangappa[1] and B. Nagesha Rao[2]

[1]Department of High Voltage Engineering
Indian Institute of Science
Bangalore - 560 012, India
[2]Central Research And Testing Laboratory
Prof.Sir.C.V.Raman Road
Bangalore - 560094, India

INTRODUCTION

A composite solid-gas insulating system is an integral part of high voltage equipments. Solid insulators are used as mechanical support and to offer electrical insulation. There have been several research investigations with primary emphasis on the understanding of initiation of charges followed by discharge development leading to surface flashover on solid dielectrics in high pressure and in high vacuum systems. Recently Wintle[1] has reviewed the present understanding of the basic insulator physics and of insulating systems.

Initial charges produced in the triple junction (solid- gas-metal) flow as surface current on insulating spacers[2]. Also, the charges produced due to localised electrical stress in the gaseous medium would drift and accumulate on spacers surface, changing the applied field conditions leading to surface flashover[3]. However, the investigations were mainly concerned with d.c. applied voltages. The present work reports on the investigations relating to a.c. current flow on the dielectric surface under normal high voltage system operating conditions. The surface currents exhibited hysteresis (charge retention) characteristics with variation in the magnitudes of a.c. applied voltages both at atmospheric pressure and at low pressure conditions. The surface currents are observed to be not just limited by the surface resistivity of the spacer, but are also influenced by the encompassed gaseous environment.

Gaseous Dielectrics VII, Edited by L.G. Christophorou
and D.R. James, Plenum Press, New York, 1994

EXPERIMENTAL DETAILS

The experimental set up used in this work is basically similar to the one used by Lee et al[4]. Briefly, the apparatus consists of uniform field brass electrodes of 95 mm dia having Rogowski profile, mounted in a vacuum chamber with view ports. The top electrode is rigidly supported at the centre of the top plate of the chamber and insulated using a high voltage bushing. The other electrode is also insulated from the body of the chamber and is supported by a movable shaft which enters the chamber through a wilson seal in the bottom plate. The two electrodes are centred and are aligned co-axially to the chamber. The inter-electrode distance may be varied from zero to 100 mm. The bottom electrode is electrically connected to the measurement system through a PTFE bushing provided in the bottom flange. The chamber could be evacuated to pressure less than micro Torr. However, the experiment was conducted in the range of 200 milli Torr and at atmospheric pressure. The insulating materials of cylindrical shape of different length and diameter were positioned symmetrically at the centre of the electrode system. It was ensured that electrode surfaces make perfect contact and coincide with machined, polished and cleaned two end faces of insulating spacers used. High voltage a.c. derived from 230 V/15 kV, 3 kVA, transformer was applied across the electrode system, and measured using a potential transformer combined with a voltmeter. The applied potential was varied across the gap by controlling input voltage to the high voltage transformer. The maximum applied voltage for any spacer between the electrodes was below the breakdown level of the gap at the operating pressure or limited by the source voltage. The surface currents were obtained by measuring the potential drop across known resistance of high tolerance (1%), shielded in an enclosure, connected in series with the bottom electrode and ground. A broad band millivoltmeter with an operating frequency range of 2Hz to 1MHz offering an input impedance of 300 MΩ, was used to measure the potential drop. The capacitance offered by the electrode gap and the spacer draws a reactive current when a.c. voltage is applied. To neutralise the reactive component of current flowing in the gap and to measure real surface current on the spacer, a compensating circuit was used[4] . The principle of the compensation circuit is to drive a current equal in magnitude and in phase opposition to the reactive component of current flowing through the sensing resistor.

RESULTS AND DISCUSSION

Surface current measurements were conducted on cylindrical insulator spacer materials placed between uniform field electrodes gap with applied power frequency (50 Hz) voltages. Surface current data excluding reactive component of the current flowing in the gap obtained for pyrex glass of length 10.5 mm and diameter 12.5 mm is plotted in the figure 1. The range of a.c. currents measured is in μA compared to that of d.c. current measured in the order of 10^{-14}A. Uncertainity of the measurements is estimated to be in the order of $\pm10\%$ in this investigation. Reactive component of the gap current measured and neutrallized is 1 μA, at the maximum voltage used, which is in close agreement with the reported measurement of 0.5 μA at 2 kV for 73 mm diameter electrode system by Lee et al[4]. It is interesting to observe that the surface current recorded with variation of the applied voltage to the electrode system from zero reaching the maximum, is lower compared to the data measured during retrogression of the applied voltage. The readings were recorded for every change in the applied voltage after allowing stabilization time of more than three minutes. The surface current on the insulating spacer exhibited a kind of retention of charge density when the applied voltage is reduced and is referred here as hysteresis (memory) effect. No attempts were made to measure charges, that might have been detained on the spacer surface after removing the applied voltage. After obtaining data under ambient laboratory pressure (680 Torr) conditions, the electrode system was shorted and left for long hours before the chamber was evacuated and the experimental procedure repeated at low pressures (1 to 2 milli Torr). The current magnitudes recorded are higher compared to that of ambient pressure condition for the range of applied voltages as evident from the figure 1, however, the hysteresis effect is not significant.

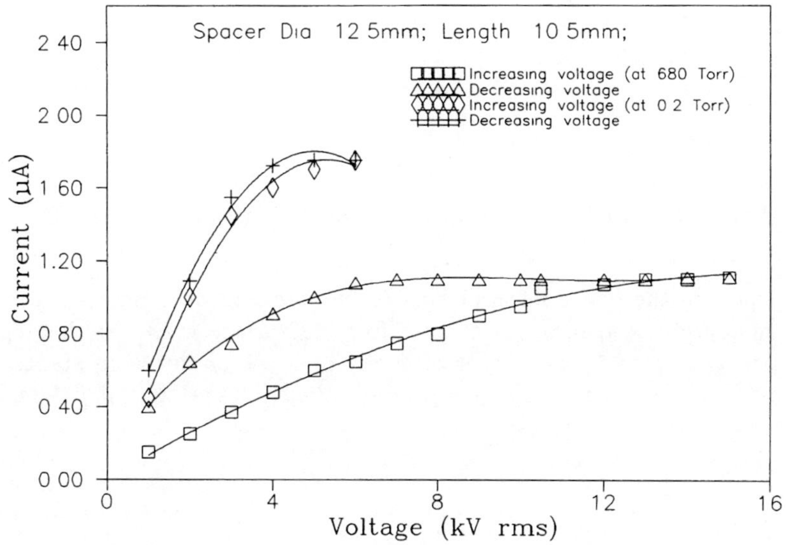

Figure 1. Surface current (a.c.) on Pyrex glass spacer versus applied voltage.

The data obtained for pyrex glass of different diameter and lengths exhibited similar characteristics. To examine this feature, experiment was extended for other insulating materials used in high voltage systems, namely Teflon, porcelain, perspex, polypropylene, FRP and Nylon. Hysteresis effects of similar nature, but of variable degrees were observed for different insulating materials. Teflon and polypropelene have shown the lower degree of hysteresis effect. The degree expressed in terms of the area bounded by two current curves corresponding to increasing and decreasing voltages. The data obtained for the Nylon spacer of length 10.5 mm and diameter 14 mm at two different pressure conditions is shown in the figure 2. The current data for low pressure with nylon spacer, also have shown an increased value with little hysteresis effect. The maximum voltage applied to the electrode gap with any spacer under low pressure conditions was within the limits of breakdown strength of the product of pressure and gap distance.

It is feasible that the current flowing over the insulating material, originate in part as charge transfer from the electrode material in addition to the charges produced due to high field concentrations at the electrode edges.

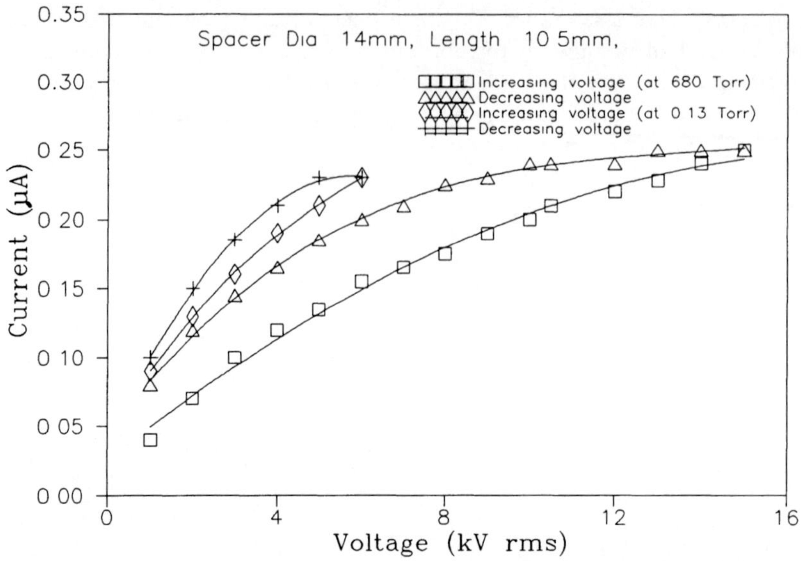

Figure 2. Surface current (a.c.) on Nylon spacer versus applied voltage.

To examine the dependence of area of metal electrode contact with the spacer, in the contribution of transfer of charges forming surface currents; the experimental data obtained for pyrex glass spacer of different diameters for a constant gap length of 5.5 mm. The surface currents for various applied voltages and at different pressure conditions are plotted in figures 3 and 4.

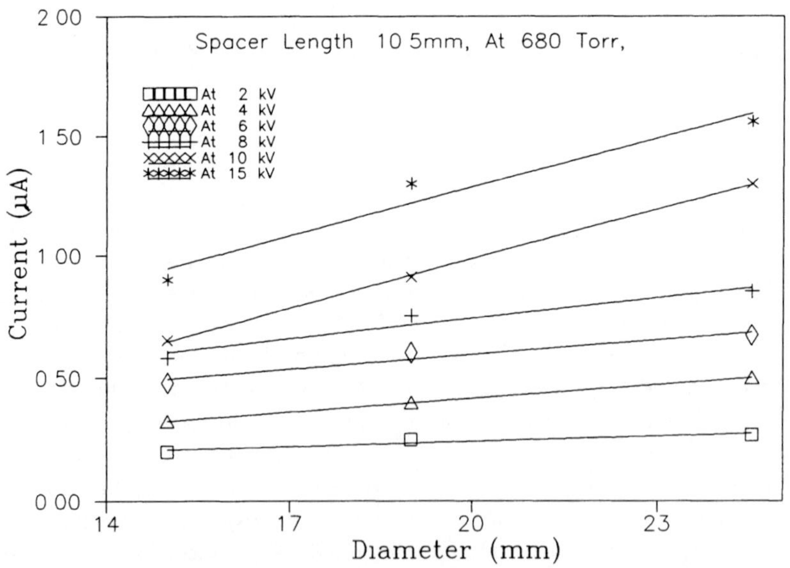

Figure 3. Surface current (a.c.) on Pyrex glass spacers of different diameters for constant applied voltages at ambient conditions.

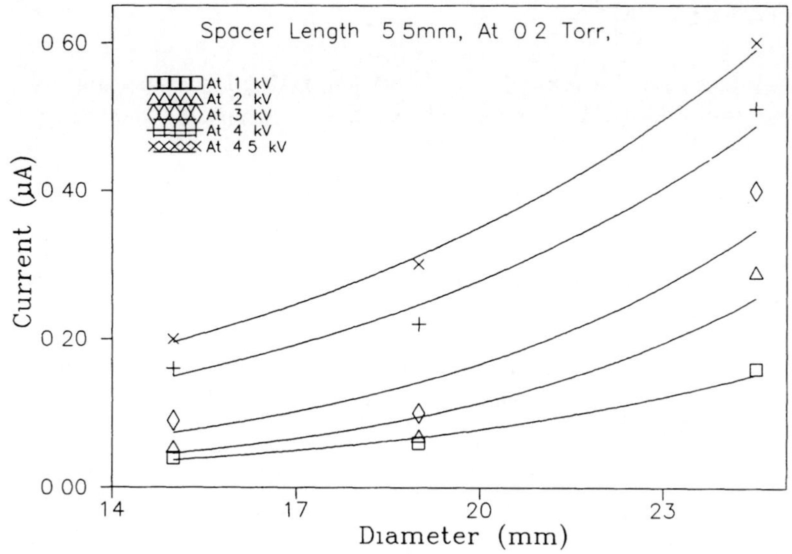

Figure 4. Surface current (a.c.) on Pyrex glass spacers of different diameters for constant applied voltages at low pressure conditions.

As the spacer diameter increases in the electrode gap of constant length, the magnitude of surface currents have increased at ambient atmospheric pressure conditions. However, it may be seen from the figure 4, that the magnitudes of current on the insulator surface shows an exponential growth at low pressures with increase in diameter compared to linear trend at atmospheric pressure conditions. Data on various insulating spacers of comparative diameters have different ranges of current flowing on the surfaces. It is probable that charge retention effect and their flow as current to electrodes are controlled by surface factors combined with gaseous pressure conditions, apart from amount of initial charges produced in the triple junction.

CONCLUSIONS

The surface currents on different insulator spacers bridging the uniform field electrode gap are investigated. Field distortion with the introduction of the spacer is not accounted. However, capacitive currents are neutralised in obtaining the data. Conclusions that may be drawn from the present work are:

a) Surface currents (a.c) data on various cylindrical spacer insulating materials have shown different degree of hysteresis characteristics.

b) The hysteresis effect was higher at ambient atmospheric pressure conditions compared to low pressure, on the insulating spacers of identical surface characteristics.

c) Thus, the present study suggests, that the current flowing on insulator spacers surface is not just limited by the resistivity of the spacer but would also be influenced by the encompassed gaseous pressure conditions.

ACKNOWLEDGEMENT

The authors wish to thank Mr.V.V.Sivaram, CMC, Hyderabad for his help during experimentation and to Mrs.Meena and Mrs.Kala in preparing the manuscript of this work.

REFERENCES

1. H.J.Wintle, "Basic Physics of insulators" IEEE Trans on Electrical Insulation EI-25 No.1, pp 27-44 (1990).

2. P.J.Atkinson and R.J. Fleming, "Surface component of vacuum absorption and resorption currents in polymers. I and II" J.Phys.D, Appl. Phys. Vol 13, pp 626-653, (1980).

3. C.M.Cooke, "Charging of insulator surfaces by ionisation and transport in gases" IEEE Trans. on Electrical insulation EI- 17, No.2, pp 172-178 (1982).

4. R.Lee, T.S.Sudarshan, J.E. Thompson and G.R.Nagabhushana, "Pre-discharge current in vacuum gaps bridged with plexi- glass insulators" IEEE Trans. on Electrical Insulation EI-18, No.3, pp 280-286 (1983).

SECTION 13: DISCUSSION PANEL: REAL TIME CONTROL DIAGNOSTICS/GASEOUS DIELECTRICS NEEDS IN HVDC

DISCUSSION PANEL: REAL TIME CONTROL DIAGNOSTICS/GASEOUS DIELECTRICS NEEDS IN HVDC*

PANELISTS: M. F. FRÉCHETTE (Chairman, Hydro Québec, Canada), O. FARISH (University of Strathclyde, U.K.), I. HERBST (ABB, Switzerland), H. D. MORRISON (Ontario Hydro Technologies, Canada), Y. MURAYAMA (Toshiba, Japan), and R. J. VAN BRUNT (NIST, U.S.A.)

M. F. FRÉCHETTE: Welcome to Session 12, Group Discussion on two topics: Real time Control Diagnostics and Gaseous Dielectrics Needs in HVDC. Group discussions have become a tradition here at this conference. The group discussions in the past have been separated into two parts covering basic research, applications, new trends, and other topics. This year, we have one group discussion of two hours. Basically, the purpose of a group discussion in this scientific forum is to try to get a vision of the future, to put things together, to see what kind of research is needed. Since we have one group discussion, we will try to ally basic research, applications, and engineering toward specific field applications in terms of new technologies, integration to existing technologies, or a positive tradeoff. So, I invite everyone to contribute depending upon where you stand with your vision - thematic basic research, engineering, developing products, developing diagnostics, or trying to integrate some of this information for applications.

The first topic to be treated is Real-time Diagnostics as referenced to gas-insulated equipment and the second topic is Gaseous Dielectrics Needs for HVDC. Short presentations, of about 10 minutes each, will be made by the panelists: Dr. Herbst from ABB, Dr. Farish from The University of Strathclyde, Dr. Van Brunt from NIST, Dr. Morrison from Ontario Hydro Technologies, and Dr. Murayama from Toshiba Corporation. We hope that these statements will provoke some reaction and discussion from you. Please try to contribute - this is your session.

There will be seven contributions - five on diagnostics and two on HVDC. After each statement we will have discussion. I would like to begin with Dr. Van Brunt from NIST on Real-time Diagnostics.

R. J. VAN BRUNT: This presentation provides an opportunity to discuss the problem of what is called technology transfer. Over the past five years or so there has been a lot of interesting laboratory research, much of it reported here at this conference already, on new techniques to investigate the physics and chemistry of partial discharges in laboratory experiments and to look at decomposition of gaseous dielectrics under controlled discharge conditions. A lot of new, highly sensitive techniques, have been developed and the advent of fast computers allows a great deal more versatility in what we can do. The application of new PD techniques has led to a significant enhancement in our understanding of the PD phenomena in electrical insulation commonly used in gas-insulated power systems. A challenge for the future is then the utilization of information gained from these studies in the design of new diagnostic and control techniques that can be applied to the testing and monitoring of large-scale power system equipment such as transformers, transmission line components and breakers, both at the time of installation and during operation. The

*Transcribed and edited by D. R. James and M. F. Fréchette

Gaseous Dielectrics VII, Edited by L.G. Christophorou and D.R. James, Plenum Press, New York, 1994

cost of instrumentation is going down, but the kind of instruments that we have developed for laboratory research often is not well suited for measurements in the field or in the factory due to factors such as high cost, transportability, noise environment, and complexity of the associated data analysis procedures. However, the measurement techniques developed for laboratory research can often provide a foundation for design of power system diagnostics given an effective technology transfer mechanism. So, I think one of the challenges for the future is to use some of the knowledge we have gained from our laboratory testing to build better instruments that can be used for field or factory testing.

I will mention four items, all of which have been discussed to some extent at this conference: (1) sensitive chemical analysis techniques for detecting by-products from PD in SF_6; (2) digital recording and subsequent stochastic analysis of PD under AC or DC conditions; (3) the definition and recognition of PD "patterns" using neural networks or other schemes; and (4) electro-optical methods for measurement of surface charge densities as well as the capacitive probe techniques discussed by Prof. Farish this morning. I mainly want to focus on the first two items since these are perhaps ripe for use by industry. The latter two items are still very much in the experimental stage and perhaps Prof. Farish will say a little more about them in his presentation.

A lot of progress has been made in chemical analysis and partial discharge data recording. One example is the GC/MS system I discussed this morning where I mentioned some of the problems about using the system. The system we have set up at NIST was a fairly expensive commercially available instrument. In fact, it is not available any more and is now somewhat obsolete. This instrument has now been modified such that you cannot buy an instrument that does what ours does - we have sort of a unique setup. But, there is no reason why you could not make a much less expensive, more portable instrument based on this idea. In fact, we are working on that right now. The cost of quadrupole mass analyzers is going down and these are now routinely used as residual gas analyzers on many discharge systems. The kind of column system we use, the interface, and forepump are fairly inexpensive so you can make a portable system at a much lower cost than what we paid for our instrument back in the 1980's. This system is an example of something where we have shown it works in the laboratory, have prototypes and would like to get it out for other laboratories to use. There is a question about what kind of laboratories should use an instrument like this. Is it really something you could put into a truck and take out to substations? We are not sure about that, but it is a device that could be installed in testing laboratories at utilities or companies that serve utilities. Also, it could be used by manufacturers in their laboratories.

There is another instrument we have been working on at NIST. This one has to do with partial discharge phenomena. There have been a lot of developments recently in measurement of digitizing PD signals and recording three dimensional plots of phase and PD amplitude, time and so on. We have been involved in recent years in looking at the stochastic properties of partial discharges, memory effects, and in trying to understand the physics of the phenomenon. But the instruments we use in these studies are very bulky. They are multichannel analyzers that are rather large and require big racks, computers, etc. At the time we started this work about six or seven years ago, those were the best instruments available. But now with the advent of 486-type computers that have hundreds of megabytes of memory at low cost (maybe $5,000 for a fairly impressive computer with large memory), you can do a lot more at lower cost in terms of data recording. What we have set out to do here is to build a device that will allow you to record data indefinitely. One example is a PD signal from a point-dielectric gap in epoxy. We have built a digitizer system that allows us to get an enormous record of this. We record every single event - the amplitude, phase, and cycle in which it occurred. We can look at aging effects, record indefinitely and play it back through the computer. We can do all kinds of sophisticated analysis to look for memory effects and correlation coefficients. We can define patterns that are far more sophisticated than those being considered for neural networks. The simple problem of recording enough data was an obstacle until recently, but we have demonstrated that it is no longer an obstacle. You can store on a magnetic tape about the size of my finger literally hundreds of megabytes of data. You can move a digitizer interfaced to a computer to any location or lab you want. We have a prototype and are interested in getting this idea out and are willing to work with others in doing this kind of data recording.

Basically, the system records a data file in hexadecimal. Every cycle is indicated and information about the phase and amplitude of all the events that occur in that cycle are stored including the sign of the discharge. This, then, becomes a small segment of an enormous file. The way the system works is as follows. We have a partial discharge detector, we do not care what kind of detector you use (it can be acoustic or electrical). You get the signal in as a stream of pulses. The signal is digitized with this system, so we can select out positive and negative pulses and record them separately; the information is stored into a huge computer file. The digitizer is fast, but not the fastest one you could make. There are fairly fast analog to digital converters so it is possible with these systems to record enormous quantities of data. This system is very compact and could be made into a card which could be plugged into a computer. It could be very inexpensive too if you manufacture it. However, as a government laboratory we are not in the business of manufacturing but we are willing to help others develop these ideas. Once you have this information, you can play it back and do all kinds of analysis. The challenge, then, is to write the software that does the analysis.

We have used this instrument to look at PD at a point dielectric gap in epoxy. The time history of the PD of the point on epoxy shows an aging effect we were able to observe with our other stochastic analysis system. This is similar to an effect that was previously reported by Dr. Bartnikas of Hydro Québec and others and is also discussed in a poster paper at this conference. We have recorded a stream of pulses and calculated for certain time intervals an integrated charge distribution for the positive and negative discharge events. The aging effect that occurs is very reproducible. The positive discharges will disappear after some period of time, but the negative discharge will continue. However, the statistical width of the distribution changes dramatically during this aging. The aging is due to the fact that when the partial discharge is occurring on the epoxy, the epoxy becomes more conductive. When this increase in conductivity takes place, it changes the nature of the partial discharge phenomenon. When you have this kind of aging going on, you really need a time history which will give you another dimension in defining a pattern. Professor Farish will talk more about neural network and pattern recognition schemes being considered now. One of the challenges in that area is to obtain a meaningful pattern that you can relate to the physics of the discharge.

We see this digitizer as another instrument that has been developed in the laboratory and shows some promise in getting out to the field. There are a number of companies that make digitizers, but they do not do exactly this. We have not actually tried to get into the software development in a big way, but sophisticated displays using windows and so on could be done. We leave it as an exercise for industry to develop software for the kind of analysis they want. What we are arguing is that if you make a permanent record of the partial discharge and have all the information stored, you can then go back and do a lot of analysis. If you use the conventional multichannel analyzer approaches, you have already prefiltered the data and excluded the possibility of doing a lot of other things with the data. With our device, there is no limit on the analysis, so you might look upon it as a multidimensional multichannel analyzer. With the advent of fast PCs there are all kinds of opportunities for data recording that are now opening up. Thank you very much.

R. BARTNIKAS: I would just like to make a comment on Dr. Van Brunt's work. We know each other very well and we discuss these things all the time. One problem we have is the sort of gulf that exists between what is being done in research and what is being done in the field. The application of the results in the field is indeed very slow. As you know, the PD measurement techniques are extremely rudimentary compared to those achievable in the lab. The present day specifications of the IEC on PD, or I should say the working group that is attempting to change some of the procedures described in the specifications, are still attempting to keep the PD test procedure as simple and as general as possible to make it applicable everywhere. Their approach is if we want to do some specific measurements, then we should still remain within that framework and modify from the general concept down to some specific measurement. So, I think this is really a very serious predicament we have in the sense that we are advancing very rapidly in the area of research but the application in the field is moving by comparison at a snail's pace.

R. J. VAN BRUNT: Do you see the development of standards such as the IEC standard on PD as an obstacle on taking in new technology and measurement techniques?

R. BARTNIKAS: No, I do not think it is an obstacle, but the problem with the IEC is that it is more or less constrained within this point-of-view in the sense that we have a whole lot of similar type of PD detectors in use. These procedures and instruments are being used universally so we cannot change very quickly - it has to be changed very slowly. The approach at the IEC is to have a specification or a standard around a method rather than around a specific instrument. Of course we have all sorts of highly specialized and very powerful instruments that you describe in the area of digital recording which can be used, but it will really take some time before they trickle down to general usage. I would like to see these methods used but due to the practical impediments, I cannot see it happening as fast as we would like it to.

G. N. TRINH: In developing real-time monitoring systems, one should bear in mind that for operating and maintenance personnel, the sophistication of the monitoring system is of secondary importance. What is required is the capability of the system to help draw rapidly adequate conclusions on the state of the equipment; i.e., the necessity of built-in interpretation of the monitored data.

M. F. FRÉCHETTE: I would now like to invite Prof. Farish to make his statement. (This contribution is co-authored by B. F. Hampton of the University of Strathclyde.)

O. FARISH: The first thing I have to do is apologize for not being Brian Hampton. As many of you know, he is the real expert in this area at Strathclyde. He unfortunately could not be with us at this conference so I have taken his place.

Our contribution is concerned with continuous monitoring of PD in GIS toward an overall strategy of control and diagnostics. To have such a system we have to be able to detect a PD of any type, identify the defect, and in addition to classify the defect in terms of severity, i.e., to make some assessment of the criticality of a defect so that we can take appropriate action. The types of defects which can exist include particles, floating shields, barrier contamination, voids, etc. They have in common the fact that they all give some form of PD prior to breakdown. Although in SF_6, of course, the warning between inception of PD and breakdown is not nearly as comfortable a margin as you would find in most insulations.

In principle these defects can be detected in a number of ways: chemical, optical, acoustical, and electrical; they include conventional methods through IEC 270 or UHF and VHF methods. We have heard about these methods on previous occasions and at this conference. The optical method is not perhaps very practical for continuous monitoring for obvious reasons. Indeed, there has been some attempt to make comparisons of these possible basic methods of monitoring. There was a series of tests done under the auspices of CIGRE 15.03 a couple of years ago in which a number of standard defects such as a needle on a busbar (also on barriers, particles and so on) were used to assess in a comparative way the various techniques. This was done in a test chamber at Strathclyde with a number of teams coming together. These results showed that the chemical method was not sensitive enough to be considered for practical purposes. The rest all gave indication of PD from these various standard defects. For continuous monitoring applications the only real candidates would be the acoustic and the UHF method. The conventional method cannot be used in that way. Indeed, the acoustic method and UHF were shown to be very effective in tests for a needle on a busbar. The acoustic method is even better for particle detection. But for a practical continuous monitor, the number of sensors that would be needed for a full substation would be rather large for the acoustic method. One is really left with the UHF method which we would suggest as the only real practical method for a complete system. UHF couplers can be situated some 10 m apart or more and, therefore, can cover a substation with a reasonable number of such sensors.

At Strathclyde in the past few years we have developed a full network for monitoring PD in practical substations. The system utilizes a token ring network, an optical bus linking the various coupler locations. At each location there is a node which collects data, processes them, and feeds them into the network. That data contain information on the coupler location, pulse magnitude and point on wave where the PD arrived. That information is processed again to provide information on the type of defect and some estimate of its criticality. One can also have real time displays and you can recall data from earlier times. Most substations are unmanned so these data can be accessed through a modem from a remote location. Again, the real choice is how you actually display the data - what you record, what you throw away. The business of handling the data is very important of course as is how you identify the defect and criticality.

We can detect PD but we have to know more than that. Take for example the case of a fixed needle on a conductor. As you raise the voltage, it will eventually develop into the dangerous condition of leaders. The PDs that occur then for this kind of defect will be on the peaks of the waves so the phase resolved recording lets you identify it as a fixed point and not as a particle. The size of the signal will tell you whether or not you are entering into a strong leader type situation or whether you have a relatively innocuous localized streamer discharge. With particles a different situation occurs. You have the particle bouncing around on the enclosure and you have random PD which is a feature that can be recognized and used to identify the PD as a particle and not as a fixed protrusion or a floating shield. The interval between the pulses can be analyzed statistically and can give you information on where the particle is locally confined or whether it is of a size and mass that it may be able to fly around and get to a dangerous condition of approaching the conductor. So, again, there is the possibility of identification and some kind of classification of severity.

We want to know how to present the data and how much time we have to collect data to give a snapshot that can be used to identify and classify the source. For a continuous monitor, we also want to have some kind of running record giving the phase-resolved peak magnitude, arriving, say, over a 10-minute interval. One should collect these data for a period of, say, 100 days or a year which is what our system does. In addition, you want to be able to have a snapshot that allows you to assess a particular defect and be triggered by the system itself. It should give a flag when a problem suddenly turns on and take enough data for analysis. For example, streamers occur every cycle, leaders less frequently, and particles, if they are flying in the gas, may last for several cycles, maybe 10 or so. Other defects may be evident every cycle. If one second of data is collected, 50 or 60 cycles, that should give enough data for analysis. That is what our system does. The display must be phase resolved and include amplitude and resolution of individual cycles so we can see cycle-to-cycle how the discharge is evolving. This will give you a snapshot display. Figure 1 is a typical one-second display with 50 odd cycles in one direction and phase resolution in the other direction. Immediately this display shows that we have scattered pulses on the negative half cycle which is typical of a fixed protrusion. If you raise the voltage, you will see more activity on the negative voltage peak and some positive streamers on the positive peak. If the voltage is raised even higher, the onset of leaders at the positive peak can be seen which clearly indicates some potential for problems in a real system. A particle bouncing randomly gives a pattern which can also be recognized. For a floating shield, very large signals give the game away here, but these are quite difficult to interpret. Discharges in the first and third quadrants are typical of a capacitive discharge, so the point on the wave tells you it is not a fixed protrusion but a floating electrode.

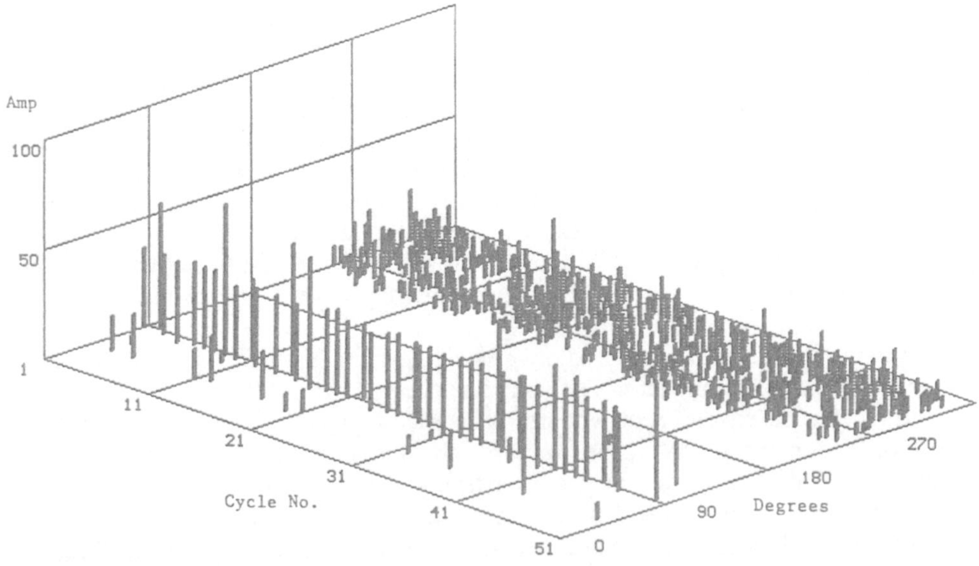

Figure 1. Snapshot of streamer and leader activity from a point on a busbar

With this kind of technique we can in fact identify the defect and have gone some way towards classifying it. We have begun to develop neural network techniques to recognize patterns but that is at its early stages. However, that is not really the first thing one should do. One has to have a good data base containing a lot of information about what kind of signals can occur from defects before you begin to worry about the artificial intelligence aspect.

These systems have in fact been brought to practical use now and there are two of them being installed in full scale GIS in the United Kingdom. They are now real working systems and are not any longer just laboratory examples. This perhaps addresses Dr. Bartnikas' point on the slowness of bringing things from the fundamental stage through to the application. This system has taken some 5 years although Brian Hampton's activity goes back many years in looking at the UHF method. Once you have such a network, you can begin readily to bring in other parameters so we will see eventually the true smart substation where you can incorporate measurements of contact time travel recording, gas density, temperature, transformer oil temperature, and so on. They can be readily added since the network has perhaps a couple of hundred spare channels to bring in other data. That whole system gives you essentially one node in what will be eventually a completely integrated protection and control system. There will be many other free standing monitoring systems that will come into this and in the long term you will see, I am sure, integration of all such systems. I think Dr. Herbst will be talking about this aspect of integration of protection and control systems. It is a long way off yet. We are still in the position of where you do need an expert with some support to see what exactly is happening. In the longer term it should be possible, once enough confidence is there, to use this system to take action to control the network. Thank you Mr. Chairman.

A. H. MUFTI: The measurement of pC in a partial discharge is very important to the utility. How can pC be measured using the UHF method?

O. FARISH: The method does not give a direct measurement of pC. Comparison with the conventional method under standard conditions can give some degree of calibration, but in a continuous monitoring situation, additional factors such as the position and orientation of the PD source will have an effect on the UHF signal recorded.

R. BARTNIKAS: It would probably be helpful instead of having just one system of measurement to have several systems. I would, therefore, favor in addition to the electrical measurements (if one decides to do the EHF measurements) also to have the acoustical measurements. Neither system is self-sufficient. There are some things the acoustical system can tell us that the electrical system cannot tell us. The location of the discharge sources could be verified between the two systems. In regard to the apparent charge measurement, I think what can be easily done, and it is being done in the field, is that the EHF system can be complemented by an ordinary low frequency PD measurement which gives the peak amplitude in charge units. These are the possibilities that one should pursue.

O. FARISH: Of course the acoustic techniques are very useful and they can be used for location but we can locate PD in other ways such as by traveling waves methods or by attenuation between couplers. The acoustic method is perhaps less suitable for a panel-installed, continuous-monitoring, multisensor system because you have to have several hundred rather than several tens of sensors which becomes much more difficult and expensive to operate.

L. NIEMEYER: As Dr. Van Brunt and Prof. Farish have demonstrated, excellent instrumentation has been developed in the last few years which provides valuable tools to be used by the expert researcher and the systems designer. However, these tools cannot serve for on-site monitoring purposes where simpler-to-handle equipment and procedures will be required. One of the key issues to be worked on in the future will be closing the gap between the flood of data provided by sensors and devices and decision making at the equipment in operation. Understanding the physics of partial discharges caused by equipment defects combined with a full exploitation of design knowledge will eventually permit us to select a relatively small number of key quantities which will have to be monitored and which will provide the input for monitoring based decisions.

O. FARISH: We would agree that present instrumentation systems for continuous monitoring of PD detection can generate much more data than can be handled by an engineer. Our aim is therefore to process the information so as to provide the basis for its interpretation as to the type and severity of any defects present in the insulation. While at present this requires "expert" assessment, our understanding of the physics of PD mechanisms is now such that (in principle) it is already possible to use an expert system to limit the need for human intervention. Neural network techniques have already been used to identify the signals associated with the various types of defects and we anticipate that, when more experience is gained, it will be possible to operate the monitoring system in a fully automatic mode including the possibility of initiating control action.

The difficult point is the next part which is to recognize not that you have a defect and it is a particle or a protrusion, but how serious it is. That is where we need the physics. Until we have the physics, there is no point in getting any more intelligence (data). To give you an example, we did a study of a particle on a barrier to be sure we could distinguish between a streamer (by definition it is locally confined and not dangerous) and a leader. There is a characteristic shape of such a discharge in the two half cycles. As you raise the voltage, you begin to get particular activity on the positive half cycle. We measured the UHF signal as a function of voltage and the signal from an infrared sensitive photomultiplier looking into the test chamber which recognizes the presence of a hot channel as in the leader channel. Streamers do not produce a signal since the PM tube is not sensitive to the UV light from them. Hence, when you begin to get leaders you will see activity from the PM. This gives the basis for recognizing the changes in the pattern that are important for an expert person to recognize. We have to have far more data on that before we can with confidence say to the operator that a particular defect is dangerous. But that's a long way ahead. It is based not on artificial intelligence shortcomings but on lack of fundamental physical data.

M. F. FRÉCHETTE: Thank you Prof. Farish. Now, I would like to invite Dr. Herbst to make his statement, co-authored by A. Jaussi, W. Lantz, and K. Pettersson of ABB.

I. HERBST: I would like to make some comments on control diagnostics of GIS in general, not only on partial discharge measurements. Both utilities and GIS manufactures are faced today with an increasing demand for lower life cycle cost and a further increased availability of GIS. What is the strategy to meet this demand?

- Improved and professional overall system concept.
- Reduction of the of the variety of different parts; this includes the primary and the secondary technology.

- Use of an overall monitoring scheme.
- Use of integrated control and protection.
- Use of reliable and cost effective sensors.
- Use of intelligent switching to reduce stresses.

An overall monitoring scheme has to fulfill several requirements:

- The overall aim of as few as possible GIS- and related parts has to be met.
- The sensitivity of the signal chain has to be good enough to ensure detection of critical defects.
- The monitoring system user has to be able to handle the available amount of information and the system should extract essential messages.
- The system must be insensitive to external noise. Here we have a small problem with GIS because any breaker or disconnect operation creates transients which are orders of magnitude stronger than what we actually want to measure. Your intelligent stations need to know all the information about the complete system in order to be made insensitive to such external signals.
- The monitoring system itself must be over an order of magnitude more reliable than the object which is monitored.
- The monitoring system must not stress the system or lead to its deterioration.
- The monitoring system should be integrated into a preventive maintenance concept.

Keeping this in mind, a hierarchy of monitoring tasks can be defined. This hierarchy should account for the system time constants of the objects to be monitored. It reveals the fact that the monitoring system must not lead to deterioration of the object which is monitored and for simplicity we want to define here only three levels:

- Level 1: Permanent control (e.g., what is continuously running).
- Level 2: Checkpoints "once-per-day or operation" (e.g., a measurement on a breaker once its operated).
- Level 3: Check-points "once-per-year or less".

Cost reasons will influence the decision whether to put a task into level one, two or three.

Level 1 is typical for self-checking electronic components: hardware checks by software or software checks by hardware. You may monitor the relay position communications links. The supervision of the contact position of switching devices is another example and also leak detection of hydraulic components. The measurement of voltage, current, gas density, or temperature falls in this category. Very basic but important checks are done. The user will only get messages if there is a problem so there is no user interaction which is necessary. However, if the user wants, he can analyze the data at any time.

Level 2 monitoring should only give messages in case of problems. Level 2 is typical of self-checking of electro-mechanical components and overall tests of electronic chains: relay checks or tests by self-generated test pulses or information pattern. The example we use is a relay check at night when everything is peaceful and you can switch off the electronics for a short time, move all relays, check that they all moved, and once everything is in order return to operation. This can be done every night. You may also generate an information pattern from the different sensors and check your software against the expected pattern. Switching devices can be monitored for time of operation, breaker contact wear (even though we have not had a single major failure due to this wear up to now), breaker speed, energy stored in the breaker drive, or dynamic contact resistance. Also in this group falls the registration of stresses which is the number of breaker operations or the integrated current during switching operations. For heavy duty breakers, current sensing equipment is usually installed in the bay. With an intelligent control and protection system it is then very easy to extract such parameters to get some important information on the status of the breaker.

Level 3 is typical for service checks. Any measurement which needs a high level of technical judgement or which does not need to be frequently performed may fall into this category (may be once a year or less). Measurements which need user interaction due to noise should not be performed so regularly. The frequently discussed PD monitoring may be Level 3 or Level 1. An acoustic PD measuring device, which is easy to use and practical for testing old substations without preinstalled sensors, may be an optimal solution for an EMF-polluted environment. Electrical PD detection, either VHF or UHF, could go into Level 3 as long as reliable warning data algorithms are missing. (Actually we propose testing at elevated voltage not only at the one per unit where you operate but maybe at the higher voltage because some defects have orders of magnitude higher level effects when you go a little bit up in voltage). Most of the breaker checks discussed in Level 2 will be most cost effective if put under Level 3 for "standard load breakers". It should be noted that the flood of information from today's PD monitoring systems may not be interesting any more after some years of problem free GIS operation!

As an example of a Level 3 instrument, the acoustic measurement technique has advantages and disadvantages as Dr. Bartnikas pointed out before. It is important to realize that you will never be able to have one system doing everything for you. So this one is very good if you have EMF-polluted areas because it is insensitive to it. It is also good for needles or sharp protrusions but not as good for voids and insulators, for example.

I would, now, like to discuss some work on breakers. There was the second CIGRE inquiry which observed 19 failures of breakers with only part of these being due to contact wear. The population was 70,700 circuit breaker years. You see that the problem is really very, very small. In fact we do not have a single failure due to contact wear now for modern SF_6 breakers. So, the message is that contact wear is important but not critical, and simple methods such as periodic checks would be good enough.

To summarize, the cost efficiency and availability of GIS can only be improved if integrated systems solutions are envisaged. The monitoring system has to be an order of magnitude more reliable than the object to be monitored which may not always be the case today. Measurements which need user interaction or which need a high level of technical judgement belong to this level three monitoring. What is called monitoring sometimes is not always monitoring in the sense that the user does not have to interact any more. The monitoring of breakers seems interesting to us only for heavy duty breakers and maybe as well only in such an intelligent substation where you have sensors already equipped so you already have parts of the puzzle and you only need some software added. It is not "what seems technically feasible" but "what seems useful" that should drive the development of monitoring. Thank you.

C. X. WANG: My comment has to do with monitoring in general. We have now very sophisticated instruments as you have shown for monitoring a lot of parameters, PD, circuit breaker conditions, etc. But it seems that we really have to target on the correct parameter to monitor, to make it cost effective - to show to the utilities that it is really a useful technique. If we go out to monitor every parameter possible, it seems that we will be overwhelmed with data that we cannot analyze or will have no time to analyze. For example, have we really established that small particles can cause problems to GIS? We know that GIS is tolerable to small particles and it is very difficult to cause a flashover in the laboratory with system voltage. This is just an example that we have to make sure the parameter that we monitor is the greatest concern to utilities.

I. HERBST: I completely agree that one has to decide what is useful and what is reasonable. I think that particles are of concern to us. At operation voltage they will usually not do any harm, but at switching overvoltages as VFT's they may become critical. Good quality assurance testing will enable us to get the problem of particles as low as possible.

H. SLOWIKOWSKA: Committee 17 IEC is preparing now the requirements for SOF_2 monitoring. The speaker did not mention about the necessity of monitoring this SF_6 by-product. Could you comment on this?

I. HERBST: To our knowledge there is no proposal to monitor SOF_2. We also do not intend to perform SOF_2 monitoring as we do not see a need for it.

G. N. TRINH: Regarding the on-site testing of GIS, how high is the test voltage you consider in terms of p.u.?

I. HERBST: We apply 1 p.u. of 3 phase to 1 phase for about 15 minutes. For $U < 800$ kV we then go to 0.8 times the power frequency withstand level. For $U = 800$ kV we have an oscillating impulse voltage testing at $0.7 \times$ BIL complementing the AC test.

T. NITTA: There is one point that is missing from the discussions of monitoring systems or occasional defect detection. That is the relation to the quality of GIS. When you apply on-line monitoring on GIS installations, how often do you expect alarms? If one alarm comes out on one installation per year, we cannot live with the system. GIS is considered to be operable for its life without opening it. To convince a user to install a monitoring system, alarms should probably not occur more than once in the lifetime of the installation. For occasional detection, if you apply the monitoring system at the installation and detect something wrong, you have to dismantle the GIS and find out what is wrong. Then, you reassemble the GIS and apply the detection technique again. If you find another defect, you have to repeat the process again. You may never complete your installation. So, before you apply monitoring, you have to improve the quality of GIS. Then, you should discuss if you need the monitoring for that high quality GIS.

M. F. FRÉCHETTE: Now I would like to go to the next statement which was prepared by Dr. Yanabu. I would like to invite Dr. Murayama to give the statement.

Y. MURAYAMA: Thank you Mr. Chairman. At the beginning, I must apologize for the absence of Dr. Yanabu and I will try to do my best for the short presentation. The following items are some typical requirements of the Japanese power utilities concerning the monitoring systems:

(1) on-line monitoring to know or to detect the instantaneous value of the status of the equipment; (2) operation simulating system; (3) fault location system; (4) automatic restoration system by using the fault location results; (5) maintenance and inspection supporting system; and, (6) automatic operation system.

In Japan the automatic restoring system or the fault location system is strongly required from the Japanese utilities because the down time of the electricity supply has a very severe impact on society. Of course the operation reliablity of the equipment is satisfactorily high but there is still a requirement for these fault location systems, sensors for detecting the beginning phenomena of the abnormalities (partial discharge, gas pressure or gas density sensor, or leakage current monitoring for the zinc oxide type surge arrestor and so on). There is a hierarchy of on-line monitoring systems. This is the equipment to be monitored and on this equipment several sensors are mounted. Data from the sensors are treated at the local control panel which is installed in the vicinity of the major equipment. The data treated by these sensors are transmitted by a local area network to the on-line monitoring system. These systems are installed in the remote control center or the control building in the substation. At this moment the Japanese client requests and also we agreed to, these monitoring systems separate from the protection and control systems because the data transmission speeds or other requirements for the duplication or other technical specification is too different from each other. For our GIS installation of 500 kV, 8,000 A we installed an on-line monitoring system. The gas pressure sensor was installed onto the GIS enclosure. In order to obtain the electromagnetic compatibility, this sensor is mounted onto the GIS and isolated from the enclosure. At the beginning stage these gas-pressure sensors are used only for monitoring the instantaneous value of the gas pressure but in order to simplify our systems we combine the two functions. This sensor can detect the instantaneous value of the gas pressure and the pressure rise after an internal fault in the GIS. In order to evaluate the accuracy of this sensor, we carried out an internal arc test. By these tests we can ensure that the gas pressure and the density sensors are very sensitive to detect the pressure rise. These sensors can also be used in the fault location system.

Another example is the sensor to detect the leakage current of the ZnO surge arrestors. The total leakage current is composed of the capacitive and resistive leakage currents. The major value of the leakage current is given by the capacitive leakage current. We assume the wave shape of the capacitive leakage current as sinusoidal waveshape so we can calculate by using this peak value the capacitive leakage current and then also the remaining value of the resistive current. The actual application of the ZnO surge arrestors in the field is very expensive so we do not want to install these costly sensors onto every surge arrestor but only on very important ones. For example, now in Japan 1100 kV transmission systems are planned and surge arrestors are used as the protective devices. In such a case the monitoring for the leakage current is a very important item and for these systems we would like to apply the leakage current sensors.

To summarize, at this moment in Japan we already supplied more than 10 systems to the actual GIS substations, but still we have some items to be improved or to be considered. Three points should be made. The first is the reliability of the sensing system itself. As you know well, the reliability of GIS is very high and further improvements in reliability will require very high technology. The second is the lifetime of the sensor system. In order to obtain the high safety for the EMC we applied fiber optics between the sensor and the microprocessor. All the data was transmitted by the fiber optics. For that system we needed a battery on the sensor which required replacement every 6 years. Recently we eliminated the battery from the sensor. The third point is the cost of the sensor system. The reliability of the sensor itself should be designed to be very high and also the lifetime should be very long without periodic maintenance. As a result, the cost of the sensor becomes relatively high. The Japanese clients judge that for the very important substations for the electricity supply, these on-line monitoring systems are used. For lower voltage, at this time, on-line systems are not required. Thank you.

M. F. FRÉCHETTE: I would, now, like to invite Dr. H. D. Morrison to present the statement on User Perspective on GIS Diagnostic Monitoring Systems prepared by Mr. N. Fujimoto.

H. D. MORRISON: Thank you Mr. Chairman:

Purpose of Monitoring

The purpose of diagnostic measurements is to provide some information concerning the condition or "state-of-health" of the equipment being measured. To be effective, this information must provide positive economic benefit to the equipment user. Dielectric diagnosis for GIS should, therefore, be capable of dealing with all defects and/or states of deterioration which might lead to premature dielectric failure of the equipment. In order for a positive economic impact to be realized, these diagnostic systems must be able to: (1) detect the presence of a harmful defect, (2) determine the location of the detected defect, (3) identify the nature of the defect, and (4) provide some information on the state of deterioration and the urgency for remedial action.

Defects in GIS

GIS is a technology whose dielectric performance is influenced mainly by the presence of defects and other abnormalities. Many of the defect types discussed in the literature, such as particles and conductor asperities, are problems primarily introduced during shipping and onsite assembly and are the focus of diagnostic measurements performed at commissioning. Inservice monitoring should deal with the defect types which result in long-term deterioration and ageing of the insulating properties of the GIS. While some of the assembly-related defects, such as particles, can be generated by moving components within the GIS during service, the main ageing-related defects fall into two categories: (1) poor/loose contact of electrostatic shields, and (2) insulator defects. Several failures resulting from both of these defect types have been experienced in operating GIS. Effective monitoring must be able to deal with both of these defect types.

Electrical connections for electrostatic shields are sometimes implemented with a low-force contact system designed for the small capacitive currents. While contact may be adequate after initial assembly, these contacts can sometimes deteriorate over time (possibly as a result of an electrochemical mechanism from microdischarges during surge events) and eventually develop into full floating components and eventually induce dielectric failure.

Defects within epoxy insulators are normally avoided by ensuring a high degree of quality during manufacture. However, the PD tests (QA testing) performed today in industrial environments are not capable of finding all the defects which could lead to premature failure. These defects can be voids or other contaminants which lead to the development of electrical trees in the epoxy matrix and eventually cause failure of the insulator.

Monitoring Strategies

While it may be always desirable to have advance warning of failures by some monitoring system, any such monitoring system must be capable of providing sufficient advance warning to have a positive economic impact on the operation and maintenance of the GIS. The cost of outages can be high, but the benefits of reliable advance warning of a failure can vary. For instance, in some cases, knowledge that a failure could occur might have little economic consequence if the actions required to mitigate the deterioration are similar to the effort required to repair a fault. However, in other cases, failure avoidance might have much greater economic impact, if sufficient lead time allows a scheduled outage for repair. With these issues, and considering the overall cost of GIS equipment, many utilities require inexpensive, easy-to-use but effective monitoring strategies. In many cases, utilities are unwilling to spend more than a few percent of the GIS equipment cost on diagnostic systems.

Gas sampling remains the most widely used monitoring scheme, because of its simplicity and low cost. The detection of decomposition by-products in the gas is a positive indication of problems. Gas sampling is normally performed only when problems are suspected, but some utilities have considered periodic sampling. Generally, trends in gas composition detected by periodic monitoring are more useful than one-time samples. The effectiveness of gas sampling is limited, however, by the size of the gas compartment and the presence of absorbing materials. Other forms of PD measurements using acoustic and electrical (UHF/VHF) methods are being considered and used by some users worldwide for occasional and/or periodic use. However, continuous monitoring is still not common, primarily because of the high cost and because few devices are available commercially. Nonetheless, a few utilities are presently using continuous monitoring schemes and others are considering its use, especially for those critical GIS at high voltage levels.

Gaps in Knowledge

Examination of the present state of technology shows that some gaps in our knowledge must be addressed before diagnostic technology can be used to its fullest advantage. While the sensitivity to many defect types may be adequate, the need for positive benefit requires that the deteriorating situations be identified as early as possible, especially for the ageing-related defects. These defects need to be studied further in order to assess the sensitivity requirements for monitoring systems and to characterize the nature of PD signals (as measured by the various techniques) in the early stages of deterioration and its time evolution towards failure. This is especially important in interpreting the data from the point of view of assessing the risk of failure, urgency for action and/or remaining life. Sensitivity limits may also dictate how useful diagnostic monitoring could be as in some cases, PD levels might remain below detectable limits until just before failure. Specifically, defects for which time evolution is important and need further study are: (1) Floating components/poor contacts - the deterioration mechanisms for low-force electrostatic contacts need to be investigated in order to understand what diagnostic information might be available to indicate a worsening situation leading to a full floating component, large magnitude discharges and eventual failure; (2) solid insulators - given that some defects have escaped detection by factory testing, the transition to electrical treeing, the required sensitivity, the associated PD signature, and the time evolution need to be fully assessed to evaluate how diagnostics could be used to a positive benefit for this type of failure.

M. F. FRÉCHETTE: I would, now, like to ask Dr. Murayama to comment on HVDC GIS. (Dr. S. Yanabu prepared a written statement which has been distributed).

Y. MURAYAMA: I will discuss the needs or future requirements for HVDC GIS. Japan has a very complicated power supplying system so it gives us very interesting technical items. Two power frequencies are used - 50 and 60 Hz - so we need power frequency converter stations. The main part of Japan is divided into 9 electric power supply companies and to obtain stability there are some back-to-back stations installed. There are some major islands in Japan so the generating power is transmitted by HVDC to this mainland operated at 250 kV HVDC, but not GIS. Recently there have been several plans for HVDC. It is planned for mountainous and seashore areas. So, in order to minimize equipment size or eliminate the influence of saline contamination conditions we are studying HVDC GIS.

The typical needs for HVDC GIS are first the insulation requirement. As already discussed in this morning's session, when metallic particles exist inside HVDC GIS, the breakdown voltage is tremendously reduced. We must know about the behavior of the metallic particles and also we must propose some counter measures. As you know well, the breakdown voltage with lightning impulse is drastically reduced when the length of metallic particles set on the surface of the high voltage conductor is increased. For HVDC, when the metallic particle is released it moves instantly to the high voltage conductor so the particle trap functions must be applied to the high voltage side. For AC GIS, the particle traps are installed at the enclosure side. Therefore, a major requirement for HVDC GIS is the insulation against the existence of metallic particles.

For gas breakers, the breaking phenomena are also determined by the dielectric properties just after current interruption. Two types of breakers are used for HVDC GIS. One is the bypass switch in parallel to the thyristor valve and the other is the metallic return transfer busbar (MRTB). We use standard AC gas type breaker for the MRTB, the AC 300 kV standard gas type breaker, and the AC 84 kV breaker. When arcing voltage is generated, some resonance phenomena occur and if this oscillating current interrupts the zero line, then this current could be interrupted at this current zero. From the point-of-view of the current interruption of the gas type breaker, when this frequency is low, the interruption of the current is easier. But in order to make the frequency lower we must supply a big capacitance. To minimize the size of the equipment, we would like to make this capacitance as small as possible. If we use 300 kV gas type breakers, only less than 20 μf capacitance is applied. Our breaker can interrupt 3500 DC A which is the specification. If we apply 84 kV gas type breakers, even if we apply 28 μf our gas type breaker cannot interrupt and at 100 μf only 900 A can be interrupted. This requirement is not only for the HVDC GIS, but also for the HVDC conventional station. (Formerly we applied the air blast breakers for this use, but these are no longer being produced. However, we are confident we can apply the gas type breaker).

Finally, I would like to address the requirement for the disconnecting switch for HVDC GIS. When the disconnector switch is making, the thyristor valve stack should be energized. If we make this stack without the pre-insertion resistor, a very steep front surge is generated and the voltage distribution at the entrance side becomes very severe. In order to design the thyristor valve appropriately for higher voltages, for example 500 kV HVDC, we would like to propose the use of a pre-insertion resistor for the disconnecting switch. These are examples for the needs for HVDC GIS. Thank you for your attention.

G. N. TRINH: When do you forsee to have the first HVDC GIS converter station in service?

Y. MURAYAMA: At the moment the plan is not sure, but I, hopefully, foresee around the year 2000 for the start of in-service operation.

L. G. CHRISTOPHOROU: Are you planning to use other insulating gases besides pure SF_6, such as SF_6/N_2 mixtures? Has anyone investigated the by-products of SF_6/N_2 mixtures?

Y. MURAYAMA: There are no plans to use mixtures for higher voltage GIS. For 72 kV cubicle-type GIS, the mixture with air is used since the cost of the gas is reduced and the on-site maintenance evacuation procedure is easier. There are no studies for the by-products of mixtures at present.

K. NAKANISHI: Are you planning on applying any special diagnostic tool different from those for AC GIS to HVDC GIS?

Y. MURAYAMA: I have no idea for the monitoring of HVDC GIS. The off-line monitoring will be used as a part of the commissioning test. The on-line monitoring will be studied.

M. F. FRÉCHETTE: I will, now, call on Dr. Herbst to make his statement on HVDC GIS.

I. HERBST: I will discuss gaseous dielectric needs in HVDC. Even though the theoretical DC breakdown voltage in SF_6 gas is, roughly, equal to the peak of the AC breakdown voltage, a lot of work was (and is) needed to design insulation systems which sustain HVDC in real applications. During the last 20 years extensive research was conducted in this field.

The main problem is the correct field steering of very weakly conducting (insulating) materials. Surface and volume irregularities which do not affect the AC performance may affect the DC performance due to distorted field configurations. Any discharge activity near insulating surfaces bears the risk that charge carriers reach the surface and subsequently disturb the potential distribution. It is, therefore, of primary importance to avoid discharge activity and to design spacers such that charge carriers will usually not reach the insulator surface. Polarity reversal tests will uncover strong field distortions as they will lead to breakdown.

In figure 2 we attempt to summarize the status of HVDC components. The different parts needed for an AC/DC converter station are listed and commented with respect to the potential of applications and the state of the art.

The application of SF_6 gas insulation is realized both in prototypes and in commercial systems. The main advantage of gas insulation comes mainly when using all components with SF_6 insulation: switchyard, connections, and filters. Substantial space demand reductions are achievable then.

POTENTIAL OF APPLYING GAS INSULATION :	AC-SIDE (PARTIALLY WITH DC COMPONENTS)				AC / DC	DC-SIDE		
		FILTERS				FILTERS		
	SWITCHYARD	PASSIVE: C&L&R	ACTIVE	CONNECTION TRAFO - VALVE HALL	VALVE HALL	PASSIVE: C&L&R	ACTIVE	SWITCHING EQUIPMENT INCL. BUSES
VERY HIGH HIGH MEDIUM LOW	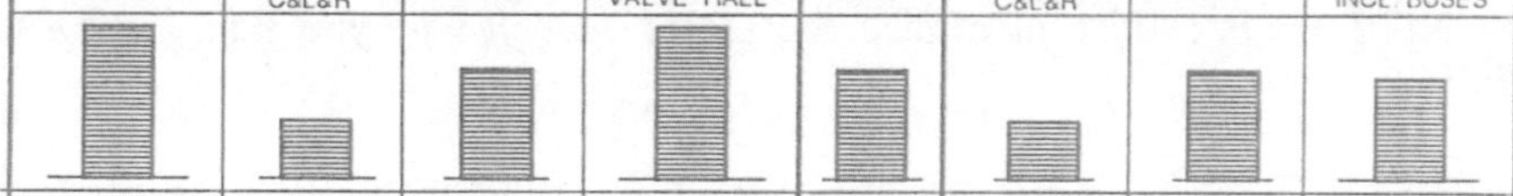							
TYPE OF EQUIPMENT	STANDARD AC-GIS	OPEN OR ENCLOSED	OPEN OR ENCLOSED	BASICALLY STD AC-GIS EXCL. SOLID INSULATORS.	OPEN OR ENCLOSED	OPEN OR ENCLOSED	OPEN OR ENCLOSED	BASICALLY STD. AC-GIS EXCL. SOLID INSULATORS
STATE OF THE ART	ESTABLISHED TECHNIQUE ("CONV." GIS)	GAS INSULATED PROTOTYPES TESTED	SOLUTIONS WITHOUT GAS INSULATION EXIST TO REPLACE PASSIVE FILTERS	IN COMMERCIAL SERVICE AT 150KV 500KV PROTOTYPE LONG TERM TESTED GAS INSULATED BUSHINGS IN SERVICE	—	GAS INSULATED PROTOTYPES TESTED	SOLUTIONS WITHOUT GAS INSULATION EXIST TO REPLACE PASSIVE FILTERS	500KV PROTO. LONG TERM TESTED -> SOLID INSUL. -> SURGE ARR. -> ..... 600KV PROTO. TESTED -> SOLID INSUL. -> OIL-TO-GAS BUSHING

Figure 2. Status of HVDC components

A first commercial installation of GIS in conjunction with AC/DC converter stations was commissioned in 1983. The 130MW monopolar cable link with a 150kV GIS connection from a transformer to the valve hall is in operation since then.

Substantial research was carried out since then to study the insulator behavior at higher voltages. The time-dependent potential distributions were evaluated. It has been found that, after a critical inspection of the design, normal GIS components are usually usable for DC operation. Extensive DC tests have been performed at BPA with two designs based on 550/800kV AC GIS. Two sizes of spacers were specially developed for use with HVDC. Furthermore, two types of field steering methods have been tested for one given size:

- Insulator with semi-conductive spacer surface
- Field steering with the insulator bulk material

Both methods have proven their reliability during all types of tests:

- Long-term DC (500kV integrated load = 10000h)
- BIL tests at 1300kV, SIL tests at 1050kV
- Voltage reversal tests
- DC tests with current/thermal load.

We can summarize the above by saying that the basic problems of gas insulated DC components are solved and a phase is reached where the market should drive a further development.

A. H. MUFTI: I would like to comment on the question raised by Dr. Christophorou on the recent work on SF_6 gas mixtures in Japan. There is a recent work by Japanese researchers to inject a fluorocarbon vapor into the GIS system under the name of "Preventive Maintenance System with a Different Gas Injecting Facility for GIS", 1993.

Y. MURAYAMA: Regarding the monitor gas, these studies have been done for the gas-insulated power transformer and GIS. But until now there is no commercial use. The control of the mixture contents is to be studied when the gas is refilled.

T. NITTA: Is reduction of size the major reason why you apply gas insulation on HVDC?

I. HERBST: Yes, but there are other reasons as well. For example, our station equipped with DC GIS bus ducts is at the coast of Sweden where contamination could be a problem.

K. NAKANISHI: In the presentation you mentioned the harmfulness of the conducting particles especially at DC fields. Do you have any countermeasures which deactivate the conducting particles in HVDC GIS? Mr. Murayama referred to this point.

I. HERBST: There are several measures to take including AC conditioning of the GIS and coating of insulators to control the potential distribution.

K. FESER: If I understood you correctly, you said there is no need for research on DC GIS?

I. HERBST: I said the principal things are done. The prototype work has been done. We can use an AC GIS and modify it a little bit. To go from an AC system with only a little modification you have to check that the component you had before will withstand the voltages under DC. The checking for every given design you want to use has to be done and its the details you have to think about not the general layout of the system because that I think is solved. It is really the small details and when you start thinking about DC, you discover problems where you would not expect them.

C. X. WANG: You have identified that for DC GIS, the insulators need to be modified and the technique is a semi-insulating material coating. Now we know for sure that even for AC GIS, periods of DC stressing exist during normal operations. Can your semi-insulator coating technique be used as a retro-fitting method?

I. HERBST: Yes, it could be used. We do not propose it though, as we do not have problems with "DC at AC". Also for cost reasons such a retrofitting is not interesting.

I. SAUERS: It would be interesting to know if anyone has considered using gas switches for converter valves in place of the solid state thyristor switches, in particular the opening switches that might allow you to eliminate a lot of the filtering that takes up so much of that space.

M. F. FRÉCHETTE: I thank the panelists and the participants and I declare the session closed.

APPENDIX: BANQUET SPEECH

WHAT DOES INDUSTRY EXPECT FROM BASIC RESEARCH?*

Tohei Nitta

President
Mitsubishi Electric Research Laboratories, Inc.
Cambridge, Massachusetts 02139

Good evening. It is nice to see many familiar faces in the audience. When I attended the last meeting of this symposium three-and-a-half years ago in this hotel, I thought it would be my last chance to participate in the symposium. By that time I had already received a new career assignment which has taken me rather far afield from the symposium topic of gaseous dielectrics. In spite of that, you kindly gave me an opportunity to return once again to speak at this banquet. It is really a pleasure and great honor for me to talk to you this evening.

BACKGROUND

The topic of the talk suggested to me is "What Does Industry Expect from Basic Research?" I know that this topic is too big, particularly for someone like me who has spent his whole career in one private company, Mitsubishi Electric Corporation (MELCO). However, there are reasons why I consider myself qualified to present my personal view on the subject. And that is why I come here and stand on this podium.

In the fall of 1989, I first learned that MELCO's Corporate R&D division intended to establish a basic research laboratory somewhere outside of Japan. Under consideration was a new research facility on electronics, though very little had been specified at that time. There was no clear vision for the lab, and key issues such as location and size were yet to be decided.

So, I volunteered for the project. First we worked to solidify a plan for approval by the Executive Board of Mitsubishi Electric Corporation. This entailed approaching all sectors in the organization, such diverse groups as Strategic Planning, Accounting, Finance, Legal, International Operations, Marketing, etc. Not an easy task — MELCO is a large company, 44th on the Fortune 500 in 1993, and a large company is normally conservative with a sizable bureaucracy, as you might expect. It meant requesting authorization of a plan which would require millions of dollars every year, not just a one-time expense. And it was understood that, most probably, there would be no valuable returns for some years to come.

While working on this project, I learned a lot about what people, particularly those without technical backgrounds, expect from research. In order to meet those expectations,

* Banquet speech at the 7th International Symposium of Gaseous Dielectrics, Knoxville, TN, April 27, 1994.

Gaseous Dielectrics VII, Edited by L.G. Christophorou and D.R. James, Plenum Press, New York, 1994

we carefully formulated the concept of our research lab and subsequently made a successful presentation to MELCO's Executive Board.

MERL, MELCO'S NEW RESEARCH LAB

Upon receiving authorization, Mitsubishi Electric Research Laboratories (MERL) was officially established, and I was named president. It was decided to locate the laboratory in the United States -- Cambridge Massachusetts, to be specific. We chose a site for the laboratory near Harvard and MIT, within walking distance of MIT. We considered the location to be a very good environment to institute an efficient network in the research community and to attract excellent researchers.

We have defined our mission as exploring entirely new functions for computers -- those not yet envisioned. While computers presently play a very important role in our society, that role is for the most part restricted to supporting the existing social systems. Our goal is to help create new social infrastructures for the 21st Century using modern computer technology. Some examples of such systems are for education, including not only school education but also life-long learning in an ever-changing society; entertainment; and new working environments which will replace today's office.

We quickly recruited researchers from all over the world and have started several research projects in the laboratory. Today we number about 30 researchers with Ph.D.'s in various fields of expertise.

I will talk about my experience with MERL in relation to the subject given to me for discussion tonight. I will also tell you about the problems I had in setting up research projects with the researchers we hired.

KEY OPERATIONS CONCEPTS FOR MERL

We set forth the following three key concepts for the operation of MERL:

(i) The laboratory will focus on long-term research;

(ii) It will be an independent lab which can set its own research direction; and

(iii) Scientific aspects of the lab will be open to the public.

These operational concepts show what a private company expects from research, as I will explain in more detail.

Long-Term Research

The laboratory focuses on long-term research. What are the implications of this concept? Some people may assume that researchers will have plenty of time to complete their research projects, but that is not true. If they keep doing research for ten years in the lab, after ten years you will not see any real impact of the research (see Figure 1).

Long-term actually has two implications. First, the lab will work on projects that may take many years before making a real impact on society. Second, MERL will restrict itself to research alone, and developing the products generated by MERL research will be handled by other organizations.

Since MERL itself will not spend major effort on development, it is MERL's responsibility to find someone to perform development outside the laboratory. Our task is not only to accomplish the research project, but also to locate a development group and capture their interest — all this has to be done as part of the research process. If we fail to find a group to carry on the development work, we consider that our project fails.

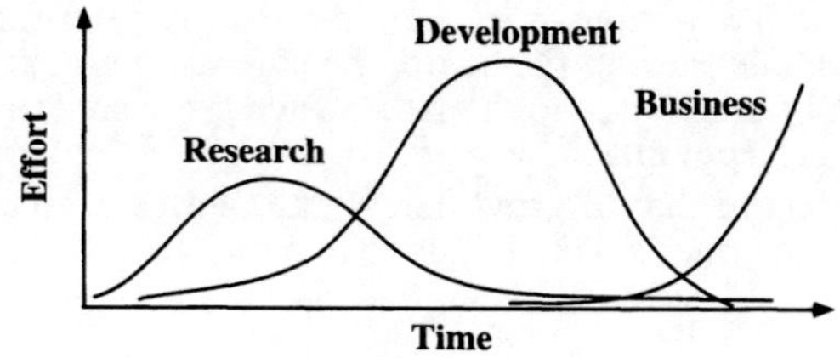

Figure l. Two implications of long-term research: 1) Period of time to impact society is long-term, but the real research term is rather limited; and 2) Merl is expected to find someone to do development outside of MERL.

Independent Lab

The second basic concept of MERL's operations is the independence of the laboratory. The laboratory has the freedom to design its own research projects; this will not be dictated by our supporter. It is the researchers and the management of MERL who will identify areas of investigation and possible application.

In exchange for this freedom, MERL is responsible for the final outcome of its research. It is necessary for MERL to find ways to sell its research for the benefit of its industrial partner (although the application of research results is not limited to MELCO). We do not have the luxury of saying, "We performed great, exciting research, but development people in industry did not recognize the importance of the research and did not take full advantage of the results."

Also, it is important to realize that an industry is not immediately and directly impressed by scientific accomplishments. In fact, I have to say that success of a research project does not have much to do with its scientific value. While contribution to the scientific community is a basic motivator for researchers, it may not be the goal of a research organization. Figure 2 illustrates the concept that the primary relationship of the research lab is with its appropriate industrial partner, not the scientific community or end-users.

At this point let me interject my observation on many of our researchers. When researchers think about application of their research, they essentially consider the benefit to end-users — a dangerous simplification. For the application of research to a product, you have to attract your industrial partner instead of end-users; if you fail to do so, the technology will never reach its targeted end-users. It is well to remember that concern for end-users is not necessarily the highest priority issue for the industrial partner who applies your research to its product.

In fact, a company is extremely sensitive to risk and competition within a given market. To sell a new technology, it must be made clear how it is appropriate and especially advantageous to that particular company. An idea favorable to IBM will not necessarily be favorable to Mitsubishi Electric. Value can rarely be generalized to a whole industry.

So, to reiterate, the freedom of independence carries with it the responsibility to identify a suitable partner.

Open Lab

MERL is designed to be an open laboratory. Researchers are encouraged to publish their findings and present them at scientific conferences and seminars, as well as to invite their colleagues for visits and technical discussions. MERL also supports projects and collaborations with a number of research institutions in the U.S. and abroad.

How can MERL benefit a corporate sponsor while remaining an open laboratory? I like to compare research with "nurturing fish in the ocean," see Figure 3. Fish in the ocean belong to everyone until caught, when they become the property of the fisherman who caught them. The fish we foster are for everybody; they are different from fish raised in our own pond. This describes the openness of our laboratory.

However, we must help our sponsor to catch the fish by determining when and where they should lower the nets. Nurturing fish is not the sole function of the laboratory, but we have an obligation to lead our sponsor to the fish. If we grow superb fish but cannot guide our sponsor to them, the project fails.

I would like to emphasize that to grow fish is one thing and to catch them is another. Industry expects a research partner to take initiative in both.

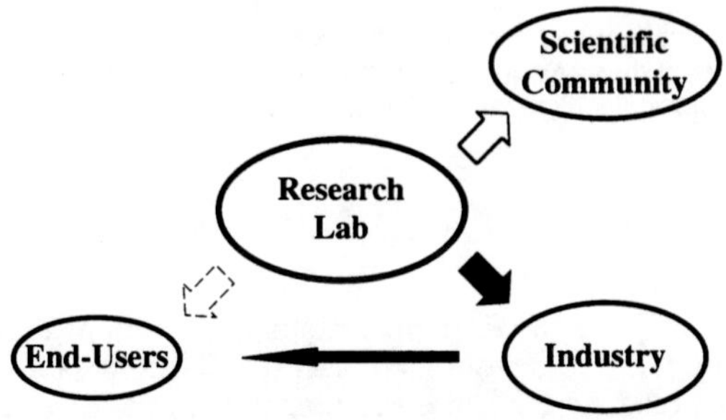

Figure 2. The primary relationship of an independent research lab is with its appropriate industrial partner, not the scientific community or end-users of the product.

INDUSTRIAL PARTNER FOR RESEARCH

MERL's industrial partner is MELCO — we work together for our mutual prosperity. To be an effective research organization, I believe, we always have to have a clear picture of our own industrial partner. This is really important, because the partner should influence the way we do research.

Start-Up Company As A Partner

The typical example of an industrial partner of a research lab is a start-up company led by an ambitious entrepreneur — a familiar American dream. Tinkering begun in the corner of a garage may eventually yield, with luck, Apple Computer or Microsoft. While conducting research with this kind of partner in mind, you can not expect too much support from industry. Eventually, there is no existing industrial partner to support your research while you are doing research. You have to convince the ambitious entrepreneur to go with your idea. Or, you have to convince yourself to change from a researcher to an entrepreneur. To accomplish this, all efforts at prototyping and a well-developed and sophisticated demonstration is required as proof of feasibility of your idea. Without this demonstration, no venture capitalist will be interested in the new technology.

Established Private Industry As A Partner

Partnership with a well established company is the other extreme. A research lab that belongs to a well established private industry has to work basically with its parent company as its industrial partner. In this case, it is critically important to attract and hold the attention of its partner so that the research activities in the lab are constantly observed and engaged in a continual exchange of information on research direction. From the lab's point of view it is crucial to know when to make proposals and what will likely result, a vulnerable position since the partner may find all kinds of reasons to reject your recommendations. You will be stuck if their response is, "Great idea, but we are busy with something else." A research lab that belongs to a private company has a tremendous handicap when compared to academia or independent research institutions that have freedom to form alliances with any development organization.

The laboratory has, however, an advantage to compensate for this handicap, and that is the proximity to the industrial partner. We are in a position to know strengths and weaknesses of the partner in terms of technical work force, marketing and servicing networks, manufacturing facilities, and so on. We have access to their future plans including when they finish with a current project, and other useful internal information. To exploit this advantage, it is essential to work very closely with the partner from a very early stage of research.

One might be tempted to wait for solid research results before approaching the industrial partner, but that tactic loses the advantage of proximity. Actually, communication at the earliest moment is much more effective. Be aware, too, your partner is existing not to apply your research to its products. They are existing for something else. All the resources, people, facilities and money in your partner's hands are allocated for other purposes. By the time work has reached the stage of a completed prototype, in most cases it is too late in the game to go with your idea; the industrial partner will have to collect hundreds of people and pour millions of dollars into the new project, a highly unlikely decision. This approach differs significantly from working with start-ups.

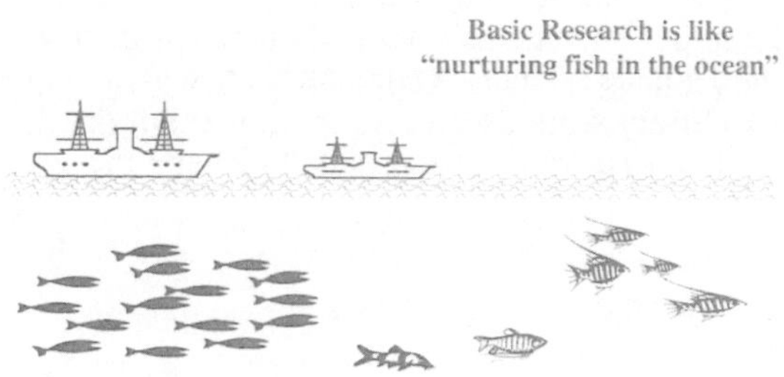

Figure 3. The independent research lab not only nurtures fish but also leads its industrial partner to the fish.

When working with an existing company, there is one other important factor that you should bear in mind. That is cooperation with the researchers in related but different fields.

Organization of the industrial partner is usually hierarchical, as shown in Figure 4, because industrial products require the coordination of a variety of technologies. Suppose you are a researcher whose field of interest is represented by one of the peaks at the left of Figure 4. When you want to talk with your industrial partner (shown by the hierarchical organization on the right of Figure 4), the specialist who understands your technology is at the bottom of the organizational chart. It is obvious that you cannot impact the organization by simply impressing a guy at the bottom of the hierarchy. To give a real impact to your industrial partner, you have to have a certain width in the spectrum of your technology, as in Figure 4. Industry expects the wide spectrum of coordinated expertise provided by a strong interdisciplinary team, not an inch-wide and mile-high technology.

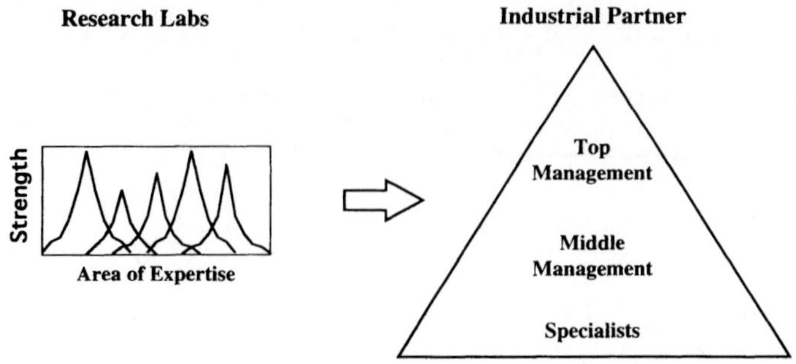

Figure 4. The strength of the independent research lab is in the wide spectrum of its echnology and how it relates to the hierarchy of its industrial partner.

CONCLUSIONS

In summary, industry expects research to:

- Accept full responsibility for the design of research projects. While it is important to work closely with your industrial partner, do not expect the partner to dictate or suggest the direction of your research.
- Make it clear to whom you must sell your results before you begin the research. Your approach will be different depending on your industrial partner.
- Recognize that to commercialize your technology it is necessary to impress your industrial partner, not the scientific community or end-user of the product.
- Expect to work closely with researchers in related but different fields. Forming a team with a wide spectrum of specialties will have the greatest impact on your industrial partner.

PARTICIPANTS

PARTICIPANTS

H. Aoyagi
Toshiba Corporation
1-6, Uchisaiwai-cho 1-Chome
Chiyoda-ku, Tokyo 100
JAPAN

R. Bartnikas
Hydro-Quebec
1800 Montee Ste-Julie
Varennes, Quebec
CANADA J3X 1S1

B. A. Berven
Oak Ridge National Laboratory
P. O. Box 2008
Oak Ridge, TN 37831-6124

P. P. M. Blom, Jr.
Eindhoven University of Technology
P. O. Box 513
5600 MB Eindhoven
THE NETHERLANDS

D. D. Bloomquist
Sandia National Laboratories
Division 1266
P. O. Box 5800
Albuquerque, NM 87185

D. W. Bouldin
Department of Electrical Engineering
The University of Tennessee
Knoxville, TN 37996

A. F. Borghesani
Universita Degli Studi di Padova
Departimento di Fisica
"Galileo Galilei"
35131 Padova-Via F.
Marzolo, 8
Padova
ITALY

D. W. Branston
Siemens AG
Postfach 3220
91050 Erlangen
GERMANY

D. Buchner
Institute of High Voltage Engineering
Technical University Munich
D 80290 München
GERMANY

F. Campbell
Tennessee Valley Authority
C9T 17A
1101 Market Street
Chattanooga, TN 37402

J. G. Carter
Oak Ridge National Laboratory
P. O. Box 2008
Oak Ridge, TN 37831-6122

J. Castonguay
'Hydro-Quebec
C.P. 1000
1800 Montee Ste-Julie
Varennes, Quebec
CANADA J3X 1S1

I. D. Chalmers
Department of Electronic
and Electrical Engineering
Royal College Building
204 George Street
University of Strathclyde
Glasgow G1 1XW
SCOTLAND, UK

J. S. Chang
Department of Engineering Physics
Nuclear Research Building
McMaster University
Hamilton, Ontario
CANADA L8S 4M1

L. G. Christophorou
Oak Ridge National Laboratory
P. O. Box 2008
Oak Ridge, TN 37831-6122

O. Claus
Technical University Darmstadt
Landgraf-Georg Stras.
64283 Darmstadt
GERMANY

A. H. Cookson
National Institute of Standards and
Technology
Bldg. 220, Room B166
Gaithersburg, MD 20899-0001

G. C. Crichton
Electric Power Engineering Department
Technical University of Denmark
Building 325
DK-2800 Lyngby
DENMARK

P. G. Datskos
Oak Ridge National Laboratory
P. O. Box 2008
Oak Ridge, TN 37831-6122

I. Dionne
IREQ
1800 Montee Ste-Julie
Varennes, Quebec
CANADA J3X 1S1

Rich Driggans
Tennessee Valley Authority
1101 Market Street
Chattanooga, TN 37402

J. De Urquijo
Instituto de Fisica, Unam
Post Office Box 139-B
62191 Cuernavaca Mor.
MEXICO

O. Farish
Department of Electronics and
Electrical Engineering
204 George Street
University of Strathclyde
Glasgow G1 1XW
SCOTLAND, UK

K. Feser
Institut für Energieübertragung und
Hochspannungstechnik
Universität Stuttgart Breitscheidstruβ 2
D-7000 Stuttgart 1
GERMANY

M. F. Fréchette
IREQ
1800 Montee Ste-Julie
Varennes, Quebec
CANADA J3X 1S1

A. Garscadden
Wright-Patterson Air Force Base
Advanced Plasma Research Group
Power Division
Aero Propulsion and Power
Laboratory
Dayton, OH 45433-6523

M. Goldman
Laboratoire de Physique
des Decharges
CNRS/ESE
Plateau du Moulon
91190 Gif-sur-Yvette-Cedex
FRANCE

G. D. Griffin
Oak Ridge National Laboratory
P. O. Box 2008
Oak Ridge, TN 37831-6101

S. Grzybowski
Mississippi State University
Drawer EE
Mississippi State, MS 39762

R. N. Hamm
Oak Ridge National Laboratory
P. O. Box 2008
Oak Ridge, TN 37831-6123

L. L. Hatfield
Department of Physics
Texas Tech University
Box 41051
Lubbock, TX 79409-1051

I. Herbst
ABB
HV Technologies
CH8050 Zürich
SWITZERLAND

R. Huang
Oak Ridge National Laboratory
Building 4500S, MS-6125
P. O. Box 2008
Oak Ridge, TN 37831-6125

S. R. Hunter
Consultec Scientific, Inc.
725 Pellissippi Parkway, Suite 110
Knoxville, TN 37932-3300

K. Ibuki
Mitsubishi Electric Co.
1-1, Tsukaguchi
Honmachi 8-chome
Amagasaki, Hyogo
JAPAN

A. Ieyasu
Nissin Electric Co., Ltd.
2121, Soja, Seji-Machi
Maebashi, Gunma, 371
JAPAN

E. Illenberger
Freie University
Takustrasse 3
D-14195 Berlin
GERMANY

D. R. James
Oak Ridge National Laboratory
P. O. Box 2008
Oak Ridge, TN 37831-6123

M. Judd
University of Strathclyde
204 George St.
Glasgow G1 1XW
UK

K. Kato
Ztech, Inc.
396 Sharon Drive
Cheshire, CT 06410

E. Kawagoe
The Kansai Electric Power Co.
3-22, 3-chome, Nakanoshima
Kita-ku, Osaka 530
JAPAN

Y. Kawaguchi
Dept. of Elec. Eng.
Kokushikan University
4-28-1, Setagaya
Tokyo
JAPAN

J. T. Kennedy
Eindhoven University of Technology
P. O. Box 513
5600 MB Eindhoven
THE NETHERLANDS

W. Kinsner
Dept. of Electrical &
Computer Engineering
University of Manitoba
Winnipeg, Manitoba
CANADA R3T 2N2

E. Kuffel
Faculty of Engineering
University of Manitoba
Winnipeg, Manitoba
CANADA R3T 2N2

R. Kurrer
IEH/University of Stuttgart
Greifscheid Str. 2
D-70174 Stuttgart
GERMANY

P. Lamp
Max-Plack-Institut f. Physik
Werner-Heisenberg-Institute
Postfach 40 12 12
8000 Munchen
GERMANY

S. M. Mahajan
Tennessee Technological University
Box 5004/Electrical Engineering
Cookeville, TN 38505

E. Marode
Laboratoire de Physique des Decharges
CNRS/ESE
Plateau du Moulon
91190 Gif-sur-Yvette
FRANCE

M. Martin
Oak Ridge National Laboratory
P. O. Box 2008
Oak Ridge, TN 37831-6122

D. L. McCorkle
Oak Ridge National Laboratory
P. O. Box 2008
Oak Ridge, TN 37831-6122

L. Ming
ABB Corporate Research
S-721 78 Västeräs
SWEDEN

H. D. Morrison
Ontario Hydro Technologies
800 Kipling Ave, KR 139
Toronto, Ontario M8Z 5S4
CANADA

A. H. Mufti
King Abdulaziz University
Electric & Computer Engineering Dept.
P. O. Box 9027
Jeddah 21413
SAUDI ARABIA

Y. Murayama
Toshiba Corporation
2-1 Ukishima-cho
Kawasaki
JAPAN

R. Nagpal
Wright Laboratory
Aerospace Power Division
Bldg. 450
Wright-Patterson AFB, OH 45433-7919

K. Nakanishi
Mitsubishi Electric Corp.
1-1 Tsukaguchi-Honmachi 8 Chome
Amagasaki, Hyogo 661
JAPAN

H. Nakazawa
Mitsubishi Electric Co.
1-1 Tsukaguchi-Honmachi 8 Chome
Amagasaki, Hyogo 661
JAPAN

A. Nasciuti
ETH-Zentrum
Physikstr. 3
CH-8092 Zürich
SWITZERLAND

L. Niemeyer
ABB Corporate Research
Dept. CRBV.3
CH-5405 Baden-Dättwil
SWITZERLAND

T. Nitta
Mitsubishi Electric Research Labs
201 Broadway
Cambridge, MA 02139

J. Olthoff
National Institute of Standards and Technology
Building 220, Room B344
Gaithersburg, MD 20899

M. O. Pace
Department of Electrical Engineering
The University of Tennessee
411 Ferris Hall
Knoxville, TN 37996-2100

D. Paul
Oak Ridge National Laboratory
P. O. Box 2008
Oak Ridge, TN 37831-6123

A. Pedersen
8 Ellenbakken
DK 3460 Birkerod
DENMARK

K. Petcharaks
ETH-Zentrum
Physikstr. 3
CH-8092 Zürich
SWITZERLAND

W. Pfeiffer
Technische Hochschule Darmstadt
Landgraf Georg Str. 4
64283 Darmstadt
GERMANY

M. Piemontesi
ETH-Zentrum
Physikstr. 3
CH-8092 Zürich
SWITZERLAND

L. A. Pinnaduwage
Oak Ridge National Laboratory
P. O. Box 2008
Oak Ridge, TN 37831-6122

Y. Qiu
Dept. of Electrical Engr.
Xi'an Jiaotong University
Xi'an 710049
CHINA

H. Rodrigo
Department of Electrical Engineering
Michigan Technological University
Houghton, MI 49931

S. W. Rowe
Merlin Gerin Company
Research Dept., USIWE A2
38050 Grenoble Cedex
FRANCE

K. Rupnik
Chemistry Department
Louisiana State Universtiy
Baton Rouge, LA 70803

I. Sauers
Oak Ridge National Laboratory
P. O. Box 2008
Oak Ridge, TN 37831-6123

R. Sawhney
College of Military Engineering
Pune-411 031
INDIA

M. Schmidt
INP
R. Blumstr. 8-10
D 17483 Greifswald
GERMANY

K. H. Schoenbach
Dept. of Electrical Engineering
Old Dominion University
Norfolk, VA 23529

D. L. Schweickart
Wright Aeronautical Laboratories
WL/POOX-2, Building 450
Wright Patterson Air Force Base
Dayton, OH 45433-6563

H. Słowikowska
Ul. Pozaryskiego 28
Warsaw 04-703
POLAND

S. J. Słowikowski
Ul. Pozaryskiego 28
Warsaw 04-703
POLAND

T. H. Teich
Swiss Federal Institute of Technology
FG Hochspannung, ETHZ
Physikstr. 3
CH-8092 Zurich
SWITZERLAND

N. G. Trinh
Cables & Isolants
Institute de recherche d' Hydro-Quebec
1800 Montee Ste-Julie
Varennes, Quebec
CANADA J3X 1S1

R. J. Van Brunt
National Institute of Standards and Technology
Building 220, Room B344
Gaithersburg, MD 20899

C. X. Wang
National Grid Company
Kelvin Avenue
Leatherhead
Surrey KT22 7ST
UK

R. J. Warmack
Oak Ridge National Laboratory
P. O. Box 2008
Oak Ridge, TN 37831-6123

A. Wasserstrom
LIMCO Mfg. Corp.
Garvies Point Road
Glen Cove, NY 11542

X. Waymel
EDF DER Laboratoire de Génie
Electrique Les Renardières
Route de Sens Ecuelles, BP N1
77250 Moret Sur Loing
FRANCE

R. Weber
Institute of High Voltage Engineering
Technical University Munich
Arcisstr. 21
D 80290 München
GERMANY

J. M. Wetzer
Eindhoven University of Technology
P. O. Box 513
5600 MB Eindhoven
THE NETHERLANDS

M. Wohlmuth
Institute of High Voltage Engineering
Technical University Munich
Arcisstr. 21
D 80290 München
GERMANY

T. Yamagiwa
Kokuba Works
Hitachi, Ltd.
1-1-1, Kukubu-cho, Hitachi-shi
Ibaraki-ken 316
JAPAN

T. Yoshikawa
Mitsubishi Electric Corporation
AKO Works
651 Tenwa, Ako City
Hyogo Pref. 678-02
JAPAN

M. Zhang-Cheng
Dept. of Electrical Engineering
Huazhong University of Sci. & Tech.
Wuhan 430074
CHINA

L. Zheng-ying
Dept. of Electrical Engineering
Huazhong University of Sci. & Tech.
Wuhan 430074
CHINA

B. Zullo
Ztech, Inc.
396 Sharon Drive
Cheshire, CT 06410

PHOTOGRAPHS OF PARTICIPANTS

Participants (left to right)

Third Row: A. Ieyasu, R. Weber, C. X. Wang, H. Aoyagi

Second Row: Y. Kawaguchi, D. Buchner, Y. Qiu, P. G. Datskos, J. Castonguay, X. Waymel, K. Nakanishi, J. G. Carter, A. H. Mufti

First Row: K. Feser, S. R. Hunter, S. J. Slowikowski, H. Slowikowska, F. Campbell, A. Pedersen, E. Illenberger, H. Rodrigo

Participants (left to right)

Third Row: T. Yoshikawa, H. Nakazawa, T. Yamagiwa, J. K. Olthoff, O. Farish, G. D. Griffin

Second Row: M. F. Fréchette, M. Judd, K. Kato, E. Kawagoe, M. Schmidt, K. Ibuki

First Row: R. Driggans, K. Petcharaks, B. Zullo, J. S. Chang, T. Nitta, Y. Murayama

Participants (left to right)

Third Row: E. Kuffel, H. D. Morrison, K. Rupnik, R. J. Van Brunt, A. Nasciuti, J. M. Wetzer, R. Bartnikas

Second Row: P. Lamp, S. W. Rowe, A. Wasserstrom, P. M. Blom, Jr., A. Piemontesi, S. Grzybowski

First Row: A. F. Borghesani, R. Kurrer, D. Paul, M. Martin, J. T. Kennedy, L. Niemeyer, W. Kinsner

Participants (left to right)

Third Row: O. Claus, I. D. Chalmers, G. C. Crichton, D. C. Crowell, B. J. Norton, W. Pfeiffer, C. M. Zhang

Second Row: L. Ming, D. Branston, A. H. Cookson, M. O. Pace, M. Goldman, Li Zheng-ying, R. Nagpal

First Row: M. Wohlmuth, I. Dionne, G. N. Trinh, N. F. Cardwell, J. A. Cripps, J. A. Carrington, D. Schweickart, L. Hatfield

Participants (left to right)

Second Row: D. R. James, E. Marode, T. H. Teich, J. de Urquijo

First Row: A. Garscadden, L. G. Christophorou, S. M. Mahajan, I. Sauers, L. A. Pinnaduwage

AUTHOR INDEX

AUTHORS INDEX

SUBJECT INDEX

SUBJECT INDEX

GPSR Compliance

The European Union's (EU) General Product Safety Regulation (GPSR) is a set of rules that requires consumer products to be safe and our obligations to ensure this.

If you have any concerns about our products, you can contact us on ProductSafety@springernature.com

In case Publisher is established outside the EU, the EU authorized representative is:

Springer Nature Customer Service Center GmbH
Europaplatz 3
69115 Heidelberg, Germany

FSC
www.fsc.org
MIX
Papier | Fördert gute Waldnutzung
FSC® C083411

Kolibri 360 GmbH
Ferdinand-Jühlke-Straße 7
99095 Erfurt, Deutschland
produktsicherheit@kolibri360.de